BIOCHEMISTRY OF LIPIDS, LIPOPROTEINS AND MEMBRANES (4TH EDN.)

# New Comprehensive Biochemistry

Volume 36

*General Editor*

## G. BERNARDI
*Paris*

**ELSEVIER**

Amsterdam · Boston · Heidelberg · London · New York · Oxford · Paris
San Diego · San Francisco · Singapore · Sydney · Tokyo

# Biochemistry of Lipids, Lipoproteins and Membranes (4th Edn.)

*Editors*

Dennis E. Vance

*CIHR Group on Molecular and Cell Biology of Lipids and Department of Biochemistry, University of Alberta, Edmonton, Canada*

and

Jean E. Vance

*CIHR Group on Molecular and Cell Biology of Lipids and Department of Medicine, University of Alberta, Edmonton, Canada*

ELSEVIER

Amsterdam · Boston · Heidelberg · London · New York · Oxford · Paris
San Diego · San Francisco · Singapore · Sydney · Tokyo

Elsevier
Radarweg 29, PO Box 211, 1000 AE Amsterdam, The Netherlands
The Boulevard, Langford Lane, Kidlington, Oxford OX5 1GB, UK

First edition 2002
Reprinted 2006

**Library of Congress Cataloging-in-Publication Data**
A catalog record for this book is available from the Library of Congress

ISBN–13: 978-0-444-51139-3 (paperback)
ISBN–10: 0-444-51139-3 (paperback)

ISBN: 0444-51138-5 (hardback)
ISBN: 0-44-80303-3 (series)

ISSN: 0167-7306 (series)

For information on all Elsevier publications
visit our website at books.elsevier.com

Transferred to digital print 2007
Printed and bound by CPI Antony Rowe, Eastbourne

Working together to grow
libraries in developing countries

www.elsevier.com | www.bookaid.org | www.sabre.org

ELSEVIER     BOOK AID
             International     Sabre Foundation

# Preface

The first edition of this textbook was published in 1985. However, research in the biochemistry and molecular biology of lipids and lipoproteins has experienced a remarkable rebirth within the past few years with the realization that lipids play important roles not only in membrane structure and the functioning of membrane proteins, but also in diseases such as heart disease, diabetes, obesity, stroke, cancer and neurological diseases. In addition, lipids are known to participate widely in signaling pathways which impact on all basic biological processes. We have therefore assembled the fourth edition of this textbook by taking account of these major advances in these fields.

The 4th edition has been written with two major objectives in mind. The first is to provide students and teachers with an advanced and up-to-date textbook covering the major areas in the fields of lipid, lipoprotein and membrane biochemistry and molecular biology. The chapters are written for students who have already taken an introductory course in biochemistry, who are familiar with the basic concepts and principles of biochemistry, and who have a general background in the area of lipid metabolism. This book should, therefore, provide the basis for an advanced course for students and teachers in the biochemistry of lipids, lipoproteins and membranes. The second objective of this book is to satisfy the need for a general reference and review book for scientists studying lipids, lipoproteins and membranes. Our goal was to provide a clear summary of these research areas for scientists presently working in, or about to enter, these and related fields. This book remains unique in that it is not a collection of exhaustive reviews on the various topics, but rather is a current, readable and critical summary of these areas of research. This book should allow scientists to become familiar with recent discoveries related to their own research interests, and should also help clinical researchers and medical students keep abreast of developments in basic science that are important for clinical advances in the future.

All of the chapters have been extensively revised since the third edition appeared in 1996. New chapters have been added on lipid modifications of proteins, bile acids, lipoprotein structure, and the relation between lipids and atherosclerosis. We have not attempted to describe in detail the structure and function of biological membranes or the mechanism of protein assembly into membranes since these topics are covered already in a number of excellent books. The first chapter, however, contains a summary of the principles of membrane structure as a basis for the subsequent chapters.

Excellent up-to-date reviews are available on all the topics included in this book and many of these reviews are cited in the relevant chapters. We have limited the number of references cited at the end of each chapter and have emphasized review articles. In addition, the primary literature is cited in the body of the text by providing the name of one author and the year in which the work was published. Using this system, readers will readily be able to find the original citation via computer searching.

The editors and contributors assume full responsibility of the content of the various chapters and we would be pleased to receive comments and suggestions for future editions of this book.

Dennis and Jean Vance
Edmonton, Alberta, Canada
January, 2002

# List of contributors [*]

Luis B. Agellon    433
*Canadian Institutes of Health Research Group in Molecular and Cell Biology of Lipids and Department of Biochemistry, University of Alberta, Edmonton, AB T6G 2S2, Canada*

Nikola A. Baumann    37
*University of Wisconsin-Madison, Department of Biochemistry, 433 Babcock Drive, Madison, WI 53706-1569, USA*

Assumpta A. Bennaars    263
*Department of Biochemistry, Molecular Biology and Biophysics, University of Minnesota, Minneapolis, MN 55455, USA*

David A. Bernlohr    263
*Department of Biochemistry, Molecular Biology and Biophysics, University of Minnesota, Minneapolis, MN 55455, USA*

Mikhail Bogdanov    1
*University of Texas-Houston, Medical School, Department of Biochemistry and Molecular Biology, Houston, TX 77030, USA*

Harold W. Cook    181
*Atlantic Research Centre, Departments of Pediatrics and Biochemistry & Molecular Biology, Dalhousie University, Halifax, Nova Scotia, B3H 4H7 Canada*

William Dowhan    1
*University of Texas-Houston, Medical School, Department of Biochemistry and Molecular Biology, Houston, TX 77030, USA*

Christopher J. Fielding    527
*Cardiovascular Research Institute, and Departments of Medicine and Physiology, University of California, San Francisco, CA 94143-0130, USA*

Phoebe E. Fielding    527
*Cardiovascular Research Institute, and Departments of Medicine and Physiology, University of California, San Francisco, CA 94143-0130, USA*

---

[*] Authors' names are followed by the starting page number(s) of their contribution(s).

Richard J. Heath    55
*St. Jude's Children's Research Hospital, Protein Science Division, Department of Infectious Diseases, 332 N. Lauderdale, Memphis, TN 38101-0318, USA*
*and*
*University of Tennessee Health Science Center, Department of Molecular Biosciences, Memphis, TN 38163, USA*

Suzanne Jackowski    55
*St. Jude's Children's Research Hospital, Protein Science Division, Department of Infectious Diseases, 332 N. Lauderdale, Memphis, TN 38101-0318, USA*
*and*
*University of Tennessee Health Science Center, Department of Molecular Biosciences, Memphis, TN 38163, USA*

Anne E. Jenkins    263
*Department of Biochemistry, Molecular Biology and Biophysics, University of Minnesota, Minneapolis, MN 55455, USA*

Ana Jonas    483
*Department of Biochemistry, College of Medicine, University of Illinois at Urbana-Champaign, 506 South Mathews Avenue, Urbana, IL 61801, USA*

Ten-ching Lee    233
*Oak Ridge Associated Universities (retired), Oak Ridge, TN 37831, USA*

Laura Liscum    409
*Department of Physiology, Tufts University School of Medicine, 136 Harrison Avenue, Boston, MA 02111, USA*

Christopher R. McMaster    181
*Atlantic Research Centre, Departments of Pediatrics and Biochemistry & Molecular Biology, Dalhousie University, Halifax, Nova Scotia, B3H 4H7 Canada*

Linda C. McPhail    315
*Department of Biochemistry, Wake Forest University School of Medicine, Winston-Salem, NC 27157, USA*

Anant K. Menon    37
*University of Wisconsin-Madison, Department of Biochemistry, 433 Babcock Drive, Madison, WI 53706-1569, USA*

Alfred H. Merrill Jr.    373
*School of Biology, Petit Institute for Bioengineering and Biosciences, Georgia Institute of Technology, Atlanta, GA 30332, USA*

Robert C. Murphy    341
*Department of Pediatrics, Division of Cell Biology, National Jewish Medical and Research Center, 1400 Jackson Street, Room K929, Denver, CO 80206-2762, USA*


John B. Ohlrogge    93
*Michigan State University, Department of Plant Biology, East Lansing, MI 48824, USA*


Vangipuram S. Rangan    151
*Children's Hospital Oakland Research Institute, 5700 Martin Luther King Jr. Way, Oakland, CA 94611, USA*


Charles O. Rock    55
*St. Jude's Children's Research Hospital, Protein Science Division, Department of Infectious Diseases, 332 N. Lauderdale, Memphis, TN 38101-0318, USA*
*and*
*University of Tennessee Health Science Center, Department of Molecular Biosciences, Memphis, TN 38163, USA*


Konrad Sandhoff    373
*Kekule-Institut für Organische Chemie und Biochemie der Rheinischen Friedrich-Wilhelms-Universität Bonn, D-53121 Bonn, Germany*


Katherine M. Schmid    93
*Butler University, Department of Biology, Indianapolis, IN 46208-4385, USA*


Wolfgang J. Schneider    553
*Institute of Medical Biochemistry, Department of Molecular Genetics, Dr. Bohr Gasse 9/2, A-1030 Vienna, Austria*


Horst Schulz    127
*City College of CUNY, Department of Chemistry, Convent Ave. at 138th Street, New York, NY 10031, USA*


Stuart Smith    151
*Children's Hospital Oakland Research Institute, 5700 Martin Luther King Jr. Way, Oakland, CA 94611, USA*


William L. Smith    341
*Department of Biochemistry and Molecular Biology, Michigan State University, East Lansing, MI 48824-1319, USA*


Fred Snyder    233
*Oak Ridge Associated Universities (retired), Oak Ridge, TN 37831, USA*

Ira Tabas    573
*Departments of Medicine and Anatomy & Cell Biology, Columbia University, New York, NY 10032, USA*

Dennis E. Vance    205
*CIHR Group on Molecular and Cell Biology of Lipids and Department of Biochemistry, University of Alberta, Edmonton, Alberta, T6G 2S2 Canada*

Jean E. Vance    505
*CIHR Group on Molecular and Cell Biology of Lipids, 328 Heritage Medical Research Centre, University of Alberta, Edmonton, Alberta T6G 2S2, Canada*

Dennis R. Voelker    449
*The Lord and Taylor Laboratory for Lung Biochemistry, Program in Cell Biology, Department of Medicine. The Natl. Jewish Center for Immunology and Respiratory Medicine, Denver, CO 80206, USA*

Moseley Waite    291
*Department of Biochemistry, Wake Forest University School of Medicine, Winston-Salem, NC 27157, USA*

David C. Wilton    291
*Division of Biochemistry and Molecular Biology, School of Biological Sciences, University of Southampton, Bassett Crescent East, Southampton SO16 7PX, UK*

Robert L. Wykle    233
*Department of Biochemistry, Wake Forest University Medical Center, Winston-Salem, NC 27517, USA*

# Contents

## *Chapter 4.* Lipid metabolism in plants
*Katherine M. Schmid and John B. Ohlrogge* . . . . . . . . . . . . . . . . 93

## Chapter 5.   Oxidation of fatty acids in eukaryotes
*Horst Schulz* . . . . . . . . . . . . . . . . . . . . . . . . . . . . . . . . .

## Chapter 6.   Fatty acid synthesis in eukaryotes
*Vangipuram S. Rangan and Stuart Smith* . . . . . . . . . . . . . . . . .

## *Chapter 7.* Fatty acid desaturation and chain elongation in eukaryotes
*Harold W. Cook and Christopher R. McMaster* . . . . . . . . . . . . . . . 181

## *Chapter 8.*   Phospholipid biosynthesis in eukaryotes
### *Dennis E. Vance* . . . . . . . . . . . . . . . . . . . . . . . . . . . . .   205

*Chapter 9.*   Ether-linked lipids and their bioactive species
*Fred Snyder, Ten-ching Lee and Robert L. Wykle*  . . . . . . . . . . . . . 233

## *Chapter 12.* Glycerolipids in signal transduction
*Linda C. McPhail* . . . . . . . . . . . . . . . . . . . . . . . **315**

## *Chapter 13.* The eicosanoids: cyclooxygenase, lipoxygenase, and epoxygenase pathways
*William L. Smith and Robert C. Murphy* . . . . . . . . . . . . . . . . . **341**

*Chapter 14.* Sphingolipids: metabolism and cell signaling
*Alfred H. Merrill Jr. and Konrad Sandhoff* . . . . . . . . . . . . . . . . . . 373

## *Chapter 15.* Cholesterol biosynthesis
*Laura Liscum* . . . . . . . . . . . . . . . . . . . . . . . . . . . . . . . . **409**

*Chapter 19.*   Assembly and secretion of lipoproteins
*Jean E. Vance* . . . . . . . . . . . . . . . . . . . . . . . . . . . . . . 505

*Chapter 20.*   Dynamics of lipoprotein transport in the human circulatory system
*Phoebe E. Fielding and Christopher J. Fielding* . . . . . . . . . . . . . . . 527

## Chapter 21.   Lipoprotein receptors
*Wolfgang J. Schneider* . . . . . . . . . . . . . . . . . . . . . . . . . . . . **553**

*Chapter 22.*   Lipids and atherosclerosis
*Ira Tabas* . . . . . . . . . . . . . . . . . . . . . . . . . . . . . . . . . .  573

# Other volumes in the series

D.E. Vance and J.E. Vance (Eds.) *Biochemistry of Lipids, Lipoproteins and Membranes (4th Edn.)*
© 2002 Elsevier Science B.V. All rights reserved

# CHAPTER 1

# Functional roles of lipids in membranes

## William Dowhan and Mikhail Bogdanov

*6431 Fannin, Suite 6.200, Department of Biochemistry and Molecular Biology, University of
Texas-Houston, Medical School, Houston, TX 77030, USA, Tel.: +1 (713) 500-6051;
Fax: +1 (713) 500-0652; E-mail: william.dowhan@uth.tmc.edu*

## 1. Introduction and overview

Lipids as a class of molecules display a wide diversity in both structure and biological
function. A primary role of lipids in cellular function is in the formation of the
permeability barrier of cells and subcellular organelles in the form of a lipid bilayer
(Fig. 1). Although the major lipid type defining this bilayer in almost all membranes
is glycerol-based phospholipid, other lipids are important components and vary in their

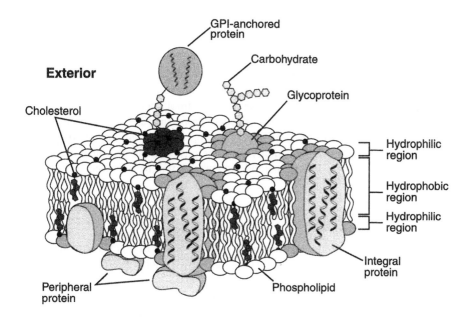

Fig. 1. Model for membrane structure. This model of the plasma membrane of a eukaryotic cell is an
adaptation of the origin model proposed by Singer and Nicholson [1]. The phospholipid bilayer is shown
with integral membrane proteins largely containing α-helical transmembrane domains. Peripheral membrane
proteins associate either with the lipid surface or with other membrane proteins. Lipid rafts (dark gray
headgroups) are enriched in cholesterol and contain a glycosylphosphatidylinositol-linked (GPI) protein.
The light gray headgroups depict lipids in close association with protein. The irregular surface and wavy
acyl chains denote the fluid nature of the bilayer.

presence and amounts across the spectrum of organisms. Sterols are present in all eukaryotic cytoplasmic membranes and in a few bacterial membranes. The ceramide-based sphingolipids are also present in the membranes of all eukaryotes. Neutral glycerol-based glycolipids are major membrane-forming components in many Gram-positive bacteria and in the membranes of plants while Gram-negative bacteria utilize a glucosamine-based phospholipid (Lipid A) as a major structural component of the outer membrane. Additional diversity results in the variety of the hydrophobic domains of lipids. In eukaryotes and eubacteria these domains are usually long chain fatty acids or alkyl alcohols with varying numbers and positions of double bonds. In the case of archaebacteria, the phospholipids have long chain reduced polyisoprene moieties, rather than fatty acids, in ether linkage to glycerol. If one considers a simple organism such as *Escherichia coli* with three major phospholipids and several different fatty acids along with many minor precursors and modified products, the number of individual phospholipid species ranges in the hundreds. In more complex eukaryotic organisms with greater diversity in both the phospholipids and fatty acids, the number of individual species is in the thousands.

If one or two phospholipids are sufficient to form a stable bilayer structure, why is the above diversity in lipid structures present in biological membranes [2]? The adaptability and flexibility in membrane structure necessitated by environment is possible only with a broad spectrum of lipid mixtures. The membrane is also the supporting matrix for a wide spectrum of proteins involved in many cellular processes. Approximately 20–35% of all proteins are integral membrane proteins, and probably half of the remaining proteins function at or near a membrane surface. Therefore, the physical and chemical properties of the membrane directly affect most cellular processes making the role of lipids dynamic with respect to cell function rather than simply defining a static barrier.

In this chapter, the diversity in structure, chemical properties, and physical properties of lipids will be outlined. Next, the various genetic approaches available to study lipid function in vivo will be summarized. Finally, how the physical and chemical properties of lipids relate to their multiple functions in living systems will be reviewed.

## 2. Diversity in lipid structure

Lipids are defined as those biological molecules readily soluble in organic solvents such as chloroform, ether, or toluene. However, some very hydrophobic proteins such as the $F_0$ subunits of ATP synthase are soluble in chloroform, and lipids with large hydrophilic domains such as lipopolysaccharide are not soluble in these solvents. Here we will consider only those lipids that contribute significantly to membrane structure or have a role in determining protein structure or function. The broad area of lipids as second messengers is covered in Chapters 12–14.

### 2.1. Glycerol-based lipids

The primary building blocks of most membranes are glycerol phosphate-containing lipids generally referred to as phospholipids (Fig. 2). The diacylglycerol backbone in

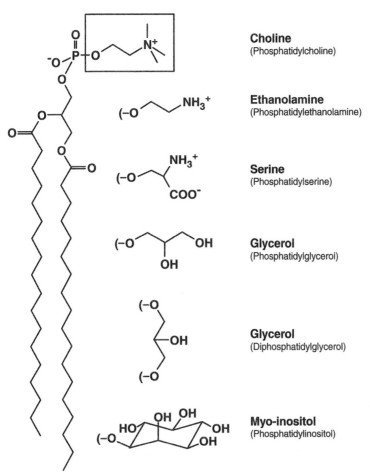

**Choline**
(Phosphatidylcholine)

**Ethanolamine**
(Phosphatidylethanolamine)

**Serine**
(Phosphatidylserine)

**Glycerol**
(Phosphatidylglycerol)

**Glycerol**
(Diphosphatidylglycerol)

**Myo-inositol**
(Phosphatidylinositol)

Fig. 2. Structure of glycerol phosphate-based lipids. The complete lipid structure shown is 1,2-distearoyl-*sn*-glycerol-3-phosphocholine or phosphatidylcholine (PC). Substitution of choline in the box with the headgroups listed below results in the other phospholipid structures. CDP-diacylglycerol has a CMP and phosphatidic acid has a hydroxyl group in place of choline (not shown). Diphosphatidylglycerol, which contains two phosphatidic acids joined by glycerol, is commonly referred to as cardiolipin (CL).

eubacteria and eukaryotes is *sn*-3-glycerol esterified at the 1- and 2-position with long chain fatty acids. In archaebacteria (Fig. 3), *sn*-1-glycerol forms the lipid backbone and the hydrophobic domain is composed of phytanyl (a saturated isoprenyl) groups in ether linkage at the 2- and 3-position (an archaeol). In addition two *sn*-1-glycerol groups are found connected in ether linkage by two biphytanyl groups (dibiphytanyldiglyc-erophosphatetraether) [3] to form a covalently linked bilayer. Some eubacteria (mainly hyperthermophiles) have dialkyl (long chain alcohols in ether linkage) glycerophosphate lipids and similar ether linkages are found in the plasmalogens of eukaryotes. The headgroups of the phospholipids (boxed area of Fig. 2) extend the diversity of lipids defining phosphatidic acid (PA, with OH), phosphatidylcholine (PC), phosphatidylserine

4

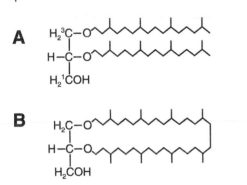

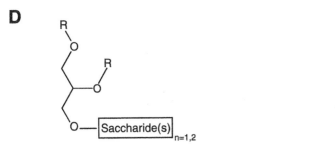

Fig. 3. Structure of dialkylglycerols in archaebacteria. Archaebacteria have phytanyl chains in ether linkage to the 2- and 3-positions of *sn*-1-glycerol (archaeol). The 1-position can be derivatized with phosphodiesters. (A) Diphytanylglycerol (C20–C20 diether) with the stereochemistry of glycerol indicated. (B) Cyclic biphytanyl (C40) diether. (C) Biphytanyl diglycerol diether. (D) A glycolipid with either a mono or disaccharide (glucose or galactose) at the 1-position of *sn*-1-glycerol. The R groups are ether-linked phytanyl chains. Similar glycolipids are found in eubacteria and plants with a *sn*-3-glycerol backbone and ester-linked fatty acid chains at the 1- and 2-positions.

(PS), phosphatidylglycerol (PG), phosphatidylinositol (PI), and cardiolipin (CL). Archaebacteria analogues exist with headgroups of glycerol and glycerolmethylphosphate as well as all of the above except PC and CL (Chapter 3). Archaebacteria also have neutral glycolipid derivatives in which mono- and disaccharides (glucose or galactose) are directly linked to *sn*-1-archaeol (Fig. 3). Plants (mainly in the thylokoid membrane) and many Gram-positive bacteria also have high levels of neutral glycolipids with mono- or disaccharides linked to the 3-carbon of *sn*-3-diacylglycerol (Chapter 4). Therefore, the diversity of glycerol-based lipids in a single organism is significant, but the diversity throughout nature is enormous. The lipid composition of various biological membranes is shown in Table 1.

Table 1
Lipid composition of various biological membranes

| Lipid | Erythrocyte[b] | Myelin[b] | Mitochondria[c] | Endoplasmic reticulum[c] | E. coli[d] |
|---|---|---|---|---|---|
| Cholesterol | 23 | 22 | 3 | 6 | – |
| PE | 18 | 15 | 35 | 17 | 70 |
| PC | 17 | 10 | 39 | 40 | – |
| Sphingomyelin | 18 | 8 | – | 5 | – |
| PS | 7 | 9 | 2 | 5 | – |
| PG | – | – | – | – | 20 |
| CL | – | – | 21 | – | 10 |
| Glycolipid[a] | 3 | 28 | – | – | – |
| Others | 13 | 8 | – | 27 | – |

The data are expressed as weight % of total lipid.
[a] Ceramide based.
[b] Human sources.
[c] Rat liver. Inner and outer mitochondrial membrane.
[d] Inner and outer membrane excluding Lipid A.

The majority of information on the chemical and physical properties of lipids comes from studies on the major phospholipid classes of eubacteria and eukaryotes with only limited information on the lipids from archaebacteria. The biosynthetic pathways and the genetics of lipid metabolism have also been extensively studied in eubacteria (Chapter 3) and eukaryotes (Chapter 8). Clearly the archaeol lipids confer some advantage with respect to the environment of archaebacteria. Many of these organisms exist in harsh environments that call for more chemically stable lipid bilayers which is afforded by the above lipids. How the physical properties of the more commonly studied lipids change with environment will be discussed later.

### 2.2. Diglucoseamine phosphate-based lipids

The outer membrane of Gram-negative bacteria (Fig. 4) contains a lipid made up of a headgroup derived from glucosamine phosphate (Chapter 3). The core lipid (Lipid A, see Fig. 5 and Chapter 3) in E. coli is a phospholipid containing two glucoseamine groups in $\beta(1–6)$ linkage that are decorated at positions 2, 3, 2′ and 3′ with R-3-hydroxymyristic acid (C14) and at positions 1 and 4′ with phosphates. Further modification at position 6′ with a KDO disaccharide (two 3-deoxy-D-manno-octulosonic acids in $\alpha(1–3)$ linkage) results in $KDO_2$-Lipid A that is further modified by an inner core, an outer core, and the O-antigen. Laboratory strains of Salmonella typhimurium and E. coli such as K-12 lack the O-antigen found in the wild-type and clinically important strains.

The complete structure either with or without O-antigen is referred to as lipopolysaccharide or LPS. The core Lipid A forms the outer monolayer of the outer membrane bilayer of Gram-negative bacteria; the inner monolayer of the outer membrane (Fig. 4) is made up of glycerophosphate-based lipids. The whole lipopolysaccharide structure defines the outer surface of Gram-negative bacteria, but only the $KDO_2$-Lipid A struc-

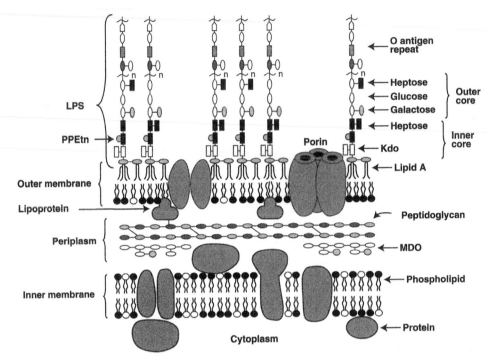

Fig. 4. *E. coli* cell envelope. The complete cell envelope of Gram-negative bacteria contains an inner membrane that is a typical phospholipid bilayer and is the permeability barrier of the cell. The outer membrane is composed of an inner monolayer of phospholipid and an outer monolayer of the Lipid A portion of lipopolysaccharide (LPS). The structure of $KDO_2$-Lipid A is shown in Fig. 5 and is connected to a polysaccharide to build up the inner core, outer core and the O antigen repeat. PPEtn is ethanolamine pyrophosphate. The outer membrane is a permeability barrier for molecules larger than 750–1000 Da that pass through various pores in the outer membrane. The periplasmic space contains many proteins and the membrane-derived oligosaccharide (MDO) that is one component of the osmolarity regulatory system. MDO is decorated with *sn*-glycerol-1-phosphate and ethanolamine phosphate derived from PG and PE, respectively. The amino acid–sugar crosslinked peptidoglycan gives structural rigidity to the cell envelope. One-third of the lipoproteins (*lpp* gene product) is covalently linked via its carboxyl terminus to the peptidoglycan and in complex with the remaining lipoproteins as trimers that associate with the outer membrane via covalently linked fatty acids at the amino terminus. The amino terminal cysteine is blocked with a fatty acid, derived from membrane phospholipids, in amide linkage and is derivatized with diacylglycerol, derived from PG, in thioether linkage. Figure is courtesy of C.R.H. Raetz.

ture is essential for viability of laboratory strains. However, the remainder of the lipopolysaccharide structure is important to survival of Gram-negative bacteria in their natural environment. This structure is modified post-assembly in response to environment including host fluids, temperature, ionic properties, and antimicrobial agents [4]. In addition, both enteric and non-enteric Gram-negative bacteria show a great diversity in all component parts of the LPS structure. Studies of Lipid A biosynthesis is of clinical importance because it is the primary antigen responsible for toxic shock syndrome caused by Gram-negative bacterial infection.

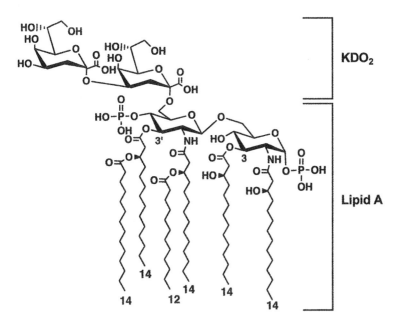

Fig. 5. Structure of KDO$_2$-Lipid A. Lipid A is a disaccharide of glucosamine phosphate that is multiply acylated in both amide and ester linkages with fatty acids of the chain lengths indicated (12 and 14). As illustrated in Fig. 4, Lipid A is attached to KDO$_2$ that is then elongated with the remainder of lipopolysaccharide structure. Figure is courtesy of C.R.H. Raetz.

## 3. Properties of lipids in solution

The matrix that defines a biological membrane is a lipid bilayer composed of a hydrophobic core excluded from water and an ionic surface that interacts with water and defines the hydrophobic–hydrophilic interface (Fig. 1). Much of our understanding of the physical properties of lipids in solution and the driving force for the formation of lipid bilayers comes from the concept of the 'hydrophobic effect' as developed by Charles Tanford [5]. The 'fluid mosaic' model for membrane structure further popularized these concepts [1]. This model, since extensively refined, envisioned membrane proteins as undefined globular structures freely moving in a homogeneous sea of lipids.

### 3.1. Why do polar lipids self-associate?

Polar lipids are amphipathic in nature containing both hydrophobic domains, which do not interact with water, and hydrophilic domains that readily interact with water. The basic premise of the hydrophobic effect is that the hydrocarbon domains of polar lipids disrupt the stable hydrogen bonded structure of water and therefore are at an energy minimum when such domains self associate to minimize the total surface area in contact with water. The polar domains of lipids interact either through hydrogen bonding or ionic interaction with water and therefore are energetically stable in an aqueous environment. The structural organization that a polar lipid assumes in water is determined by its concentration and the law of opposing forces, i.e. hydrophobic forces driving self-association versus steric and ionic repulsive forces of the polar domains in opposing self-association. At low concentrations, amphipathic molecules exist as monomers in solution. As the concentration of the molecule increases, its stability in solution as a monomer decreases until the favorable interaction of the polar domain with water is outweighed by the unfavorable interaction of the hydrophobic domain with water. At this point, a further increase in concentration results in the formation of increasing amounts of self-associated monomers in equilibrium with a constant amount of free monomer. This point of self-association and the remaining constant free monomer concentration is the critical micelle concentration [6]. The larger the hydrophobic domain, the lower the critical micelle concentration due to the increased hydrophobic effect. However, the larger the polar domain, either because of the size of neutral domains or charge repulsion for ionic domains, the higher the critical micelle concentration due to the unfavorable steric hindrance in bringing these domains into close proximity. The critical micelle concentration of amphipathic molecules with a net charge is influenced by ionic strength of the medium due to dampening of the charge repulsion effect. Therefore, the critical micelle concentration of the detergent sodium dodecyl sulfate is reduced ten-fold when the NaCl concentration is raised from 0 to 0.5 M.

These physical properties and the shape of amphipathic molecules define three supramolecular structural organizations of polar lipids and detergents in solution (Fig. 6). Detergents, lysophospholipids (containing only one alkyl chain), and phospholipids with short alkyl chains (eight or fewer carbons) have an inverted cone-shape (large head group relative to a small hydrophobic domain) and self associate above the critical micelle concentration with a small radius of curvature to form micellar structures with a hydrophobic core excluding water. The micelle surface, rather than being a smooth spherical or elliptical structure with the hydrophobic domains completely sequestered inside a shell of polar residues that interact with water, is a very rough surface with many of the hydrophobic domains exposed to water. The overall structure reflects the packing of amphipathic molecules at an energy minimum by balancing the attractive force of the hydrophobic effect and the repulsive force of close headgroup association. The critical micelle concentration for most detergents ranges from micromolar to millimolar. Lysophospholipids also form micelles with critical micelle concentrations in the micromolar range. However, phospholipids with chain lengths of 14 and above self associate at a concentration around $10^{-10}$ M due to the hydrophobic driving force contributed by two alkyl chains. Phospholipids with long alkyl chains do not form micelles but organize into bilayer structures, which allow tight packing of adjacent side chains with the maximum exclusion of water from the hydrophobic domain. In living cells, phospholipids are not found free as monomers in solution, but are organized into either membrane bilayers or protein complexes. When long chain phospholipids are first dried to a solid from organic solvent and then hydrated, they spontaneously form large multilamellar bilayer sheets separated by water. Sonication disperses these sheets into smaller unilamellar bilayer structures that satisfy the hydrophobic nature of the ends of the bilayer by closing into sealed vesicles (also termed liposomes) defined by

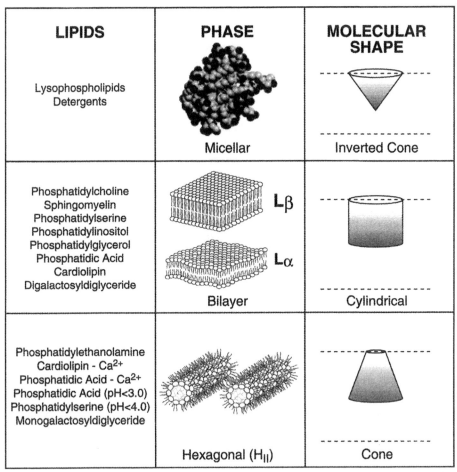

| LIPIDS | PHASE | MOLECULAR SHAPE |
|---|---|---|
| Lysophospholipids Detergents | Micellar | Inverted Cone |
| Phosphatidylcholine Sphingomyelin Phosphatidylserine Phosphatidylinositol Phosphatidylglycerol Phosphatidic Acid Cardiolipin Digalactosyldiglyceride | L$\beta$ L$\alpha$ Bilayer | Cylindrical |
| Phosphatidylethanolamine Cardiolipin - Ca$^{2+}$ Phosphatidic Acid - Ca$^{2+}$ Phosphatidic Acid (pH<3.0) Phosphatidylserine (pH<4.0) Monogalactosyldiglyceride | Hexagonal (H$_{II}$) | Cone |

Fig. 6. Polymorphic phases and molecular shapes exhibited by lipids. The space filling model for the micellar phase is of β-D-octyl glucoside micelle (50 monomers). The polar portions of the detergent molecules (oxygen atoms are black) do not cover completely the micelle surface (hydrocarbons in gray) leaving substantial portions of the core exposed to bulk solvent. Inverted cone-shaped molecules form micelles. Model adapted with permission from Garavito and Ferguson-Miller [6]. Copyright 2001 The American Society for Biochemistry and Molecular Biology. Polar lipids with two long alkyl chains adopt a bilayer or non-bilayer (H$_{II}$) structure depending on the geometry of molecule (cylindrical or cone-shaped, respectively) and environmental conditions. The L$_\beta$ (order gel) and L$_\alpha$ (liquid crystalline) bilayer phases differ in the order within the hydrophobic domain and in mobility of the individual molecules.

a continuous single bilayer and an aqueous core much like the membrane surrounding cells. Liposomes can also be made by physical extrusion of lamellar structures through a small orifice or by dilution of a detergent–lipid mixture below the critical micelle concentration of the detergent.

Cylindrical shaped lipids (head group and hydrophobic domains of similar diameter) such as PC form lipid bilayers. Cone-shaped lipids (small head groups relative to a large hydrophobic domain) such as PE (unsaturated fatty acids) favor an inverted micellar structure where the headgroups sequester an internal aqueous core and the hydrophobic domains are oriented outward and self-associate in non-bilayer structures. These are denoted as the hexagonal II ($H_{II}$) and cubic phases (a more complex organization similar to the $H_{II}$ phase). The ability of lipids to form multiple structural associations is referred to as lipid polymorphism. Lipids such as PE, PA, CL, and monosaccharide derivatives of diacylglycerol can exist in either bilayer or the $H_{II}$ phase, depending on solvent conditions, alkyl chain composition, and temperature.

Both cone-shaped and inverted cone-shaped lipids are considered as non-bilayer-forming lipids and when mixed with the bilayer-forming lipids change the physical properties of the bilayer and introduce stress or strain in the bilayer structure. When bilayer-forming lipids are spread as a monolayer at an aqueous–air interface, they have no tendency to bend away from or toward the aqueous phase due to their cylindrical symmetry. In such a system, the hydrophobic domain orients toward the air. Monolayers of the asymmetric cone-shaped lipids ($H_{II}$-forming) tend to bend toward from the aqueous interface (negative radius of curvature) while monolayers of asymmetric inverted cone shaped lipids (micelle-forming) tend to bend away from the aqueous phase (positive radius of curvature). The significance of shape mis-match in lipid mixtures will be covered later.

### 3.2. Physical properties of membrane bilayers

The organization of diacylglycerol-containing polar lipids in solution (Fig. 6) is dependent on the nature of the alkyl chains, the headgroups, and the solvent conditions (i.e., ion content, pH, and temperature). The transition between these phases for pure lipids in solution can be measured by various physical techniques such as $^{31}$P-NMR and microcalorimetry. The difference between the ordered gel ($L_\beta$) and liquid crystalline ($L_\alpha$) phases is the viscosity or fluidity of the hydrophobic domains of the lipids which is a function of temperature and the alkyl chain structure. At any given temperature the 'fluidity' (the inverse of the viscosity) of the hydrocarbon core of the bilayer increases with increasing content of unsaturated or branched alkyl chain or with decreasing alkyl chain length. Due to the increased mobility of the fatty acid chains with increasing temperature, the fluidity and also space occupied by the hydrophobic domain of lipids also increases. A bilayer-forming lipid such as PC assumes a cylindrical shape over a broad temperature range and with different alkyl chain compositions. When analyzed in pure form, PC exists in either the $L_\beta$ or $L_\alpha$ phase mainly dependent on the alkyl chain composition and the temperature. Non-bilayer-forming lipids such as PE exist at low temperatures in the $L_\beta$ phase, at intermediate temperatures in the $L_\alpha$ phase, and at elevated temperatures in the $H_{II}$ or cubic phase (Fig. 7). The last transition is temperature dependent but also depends on the shape of the lipid. The shape of lipids with relatively small head groups can change from cylindrical to conical ($H_{II}$ phase) with increasing unsaturation or length of the alkyl chains or with increasing temperature. As can be seen from Fig. 7, the midpoint temperature ($T_m$) of the transition from the $L_\beta$ to $L_\alpha$ phase increases with an increase in the length of the fatty acids, but the midpoint of the transition temperature ($T_{LH}$) between the $L_\alpha$ and $H_{II}$ phases decreases with increasing chain length (or increasing unsaturation, not shown).

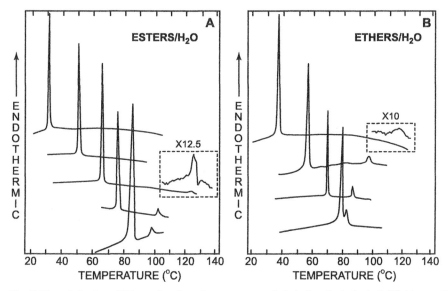

Fig. 7. Phase behavior of PE as a function of temperature and chain length. As hydrated lipids pass through a phase transition heat is absorbed as indicated by the peaks. The large peaks at the lower temperatures are due to the $L_\beta$ to $L_\alpha$ transition and the smaller peaks at higher temperatures are due to the $L_\alpha$ to $H_{II}$ transition. (A) Even numbered diacyl-PEs ranging from C12 to C20 top to bottom. (B) Even numbered dialkyl-PEs in ether linkage ranging from C12 to C18 top to bottom. The inserts indicate an expanded scale for the transition to $H_{II}$. Figure adapted with permission from Seddon et al. [7]. Copyright 1983 American Chemical Society.

Similar transition plots as well as complex phase diagrams have been generated with mixtures of lipids. The physical property of a lipid mixture is a collective property determined by each of the component lipids. A large number of studies indicate that the $L_\alpha$ state of the membrane bilayer is required for cell viability and cells adjust their lipid composition in response to many environmental factors so that the collective property of the membrane exhibits the $L_\alpha$ state. Addition of non-bilayer-forming lipids to bilayer-forming lipids can result in non-bilayer formation, but at a higher temperature than for the pure non-bilayer-forming lipid. Addition of non-bilayer-forming lipids also adds another parameter of tension between the two monolayers. These lipids in each half of the bilayer tend to reduce the radius of curvature of each monolayer that results in a tendency to pull the bilayer apart by curving the monolayers away from each other (see the end of Section 3.1). This process results in potential energy residing in the bilayer that is a function of the presence of non-bilayer lipids. Forcing non-bilayer-forming lipids into a bilayer structure also exposes the hydrophobic core to the aqueous phase. Mixtures of lipids with dissimilar phase properties can also generate phase separations with local domain formation. Such discontinuities in the bilayer structure may be required for many structural organizations and cellular processes such as accommodation of proteins into the bilayer, movement of macromolecules across the bilayer, cell division, and membrane fusion and fission events. The need for bilayer discontinuity may be the reason that all natural membranes contain a significant proportion of non-bilayer-forming lipids even though the membrane under physiological conditions is in the $L_\alpha$ phase.

Addition of cholesterol to lipid mixtures has a profound effect on the physical properties of a bilayer. Increasing amounts of cholesterol inhibit the organization of lipids into the $L_\beta$ phase and favor a less fluid but more ordered structure than the $L_\alpha$ phase resulting in the lack of a phase transition normally observed in the absence of cholesterol. The solvent surrounding the lipid bilayer also influences these transitions primarily by affecting the size of the headgroup relative to the hydrophobic domain. $Ca^{2+}$ and other divalent cations ($Mg^{2+}$, $Sr^{2+}$, but not $Ba^{2+}$) reduce the effective size of the negatively charged headgroups of CL and PA allowing organization into the $H_{II}$ phase. Low pH has a similar effect on the headgroup of PS. Since $Ca^{2+}$ is an important signaling molecule that elicits many cellular responses, it is possible that part of its effects may be transmitted through changes in the physical properties of membranes. In eukaryotes, CL is found almost exclusively in the inner membrane of the mitochondria where $Ca^{2+}$ fluxes play important regulatory roles.

### 3.3. Special properties of cardiolipin

CL has the unique property of being both a bilayer and non-bilayer lipid depending on the absence or presence of divalent cations. CL is found almost exclusively in eukaryotic mitochondria and in bacteria that utilize oxidative phosphorylation for proton pumping across the membrane. A property of CL that has gone largely unrecognized is the ionization constants of its two phosphate diesters. Rather than displaying two $pK$ values in the range of 2–4, $pK_2$ of CL is >8.5 [8] indicating that CL is protonated at physiological pH (Fig. 8). This property may make CL a proton sink or a conduit for protons in transfer processes. Although PG appears to substitute for CL in many processes in both bacteria and yeast, lack of CL results in a reduction in cell growth dependent on oxidative processes. Therefore, CL is not absolutely essential, but it appears to be required for optimal cell function.

### 3.4. What does the membrane bilayer look like?

The functional properties of natural fluid bilayers not only include the hydrophobic core and the hydrophilic surface but the interfacial region containing bound water and ions. Fig. 9A shows the distribution of the component parts of dioleoylphosphatidylcholine across the bilayer [9] and illustrates the dynamic rather than static nature of the membrane. The bilayer thickness of 30 Å is defined by the length of the fatty acid chains. However, the thickness is not a static number as indicated by the probability of finding $CH_2$ residues outside of this limit. Bilayer thickness can vary over the surface of a membrane if microdomains of lipids are formed with different alkyl chain lengths. What is generally not appreciated is the width (15 Å on either side of the bilayer) of the interface region between the hydrocarbon core and the free water phase of the cytosol. This region contains a complex mixture of chemical species defined by the ester linked glycerophosphate moiety, the variable headgroups, and bound water and ions. Many biological processes occur within this interface region and are dependent on its unique

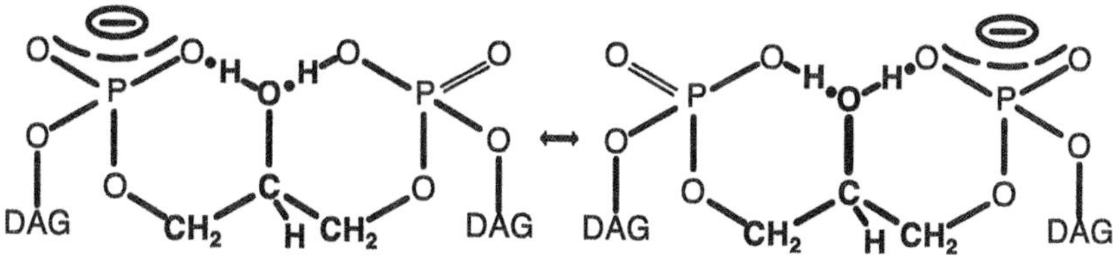

CARDIOLIPIN ACID-ANION STRUCTURE

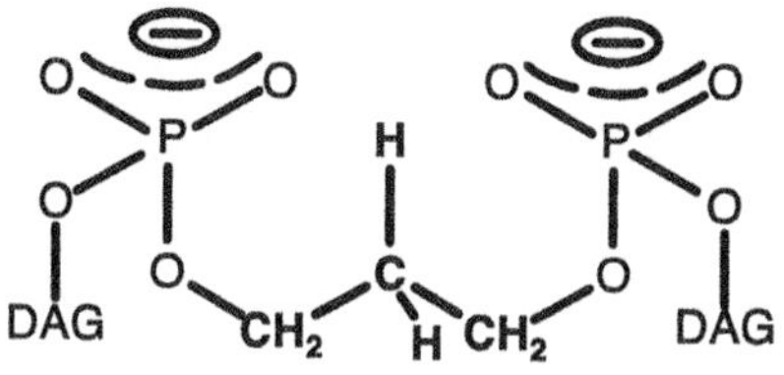

DEOXYCARDIOLIPIN ANION STRUCTURE

Fig. 8. Ionization state of CL and deoxy-CL at physiological pH. CL is only partially ionized under these conditions ($pK_2 > 8.5$) and therefore can trap a proton by hydrogen bonding with the *sn*-2 hydroxyl of glycerol that joins the two phosphatidic acids in the CL structure. Deoxy-CL lacking the *sn*-2 hydroxyl is fully ionized and cannot trap a proton.

properties including the steep polarity gradient (Fig. 9B) within which surface bound cellular processes occur.

## 4. Engineering of membrane lipid composition

Given the diversity in both lipid structure and function, how can the role of a given lipid be defined at the molecular level? Unlike proteins, lipids have neither inherent catalytic activity nor obvious functions in isolation (except for their physical organization). Many functions of lipids have been uncovered serendipitously based on their effect on catalytic processes or biological functions studied in vitro. Although considerable information has accumulated with this approach, such studies are highly prone to artifacts. The physical properties of lipids are as important as their chemical properties in determining function. Yet there is little understanding of how the physical properties of lipids measured in vitro relate to their in vivo function. In addition, the physical properties of lipids have been ignored in many in vitro studies. Genetic approaches are generally the most useful in studying in vivo function, but this approach has considerable limitations when applied to lipids. First, genes do not encode lipids, and in order to make mutants with altered lipid composition, the genes encoding enzymes along a biosynthetic pathway must be targeted. Therefore, the results of genetic mutation are indirect and many times far removed from the primary lesion. Second, a primary

14

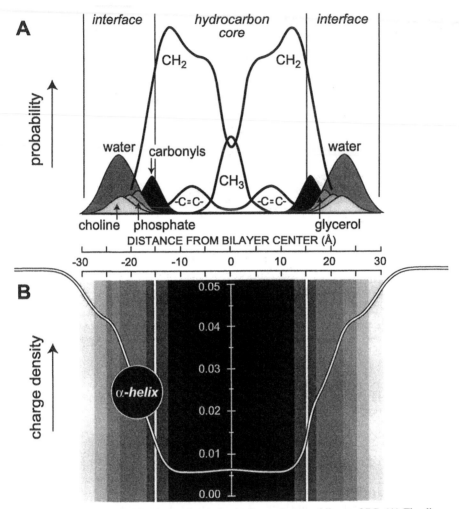

Fig. 9. The probability distribution for chemical constituents across a bilayer of PC. (A) The diagram was generated from X-ray and neutron diffraction data. The interface region between the hydrocarbon core and the free solvent region extends for approximately 15 Å on either side of the 30-Å-thick bilayer. The width of each peak defines the mobility of each constituent of PC. (B) As an α-helical peptide moves from either side of the bilayer towards the center, the charge density of the environment steeply declines as indicated by the line. Figure adapted with permission from White et al. [9]. Copyright 2001 American Society for Biochemistry and Molecular Biology.

function of most membrane lipids is to provide the permeability barrier of the cell. Therefore, alterations in lipid composition may compromise cell permeability before other functions of a particular lipid are uncovered. One may learn from genetics that a lipid is essential for cell viability but never learn the molecular bases for other requirements. Over the past 20 years, genetic approaches have been successfully used to establish the biosynthetic pathways of most of the common lipids. The challenge is to use this genetic information to manipulate the lipid composition of cells without severely compromising cell viability. In those cases where this has been possible, the combination of the genetic approach to uncover phenotypes of cells with altered lipid composition and the dissection in vitro of the molecular basis for the phenotype has proven to be a powerful approach to defining lipid function. The more complex the organelle content and accompanying membrane structure of a cell the more difficult is the application of the genetic approach. Therefore, the most useful information to date has come from genetic manipulation of prokaryotic and eukaryotic microorganisms. However, the basic molecular principles underlying lipid function will be generally applicable to more complex mammalian systems.

*4.1. Alteration of lipid composition in bacteria*

The pathways for formation of the major phospholipids (PE, PG, and CL) of *E. coli* were biochemically established mainly by Eugene Kennedy and coworkers and subsequently verified using genetic approaches as described in Chapter 3. The design of strains in which lipid composition can be genetically altered in a systematic manner has been very important in defining new roles for lipids in cell function [2]. Unlike many other mutations affecting the metabolic pathways in *E. coli*, mutants in phospholipid biosynthesis cannot be bypassed by supplementation of the growth media with phospholipids due to the barrier function of the outer membrane. Therefore, the isolation and study of *E. coli* phospholipid auxotrophs has not been possible.

With the exception of the synthesis of CL, mutants in all steps of phospholipid biosynthesis were thought to be lethal even under laboratory conditions. To date, no growth conditions have been established for cells unable to synthesize CDP-diacylglycerol. Null mutants in the *pgsA* gene (encodes phosphatidylglycerophosphate synthase) that cannot synthesize PG and CL are lethal, but a suppressor of this mutation has been identified [10]. In such mutants, the major outer membrane lipoprotein precursor (see Fig. 4), that depends on PG for its lipid modification, accumulates in the inner membrane and apparently kills the cell. Cells unable to make this lipoprotein are viable but are temperature sensitive for growth indicating that PG and CL are not absolutely required for viability, only for optimal growth. However, the anionic nature of these lipids (apparently substituted by increased levels of PA) is necessary for the proper membrane association and function of peripheral membrane proteins as discussed in Sections 5.4 and 5.5.

The amine-containing lipids, PS and PE, were also thought to be essential based on initial mutants carrying temperature sensitive alleles of the genes (*pssA* and *psd*) encoding their respective biosynthetic enzymes. However, the growth phenotype of these mutants (as well as *pssA* null strains) with reduced amine-containing lipids could be suppressed by adding $Ca^{2+}$, $Mg^{2+}$, and $Sr^{2+}$ in millimolar concentrations to the growth medium. These mutants, although viable, have a complex mixture of defects in cell division, reduced growth rate, loss of outer membrane barrier function, defects in energy metabolism, mis-assembly of membrane proteins, and defects in sugar and amino acid transport.

The key to defining new functions for the anionic and zwitterionic phospholipids of *E. coli* was the design of strains in which the content of PG/CL and PE could be regulated in a systematic manner in viable cells. The *pgsA* gene (encoding the phosphatidylglycerophosphate synthase) was placed under the control of the exogenously regulated promoter *lacOP* (promoter of the *lac* operon) that is controlled by isopropyl-β-thiogalactoside levels in the growth media. Variation in PG plus CL levels were correlated in a dose–response manner with the functioning of specific cellular processes both in vivo and in vitro to determine lipid function. Similarly, the involvement of PE in function was uncovered by comparing phenotypes of cells with and without PE or by placing the *pssA* gene (encoding PS synthase) under exogenous regulation. Therefore, these genetically altered strains have been used as reagents to define potential lipid involvement in cellular process in vivo that can be verified by biochemical studies in vitro.

### 4.2. Alteration of lipid composition in yeast

The pathways of phospholipid synthesis and the genetics of lipid metabolism in yeast *Saccharomyces cerevisiae* [11] are as well understood as in *E. coli*. Yeast have pathways (see Chapters 3 and 8) similar to those in the *E. coli* for PE and PG synthesis. CL synthesis in all eukaryotes involves transfer of a phosphatidyl moiety from CDP-diacylglycerol to PG rather than from one PG to another PG as in bacteria. In addition, yeast utilize the mammalian pathways for synthesis of PI, PE, and PC including the methylation of PE to form PC (Chapter 8).

All gene products necessary for the synthesis of diacylglycerol, CDP-diacylglycerol, and PI in yeast are essential for viability. PS synthesis is not essential if growth medium is supplemented with ethanolamine in order to make PE and PC. However, PE is definitely required since *pss1* (encodes PS synthase) mutants also lacking a sphingolipid degradative enzyme that generates ethanolamine internally, require ethanolamine supplementation [12].

No gene products involved in lipid metabolism are encoded by the mtDNA which in *Saccharomyces cerevisiae* encodes eight proteins (subunits I, II, and III of cytochrome *c* oxidase, cytochrome *b*, the 3 subunits that make up the $F_0$ component of ATP synthase, and the *VAR1* gene product which is part of the mitochondrial ribosome). The enzymes necessary for synthesis of PE from PS, and for PG and CL, are all encoded by nuclear genes and imported into the mitochondria. Null mutants of *crd1* (encodes CL synthase) grow normally on glucose for which mitochondrial function is not required. However, on non-fermentable carbon sources such as glycerol or lactate, they grow slower. Therefore, CL appears to be required for optimal mitochondrial function but is not essential for viability. However, lack of PG and CL synthesis due to a null mutation in the *PGS1* gene (encodes phosphatidylglycerophosphate synthase) results in the inability to utilize non-fermentable carbon sources for growth. Mitochondrial membrane potential is reduced to near undetectable levels although remains sufficient to support the import of all nuclear encoded proteins thus far investigated. Similar effects are seen in mammalian cells with a mutation in the homologous *PGS1* gene. The surprising consequence of lack of PG and CL in yeast is the lack of translation of mRNAs of four mitochondria-encoded proteins (cytochrome *b* and cytochrome *c* oxidase subunits I–III) as well as cytochrome *c* oxidase subunit IV [13] that is nuclear encoded. These results would indicate that either some aspects of translation of a subset of mitochondrial proteins (those associated with electron transport complexes in the inner membrane but not ATP metabolism) require PG and/or CL or the lack of these lipids indirectly affects both mitochondrial and cytoplasmic mRNA translation.

# 5. Role of lipids in cell function

There are at least two ways by which lipids can affect protein structure and function and thereby cell function. Protein function is influenced by specific protein–lipid interactions that are dependent on the chemical and structural anatomy of lipids (headgroup, backbone, alkyl chain length, degree of unsaturation, chirality, ionization and chelating properties). However, protein function is also influenced by the unique self-association properties of lipids that result from the collective properties (fluidity, thickness, shape, packing properties) of the lipids organized into membrane structures.

## 5.1. The bilayer as a supramolecular lipid matrix

Biophysical studies on membrane lipids coupled with biochemical and genetic manipulation of membrane lipid composition have established that the $L_\alpha$ state of the membrane bilayer is essential for cell viability. However, membranes are made up of a vast array of lipids that have different physical properties, can assume individually different physical arrangements, and contribute collectively to the final physical properties of the membrane. Animal cell membranes are exposed to a rather constant temperature, pressure, and solvent environment and therefore do not change their lipid make up dramatically. The complex membrane lipid composition including cholesterol stabilizes mammalian cell membranes in the $L_\alpha$ phase over the variation in conditions they encounter. Microorganisms are exposed to a broad range of environmental conditions so have developed systems for changing membrane lipid composition in order to exist in the $L_\alpha$ phase. Yet all biological membranes contain significant amounts of non-bilayer-forming lipids.

### 5.1.1. Physical organization of the bilayer

As the growth temperature of *E. coli* is lowered, the content of unsaturated fatty acids in phospholipids increases. Genetic manipulation of phospholipid fatty acid composition in *E. coli* is possible by introducing mutations in genes required for the synthesis of unsaturated fatty acids (Chapter 3). The mutants require supplementation of unsaturated fatty acids from the growth medium and incorporate these fatty acids to adjust membrane fluidity in response to growth temperature. Mutants grown at low temperature with high unsaturated fatty acid content lyse when raised rapidly to high temperature probably due to the increased membrane permeability of fluid membranes and a transition from the $L_\alpha$ to $H_{II}$ phase of the lipid bilayer. Conversely, growth at high temperatures with high saturated fatty acid content results in growth arrest after a shift to low temperature due to the reduced fluidity of the membrane. Wild-type cells, that do not normally contain such extremes in fatty acid content as can be generated with mutants, arrest growth after a temperature shift until fatty acid composition is adjusted to provide a favorable membrane fluidity.

Bacterial cells also regulate the ratio of bilayer to non-bilayer-forming lipids in response to growth conditions [2]. Bacterial non-bilayer-forming lipids are PE with unsaturated alkyl chains, CL in the presence of divalent cations, and monoglucosyl diacylglycerol (MGDG). Extensive studies of lipid polymorphism have been carried out on *Acholeplasma laidlawii* because this organism alters its ratio of MGDG (capable of assuming the $H_{II}$ phase) to DGDG (diglucosyl diacylglycerol, which only assumes the $L_\alpha$ or $L_\beta$ phase) in response to growth conditions. High temperature and unsaturation in the fatty acids favor the $H_{II}$ phase for MGDG. At a given growth temperature, the MGDG to DGDG ratio is inversely proportional to the unsaturated fatty acid content of MGDG. As growth temperature is lowered, *A. laidlawii* either increases the incorporation into MGDG of unsaturated fatty acids from the medium or increases the ratio of MGDG to DGDG to adjust the $H_{II}$ phase potential of its lipids to remain just below the transition from bilayer to non-bilayer. Therefore, the cell maintains the physical properties of the membrane well within that of the $L_\alpha$ phase but with a constant potential to undergo transition to the $H_{II}$ phase.

Contrary to *A. laidlawii*, *E. coli* maintains its non-bilayer lipids, CL (in the presence of divalent cations) and PE, within a narrow range and in wild-type cells adjusts the fatty acid content of PE to increase or decrease its non-bilayer potential. The unsaturated fatty acid content of inner membrane PE is higher than that of the PE on the inner leaflet of the outer membrane (which is 90% PE). The result is that the $L_\alpha$ to $H_{II}$ transition for the inner membrane pool is only 10–15°C above the normal growth temperature of 37°C while this transition for the outer membrane phospholipids is 10°C higher than the inner membrane phospholipids. This increased potential for the inner membrane lipids to form non-bilayer structures is believed to be biologically significant to the function of the inner membrane. In mutants completely lacking PE, the role of non-bilayer lipid appears to be filled by CL. The growth defect of mutants lacking PE is suppressed by divalent cations in the same order of effectiveness ($Ca^{2+} < Mg^{2+} < Sr^{2+}$) as these ions induce the formation of non-bilayer phases of CL. Neither growth of the mutant nor the $H_{II}$ phase for CL is supported by $Ba^{2+}$. The CL content of these mutants varies with the divalent cation used during growth. However, the $L_\alpha$ to $H_{II}$ transition for the extracted lipids (in the presence of the divalent cation) is always the same as that of lipids from wild-type cells (containing PE) grown in the absence of divalent cations. Therefore, even though *E. coli* normally does not alter its PE or CL content to adjust the physical properties of the membrane, these mutants are able to adjust CL levels to maintain the optimal physical properties of the membrane bilayer.

### 5.1.2. Biological importance of non-bilayer lipids

It is obvious why prevention of formation of large amounts of non-bilayer phase would be important to maintaining cell integrity. However, why is there a need for non-bilayer-forming lipids? There are numerous biological processes that can be envisioned to require discontinuity in the membrane bilayer. Integration of proteins into the bilayer might require 'annular lipids' (those in close proximity to the protein) to interface with the more regular structure of the bilayer. Movement of proteins or other macromolecules through the bilayer might also require such discontinuity. The process of membrane vesicle fusion and fission requires a transition state that is not bilayer in nature. Finally, the tension resulting from the pulling apart of the two halves of the bilayer induced by either one or both monolayers containing non-bilayer lipids may be of biological importance.

Since cells homeostatically adjust their mixture of bilayer and non-bilayer-forming lipids, some proteins must be sensitive to the intrinsic curvature of the composite membrane lipids. There is a correlation between the spontaneous curvature of the membrane and the performance of embedded proteins [14]. Protein kinase C is a peripheral membrane protein that binds to the membrane and is activated by a complex of PS (probably at least six molecules), one molecule of diacylglycerol, and one molecule of $Ca^{+2}$. In the presence of diacylglycerol protein kinase C is highly specific for PS but in the absence of diacylglycerol the kinase will bind to any anionic lipid. Stereoselectivity for the 1,2-diacyl-sn-glycerol is not absolute, but protein kinase C is stereospecific for the natural L-serine isomer of PS independent of whether or not other non-bilayer-forming lipids are present. However, this stereoselectivity appears to be related to the fact that in the presence of $Ca^{2+}$ the natural isomer of PS undergoes the $L_\alpha$ to $H_{II}$ transition at a lower temperature than the D-serine isomer [15]. Diacylglycerol is highly non-bilayer promoting and might selectively partition to a non-bilayer domain formed by the natural isomer of PS. The specific interaction of these two lipids may provide the unique allosteric switch regulating protein kinase C activity.

Phospholipase C activity is not directly influenced by the formation of non-bilayer structures. However, the presence of lipids (e.g. PE) with a tendency to form such structures, stimulates the enzyme even under conditions at which purely bilayer phases exist. Conversely sphingomyelin, a well-known stabilizer of the bilayer phase, inhibits the enzyme. Thus phospholipase C appears to be regulated by the overall geometry and composition of the bilayer [16] supporting the hypothesis that the collective physical properties of the lipid bilayer can modulate the activities of membrane-associated proteins.

Therefore, it is not always clear from initial studies which property of lipids, i.e. chemical or physical, is required for optimum function. The complex interplay between chemical and physical properties of lipids exemplifies the difficulty in understanding how lipids affect biological processes at the molecular level.

## 5.2. Selectivity of protein–lipid interactions

A specific phospholipid requirement has been determined for optimal reconstitution of function in vitro for more than 50 membrane proteins. If one considers specific lipid requirements for membrane association and activation of peripheral membrane proteins, the number is in the hundreds. Integral membrane proteins fold and exist in a very complex environment and have three modes of interaction with their environment. The extramembrane parts are exposed to the water milieu, where they interact with water, solutes, ions and water-soluble proteins. Part of the protein is exposed to the hydrophobic–aqueous interface region (see Fig. 9). The remainder of the protein is buried within the approximately 30 Å thick hydrophobic interior of the membrane.

Peripheral membrane proteins may spend part of their time completely in the cytosol and are recruited to the membrane surface, or even partially inserted into the membrane, in response to various signals.

Much of what is known about these protein–lipid interactions has come from protein purification and reconstitution of function dependent on lipids. Genetic approaches coupled with in vitro verification of function has uncovered new roles for lipids. Most exciting have been recent results from X-ray crystallographic analysis of membrane proteins which have revealed lipids in specific and tight association with proteins [17]. The predominant structural motif for the membrane spanning domain of membrane proteins is an α-helix of 20–25 amino acids which is sufficient to span the 30 Å core of the bilayer. A β-barrel motif is also found to a lesser extent.

### 5.2.1. Lipid association with α-helical proteins

CL is found aligned with a high degree of structural complementarity within a high-resolution structure of the light harvesting photosynthetic reaction center from *Rhodobacter sphaeroides*. The head group of CL is located on the surface of the reaction center, is in close contact with residues from all three of the reaction center subunits, and is engaged in hydrogen bond interactions with polar residues in the membrane interfacial region (at the cytoplasmic side of the membrane) (Fig. 10A). Bonding interactions between CL and the protein involve either direct contacts of the phosphate oxygens of the lipid headgroup with basic amino acids and backbone amide groups exposed at the protein surface, or indirect contacts with amino acid side chains that are mediated by bound water molecules. A striking observation was that the acyl chains of CL lie along grooves in the α-helices that form the hydrophobic surface of the protein and are restricted in movement by van der Waals interactions. A PE molecule was resolved in the X-ray structure of the photosynthetic reaction center from *Thermochromatium tepidum*. The phosphate group of PE is bound to Arg and Lys by electrostatic interaction, and to Tyr and Gly by hydrogen bonds. PE acyl chains fit into the hydrophobic clefts formed between α-helices of three different subunits of the complex.

Bacteriorhodopsin is a light-driven ion pump that is found in the purple membrane of the archaebacterium *Halobacterium salinarum*. Bacteriorhodopsin monomers consist of a bundle of seven transmembrane α-helices that are connected by short inter-helical loops, and enclose a molecule of retinal that is buried in the protein interior, approximately half way across the membrane. Proton pumping by bacteriorhodopsin is linked to photoisomerization of the retinal and conformational changes in the protein, in a series of changes called a photochemical cycle. Specific lipids can influence the steps in this cycle. A combination of squalene (a reduced isoprenoid) and the methyl ester of phosphatidylglycerophosphate is required to maintain normal photochemical cycle behavior. In a high-resolution (1.55 Å) structure of bacteriorhodopsin, 18 full or partial lipid acyl chains per monomer were resolved (Fig. 10C), four pairs of which were linked with a glycerol backbone to form diether lipids identified as native archaeol-based lipids. One of the lipid alkyl chains buried in the center of the membrane appears to be squalene. Lipid chains were also observed in the hydrophobic crevices between the ends of the monomers in the trimeric structure and probable hold the complex together.

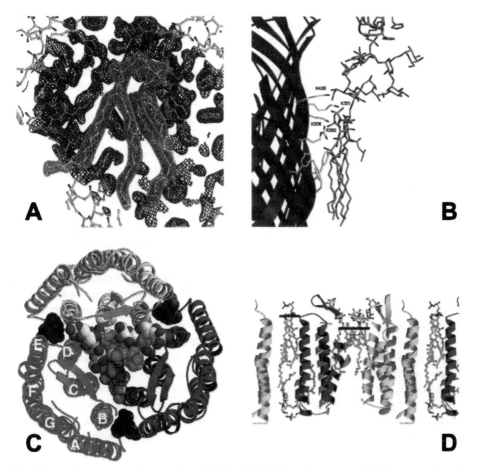

Fig. 10. Atomic structure of protein bound lipids. (A) Model of CL (green) tightly bound to the surface of photosynthetic reaction center (blue) from *R. sphaeroides*. The space-fill model was derived from X-ray crystallographic data that resolved between 9 and 15 carbons of the acyl chains of CL. *R. sphaeroides* contains mainly $18:0$ and $18:1$ fatty acids. Figure adapted with permission from McAuley et al. [18]. Copyright 1999 National Academy of Sciences, USA. (B) Crystal structure of FhuA complexed with lipopolysaccharide. The ribbon structure (blue) represents the outside surface of the β barrel of FhuA with extended chains (yellow) of amino acids. The amino acids of the aromatic belt interact with the acyl chains (gray) and the basic amino acids interact with the phosphate (green and red) groups of Lipid A. The remainder of the lipopolysaccharide structure extends upward into the periplasm. Figure adapted with permission from Ferguson et al. [19]. Copyright 1999 Elsevier Science Ltd. (C) Lipid packing in crystals of bacteriorhodopsin. Top view of the trimer in three different colors (domains A–E noted in one of the trimers) in complex with lipid (space-fill models) as viewed from the extracellular side. Three phytanyl chains of lipid (gray) lie in the crevices formed between the A and B domain of one monomer and the D and E domain of the adjacent monomer. The central core of the trimer is filled with a sulfated triglycoside attached to archaeol (glycolipid S-TGA-1). Red denotes the oxygen molecules of the sugars in white. (D) Cross section of the lipid bilayer showing phytanyl archaeol molecules (blue hydrophobic domain with red and white for the headgroups) extending from both sides of the bilayer and interacting with protein monomers (red and yellow helices) of bacteriorhodopsin. In the central core of the protein are two glycolipid S-TGA-1 molecules on the extracellular side of the membrane with a the gap on the cytosolic side of the membrane. The black lines indicate the boundary of the interface region. Figure adapted with permission from Essen et al. [20]. Copyright 1998 National Academy of Sciences, USA.

This organization explains the requirement for the natural archaeol lipids to maintain structure and function of the protein. The X-ray data demonstrate a good match between the hydrophobic face of membrane proteins and selective highly ordered surrounding lipids. The positioning of two glycolipid S-TGA-1 molecules to the extracellular side of the central hole in the bacteriorhodopsin trimer (Fig. 10D) results in a 5 Å 'membrane thinning' relative to the surrounding bilayer. This may cause a steeper electric field gradient across the central core than in the bulk lipid phase that might affect the proton pathway in bacteriorhodopsin.

### 5.2.2. Lipid association with β-barrel proteins

The pore-forming proteins of the outer membrane of *E. coli* are organized as antiparallel β chains forming a barrel structure with an aqueous pore on one side and an interface with the membrane bilayer on the other (Fig. 10B). X-ray crystal structure of *E. coli* outer membrane ferric hydroxamate uptake receptor (FhuA) contains a bound lipopolysaccharide in 1 : 1 stoichiometric amounts. The acyl chains of the lipopolysaccharide are ordered on the protein surface approximately parallel to the axis of the β-barrel along the half of the hydrophobic belt oriented toward the extracellular surface of the outer membrane. Numerous van der Waals interactions with surface-exposed hydrophobic residues are observed. The large polar headgroup of lipopolysaccharide makes extensive interactions with charged and polar residues of the protein near the outer surface of the membrane. Most of the favorable interactions are contributed by a cluster of 8 positively charged residues on the surface of the barrel, which interact by hydrogen bonding at distances around 3 Å or electrostatically at longer distances. In the interface region of the membrane there are clusters of aromatic amino acid residues positioned as belts around the protein. Similar organization of aromatic amino acids near the membrane interface region has been observed in other membrane proteins and may be involved in π-bonding interactions with the headgroups of lipids.

### 5.2.3. Organization of protein complexes

Rather than being associated with the exterior surface of membrane proteins, many phospholipids, particularly anionic ones like PG and CL, are found wedged between the subunits of monotropic and heterotropic oligomeric complexes. Anionic phospholipids have a particularly important function in energy-transducing membranes such as the bacterial cytoplasmic membrane and the inner mitochondrial membrane. In particular, CL has been shown to be a key factor in the maintenance of the optimal activity of the major integral proteins of the inner mitochondrial membrane, including NADH dehydrogenase, the cytochrome $bc_1$ complex, ATP synthase, cytochrome $c$ oxidase, and the ATP/ADP translocase.

The crystal structure of bovine cytochrome $c$ oxidase has been determined to a resolution of 2.3 Å. This integral membrane protein (a homodimer with 13 different subunits in each monomer) is responsible for the reduction of molecular oxygen to water during aerobic respiration, with concomitant proton pumping across the mitochondrial inner membrane. Several lipids were resolved in this structure: three PEs, seven PGs, one PC and two CLs. One of the CLs is located at the interface between the two monomers of the dimer and the remaining phospholipids are located between subunits within each monomer. The resolution of CL in the structure of the bovine cytochrome *c* oxidase is particularly intriguing because it is well documented that CL cannot be removed from cytochrome *c* oxidase without a loss of enzyme activity.

In the above case, specific lipids mediate protein–protein contacts within a multimeric complex and may be very important for structural and functional integrity of complex membrane proteins. The advantage of using lipid molecules to form a significant part of the contact surface between adjacent protein subunits is that they have a high degree of conformational flexibility, and are usually available in a range of molecular shapes and sizes. Using lipids as interface material reduces the need for highly complementary protein–protein interactions and provides for flexible interactions between subunits.

### 5.2.4. Binding sites for peripheral membrane proteins

A common mechanism of cellular regulation is to organize functional complexes on demand from existing components. In many cases, the components are initially distributed between membranes and the cytosol. Post-translational modification of a protein, or the appearance of a cytoplasmic or membrane component, signals the organization of a functional complex on the membrane surface. There are many examples of primarily anionic phospholipids being either the signals or the organization site for such protein complexes at the membrane surface. Three structure-specific domains have been recognized, mostly in eukaryotes, that serve as specific lipid binding domains.

The C1 lipid clamp is a conserved cysteine-rich protein domain that binds lipids and is found in protein kinases C and other enzymes regulated by the second messenger diacylglycerol. This receptor domain interacts with one molecule of diacylglycerol and recruits the protein kinase C to specific membrane sites [21]. The C1 domain adopts a β-sheet structure with an open cavity.

The C2 domain generally binds anionic phospholipids such as PS in a $Ca^{2+}$-dependent manner and is conserved among phospholipase C, phospholipase $A_2$, PI-3-phosphate kinase and calcium-dependent protein kinase C [22]. The crystal structure of C2 domain of protein kinase Cα in complex with PS reveals that the recognition of PS involves a direct interaction with two $Ca^{2+}$ ions.

The PH, or pleckstrin homology, domain is shared by protein kinase Cβ and some phospholipases C. This domain is responsible for associating peripheral membrane proteins with the membrane via the phosphoinositide head group of polyphosphorylated PIs in an enantiomer-specific manner [23]. PH domains consist of 7-stranded β-sheets with positively charge pockets that attract the negatively charged PI head group.

In prokaryotic cells, the protein structural features defining lipid-binding domains is less well conserved than in eukaryotes, and the membrane ligand appears to be an anionic lipid-rich domain with little selectivity for the chemical species of lipid. DnaA protein and SecA protein (see Section 5.3) are peripheral membrane proteins in *E. coli* that carry out different functions but become membrane associated and activated by similar mechanisms. The involvement of anionic lipids in the function of these proteins was discovered through the use of *E. coli* mutants in which anionic lipid content could be controlled [2]. DnaA protein is required for initiation of DNA replication and is active in its ATP- but not ADP-bound form. In vitro, the exchange of ATP for ADP in the complex is greatly stimulated by almost any anionic phospholipid including non-*E.*

*coli* lipids like PI. In vivo DnaA function is compromised in mutants with reduced PG/CL levels. This phenotype can be suppressed by mutations that bypass the need for DnaA. An anionic specific membrane binding domain has been identified that appears to direct initial membrane association followed by partial insertion of the protein into the bilayer. The resulting conformational changes alter the ATP/ADP binding properties.

What was once thought to be a specific interaction of SecA and DnaA with either PG or CL, is actually an interaction with an anionic surface charge on the membrane. Mutants completely lacking PG and CL but with highly elevated levels of phosphatidic acid still initiate DNA replication (DnaA) and export protein (SecA) [10]. Both of these proteins can be activated in in vitro reconstituted systems with a wide range of anionic lipids including those not found in *E. coli*. It appears that both of these proteins recognize, via positively charged amphipathic helices, clusters or domains of negative charge rather than specific lipids on the membrane surface.

CTP: phosphocholine cytidylyltransferase is responsible for the synthesis of CDP-choline, an early precursor to PC synthesis in mammalian cells (Chapter 8). The enzyme has affinity for membranes depleted of PC that leads to activation of the enzyme and increased synthesis of PC. A complex mixture of factors including anionic lipids and non-bilayer-forming lipids stimulates membrane association. When binding occurs via two positively charged amphipathic helices, a large structural change occurs leading to enzyme activation. Affinity for an anionic membrane surface is understandable, but the role of other lipids such as PE and diacylglycerol in activation and the negative effect of fatty acids have only been recently clarified. The former two non-bilayer-forming lipids induce a negative curvature of the two halves of the bilayer toward the aqueous domain and away from the hydrophobic domain. Incorporation of free fatty acids (micelle forming) has the opposite effect. Surface association of the amphipathic helices would be favored by the decrease in hydration of the interfacial region due to the induced negative curvature, and penetration into the surface of the bilayer would reduce the stress imposed by negative curvature [24].

## 5.3. Translocation of proteins across membranes

Movement of proteins across membranes is required to transfer a protein from its site of synthesis to its site of function. The process involves the transfer of hydrophobic and hydrophilic segments of proteins through the hydrophobic core of the membrane. The role of lipids in this process has only recently received considerable attention. The in vivo evidence for the participation of anionic phospholipids in protein translocation was obtained from experiments with *E. coli* mutant strains defective in the biosynthesis of PG and CL [2]. The in vivo translocation of the outer membrane precursor proteins, prePhoE and proOmpA, was severely hampered in these cells. When the expression of the *pgsA* gene (encodes phosphatidylglycerophosphate synthase) placed under control of the *lac* promoter/operator was used to fine-tune the level of PG in the membrane, the translocation rate of the proteins was directly proportional to the amount of PG. The molecular basis for this anionic lipid requirement is for the function of SecA. SecA is a peripheral membrane protein that acts as a translocation ATP-driven motor which moves secreted proteins through the membrane translocation pore composed of SecY and two other membrane proteins, SecE and SecG. SecA requires both anionic phospholipids and pore component SecY for high affinity binding to the membrane, for membrane penetration, and for high level ATPase-dependent function. Functional reconstitution of purified and delipidated SecYEG complex from *E. coli* and *Bacillus subtilis* into liposomes of defined lipid composition revealed an absolute requirement for PG [25]. Translocation activity was proportional to the amount of PG in reconstituted proteoliposomes and optimum activity was obtained only with the specific lipid composition of each organism.

The N-terminal signal peptides of *E. coli* secreted proteins possess at least one positively charged amino acid. Protein translocation efficiency is dependent on both the number of positive charges and the anionic phospholipid content of the membrane. Photocrosslinking of the secreted proteins with the fatty acid chains of membrane lipids demonstrated direct contact between the signal peptide and lipids during early stages of protein translocation.

Non-bilayer-forming lipids are also required for protein translocation of proteins across the membrane of *E. coli*. The only non-bilayer-forming lipid in *E. coli* mutants lacking PE is CL in the presence of $Mg^{2+}$ or $Ca^{2+}$. Protein translocation into inverted membrane vesicles prepared from PE-deficient cells is reduced with divalent cation-depletion but can be enhanced by inclusion of $Mg^{2+}$ or $Ca^{2+}$ [2]. Translocation in the absence of divalent cations is restored by incorporation of non-bilayer PE (18 : 1 acyl chains) but not by bilayer-prone PE (14 : 0 acyl chains). These results would indicate that lipids with a tendency to form non-bilayer structures provide a necessary environment for the translocase of proteins across the membrane.

## 5.4. Assembly of integral membrane proteins

Much less is known about the role of phospholipids in insertion and organization of integral membrane proteins. Most such proteins are organized with several α-helical transmembrane domains spanning the membrane bilayer. These helices are connected by extramembrane loops alternately exposed on either side of the membrane. In bacteria these proteins generally obey the 'positive inside' rule in which those extramembrane loops on the cytoplasmic side of the membrane have a net positive charge and the loops exposed to the exterior are either neutral or negatively charged. Variation in the positive charge density of these loops as well as the anionic phospholipid content of the membrane indicates that these positive loops are anchored to the inside of the cell by interaction with anionic phospholipids.

### 5.4.1. Lipid-assisted folding of membrane proteins
The membrane clearly serves as the solvent within which integral membrane proteins fold and function [9]. However, do lipids act in more specific ways to guide and determine final membrane protein structure and organization? Recent evidence supports a role for lipids analogous to that of protein molecular chaperones in the folding of membrane proteins. Molecular chaperones facilitate the folding of substrate proteins by interacting with non-native folding intermediates but do not interact with native or totally unfolded proteins, and are not required to maintain native conformation. Lipids

that fulfill these requirements in assisting the folding of specific membrane proteins have been termed 'lipochaperones' to distinguish their function from simply providing a solvent for the folding process.

The major evidence for the existence of lipochaperones comes from studies on the requirement for PE in the assembly and function of lactose permease (LacY) of *E. coli* [26]. LacY is a polytopic membrane protein with 12 transmembrane-spanning domains connected by alternating cytoplasmic and periplasmic loops (Fig. 11). LacY carries out transport of lactose either in an energy-independent mode to equilibrate lactose across the membrane (facilitated transport) or by coupling uphill movement of lactose against a concentration gradient with downhill movement of a proton coupled to the proton electrochemical gradient across the membrane (active transport). Uncovering a lipochaperone role for PE in the assembly of LacY came about by the fortuitous availability of reagents and techniques. A large number of biochemical and molecular genetic resources are available for studying LacY including antibodies and molecularly engineered derivatives of LacY. The availability of viable *E. coli* strains that either lack PE, or in which the level of PE can be regulated, provided reagents to study the requirement for PE in the assembly of LacY in vivo and in isolated membranes. The development of a blotting technique termed an 'Eastern–Western' made possible the screening for lipids affecting the refolding of LacY in vitro or the conformation of LacY made in vivo.

In the Eastern–Western procedure, lipids are first applied to a solid support such as nitrocellulose. Next, proteins subjected to sodium dodecyl sulfate polyacrylamide gel electrophoresis are transferred by standard Western blotting techniques to the solid support in such a manner that the protein of interest is transferred to the lipid patch. During electrotransfer of protein to the solid support, protein, lipid, and sodium dodecyl sulfate mix and as transfer continues the sodium dodecyl sulfate is removed leaving behind the protein to refold in the presence of lipid. Attachment of the refolded protein to a solid support allows one to probe protein structure using conformation-sensitive antibodies or protein function by direct assay. This combined blotting technique allows the detection of membrane protein conformational changes as influenced by individual lipids during refolding. Since many membrane proteins including LacY retain significant amounts of secondary structure even in sodium dodecyl sulfate, refolding from sodium dodecyl sulfate detects intermediate and late steps of folding.

The initial observation that PE was required for LacY function was concluded from studies of reconstitution of transport function in sealed vesicles made of purified LacY and lipid. When reconstituted in lipid mixtures containing PE, both active and facilitated transport was restored. In mixtures containing only PG and/or CL or even PC only facilitated transport occurred. The physiological importance of PE for LacY function was established using mutants lacking PE. LacY expressed in PE-containing cells had full transport function, but cells lacking PE only displayed facilitated transport even though bioenergetic parameters of the membrane were normal. Using Western and Eastern–Western blotting techniques and a conformation sensitive antibody, it was established that LacY assembled in the presence, but not in the absence, of PE displays 'native' structure. LacY maintains its native structure even when PE is completely removed, and LacY originally assembled in the absence of PE is restored to native structure by partial denaturation in sodium dodecyl sulfate followed by renaturation in

**A**

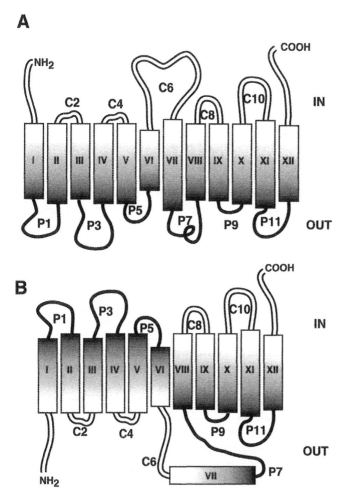

Fig. 11. Topological organization of LacY in the *E. coli* inner membrane. (A) The topological organization of LacY when assembled in membranes containing PE (normal configuration). The 12 hydrophobic membrane spanning α-helices are numbered in Roman numerals from the amino (NH₂) to the carboxyl (COOH) terminus. The even-numbered hydrophilic loops ('C' and open) connect the transmembrane domains on the cytoplasmic side (IN) of the membrane. The odd-numbered hydrophilic loops ('P' and filled) connect the transmembrane domains on the periplasmic side (OUT) of the membrane. (B) The topological organization of lactose permease when assembled in the absence of PE. Note the first six transmembrane domains along with their connected hydrophilic loops are inverted with respect to the plane of the membrane when compared to A. Transmembrane domain VII is very hydrophilic and is postulated to reside outside of the membrane in the altered structure.

the presence of specifically PE (or other primary amine-containing phospholipids such as PS). LacY assembled (either in vivo or in vitro) in membranes lacking PE is restored to native structure by post-assembly addition of PE to the membranes. Furthermore, LacY extensively denatured in urea–sodium dodecyl sulfate (which eliminates most secondary structure) cannot be renatured by simple exposure to PE. Taken together these data strongly suggest that PE assists in the folding of LacY by a transient non-covalent interaction with a late-folding and non-native intermediate thereby fulfilling the minimum requirements of a molecular chaperone [26].

The molecular basis for the loss of native structure and function of LacY assembled in the absence of PE is a topological mis-assembly of the protein [27]. In the absence of PE the first six transmembrane domains and the loops that connect them assume an inverted topology with respect to the plane of the membrane bilayer (Fig. 11). Cytoplasmic loops become periplasmic, and periplasmic loops become cytoplasmic. If LacY is first assembled in the absence of PE and then post-assembly PE synthesis is initiated, there is correction of topology and regain of full transport function. These results dramatically illustrate the specific effects of membrane lipid composition on both structure and function of membrane proteins. The ability of changes in lipid composition to effect such large changes in protein structure has important implications for regulatory roles of lipids in cell processes. For example, as eukaryotic proteins move through the secretory pathway, they encounter different membrane lipid compositions that might affect protein structure in dramatic ways to turn on or turn off function. Local changes in lipid composition may also result in similar changes in structure and function.

### 5.4.2. Scope of lipochaperone function
Secondary transport proteins (those that couple active transport to proton uptake) in *E. coli* show high overall structural homology to LacY. Amino acid permeases specific for Phe, Lys, Pro, and aromatic amino acids are defective in active transport in cells lacking PE. The melibiose permease requires PE in order to reconstitute transport function in vitro. Therefore, the lipochaperone function of PE may be a general requirement for proper assembly of this family of secondary transport proteins. A lipochaperone requirement in *E. coli* has been reported for the proper folding of the β-barrel outer membrane protein PhoE. During passage of the protein through the inner membrane of *E. coli*, interaction with precursors to lipopolysaccharide is required to obtain assembly competent protein upon arrival at the outer membrane [28]. Therefore, a role for lipids to act in membrane protein folding in a more specific manner than providing the solvent has been established.

### 5.5. Lipid domains

Compartmentalization of many biological processes such as biosynthesis, degradation, energy production, and metabolic signaling plays an important role in cell function. Subcellular organelles, multiple membrane structures, cytosol versus membrane are all utilized to compartmentalize functions. The original fluid mosaic model envisioned the membrane bilayer as a homogenous sea of lipid into which proteins are dispersed (Fig. 1). The current view of biological membranes is that they contain microdomains of different lipid and protein composition than is reflected by the whole membrane and that these domains serve to further compartmentalize cellular processes.

Lipid mixtures made up of defined lipids undergo phase separations due to lipid polymorphism and differences in steric packing of the acyl chains. Mixtures of bilayer and non-bilayer lipids undergo multiple phase transitions supporting the existence of segregated domains within the bilayer. In model systems amphipathic polar lipid analogues self associate into domains if their hydrophobic domains are the same even if their polar domains carry the same net charge. Therefore, headgroup repulsive forces can be overcome by orderly packing of the hydrophobic domains. There is considerable acyl chain mismatch between phospholipids and sphingolipids (see Chapter 14 for structures), i.e., phospholipids tend to have shorter acyl chains (16–18) with higher degrees of unsaturation as compared to the longer (20–24 for the acyl group) saturated chains of sphingolipids. Naturally occurring sphingolipids undergo the $L_\beta$ to $L_\alpha$ transition near the physiological temperature of 37°C while this transition for naturally occurring phospholipids is near or below 0°C. Therefore, the more laterally compact hydrophobic domains of sphingolipids can readily segregate from the more disordered and expanded domains of unsaturated acyl chains of phospholipids. Lipid segregation can also be facilitated by specific polar headgroup interactions, particularly intermolecular hydrogen bonding to other lipids and to protein networks involving hydroxyls, phosphates, amines, carbohydrates, and alcohols. The hydrogen bonding properties of CL due to its high $pK_2$, as noted earlier, may be the basis for the formation of clusters of CL in natural and artificial membranes [29].

### 5.5.1. Lipid rafts

Lipid rafts are liquid ordered phases of cholesterol, glycosphingolipids (gangliosides), sphingomyelin, and proteins that exist as microdomains within the more dispersed liquid crystalline bilayer. Lipid rafts are operationally defined as the membrane fraction of eukaryotic cells that is resistant to solubilization in the cold by the detergent Triton X-100 (detergent-resistant membrane fraction). This fraction is greatly enriched in cholesterol, glycosphingolipids, sphingomyelin and a subset of membrane proteins. The proteins co-clustered in lipid rafts are soluble globular proteins tethered to the raft lipids via covalent linkage to fatty acids, cholesterol, or phosphatidylinositol (Chapter 2) [30]. The latter glycosylphosphatidylinositol-linked proteins are attached directly to the amino group of ethanolamine phosphate which in turn is linked to a trisaccharide and then to the inositol of PI (Fig. 12). The sphingolipids and glycosylphosphatidylinositol-linked proteins occupy the outer surface monolayer of the plasma membrane bilayer, and the acyl chains of these lipids are generally more saturated and longer than those of the plasma membrane phospholipids. The similarity in the structure of the more ordered hydrophobic domains of the raft lipids and their dissimilarity with the surrounding more fluid phospholipids favor a self-association of the raft lipids and the glycosylphosphatidylinositol-linked proteins. The hydrogen-bonding properties of the glycosphingolipids with themselves and with the constituents of the glycosylphosphatidylinositol-linked proteins stabilizes the complexes. Finally, the planar shape of cholesterol favors its intercalation parallel to the ordered acyl chains of the raft lipids with its single hydroxyl group facing the surface. The stability of this structure appears to be why it is not dissipated by detergent extraction.

Lipid rafts appear to be a mechanism to compartmentalize various processes on the cell surface by bringing together various receptor-mediated and signal transduction processes. A general phenomenon is that when glycosylphosphatidylinositol-linked proteins aggregate on the surface, they also become enriched in the detergent-resistant

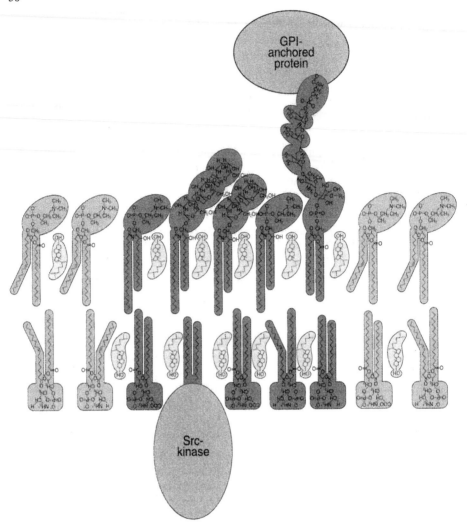

Fig. 12. Model of lipid raft. A glycosylphosphatidylinositol-linked protein is attached to the exterior monolayer of the membrane and a Src-kinase to the interior monolayer of the membrane by their respective covalently attached lipids. The mechanism for clustering and coupling Src-kinase with a glycosylphosphatidylinositol-linked protein is hypothetical. Clustered (dark gray) around the glycosylphosphatidylinositol are ordered (straight alkyl chains) glycosphingolipids, sphingomyelin, and PC with intercalated cholesterol. The phospholipids with kinked (unsaturated) chains indicate the more disordered liquid crystalline state of the surrounding bilayer. Reprinted (abstracted/excerpted) with permission from Simons and Ikonen [30]. Copyright 2000 American Association for the Advancement of Science.

membrane fraction as well as phosphorylated. For example, clustering of the IgE receptor, FcεRI, by binding its ligand on the cell surface results in its phosphorylation by a Src-family kinase that activates a signaling cascade. Only the receptor associated with the detergent-resistant membrane fraction is phosphorylated. The Src-family of kinases is believed to localize to lipid rafts from the cytosolic side of the membrane via covalently attached fatty acids that insert into the membrane (Fig. 12).

A class of lipid domains related to rafts are caveolae which are invaginations on the surface of the plasma membrane of eukaryotic cells (Chapter 20). They contain the protein caveolin which when complexed with cholesterol forms a hairpin structure believed to induce curvature in the bilayer to form the caveolae. Caveolin is an integral part of the intracellular cholesterol transport machinery and caveolae are involved in regulating influx and efflux of cellular cholesterol. Caveolae are also resistant to detergent solubilization and also contain several proteins responsible for signal transduction suggesting a regulatory role for these structures in response to cholesterol.

The existence of lipid rafts and their function is still an evolving area. Isolation and characterization of detergent-resistant membrane fractions, studies in model systems, and studies with whole cells all support the concept of lipid rafts. However, they have never been observed in the native state, presumably due to their small size and possibly their loose association and dynamic properties. The raft components may normally have low affinity for each other, but increase their affinity when clustering is induced by other processes. It still is not clear how glycosylphosphatidylinositol-linked proteins on the exterior of the cell interact with acylated proteins that appear to be associated with rafts on the interior surface of the membrane.

### 5.5.2. Lipid domains in bacteria

One of the few examples of lipid domains being observed in living cells comes from the staining of CL in the membrane of E. coli by the fluorescent dye 10-N-nonyl acridine orange [31]. This reagent is highly specific for CL and has little or no interaction with other lipids or cell components. In wild-type cells, fluorescence is observed at the poles of the cell and at the division septum. In filamentous cells of E. coli with multiple genomes distributed along their length, the fluorescence is localized between the genomes. These fluorescence-enriched domains may be due either to regions enriched in CL (or possibly anionic lipids including CL) or areas enriched in lipid relative to protein. Initiation of DNA replication and the formation of the septal domain of E. coli have been postulated to be dependent on anionic phospholipids that would be consistent with the localization of CL domains.

### 5.6. Cytokinesis

The function of cytokinesis is to divide one cell into two by building a membrane barrier between the two daughter cells. In eukaryotic cells, the interaction of actin filaments with myosin filaments applies tension to the membrane to form a cleavage furrow, which gradually deepens until it encounters the narrow remains of the mitotic spindle between the two nuclei. Phospholipids play an essential role in the division processes in eukaryotic cells.

Phospholipids in biological membranes are distributed asymmetrically between the inner and outer leaflets of the lipid bilayer [32]. In the plasma membrane of eukaryotic cells PE and PS are localized to the inner leaflet and PC and sphingomyelin are enriched in the outer leaflet. Using a cyclic peptide highly specific for binding to PE, it was demonstrated that PE is exposed on the cell surface of the cleavage furrow of eukaryotic cells at the final stage of cytokinesis. Immobilization of cell surface PE by the PE-binding peptide inhibited disassembly of the contractile ring resulting in formation of a long cytoplasmic bridge between the daughter cells. Removal of the peptide from the surface of arrested cells allowed cell division to proceed with disappearance of exposed PE. Furthermore, a mutant cell line defective in PE biosynthesis was isolated as a variant resistant to the cytotoxicity of the PE-binding peptide [33]. This cell line required either PE or ethanolamine for normal growth and cell division. In ethanolamine-deficient medium these mutant cells arrested with a cytoplasmic bridge between the two daughter cells. Addition of PE or ethanolamine restored normal cytokinesis. These findings provide the evidence that transbilayer movement of PE at the cleavage furrow contributes to regulation of cytokinesis.

In *E. coli*, cell division is initiated after genome duplication by organization of FtsZ protein monomers midway between the poles of the cell. This protein ring is the scaffold that recruits a series of proteins to the division site that brings about constriction and eventually cell division. An *E. coli* mutant completely lacking PE propagates as long filamentous cells. The FtsZ protein complex is recruited to the division site, but the FtsZ ring fails to constrict [34]. This phenotype is not observed in strains with specific defects in other steps of phospholipid biosynthesis. Although not firmly documented, prokaryotic membranes also appear to have an asymmetric enrichment of PE on the inner leaflet of the cytoplasmic membrane [35]. It is likely that PE is essential for cytoskeletal organization in the completion of cytokinesis in prokaryotic cells as well as mammalian cells.

## 6. Future directions

The roles lipids play in cellular processes is as diverse as the chemical structures of lipids found throughout nature. Although a single phospholipid can form a sealed bilayer vesicle in solution, a diversity of lipid structure and physical properties is necessary to fill the broad range of roles that lipids play in cells. Lipid structures vary greatly from the archaebacteria, with their often-harsh environments, to the eubacteria that also must carry out a diversity of processes in one or two membrane structures, and to eukaryotic cells that have specialized organelles with different lipid compositions tailored to their function.

Defining lipid function is a challenging undertaking because of the diversity of chemical and physical properties of lipids and the fact that each lipid type potentially is involved at various levels of cellular function. Biological membranes are flexible self-sealing boundaries that form the permeability barrier for cells and organelles and provide the means to compartmentalize functions, but at the same time they perform many other duties. As a support for both integral and peripheral membrane processes, their physical properties directly affect these processes in ways that are often difficult to assess. Each specialized membrane has a unique structure, composition and function. Also within each membrane exist subdomains such as lipid rafts, lipid domains, and organizations of membrane associated complexes with their own unique composition. These complexes can be made up of specific lipids as the organization site for integral and peripheral membrane proteins and many times are transient responding to cellular signals that can themselves be lipids. Lipids are integral components of stable complexes and serve specific structural roles by affecting protein conformation, by serving as the 'glue' that holds complexes together, or by providing the flexible interface between protein subunits. Lipids provide the complex hydrophobic–hydrophilic solvent within which membrane proteins fold and function, but they can also act in a more specific manner as molecular chaperones directing the attainment of final membrane protein organization. These diverse functions of lipids are made possible by a family of low molecular weight molecules that are physically fluid and deformable to enable interaction in a flexible and specific manner with macromolecules. At the same time they can organize into the very stable but highly dynamic supramolecular structures we know as membranes.

The challenge for the future will be to determine the function at the molecular level for the many lipid species already discovered. Coupling genetic and biochemical approaches has been historically a very powerful approach to defining structure–function relationships of physiological importance. Using this approach in microorganisms has proven to be very fruitful. As the sophistication of mammalian cell and whole animal genetics evolves, genetic manipulation coupled with biochemical characterization will begin to yield new and useful information on the function of lipids in more complex organisms. The interest in understanding biodiversity through the detailed characterization of the vast number of microorganisms will yield additional novel lipids that must be characterized structurally and functionally. Finally, as we discover more about the role of lipids in normal cell function, the role lipids play in disease will become more evident.

## Abbreviations

CL      cardiolipin
DGDG    diglucosyl or digalactosyl diacylglycerol
MGDG    monoglucosyl or monogalactosyl diacylglycerol
PA      phosphatidic acid
PC      phosphatidylcholine
PE      phosphatidylethanolamine
PG      phosphatidylglycerol
PI      phosphatidylinositol
PS      phosphatidylserine

## References

1. Singer, S.J. and Nicolson, G.L. (1972) The fluid mosaic model of the structure of cell membranes. Science 175, 720–731.
2. Dowhan, W. (1997) Molecular basis for membrane phospholipid diversity: why are there so many phospholipids. Annu. Rev. Biochem. 66, 199–232.

3. Kates, M. (1993) Biology of halophilic bacteria, Part II. Membrane lipids of extreme halophiles: biosynthesis, function and evolutionary significance. Experientia 49, 1027–1036.

4. Raetz, C.R. (2001) Regulated covalent modifications of lipid A. J. Endotoxin Res. 7, 73–78.

5. Tanford, C. (1980) The Hydrophobic Effect: Formation of Micelles and Biological Membranes. Wiley, New York.

6. Garavito, R.M. and Ferguson-Miller, S. (2001) Detergents as tools in membrane biochemistry. J. Biol. Chem. 276, 32403–32406.

7. Seddon, J.M., Cevc, G. and Marsh, D. (1983) Calorimetric studies of the gel-fluid (Lβ-Lα) and lamellar-inverted hexagonal (Lα-H$_{II}$) phase transitions in dialkyl- and diacylphosphatidylethanol-amines. Biochemistry 22, 1280–1289.

8. Kates, M., Syz, J.Y., Gosser, D. and Haines, T.H. (1993) pH-dissociation characteristics of cardiolipin and its 2'-deoxy analogue. Lipids 28, 877–882.

9. White, S.H., Ladokhin, A.S., Jayasinghe, S. and Hristova, K. (2001) How membranes shape protein structure. J. Biol. Chem. 276, 32395–32398.

10. Matsumoto, K. (2001) Dispensable nature of phosphatidylglycerol in Escherichia coli: dual roles of anionic phospholipids. Mol. Microbiol. 39, 1427–1433.

11. Carman, G.M. and Henry, S.A. (1999) Phospholipid biosynthesis in the yeast Saccharomyces cerevisiae and interrelationship with other metabolic processes. Prog. Lipid Res. 38, 361–399.

12. Birner, R., Burgermeister, M., Schneiter, R. and Daum, G. (2001) Roles of phosphatidylethanolamine and its several biosynthetic pathways in Saccharomyces cerevisiae. Mol. Biol. Cell 12, 997–1007.

13. Ostrander, D.B., Zhang, M., Mileykovskaya, E., Rho, M. and Dowhan, W. (2001) Lack of mitochon-drial anionic phospholipids causes an inhibition of translation of protein components of the electron transport chain. A yeast genetic model system for the study of anionic phospholipid function in mitochondria. J. Biol. Chem. 276, 25262–25272.

14. Bezrukov, S.M. (2000) Functional consequences of lipid packing stress. Curr. Opin. Coll. Interface Sci. 5, 237–243.

15. Epand, R.M., Stevenson, C., Bruins, R., Schram, V. and Glaser, M. (1998) The chirality of phos-phatidylserine and the activation of protein kinase C. Biochemistry 37, 12068–12073.

16. Ruiz-Arguello, M.B., Goni, F.M. and Alonso, A. (1998) Phospholipase C hydrolysis of phospholipids in bilayers of mixed lipid compositions. Biochemistry 37, 11621–11628.

17. Fyfe, P.K., McAuley, K.E., Roszak, A.W., Isaacs, N.W., Cogdell, R.J. and Jones, M.R. (2001) Probing the interface between membrane proteins and membrane lipids by X-ray crystallography. Trends Biochem. Sci. 26, 106–112.

18. McAuley, K.E., Fyfe, P.K., Ridge, J.P., Isaacs, N.W., Cogdell, R.J. and Jones, M.R. (1999) Structural details of an interaction between cardiolipin and an integral membrane protein. Proc. Natl. Acad. Sci. USA 96, 14706–14711.

19. Ferguson, A.D., Welte, W., Hofmann, E., Lindner, B., Holst, O., Coulton, J.W. and Diederichs, K. (2000) A conserved structural motif for lipopolysaccharide recognition by procaryotic and eucaryotic proteins. Struct. Fold. Des. 8, 585–592.

20. Essen, L., Siegert, R., Lehmann, W.D. and Oesterhelt, D. (1998) Lipid patches in membrane protein oligomers: crystal structure of the bacteriorhodopsin–lipid complex. Proc. Natl. Acad. Sci. USA 95, 11673–11678.

21. Hurley, J.H., Newton, A.C., Parker, P.J., Blumberg, P.M. and Nishizuka, Y. (1997) Taxonomy and function of C1 protein kinase C homology domains. Protein Sci. 6, 477–480.

22. Rizo, J. and Sudhof, T.C. (1998) Mechanics of membrane fusion. Nat. Struct. Biol. 5, 839–842.

23. Rebecchi, M.J. and Scarlata, S. (1998) Pleckstrin homology domains: a common fold with diverse functions. Annu. Rev. Biophys. Biomol. Struct. 27, 503–528.

24. Davies, S.M., Epand, R.M., Kraayenhof, R. and Cornell, R.B. (2001) Regulation of CTP: phospho-choline cytidylyltransferase activity by the physical properties of lipid membranes: an important role for stored curvature strain energy. Biochemistry 40, 10522–10531.

25. van der Does, C., Swaving, J., van Klompenburg, W. and Driessen, A.J. (2000) Non-bilayer lipids stimulate the activity of the reconstituted bacterial protein translocase. J. Biol. Chem. 275, 2472–2478.

26. Bogdanov, M. and Dowhan, W. (1999) Lipid-assisted protein folding. J. Biol. Chem. 274, 36827–36830.

27. Bogdanov, M., Heacock, P. and Dowhan, W. (2002) A polytopic membrane protein displays a reversible topology dependent on membrane lipid composition. EMBO J. 21, 2107–2116.

28. De Cock, H., Brandenburg, K., Wiese, A., Holst, O. and Seydel, U. (1999) Non-lamellar structure and negative charges of lipopolysaccharides required for efficient folding of outer membrane protein PhoE of *Escherichia coli*. J. Biol. Chem. 274, 5114–5119.

29. Mileykovskaya, E., Dowhan, W., Birke, R.L., Zheng, D., Lutterodt, L. and Haines, T.H. (2001) Cardiolipin binds nonyl acridine orange by aggregating the dye at exposed hydrophobic domains on bilayer surfaces. FEBS Lett. 507, 187–190.

30. Simons, K. and Ikonen, E. (2000) How cells handle cholesterol. Science 290, 1721–1726.

31. Mileykovskaya, E. and Dowhan, W. (2000) Visualization of phospholipid domains in *Escherichia coli* by using the cardiolipin-specific fluorescent dye 10-*N*-nonyl acridine orange. J. Bacteriol. 182, 1172–1175.

32. Sprong, H., van der Sluijs, P. and van Meer, G. (2001) How proteins move lipids and lipids move proteins. Nat. Rev. Mol. Cell. Biol. 2, 504–513.

33. Emoto, K. and Umeda, M. (2000) An essential role for a membrane lipid in cytokinesis. Regulation of contractile ring disassembly by redistribution of phosphatidylethanolamine. J. Cell. Biol. 149, 1215–1224.

34. Mileykovskaya, E., Sun, Q., Margolin, W. and Dowhan, W. (1998) Localization and function of early cell division proteins in filamentous *Escherichia coli* cells lacking phosphatidylethanolamine. J. Bacteriol. 180, 4252–4257.

35. Rothman, J.E. and Kennedy, E.P. (1977) Asymmetrical distribution of phospholipids in the membrane of *Bacillus megaterium*. J. Mol. Biol. 110, 603–618.

D.E. Vance and J.E. Vance (Eds.) *Biochemistry of Lipids, Lipoproteins and Membranes (4th Edn.)*

# CHAPTER 2

# Lipid modifications of proteins

Nikola A. Baumann and Anant K. Menon

*Department of Biochemistry, University of Wisconsin-Madison, 433 Babcock Drive,
Madison, WI 53706-1569, USA*

## 1. Preamble

Lipid modifications of proteins (Fig. 1) are widespread and functionally important in eukaryotic cells. For example, many intracellular proteins such as the signal-transducing heterotrimeric GTP-binding proteins (G proteins) and the Ras superfamily of G proteins are modified by 14- or 16-carbon fatty acids and/or 15- or 20-carbon isoprenoids. Also, a variety of cell surface proteins such as acetylcholinesterase, the T lymphocyte surface antigen Thy-1, and members of the cell adhesion protein family are modified by glycosylphosphatidylinositol (GPI) anchors. In most cases, the lipid moiety is crucial to protein function as it allows an otherwise water-soluble protein to interact strongly with membranes. The lipid moiety may also aid in the sorting of the protein to membrane domains that promote lateral and transbilayer protein–protein interactions that are critical for cell function. In some instances, the covalent lipid acts as a functional switch resulting in functional membrane association of certain protein conformations but not of others.

The covalent attachment of lipid to protein was first described in a study of myelin protein in 1951, but only clearly documented as important for protein biosynthesis and function in a study of the outer membrane murein lipoprotein of *Escherichia coli* by Braun and Rehn in 1969. These early discoveries were followed, in the 1970s, by the identification of fatty acids linked to viral glycoproteins and of isoprenoids covalently attached to fungal mating factors and to GTP-binding proteins. The 1980s saw the identification and characterization of *N*-myristoylated proteins and GPI-anchored proteins, and work on tissue patterning factors in the 1990s revealed a new class of autoprocessed proteins modified by cholesterol. Our purpose in this chapter is to document the structure of these various lipid modifications, describe their biosynthesis, and survey their functional significance. The chapter does not cover the structure and biosynthesis of diacylglycerol-modified proteins found in *E. coli* and other bacteria; information on this may be found in articles by Wu and colleagues (Wu, 1993).

## 2. Protein prenylation

Prenylated proteins constitute approximately 0.5–2% of all proteins in mammalian cells. They contain a farnesyl (15-carbon) or geranylgeranyl (20-carbon) isoprenoid attached via a thioether linkage to a cysteine residue at or near the carboxy-terminus of the protein. Protein prenylation was originally discovered when certain fungal

38

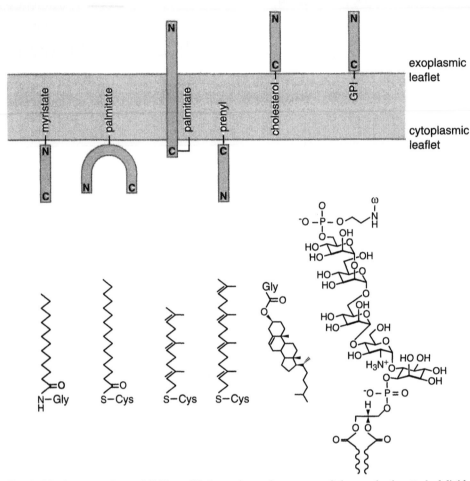

Fig. 1. Membrane topology of lipid modified proteins and structures of the covalently attached lipids. The structures (left to right) represent *N*-myristoyl glycine, palmitate thioester-linked to cysteine, farnesyl thioether-linked to cysteine, geranylgeranyl thioether-linked to cysteine, cholesterol ester-linked to glycine, and a minimal GPI anchor linked to the ω amino acid in a GPI-anchored protein. The GPI structure is shown with a diacylglycerol moiety containing two, ester-linked fatty acids. Other GPI anchors are based on ceramide, while yet others have monoacylglycerol, a fatty acid in ether-linkage to glycerol, and/or a fatty acid ester-linked to inositol.

peptide mating factors were shown to contain a carboxy-terminal cysteine modified by a thioether-linked farnesyl group. However, it was not until Glomset and colleagues (Schmidt, 1984) subsequently showed that animal cell proteins could be metabolically radiolabeled with radioactive mevalonate, an isoprenyl group precursor, that protein prenylation was more widely appreciated. Glomset and colleagues initially observed that the growth arrest of mammalian cells induced by compactin, an inhibitor of isoprenoid biosynthesis, could not be reversed by exogenously added sterols including cholesterol, the major product of the isoprenoid pathway (Brown, 1980). The compactin-induced growth arrest could, however, be alleviated by small amounts of mevalonate, suggesting that mevalonate itself or a non-sterol metabolite of mevalonate played an important role in the growth cycle of cells. This result prompted studies in which cells were metabolically labeled with radioactive mevalonate and led to the discovery that almost 50% of the cell-associated radioactive mevalonate could not be extracted into lipid solvents as a result of post-translational (cycloheximide-insensitive) covalent association with proteins.

Protein prenylation is catalyzed by one of three different multisubunit prenyltransferases located in the cytoplasm of cells [1]. The majority of prenylated proteins, including most members of the Ras family of G proteins [2], contain a carboxy-terminal CaaX motif (CaaX box) composed of a conserved cysteine residue, two aliphatic amino acids (a) and a variable carboxy-terminal residue (X). The CaaX box is recognized by CaaX prenyltransferases that catalyze the attachment of a farnesyl or geranylgeranyl group from the corresponding isoprenyl pyrophosphates to the cysteine residue (Fig. 2). The CaaX prenyltransferases involved in these reactions are protein farnesyltransferase (FTase) and protein geranylgeranyltransferase type I (GGTase-I). FTase recognizes CaaX boxes where X = M, S, Q, A, or C, whereas GGTase-I recognizes CaaX boxes with X = L or F. Other prenylated proteins, such as the Rab proteins involved in vesicular transport, terminate in a CC or CXC motif; these proteins are substrates for protein geranylgeranyltransferase type II (GGTase-II) [3].

Subsequent to prenyl modification, Ras and most other CaaX proteins are further processed by two ER-localized, membrane-bound enzymes. The first prenyl-dependent processing step is the proteolytic removal of the –aaX tripeptide by the CaaX protease

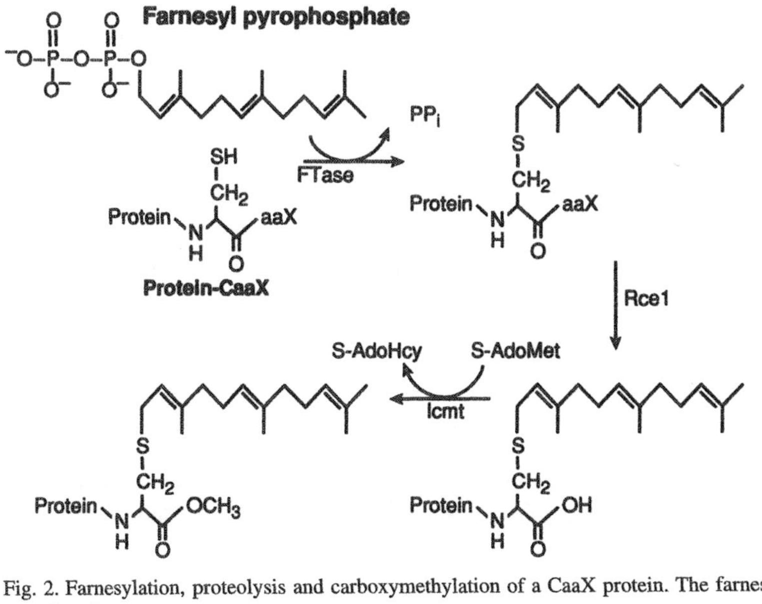

Fig. 2. Farnesylation, proteolysis and carboxymethylation of a CaaX protein. The farnesyl donor is farnesyl pyrophosphate.

Rce1; this is followed by carboxymethylation of the now C-terminal prenylcysteine residue by the methyltransferase Icmt (Fig. 2). The result of these modifications is to produce a protein that exhibits a high affinity for cellular membranes and also to impart a unique structure at the C-terminus that can serve as a specific recognition motif in certain protein–protein interactions [4].

The importance of prenylation in CaaX protein function, most notably as a regulator of the oncogenic potential of the Ras proteins [2], has led to considerable efforts to identify inhibitors of the prenyltransferases involved for evaluation as therapeutic agents [5,6]. The majority of these studies have focused on FTase, since this enzyme modifies Ras proteins, and early preclinical studies indicated significant anticancer potential for FTase inhibitors. A wide variety of FTase inhibitors have been developed, including some very promising ones that possess antitumor activity in animal models and are now in clinical development [6]. In addition, the success of FTase inhibitors in preclinical models of tumorigenesis, the increasing realization that proteins modified by GGTase-I play important roles in oncogenesis, and the finding that post-prenylation processing by Rce1 is important in the function of Ras and other CaaX proteins has led to the current situation in which all of the enzymes involved in CaaX protein processing are viewed as potential therapeutic targets.

## 2.1. The CaaX prenyltransferases FTase and GGTase-I

FTase is a heterodimer consisting of 48-kDa ($\alpha$) and 46-kDa ($\beta$) subunit polypeptides. GGTase-I also consists of two subunits, a 48-kDa $\alpha$ subunit shared with FTase, and a 43-kDa $\beta$ subunit. The isoprenoid substrates for the two enzymes are farnesyl pyrophosphate and geranylgeranyl pyrophosphate. Protein substrates for FTase in mammalian cells include Ras GTPases, lamin B, several proteins involved in visual signal transduction and at least three protein kinases and phosphatases. Known targets of GGTase-I include most $\gamma$ subunits of heterotrimeric G proteins and Ras-related GTPases such as members of the Ras and Rac/Rho families. Both FTase and GGTase-I recognize short peptides containing appropriate CaaX motifs, and tetrapeptide substrates were instrumental in purifying the enzymes to homogeneity.

Both FTase and GGTase-I are zinc metalloenzymes in which the single bound zinc atom participates directly in catalysis. FTase additionally requires high concentrations (>1 mM) of magnesium for catalysis. FTase proceeds via a functionally ordered kinetic mechanism, with farnesyl pyrophosphate binding first to create an FTase–farnesyl pyrophosphate binary complex that then reacts rapidly with a CaaX substrate to form a prenylated product. In the absence of excess substrate, the dissociation rate is so slow that FTase–product complexes can be isolated. A wealth of structural information has emerged for FTase in the past several years beginning with the first X-ray crystal structure of unliganded FTase solved at 2.2 Å resolution (Park, 1997), greatly enhancing the ability of investigators to conduct structure–function analyses on the enzyme and investigating the roles of specific residues in substrate binding and catalysis.

## 3. Myristoylation

N-Myristoylated proteins comprise a large family of functionally diverse eukaryotic and viral proteins. Myristate, a relatively rare 14-carbon, saturated fatty acid, is transferred from myristoyl-CoA and linked via an amide bond to the N-terminal glycine residue of the target protein. The reaction is catalyzed by the enzyme myristoyl-CoA: protein N-myristoyltransferase (NMT) which recognizes and modifies N-terminal glycine residues presented in a particular sequence context. Although myristate can be attached post-translationally to N-terminal glycine in synthetic peptides of the appropriate sequence, in vivo myristoylation is an early co-translational event occurring in the cytoplasm as soon as ~60 amino acids of the nascent peptide emerge from the ribosomal tunnel and after the N-terminal glycine residue is made available by cellular methionyl-aminopeptidases that remove the initiator methionine residue. The myristoyl-CoA pools used by NMT appear to be supplied by de novo synthesis and by activation of exogenous myristate. Of 6220 open reading frames surveyed in the *Saccharomyces cerevisiae* genome data base for an appropriately positioned glycine residue, 70 (1.1%) are known to be or predicted to be N-myristoylated proteins. Some of these proteins play critical roles in cell survival since NMT is an essential enzyme in *S. cerevisiae*. NMT is also essential in *Candida albicans*, and *Cryptococcus neoformans*, the most common causes of systemic fungal infections in immunocompromised individuals.

### 3.1. N-Myristoyltransferase (NMT)

NMT was first purified from *S. cerevisiae* and characterized as a ~55-kDa monomeric protein with no apparent cofactor requirements. The crystal structures of the *S. cerevisiae* and *C. albicans* NMTs have been determined, the former as a co-crystal with a non-hydrolyzable myristoyl-CoA derivative and a dipeptide inhibitor of the enzyme [7].

NMT uses an ordered Bi–Bi reaction mechanism: myristoyl-CoA binds first, followed by the peptide substrate. After catalysis, CoA is discharged first, followed by the N-myristoyl peptide. The enzyme is highly selective for myristoyl-CoA and for polypeptide substrates with an N-terminal glycine, an uncharged residue at position 2, and neutral residues at positions 3 and 4. Serine is found at position 5 of all known yeast N-myristoyl proteins, while lysine is commonly found at position 6. In vitro analyses of human and fungal NMTs indicate that while their reaction mechanism and acyl-CoA substrate specificities are the same, their peptide specificities are different — this difference has been exploited to develop species-selective NMT inhibitors that act as fungicidal agents.

### 3.2. Myristoyl switches to regulate protein function

The myristoyl group in N-myristoylated proteins frequently acts as a key regulator of protein function. In some cases, the myristate residue provides a constitutive source of membrane affinity that needs to be supplemented by a second interaction between the protein and the membrane in order for the protein to stay membrane associated. For the MARCKS protein (myristoylated alanine-rich C-kinase substrate) as well as the tyrosine kinase Src, this second interaction is provided by electrostatic affinity between a polybasic region of the protein and the negatively charged headgroups of phospholipids in the cytoplasmic leaflet of cell membranes. When serine residues in the polybasic region of MARCKS or Src are phosphorylated, the electrostatic contribution to membrane binding is reduced, and the protein moves off the membrane into the cytoplasm.

Myristate can also provide a regulated source of membrane affinity. Some proteins such as ARF (ADP ribosylation factor) and recoverin exist in alternate conformations in which the myristoyl group is exposed and available for membrane binding, or sequestered within a hydrophobic pocket in the protein. On ligand binding (GTP for ARF, and $Ca^{2+}$ for recoverin), the myristoyl group is exposed and becomes available to promote interactions with target membranes and or protein partners.

## 4. Protein thioacylation

Thioacylated proteins contain fatty acids in thioester linkage to cysteine residues [8,9]. This class of lipid modification of proteins was first identified in studies of brain myelin protein, but only firmly established in the late 1970s when Schlesinger, Schmidt and co-workers reported the palmitoylation of Sindbis virus and vesicular stomatitis virus glycoproteins. Protein thioacylation is frequently referred to as palmitoylation, although fatty acids other than palmitate may be found on thioacylated proteins. Membrane proteins as well as hydrophilic proteins are thioacylated, the latter, in many cases, acquiring the modification when they become associated with a membrane compartment as a result of N-myristoylation or prenylation. Thioacylated cysteine residues are found in a variety of sequence contexts and are invariably located in portions of the protein that are cytoplasmic or within a predicted transmembrane domain. Unlike the other known lipid modifications of proteins, thioacylation is reversible: the protein undergoes cycles of acylation and deacylation, and as a result, the half-life of the acyl group is much shorter than that of the polypeptide (~20 min for the acyl group versus ~1 day for the polypeptide in the case of N-Ras (Magee, 1987)). Several protein acyltransferases have been isolated, but it is unclear whether they are required for thioacylation in living cells: non-enzymatic thioacylation has been observed in vitro suggesting that acyl transfer may occur through an autocatalytic process. In contrast, a number of thioacyl protein thioesterases have been identified and these appear to be responsible for the deacylation of thioacylated proteins.

### 4.1. Examples of thioacylated proteins

There are three classes of thioacylated proteins [9]: polytopic membrane proteins such as some G-protein coupled receptors (β-adrenergic receptor, rhodopsin), monotopic membrane proteins including viral glycoproteins, the transferrin receptor, and the cation-dependent mannose-6-phosphate receptor, and hydrophilic proteins such as members of the Src family of protein tyrosine kinases (e.g., p59[fyn] and p56[lck]) [10], as well as H-Ras, N-Ras, and the synaptic vesicle protein SNAP-25. The functional significance of thioacylation of polytopic and monotopic membrane proteins is unclear, but the acyl modification may dictate how the protein is trafficked between membranes and whether it partitions into sterol/sphingolipid-rich membrane domains (lipid rafts) (Chapter 1) (Melkonian, 1999). For intrinsically hydrophilic proteins such as p59$^{fyn}$ and SNAP-25, the function of thioacylation is clear — the acyl group functions together with other acyl groups, or other lipid modifications, to provide membrane anchoring in the cytoplasmic leaflet of the membrane bilayer for an otherwise soluble protein.

## 4.2. Membrane anchoring of thioacylated proteins: the need for multiple lipid modifications and the role of dynamic thioacylation

Thioacylated proteins that lack transmembrane spans must have more than one co-valently bound lipid chain in order to be stably associated with membranes. This is also true for N-myristoylated proteins and prenylated CaaX proteins that, when newly synthesized, are modified by only a single lipid moiety. The association of monoacylated or monoprenylated proteins with the lipid bilayer appears to be rapidly reversible and thus, for stable membrane association, lipid modified proteins require at least two lipid chains or must rely on some other interaction with membranes such as recognition by a membrane-bound receptor or electrostatic interaction between charged amino acids in the protein and charged phospholipids in the membrane [10].

The N-myristoylated Src family protein tyrosine kinases are frequently thioacylated at one or more cysteine residues near the myristoylated glycine, and it is these doubly or triply lipid modified proteins that are found associated with the cytoplasmic face of the plasma membrane [10] (the N-terminal sequences of p59$^{fyn}$ and p56$^{lck}$ are [1]MGCVC– and [1]MGCVQC–, respectively, where the N-myristoylated glycine is shown in bold and the thioacylated cysteine residues are underlined). A similar situation is seen for the prenylated Ras proteins which must be thioacylated before they associate firmly with membranes (the C-terminal sequences of H-Ras and N-Ras are –GCMSCKCVLS-COOH and –GCMGLPCVVM-COOH, respectively (the farnesylated cysteine is shown in bold and the thioacylated cysteines are underlined)). A third example is provided by proteins such as SNAP-25 which are exclusively thioacylated, but display at least four thioacyl chains through which they become stably associated with the synaptic vesicle membrane.

An interesting and persuasive model for the targeting of lipid modified proteins to particular membranes was suggested by Shahinian and Silvius [11] (Fig. 3), based on the notion that single lipid modifications allow the protein to undergo transient interactions with a variety of intracellular membranes whereas tandem modifications promote stable membrane association. Thus, a protein with a single lipid modification such as the cytoplasmically synthesized N-myristoylated p59$^{fyn}$ becomes stably associated with the cytoplasmic face of the plasma membrane only when it becomes thioacylated at this location. In this scenario, thioacylation would not only provide for stable membrane association of a protein with a single lipid modification, but it would also ensure targeting of that protein to the membrane where thioacylation occurs. Thus, within 5 min of the completion of peptide synthesis, p59$^{fyn}$ becomes N-myristoylated, thioacylated and located to the cytoplasmic face of the plasma membrane (van 't Hof, 1997). Removal of the palmitoylation sites slows the kinetics of membrane association,

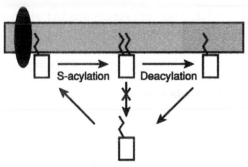

Fig. 3. Bilayer trapping mechanism for membrane targeting of lipid-anchored proteins lacking transmembrane spans (redrawn from [11]). A singly lipid-modified protein associates transiently with a membrane containing a membrane targeting receptor, possibly a thioacyltransferase. Thioacylation at the membrane yields a dual-anchored species that is stably associated with the membrane until it is deacylated.

and reduces the proportion of p59$^{fyn}$ that is membrane associated. The dynamic nature of thioacylation would suggest that the duration of association of p59$^{fyn}$ with the plasma membrane would be dictated by the half-life of the thioacyl chain. However, since p59$^{fyn}$ is doubly thioacylated, it is unlikely that it would revert to its solely $N$-myristoylated state and re-enter the cytoplasm.

Thioacylation frequently dictates plasma membrane targeting of proteins lacking transmembrane spans. In the case of p59$^{fyn}$, targeting occurs directly, with the $N$-myristoylated protein becoming thioacylated and plasma membrane associated rapidly upon completion of synthesis. In contrast, p56$^{lck}$ appears to be thioacylated on intracellular membranes and arrive at the plasma membrane via vesicular transport (bound to the cytoplasmic face of secretory vesicles) (Bijlmakers, 1999). In yet another targeting variation, newly synthesized $N$-myristoylated G$_{z\alpha}$, a dually acylated trimeric G protein $\alpha$ subunit, associates with all cellular membranes but accumulates eventually at the plasma membrane: the plasma membrane form is the only one that is both $N$-myristoylated and thioacylated.

### 4.3. Thioesterases

Three thioacyl protein thioesterases have been identified [9]. Two of these, PPT1 (Camp, 1994) and PPT2 (Soyombo, 1997) (palmitoyl–protein thioesterases 1 and 2) are localized to the lysosomes and are thus likely to be involved in the catabolism of thioacylated proteins or peptides. The thioacylated molecules are presumed to gain access to the lysosomal lumen by an autophagic pathway in which membrane fragments are captured into a vacuole that subsequently fuses with lysosomes. A defect in PPT1 leads to a severe neurodegenerative disorder termed infantile neuronal ceroid lipofuscinosis characterized by the accumulation of autofluorescent material (including lipid thioesters) in all tissues. No diseases have been linked to PPT2.

A third thioesterase was purified from rat liver cytosol using palmitoylated G-protein $\alpha$ subunit as a substrate (Duncan, 1998). This thioesterase, a 25-kDa monomeric protein, is likely to be the one involved in turnover of thioacyl groups on proteins. It displays both acylprotein thioesterase activity as well as lysophospholipase activity, but thioacylproteins are by far the preferred substrate.

## 5. Cholesterol modification

In addition to its numerous roles in membrane architecture and steroid and bile acid synthesis (Chapters 15 and 16), cholesterol has recently been discovered as a post-translational protein modification on a family of signaling proteins referred to as hedgehog (Hh) proteins [12]. This modification was identified in studies on the processing of the *Drosophila* Hh protein (Porter, 1996). During biosynthesis, the Hh protein undergoes an autocatalytic cleavage event during which the carboxy-terminal domain is removed and cholesterol is covalently attached to the amino-terminal domain yielding an active signaling molecule (Fig. 4). The Hh proteins, found in insects, vertebrates, and other multicellular organisms, are part of a family of secreted signaling molecules involved in the patterning of diverse tissues during development. These proteins function

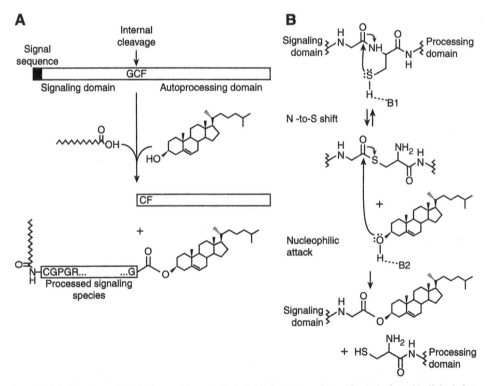

Fig. 4. Modification of hedgehog with ester-linked cholesterol and an N-terminal amide-linked fatty acid (redrawn from [12]). (A) Overall processing steps leading to lipid-modified Hh. (B) Mechanism of cholesterol addition. Residues indicated as B1 and B2 in panel B are catalytic bases presumably contained in the carboxy-terminal domain of the Hh precursor.

in a graded manner, emanating from cells in which the protein is produced and acting upon target cells several cell diameters away, thus specifying different cell fates by altering programs of gene expression. Cholesterylation of the Hh signaling protein is crucial for its proper targeting to specific tissues during development [13,14].

## 5.1. Addition of cholesterol to hedgehog proteins

Newly synthesized Hh precursor proteins of approximately 45 kDa contain an N-terminal signal sequence which targets the protein to the ER and is subsequently trimmed, followed by an N-terminal signaling domain (~19 kDa), an absolutely conserved Gly–Cys–Phe tripeptide motif, and a C-terminal processing domain which is removed to yield the active signaling molecule (Fig. 4). Addition of cholesterol to Hh proteins proceeds via an autoproteolytic internal cleavage reaction at the Gly–Cys–Phe tripeptide sequence. The cleavage is initiated through a nucleophilic attack by the thiol side chain of Cys on the Gly carbonyl, replacing the Gly–Cys peptide bond with a thioester linkage. A second nucleophilic attack on the same carbonyl by the hydroxyl group of cholesterol results in removal of the C-terminal processing domain and produces the active Hh signaling molecule containing a C-terminal cholesterol modification (Fig. 4). Although other sterols have been shown to be able to substitute for cholesterol in in vitro Hh processing assays, the free 3β-hydroxyl group is essential for processing [12].

The autocleavage reaction is mediated by the C-terminal processing domain of the Hh precursor polypeptide. The first 145 amino-terminal residues of C-terminal processing domain are sufficient for thioester formation as indicated by the ability of Hh proteins truncated after this point to undergo cleavage in the presence of dithiothreitol. At least some part of the remaining carboxy-terminal residues is required for cholesterol addition and contains a sterol recognition region (Hall, 1997).

In addition to the cholesterol modification, Hh proteins can also be palmitoylated at the N-terminal Cys residue (part of a CGPGR motif) after cholesterol is added. An acyltransferase, named Skinny Hh, was discovered in *Drosophila*. Skinny Hh mutant cells express Hh proteins containing the cholesterol modification but lacking aminoterminal palmitate (Chamoun, 2001). Available information indicates that the palmitate residue is amide-linked to the N-terminus of the protein and not thioester-linked to the cysteine side chain.

## 5.2. Biological significance of cholesterol modification

The role of cholesterol-anchoring of the hedgehog signaling molecule is not fully understood. Cholesterylated Hh must travel to other cells to elicit its effect, even though the hydrophobic cholesterol molecule is presumed to act as a membrane anchor. Hh proteins truncated at the cleavage site are not modified by cholesterol, but nevertheless produce active signaling molecules; however, studies in *Drosophila* have shown that these proteins are not targeted properly and cause mispatterning and lethality in embryos. The subcellular location of the autocatalytic cleavage and processing events is not clear. Cholesterol-modified Hh in *Drosophila* partitions into detergent insoluble membranes, or rafts, which are enriched in specific molecules such as sterols, sphingolipids, thioacylated proteins and GPI-anchored proteins and usually considered predominantly plasma membrane domains (Chapter 1) (Rietveld, 1999). Several proteins, each containing a sterol-sensing domain, have been identified that function in the release of Hh from Hh-producing cells and proper sequestering of Hh in target cells. For example, the proteins Patched and Dispatched are involved in sequestration of the Hh signal within tissues and release of functional signal from Hh-producing cells, respectively. A soluble form of mammalian Hh (sonic Hh) that is cholesterol-modified, multimeric and biologically potent has been isolated from developing chick limb (Zeng, 2001). The protein Patched is required for release of this soluble complex. A proposed model is that cholesterol-modified Hh protein is recruited to lipid rafts (exoplasmic leaflet of the plasma membrane) where multimerization occurs and the soluble complex can then travel to target cells and elicit its effect.

# 6. GPI anchoring of proteins

Glycosylphosphatidylinositols (GPIs) are a ubiquitous family of eukaryotic glycolipids. GPIs were originally discovered covalently linked to cell surface glycoproteins and recognized to be an important alternative mechanism for anchoring proteins to the cell surface [15]. After synthesis of GPI anchors and attachment to protein in the endoplasmic reticulum (ER), GPI-anchored proteins are transported via the secretory pathway to the cell surface. GPI anchors are found on a variety of functionally diverse proteins and glycoconjugates including cell surface receptors (folate receptor, CD14), cell adhesion molecules (NCAM isoforms, carcinoembryonic antigen variants), cell surface hydrolases (5′-nucleotidase, acetylcholinesterase), complement regulatory proteins (decay accelerating factor), and protozoal surface molecules (*Trypanosoma brucei* variant surface glycoprotein, *Leishmania* lipophosphoglycan). Along with serving to attach proteins to the cell surface, GPI-anchored proteins appear to be markers and major constituents of 'detergent-resistant' lipid rafts, the sphingolipid- and sterol-rich domains in membranes that are postulated to play an important role in the activation of signaling cascades [16].

## 6.1. Biosynthesis of GPI

GPIs are assembled via a biosynthetically and topologically complex metabolic pathway that is comprised of at least 10 steps and requires the participation of at least 20 distinct gene products (reviewed in [17–19]). In basic terms, synthesis occurs in the ER and involves the sequential addition of monosaccharides to phosphatidylinositol (PI) yielding the core GPI structure of ethanolamine-$P$-6Man$\alpha$1–2Man$\alpha$1–6Man$\alpha$1–4GlcN$\alpha$1–6$myo$-inositol-P-lipid which is then added to proteins. Many of the genes encoding enzymes involved in synthesis have been identified. Fig. 5 illustrates the sequence of reactions involved in the assembly of a GPI protein anchor precursor in mammalian cells; variations of this general sequence are found in all organisms studied thus far.

48

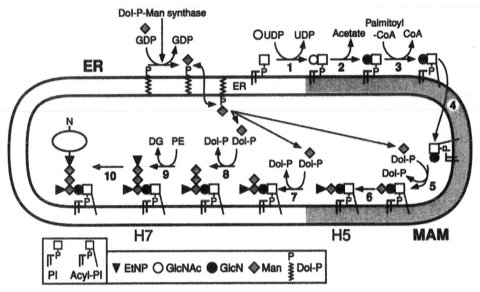

Fig. 5. GPI biosynthesis in the ER (redrawn from [17]). The pathway is initiated on the cytoplasmic face of the ER and proceeds clockwise in the figure to yield the lumenally oriented mature GPI structure H7 (mammalian cells synthesize another mature GPI termed H8 (not depicted) which is identical to H7 except that it contains an additional EtNP residue linked to the second mannose). H7 is attached to ER-translocated proteins bearing a C-terminal, GPI-directing signal sequence. MAM, mitochondria-associated membrane. See text for other abbreviations.

Initiation of GPI biosynthesis occurs on the cytoplasmic leaflet of the ER via the addition of *N*-acetylglucosamine (GlcNAc) from UDP-GlcNAc to PI yielding GlcNAc-PI (Fig. 5, step 1). The enzyme which mediates this first step is a complex consisting of at least six proteins in mammalian cells: PIG-A, PIG-C, PIG-H, GPI1, PIG-P, and DPM2; however, the subunit stoichiometry and functions of most of the subunits are not yet known. PIG-A which has homology to many glycosyltransferases is most likely the catalytic component. PIG-C, PIG-H, and PIG-P are essential components of the complex but their precise functions are unknown. GPI1 is important but not essential for activity and DPM2 appears to be a positive regulator stimulating GlcNAc transferase activity approximately 3-fold.

GlcNAc-PI is subsequently de-*N*-acetylated to generate GlcN-PI by the GlcNAc-PI de-N-acetylase, PIG-L (Fig. 5, step 2). The third step in biosynthesis is the acylation, predominantly palmitoylation, of the inositol residue of GlcN-PI at the 2 position (Fig. 5, step 3). The acylation step is required prior to the addition of three mannose residues (four in yeast) to GlcN-(acyl)PI and modification of the mannose residues with phosphoethanolamine (EtNP) side chains. In contrast to the first two steps of synthesis which occur on the cytoplasmic leaflet of the ER, the mannosylation reactions are thought to occur lumenally based on the predicted topology of the mannosyl transferases. This indicates that the substrate (GlcN-PI or GlcN-(acyl)PI must be flipped across the ER membrane bilayer (Fig. 5, step 4).

Dolichol-P-mannose serves as the mannose donor for all three mannosylation steps. The first mannosylation is catalyzed by PIG-M, the GPI α1–4-mannosyltransferase (Fig. 5, step 5). Phosphoethanolamine is then added to the 2 position of the first mannose residue (Fig. 5, step 6); PIG-N is involved in this reaction and is probably the EtNP transferase. A second mannose residue is then added by an α1–6-mannosyltransferase that has not yet been identified (Fig. 5, step 7). An α1–2-mannosyltransferase, most likely the protein PIG-B, mediates transfer of the third mannose residue to the GPI moiety (Fig. 5, step 8). EtNP residues can be added to both the second and third mannose residues as well (Fig. 5, steps 9 and 10). PIG-O and PIG-F are the gene products responsible for transferring EtNP from phosphatidylethanolamine (PE) to the 6 position on the third mannose (Fig. 5, step 10). These proteins form a complex, with PIG-O most likely serving the catalytic role. This modification is crucial as it is this EtNP residue on the third, or terminal mannose that is involved in the linkage of GPI to the carboxy-terminus of proteins destined to be GPI-anchored (Fig. 5, step 11). Fully assembled GPI structures, or the GPI moiety in GPI-anchored proteins, are frequently subject to lipid re-modeling reactions in which fatty acids or the entire lipid structure are replaced with different fatty acids or lipids.

### 6.2. Subcellular location and membrane topology of GPI biosynthesis

GPI biosynthesis occurs in the ER, however the ER is an extremely heterogeneous organelle and GPI biosynthetic activities are not uniformly distributed throughout the bulk ER [19]. The capacity to synthesize GlcNAc-PI appears to be uniformly distributed in the ER in mammalian cells while subsequent biosynthetic steps leading to the formation of (Etn-P)Man-GlcN-PI are highly enriched in an ER domain that is associated with mitochondria (mitochondria-associated ER membrane domain) (Fig. 5). The reason for segregation of some GPI biosynthetic reactions to the mitochondria-associated membrane is not known.

In addition to spatial segregation of GPI biosynthetic activities within subdomains of the ER, the biosynthetic pathway is also topologically complex with the first two steps occurring in the cytoplasmic leaflet of the membrane and later mannosylation steps and attachment of GPI to protein occurring in the lumenal leaflet (Vidugiriene, 1993, 1994). Thus, for GPI synthesis to go to completion GPI precursors must be flipped across the ER bilayer to access the enzymes involved in the late steps of synthesis [19]. The point at which this flipping occurs as well as the mechanism for flipping GPI precursors remain to be found.

### 6.3. Attachment of GPIs to proteins

Newly synthesized proteins are attached to pre-existing GPIs in the lumenal leaflet of the ER by a GPI: protein transamidase complex [20,21]. Proteins destined to be GPI-anchored contain an N-terminal signal sequence for targeting to the ER and a C-terminal signal sequence that directs GPI-anchoring. The N-terminal signal sequence is cleaved by signal peptidase during or after translocation of the nascent polypeptide into the ER lumen. The C-terminal GPI signal sequence is then

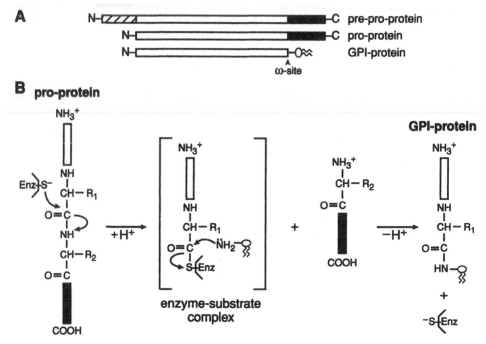

Fig. 6. GPI anchoring of proteins. (A) The processing steps involved in the conversion of a pre-pro-protein to a GPI-anchored protein. (B) The transamidation reaction in which a GPI anchor is attached to a pro-protein. The transamidase enzyme, which contains an enzymatically critical cysteine residue, is depicted as Enz-S$^-$. The $\omega$ amino acid is depicted with a side chain labeled R1; the $\omega + 1$ amino acid has a side chain labeled R2.

cleaved and replaced with a GPI anchor through the action of GPI transamidase (Fig. 6).

The C-terminal GPI-signal peptide, which is cleaved at what is referred to as the $\omega$ site, is necessary and sufficient for designating that a protein becomes GPI-anchored. The sequence motif for GPI-anchoring includes four sequence elements: (1) an unstructured linker region of about 11 residues ($\omega - 11$ to $\omega - 1$); (2) four preferably small amino acids ($\omega - 1$ to $\omega + 2$) including the cleavage site; (3) a moderately polar spacer region ($\omega + 3$ to $\omega + 9$); and (4) a hydrophobic tail from $\omega + 9$ or $\omega + 10$ to the C-terminal end [21].

The GPI transamidase complex responsible for attaching the protein to the GPI-anchor consists of at least five subunits: Gpi8p, Gaa1p, PIG-S, PIG-T and PIG-U (Vidugiriene, 2001; Ohishi, 2001; Fraering, 2001; Vainauskas, 2002). The enzyme recognizes a pro-protein containing a GPI-anchoring signal sequence, cleaves the signal sequence and attaches the new carboxy terminus to the terminal EtNP of a pre-existing GPI. Gpi8p is a membrane protein with a large lumenal domain and has homology to a family of cysteine proteases, one of which is a plant endopeptidase with transamidase activity. Cysteine and histidine residues conserved between the cysteine proteases and Gpi8p are essential for GPI transamidase activity, thus it is proposed that Gpi8p cleaves the signal sequence and functions in the transamidation reaction. The functions of the remaining subunits are unknown.

## 6.4. GPI-anchoring in mammals, parasitic protozoa and yeast

GPI-deficient mammalian cells are viable in tissue culture, but a GPI defect has clear consequences for multicellular organisms. Transgenic mouse embryos lacking the ability to initiate GPI biosynthesis (defective in PIG-A) do not develop beyond the ninth day of gestation [18]. The inability of certain blood cells to express GPI-anchored proteins results in the rare human disease, paroxysmal nocturnal hemoglobinuria (PNH), characterized by intravascular hemolysis, thrombosis and bone marrow failure. The disease is caused by a somatic mutation of the X-linked PIG-A gene (encoding the catalytic component of the enzyme responsible for the first step in GPI biosynthesis, see above) in hematopoietic stem cells. Cells defective in PIG-A are either unable to synthesize GPIs or exhibit a significant decrease in the synthesis of GPIs and thus decreased expression of GPI-anchored proteins. Red blood cells no longer expressing GPI-anchored complement regulatory proteins (e.g. CD55 and CD59) become susceptible to complement-mediated lysis, resulting in the release of heme and hemoglobin into the blood, filtering by the kidney, and excretion in the urine. Interestingly, while the GPI-deficient phenotype of hematopoietic stem cells makes the cells more sensitive to complement-mediated lysis, the defective PNH clone still persists. This implies that there must be other factors which select for clonal expansion and maintenance of GPI-deficient blood cells. One proposal is that PNH patients possess autoreactive T cells that target GPI on the surface of hematopoietic stem cells, thus PNH cells would evade damage because they lack surface GPI molecules [22].

GPI anchoring is the most prominent mode of attachment for cell surface proteins and glycans in parasitic protozoa [23]. Pathogenic protozoa, including species of the genera *Trypanosoma*, *Leishmania*, and *Plasmodium*, display abundant GPI-anchored cell surface macromolecules that play crucial roles in parasite infectivity and survival. An example is the GPI-anchored variant surface glycoprotein (VSG) of bloodstream forms of *Trypanosoma brucei*, the causative agent of African sleeping sickness.

GPIs serve a unique function in yeast (*S. cerevisiae*). In addition to anchoring secretory proteins to the cell surface, GPIs play an additional role in yeast cell wall biosynthesis [24]. The yeast cell wall consists of a fibrous lattice of mannoproteins, $\beta 1,3$-glucan, $\beta 1,6$-glucan and chitin. During cell wall biosynthesis GPI-anchored mannoproteins are transported through the secretory pathway to the cell surface. After arrival at the plasma membrane, a transglycosylation reaction (catalyzed by an unknown enzyme) results in cleavage of the GPI moiety between GlcN and the first mannose residue and formation of a glycosidic linkage between the mannoprotein-GPI-remnant and $\beta 1,6$-glucan. Mutations in GPI biosynthesis are lethal in yeast and decreased levels of GPI biosynthesis cause growth defects and aberrant cell wall biogenesis. However, not all GPI-anchored proteins become crosslinked to $\beta 1,6$-glucan; the amino acids within four or five residues upstream of the $\omega$-site determine whether the protein becomes incorporated into the cell wall or remains anchored to the plasma membrane. A unique characteristic of yeast GPIs is the addition of a fourth mannose residue to the core GPI structure (via an $\alpha 1-2$

linkage to the third mannose). Most GPI-anchored proteins in yeast also undergo lipid remodeling replacing the glycerolipid backbone with ceramide. The remodeling occurs after the attachment of protein to the GPI and can occur in both the ER and the Golgi.

### 6.5. Functions of GPI anchors

GPIs function as membrane anchors for secretory proteins and are believed to provide targeting signals that influence the intracellular trafficking of these proteins. Recent results indicate that GPI-proteins are packaged into unique transport vesicles, distinct from those carrying other secretory proteins, for export from the ER (Muñiz, 2001). GPI-anchored proteins have been shown to coalesce with sphingolipids and cholesterol into detergent insoluble membrane domains, or lipid rafts. Association of molecules in rafts at the plasma membrane, including GPI-anchored proteins (in the exoplasmic leaflet) and acylated signaling molecules (in the cytoplasmic leaflet), is postulated to play an important role in the activation of signaling cascades [16]. In some polarized epithelial cells, GPI-anchored proteins and many glycosphingolipids are sorted in the *trans* Golgi network and specifically targeted to the apical membrane. Once at the plasma membrane GPI-anchored proteins can undergo endocytosis and recycling back to the cell surface. However, uptake of GPI-anchored proteins is approximately five-times slower than that of receptor-mediated endocytosis and recycling to the plasma membrane is about three-times slower than that of recycling receptors (Sabharanjak, 2002). The current view is that GPI-anchoring of proteins directs their segregation into lipid rafts and affects their sorting in both the exocytic and endocytic pathways.

A GPI-anchor may also allow a protein to be selectively released from the cell surface upon hydrolysis by a GPI-specific phospholipase (e.g., PI-PLC or GPI-PLD). This has been shown to occur for certain GPI-anchored proteins in mammalian cell culture. One example is GPI-anchored membrane dipeptidase which is released from the adipocyte cell surface by a phospholipase C in response to insulin (Movahedi, 2000). Interestingly, other GPI-anchored proteins are not released indicating a level of regulation in insulin-stimulated hydrolysis of GPI-anchored proteins. GPI-anchored molecules have also been to shown to transfer between cells and stably insert in the external leaflet of the acceptor cell's plasma membrane (Low, 1998). The biological significance of this is unclear; however, the ability of GPI-anchored proteins to transfer between cells has implications for the expression of foreign proteins on the cell surface.

## 7. Future directions

This chapter describes post-translational lipid modifications of proteins that represent functionally critical elaborations of protein structure. These modifications, with the possible exception of thioacylation, can be anticipated at the level of primary protein sequence, enabling predictions about the localization and likely behavior of the modified proteins within cells. Although the functional diversity of lipid-modified proteins makes it difficult to arrive at generalizations about the evolutionary impetus for lipid anchoring compared to the use of 'conventional' protein transmembrane domains, the observation that many lipid modified proteins are associated with cholesterol and sphingolipid-rich membrane domains (rafts) is suggestive. Analyses of the biophysical characteristics of membrane association of lipid modified proteins, the role of raft-associated proteins in cell signaling, and the regulation of protein function by lipid modification are likely to remain fruitful areas of investigation.

Although many biochemical and cell biological aspects of the various lipid modifications remain to be elucidated, it is worth highlighting the scope for investigation offered by the GPI biosynthetic pathway. GPI anchoring is arguably the most biosynthetically and topologically complex of the lipid modifications described in this chapter. Despite the identification of numerous genes and gene products associated with the GPI biosynthetic pathway, the enzymology of GPI biosynthesis, including analyses of enzyme structure, basis for enzyme localization to the ER and ER domains, and transbilayer distribution of the pathway remains open to new investigation.

## Abbreviations

| | |
|---|---|
| ER | endoplasmic reticulum |
| EtN | ethanolamine |
| FTase | protein farnesyltransferase |
| FTI | FTase inhibitor |
| GGTase | protein geranylgeranyltransferase |
| GlcN | glucosamine |
| GlcNAc | $N$-acetylglucosamine |
| GPI | glycosylphosphatidylinositol |
| G proteins | GTP-binding proteins |
| Hh | hedgehog protein |
| Man | mannose |
| NMT | $N$-myristoyltransferase |
| PE | phosphatidylethanolamine |
| PI | phosphatidylinositol |
| PNH | paroxysmal nocturnal hemoglobinuria |

## References

1. Zhang, F.L. and Casey, P.J. (1996) Protein prenylation: molecular mechanisms and functional consequences. Annu. Rev. Biochem. 65, 241–269.
2. Magee, T. and Marshall, C. (1999) New insights into the interaction of ras with the plasma membrane. Cell 98, 9–12.
3. Pereira-Leal, J.B., Hume, A.N. and Seabra, M.C. (2001) Prenylation of rab GTPases: molecular mechanisms and involvement in genetic disease. FEBS Lett. 498, 197–200.
4. Gelb, M.H. (1997) Protein prenylation — signal transduction in two dimensions. Science 275, 1750–1751.
5. Gelb, M.H., Scholten, J.D. and Sebolt-Leopold, J.S. (1998) Protein prenylation: from discovery to prospects for cancer treatment. Curr. Opin. Chem. Biol. 2, 40–48.

6. Johnston, S.R. (2001) Farnesyltransferase inhibitors: a novel targeted therapy for cancer. Lancet Oncol. 2, 18–26.

7. Farazi, T.A., Waksman, G. and Gordon, J.I. (2001) The biology and enzymology of protein N-myristoylation. J. Biol. Chem. 276, 39501–39504.

8. Mumby, S.M. (1997) Reversible palmitoylation of signaling proteins. Curr. Opin. Cell Biol. 9, 148–154.

9. Linder, M.E. (2000) Reversible modification of proteins with thioester-linked fatty acids. In: F. Tamanoi and D.S. Sigman (Eds.), The Enzymes, Volume XXI 'Protein Lipidation'. Academic Press, New York, pp. 215–240.

10. Resh, M.D. (1999) Fatty acylation of proteins: new insights into membrane targeting of myristoylated and palmitoylated proteins. Biochim. Biophys. Acta 1451, 1–16.

11. Shahinian, S. and Silvius, J.R. (1995) Doubly-lipid-modified protein sequence motifs exhibit long-lived anchorage to lipid bilayer membranes. Biochemistry 34, 3813–3822.

12. Mann, R.K. and Beachy, P.A. (2000) Cholesterol modification of proteins. Biochim. Biophys. Acta 1529, 188–202.

13. Incardona, J.P. and Eaton, S. (2000) Cholesterol in signal transduction. Curr. Opin. Cell Biol. 12, 193–203.

14. Ingham, P.W. (2000) Hedgehog signaling: how cholesterol modulates the signal. Curr. Biol. 10, R180–R183.

15. Low, M.G., Ferguson, M.A.J., Futerman, A.H. and Silman, I. (1986) Covalently attached phosphatidylinositol as a hydrophobic anchor for membrane proteins. Trends Biochem. Sci. 11, 212–215.

16. Simons, K. and Toomre, D. (2000) Lipid rafts and signal transduction. Nat. Rev. (Mol. Cell Biol.) 1, 31–39.

17. Kinoshita, T. and Inoue, N. (2000) Dissecting and manipulating the pathway for glycosylphosphatidylinositol-anchor biosynthesis. Curr. Opin. Chem. Biol. 4, 632–638.

18. Tiede, A., Bastisch, I., Schubert, J., Orlean, P. and Schmidt, R.E. (1999) Biosynthesis of glycosylphosphatidylinositols in mammals and unicellular microbes. Biol. Chem. 380, 503–523.

19. McConville, M.J. and Menon, A.K. (2000) Recent developments in the cell biology and biochemistry of glycosylphosphatidylinositol lipids. Mol. Membr. Biol. 17, 1–16.

20. Udenfriend, S. and Kodukula, K. (1995) How glycosyl-phosphatidylinositol-anchored proteins are made. Annu. Rev. Biochem. 64, 563–591.

21. Eisenhaber, B., Bork, P. and Eisenhaber, F. (2001) Post-translational GPI lipid anchor modification of proteins in kingdoms of life: analysis of protein sequence data from complete genomes. Protein Eng. 14, 17–25.

22. Karadimitris, A. and Luzatto, L. (2001) The cellular pathogenesis of paroxysmal nocturnal hemoglobinuria. Leukemia 15, 1148–1152.

23. Ferguson, M.A.J. (2000) Glycosylphosphatidylinositol biosynthesis validated as a drug target for African sleeping sickness. Proc. Natl. Acad. Sci. USA 97, 10673–10675.

24. Lipke, P.N. and Ovalle, R. (1998) Cell wall architecture in yeast: new structure and new challenges. J. Bacteriol. 180, 3735–3740.

D.E. Vance and J.E. Vance (Eds.) *Biochemistry of Lipids, Lipoproteins and Membranes (4th Edn.)*
© 2002 Elsevier Science B.V. All rights reserved

# CHAPTER 3

# Fatty acid and phospholipid metabolism in prokaryotes

Richard J. Heath [1], Suzanne Jackowski [1,2] and Charles O. Rock [1,2]

[1] *Protein Science Division, Department of Infectious Diseases, St. Jude Children's Research Hospital, Memphis, TN 38105, USA, Tel.: +1 (901) 495-3495; Fax: +1 (901) 525-8025; E-mail: charles.rock@stjude.org*
[2] *Department of Molecular Biosciences, University of Tennessee Health Science Center, Memphis, TN 38163, USA*

## 1. Bacterial lipid metabolism

Bacteria are a versatile tool for the study of metabolic pathways. This is especially true for the Gram-negative *Escherichia coli*. These bacteria are easy to grow, and the growth conditions can be controlled and manipulated by the investigator. Most importantly, they are suitable for genetic manipulation and their genome sequence is available [1]. *E. coli* have been studied extensively, and the genes and enzymes of phospholipid metabolism were first delineated using this organism [2]. However, *E. coli* is not typical of all Gram-negative bacteria, let alone all eubacteria. Certain key differences between this model system and other organisms will be highlighted in the relevant sections.

Phospholipids in bacteria comprise about 10% of the dry weight of the cell, and each mole of lipid requires about 32 mole of ATP for its synthesis. Thus, phospholipid synthesis requires significant investment by the cell, and the advantages of maintaining fine control over the pathway are obvious. The pathway in most bacteria is catalyzed by a series of discrete proteins: the enzymes of fatty acid synthesis are cytosolic, while those of membrane lipid synthesis are mainly integral inner membrane proteins. The differences between the bacterial and mammalian enzymes offer attractive targets for novel antimicrobial drugs, and this has been a driving force behind much of the recent research.

The study of phospholipid enzymology in *E. coli* dates back to the early 1960s, when work in the laboratory of Vagelos discovered that the intermediates in fatty acid synthesis are bound to a heat stable cofactor termed acyl carrier protein (ACP) [3]. The enzymes of fatty acid synthesis are soluble proteins whose individual activities can be assayed in crude cell extracts or purified preparations. This is markedly different from the mammalian fatty acid synthase, a large multi-functional polypeptide with intermediates covalently attached (Chapter 6). Thus, the bacterial enzymes became a focus of intensive study, especially in the laboratories of Vagelos, Bloch and Wakil. Soon, the structures of all of the intermediates were known, and the basic chemical reactions required had been described [4]. During the late 1960s, work on the enzymes of phospholipid synthesis in bacteria flourished, and, mainly through classical identification experiments in the Kennedy laboratory, the intermediates in that pathway were established [5].

A second phase of bacterial phospholipid research, during the 1970s and early 1980s, was the identification of mutants in the pathway [6]. The genetic manipulation of *E. coli* is relatively facile. Mutants in many specific enzymes were generated by employing mutagens in combination with a battery of clever selection and screening techniques [7]. Such mutations generally fall into one of two classes. Firstly, they may confer an auxotrophy on a strain, such as a requirement for unsaturated fatty acids or glycerol phosphate. Such mutants have generally lost the ability to produce a key biosynthetic enzyme (e.g. *fabA* mutants, which require supplementation with unsaturated fatty acids), or they may be more complex (e.g. *plsB* mutants require high glycerol phosphate concentrations due to a $K_m$ defect in the enzyme and a second site mutation). Secondly, they may be conditionally defective, usually at elevated temperatures. For example, strains with the *fabI*(Ts) mutation grow at 30°C, but are not viable at 42°C. This type of defect is usually ascribed to an amino acid change in an essential enzyme that renders it unstable at higher temperatures. Techniques using the bacteriophage P1 are available for the movement of these alleles into other host strains, thus allowing for the mapping of the genes to specific regions of the chromosome, or the generation of strains with particular combinations of mutations. Regulatory mutants were identified, affecting multiple enzymes with a single mutation, that allowed for regulatory networks to be investigated. The membrane-bound enzymes of phospholipid synthesis were not amenable to analysis using standard biochemical approaches, and the genetic approach allowed these enzymes to be identified.

Next came the cloning and detailed study of the enzymes of lipid metabolism during the late 1980s and 1990s [2]. Plasmid-based expression systems were used to examine overexpression of enzymes on pathway regulation, and purified enzymes could be more easily obtained for biochemical analysis. With the availability of the sequence data generated by these clones, more precise methods for the construction of specific mutations also became available. Specific genes, or portions of genes, could be 'knocked-out' by targeted replacement based on sequence information, as opposed to random insertion of phage DNA. In the late 1990s, the genomic sequences of a broad spectrum of bacteria started to become available [8]. These data fuel the current phase of bacterial lipid metabolic research and comparative enzymology.

With the advent of genome information, many of the genes and gene products first identified and characterized in *E. coli* (Table 1) were identified in other bacteria, thus these genes can easily be cloned and their properties compared to those of *E. coli*. The pathways, especially fatty acid biosynthesis, are generally highly conserved amongst bacteria. Coupled with recent findings about the efficacy of certain inhibitors of the pathway, considerable attention is now being focused on fatty acid synthesis as a target for novel drug design [9]. The availability of pure enzymes has also stimulated progress in structural biology, and many new 3-dimensional structures of enzymes related to lipid metabolism have been recently solved.

## 2. Membrane systems of bacteria

Phospholipids in *E. coli* and other Gram-negative bacteria are used in the construction of the inner and outer membranes (Fig. 4 in Chapter 1). The inner membrane is impermeable to solutes unless specific transport systems are present. The outer membrane contains pores that allow the passage of molecules having a molecular weight less than 600, and is rich in structural lipoproteins and proteins involved in the transport of high molecular weight compounds. The outer layer of the outer membrane is composed primarily of lipopolysaccharides rather than phospholipid. Between the inner and outer membranes is an osmotically active compartment called the periplasmic space. Membrane-derived oligosaccharides, peptidoglycan, and binding proteins involved with metabolite transport are found in this compartment. Gram-positive bacteria do not possess an outer membrane. Instead, they have a membrane bilayer surrounded by a thick layer of peptidoglycan decorated with proteins, carbohydrates and, often, teichoic and lipoteichoic acid.

## 3. Bacterial fatty acid biosynthesis

### 3.1. Acyl carrier protein

A unique feature of fatty acid synthesis in bacteria is the presence of the small (8.86 kDa), acidic and highly soluble ACP, the product of the *acpP* gene. ACP is one of the most abundant proteins in *E. coli*, constituting about 0.25% of the total soluble protein ($\sim 6 \times 10^4$ molecules/cell). The acyl intermediates of fatty acid biosynthesis are bound to the protein through a thioester linkage to the terminal sulfhydryl of the 4'-phosphopantetheine prosthetic group. The prosthetic group sulfhydryl is the only thiol group of ACP and is attached to the protein via a phosphodiester linkage to Ser-36. ACP must interact specifically and transiently with all of the enzymes of fatty acid biosynthesis (except acetyl-CoA carboxylase), and does so through interactions with exposed negative residues on ACP with a patch of positive residues on the surfaces of the *fab* enzymes.

The ACP pool in normally growing cells is approximately one-eighth the coenzyme A (CoA) pool, the other acyl group carrier in cells. The prosthetic group of ACP is produced from CoA, and a common feature of both is the pantetheine arm for thioester formation. Virtually all of the ACP is maintained in the active, holo-form in vivo indicating that the supply of prosthetic group does not limit fatty acid biosynthesis. During logarithmic growth, a significant pool of ACP is unacylated. The ACP pool must be severely depleted before an effect on fatty acid and phospholipid synthesis can be detected. Overproduction of ACP generally yields high levels of apo-ACP, which is toxic to the cell by inhibition of the glycerol phosphate acyltransferase. The 4'-phosphopantetheine prosthetic group is transferred from CoA to apo-ACP by the 14-kDa monomeric [ACP]synthase. The [ACP]synthase from *Bacillus subtilis* has been crystallized in complex with ACP to give the first detailed look at ACP–protein interactions. ACP plays other roles in cell physiology, donating acyl chains to membrane-derived oligosaccharides, lipoic acid and quorum sensors.

Table 1

Genes of lipid metabolism

| Gene | Protein |
|------|---------|
| *aas* | 2-acyl-GPE acyltransferase |
| *accA* | Carboxyltransferase subunit |
| *accB* | Biotin carboxy carrier protein |
| *accC* | Biotin carboxylase |
| *accD* | Carboxyl transferase subunit |
| *acpP* | Acyl carrier protein |
| *acpD* | Azoreductase |
| *acpS* | Acyl carrier protein synthase |
| *cdh* | CDP-diacylglycerol hydrolase |
| *cdsA* | CDP-diacylglycerol synthase |
| *cdsS* | Stabilizes mutant CDP-diacylglycerol synthase |
| *cfa* | Cyclopropane fatty acid synthase |
| *cls* | Cardiolipin synthase |
| *desA* | Desaturase |
| *dgk* | Diacylglycerol kinase |
| *fabA* | β-Hydroxydecanoyl-ACP dehydrase |
| *fabB* | β-Ketoacyl-ACP synthase I |
| *fabD* | Malonyl-CoA : ACP transacylase |
| *fabF* | β-Ketoacyl-ACP synthase II |
| *fabG* | β-Ketoacyl-ACP reductase |
| *fabH* | β-Ketoacyl-ACP-synthase III |
| *fabI* | Enoyl-ACP reductase I |
| *fabK* | Enoyl-ACP reductase II |
| *fabL* | Enoyl-ACP reductase III |
| *fabZ* | β-Hydroxyacyl-ACP dehydrase |
| *fadA* | β-Ketoacyl-CoA thiolase |
| *fadB* | 4-Function enzyme of β-oxidation: β-hydroxyacyl-CoA dehydrogenase and epimerase; *cis*-β-*trans*-2-enoyl-CoA isomerase and enoyl-CoA hydratase |
| *fadD* | Acyl-CoA synthetase |
| *fadE* | Electron transferring flavoprotein |
| *fadF* | Acyl-CoA dehydrogenase |
| *fadG* | Acyl-CoA dehydrogenase? |
| *fadH* | 2,4-dienoyl-CoA reductase |
| *fadL* | Long-chain fatty acid transport protein precursor |
| *fadR* | Transcriptional regulator |
| *fatA* | Unknown, possible transcription factor |
| *gpsA* | Glycerol phosphate synthase |
| *htrB* | $KDO_2$-lipid $IV_A$ acyloxy lauroyltransferase |
| *kdtA* | KDO transferase |
| *lpxA* | UDP-GlcNAc β-hydroxymyristoyl-ACP acyltransferase |
| *lpxB* | Disaccharide-1-P synthase |
| *lpxC* | UDP-β-*O*-hydroxymyristoyl-GlcNAc deacetylase |
| *lpxD* | UDP-β-*O*-hydroxymyristoyl-GlcN *N*-acyltransferase |
| *lpxK* | Disaccharide-1-P 4'-kinase |
| *lpxP* | $KDO_2$-lipid $IV_A$ acyloxy palmitoyltransferase |
| *mdoB* | Phosphatidylglycerol : membrane-oligosaccharide glycerophosphotransferase |
| *msbA* | Lipid flippase |
| *msbB* | $KDO_2$-lipid $IV_A$ acyloxy myristoyltransferase |

Table 1 (continued)

| Gene | Protein |
| --- | --- |
| *pgpA* | PGP phosphatase |
| *pgpB* | PGP phosphatase |
| *pgsA* | PGP synthase |
| *pldA* | Detergent-resistant phospholipase A |
| *pldB* | Inner membrane lysophospholipase |
| *plsB* | Glycerol phosphate acyltransferase |
| *plsC* | 1-Acylglycerol phosphate acyltransferase |
| *plsX* | Unknown |
| *plsD* | Glycerol phosphate acyltransferase |
| *psd* | PS decarboxylase |
| *pssA* | PS synthase |
| *tesA* | Thioesterase I |
| *tesB* | Thioesterase II |

## 3.2. Acetyl-CoA carboxylase

Acetyl-CoA carboxylase catalyzes the first committed step of fatty acid synthesis, the conversion of acetyl-CoA to malonyl-CoA. Acetyl-CoA is a key intermediate in many pathways, and forms the majority of the CoA species within the cell at concentrations of about 0.5–1.0 mM during logarithmic growth on glucose [10]. Malonyl-CoA is normally present at 0.5% of this level, and is used exclusively for fatty acid biosynthesis. The overall carboxylation reaction is composed of two distinct half reactions: the ATP-dependent carboxylation of biotin with bicarbonate to form carboxybiotin; and transfer of the carboxyl group from carboxybiotin to acetyl-CoA, forming malonyl-CoA (Fig. 1).

Each acetyl-CoA carboxylase half reaction is catalyzed by a different protein subcomplex. The vitamin biotin is covalently coupled through an amide bond to a lysine residue on biotin carboxyl carrier protein (BCCP; a homodimer of 16.7-kDa

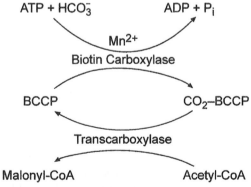

Fig. 1. The acetyl-CoA carboxylase reaction is performed in two steps. Biotin, covalently attached to BCCP (biotin carboxyl carrier protein, *accB*), is carboxylated by the carboxylase subunit (*accC*). The heterodimeric transcarboxylase (*accA* and *accD*) then transfers the $CO_2$ to acetyl-CoA, forming malonyl-CoA.

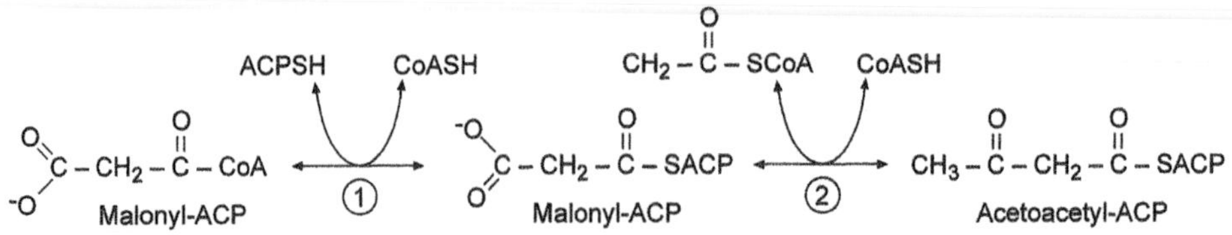

Fig. 2. Initiation of fatty acid synthesis. (1) Malonyl-CoA : ACP transacylase (FabD) transfers the malonyl group from CoA to ACP and then (2) β-ketoacyl-ACP synthase III (FabH) catalyzes the initial irreversible condensation of malonyl-ACP with acetyl-CoA to form acetoacetyl-ACP.

monomers encoded by *accB*) by a specific enzyme, biotin-apoprotein ligase (encoded by *birA*), and is essential to activity. The crystal and solution structures of the biotinyl domain of BCCP have been determined, and reveal a unique 'thumb' required for activity. Carboxylation of biotin is catalyzed by biotin carboxylase (encoded by *accC*), a homodimeric enzyme composed of 55-kDa subunits that is copurified complexed with BCCP. The *accB* and *accC* genes form an operon. The 3-dimensional structure of the biotin carboxylase subunit has been solved by X-ray diffraction revealing an 'ATP-grasp' motif for nucleotide binding. The mechanism of biotin carboxylation involves the reaction of ATP and $CO_2$ to form the short-lived carboxyphosphate, which then interacts with biotin on BCCP for $CO_2$ transfer to the 1'-nitrogen.

The carboxyltransferase enzyme that transfers the carboxy group from the biotin moiety of BCCP to acetyl-CoA is a heterotetramer composed of two copies of two dissimilar subunits, α (35 kDa) and β (33 kDa) (encoded by *accA* and *accD*, respectively). Sequence analysis suggests that the acetyl-CoA binding site lies within the AccA subunit. Strains with mutations in *accB* and *accD* have been obtained that are temperature sensitive for growth, indicating that this reaction is essential. It is thought that the enzyme present in vivo is composed of one copy of each subcomplex, with a combined molecular weight of 280 kDa.

### 3.3. Initiation of fatty acid biosynthesis

For the malonate group to be used for fatty acid synthesis, it must first be transferred from malonyl-CoA to malonyl-ACP by the 32.4-kDa monomeric malonyl-CoA : ACP transacylase, the product of the *fabD* gene (Fig. 2). A stable malonyl-serine enzyme intermediate is formed during the course of the FabD reaction, and subsequent nucleophilic attack on this ester by the sulfhydryl of ACP yields malonyl-ACP. The high reactivity of the serine in malonyl-ACP transacylase is due to the active site being composed of a nucleophilic elbow as observed in α/β hydrolases. The serine is hydrogen bonded to His-201 in a fashion similar to serine hydrolases.

The last two carbons of the fatty acid chain (i.e. those most distal from the carboxylate group) are actually the first introduced into the nascent chain, and acetyl-CoA can be thought of as the 'primer' molecule of fatty acid synthesis in *E. coli*. The initial condensation reaction, catalyzed by β-ketoacyl-ACP synthase III (FabH), utilizes acetyl-CoA and malonyl-ACP to form the four carbon acetoacetyl-ACP with concomitant loss of $CO_2$ (Fig. 2). FabH also possesses acetyl-CoA : ACP transacylase

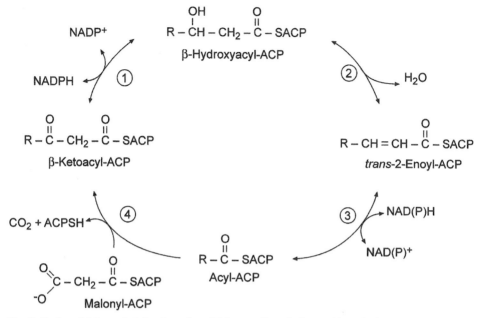

Fig. 3. Cycles of fatty acyl chain elongation. All intermediates in fatty acid synthesis are shuttled through the cytosol as thioesters of the acyl carrier protein (ACP). (1) β-Ketoacyl-ACP reductase (FabG). (2) β-Hydroxyacyl-ACP dehydrase (FabA or FabZ). (3) *trans*-2-Enoyl-ACP reductase I (FabI). (4) β-Ketoacyl-ACP synthase I or II (FabB or FabF).

activity, and for many years it was thought that acetyl-ACP was the actual primer. However, acetyl-ACP appears to be a product of a side reaction, and plays no direct role in fatty acid synthesis.

### 3.4. Elongation of acyl chains

Four enzymes participate in each iterative cycle of chain elongation (Fig. 3). First, β-ketoacyl-ACP synthase I or II (*fabB* or *fabF*) adds a two-carbon unit from malonyl-ACP to the growing acyl-ACP. The resulting ketoester is reduced by a NADPH-dependent β-ketoacyl-ACP reductase (*fabG*), and a water molecule is then removed by a β-hydroxyacyl-ACP dehydrase (*fabA* or *fabZ*). The last step is catalyzed by enoyl-ACP reductase (*fabI*) to form a saturated acyl-ACP, which in turn serves as the substrate for another condensation reaction.

#### 3.4.1. The β-ketoacyl-ACP synthases

Three *E. coli* enzymes catalyze the Claisen condensation that is the β-ketoacyl-ACP synthase reaction. These enzymes are the products of the *fabB*, *fabF*, and *fabH* genes. β-Ketoacyl-ACP synthase I, or FabB, is composed of two identical 42.6-kDa subunits, and has both malonyl-ACP and fatty acyl-ACP binding sites. In the condensation reaction, the acyl group is covalently linked to the active site cysteine. The acyl-enzyme undergoes condensation with malonyl-ACP to form β-ketoacyl-ACP, $CO_2$, and free enzyme.

Overproduction of FabB has two effects: an increased amount of *cis*-vaccenic acid in phospholipids; and resistance to the antibiotics thiolactomycin and cerulenin (Section 12.2). β-Ketoacyl-ACP synthase II (FabF) is very similar to FabB (38% identical at the amino acid level). Like FabB, FabF has a dimeric structure of 43-kDa subunits, and is inhibited by cerulenin and thiolactomycin. FabF is not essential to growth in *E. coli*, but is essential for the regulation of fatty acid composition in response to temperature fluctuations. Mutants lacking FabF activity, unlike wild-type *E. coli*, do not produce increased amounts of the long, unsaturated fatty acid *cis*-vaccenate at lower temperatures. β-Ketoacyl-ACP synthase III (FabH) is a dimeric protein of identical 33.5-kDa subunits first detected as a condensation activity resistant to cerulenin. The FabH reaction is characterized by the preference for a CoA-linked primer, rather than acyl-ACP.

The crystal structures of all three synthases have been determined, and all share a common thiolase fold. Structures of FabB and FabF are virtually identical. Both utilize a conserved catalytic triad of Cys–His–His. FabH has a much more closed active site, reflecting its use of less hydrophobic, shorter chain, substrates than FabB and FabF. FabH also differs in that it contains a catalytic triad composed of Cys–His–Asn. This difference renders FabH more resistant to thiolactomycin and cerulenin than FabB or FabF. The exact reason for this discrepancy is not clear at this time, since the key nitrogens of the His or Asn residues of the respective synthases occupy equivalent space in the respective structures.

### 3.4.2. β-Ketoacyl-ACP reductase

The β-ketoacyl-ACP reductase gene (*fabG*) is located within the *fab* gene cluster between the *fabD* and *acpP* genes and is cotranscribed with *acpP*. Insertional mutants that prevent *fabG* transcription while allowing ACP to be produced were generated in Cronan's laboratory and suggest that *fabG*-encoded reductase activity is essential in *E. coli*. The *fabG*-encoded NADPH-specific β-ketoacyl-ACP reductase is a homotetrameric protein of 25.6-kDa monomers. The protein functions with all chain lengths in vitro and exhibits cooperative binding of NADPH. A dramatic conformational change occurs on cofactor binding, as evidenced by the crystal structures of the free and NADPH-bound protein.

### 3.4.3. β-Hydroxyacyl-ACP dehydrase

*E. coli* possesses two β-hydroxyacyl-ACP dehydrases (more properly termed dehydratases). One is encoded by *fabZ*, and is active on all chain lengths of saturated and unsaturated intermediates. This enzyme is distinct from the dual-function β-hydroxydecanoyl-ACP dehydrase/isomerase (encoded by *fabA*) first described by Bloch and coworkers. The FabA enzyme dehydrates saturated, but not unsaturated, fatty acid intermediates and catalyzes a key isomerization reaction at the point where the biosynthesis of unsaturated fatty acids diverges from saturated fatty acids (Fig. 4). Both enzymes share weak overall homology (28% identity and 50% similarity at the amino acid level). The mono-functional FabZ protein (17 kDa) is somewhat smaller than the *fabA*-encoded bifunctional enzyme (19 kDa). The FabA dehydrase/isomerase has been crystallized, and its structure solved. The active site His is located in a long tunnel which acts a molecular ruler to ensure that only 10-carbon intermediates are isomerized.

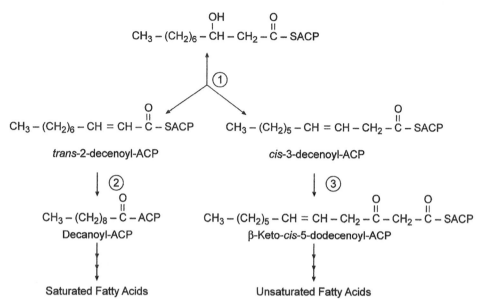

Fig. 4. Branch-point in unsaturated fatty acid synthesis. (1) FabA catalyzes the interconversion of β-hydroxydecanoyl-ACP, *trans*-2-decenoyl-ACP and *cis*-3-decenoyl-ACP. (2) *trans*-2-Decenoyl-ACP is a substrate for enoyl-ACP reductase (FabI), while (3) *cis*-3-decenoyl-ACP is elongated by β-ketoacyl-ACP synthase I (FabB). Competition between FabI and FabB is partly responsible for the ratio of unsaturated fatty acids made.

### 3.4.4. Enoyl-ACP reductase

The final step in each round of fatty acyl elongation in *E. coli* is the NADH-dependent reduction of the *trans* double bond, catalyzed by the homotetrameric (subunit mass of 29 kDa) NADH-dependent enoyl-ACP reductase I (encoded by *fabI*). The FabI amino acid sequence is similar (34% identical) to the product of a gene (called *inhA*) from Mycobacteria. InhA is involved in mycolic acid biosynthesis. The synthesis of these unusual 70–80 carbon mycobacterial acids requires a pathway composed of enzymes essentially identical to those of fatty acid synthesis. Missense mutations within the *inhA* gene result in resistance to the anti-tuberculosis drugs, isoniazid and ethionamide. The crystal structures of FabI and InhA have been solved, and are virtually superimposable for most of the protein. FabI has a flexible substrate binding loop that becomes ordered on binding of the specific inhibitors, diazaborine or triclosan (and presumably substrate), while InhA has two short helices in this region that move 4 Å on drug or substrate binding. Novel enoyl-ACP reductase isoforms II (FabK) and III (FabL) have been discovered recently in Gram-positive bacteria by bioinformatics (Section 11).

### 3.5. Synthesis of unsaturated fatty acids

The pathway described above suffices to produce the straight chain saturated fatty acids found in the membrane phospholipids, mainly palmitic acid (16:0) in *E. coli*.

Gram-negative bacteria also contain unsaturated fatty acids, and the ratio of saturated:unsaturated fatty acids in the membrane phospholipids is a key determinant in membrane fluidity, and changes according to temperature [11]. The *fabA*-encoded dehydrase/isomerase (Section 3.4.3) catalyzes a vital reaction at the branch point of the two pathways: the isomerization of *trans*-2-decenoyl-ACP to *cis*-β-decenoyl-ACP (Fig. 4). The *cis*-3 compound is not a substrate for the enoyl-ACP reductase, but instead is rapidly condensed in a reaction requiring FabB, but not FabF. Strains with mutations in either *fabA* or *fabB* require supplementation with unsaturated fatty acids for growth, showing the specific requirement for both enzymes. As their names suggest, these were the first two mutants in fatty acid biosynthesis identified by classical genetic techniques [12,13].

Both FabB and FabF are capable of participating in saturated and unsaturated fatty acid synthesis and the enzymes have been shown, in vitro, to function similarly with all long-chain acyl-ACPs except palmitoleoyl-ACP. Palmitoleoyl-ACP is an excellent substrate for FabF, but not for FabB. In vivo, the reactivity of FabF towards this substrate increases at lower temperatures, leading to increased amounts of the more fluid *cis*-vaccenate ($18:1\Delta11$) in membrane phospholipids.

### 3.6. Afterword: dissociable or dissociated enzymes?

Historically, scientists in the field have debated whether the enzymes of bacterial fatty acid synthase system are 'dissociable' or 'dissociated'. This seemingly minor semantic distinction has larger ramifications for the in vivo physiology of the cell. The implication of the use of the word *dissociable* is that the enzymes form a complex in vivo, and only became separated (or dissociated) on cell disruption; whereas in a *dissociated* system, the enzymes do not form a complex in vivo. The concept of a large complex mimicking the multifunctional type I enzyme would support the notion of substrate channeling between the active sites, increasing the catalytic efficiency of the pathway as a whole. However, there are no data to support the existence of either a large complex or substrate channeling. A minimal fatty acid synthetic unit must consist of at least six separate activities, encoded by the *acpP*, *fabD*, *fabH*, *fabG*, *fabI*, *fabZ* and *fabB* genes. In vitro, these enzymes appear as monomers (ACP, FadD), dimers (FabH, FabZ and FabB) or tetramers (FabG and FabI). Thus, a complex of these proteins in vivo would possess 16 subunits (assuming only one of each individual complex was present) with a combined mass of over 440 kDa. With our current knowledge of the 3-dimensional structures of these proteins (only FabZ has not been solved to date), it is hard to envision how the pieces of such a puzzle would fit together. Many of the enzymes have active sites located in narrow tunnels, leaving no opportunity for the prosthetic group to swing between them. The acyl-ACP intermediate must instead completely dissociate from the enzyme to interact with the next enzyme in the pathway. Studies with the yeast two-hybrid system have also failed to detect interactions between different enzymes. Thus, type II fatty acids synthases should be considered dissociated and not dissociable.

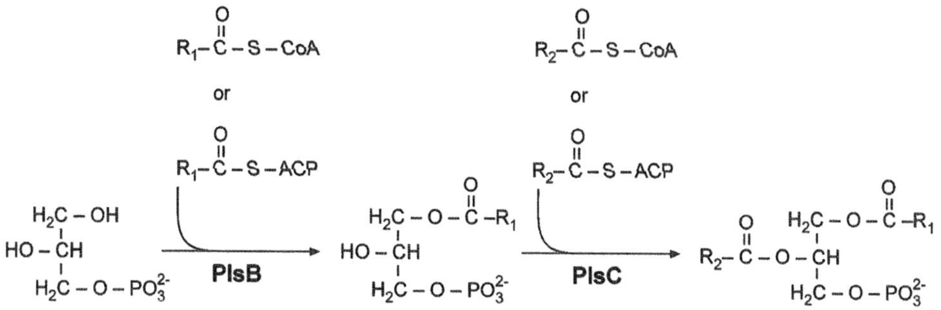

$$R_1-\overset{\overset{\text{O}}{\|}}{C}-S-CoA \qquad\qquad R_2-\overset{\overset{\text{O}}{\|}}{C}-S-CoA$$

or                                                  or

$$R_1-\overset{\overset{\text{O}}{\|}}{C}-S-ACP \qquad\qquad R_2-\overset{\overset{\text{O}}{\|}}{C}-S-ACP$$

$$\begin{array}{c}H_2C-OH\\HO-CH\\H_2C-O-PO_3^{2-}\end{array} \xrightarrow{\textbf{PlsB}} \begin{array}{c}H_2C-O-\overset{\overset{\text{O}}{\|}}{C}-R_1\\HO-CH\\H_2C-O-PO_3^{2-}\end{array} \xrightarrow{\textbf{PlsC}} \begin{array}{c}H_2C-O-\overset{\overset{\text{O}}{\|}}{C}-R_1\\R_2-\overset{\overset{\text{O}}{\|}}{C}-O-CH\\H_2C-O-PO_3^{2-}\end{array}$$

Fig. 5. Transfer of fatty acyl groups to the membrane. A (primarily saturated) fatty acid is transferred from acyl-ACP or acyl-CoA to the 1 position of *sn*-glycerol-3-phosphate by glycerol phosphate acyltransferase (PlsB). A second fatty acid is transferred to the 2-position by the 1-acylglycerol phosphate acyltransferase (PlsC).

## 4. Transfer to the membrane

Fatty acid biosynthesis in *E. coli* normally ends when the acyl chain is 16 or 18 carbons in length. These acyl-ACPs are now substrates for the acyltransferases that will transfer the fatty acyl chain into the membrane phospholipids (Fig. 5). Alternatively, *E. coli* can incorporate exogenous fatty acids, following esterification to CoA. The first enzyme (the *plsB* gene product) transfers fatty acids from either the soluble acyl-ACP or acyl-CoA to the 1-position of glycerol phosphate. The product of the reaction, 1-acylglycerol phosphate, partitions into the membrane. The PlsB protein is an integral inner membrane protein of 91 kDa, and has a preference for saturated fatty acids. The second acyltransferase (the *plsC* gene product), a membrane protein of 27 kDa, esterifies the 2-position of the glycerol backbone and prefers unsaturated acyl chains. Thus, bacterial phospholipids have an asymmetric distribution of fatty acids between the 1- and 2-positions of the glycerol phosphate backbone. The glycerol phosphate acyltransferase system influences the chain length of the fatty acids incorporated into the phospholipids by competition with the elongation condensing enzymes for acyl-ACP, and the rate of fatty acid biosynthesis via modulation of its activity by ppGpp (Section 10.5).

The isolation of *E. coli* mutants with defective acyltransferase activity (*plsB*) by Bell's laboratory heralded a major advance in the study of the acyltransferases. These mutants were glycerol phosphate auxotrophs and exhibited an increased Michaelis constant for glycerol phosphate in in vitro acyltransferase assays. The increased $K_m$ was subsequently shown to arise from a single missense mutation in the open reading frame. Therefore, *plsB* mutants require an artificially high intracellular concentration of glycerol phosphate for activity. Complementation of these mutants facilitated the cloning of the glycerol phosphate acyltransferase. Plasmids that suppressed the glycerol phosphate requirement of *plsB* strains overexpressed glycerol phosphate acyltransferase activity 10-fold. The protein possesses a catalytic His–Asp dyad common to all glycerolipid acyltransferases. The glycerol phosphate acyltransferase is specifically activated by acidic phospholipids, phosphatidylglycerol and cardiolipin, as shown by micelle assays containing detergents and phospholipid, and is active as a monomer. Glycerol phosphate acyltransferase also exhibits negative cooperativity with respect to glycerol phosphate binding, a property that may account in part for the finding that dramatic increases in the intracellular glycerol phosphate concentration do not increase the amount of phospholipid in *E. coli*.

The interpretation of the *plsB* mutants is complicated by the finding that the *plsB* growth phenotype depends on mutations in two unlinked genes. One mutation is in the *plsB* gene discussed above, and the second is in a gene called *plsX*. Both mutations are required for a strain to exhibit a requirement for glycerol phosphate since strains harboring either the *plsB* or *plsX* lesion do not have a defective growth phenotype. The *plsX* gene is located in the *fab* cluster next to *fabH* and predicted to encode a protein of 37.1 kDa. Despite extensive study, no enzymatic activity has been described for the PlsX protein. It is possible that the native PlsB protein exists in a complex with PlsX. Analysis of the genomic databases reveals many bacteria contain *plsX* homologues, but provides no further clues as to its biochemical function.

The next step in phospholipid biosynthesis is catalyzed by 1-acylglycerol phosphate acyltransferase (the *plsC* gene product) which acylates the product of the PlsB step to form phosphatidic acid (Fig. 5). Phosphatidic acid comprises only about 0.1% of the total phospholipid in *E. coli* and turns over rapidly, a property consistent with its role as an intermediate in phospholipid synthesis. The 1-acyl-glycerol phosphate acyltransferase is thought to transfer unsaturated fatty acids selectively to the 2-position.

Certain bacteria, such as Clostridia, do not possess a *plsB* homologue. Instead, the *plsD* gene was isolated from a *Clostridium difficile* genomic library by functional complementation of the *plsB* phenotype of *E. coli*. The protein encoded by *plsD* is a glycerol phosphate acyltransferase, but shares no homology with the *plsB*-encoded enzyme, except for a predicted His–Asp dyad active site. Overall, it appears more similar to the *plsC*-encoded 1-acylglycerol phosphate acyltransferase of *E. coli* in both amino acid composition (27% identity) and size (26.5 kDa). However, there are many bacterial species that lack both *plsB* and *plsD* genes, thus the enzyme(s) that acylates glycerol phosphate in these organisms remains unknown.

# 5. Phospholipid biosynthesis

*E. coli* possesses only three major phospholipid species in its membranes, making it one of the simplest organisms to study with regard to phospholipid biosynthesis. Phosphatidylethanolamine (PE) comprises the bulk of the phospholipids (75%), with phosphatidylglycerol (PG) and cardiolipin (CL) forming the remainder (15–20% and 5–10%, respectively). The scheme for the synthesis of membrane phospholipids follows the classic Kennedy pathway (Fig. 6).

## 5.1. Phosphatidate cytidylyltransferase

The key activated intermediate in bacterial phospholipid synthesis, CDP-diacylglycerol, comprises only 0.05% of the total phospholipid pool. The 27.6-kDa enzyme phosphati-

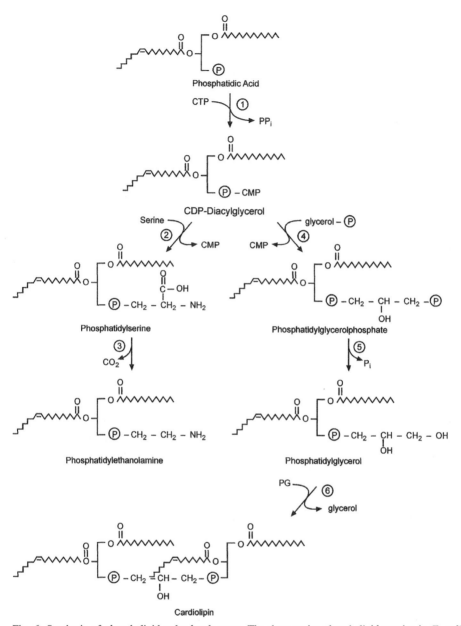

Fig. 6. Synthesis of phospholipid polar headgroups. The three major phospholipid species in *E. coli* are synthesized by a total of six different enzymatic activities: (1) phosphatidate cytidylyltransferase (Cds); (2) phosphatidylserine synthase (Pss); (3) phosphatidylserine decarboxylase (Psd); (4) phosphatidylglycerolphosphate synthase (PgsA); (5) phosphatidylglycerolphosphate phosphatase (PgpA or PgpB); and (6) cardiolipin synthase (Cls).

date cytidylyltransferase (or CDP-diacylglycerol synthase) catalyzes the conversion of phosphatidic acid to a mixture of CDP-diacylglycerol and dCDP-diacylglycerol. Strains of *E. coli* with mutations in the *cds* gene retain 5% of the normal levels of CDP-diacylglycerol synthase and grow normally under standard laboratory conditions, although are hypersensitive to erythromycin and elevated pH. Thus, CDP-diacylglycerol synthase is present in large excess of the minimum amount of enzyme required to sustain phospholipid synthesis. These mutants accumulate substantial amounts of phosphatidic acid (up to 5% of the total phospholipid). Null mutations in *cds* have not been reported, and would presumably be non-viable due to the complete lack of phospholipid synthesis from this point.

CDP-diacylglycerol stands at the branch point between PE synthesis and PG and CL synthesis (Fig. 6). It has been hypothesized that the presence of both ribo and deoxyribo forms of the liponucleotide could play a role in determining the relative amount of intermediate entering these two arms of the pathway. For this to be true, the respective synthases that utilize this compound would have to be selective toward either dCDP- or CDP-diacylglycerol. In vivo the ratio of dCDP- to CDP-diacylglycerol is 0.88. A change in this ratio to 3.1 has no effect on the relative rates of PE and PG synthesis in vivo, arguing against selectivity of the subsequent enzymes for one form of the other. Further, both ribo- and deoxyribo-liponucleotides are substrates for PS synthase in vitro, and thus, the significance, if any, of the two forms of liponucleotide remains to be determined.

### 5.2. Phosphatidylethanolamine production

#### 5.2.1. Phosphatidylserine synthase

The first step in the synthesis of PE is the condensation of CDP-diacylglycerol with serine catalyzed by PS synthase to form PS. During cell disruption, the 58-kDa PS synthase appears associated with ribosomes, but reattaches to the membrane vesicles once substrate is added. PS is a minor membrane constituent of *E. coli* since it is rapidly converted to PE by PS decarboxylase. Mutants in the *pss* gene encoding PS synthase are viable only when supplemented with divalent metal ions. PE is capable of forming the hexagonal (non-bilayer) $H_{II}$ lipid phase, and Dowhan has demonstrated that the divalent cations interact with CL to replace the function of PE in the formation of an $H_{II}$ phase (Chapter 1). The cells lack PS synthase activity, and thus contain no PS or PE in their membranes. There are also perturbations in the function of permeases, electron transport, motility and chemotaxis.

#### 5.2.2. Phosphatidylserine decarboxylase

PS is decarboxylated by PS decarboxylase to yield the zwitterionic PE. This inner membrane enzyme has a subunit molecular mass of 36 kDa. PS decarboxylase has a pyruvate prosthetic group that participates in the reaction by forming a Schiff base with PS. Overproduction of the enzyme 30–50-fold by plasmid-borne copies of the *psd* gene has no effect on membrane phospholipid composition indicating that the level of this enzyme does not regulate the amount of PE in the membrane. The majority of the PE is found in the periplasmic leaflet of the inner membrane, and there is a rapid flipping from the inner to outer leaflet by the MsbA lipid flippase (Section 7).

Mutants with a temperature-sensitive decarboxylase accumulate PS at the non-permissive temperature. The mutants continue to grow for several hours after the shift to the non-permissive temperature, despite the reduced levels of PE and the concomitant increase in PS. Complete inactivation of *psd* by insertional mutagenesis has the same divalent cation-requiring phenotype as the *pss* mutants described above. The requirement for CL is consistent with the inability to introduce a null *cls* allele into *pss* strains. Thus, PE is essential for the polymorphic regulation of lipid structure. Evidence from the Dowhan laboratory demonstrates that PE is a molecular chaperone that is essential for the proper folding of integral membrane proteins (Chapter 1). All of the physiological processes dependent on the formation of local regions of non-bilayer structure or that specifically require PE remain to be elucidated, but the process of cell division, the formation of contacts between inner and outer membranes, and the translocation of molecules across the membrane are viable candidates.

## 5.3. Phosphatidylglycerol synthesis

### 5.3.1. Phosphatidylglycerolphosphate synthase
CDP-diacylglycerol is condensed with glycerol phosphate to form phosphatidylglyc-erolphosphate (PGP), an intermediate in the production of the acidic phospholipids PG and CL (Fig. 6). The reaction is analogous to the synthesis of PS, with the product CMP being released. Mutants (*pgsA*) defective in PGP synthesis contain less than 5% of normal PGP synthase activity in vitro, however, there is no growth phenotype associated with these mutants. The PgsA protein is predicted to be a 20.7-kDa integral membrane protein. It has long been thought that PGP synthase is essential, and that cells cannot survive without acidic phospholipids. There are many important cellular functions that are affected by reduced PG and/or CL content of the membrane. PG is required for protein translocation across the membrane, and acidic phospholipids are required for channel activity of bacterial colicins and the interaction of antibiotics with the membrane. Cell division proteins such as FtsY also apparently require acidic phospholipids for activity, as does the DnaA protein involved in chromosome segregation. However, Matsumoto has recently inactivated the *pgsA* gene with a kanamycin cassette. This *pgsA::kan* strain has no detectable PG or CL, is not viable above 40°C, and contains increased concentrations of PA, which may at least partially compensate for the absence of PG and CL.

### 5.3.2. PGP phosphatases
The second step in the synthesis of PG is the dephosphorylation of PGP (Fig. 6). Two independent genes have been identified, *pgpA* and *pgpB*, that encode PGP phosphatases based on an in vitro assay. Both proteins are small (19.4 and 29 kDa, respectively) but share no sequence homology. In vitro, the *pgpA*-encoded phosphatase specifically hydrolyzes PGP, whereas the PgpB phosphatase also hydrolyzes phosphatidic acid. However, disruption of both of these genes in a single strain did not impair PG synthesis, although the respective phosphatase activities were reduced. Thus, neither of these phosphatases is required for PG synthesis suggesting that another phosphatase capable of operating in the PG biosynthetic pathway remains to be discovered.

## 5.4. Cardiolipin biosynthesis

Unlike in mammalian mitochondria, where CL is synthesized by the reaction of CDP-diacylglycerol with PG (Chapter 8), CL is produced in bacteria by the condensation of two PG molecules (Fig. 6). CL accumulates as the cells enter the stationary phase of growth, and is required for prolonged survival of the bacteria. CL synthase is post-translationally processed from a 55-kDa precursor to a 45–46-kDa form. Mutants deficient in CL synthase (*cls*) possess very low levels of CL and lose viability in stationary phase. The mutants also grow at a slower rate and to a lower density than the corresponding wild-type cells. Low, residual concentrations of CL are present in the *cls* null mutants, hampering efforts to isolate the role of this lipid in cell physiology. A second gene has recently been described that can catalyze the formation of CL in vitro, but does not appear to do so in vivo. Amplification of CL synthase leads to the overproduction of CL, a decrease in membrane potential, and loss of viability. Therefore, *E. coli* can tolerate changes in the overall CL content but the elimination or significant overproduction of CL leads to significant physiological imbalance.

## 5.5. Cyclopropane fatty acids

Fatty acids attached to membrane phospholipids can be post-synthetically converted to their cyclopropane derivatives during the stationary phase of bacterial growth. Their biosynthesis and function have been elucidated by the Cronan laboratory. *E. coli* mutants that completely lack cyclopropane fatty acid synthase activity (owing to null mutations in the *cfa* gene) grow and survive normally under virtually all conditions, except that *cfa* mutant strains are more sensitive to freeze–thaw treatment and acid shock than are isogenic *cfa*+ strains. Thus, the stable cyclopropane derivative protects the reactive double bond from adverse reactions during stationary phase. Cyclopropanation involves a significant energy commitment by the cell: the reaction uses *S*-adenosylmethionine, which requires three molecules of ATP are required for regeneration. It is not known how the soluble enzyme and substrate gain access to the phospholipids of the inner and outer membranes.

The 44-kDa cylcopropane synthase protein is metabolically unstable, but protein levels peak sharply due to increased *cfa* transcription as cultures enter the stationary phase. Cfa levels drop in late stationary phase cultures as the enzyme is destroyed by proteolysis, probably by a protease of the heat shock response. The *cfa* gene possesses two promoters of approximately equal strengths, with the more distal promoter functioning through-out the growth cycle. The proximal promoter requires the specialized sigma factor, $\sigma^S$ (encoded by *rpoS*), for transcription, and is thus active only as cultures enter stationary phase. Indeed, the cyclopropane fatty acid content of *rpoS* strains is low and transcription from the proximal promoter is absent in these strains. As cells remain in stationary phase and phospholipid biosynthesis ceases, the low levels of Cfa that do persist no longer en-counters an expanding substrate pool. Thus, an increasing amount of the fatty acyl chains are converted to their cyclopropane derivatives over time. The instability of Cfa results in little carry-over of synthetic capacity when exponential growth resumes, and the existing cyclopropane fatty acids are quickly diluted by de novo phospholipid synthesis.

## 6. Lipid A biosynthesis

Lipopolysaccharides (LPS) form the majority of the outermost leaflet of the membrane in most Gram-negative bacteria (Fig. 4 in Chapter 1), and display a tremendous amount of structural variability [14]. LPS is essential to the growth of Gram-negative bacteria, and provides an effective hydrophobic barrier to toxic compounds. LPS are comprised of three components: the O-antigen, a core polysaccharide and lipid A. The O-antigen is a polysaccharide that extends from the cell surface. O-antigens are constructed from 10 to 30 repeats of specific β-6 sugar oligosaccharide units, and each is essentially unique to a given serotype of bacteria. The O-antigen is linked to the core polysaccharide region, which is common to groups of bacteria. The membrane associated portion of LPS is lipid A. The core polysaccharide is attached to lipid A by a 2-keto-β-deoxyoctonate (KDO) disaccharide. Lipid A anchors the LPS to the outer membrane and functions as an endotoxin and a mitogen during bacterial infections. The lipid A is synthesized and ligated to the oligosaccharide core on the cytoplasmic face of the inner membrane, while the O-antigen is added in the periplasm. O-antigen is not essential for the viability of *E. coli*, and is in fact missing from *E. coli* K12, making it safe for laboratory use. Details of the synthesis of the O-antigen and core region are outside the scope of this discussion and can be found in a review by Raetz [14].

The pathway of lipid A synthesis has been determined mainly in the laboratory of Raetz (Fig. 7). The first step in lipid A synthesis is the reversible transfer of the β-hydroxymyristoyl group from ACP to UDP-*N*-acetylglucosamine (UDP-GlcNAc) by the UDP-GlcNAc acyltransferase (*lpxA*). Competition between LpxA and the FabZ dehydratase of fatty acid synthesis help determine the rate of lipid A synthesis. The second step in the pathway is the deacetylization of UDP-3-Acyl-GlcNAc by the zinc-dependent 34-kDa UDP-β-*O*-acyl-GlcNAc deacetylase (*lpxC*). The *lpxC* gene was first described as *envA*, an essential gene involved in envelope production. Mutations in *lpxC* cause a plethora of effects, including increased sensitivity to antibiotics, increased dye permeability and defects in cell division. Null mutations in *lpxC* are lethal. The LpxA-catalyzed acyltransfer step is thermodynamically unfavorable, thus the irreversible deacetylase reaction is the first committed step in the pathway. The essential nature of LpxC has driven the development of a novel group of antimicrobial compounds active against this step (Section 12). These compounds are effective against a wide range of Gram-negative bacteria in vivo.

The third step is a second β-hydroxymyristoyl-ACP acyltransferase, catalyzed by UDP-β-*O*-[β-hydroxymyristoyl]GlcN acyltransferase (*lpxD*). Like the LpxA acyltransferase, the 36-kDa LpxD possesses repeating hexapeptide units and will presumably have a fold and trimeric structure similar to LpxA. The product of the LpxD reaction is UDP-2,3-diacyl-GlcN. UMP is removed from this compound to form 2,3-diacyl-GlcN-1-phosphate (lipid X), but the enzyme responsible for this reaction is not known. There is an approximately 10-fold excess of lipid X over UDP-2,3-diacyl-GlcN in wild-type cells. Mutations in *lpxB*, cause a 500-fold increase in the amount of lipid X in the membrane, although LpxB is not a pyrophosphatase. LpxB, a dimer of 42-kDa monomers, catalyzes the condensation of lipid X with UDP-2,3-diacyl-GlcN to form the lipid A disaccharide-1-phosphate. *lpxB* is present in a complex cluster with *lpxA*, *lpxD*

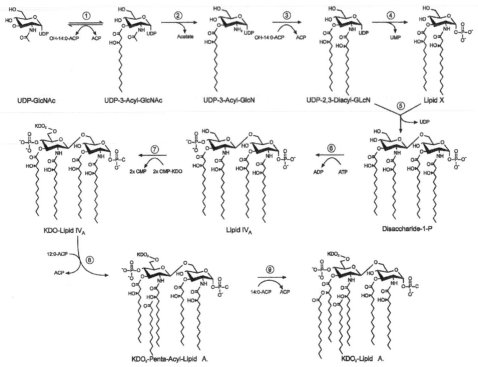

Fig. 7. Biosynthesis of endotoxin in *E. coli*. The first step (1) in the pathway is catalyzed by UDP-N-acetylglucosamine (UDP-GlcNAc) acyltransferase (LpxA). (2) The committed step is catalyzed by the LpxC deacetylase, followed by (3) a second acyltransferase (LpxD). (4) Lipid X is generated by the removal of UMP from UDP-2,3-diacyl-GlcNac by an unknown enzyme. (5) Lipid X and UDP-2,3-diacyl-GlcNac are then condensed together by LpxD to form lipid IV$_A$. (6) A 4'-kinase phosphorylates the disaccharide to produce lipid IV$_A$. (7) Two consecutive additions of KDO by KdtA, and two O-acylations by (8) HtrB and (9) MsbB yield KDO$_2$-lipid A. Subsequent addition of core sugars and O-antigen chains (not shown) yield the mature lipopolysaccharide.

and *fabZ*, and other genes encoding enzymes of phospholipid synthesis, outer membrane proteins, DNA synthesis and fatty acid synthesis.

Strains mutated in *lpxK* accumulate lipid A disaccharide-1-P, which led to the cloning and characterization of lipid A disaccharide kinase, a Mg$^{2+}$-dependent 4'-kinase activity stimulated by CL, that makes the key intermediate, lipid IV$_A$. The presence of the 4'-phosphate is essential for the recognition of lipid A by mammalian cells during endotoxin stimulation. Two subsequent additions of KDO are catalyzed by the KDO transferase, the 47-kDa KdtA protein. A single protein catalyzes both additions, since overexpression of *kdtA* causes a large increase in both transfers. KDO transferase activity is essential for growth, and conditional mutants accumulate massive amounts of lipid IV$_A$ prior to cessation of growth at the non-permissive temperature.

In the final steps of lipid A synthesis, two fatty acids are transferred to the hydroxyl groups of the β-hydroxymyristate on the distal unit. The first of these is usually laurate (12 carbons) added from lauroyl-ACP by the enzyme HtrB, and then MsbB adds myristate from myristoyl-ACP. During cold shock (on shifting *E. coli* from 30 to 12°C), palmitoleate appears in the lipid A at the expense of laurate. A novel gene, *lpxP*, is induced for this reaction, and null mutants in HtrB are not defective in this adaptation. Double null *htrB msbB* mutants can also be generated. Thus, the extensive acylation of the lipid A is not absolutely required for its proper insertion into the outer membrane or the formation of a hydrophobic barrier.

## 7. Phospholipid flippase

All of the enzymes of phospholipid and lipid A biosynthesis are either cytosolic or located on the inner aspect of the inner membrane. How then do the lipids get to the outer face of the inner membrane, or to the outer membrane? Recent work from Raetz's laboratory has shown that the product of the *msbA* gene is a lipid 'flippase' required for the trafficking of lipids across the inner membrane. A temperature-sensitive mutant, *msbA*(Ts), with a A270T substitution in the MsbA protein, shows a rapid and dramatic reduction in the export of all major lipid classes, including PE and core-lipid A, to the outer membrane. Lipid export is inhibited by >90% after 30 min at the non-permissive temperature, while protein transport is not affected. *E. coli* harboring this mutation possess duplicated inner membranes at the elevated temperature. Null mutations in *msbA* are lethal, and this is the only bacterial transporter that has been shown to be essential. The MsbA protein is a member of the ATP-binding cassette (ABC) family of transporters, and has most similarity to mammalian P-glycoprotein multiple-drug resistance ABC transporters (>30% identity). MsbA exists as a homodimer of 64.6-kDa monomers, each of which has a single membrane spanning region (comprised of six transmembrane helices) and a nucleotide binding domain. This arrangement is distinct from the mammalian multiple-drug resistance pumps, which have two membrane spanning regions and two nucleotide binding domains fused into a single polypeptide. Thus, MsbA is a 'half-transporter', and is actually homologous to both amino- and carboxy-terminal halves of the mammalian multiple-drug resistance protein.

The crystal structure of the integral membrane MsbA has been determined by Chang and Roth to 4.5 Å [15]. The protein traverses the entire inner membrane and is cone-shaped, with an opening of about 25 Å on the cytoplasmic side, leading into a chamber of sufficient size for core-lipid A. The nucleotide binding domains are on the cytoplasmic face, and share no contacts. Dimer contacts are made in the half of the protein in the outer membrane. A model for transport has been suggested in which the lipid A or phospholipid enters the chamber through the cytoplasmic membrane, the chamber then closes allowing for flip-flop of the molecule, and then the chamber re-opens, and the lipid is ejected. Presumably, the lipid is released into the outer leaflet of the inner membrane, since MsbA does not extend across the periplasm. Thus, lipids that are required in the outer membrane must also traverse the periplasm. LPS molecules must also be flipped across the outer membrane to be displayed on the surface of the bacteria. Currently, nothing is known about these processes.

# 8. Degradation of fatty acids and phospholipids

## 8.1. β-Oxidation of fatty acids

### 8.1.1. Transport of fatty acids across the membrane

Exogenous long chain fatty acids are utilized by *E. coli* in two ways. Firstly, they can be incorporated into the membrane phospholipids by the acyltransferase system (PlsB and PlsC; Section 4). Secondly, they can be used as the sole carbon source for growth, and are in fact an important source of energy for *E. coli* in their normal habitat, the intestine [16]. The CoA thioester of the fatty acid is the substrate for both of these pathways. Fatty acids greater than 10 carbons in length require the *fadL* gene product to be taken up from the growth medium in sufficient quantities to support growth. FadL is a 46-kDa outer membrane protein produced following the cleavage of a 28-residue signal peptide from the propeptide. Fatty acid uptake is closely coupled to acyl-CoA formation, since very low levels of free fatty acid are found in the cells. The acyl-CoA synthetase, encoded by *fadD*, is a homodimer of 62-kDa subunits, and associates with the cytoplasmic leaflet of the inner membrane. Strains mutated in *fadD* cannot produce acyl-CoA and thus cannot grow on exogenous fatty acids, nor incorporate them into their membrane phospholipids. The esterification of the free fatty acid to CoA traps the fatty acid inside the cell, driving its transport across the inner membrane, and the net accumulation of fatty acid from the medium. Medium chain fatty acids do not require FadL to enter the cells, and may traverse the outer membrane by passive diffusion.

### 8.1.2. Degradation of fatty acids

Degradation of fatty acids proceeds via an inducible set of enzymes that catalyze the pathway of β-oxidation [16]. β-Oxidation occurs via repeated cycles of reactions that are essentially the reverse of the reactions of fatty acid synthesis (Fig. 8). However, three major differences distinguish the two pathways. Firstly, β-oxidation utilizes acyl-CoA thioesters, and not acyl-ACPs. Secondly, the β-hydroxy intermediates have the opposite stereochemistry (L in β-oxidation and D in synthesis). Finally, the enzymes of β-oxidation share no homology with those of synthesis.

The first step in the pathway is the dehydrogenation of acyl-CoA by the enzyme acyl-CoA dehydrogenase. While other organisms have several dehydrogenases with different chain length specificities (i.e. for short, medium or long acyl chains), it has been reported that *E. coli* has one enzyme active on all chain lengths. The dehydrogenase has been linked to two mutations in the 5 min region of the *E. coli* chromosome, *fadF* and *fadG*. However, the genome sequence suggests that a single gene in this region, *yahF*, encodes for a 92-kDa dehydrogenase. Thus, *fadF* and *fadG* probably represent different mutations in the same YahF dehydrogenase, and do not encode two distinct proteins. The acyl-CoA dehydrogenase is a flavoprotein, and is linked to a electron transferring flavoprotein (*fadE*). Further confusion exists in the literature as many acyl-CoA dehydrogenase (YafH) homologues are annotated as FadE homologues. Mutant strains of *E. coli* blocked in β-oxidation with *fadE* or *fadF* (*yahF*) mutations can accumulate acyl-CoA species, but cannot degrade them.

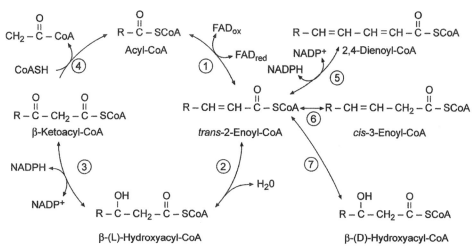

Fig. 8. β-Oxidation of fatty acids in *E. coli*. Long chain fatty acids are transported into the cell by FadL and converted to their CoA thioesters by FadD (not shown). The acyl-CoAs are substrates for the (1) acyl-CoA dehydrogenase (YafH) to form a *trans*-2-enoyl-CoA. The double bond is reduced by (2) trans-2-enoyl-hydratase (crotonase) activity of FadB. The β-hydroxyacyl-CoA is then a substrate for the NADP+-dependent dehydrogenase activity of FadB (3). A thiolase, FadA (4), releases acetyl-CoA from the β-ketoacyl-CoA to form an acyl-CoA for subsequent cycles. (5) Polyunsaturated fatty acids are reduced by the 2,4-dienoyl-CoA reductase (FadH). (6) FadB also catalyzes the isomerization of *cis*-unsaturated fatty acids to *trans*. (7) The epimerase activity of FadB converts D-β-hydroxy thioesters to their L-enantiomers via the *trans*-2-enoyl-CoA.

The second step in the cycle is enoyl-CoA hydratase, an activity commonly referred to as 'crotonase'. Traditionally, in vitro measurements of this activity utilize crotonoyl-CoA (*trans*-2-butenoyl-CoA) as the substrate. Crotonase activity in *E. coli* is one function present in a multifunctional protein encoded by *fadB*. The next step in the cycle is β-hydroxyacyl-CoA dehydrogenase, another function of the FadB enzyme. The β-ketoacyl-CoA produced in this reaction is a substrate for the monofunctional *fadA*-encoded β-ketoacyl thiolase, which cleaves acetyl-CoA from the acyl-CoA to produce an acyl-chain two carbons shorter than when it entered the cycle. The cycle is then repeated until the fatty acid is metabolized. The FadB protein is a homodimer of 78-kDa subunits, and is purified in complex with the homodimeric 42-kDa *fadA* gene product. The total complex is thus an $\alpha_2\beta_2$ heterotetramer with an apparent mass of about 260 kDa. The *fadA* gene encodes the β-subunit, while the *fadB* gene gives the α-subunit, the confusing nomenclature a remnant of the days of classical genetics.

Unsaturated fatty acids can also be degraded by the β-oxidation pathway. The FadB protein possesses *cis*-β-enoyl-CoA isomerase activity, which converts *cis*-3 double bonds to *trans*-2 (Fig. 8). A 2,4-dienoyl-CoA reductase encoded by *fadH* is also required for the metabolism of polyunsaturated fatty acids (Fig. 8). This protein is a 73-kDa monomeric, NADP+-dependent, 4Fe–4S flavoprotein. The FadH protein can utilize compounds with either *cis* or *trans* double bonds at the 4-position. An epimerase activity of FadB allows for the utilization of D-hydroxy fatty acids. The epimerase is actually a combination of a D-β-hydroxyacyl-CoA dehydratase and the

Table 2

Phospholipid degradative activities in *E. coli*

| Enzyme | Gene | Location | Substrates |
|---|---|---|---|
| Phospholipase A1 | *pldA* | Outer membrane | Phosphatidylethanolamine, phosphatidylglycerol, cardiolipin and lyso derivatives |
| Phospholipase A | | Cytoplasm | Phosphatidylglycerol |
| Lysophospholipase L2 | *pldB* | Inner membrane | Lyso-phosphatidylethanolamine |
| Lysophospholipase | | Cytoplasm | Lyso-phosphatidylethanolamine, lyso-phosphatidylglycerol |
| Phospholipase C | | Unknown | Phosphatidylethanolamine |
| Phospholipase D | | Cytoplasm | Cardiolipin |
| Phospholipase D | | Cytoplasm | Phosphatidylserine |
| Lipase | | Membrane | Triacylglycerol |
| CDP-diacylglycerol hydrolase | | Inner membrane | CDP-diacylglycerol |
| Phosphatidic acid phosphatase | | Membrane | Phosphatidic acid |
| Thioesterase I | *tesA* | Periplasm | Acyl-CoA |
| Thioesterase II | *tesB* | Cytoplasm | Acyl-CoA |

crotonase (hydratase) activities, resulting in the conversion of the D to the L enantiomer (Fig. 8).

The substrate specificities of the enzyme complex in vitro suggests that all of the enzymes can utilize all chain lengths of substrates, with the possible exception of the crotonase activity. This function of FadB appears somewhat limited to short chain substrates, and it has been suggested that a separate long chain enoyl-CoA hydratase may exist in *E. coli*. Two open reading frames in the *E. coli* genome, discovered by bioinformatics, are predicted to encode homologs of FadA and FadB. Thus, two complexes may be present with preferences for long or short chain acyl-CoAs.

### 8.2. Phospholipases

Based mainly on cell free assays, 10 enzymatic activities that degrade phospholipids, intermediates in the phospholipid biosynthetic pathway, or triacylglycerol have been reported (Table 2). The detergent-resistant phospholipase A$_1$ (encoded by *pldA*) of the outer membrane, characterized by Nojima and colleagues, is the most studied of these enzymes. This enzyme is unusually resistant to inactivation by heat and ionic detergents and requires calcium for maximal activity. The mature phospholipase has a subunit molecular mass of 31 kDa. Hydrolysis of fatty acids from the 1-position of phospholipids is the most rapid reaction, but the enzyme will also hydrolyze 2-position fatty acids, as well as both isomeric forms of lysophosphatides and mono- and di-acylglycerols. A detergent-sensitive phospholipase A$_1$ has also been described, although this activity has not been designated to a gene. This enzyme differs from the detergent-resistant protein in that it is located in the soluble fraction of the cell, is inactivated by heat and ionic detergents, and has a high degree of specificity for PG. The cytoplasmic phospholipase A also requires calcium for activity. There are also inner membrane and cytoplasmic lysophospholipases. The best characterized of these is the inner membrane lysophospholipase L2 (*pldB*) which hydrolyzes 2-acyl-glycerophosphoethanolamine efficiently, but is barely active on the 1-acyl isomer. This lysophospholipase also catalyzes the transfer of fatty acids from 2-acyl-glycerophosphoethanolamine to PG to form acyl-PG.

The physiological role of these degradative enzymes remains unknown. Mutants lacking the detergent-resistant phospholipase (*pldA*), lysophospholipase L2 (*pldB*), or both enzymes do not have any obvious defects in growth, phospholipid composition, or turnover. Moreover, strains that overproduce the detergent-resistant enzyme also grow normally. It has been established that the detergent-resistant phospholipase is responsible for the release of fatty acids from phospholipids that occurs during infection with T4 and λ phages. However, phospholipid hydrolysis is not essential for the life cycle of these bacteriophages. One possible function for the hydrolytic activities with unassigned genes is that they are actually biosynthetic proteins (acyltransferases) that act as lipases in the absence of suitable acceptor molecules in the assay systems employed. An example of such an enzyme is PS synthase, which catalyzes both phospholipase D and CDP-diacylglycerol hydrolase reactions. PS synthase appears to function via a phosphatidyl–enzyme intermediate, and in the absence of a suitable acceptor such as serine or CMP, the phosphatidyl–enzyme complex is hydrolyzed by water; thus, the enzyme acts as a phospholipase D. Also, some of these enzyme activities may reflect a broad substrate specificity of a single enzyme rather than the presence of several distinct protein species. For example, the observed lipase activity that cleaves the 1-position fatty acids from triacylglycerols (a lipid usually not found in *E. coli*) may arise from the detergent-resistant phospholipase $A_1$ acting on triacylglycerol as an alternate substrate. Further, a lysophospholipase L(1) activity has been attributed to thioesterase I. Three open reading frames (*ybaC*, *yhjY*, and *yiaL*) present in the *E. coli* genome potentially encode proteins with lipase activity, but have not been studied to date.

## 8.3. Thioesterases

Thioesterases preferentially cleave the thioester bond of acyl-CoA molecules to produce CoA and free fatty acid. *E. coli* contains two well characterized thioesterases. Thioesterase I (encoded by *tesA*) is a periplasmic enzyme of 20.5 kDa, with a substrate specificity for acyl chains >12 carbon atoms. Thioesterase I hydrolyzes synthetic substrates used in the assay of chymotrypsin, which led to the initial conclusion that TesA was a protease ('protease I'). However, the purified protein does not cleave peptide bonds. Thioesterase I also appears to possess lysophospholipase L(1) activity. Thioesterase II is a cytosolic tetrameric protein composed of 32-kDa subunits encoded by the *tesB* gene. Thioesterase II cleaves acyl-CoAs of >6 carbons and β-hydroxyacyl-CoAs, but is unable to cleave acyl-pantetheine thioesters. The physiological function of thioesterases I and II is unknown. Null mutants have been constructed in both *tesA* and *tesB*, and the double mutant strain generated. None of these strains has an observable growth phenotype indicating that neither protein is essential. However, the *tesAB* double-null mutant still retains about 10% of the total wild-type thioesterase activity indicating the existence of a third unidentified thioesterase in *E. coli*.

## 9. Phospholipid turnover

### 9.1. The diacylglycerol cycle

The polar head group of PG is rapidly lost in a pulse-chase experiment, whereas that of PE is stable. The conversion of PG to CL, catalyzed by CL synthase, does not account for all the loss of $^{32}$P-labeled PG observed in the pulse-chase experiments. A phosphate-containing non-lipid compound derived from the head group of PG was sought, which led to the discovery of the membrane-derived oligosaccharides (MDO) by Kennedy's group [17]. These compounds are composed of sn-glycerol-1-phosphate (derived from PG), glucose and (usually) succinate moieties, and have molecular weights in the range of 4000–5000. They are found in the periplasm of Gram-negative bacteria, an osmotically sensitive compartment. The synthesis of the MDO compounds is regulated by the osmotic pressure of the growth medium and decreased osmotic pressure gives an increased rate of MDO synthesis. Thus, MDO compounds seem to be involved in osmotic regulation.

In the synthesis of MDO, the sn-glycerol-1-phosphate polar headgroup of PG is transferred to the oligosaccharide, with 1,2-diacylglycerol as the other product (Fig. 9). The 85-kDa transmembrane protein that catalyzes this reaction is encoded by mdoB. Diacylglycerol kinase phosphorylates the diacylglycerol to phosphatidic acid, which reenters the phospholipid biosynthetic pathway (Fig. 6) to complete the diacylglycerol cycle (Fig. 9). In the overall reaction only the sn-glycerol-1-phosphate portion of the PG molecule is consumed; the lipid portion of the molecule is recycled back into phospholipid. MDO synthesis is responsible for most of the metabolic instability of the polar group of PG, since blocking MDO synthesis at the level of oligosaccharide synthesis by lack of UDP-glucose greatly reduces PG turnover. Moreover, the rate of accumulation of diacylglycerol in strains lacking diacylglycerol kinase (dgk) correlates with the presence of both the oligosaccharide acceptor and the osmolarity of the growth medium. dgk is immediately upstream of plsB in E. coli. The protein is a trimer of identical 13-kDa subunits, each with three predicted trans-membrane helices. Activity appears to be limited by diffusion of substrate across the membrane to the cytoplasmic active site. It should be noted that some species of MDO contain phosphoethanolamine. Although direct proof is lacking, it is likely that the ethanolamine moiety is derived from PE, as this is the only known source of ethanolamine.

### 9.2. The 2-acylglycerolphosphoethanolamine cycle

2-Acylglycerolphosphoethanolamine is a minor membrane lipid in E. coli generated from fatty acid transfer of the acyl moiety at the 1-position of PE to the outer membrane lipoprotein. 2-Acylglycerolphosphoethanolamine acyltransferase is an inner membrane enzyme that esterifies the 1-position of 2-acylglycerolphosphoethanolamine utilizing acyl-ACP (and not acyl-CoA) as the acyl donor. The acyltransferase was first recognized as a protein called acyl-ACP synthetase that catalyzes the ligation of fatty acids to ACP, hence the gene designation aas. ACP acts as a bound subunit for accepting the acyl intermediate in the normal acyltransferase reaction and high salt

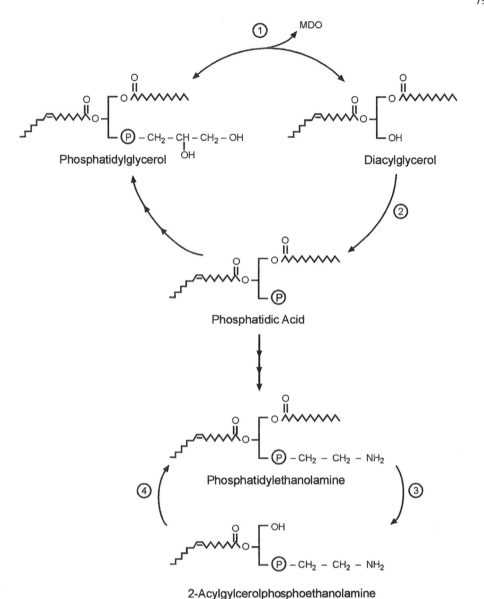

Fig. 9. Phospholipid turnover. The 1,2-diacylglycerol kinase cycle involves (1) the transfer of the *sn*-1-glycerol-phosphate moiety from phosphatidylglycerol to membrane derived oligosaccharides (MDO) by the enzyme MdoB. (2) Diacylglycerol kinase converts the diacylglycerol to phosphatidic acid, which can regenerate the phosphatidylglycerol (see Fig. 6). Phosphatidylethanolamine cycling involves (3) the transfer of an acyl chain to membrane lipoprotein and (4) re-esterification of the 1-position by 2-acylglycerophosphoethanolamine (Aas).

concentrations are required to dissociate the acyl-ACP intermediate from the enzyme in vitro. However, 2-acylglycerolphosphoethanolamine acyltransferase is the only reaction catalyzed by Aas in vivo. *aas* mutants are defective in both acyl-ACP synthetase and 2-acylglycerolphosphoethanolamine acyltransferase activities in vitro. They do not accumulate 2-acylglycerolphosphoethanolamine in vivo unless they are also defective in the *pldB* gene which encodes a lysophospholipase that represents a second pathway for 2-acylglycerolphosphoethanolamine metabolism. The acyl-ACP synthetase reaction has proven extremely valuable in the preparation of acyl-ACPs for use as substrates and inhibitors for the enzymes of fatty acid synthase.

## 10. Regulation of lipid metabolism

### 10.1. Regulation of fatty acid chain length

Fatty acyl chains in the membrane phospholipids of *E. coli* are normally 16 or 18 carbons in length. This specificity is a result of a combination of two factors: the poor reactivity of the β-ketoacyl-ACP synthases for longer chains; and the high specificity of the acyltransferases for 16- and 18-carbon products. Overexpression of FabB leads to the overproduction of *cis*-vaccenate, which is incorporated into the membranes. Overexpression of FabH causes a decrease in the average fatty acid chain length and the appearance of significant amounts of myristic acid (14:0) in the phospholipids. This effect is attributed to an increased rate of fatty acid initiation, which leads to a deficiency in malonyl-ACP for the terminal elongation reactions. The fatty acid biosynthetic machinery has the capacity to produce longer chains fatty acids. Under normal conditions, the 16 or 18 carbon chains are removed from the cytoplasm by the action of the acyltransferases. However, when phospholipid synthesis is blocked at the acyltransferase step, the fatty acids that accumulate have abnormally long chain lengths (e.g. 20 and 22 carbons). Conversely, overproduction of the acyltransferase results in a somewhat decreased average chain length, represented mainly by an increase in myristic acid. Thus, competition among the elongation synthases, the supply of malonyl-ACP, and the utilization of acyl-ACPs by the acyltransferase are the most significant determinants of fatty acid chain length.

### 10.2. Temperature modulation of fatty acid composition

All organisms regulate the fluidity of their membranes to maintain a membrane bilayer in a largely fluid state. As temperatures are lowered, membranes undergo a reversible change from a fluid (disordered) to a non-fluid (ordered) state. In *E. coli*, the temperature of the transition point depends on the fatty acid composition of the membrane phospholipids [11]. At lower temperatures, the amount of *cis*-vaccenic acid is rapidly (within 30 s) increased due to the increased activity of FabF. Synthesis of mRNA and protein are not required. Overexpression of FabB also increases the amount of this fatty acid in the membrane, although in a temperature-independent manner. Mutants that lack FabF are unable to modulate their fatty acid composition in a temperature-dependent manner. Thus FabF, and not FabB, is involved in the thermal regulation of the fatty acid composition of the membranes.

As a result of the specificity of the acyltransferases, palmitic acid occupies position 1 of the phospholipid backbone at 37°C, whereas palmitoleic and *cis*-vaccenic acids are found at position 2. As the growth temperature is lowered, *cis*-vaccenic acid competes with palmitic acid for position 1 of the newly synthesized phospholipids. This indicates that the specificity of the acyltransferases changes during this temperature shift also, although how this happens is not yet understood.

## 10.3. Transcriptional regulation of the genes of fatty acid synthesis and degradation

The known genes of fatty acid synthesis are scattered along the genome with only two clusters, the minimal *accBC* operon and the *fab* cluster. The *fab* cluster contains the *fabH*, *fabD*, *fabG*, *acpP*, and *fabF* genes and may have functional significance, since the work of Cronan has demonstrated that several genes are cotranscribed. However, most genes also appear to have a unique promoter, and the full relevance of this cluster is not yet fully understood.

For balanced production of each member of the acetyl-CoA carboxylase complex, one might expect each gene to be regulated in the same manner. However, while transcription of all four *acc* genes is under growth rate control, with the rate of transcription decreasing with decreased growth rate, the *accBC* operon seems to be regulated by a mechanism that differs from the regulation of the *accA* and *accD* genes. The *accBC* operon is transcribed from a promoter located unusually far upstream of the *accB* gene. The major *accA* promoter lies within the coding sequence of the upstream *polC* (*dnaE*) gene, although transcription through *polC* and perhaps other upstream genes also reads through the *accA* sequence. The *accD* gene is transcribed from a promoter located within the upstream *dedA* gene.

Transcriptional regulation of the other genes of fatty acid synthesis is no less complicated. The FadR protein, which was first identified as a repressor of transcription of genes in the *fad* regulon of β-oxidation and fatty acid transport, also positively regulates *fabA* transcription. FadR binds to DNA in the absence of acyl-CoA, to repress the β-oxidation regulon and activate *fabA*. Acyl-CoA (formed from exogenous fatty acids transported into the cell), binds FadR and the protein is released from the DNA. The molecular details of these interactions have been examined by the crystallization of FadR and the FadR–acyl-CoA and FadR–DNA complexes. Whether FadR activates or represses transcription depends on the location of its binding site within the promoter region. For repression, FadR binds in the $-30$ to $+10$ region of the promoter and prevents binding of DNA polymerase. For activation, the FadR operator site is located in a 17-bp region at $-40$ of the *fabA* promoter, and FadR binding promotes DNA polymerase binding. In *fadR* null mutants, the *fabA* gene is transcribed from two weak promoters of about equal strength whereas in wild-type strains, a 20-fold increase in transcription from the proximal promoter is seen in the absence of acyl-CoA. Thus, FadR monitors the intracellular concentration of acyl-CoA and coordinately regulates fatty acid synthesis and β-oxidation in response to these compounds.

The *fabB* gene, encoding β-ketoacyl-ACP synthase I, possesses a nucleotide sequence in its −40 region that matches perfectly with the highest affinity FadR operator sites. Accordingly, *fabB* is also positively regulated by FadR; however, the changes in *fabB* mRNA levels due to FadR regulation are much lower than for *fabA*.

## 10.4. Regulation of phospholipid headgroup composition

Within a given strain of *E. coli*, the phospholipid ratio (PE : PG : CL) is maintained under a variety of growth rates and conditions. The exception to this is the increased conversion of PG to CL during the stationary phase (Section 5.4). Thus, a mechanism must exist to maintain phospholipid homeostasis. Regulatory mutants resulting in the overexpression of PS synthase (*pssR*) and diacylglycerol kinase (*dgkR*) have been identified, suggesting the existence of *trans* acting factors that control the expression of these key enzymes, but their significance is unclear. The hypothetical 32-kDa PssR protein is similar to the LysR family of transcriptional regulators. Overexpression of PS synthase, PGP synthase or CL synthase in plasmid-based systems do not lead to dramatic changes in the membrane phospholipid composition. Thus, modulation of protein level is unlikely to have a role in the regulatory scheme.

So how is phospholipid homeostasis maintained? Control of the individual enzymes at the level of activity by feedback-regulation appears a more probable mechanism. Perturbations in the ratio of phospholipids were attempted experimentally by the activation of phosphoglycerol transferase I (*mdoB*). This enzyme, involved in MDO synthesis, catalyzes the transfer of glycerol phosphate from PG to the extracellular arbutin (4-hydroxyphenyl-*O*-β-D-glucoside). Treatment with arbutin (a MDO substrate analogue) causes a 7-fold increase in the rate of PG synthesis without a concomitant increase in PGP synthase proteins levels or significant changes in membrane phospholipid composition. Thus, PS synthase and PGP synthase are independently regulated by phospholipid composition. Similarly, purified CL synthase is strongly feedback inhibited by CL, and this inhibition is partially relieved by PE. Thus, the regulation of phospholipid content in *E. coli* appears to be an intrinsic property of the enzymes.

## 10.5. Coordinate regulation of fatty acid and phospholipid synthesis with macromolecular biosynthesis

Fatty acid biosynthesis is coordinately regulated with phospholipid synthesis since, in growing cultures of *E. coli*, there is no significant accumulation of any of the intermediates in fatty acid synthesis. Following inhibition of phospholipid biosynthesis, fatty acid biosynthesis carries on at about 10–20% of the uninhibited rate. Measurements of this rate were severely hampered by the requirement to first prevent β-oxidation of newly synthesized fatty acids, and second to use a strain in which acetate was solely channeled into fatty acid synthesis, so that incorporation of [14]C-label from the acetate could be measured.

Labeling the ACP moiety of the fatty acid intermediates by growth of a *panD* strain on medium containing tritiated β-alanine, a precursor of 4′-phosphopantetheine, also shows that long chain acyl-ACPs accumulate for a short period following the cessation

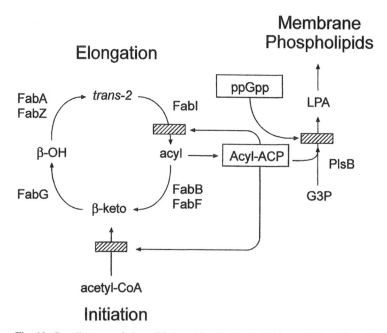

Fig. 10. Coordinate regulation of fatty acid and phospholipid metabolism. The pleiotropic regulator ppGpp regulates transfer of fatty acids to the membrane via inhibition of the PlsB acyltransferase step, coordinating phospholipid synthesis with macromolecular synthesis. Long chain acyl-ACPs accumulate for a period, then feedback inhibit their own synthesis at the point of initiation (inhibition of acetyl-CoA carboxylase and/or FabH) and elongation, by inhibition of FabI.

of phospholipid synthesis. This accumulation does not continue indefinitely, however, and reaches a plateau after about 20 min following inhibition of phospholipid synthesis. Thus, de novo fatty acid synthesis ceases, probably by a feedback inhibition mechanism involving long chain acyl-ACPs inhibiting early steps in the fatty acid biosynthesis pathway (Fig. 10). A significant finding in support of this idea is that overexpression of a thioesterase (which prevent the accumulation of acyl-ACP by cleavage of the thioester linkage and release of the acyl chain), allows continued fatty acid synthesis following cessation of phospholipid synthesis. This also further suggests that acyl-ACP and not free fatty acids mediate the inhibition. A reduction in total ACP is not responsible for the inhibition of fatty acid synthesis, since the free ACP pools of the glycerol-starved *plsB* mutants are not significantly depleted, and overproduction of ACP fails to relieve inhibition of fatty acid synthesis. A *fadD* mutant strain, which cannot produce acyl-CoA, overexpressing a thioesterase gave the same results as strains blocked elsewhere in β-oxidation or wild-type strains, thus ruling out a role for acyl-CoA.

As *E. coli* enter the stationary phase, levels of an unusual nucleotide, guanosine 5′-diphosphate-3′-diphosphate (ppGpp), rise [18]. Wild-type strains of *E. coli* undergo the so-called 'stringent response' following starvation for a required amino acid, an effect also mediated by increased intracellular ppGpp. Increased levels of ppGpp cause a strong inhibition of stable RNA synthesis, and inhibition of protein and phospholipid synthesis. Mutant strains (relA) do not undergo the stringent response, due to the lack of ppGpp synthase I, a ribosomal protein that produces ppGpp in response to uncharged tRNA. The interaction of ppGpp with RNA polymerase mediates the inhibitory effects on stable RNA synthesis. ppGpp directly inhibits phospholipid biosynthesis by inhibition of the glycerol phosphate acyltransferase (PlsB) and causes an accumulation of long chain acyl-ACPs, which in turn lead to the inhibition of fatty acid biosynthesis. Overexpression of the acyltransferase relieves the inhibition on both fatty acid and phospholipid synthesis.

The target enzyme of the acyl-ACP feedback inhibition is not clear, and indeed several enzymes are still under consideration (Fig. 10). An obvious target is acetyl-CoA carboxylase, the first enzyme of the pathway. Inhibition of malonyl-CoA production would prevent fatty acid synthesis and elongation. Cronan has shown that acetyl-CoA carboxylase appears to be a rate-limiting step for fatty acid synthesis and that the enzyme is inhibited by acyl-ACP in vitro. However, β-ketoacyl-ACP synthases I and II (FabB and FabF) are potential regulators due to their ability to degrade malonyl-ACP to acetyl-ACP, an activity stimulated by the presence of long chain acyl-ACP, and hence attenuate cycles of fatty acid elongation. FabH catalyzes the first step in the pathway, and inhibition of this enzyme would halt initiation of new acyl chains, but would allow the elongation of existing fatty acid intermediates. Inhibition of FabH by physiologically relevant concentrations of long chain acyl-ACPs has been demonstrated in vitro. Finally, enoyl-ACP reductase (FabI) is a potential target since the activity of this enzyme is a determining factor in completing rounds of fatty acid elongation and acyl-ACP could act as product inhibitors. Accumulation of the precursors of the enoyl-ACP step can be seen in vivo following inhibition of the acyltransferase. It remains possible that multiple of steps may be targeted to different degrees. The identification of the inhibited enzyme(s) and their relative contribution to the regulation will require an in vitro system including acetyl-CoA carboxylase that accurately reflects in vivo metabolism, and the isolation of mutants refractory to inhibition.

The inhibition of fatty acid biosynthesis triggers the stringent response. The accumulation of ppGpp is in this case dependent on the activity of the spoT gene product (ppGpp synthase II). How this regulatory system operates is unknown and it will be important to determine whether intermediates, such as acyl-ACP or malonyl-CoA, are the intracellular metabolites that mediate ppGpp synthase II activity.

## 11. Lipid metabolism in other bacteria

### 11.1. Analysis of lipid metabolism by genomic inference

The availability of genomic sequences for a variety of bacteria [8] allows for the rapid assessment of the lipid metabolic pathways present. Open reading frames encoding species specific isoforms of known genes can be amplified by the polymerase chain reaction, the protein expressed in E. coli, and the properties of the enzyme compared to the E. coli or other known proteins. This approach has been used extensively in Rock's laboratory for the genes of fatty acid synthesis, with isoforms from pathogenic bacteria being isolated to assess their unique biochemical characteristics and for use in drug-screening programs. For example, the enoyl-ACP reductase isoforms from *E. coli*, *Staphylococcus aureus* and *B. subtilis* have been compared. Subtle differences in cofactor specificity could be detected in this way that are not obvious from the primary sequences.

Novel proteins have been identified using genomic information [19]. The *fabI*-encoded enzyme is the sole reductase present in *E. coli*, and it was assumed that this was probably true for all bacteria. However, analysis of the genome of *Streptococcus pneumoniae* did not reveal a *fabI* homologue, while genes for all of the other enzymes required for saturated fatty acid biosynthesis could be identified in a gene cluster. Thus, a novel enoyl-ACP reductase isoform was sought, and the *fabK*-encoded enoyl-ACP reductase II identified. The *fabK*-encoded protein possesses no homology to the FabI protein, and utilizes flavin mononucleotide as a cofactor, and is resistant to inhibition by the antibiotic triclosan (Section 12.3).

A third enoyl-ACP reductase isoform was discovered in *B. subtilis*, which also possesses *fabI*. A gene, *fabL*, was identified with low overall homology to *fabI*, although it contained lysine and tyrosine residues in the distinct motif of the catalytic residues in the enoyl-ACP reductase I. Heath and coworkers cloned the gene and demonstrated that the product possesses enoyl-ACP reductase activity in vitro. Insertional inactivation of either *fabI* or *fabL* in *B. subtilis* results in no obvious growth phenotype, but the double null strain could not be constructed. Thus, *fabL* encodes enoyl-ACP reductase III.

## 11.2. Branched chain fatty acid biosynthesis

Not all bacteria regulate membrane fluidity through the production of straight chain unsaturated fatty acids. In fact, Gram-positive bacteria often use branched chain fatty acids to modulate membrane fluidity [20]. The branch is a methyl group in the iso- or anteiso-position in the chain (i.e. the second or third carbon from the distal end of the chain). Based on the concept of acetyl-CoA as a primer for straight chain fatty acid synthesis in *E. coli*, it can be seen that the methyl group could be introduced using a 'branched-chain' primer. Indeed, isotope labeling and biochemical analysis identifies precursors of the branched chain amino acids, valine and isoleucine (isobutyryl-CoA or 2-methylvaleryl-CoA, respectively), as the primers for branched chain fatty acid synthesis. As in straight chain synthesis, the primer is condensed with malonyl-ACP by the action of FabH. The substrate specificities of the FabH enzyme(s) present in the bacteria determine the relative amounts of the respective fatty acids produced. *B. subtilis* contains a high proportion of branched chain fatty acids in its membranes, and has two FabH enzymes, each of which prefer the branched chain substrates over acetyl-CoA. *E. coli* FabH cannot use branched chain primers. Why *B. subtilis* possesses two FabH enzymes, with only minor differences in substrate specificity, is not understood.

## 11.3. Other ways to make unsaturated fatty acids

The *fabA* gene encodes the dehydratase/isomerase specifically required for the production of unsaturated fatty acids in Gram-negative bacteria. Gram-positive bacteria do not contain an identifiable *fabA* homologue in their genome, but do possess unsaturated fatty acids. In *Bacillus*, it has been shown that a cold-shock induced gene (*desA*) encodes a desaturase that is active on the existing fatty acids present in membrane phospholipids. Desaturase activities are dependent on oxygen, and thus an aerobic lifestyle. This mechanism is analogous to that observed in plants (Chapter 4). *Streptococcus pneumoniae*, on the other hand, has a fatty acid profile similar to *E. coli*, but possesses neither an identifiable isomerase nor a desaturase. Thus, these facultatively anaerobic bacteria must contain a novel gene that encodes an isomerase with no homology to *fabA*. Gram-positive bacteria possess a *fabZ* homologue that performs all of the dehydration steps.

## 11.4. Bacteria with other phospholipid headgroups

*E. coli* have a very simple phospholipid composition with just three major forms, PE, PG and CL. However, the prokaryotic kingdom possesses a wide array of headgroups that defy adequate description in this short space; hence the reader is referred to Goldfine's review [21] for a comprehensive treatment of bacterial phospholipid structures. The phosphocholine headgroup stands worthy of mention for its uniqueness and distinct mechanisms of synthesis. Phosphatidylcholine (PC) had long been considered a eukaryotic phospholipid, where it is synthesized by transfer of the choline from CDP-choline to diacylglycerol, or by methylation of PE (Chapter 8).

*Rhodospeudomonas spheroides*, *Bradyrhizobium japonicum* and a few other specialized photosynthetic or nitrogen-fixing bacteria synthesize PC by three subsequent methylations of PE. The first methylase, encoded by *pmtA*, has been disrupted in *Bradyrhizobium japonicum*, and the mutants, which contain significantly reduced PC content, are less able to fix nitrogen in colonization assays. Thus, PC seems to be involved in host : bacteria interactions to establish symbiosis. The prokaryotic PE methyltransferases share weak homology to other bacterial methyltransferases, but no homology with their eukaryotic counterparts. *Rhodospeudomonas spheroides* are also somewhat unique amongst bacteria in that they contain intracellular membranes that hold the photosynthetic machinery. The amount of the intracellular membrane correlates to the amount of incident light, indicating a light-specific regulation of phospholipid synthesis in these organisms.

*Sinorhizobium meliloti* has been shown to synthesize PC by direct condensation of choline with CDP-diacylglycerol, as well as by the methyltransferase pathway. The *pcs* gene was identified and expression in *E. coli* demonstrates that it does code for a PC synthase [22]. The genomes of *Pseudomonas aeruginosa* and *Borrelia burgdorferi* contain similar genes, and have been reported to possess PC in their membranes. The PC synthase protein shares weak homology with PS synthase (a CDP-diacylglycerol : serine *O*-phosphatidyltransferase) from other bacteria, but not to any eukaryotic proteins.

A genus of bacteria, termed the Sphingobacterium, produce sphingolipids by a pathway similar to that in mammals. However, little is known about the enzymes involved at this time. Clostridia produce plasmalogens (1-alk-1'-enyl lipids) by an anaerobic pathway clearly different to the $O_2$-dependent pathway in mammals (Chapter 9). Branched chains fatty acids are also found in which the methyl group is inserted post-synthetically into the middle of the chain, in a manner analogous to cyclopropane fatty acid synthesis. S-Adenosylmethionine is also the methyl donor for these reactions.

## 11.5. Bacteria with a type I fatty acid synthase

A general distinction between prokaryotic and eukaryotic fatty acids synthases is that bacteria possess the dissociated enzymes described above (type II), while higher organisms have a single, multifunctional, protein (type I) that catalyses all of the reactions. There are exceptions to this rule, however. Mycobacteria, for example, possess a type I fatty acid synthase for the production of their membrane fatty acids. This enzyme is a homohexamer of 290-kDa subunits. Each subunit possesses the six different active sites required to generate a fatty acid. Unlike the type II system, the products of a type I enzyme are acyl-CoAs. For the mycobacterial enzyme, the saturated acyl chains produced are between 16 and 24 carbons in length. Unsaturations must be added post-synthetically by a desaturase. Even more unusual is that the Mycobacterium possesses a type II synthase system for the further elongation of the fatty acyl products of the type I system into the 70–80 carbon mycolic acids. *Brevibacterium ammoniagenes*, a highly developed bacteria thought to be a progenitor of the fungi, possesses a type I fatty acids synthase that is capable of producing both saturated and unsaturated fatty acids anaerobically.

## 11.6. Lipid synthesis in Archea

Archea are a group of organisms, previously classified as bacteria, from which eubacteria and other life may have evolved. A melavonic acid (6 carbon) building block is used for synthesis instead of acetic acid. The generated phytanyl chains are attached to glycerol moieties of complex lipids by ether linkages. Thus, these lipids are unlike anything found in eubacteria or eukaryotes today.

## 11.7. Other organisms with a bacterial-like fatty acid synthase system

The dissociated enzymes that form the fatty acid synthesis system of most bacteria are not limited to the prokaryotic kingdom [23]. Plants utilize a homologous series of enzymes for synthesis of their fatty acids (Chapter 4). Although the genes are present on the nuclear chromosomes, fatty acid biosynthesis occurs in the chloroplasts of plants. It is thus hypothesized that the pathway evolved from the endosymbiont bacteria that became the chloroplast in plants. Recently, is has also been shown that the apicomplexans, a group of intracellular parasites including *Plasmodium falciparum*, the causative organism of malaria, possess a bacterial-like fatty acid synthase system [24]. These eukaryotic organisms possess an organelle described as a vestigial chloroplast (the apiplast), and thus are presumably evolved from chloroplast-containing algae. Plasmodium have been shown to be sensitive to the antibiotics thiolactomycin and triclosan, indicating that this pathway could be exploited for the development of novel antimalaria drugs.

## 12. Inhibitors of lipid metabolism

Much of the work on prokaryotic membrane biosynthesis in recent years has been directed towards analysis of the pathway enzymes as potential targets for the discovery on new antimicrobial agents [9]. It should be remembered that bacterial fatty acid biosynthesis is catalyzed on a series of small, discrete proteins and utilizes the soluble ACP to shuttle the soluble intermediates within the cytosol. This setup is quite different from mammalian fatty acid synthesis, and thus should allow for specific inhibitors to be designed with minimal mammalian toxicity.

### 12.1. β-Decynoyl-N-acetylcysteamine

The β-hydroxydecanoyl-ACP dehydrase (the *fabA* gene product) is specifically and irreversibly inhibited by the synthetic acetylenic substrate analog β-decynoyl-*N*-acetylcysteamine. A covalent adduct is formed with the active site histidine and results in the loss of both dehydrase and isomerase reactions. The description of this mode of inhibition was the first report of a suicide or mechanism-based inhibitor [25]. β-Decynoyl-*N*-acetylcysteamine concentrations between 10 and 50 μM are sufficient to completely inhibit unsaturated fatty acid synthesis and bacterial growth. Growth inhibition is relieved by the addition of unsaturated fatty acids to the medium. Saturated fatty acid synthesis continues normally in the presence of β-decenoyl-*N*-acetylcysteamine due to the activity of the *fabZ*-encoded dehydrase. Since most Gram-positive bacteria do not contain unsaturated fatty acids, inhibitors of this step are not being intensively developed.

### 12.2. Cerulenin and thiolactomycin

Cerulenin, (2R)(3S)-2,β-epoxy-4-oxo-7,10-dodecadienolyamide, is a fungal product that irreversibly inhibits β-ketoacyl-ACP synthases I and II. It is extremely effective in blocking the growth of a large spectrum of bacteria. Cerulenin forms a covalent bond with the active site cysteine of the synthases. Cerulenin can be seen in the hydrophobic acyl binding tunnel in the crystal structures of both FabB and FabF complexed with this drug. Although cerulenin is a versatile biochemical tool, it is not a suitable antibiotic for clinical use because it is also a potent inhibitor of the multifunctional mammalian fatty acid synthase, which contains similar active site residues and catalyzes the same chemical reaction.

Thiolactomycin, (4S)(2E,5E)-2,4,6-trimethyl-β-hydroxy-2,5,7-octatriene-4-thiolide, inhibits all three β-ketoacyl-ACP synthases of bacterial fatty acid synthesis in vivo and in vitro, but not the multifunctional fatty acid synthases. Malonyl-ACP protects the synthases from thiolactomycin inhibition, indicating that this antibiotic targets a different site on the condensing enzyme from that targeted by cerulenin. The crystal structure of FabB in complex with thiolactomycin reveals that the drug does in fact bind in the malonyl-ACP binding pocket [26], which is presumably distinct from the malonyl-pantotheine site of mammalian type I fatty acid synthase. The structure of the mammalian enzyme has not been solved to date, due in part at least to the large size of the enzyme. Overproduction of FabB but not FabH imparts thiolactomycin resistance.

Since FabF is not essential, these data suggest that FabB is the relevant thiolactomycin target in vivo. Thiolactomycin resistance phenotype maps to the *emrR* locus of the *E. coli* chromosome and results in the inactivation of a repressor that governs the expression of the *emrAB* multidrug resistance pump. The subtly different active site architecture of FabH compared to FabB and FabF makes it resistant to both cerulenin and thiolactomycin [26].

## 12.3. Diazaborines, isoniazid and triclosan

Diazaborines, a group of boron-containing antibacterial heterocyclic compounds, inhibit fatty acid synthesis in *E. coli* by inhibition of the enoyl-ACP reductase I. Inhibition by diazaborines requires the presence of NAD or NADH, and the drug mimics the binding of the substrate. The discovery that the FabI analog in *M. tuberculosis* (InhA) is the target for isoniazid and ethionamide, drugs used to treat tuberculosis, illustrates the potential importance of enoyl-ACP reductase I as an antibiotic target. Isoniazid is first activated by the KatG catalase/peroxidase, and then forms a covalent adduct with the nictinamide cofactor [27]. It was recently discovered that the enoyl-ACP reductase was also targeted by another antimicrobial compound in widespread use. Triclosan, the active component in a plethora of household items including toothpaste, hand soaps, cutting boards, mattress pads and undergarments, is a very effective inhibitor of the enoyl-ACP reductase I. Its ubiquitous use has been justified by the claim, based on improperly analyzed data, that it caused non-specific membrane disruption, a delayed leakage of cytoplasmic material and finally cell death. This phenotype is essentially identical to that observed by Egan and Russell in 1973 [28] for the original temperature-sensitive *fabI* mutant at the non-permissive temperature. Mutants of *E. coli* can be generated that are resistant to triclosan, and contain a Gly to Val substitution in the active site of enoyl-ACP reductase I. The mutant protein is resistant to triclosan in vivo and in vitro. Definitive proof that triclosan does not disrupt the membranes of Gram-negative bacteria in a non-specific manner comes from the overexpression of the triclosan-resistant enoyl reductase II (from a plasmid-borne copy of the *fabK* gene) in *E. coli*, which increases resistance by over 10,000-fold.

A key feature of triclosan's mode of action is the slow but complete inactivation of the enoyl reductase. The drug binds to the active site of the enzyme and forms very strong, non-covalent interactions with active site residues and the nicotinamide ring of the cofactor. The half life of the ternary enzyme–$NAD^+$–triclosan complex is over 1 h. The mutation that leads to triclosan resistance is the same as that for diazaborine resistance, and adjacent to the mutation required for isoniazid resistance in *M. tuberculosis*. There is some concern that the unregulated use of triclosan in the household setting, where it has not been proven to be effective in promoting public health, will lead to the generation of bacteria resistant to all enoyl-ACP reductase I-directed inhibitors.

## 12.4. Lipid A biosynthesis inhibitors

A new class of antibiotics was recently discovered that inhibit lipid A biosynthesis [29]. These compounds are still in their infancy, but could prove useful additions to our antibacterial arsenal. The first generation of compounds are effective inhibitors of bacterial growth in vivo, and potent inhibitors of the LpxC deacetylase, the second step in the pathway, in vitro. Optimization of this family of compounds may lead to effective agents against Gram-negative bacteria.

## 13. Future directions

Much progress has been made over the last several years in elucidating the details of the enzymes of fatty acid biosynthesis. This pathway has become a major target for therapeutic intervention in bacterial-mediated disease. Significant inroads have also been made into the molecular mechanisms that regulate fatty acid synthesis. The availability of genomic data has also allowed for the facile translation of many of the findings made in *E. coli* to other bacteria facilitating the discovery of novel genes involved in lipid biosynthesis.

Many of the details of the regulation of fatty acid biosynthesis are still to be worked out, including the probable discovery of new transcription factors and new effector molecules. The fine details of the enzymatic mechanisms, and the comparison of the biochemical properties and functions of different isoforms, will continue in an effort, in part, to elucidate probable activity spectrums of next generation antibiotics that will surely be generated against this pathway. The rush to study fatty acid synthesis has somewhat overshadowed bacterial lipid synthesis in recent years, and discoveries in this area are more difficult since the enzymes and substrates involved are membrane-associated. However, techniques to study the structure and function of membrane proteins are rapidly evolving and will certainly be applied to resolving outstanding issues in bacterial lipid biogenesis.

## Abbreviations

| | |
|---|---|
| ACP | acyl carrier protein |
| CoA | coenzyme A |
| *fabA*, FabA | lowercase italics indicates gene, while uppercase Roman type indicates the protein product of the gene |
| PE | phosphatidylethanolamine |
| PG | phosphatidylglycerol |
| PS | phosphatidylserine |
| PGP | phosphatidylglycerolphosphate |
| CL | cardiolipin |
| LPS | lipopolysaccharides |
| ABC | ATP-binding cassette |
| MDO | membrane-derived oligosaccharides |

# *References*

1. Blattner, F.R., Plunkett, G.I., Bloch, C.A., Perna, N.T., Burland, V., Riley, M., Collado-Vides, J., Glasner, J.D., Rode, C.K., Mayhew, G.F., Gregor, J., Davis, N.W., Kirkpatrick, H.A., Goeden, M.A., Rose, D.J., Mau, B. and Shao, Y. (1997) The complete genome sequence of *Escherichia coli* K-12. Science 277, 1453–1474.
2. Cronan, J.E., Jr. and Rock, C.O. (1996) Biosynthesis of membrane lipids. In: F.C. Neidhardt, R. Curtis, C.A. Gross, J.L. Ingraham, E.C.C. Lin, K.B. Low, B. Magasanik, W. Reznikoff, M. Riley, M. Schaechter and H.E. Umbarger (Eds.), *Escherichia coli* and *Salmonella typhimurium*: Cellular and Molecular Biology. American Society for Microbiology, Washington, DC, Chapter 37.
3. Majerus, P.W., Alberts, A.W. and Vagelos, P.R. (1969) Acyl carrier protein from *Escherichia coli*. Methods Enzymol. 14, 43.
4. Bloch, K. and Vance, D.E. (1977) Control mechanisms in the synthesis of saturated fatty acids. Annu. Rev. Biochem. 46, 263–298.
5. Raetz, C.R.H. (1978) Enzymology, genetics, and regulation of membrane phospholipid synthesis in *Escherichia coli*. Microbiol. Rev. 42, 614–659.
6. Clark, D.P. and Cronan Jr., J.E. (1981) Bacterial mutants for the study of lipid metabolism. Methods Enzymol. 72, 693–707.
7. Raetz, C.R.H. (1982) Genetic control of phospholipid bilayer assembly. In: J.N. Hawthorne and B. Ansell (Eds.), Comprehensive Biochemistry. Vol. 4, Elsevier, Amsterdam, Chapter 11.
8. Fraser, C.M., Eisen, J., Fleischmann, R.D., Ketchum, K.A. and Peterson, S. (2000) Comparative genomics and understanding of microbial biology. Emerg. Infect. Dis. 6, 505–512.
9. Heath, R.J., White, S.W. and Rock, C.O. (2001) Lipid Biosynthesis as a target for antibacterial agents. Prog. Lipid Res., in press.
10. Jackowski, S. (1996) Biosynthesis of pantothenic acid and coenzyme A. In: F.C. Neidhardt, R. Curtiss, C.A. Gross, J.L. Ingraham, E.C.C. Lin, K.B. Low, B. Magasanik, W. Reznikoff, M. Riley, M. Schaechter and H.E. Umbarger (Eds.), *Escherichia coli* and *Salmonella typhimurium*: Cellular and Molecular Biology. American Society for Microbiology, Washington, DC, Chapter 46.
11. de Mendoza, D. and Cronan Jr., J.E. (1983) Thermal regulation of membrane lipid fluidity in bacteria. Trends Biochem. Sci. 8, 49–52.
12. Cronan Jr., J.E., Silbert, D.F. and Wulff, D.L. (1972) Mapping of the *fabA* locus for unsaturated fatty acid biosynthesis in *Escherichia coli*. J. Bacteriol. 112, 206–211.
13. Rosenfeld, I.S., D'Agnolo, G. and Vagelos, P.R. (1973) Synthesis of unsaturated fatty acids and the lesion in *fabB* mutants. J. Biol. Chem. 248, 2452–2460.
14. Raetz, C.R.H. (1996) Bacterial lipopolysaccharides: A remarkable family of bioactive macroamphiphiles. In: F.C. Neidhardt, R. Curtiss, C.A. Gross, J.L. Ingraham, E.C.C. Lin, K.B. Low, B. Magasanik, W. Reznikoff, M. Riley, M. Schaechter and H.E. Umbarger (Eds.), *Escherichia coli* and *Salmonella typhimurium*. American Society for Microbiology, Washington, DC, Chapter 69.
15. Chang, G. and Roth, C.B. (2001) Structure of MsbA from *E. coli*: a homolog of the multidrug resistance atp BINDING cassette (ABC) transporters. Science 293, 1793–1800.
16. Clark, D.P. and Cronan, J.E., Jr. (1996) Two-carbon compounds and fatty acids as carbon sources. In: F.C. Neidhardt, R. Curtiss, J.L. Ingraham, E.C.C. Lin, K.B. Low, B. Magasanik, W.S. Reznikoff, M. Riley, M. Schaechter and H.E. Umbarger (Eds.), *Escherichia coli* and *Salmonella*: Cellular and Molecular Biology. American Society for Microbiology, Washington, DC, Chapter 21.
17. Kennedy, E.P. (1996) Membrane-derived oligosaccharides (periplasmic Beta-D-glucans) of *Escherichia coli*. In: F.C. Neidhardt, R. Curtiss, J.L. Ingraham, E.C.C. Lin, K.B. Low, B. Magasanik, W.S. Reznikoff, M. Riley, M. Schaechter and H.E. Umbarger (Eds.), *Escherichia coli* and *Salmonella*: Cellular and Molecular Biology. American Society for Microbiology, Washington DC, Chapter 70.
18. Cashel, M., Gentry, D.R., Hernadez, V.J. and Vinella, D. (1996) The stringent response. In: F.C. Neidhardt, R. Curtis, E.C.C. Lin, J.L. Ingraham, B.L. Low, B. Magasanik, W.S. Reznikoff, M. Riley, M. Schaechter and H.E. Umbarger (Eds.), *Escherichia coli* and *Salmonella*: Cellular and Molecular Biology. American Society for Microbiology, Washington, DC, Chapter 92.
19. Heath, R.J. and Rock, C.O. (2000) A triclosan-resistant bacterial enzyme. Nature 406, 145–146.
20. Kaneda, T. (1963) Biosynthesis of branched chain fatty acids. J. Biol. Chem. 238, 1229–1235.

21. Goldfine, H. (1982) Lipids of prokaryotes: structure and distribution. In: S. Razin and S. Rottem (Eds.), Current Topics in Membranes and Transport, Academic Press, New York.

22. Sohlenkamp, C., de Rudder, K.E., Rohrs, V., Lopez-Lara, I.M. and Geiger, O. (2000) Cloning and characterization of the gene for phosphatidylcholine synthase. J. Biol. Chem. 275, 18919–18925.

23. Rock, C.O. and Cronan Jr., J.E. (1996) *Escherichia coli* as a model for the regulation of dissociable (type II) fatty acid biosynthesis. Biochim. Biophys. Acta 1302, 1–16.

24. Waller, R.F., Keeling, P.J., Donald, R.G., Striepen, B., Handman, E., Lang-Unnasch, N., Cowman, A.F., Besra, G.S., Roos, D.S. and McFadden, G.I. (1998) Nuclear-encoded proteins target to the plastid in *Toxoplasma gondii* and *Plasmodium falciparum*. Proc. Natl. Acad. Sci. USA 95, 12352–12357.

25. Helmkamp Jr., G.M., Brock, D.J.H. and Bloch, K. (1968) β-Hydroxydecanoyl thioester dehydrase. Specificity of substrates and acetylenic inhibitors. J. Biol. Chem. 243, 3229–3231.

26. Price, A.C., Choi, K.H., Heath, R.J., Li, Z., White, S.W. and Rock, C.O. (2001) Inhibition of β-ketoacyl-acyl carrier protein synthases by thiolactomycin and cerulenin. Structure and mechanism. J. Biol. Chem. 276, 6551–6559.

27. Rozwarski, D., Grant, G., Barton, D., Jacobs, W. and Sacchettini, J.C. (1998) Modification of NADH of the isoniazid target (InhA) from *Mycobacterium tuberculosis*. Science 279, 98–102.

28. Egan, A.F. and Russell, R.R.B. (1973) Conditional mutants affecting the cell envelope of *Escherichia coli*. Genet. Res. 21, 3603–3611.

29. Wyckoff, T.J., Raetz, C.R. and Jackman, J.E. (1998) Antibacterial and anti-inflammatory agents that target endotoxin. Trends Microbiol. 6, 154–159.

D.E. Vance and J.E. Vance (Eds.) *Biochemistry of Lipids, Lipoproteins and Membranes (4th Edn.)*
© 2002 Elsevier Science B.V. All rights reserved

# CHAPTER 4

# Lipid metabolism in plants

## Katherine M. Schmid [1] and John B. Ohlrogge [2]

[1] *Department of Biological Sciences, Butler University, Indianapolis, IN 46208-3485, USA,*
*Tel.: +1 (317) 940-9956; Fax: +1 (317) 940-9519; E-mail: kschmid@butler.edu*
[2] *Department of Plant Biology, Michigan State University, East Lansing, MI 48824, USA,*
*Tel.: +1 (517) 353-0611; Fax: +1 (517) 353-1926; E-mail: ohlrogge@pilot.msu.edu*

## 1. Introduction

Plants produce the majority of the world's lipids, and most animals, including humans, depend on these lipids as a major source of calories and essential fatty acids. Like other eukaryotes, plants require lipids for membrane biogenesis, as signal molecules, and as a form of stored carbon and energy. In addition, soft tissues and bark each have distinctive protective lipids that help prevent desiccation and infection. To what extent does the biochemistry of plant lipid metabolism resemble that in other organisms? This chapter mentions a number of similarities, but emphasizes aspects unique to plants. Major differences between lipid metabolism in plants and other organisms are summarized in Table 1.

The presence of chloroplasts and related organelles in plants has a profound effect on both gross lipid composition and the flow of lipid within the cell. Fatty acid

Table 1
Comparison of plant, mammalian, and bacterial lipid metabolism

|  | Higher plants | Mammals | E. coli |
|---|---|---|---|
| *Fatty acid synthase* |  |  |  |
| Structure | Type II (multicomponent) | Type I (multifunctional) | Type II (multicomponent) |
| Location | Plastids | Cytosol | Cytosol |
| Acetyl-CoA carboxylase(s) | Multisubunit and multifunctional | Multifunctional | Multisubunit |
| *Primary desaturase substrates* |  |  |  |
| $\Delta^9$ | 18:0-ACP | 18:1-CoA | none |
| ω-6 | 18:1 on glycerolipids | none | none |
| ω-3 | 18:2 on glycerolipids | none | none |
| Primary substrate(s) for phosphatidic acid synthesis | acyl-ACP and acyl-CoA | acyl-CoA | acyl-ACP |
| Prominent bilayer lipids | Galactolipid > phospholipid | Phospholipid | Phospholipid |
| Main β-oxidation function | Provides acetyl-CoA for glyoxylate cycle | Provides acetyl-CoA for TCA cycle | Provides acetyl-CoA for TCA cycle |

synthesis occurs not in the cytosol as in animals and fungi, but in the chloroplast and other plastids. Acyl groups must then be distributed to multiple compartments, and the complex interactions between competing pathways are a major focus of plant lipid biochemists. It is also significant that the lipid bilayers of chloroplasts are largely composed of galactolipids rather than phospholipids. As a result, galactolipids are the predominant acyl lipids in green tissues and probably on earth.

Plant lipids also have a substantial impact on the world economy and human nutrition. More than three-quarters of the edible and industrial oils marketed annually are derived from seed and fruit triacylglycerols. These figures are particularly impressive given that, on a whole organism basis, plants store more carbon as carbohydrate than as lipid. Since plants are not mobile, and since photosynthesis provides fixed carbon on a regular basis, plant requirements for storage lipid as an efficient, light weight energy reserve are less acute than those of animals.

Finally, hundreds of genes required for plant lipid biosynthesis, utilization and turnover have now been cloned. In addition to providing valuable information on enzyme structure and function, these genes are being exploited to design new, more valuable plant oils. The coordination of lipid metabolic genes with each other and with their potential regulators may also become better understood, as DNA microarray and other genomic technologies mature.

## 2. Plant lipid geography

### 2.1. Plastids

Although all eukaryotic cells have much in common, the ultrastructure of a plant cell differs from that of the typical mammalian cell in three major ways. The plasma membrane of plant cells is shielded by the cellulosic cell wall, preventing lysis in the naturally hypotonic environment but making preparation of cell fractions more difficult. The nucleus, cytosol and organelles are pressed against the cell wall by the tonoplast, the membrane of the large, central vacuole that can occupy 80% or more of the cell's volume. Finally, all living plant cells contain one or more types of plastid.

The plastids are a family of organelles containing the same genetic material, a circular chromosome present in multiple copies. Young or undifferentiated cells contain tiny proplastids that, depending on the tissue, may differentiate into photosynthetic chloroplasts, carotenoid-rich chromoplasts, or any of several varieties of colorless leucoplasts, including plastids specialized for starch storage [1]. These different types of plastids, which may be interconverted in vivo, have varying amounts of internal membrane but invariably are bounded by two membranes. The internal structure of chloroplasts is dominated by the flattened green membrane sacks known as thylakoids. The thylakoid membranes contain chlorophyll and are the site of the light reactions of photosynthesis.

As noted above, chloroplasts and other plastids are enriched in galactolipids (Fig. 1). They also contain a unique sulfolipid, sulfoquinovosyldiacylglycerol, whose head group is a modified galactose. The phospholipid components of plastids are less

**Monogalactosyldiacylglycerol**

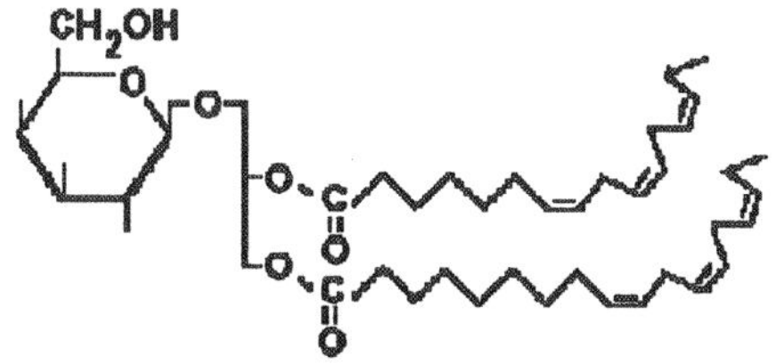

| Pea Chloroplasts | % |
|---|---|
| Thylakoids | 56 |
| Inner membrane | 64 |
| Outer membrane | 1 |
| Daffodil chromoplasts | 63 |
| Cauliflower proplastids | 31 |
| Potato leucoplasts | 14 |

**Digalactosyldiacylglycerol**

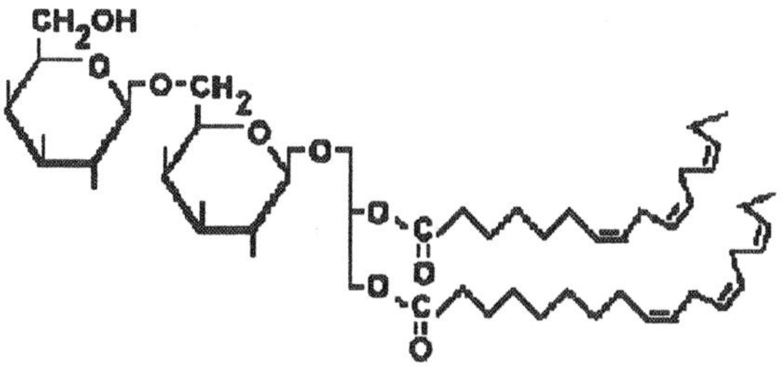

| Pea Chloroplasts | % |
|---|---|
| Thylakoids | 32 |
| Inner membrane | 31 |
| Outer membrane | 40 |
| Daffodil chromoplasts | 18 |
| Cauliflower proplastids | 29 |
| Potato leucoplasts | 45 |

**Sulfoquinorosyldiacylglycerol**

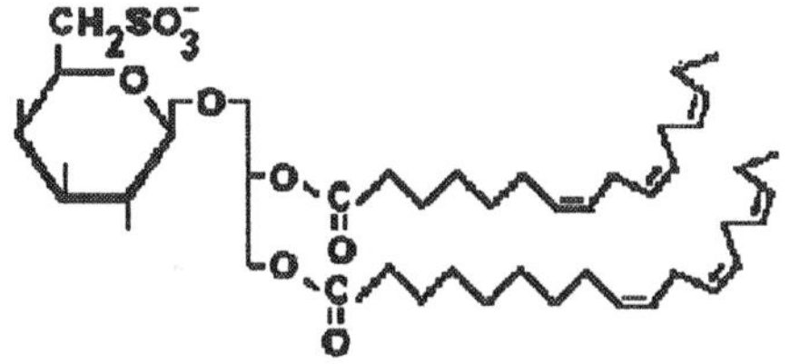

| Pea Chloroplasts | % |
|---|---|
| Thylakoids | 4 |
| Inner membrane | 4 |
| Outer membrane | 4 |
| Daffodil chromoplasts | 5 |
| Cauliflower proplastids | 6 |
| Potato leucoplasts | 5 |

Fig. 1. Composition of plastid membranes. Figures given are percentages (as % of total lipid) of the pictured lipid in the membranes specified. Data from Harwood, J.L. (1980) Plant acyl lipids: structure, distribution and analysis. In: P.K. Stumpf (Ed.) The Biochemistry of Plants, Vol. 4., Academic Press, pp. 2–56 and Sparace, S.A., Kleppinger-Sparace, K.F. (1993) Metabolism in non-photosynthetic, non-oilseed tissues. In: T.S. Moore Jr. (Ed.) Lipid Metabolism in Plants, Boca Raton, FL, CRC Press, pp. 569–589.

abundant. Phosphatidylglycerol is the most prominent phospholipid contributor to the thylakoid membrane system of chloroplasts (but < 10% of the glycerolipids), whereas most of the limited phosphatidylcholine of chloroplasts is associated with their outer membrane.

The synthetic capabilities of plastids and other plant organelles are summarized in Table 2. Representatives of each type of plastid have been isolated and found to incorporate acetate into long-chain fatty acids, to desaturate $18:0$ to $18:1$, and to assemble phosphatidic acid and galactolipids. Chloroplasts have also been shown to synthesize phosphatidylglycerol, including molecular species containing the unusual *trans*-3-hexadecenoic acid at the 2-position. In addition to the components normally retained within the plastids, large quantities of fatty acids, particularly $18:1$ and $16:0$,

Table 2

Compartmentation of lipids and lipid biosynthesis in plant cells [a]

| Membrane or organelle | Activities | Prominent lipids (not listed if <5%) |
|---|---|---|
| Plastids | Acetyl-CoA carboxylase, fatty acid synthase, 18 : 0-ACP desaturase, ω3 and ω6 desaturases, glycerol-3-phosphate acyltransferase, lysophosphatidic acid acyltransferase, phosphatidic acid phosphohydrolase, CTP : phosphatidate cytidylyltransferase, phosphatidylglycerol phosphate synthase and phosphatase, galactolipid and sulfolipid synthesis, diacylglycerol acyltransferase (minor), acyl-CoA synthetase, CTP-dependent lipid kinase | Digalactosyldiacylglycerol, monogalactosyldiacylglycerol, phosphatidylglycerol, sulfoquinovosyldiacylglycerol, phosphatidylcholine |
| Endoplasmic reticulum | Fatty acid elongase, ω3 and ω6 desaturases, other fatty acid modifying reactions, acyl-CoA thioesterase, glycerol-3-phosphate acyltransferase, lysophosphatidic acid acyltransferase, phosphatidic acid phosphatase, CTP : phosphatidate cytidylyltransferase, phosphatidylglycerol phosphate synthase and phosphatase, CDP-choline : diacylglycerol cholinephosphotransferase, phosphatidylmono- and dimethylethanolamine methyltransferases, CDP-ethanolamine : diacylglycerol ethanolaminephosphotransferase, phosphatidylserine decarboxylase, phosphatidylethanolamine : serine phosphatidyltransferase, CDP-diacylglycerol : serine phosphatidyltransferase, CDP-diacylglycerol : myo-inositol phosphatidyltransferase, diacylglycerol acyltransferase, diacylglycerol : diacylglycerol acyltransferase, some enzymes of sterol and sphingolipid synthesis, N-acylphosphatidylethanolamine synthase | Phosphatidylcholine, phosphatidylethanolamine, phosphatidylinositol, phosphatidylglycerol |
| Golgi bodies | Glycerol-3-phosphate acyltransferase, lysophosphatidic acid acyltransferase, CDP-choline : diacylglycerol cholinephosphotransferase, CDP-ethanolamine : diacylglycerol ethanolaminephosphotransferase | Phosphatidylcholine, phosphatidylethanolamine, phosphatidylserine |
| Lipid bodies | Fatty acid elongase, diacylglycerol acyltransferase, lipase | Triacylglycerol |
| Mitochondria | Fatty acid synthesis (minor), glycerol-3-phosphate acyltransferase, lysophosphatidic acid acyltransferase, CTP : phosphatidate cytidylyltransferase, phosphatidylglycerol phosphate synthase and phosphatase, phosphatidylglycerol : CDP diacylglycerol phosphatidyltransferase, β-oxidation | Phosphatidylcholine, phosphatidylethanolamine, cardiolipin, phosphatidylinositol |
| Nuclear membranes | None reported | Phosphatidylcholine, phosphatidylethanolamine, phosphatidylinositol, phosphatidylglycerol |

| | Enzymes | Lipids |
|---|---|---|
| Plasma membrane | Phosphatidylinositol and phosphatidylinositol phosphate kinases, serine exchange enzyme, sterol glucosyltransferase, glucosylceramide synthase | Phosphatidylethanolamine, phosphatidylcholine, phosphatidylinositol, sphingolipids, sterols and derivatives |
| Protein body membranes | CTP : phosphocholine cytidylyltransferase, diacylglycerol kinase, phosphatidylinositol and phosphatidylinositol phosphate kinases | Phosphatidic acid, phosphatidylcholine, phosphatidylethanolamine, phosphatidylglycerol, phosphatidylinositol |
| Tonoplast | None reported | Phosphatidylcholine, phosphatidylethanolamine, phosphatidylinositol, sphingolipids, sterols and derivatives |
| Glyoxysomes | Lipase, $\beta$-oxidation, glyoxylate cycle | Phosphatidylcholine, phosphatidylethanolamine, phosphatidylinositol |
| Peroxisomes | $\beta$-Oxidation, acyl-CoA thioesterase | Phosphatidylcholine, phosphatidylethanolamine, phosphatidylglycerol, phosphatidylinositol |

[a] Information collated primarily from T.S. Moore Jr., (Ed.) (1993) Lipid Metabolism in Plants, CRC Press, Boca Raton, FL; J.L. Harwood (1989) Crit. Rev. Plant Sci. 8, 1–43. Lipids that comprise less than 5% are not listed as prominent lipids. Note that many of the fractions cited were not analyzed for sterols, sterol esters and glycosides, and sphingolipids.

are produced for export to the rest of the cell. An acyl-CoA synthetase identified on the outer membrane of plastids is thought to facilitate release of acyl groups into the cytosol. It should also be noted that, although net lipid traffic is from the plastids, this organelle can likewise be on the receiving end. In addition to small quantities of plastidial phospholipids whose head groups are not known to arise in that compartment, there may be considerable flow of extraplastidially constructed diacylglycerol backbones into the galactolipid synthesis pathway. The quantitative significance of this backflow depends on the plant species, as will be discussed in Section 5.3.

## 2.2. Endoplasmic reticulum and lipid bodies

The endoplasmic reticulum has traditionally been viewed as the primary source of phospholipids in plant cells. With the exception of cardiolipin, all of the common phospholipids can be produced by microsomal fractions. The endoplasmic reticulum also serves as the major site of fatty acid diversification. Although plastids do have the ability to synthesize polyunsaturated fatty acids, they are formed on acyl lipid substrates and are not typically exported. Thus, the endoplasmic reticulum desaturation pathways are of particular importance for developing seeds that store large quantities of $18:2$ and $18:3$. Pathways for the production of unusual fatty acids found primarily in seed oils have likewise been described in microsomes. Not surprisingly, the endoplasmic reticulum also appears to be instrumental in the formation of the triacylglycerols themselves and the lipid bodies in which they are stored (Section 7).

## 2.3. Mitochondria

Next to plastids and the endoplasmic reticulum, the plant mitochondrion is probably the organelle investigated the most thoroughly with respect to lipid metabolism. Its ability to synthesize phosphatidylglycerol and cardiolipin is well established. Although most fatty acids for mitochondrial membranes are imported from the plastids or the ER, recently mitochondria have been shown to synthesize low levels of fatty acids from malonate. Octanoate is a major product of this pathway and serves as a precursor for the lipoic acid cofactor needed by glycine decarboxylase and pyruvate dehydrogenase [2].

## 2.4. Glyoxysomes and peroxisomes

A discussion of the compartmentation of lipids and their metabolism would be incomplete without reference to the organelles responsible for fatty acid oxidation. As in mammals, there is evidence for both mitochondrial and peroxisomal β-oxidation systems. In plants, the peroxisomal system appears to be the more significant [3]. Unlike mammals, plants can use the peroxisomal enzymes to catabolize long-chain fatty acids all the way to acetyl-CoA. Under certain conditions, such as oilseed germination, plants also differentiate specialized peroxisomes called glyoxysomes. In addition to the β-oxidation pathway, glyoxysomes contain the enzymes of the glyoxylate cycle, a pathway absent from animals. Plants are able to use the glyoxylate cycle to feed the acetyl-CoA produced by β-oxidation into carbohydrate synthesis. Since plants cannot transport fatty acids over long distances, only this conversion of acetate to sucrose, which can be transported by the plant's vascular system, makes lipid a practical carbon reserve for the growing shoots and roots of seedlings.

## 3. Acyl-ACP synthesis in plants

Fatty acid synthases may be classified into two groups. 'Type I' fatty acid synthases are characterized by the large, multifunctional proteins typical of yeast and mammals (Chapter 6), while 'Type II' synthases of most prokaryotes are dissociable into components that catalyze individual reactions (Chapter 3). Plants, while certainly themselves eukaryotic, appear to have inherited a Type II fatty acid synthase from the photosynthetic prokaryotes from which plastids originated.

The ground-breaking studies of Overath and Stumpf in 1964 [4] established not only that the constituents of the avocado fatty acid synthesis system could be dissociated and reconstituted, but also that the heat stable fraction from *E. coli* we now know as acyl carrier protein (ACP) could replace the corresponding fraction from avocado. Plant ACPs share both extensive sequence homology and significant elements of three-dimensional structure with their bacterial counterparts. In plants, this small, acidic protein not only holds the growing acyl chain during fatty acid synthesis, but also is required for synthesis of monounsaturated fatty acids and plastidial glycerolipids.

### 3.1. Components of plant fatty acid synthase

Fatty acid synthase is generally defined as including all polypeptides required for the conversion of acetyl and malonyl-CoA to the corresponding ACP derivatives, the acyl-ACP elongation cycle diagrammed in Chapter 3, and the cleavage of ACP from completed fatty acids by enzymes termed thioesterases or acyl-ACP hydrolases [5]. All components of fatty acid synthase occur in plastids, although they are encoded in the nuclear genome and synthesized on cytosolic ribosomes. Most of the 8–10 enzymes of the pathway are soluble when isolated from homogenates. Nevertheless, some evidence suggests that at least ACP and some subunits of acetyl-CoA carboxylase may be associated with the plastid membranes.

Despite the presence of acetyl-CoA : ACP acyltransferase activity in plant fatty acid synthase preparations, acetyl-ACP does not appear to play a major role in plant fatty acid synthesis (Jaworski, 1992). Instead, the first condensation takes place between acetyl-CoA and malonyl-ACP. This reaction is catalyzed by β-ketoacyl-ACP synthase III, one of three ketoacyl synthases in plant systems (Fig. 2). The acetoacetyl-ACP product then undergoes the standard reduction–dehydration–reduction sequence to produce 4 : 0-ACP, the initial substrate of ketoacyl-ACP synthase I. KAS I is responsible for the condensations in each elongation cycle up through that producing 16 : 0-ACP. The third ketoacyl synthase, KAS II, is dedicated to the final plastidial elongation, that of 16 : 0-ACP to 18 : 0-ACP.

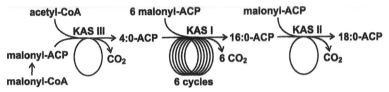

Fig. 2. Contribution of the three ketoacyl synthases (KASI, II and III) to fatty acid elongation. Each circle represents one round of the elongation cycle catalyzed by ketoacyl-ACP synthase, enoyl-ACP reductase, hydroxyacyl-ACP dehydrase, and acyl-ACP reductase.

## 3.2. Desaturation of acyl-ACPs

The major components of the long-chain acyl-ACP pool in most plant tissues are $16:0$-ACP, $18:0$-ACP and $18:1$-ACP. This finding highlights the importance of stearoyl-ACP desaturase, the plastidial enzyme responsible for $\Delta^9$-desaturation in plants. In contrast to the desaturation system of *Escherichia coli* (Chapter 3), the plant enzyme introduces the double bond directly to the $\Delta^9$-position. Unlike yeast and mammalian $\Delta^9$-desaturases, it is a soluble enzyme and is specific for acyl-ACPs rather than acyl-CoAs (McKeon, 1982). In recent years, work on stearoyl-ACP desaturase has progressed rapidly and is now providing a more detailed understanding of the fundamental mechanisms of oxygenic fatty acid desaturation. Genes for the enzyme have been cloned from a number of species, and the structure of the castor bean $\Delta^9$-desaturase has been determined to 2.4 Å resolution. A combination of the crystal structure and spectroscopic methods has revealed two identical monomers, each with an active site containing a diiron-oxo cluster. Reduction of the iron by ferredoxin leads to its binding of molecular oxygen. The resulting complex ultimately removes electrons at the $\Delta^9$-position, resulting in double bond formation [6].

Although the most common unsaturated fatty acids in plants are derived from oleic acid, a wide range of unusual fatty acids are found in the seed oils of different species. Divergent plastid acyl-ACP desaturases have been shown to account for some of this diversity. For example, *Coriandrum sativum* achieves seed oils rich in $\Delta^6$-$18:1$ (petroselinic acid) by desaturation of $16:0$ at the $\Delta^4$-position followed by elongation, while *Thunbergia alata* attains a similar oil by direct $\Delta^6$-desaturation of $18:0$. Several structure–function relationships suggested by unusual desaturases have been tested in the acyl-ACP desaturase system. Shortening the acyl binding pocket of the $\Delta^9$-$18:0$ desaturase by altering a single amino acid as in *Doxantha unguis-cati* shifts substrate specificity in favor of $16:0$-ACP (Cahoon, 1998). In addition, a set of five specific amino acids suggested by the *Thunbergia* gene transforms the castor bean $\Delta^9$-desaturase to a $\Delta^6$-desaturase, while enzymes with certain subsets of the five amino acids can desaturate at either position [6].

## 3.3. Acyl-ACP thioesterases

Among prokaryotes, all acyl groups exiting the dissociable fatty acid synthase are transferred directly from ACP to polar lipids. However, plants must also release sufficient fatty acid from ACP to supply the extraplastidial compartments. Since the typical chloroplast exports primarily 18 : 1 and 16 : 0, the same fatty acids that comprise the greatest fraction of long-chain acyl-ACPs, it might be assumed that a relatively non-specific thioesterase releases 16- and 18-carbon fatty acids from ACP. However, molecular and biochemical analyses of cloned plant thioesterases suggest that plants possess individual thioesterases with specificity either for 18 : 1 or for one or more saturated fatty acids [7]. The most prominent thioesterase in most plants has a strong preference for 18 : 1-ACP, making 18 : 1 the fatty acid most available for extraplastidial glycerolipid synthesis. In contrast, mangosteen, a plant with seed oil particularly high in 18 : 0, contains an 18 : 0-ACP thioesterase gene that has been used to engineer rapeseed with high 18 : 0 content (Hawkins, 1998).

Plants that synthesize certain unusual fatty acids have additional or modified thioesterases. For example, several plant species that produce storage oils containing large amounts of 8- to 14-carbon acyl chains contain thioesterases specific for those chain lengths. By removing acyl groups from ACP prematurely, the medium-chain thioesterases simultaneously prevent their further elongation and release them for triacylglycerol synthesis outside the plastids. In addition, both the standard $\Delta^9$-18 : 1 thioesterase and a $\Delta^6$-18 : 1 thioesterase have been purified from the $\Delta^6$-18 : 1 accumulating coriander plant. Thus plants, by regulating expression of different thioesterases, can both fine tune and radically modify the exported fatty acid pool.

# 4. Acetyl-CoA carboxylase and control of fatty acid synthesis

## 4.1. Two forms of acetyl-CoA carboxylase

The malonyl-CoA that supplies all but two carbons per fatty acid is produced from acetyl-CoA and carbon dioxide by acetyl-CoA carboxylase (ACC). In plants, malonyl-CoA for fatty acid synthesis is apparently provided by a plastid ACC, while a cytosolic ACC contributes malonyl units for fatty acid elongation as well as synthesis of flavonoids, polyketides, and other metabolites. As with fatty acid synthase, ACC forms may be categorized either as 'eukaryotic' enzymes, which are dimers of a multifunctional polypeptide (Chapter 6), or 'prokaryotic' enzymes, which are heteromers of biotin carboxyl carrier protein, biotin carboxylase, and two subunits of carboxyltransferase (Chapter 3). In the grass family, both plastids and cytosol house 'eukaryotic' enzymes. However, dicots and monocots other than grasses appear to have both forms, with the 'eukaryotic' form limited primarily to the cytosol, and 'prokaryotic' enzymes dominating in the plastids [8]. Assembly of the 'prokaryotic' form requires participation of both the nuclear genome, which encodes biotin carboxyl carrier protein, biotin carboxylase, and the α-subunit of carboxyltransferase, and the plastid genome, which has retained the gene for the carboxyltransferase β-subunit, perhaps due to a requirement for RNA editing (Sasaki, 2000).

## 4.2. Acetyl-CoA carboxylase as control point

In other kingdoms, ACC is a major control point for fatty acid biosynthesis. Although the mechanisms acting in plants are incompletely characterized, there is evidence that plant ACCs are also tightly regulated [9]. For example, both redox regulation via thioredoxin and phosphorylation of the carboxyltransferase have been implicated in up-regulation of the 'prokaryotic' ACC by light. Conversely, feedback inhibition is observed at the level of ACC when tobacco cell cultures are given exogenous fatty acids. Due to its impact on the rate of fatty acid synthesis, ACC is considered a promising target in oilseed improvement programs, and some increases in oil content have been obtained by engineering a cytosolic ACC gene to be expressed in rapeseed plastids.

## 5. Phosphatidic acid synthesis: 'prokaryotic' and 'eukaryotic' acyltransferases

Since phosphatidic acid serves as a precursor of phospholipids, galactolipids and triacylglycerols, it is not surprising that its own synthesis has been reported in four plant compartments: plastids, endoplasmic reticulum, mitochondria, and Golgi bodies. In each case, esterification of the first acyl group to the sn-1 position of glycerol-3-phosphate is catalyzed by glycerol-3-phosphate acyltransferase. Lysophosphatidic acid acyltransferase then completes the synthesis by acylating the sn-2 position. However, plastidial and extraplastidial acyltransferases show distinct differences in structure and specificity.

### 5.1. Plastidial acyltransferases

In the plastids, acyltransferases provide a direct route for acyl groups from ACP to enter membrane lipids. Since this is the standard pathway in E. coli and cyanobacteria, both the enzymes of phosphatidic acid synthesis in plastids and the glycerolipid backbones they produce are termed 'prokaryotic.' In both chloroplasts and non-green plastids, the glycerol-3-phosphate acyltransferase is a soluble enzyme that, unlike the E. coli enzyme, shows preference for 18:1-ACP over 16:0-ACP. The lysophosphatidic acid acyltransferase, which is a component of the inner envelope of plastids, is extremely selective for 16:0-ACP. The presence of a 16-carbon fatty acid at the 2-position is therefore considered diagnostic for lipids synthesized in the plastids.

### 5.2. Extraplastidial acyltransferases

At least superficially, the mitochondrial and Golgi acyltransferase activities resemble those of the better studied endoplasmic reticulum system. All three compartments have membrane-bound glycerol-3-phosphate and lysophosphatidic acid acyltransferases that utilize acyl-CoA substrates. In the endoplasmic reticulum, which is quantitatively the most significant of the extraplastidial sites for phosphatidic acid synthesis, saturated fatty acids are almost entirely excluded from the sn-2 position. The glycerol-3-phosphate acyltransferase is less selective, but, due to substrate availability, more often fills the sn-1 position with 18:1 than with 16:0. It is therefore possible to judge the relative contributions of the prokaryotic and eukaryotic pathways by comparing the proportions of eukaryotic 18/18 or 16/18 glycerolipids with prokaryotic 18/16 or

16/16 glycerolipids [5]. Mitochondrial lysophosphatidic acid acyltransferase, which shows little selectivity for chain length or unsaturation, can usually be ignored in discussions of lipid flow.

### 5.3. 16 : 3 and 18 : 3 plants

Relative fluxes through the prokaryotic and eukaryotic pathways vary between organisms and among tissues. Plastids have the potential to use phosphatidic acid from the prokaryotic pathway for all of their glycerolipid syntheses. However, not all plants do so; in some cases, the prokaryotic acyl chain arrangement is found only in plastidial phosphatidylglycerol, whereas galactolipids are derived from diacylglycerol imported to the plastids from the ER. As indicated above, the eukaryotic acyltransferases of the endoplasmic reticulum produce substantially more 18/18 than 16/18 lipids, and it is chiefly the 18/18 units that are assembled into galactolipids by plants with a minor prokaryotic pathway. Because galactolipids become highly unsaturated, plants that import diacylglycerol for galactolipids are rich in 18 : 3 and are called 18 : 3 plants. Species in which most galactolipid is derived from the prokaryotic 18/16 or 16/16 diacylglycerol contain substantial 16 : 3 and are known as 16 : 3 plants.

Kunst et al. [10] have demonstrated that a 16 : 3 plant, *Arabidopsis thaliana*, may be converted to a de facto 18 : 3 plant by a single mutation in plastidial glycerol-3-phosphate acyltransferase. Under these conditions, 16 : 3 content is reduced dramatically, and when isolated chloroplasts are labeled with glycerol-3-phosphate, only phosphatidylglycerol is labeled. Nevertheless, the percentage of galactolipids in mutant plants is practically identical to that in wild-type plants, emphasizing the ability of plants to compensate for reduction of the prokaryotic pathway. Other studies in mutants have confirmed that plants have an amazing capacity to adapt to many, but not all, perturbations of lipid metabolism (Section 11).

## 6. Glycerolipid synthesis pathways

In plants, glycerolipid biosynthesis involves a complex web of reactions distributed among multiple compartments [11,12]. As in mammals, the synthesis of individual glycerolipids is initiated either by the formation of CDP-diacylglycerol from phosphatidic acid and CTP, or by cleavage of phosphate from phosphatidic acid to produce diacylglycerol.

CTP : phosphatidate cytidylyltransferase has been observed in plastids, mitochondria and endoplasmic reticulum. In all three compartments, the CDP-diacylglycerol derived from phosphatidic acid is used in the synthesis of phosphatidylglycerol; in mitochondria, the reaction of phosphatidylglycerol with a second CDP-diacylglycerol then produces cardiolipin. The endoplasmic reticulum can also incorporate CDP-diacylglycerol into phosphatidylinositol and phosphatidylserine.

Phosphatidic acid phosphatase is present in the same three compartments. In the endoplasmic reticulum and mitochondria, diacylglycerol combines with CDP-ethanolamine or CDP-choline to produce phosphatidylethanolamine or phosphatidylcholine, respectively. Although separate enzymes catalyze ethanolamine and choline transfer in animals and yeast, there are indications that a single aminoalcoholphosphotransferase may be responsible in plants. Flux into phosphatidylcholine is at least partially determined by regulation of the phosphocholine cytidylyltransferase that generates CDP-choline. The production of CDP-ethanolamine from phosphoethanolamine is less well studied, but is also considered a probable regulatory step. In addition, there is clear evidence that phosphothanolamine can be methylated to monomethylphosphoethanolamine, dimethylphosphoethanolamine, and phosphocholine, and that this pathway is inhibited by exogenous choline at the initial methylation step. The methylation of phosphoethanolamine in plants is frequently contrasted with the methylation of phosphatidylethanolamine to phosphatidylcholine in animals and yeast. In general, no significant methylation of phosphatidylethanolamine itself occurs in plants. However, phosphatidylmonomethylethanolamine is synthesized and converted to phosphatidylcholine [11].

The diacylglycerol released in plastids reacts either with UDP-galactose or with UDP-sulfoquinovose to generate sulfolipid or monogalactosyldiacylglycerol. Synthesis of digalactosyldiacylglycerol from the latter may use either UDP-galactose or a second monogalactosyldiacylglycerol as the galactose donor [12] (Kelly, 2002). There is also evidence for some extraplastidial galactolipid synthesis in plants suffering from phospholipid deficiency (Hartel, 2000).

## 6.1. Glycerolipids as substrates for desaturation

In addition to the soluble acyl-ACP desaturases, plants contain a number of membrane-bound enzymes that desaturate fatty acids while they are esterified within glycerolipids [5,6]. The recent cloning and characterization of these desaturases is of great interest to the scientific community because the products of the membrane-bound systems include $\Delta^{9,12}$-18 : 2 and $\Delta^{9,12,15}$-18 : 3, both of which are essential to the human diet, and thought to play a major role in human health and disease.

Once again, separate pathways occur in plastids and endoplasmic reticulum, although, as should be evident from the discussion above, fatty acids from the endoplasmic reticulum may make their way back to the plastids. Clarification of the number of desaturases involved in plant lipid metabolism and isolation of their genes has been greatly assisted by the isolation of a large number of mutants in *A. thaliana*, a small weed of the mustard family used as a model organism by plant geneticists and molecular biologists. Briefly, three membrane-bound desaturation sequences are evident in *Arabidopsis* [5,6].

(1) In chloroplasts, 16 : 0 at the 2-position of phosphatidylglycerol is desaturated to *trans*-3-16 : 1. This desaturase is most likely encoded by the *FAD4* gene.

(2) Plastids are able to convert 18 : 1 to 18 : 3 and 16 : 0 to 16 : 3 using a combination of three membrane-bound desaturases. One of them, encoded by *FAD5*, is relatively specific for the conversion of 16 : 0 on monogalactosyldiacylglycerol to $\Delta^7$-16 : 1. This 16 : 1 and $\Delta^9$-18 : 1 may then be given a second and third double bond by the *FAD6* and the *FAD7* or *FAD8* gene products, respectively. The latter two desaturases are less selective in their choice of glycerolipid substrate, and will

accept appropriate fatty acids on phosphatidylglycerol, sulfolipid, or either of the major galactolipids.

(3) In the endoplasmic reticulum, 18 : 1 esterified to phosphatidylcholine or occasionally phosphatidylethanolamine may be desaturated to 18 : 2 by *FAD2* and to 18 : 3 by *FAD3*.

It should be noted that fatty acids entering one of the multistep desaturation pathways listed above are not necessarily committed to completing that set of reactions. It is particularly common for 18 : 2 to be an end product of endoplasmic reticulum desaturation. This 18 : 2 may remain in phospholipid, be incorporated into triacylglycerol, or enter the galactolipid pathway and receive a third double bond in the chloroplast.

# 7. Lipid storage in plants

A plant stores reserve material in its seeds in order to allow seedling growth of the next generation until photosynthetic capacity can be established. The three major storage materials are oil, protein and carbohydrate, and almost all seeds contain some of each. However, their proportions vary greatly. For example, the amount of oil in different species may range from as little as 1–2% in grasses such as wheat, to as much as 60% of the total dry weight of the castor seed. With the exception of the jojoba plant, which accumulates wax esters in seeds, plants store oil as triacylglycerol (TAG).

## 7.1. Lipid body structure and biogenesis

In the mature seed, TAG is stored in densely packed lipid bodies, which are roughly spherical in shape with an average diameter of 1 μm (Fig. 3) [13]. This size does not change during seed development, and accumulation of oil is accompanied by an increase in the number of lipid bodies. The very large number of lipid bodies in an oilseed cell (often >1000) contrasts strikingly with animal adipose tissue where oil droplets produced in the cytosol can coalesce into a few or only one droplet. The plant lipid bodies appear to be surrounded by a lipid monolayer in which the polar headgroups face the cytosol, while the non-polar acyl groups are associated with the non-polar TAG within. The membranes of isolated lipid bodies, which comprise less than 5% of a lipid body's weight, contain both phospholipids and characteristic proteins known as caleosins and oleosins. The recently discovered caleosins are calcium binding proteins whose function is still unknown. Oleosins are small (15–26 kDa) proteins that are believed to preserve individual lipid bodies as discrete entities. Desiccated seeds lacking oleosins undergo lipid body fusion and cell disruption when rehydrated (Leprince, 1998). The cDNAs encoding many oleosins have been cloned and each has a sequence encoding a totally hydrophobic domain of 68–74 amino acids which is likely to be the longest hydrophobic sequence found in any organism. Structurally, oleosins are roughly analogous to the animal apolipoproteins which coat the surface of lipid droplets during their transport between tissues.

When a seed germinates, the TAG stored in the lipid bodies becomes the substrate for lipases. In at least some cases, peroxidation of polyunsaturated fatty acids by a lipid

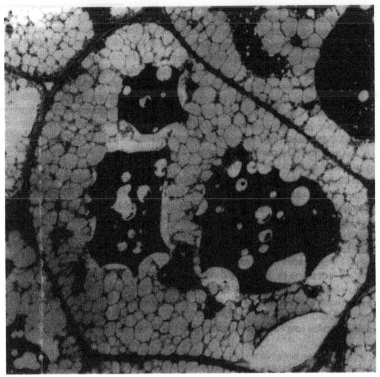

Fig. 3. Thin-sectional view of cells in a cotyledon of a developing cotton embryo harvested 42 days after anthesis. The cells are densely packed with lipid bodies and several large storage protein bodies (dark). Magnification ×9000. Photo courtesy of Richard Trelease, Arizona State University.

body lipoxygenase precedes the release of fatty acids from TAG [14]. Typically lipases and lipid body lipoxygenase are active only after germination is triggered by imbibition and other environmental signals. Fatty acids released by the lipid bodies are further metabolized through the β-oxidation pathway and glyoxylate cycle in the glyoxysomes (Section 2.4).

### 7.2. Seed triacylglycerols often contain unusual fatty acids

The structural glycerolipids of all plant membranes contain predominantly 5 fatty acids (18:1, 18:2, 18:3, 16:0, and in some species, 16:3). However, the fatty acid composition of storage oils varies far more than in membrane glycerolipids. Altogether more than 300 different fatty acids are known to occur in seed TAG [15]. Chain length may range from less than 8 to over 22 carbons. The position and number of double bonds may also be unusual, and hydroxy, epoxy or other functional groups can modify the acyl chain. Many of the different fatty acid structures, including hydroxy, epoxy, acetylenic and conjugated varieties, are now known to originate from minor modifications in the amino acid sequence of the oleate desaturase. For example, only

Table 3
Some unusual fatty acids produced in plant seeds

| Fatty acid type | Example | Major sources | Major or potential uses | Approx. US market size, $10^6$ |
|---|---|---|---|---|
| Medium chain | Lauric acid (12:0) | Coconut, palm kernal | Soaps, detergents, surfactants | 350 |
| Long chain | Erucic acid (22:1) | Rapeseed | Lubricants, anti-slip agents | 100 |
| Epoxy | Vernolic acid $18:1\Delta^9$epoxy12,13 | Epoxidized soybean oil, *Vernonia* | Plasticizers | 70 |
| Hydroxy | Ricinoleic acid $18:1\Delta^9$,12OH | Castor bean | Coatings, lubricants | 80 |
| Acetylenic | Crepenynic acid $18:2\Delta^9$,12yne | *Crepis foetida* | Polymers | – |
| Cyclopropene | Sterculic acid 19:1 | *Sterculia foetida* | Lubricants, polymers | – |
| Conjugated | Parinaric $18:4\Delta^9$c11t13t15c | *Impatiens balsamina* | Coatings | – |
| Trienoic | Linolenic acid ($\alpha$18:3) | Flax | Paints, varnishes, coatings | 45 |
| Wax esters | Jojoba oil | Jojoba | Lubricants, cosmetics | 10 |

4 amino acid changes have been shown to convert a desaturase into a hydroxylase [16].

The reason for the great diversity in plant storage oils is unknown. However, the special physical or chemical properties of the 'unusual' plant fatty acids have been exploited for centuries. In fact, approximately one-third of all vegetable oil is used for non-food purposes (Table 3). Reading the ingredients of a soap or shampoo container reveals one of the major end uses of high lauric acid specialty plant oils. Other major applications include the use of erucic acid (22:1) derivatives to provide lubricants and as a coating for plastic films. Hydroxy fatty acids from the castor bean have over 100 industrial applications including plastic and lubricant manufacture. As discussed further below, the ability of genetic engineering to transfer genes for some unusual fatty acid production from exotic wild species to high yielding oil crops is now providing the ability to produce new renewable agricultural products and to replace feedstocks derived from petroleum.

### 7.3. The pathway of triacylglycerol biosynthesis

As in animal tissues, it has been suggested that TAGs are produced by a relatively simple four reaction pathway. According to this model, phosphatidic acid is synthesized by the extraplastidial pathway (Section 5) and dephosphorylated to diacylglycerol. A third fatty acid is then transferred from CoA to the vacant third hydroxyl of the diacylglycerol, producing TAG. This last and single committed step is catalyzed by diacylglycerol acyltransferase (Fig. 4, reaction 6). Although plants possess all of the enzymes for the reactions above, the assembly of three fatty acids onto a glycerol backbone is not always

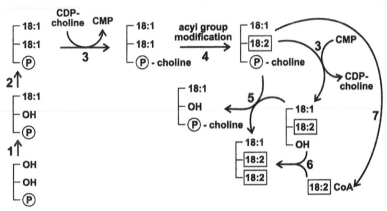

Fig. 4. Pathway depicting how flux through phosphatidylcholine (product of reaction 3) can promote acyl group diversity in plant triacylglycerols. Production of 18:2 (boxed) at the *sn*-2 position and its transfer to TAG is used as a sample modification. Other fatty acid alterations may be substituted. Enzymes: 1 = glycerol-3-phosphate:acyl-CoA acyltransferase; 2 = lysophosphatidic acid:acyl-CoA acyltransferase; 3 = CTP:phosphatidate cytidylyltransferase; 4 = T6 18:1-desaturase or other fatty acid modifying enzyme; 5 = phospholipid:diacylglycerol acyltransferase; 6 = diacylglycerol acyltransferase; 7 = acyl-CoA:phosphatidylcholine acyltransferase or phospholipase plus acyl-CoA synthetase.

as straightforward as suggested by the above pathway. In many oilseeds, pulse-chase labeling has revealed that fatty acids reach TAG only after passing through phosphatidylcholine, (or phosphatidylethanolamine to a lesser extent). Given the range of desaturation and other modification reactions that can take place on phosphatidylcholine, transit through this phospholipid helps to explain some of the fatty acid diversity in TAG.

Fatty acids from phosphatidylcholine may become available for TAG synthesis in several ways [17]. In some plants, a phospholipid:diacylglycerol acyltransferase produces TAG by direct transfer of a fatty acid from the 2-position of phospholipid to diacylglycerol (Fig. 4, reaction 5) [18]. The second mechanism by which phosphatidylcholine can participate in TAG synthesis is by donation of its entire diacylglycerol unit. In many plants, the synthesis of phosphatidylcholine from diacylglycerol and CDP-choline appears to be rapidly reversible. As shown in Fig. 4 (reaction 3), this activity of the choline phosphotransferase can allow diacylglycerol moieties modified on phosphatidylcholine to be incorporated into TAG. Finally, a fatty acid removed from phosphatidylcholine may subsequently be used for TAG synthesis. Such an 'acyl exchange' may provide acyl-CoA either by the combined reverse and forward reactions of an acyl-CoA:phosphatidylcholine acyltransferase, or by a combination of phospholipase and acyl-CoA synthase.

## 7.4. Challenges in understanding triacylglycerol synthesis

Although the basic reactions of TAG biosynthesis have been determined, several fundamental and potentially related questions persist. As highlighted above, TAG and membrane lipids frequently have radically different fatty acid compositions. How do plants control which fatty acids are stored in TAG as opposed to which fatty acids are restricted to membranes? Are unusual fatty acids excluded from membranes because their physical and chemical idiosyncracies would perturb membrane fluidity or other physical characteristics? Is TAG synthesis spatially distinct from the synthesis of membrane lipids, or do enzyme specificities dictate the partitioning of fatty acid species among glycerolipids? Although all of these factors may be significant, selectivity by enzymes such as phospholipid:diacylglycerol acyltransferase for unusual fatty acids, and editing of unusual fatty acids from phospholipids, are currently the best documented [17,18].

## 8. Protective lipids

In plants, tissues are protected against desiccation and pathogens by both a cuticle and epicuticular wax. The cuticle itself contains some wax, but is anchored to the plant cell wall by cutin, a complex polyester of fatty acid derivatives with a wide range of oxygen-containing functional groups (Fig. 5). 16- and 18-carbon dicarboxylic acids with one or more hydroxyl groups are particularly common [19]. *Arabidopsis* transformed with cutinase from a fungal plant pathogen not only develops a leaky cuticle, but also suffers from fusions between leaves and flower parts (Sieber, 2000).

Surface waxes are complex mixtures including a range of very long-chain alkanes, aldehydes and ketones as well as wax esters and their building blocks. Although only one mutation in cutin formation has been identified (Wellesen, 2001), visual screening of plant surfaces has allowed isolation of several wax mutants blocked in malonyl-CoA-dependent elongation of fatty acids, decarbonylation and reduction (Post-Beittanmiller, 1996). The cDNA of an elongase required for extension of wax acyl units beyond 24 carbons has been cloned, and resembles the condensing enzymes involved in synthesis of erucic acid in seed oil and wax ester precursors in jojoba seeds (Millar, 1999). A cDNA encoding the wax synthase of jojoba seeds has been cloned and successfully expressed in *Arabidopsis* (Lardizabal, 2000), and may provide clues to the corresponding genes for surface waxes.

Bark, wound callus, and specialized tissues such as the endodermis that controls entry into the root vascular system, have walls lined with suberin. Suberin, like cutin, is a polyester incorporating fatty acids enriched in carboxyl and hydroxyl groups. In addition to placement on the inner surface of cell walls rather than outside, the tough, waterproof suberin differs from cutin in its preference for longer fatty acids and in its incorporation of large amounts of phenylpropanoids [19].

## 9. Sterol, isoprenoid and sphingolipid biosynthesis

In the plant kingdom, isoprenoids represent the most diverse range of natural products with over 25,000 lipophilic structures known, ranging from small, volatile compounds to rubber. Quantitatively, the photosynthetic apparatus is probably the primary consumer of isoprenoids, since carotenoids, plastoquinone and the phytol tail of chlorophyll

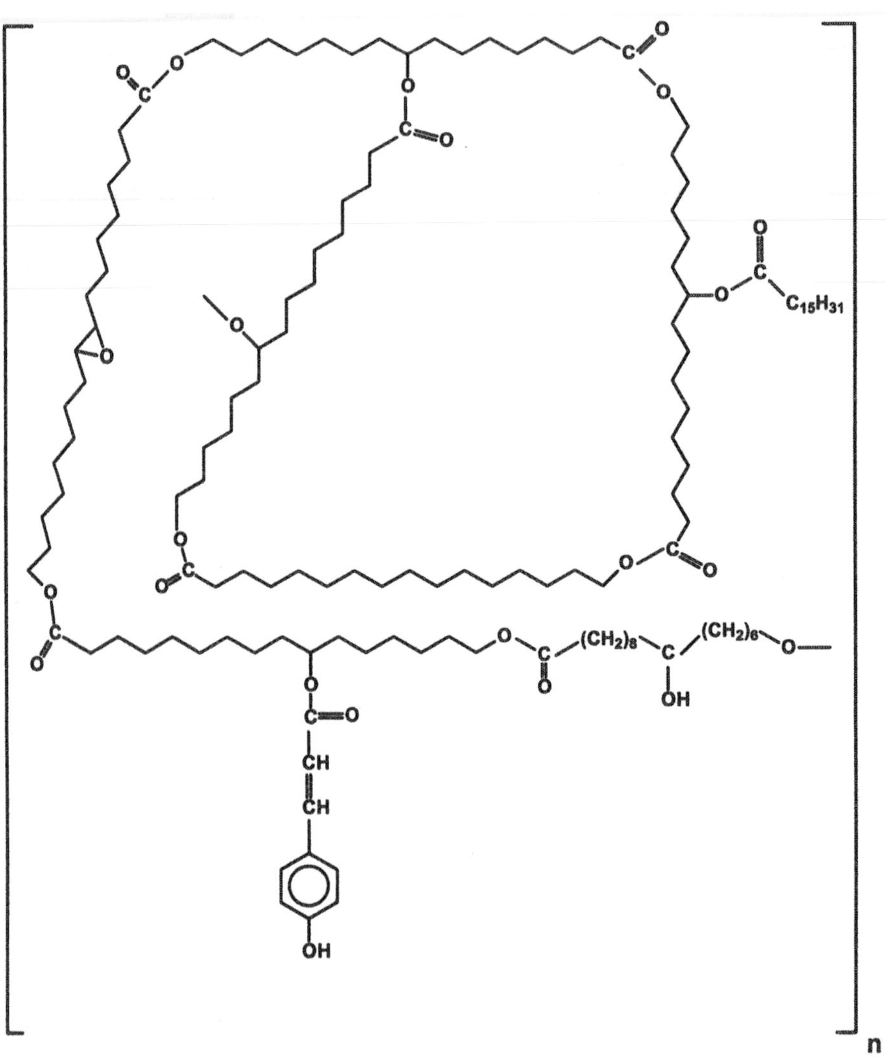

110

Fig. 5. Model showing some of the linkages in cutin, after Kolattukudy [19]. Note the cross-linking made possible by mono- and dihydroxy-fatty acids.

all belong to this group. Given that vital plant hormones such as gibberellin and abscisic acid, plus many defensive compounds, are isoprenoids, the early steps of this pathway have been studied intensely. However, surprisingly, it was not until the late 1990s that researchers realized that plants have two very different pathways for production of isopentenyl pyrophosphate, the five-carbon central precursor of all isoprenoids [20]. For several decades it was known that, as in other organisms, plants join three molecules of acetyl-CoA to form hydroxymethylglutaryl-CoA followed by the highly regulated reduction of that compound to mevalonic acid. Furthermore, plants contain multiple well-studied hydroxymethylglutaryl-CoA reductase genes that are differentially expressed during development and in response to such stimuli as light, wounding and infection. It was incorrectly suspected that this 'mevalonate' pathway was localized in both cytosol and plastids and produced all classes of isoprenoids. The story has now been clarified [20] with the discovery that plastids produce isopentenyl pyrophosphate by a 'non-mevalonate' pathway that begins with the condensation of pyruvate with glyceraldehyde-3 phosphate to produce 1-deoxy-D-xylulose-5-P. At least three additional enzymes are required to produce isopentenyl pyrophosphate in the plastids. In parallel to work in plants, this pathway has also been demonstrated in bacteria and algae. The non-mevalonate pathway in plastids is responsible for production of the classic plant photosynthetic isoprenoids such as phytols and carotenoids, as well as mono- and diterpenes.

The mevalonate pathway in the cytosol is responsible for biosynthesis of sterols, sesquiterpenes and triterpenoids. After conversion of mevalonic acid to isopentenyl pyrophosphate, three C5 units can be joined head to tail to produce a C15 compound, farnesyl pyrophosphate. Two farnesyl pyrophosphates are then united head to head to form squalene, the progenitor of the C30 isoprenoids from which sterols are derived. The plant squalene synthetase, like its mammalian homologue, is found in the endoplasmic reticulum and the reaction proceeds via a presqualene pyrophosphate intermediate. In the last step prior to cyclization, squalene is converted to squalene 2,3-epoxide.

It is also in the cyclization step that photosynthetic and non-photosynthetic organisms diverge. Whereas animals and fungi produce lanosterol, organisms with a photosynthetic heritage produce cycloartenol. Despite the differences in the cyclization product, there is substantial conservation between the enzymes responsible, with 34% identity between an *Arabidopsis* cycloartenol synthase and lanosterol synthase.

A complex series of reactions including opening of the cyclopropane ring, double bond formation and isomerization, demethylation of ring carbons, and methylation of the side chain result in formation of a number of different plant sterols. Sitosterol is the most common plant sterol (Fig. 6); however, plants normally contain mixtures of sterols whose proportions differ from tissue to tissue. In addition, sterol esters, sterol glycosides, and acylated sterol glycosides are common plant constituents whose physiological significance is under scrutiny. Both cold adaptation and pathogenesis drastically alter free and derivatized sterol pools. Plants also produce a steroid hormone, brassinolide, required for both light-induced development and fertility. Interestingly, the gene for a 5α-reductase in the brassinolide pathway can complement the corresponding reductase in the testosterone pathway [21].

Sphingolipids are usually considered minor constituents of plant lipids, accounting for 5% or less of most lipid extracts. This fact, and the more complex methods needed for their identification and characterization have resulted in a comparative lack of information on plant sphingolipid biosynthesis and function. Nevertheless, sphingolipids make up a substantial proportion (25% or more) of the composition of plasma and tonoplast membranes, with the glucosylceramides constituting the largest fraction. In addition, difficulties in extraction and analysis may have led to underestimates of sphingolipid contents. As in animals (Chapter 14), sphingolipid biosynthesis begins with condensation of palmitoyl-CoA with serine to form 3-keto-sphinganine, with the enzyme from *A. thaliana* showing 68% similarity to the human homologue (Tamura,

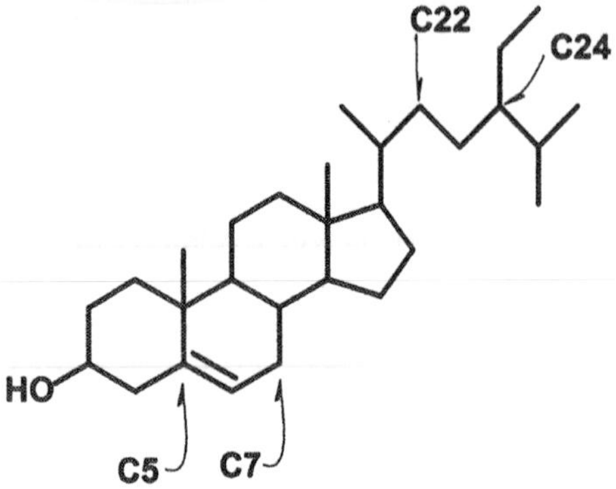

Fig. 6. Sitosterol (24α-ethylcholesterol), shown here, is the most common plant sterol, but plants generally contain complex mixtures of sterols. Other prominent phytosterols differ from sitosterol as follows. Campesterol, 24α-methyl; stigmasterol, C22 double bond; dihydrospinasterol, move double bond from C5 to C7; spinasterol, move C5 double bond to C7, add C22 double bond; dihydrobrassicasterol, 24β-methyl; brassicasterol, 24β-methyl, add C22 double bond.

2000). Reduction of 3-keto-sphinganine by an NADPH-dependent reaction yields sphinganine, and cDNAs encoding enzymes for the subsequent C4-hydroxylation and $\Delta^8$-desaturation of sphinganine have recently been cloned [22]. Less information is available on the synthesis of ceramides and glucosylceramides, although there is evidence that either sterol glucoside or UDP-glucose could serve as the glucose donor for the latter [23]. The finding that sphingolipids serve as signal molecules in plants, particularly in the defensive apoptosis induced by some plant pathogens, has begun to stimulate interest in this relatively neglected area of plant lipid biochemistry.

## 10. Oxylipins as plant hormones

Jasmonate is one of several lipid-derived plant growth regulators referred to as oxylipins [24]. The structure and biosynthesis of jasmonate have intrigued plant biologists because of the parallels to some eicosanoids (Chapter 13), which are central to inflammatory responses and other physiological processes in mammals. In plants, jasmonate derives from α-linolenic, presumably released from membrane lipids by a phospholipase A. The linolenic acid is oxidized by lipoxygenase and the resulting products, 9-hydroperoxylinolenic acid or 13-hydroperoxylinolenic acid may be further metabolized by one of three routes to produce a wide variety of oxylipin (Fig. 7). The pathways by which 13-hydroperoxylinoleic acid may be metabolized include hydroperoxide lyase catalyzed scission of the *trans*-11,12-double bond to produce a C6 aldehyde, *cis*-3-hexenal, and the C12 compound, 12-oxo-*cis*-9-dodecenoic acid. The acid is subsequently metabolized to 12-oxo-*trans*-10-dodecenoic acid, the wound hormone traumatin. The enzyme hydroperoxide dehydratase (allene oxide synthase) catalyzes the dehydration of hydroperoxides to unstable allene oxides that readily decompose to form a 9,12-ketol or a 12,13-ketol. The allene oxide of 13-hydroperoxylinolenic acid may also be converted by allene oxide cyclase to 12-oxo-phytodienoic acid which can

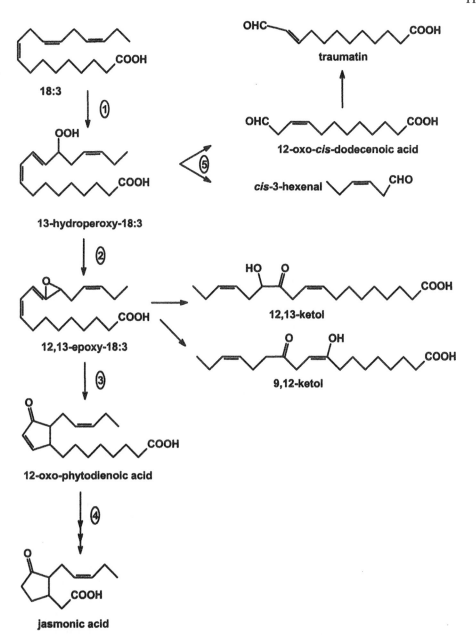

Fig. 7. Metabolism of 18:3 to oxylipins. 1, lipoxygenase; 2, allenoxide synthase; 3, allene oxide cyclase; 4, 12-oxo-phytodienoic acid reductase, β-oxidation; 5, hydroperoxide lyase.

be further metabolized to 7-iso-jasmonic acid. In the last few years, it has become clear that jasmonate is a key component of a wound-signaling pathway that allows plants to protect themselves against insect attack. When experimentally applied to plants at low concentrations, jasmonate leads to the induction of protease inhibitors and other defense compounds. Furthermore, mutants of tomato and *Arabidopsis* that are deficient in jasmonate synthesis are much more susceptible to insect damage. In addition to jasmonate, a number of the other oxylipins have been reported to function as signal molecules. In particular, the oxylipin traumatin has been suggested to trigger cell division at the site of wounds, leading to the development of a protective callus. The lipoxygenase product 13-hydroxylinolenic acid triggers phytoalexin production. Similarly, C6–C10 alkenals act as volatile elicitors of a defense response in cotton.

## 11. Progress in plant lipid research: the value of mutants

Biochemical approaches toward understanding plant lipid biosynthesis and function provided much of the information summarized above. However, in recent years, the isolation of mutants in plant lipid metabolism has been extremely fruitful in providing new insights and new methods for gene isolation. Much of the progress in the genetic dissection of plant lipid metabolism has come from the extensive studies of Somerville, Browse and coworkers with *A. thaliana*, which has one of the smallest genomes (113 megabases) of higher plants. By using gas chromatography to screen several thousand randomly selected plants from a mutagenized population, Somerville and Browse were able to obtain an extensive collection of mutants showing altered leaf or seed fatty acid compositions. As described above, these mutants included isolates instrumental in confirming the relationships between prokaryotic and eukaryotic phosphatidic acid synthesis and in the analysis and cloning of membrane-bound desaturases.

### 11.1. Mutants in lipid metabolism have helped link lipid structure and function

Two other major benefits have been derived from the *Arabidopsis* lipid mutants. First, the physiological effects of the mutations have provided the opportunity to evaluate the relationships between lipid structure and function. There has been a long-term assumption, based on the strong association of high levels of polyunsaturated fatty acids with photosynthetic membranes and the conservation of this property among higher and lower plant species, that these fatty acids must be essential to photosynthesis. However, many attempts to understand the relationships between membrane fatty acid composition and cell physiology or photosynthesis have led to equivocal results. The isolation of mutants totally lacking certain unsaturated fatty acids has now provided much more convincing evaluations of their function and indeed, the results have forced re-evaluation of several previous hypotheses. For example, $\Delta^3$-*trans*-hexadecenoate is an unusual plant fatty acid which is associated with phosphatidylglycerol of chloroplast membranes, is evolutionarily conserved, and is synthesized in coordination with the assembly of the photosynthetic apparatus. These observations led to the suggestion that $\Delta^3$-*trans*-hexadecenoate is a highly essential component of photosynthesis. However, mutants which contain no detectable $\Delta^3$-*trans*-hexadecenoate grow as well as wild type plants, and all photosynthetic parameters examined appear normal (Browse, 1985). A minor difference in stability of some components of the photosystem can be detected by polyacrylamide gel electrophoresis. It has been concluded from such analyses that, although $\Delta^3$-*trans*-hexadecenoate may facilitate assembly of the light harvesting complex into thylakoids, a more obvious phenotype could be restricted to certain unusual environmental conditions.

As mentioned above, a number of mutants blocked in the production of polyunsaturated fatty acid biosynthesis have also been isolated (Table 4). Because leaves have desaturases both in chloroplasts and in the endoplasmic reticulum, single mutations lead only to partial reduction of polyunsaturated fatty acid levels. Again, these mutants grow normally under most conditions and have normal photosynthetic parameters. However, several alterations in physiology are observed including changes in chloroplast ultrastructure, a reduction in the cross-sectional area of chloroplasts, and increased stability to thermal disruption of photosynthesis. Moreover, whereas wild-type *Arabidopsis* plants are chilling resistant and can reproduce normally at temperatures as low as 6°C, the mutants blocked in plastidial $\Delta^7$ (*fad5*) and ω6 (*fad6*) desaturation become chlorotic at 6°C and show a 20–30% reduction in growth rate relative to the wild-type plant. The *fad2* mutants, in which the endoplasmic reticulum ω-6 desaturase is blocked, are even more sensitive to 6°C and die if left at this temperature for several days. These results demonstrate that polyunsaturated fatty acids are essential for maintaining cellular function and plant viability at low temperatures.

While most mutants which are reduced only in polyunsaturated fatty acid synthesis grow and develop normally at 22°C, a high-stearate mutant with 14% 18 : 0 is strikingly abnormal (Fig. 8) [26]. Many cell types fail to expand, resulting in mutant plants growing to less than one-tenth the size of wild-type. At higher growth temperatures (36°C), the effects are less dramatic, suggesting that the physical properties or fluidity of highly saturated membranes are less impaired under these conditions.

Other large scale alterations in membrane fatty acid composition and phenotypes have been obtained by creation of multiple-mutant lines. When mutants defective in endoplasmic reticulum ω6-desaturase were crossed with plants defective in the plastid ω6-desaturase, double mutants could be recovered only on sucrose-supplemented media. The sucrose grown plants, which contained less than 6% polyunsaturated fatty acids, were chlorotic and unable to carry out photosynthesis but otherwise remarkably normal. These results, while confirming the significance of polyunsaturated fatty acids to photosynthesis, indicate that the vast majority of membrane functions can proceed despite drastically reduced levels of polyunsaturates [27].

Triunsaturated fatty acids normally dominate chloroplast membranes and thus are the most abundant fatty acids on the planet. By constructing a triple mutant of *fad3*, *fad7* and *fad8*, it has been possible to eliminate triunsaturated fatty acids from *Arabidopsis* without affecting 16 : 2 and 18 : 2 production (McConn, 1996). Surprisingly, these plants are able to grow, photosynthesize, and even flower. However, they are male sterile and therefore cannot produce seeds. This observation led to the discovery of a very different role for jasmonate. This mutant cannot synthesize jasmonate because it lacks the 18 : 3 precursor, and the plants are male-sterile because pollen does not mature properly and is not released from the anthers. Application of jasmonate or linolenic acid to the anthers restores fertility, demonstrating that jasmonate is a key signal in pollen development. This result is a dramatic example of a change in fatty acid composition having a very specific effect on an essential developmental and tissue specific reproductive process.

Table 4

Biochemical and physiological responses of selected *Arabidopsis* lipid mutants [a]

| Mutant | Enzyme blocked [a] | Fatty acid or lipid phenotype | Physiological response |
|---|---|---|---|
| fab1 | 3-Ketoacyl-ACP synthase II | 16:0↑ | Death of plants after prolonged exposure to 2°C |
| fab2 | 18:0-ACP $\Delta^9$-desaturase | 18:0↑ | Dwarf at 22°C |
| fad4 | t$\Delta^3$, 16:0 desaturase? | t$\Delta^3$, 16:1↓ | Altered stability of photosystem? |
| fad5 | 16:0 $\Delta^7$-desaturase | 16:0↑; 16:3↓ | Enhanced growth rate at high temperatures. Leaf chlorosis, reduced growth rate and impaired chloroplast development at low temperature. |
| fad6 | Plastid ω-6 desaturase | 16:1↑; 16:3↓ | Leaf chlorosis, reduced growth rate and impaired chloroplast development at low temperature. Enhanced thermotolerance of photosynthetic electron transport at high temperatures. |
| fad7 | Plastid ω-3 desaturase | 16:2↑; 18:2↑; 16:3↓; 18:3↓ | Reduced chloroplast size and altered chloroplast ultrastructure. |
| fad2 | Cytosolic ω-6 desaturase | 18:1↑; 18:2↓ | Greatly reduced stem elongation at 12°C. Death at 6°C. |
| fad2/fad6 | Plastid and cytosolic ω-6 desaturase | <6% Polyunsaturated | Loss of photosynthesis |
| fad3/fad7/fad8 | Plastid and cytosolic ω-3 desaturase | <1% Trienoic | Male sterile, insect resistance decreased |
| dgd1 | Digalactosyldiacylglycerol synthase | Digalactosyldiacylglycerol↓ | Dwarfism, abnormal chloroplast size |
| act1 | Plastid acyl-ACP : G3P acyltransferase | Phosphatidylglycerol↓, 16:3↓ | Altered chloroplast structure |
| mgd1 | Diacylglycerol glycosyltransferase | Monogalactosyldiacylglycerol↓ | Abnormal chloroplast development |
| AS11 | Diacylglycerol acyltransferase | 50% Reduction in seed TAG | Slow germination |
| wri1 | Unknown; glycolysis impaired | 80% Reduction in seed TAG | Slow germination |

[a] In some cases, the enzyme defect in the mutation is not known and this table lists the most likely possibility.

Fig. 8. Increased stearic acid causes severe dwarfing of *Arabidopsis*. A wild-type *Arabidopsis* plant (left) is compared to the *fab2* mutant (right) in which leaf stearic acid content has increased from 1 to 14%. Photo courtesy of John Browse, Washington State University.

## 11.2. Arabidopsis mutants have allowed cloning of desaturases and elongases

*A. thaliana* mutants have also provided a means of cloning genes difficult to isolate by other methods. As in other kingdoms, the membrane bound enzymes of plants have been notoriously difficult to purify and characterize. However, cDNAs or genes encoding a number of these enzymes have now been isolated using molecular genetic strategies based on mutations. Several approaches have been successful. A cDNA encoding the ω-3 desaturase which converts linoleic to linolenic acid was isolated in 1992 by Arondel et al. [28] after a mutation leading to the loss of function was genetically mapped and a yeast artificial clone corresponding to this region was selected. The genomic region was then used to screen a cDNA library, and some of the clones which hybridized to the yeast artificial chromosome had sequence similarity to cyanobacterial desaturases. These clones subsequently were shown to complement the loss of 18 : 3.

Gene 'tagging' strategies have also proved enormously valuable in identifying clones of membrane bound enzymes. *Arabidopsis* genes can be 'tagged' by insertional inactivation when T-DNA from *Agrobacterium tumefaciens* inserts randomly in the genome. When a promising phenotype is observed, the inactivated gene can be identified by probing with the T-DNA sequence. This method was used to identify the ω-6 desaturase required for the 18 : 1 to 18 : 2 conversion in endoplasmic reticulum [25]. In addition, transposon tagging led to the cloning of a gene which controls elongation of oleic acid to 20 : 1 and 22 : 1 in developing *Arabidopsis* seeds [29]. Since no membrane-bound fatty acid elongase had ever been completely purified or cloned from a eukaryotic organism, the finding that this gene encodes a 60-kDa condensing enzyme provided the first direct evidence that membrane fatty acid elongation is not catalyzed by a type I multifunctional fatty acid synthase.

## 12. Design of new plant oils

In recent years, tremendous progress has occurred not only in the isolation of many plant lipid biosynthetic genes, but also in the use of these genes to manipulate plant oil composition. As shown in Table 5, both substantial changes in seed oil composition and introduction of unusual fatty acids to heterologous species have been achieved. Progress in this area has been accelerated by several industrial laboratories whose goal has been the production of higher value oilseeds.

Table 5
Some examples of genetic engineering of plant lipid metabolism

| Modification achieved | Enzyme engineered | Source of gene | Plant transformed |
| --- | --- | --- | --- |
| Lauric acid production | Acyl-ACP thioesterase | California Bay | Rapeseed |
| Increased stearic acid | Antisense of stearoyl-ACP desaturase | Rapeseed | Rapeseed |
| Reduced saturated fatty acids | Stearoyl-CoA desaturase | Rat, yeast | Tobacco |
| | Acyl-ACP thioesterase | Soybean | Soybean |
| Reduced 18 : 3 | ω-3 Desaturase | Soybean, canola | Soybean, rapeseed |
| Altered lauric acid distribution in TAG | 1-Acyl-glycerol-3-phosphate acyltransferase | Coconut | Rapeseed |
| Altered cold tolerance | Acyl-ACP : glycerol-3-phosphate acyltransferase | *E. coli*, squash, *Arabidopsis* | Tobacco, *Arabidopsis* |
| Petroselinic acid production | Acyl-ACP desaturase | Coriander | Tobacco |
| Increased linolenic acid | ω-3 Desaturase | *Arabidopsis* | *Arabidopsis* |
| Cyclopropane fatty acid production | Cyclopropane synthase | *E. coli*, *Sterculia* | Tobacco |
| γ-Linolenic acid production | Linolenic Δ⁶-desaturase | *Synechocystis* | Tobacco |
| Increased long-chain fatty acids | Fatty acid elongase | Jojoba | Rapeseed |
| Wax ester synthesis | Wax ester synthase, fatty acid reductase, fatty acid elongase | Jojoba | *Arabidopsis* |

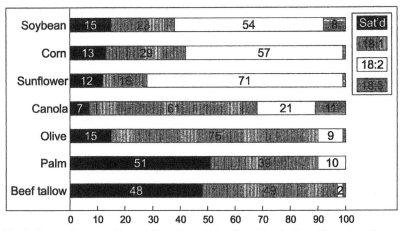

Fig. 9. Fatty acid composition of dietary vegetable oils and beef tallow. The values shown represent typical compositions of varieties grown commercially. Lines modified substantially through breeding or genetic engineering are available for soybean, canola, corn and sunflower.

## 12.1. Design of new edible oils

### 12.1.1. Improvements in nutritional value and stability of vegetable oils

Vegetable oils have gradually replaced animal fats as the major source of lipids in human diets and now constitute 15–25% of total caloric intake by industrialized nations. As shown in Fig. 9, vegetable oils display a wide range in the relative proportions of saturated and unsaturated fatty acid acids although in the United States, up to 80% of vegetable oil consumed is soybean oil making it the single largest source of calories in American diets. Most nutritionists recommend a reduction in saturated fat content in diets, and genetic engineering of plant oils can substantially help achieve this goal. Most of the saturated fatty acid in common plant oils is palmitic acid, and its occurrence is largely related to the action of a palmitoyl-ACP thioesterase. Reduction of the expression of this activity in transgenic soybean using co-suppression has led to decreases of the palmitic acid content from 15 to 6%.

A very high stability liquid soybean oil with low saturated fatty acids was also produced in soybean by suppression of the oleoyl-desaturase (Kinney, 1996). Oleic acid content was increased up to 89%, 18 : 2 content reduced from 57% to less than 3%, and saturated fatty acids reduced to less than 7%. This oil has been produced commercially and is extremely stable for high-temperature frying applications. In addition, its stability matches that of mineral oil derived lubricants, and therefore such non-food uses as biodegradable lubricants are underway. One added consumer benefit to wide future use of the engineered high-oleic acid oils may be reduction in the pathologies associated with high ω-6 consumption.

Besides reducing the amounts of *trans* and ω-6 unsaturated fatty acids in vegetable oils, efforts directed at increasing the amount of health beneficial fatty acids in vegetable oils appear to be promising. Two potential health-promoting fatty acids are stearidonic, $(18:4-\Delta^{6,9,12,15})$, an ω-3 fatty acid precursor of the ω-3 structures found in fish oils, and γ-linolenic acid ($18:3\text{-}\Delta^{6,9,12}$), which is implicated in relieving arthritis and other conditions [30]. γ-Linolenic acid is produced in some fungi and the seeds of some plants by the desaturation of linoleic acid to γ-linolenic acid by an endoplasmic reticulum-localized $\Delta^6$-desaturase. Identification of cDNAs encoding this desaturase from borage and the fungus *Mortierella alpina* has led to their heterologous expression in plants. Expression of the *Mortierella* gene in *Brassica napus* seed resulted in 47% γ-linolenic acid production (Ursin, 2000). Expression of the fungal $\Delta^6$-desaturase in a low α-linolenic acid canola line further enhanced γ-linolenic acid production to 68%. When canola lines with high α-linolenic content were crossed with high γ-linolenic acid lines, up to 17% stearidonic acid was produced. These examples illustrate that the level of desired end-product can sometimes be substantially increased, although often unpredictably, by the choice of host and/or enzyme source.

### 12.1.2. Alternatives to hydrogenated margarines and shortenings

Since most vegetable oils are liquids at room temperature, the production of margarines and shortenings from such oils requires alteration of their physical properties. This is most frequently achieved by catalytic hydrogenation of the oil, a process which reduces the double bonds and thereby increases the melting point of the oil. However, hydrogenation substantially increases the saturated fat content of the oil. An additional side reaction which occurs during hydrogenation is the conversion of much of the naturally occurring *cis* double bonds to the *trans* configuration. Typically, hydrogenated oils used in margarine or shortening contain up to 25–40% saturated or *trans* fatty acids. Although convincing evidence for a deleterious effect of *trans*-isomers in the diet is lacking, some nutritionists consider reduction of *trans*-double bonds in the diet advantageous. An added disadvantage to hydrogenation is the 2–3 cents per pound cost which it adds to the price of the oil. Thus, for several reasons, an alternative to vegetable oil hydrogenation is desirable for manufacture of margarines and shortenings.

Further progress toward reducing the need for hydrogenation has been made using a molecular genetic approach to increase the melting point of rapeseed oil. As described in Section 3.2, the introduction of the first double bond into plant fatty acids occurs by the action of stearoyl-ACP desaturase. An obvious route to alter the activity of this enzyme in oilseeds is the use of antisense RNA. This objective was achieved in *Brassica napus* and other plants, by suppression of stearoyl-ACP desaturase messenger RNA, which reduced enzyme activity and desaturase protein to barely detectable levels (Knutzon, 1992). As a result, the content of stearic acid in the seed oil was increased from 2 to 40%. Due to its high saturated fatty acid content, the oil from these plants has a high melting point and may be directly suitable for margarine or shortening manufacture without added hydrogenation.

An alternative fatty acid modification that might simultaneously increase unsaturation in diets and reduce the need for hydrogenation is the production of petroselinic acid rich vegetable oils. Petroselinic acid is an isomer of oleic acid which has a *cis* double bond at the 6th carbon from the carboxyl end of the molecule rather than the 9th carbon. This minor modification of the structure alters the melting point of the fatty acid such that petroselinic acid melts at 33°C whereas oleic acid melts at 5°C. Thus petroselinic acid might provide the means to produce an unsaturated vegetable oil which is also a solid at room temperatures and therefore ideal for margarine and shortening manufacture. Although petroselinic acid is a major component of seed oils in species such as coriander and carrot, these crops have a low yield of oil per hectare. It is therefore hoped that an oilseed crop can be engineered to become a commercially viable source of petroselinic acid. When a cDNA encoding the acyl-ACP desaturase involved in petroselinic acid synthesis was cloned from coriander and used to transform other plants, petroselinic acid was produced but only at low levels. Therefore, to achieve economic production of petroselinic acid, it may be necessary to introduce additional genes. It is now known that coriander and its relatives possess several proteins devoted to production of this special fatty acid, including ACP isoforms, a modified condensing enzyme (3-ketoacyl-ACP synthase) specific for the elongation of $cis$-4-16:1-ACP to $cis$-6-18:1-ACP and a novel acyl-ACP thioesterase specific for petroselinoyl-ACP.

## 12.2. Design of new industrial oils

As described above, plants have evolved the ability to produce a diverse range of fatty acid structures. A number of specialty fatty acids have already been extensively exploited for industrial uses such as lubricants, plasticizers and surfactants (Table 3). In fact, approximately one-third of all vegetable oil is now used for non-food products, and this figure is expected to increase as petroleum reserves are depleted. Thus, in addition to providing food, oilseed crops can be considered efficient, low polluting chemical factories which are able to harness energy from sunlight and transform it into a variety of valuable chemical structures with a multitude of non-food uses.

### 12.2.1. High laurate and caprate oils
One of the major non-food uses of vegetable oils, consuming approximately 500 million pounds of oil per annum in the US, is the production of soaps, detergents and other surfactants. The solubility and other physical properties of medium-chain fatty acids and their derivatives make them especially suited for surfactant manufacture. Coconut and palm kernel oils, which contain 40–60% lauric acid (12:0), are the current major feedstocks for the surfactant industry. The mechanism for synthesis of lauric and other medium-chain fatty acids in plants involves the action of a medium-chain acyl-ACP thioesterase which terminates fatty acid synthesis after a 10 or 12 carbon chain has been assembled (Pollard, 1991). A cDNA encoding such a thioesterase was isolated from seeds of the California Bay tree and transformed into rapeseed. As shown in Fig. 10, the introduction of this specialized thioesterase resulted in transgenic seeds that produce up to 60 mol% lauric acid. The plants grow normally and oil yields are very similar to those of the untransformed cultivars. Commercial production of high lauric rapeseed oil began in 1995, and this crop has the potential to provide a new, non-tropical source of lauric oils for the surfactant industry.

Since current oil crops do not produce significant amounts of caprate (10:0) and caprylate (8:0), the cloning of thioesterases from plants accumulating these species raised hopes for development of commercial sources of 8:0 and 10:0. Surprisingly, medium-chain thioesterase from elm, a 10:0 accumulator, recognized either 10:0 or 16:0, and a thioesterase from a $C_{8-10}$-rich *Cuphea* species, although showing strong

122

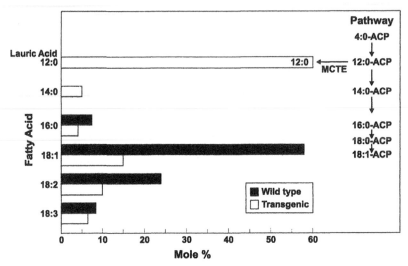

Fig. 10. Genetic engineering of rapeseed oil to produce lauric acid. mol% of major fatty acids in a typical canola cultivar are compared to the composition achieved through genetic engineering using a California Bay medium chain acyl-ACP thioesterase (MCTE) under control of a napin promoter.

selectivity for medium-chain fatty acids, gave disappointing results in transformed plants. Part of this discrepancy was resolved with the discovery that medium-chain production in *Cuphea* also involves a specific keto-acyl synthase. Cotransformation with the medium-chain thioesterase and the keto-acyl synthase IV give plants with substantial caprate and caprylate [31], and further improvements may be possible with addition of acyltransferases better able to introduce 8 : 0 and 10 : 0 to underrepresented positions in triacylglycerol [32].

### 12.2.2. Production of waxes

Long-chain wax esters were once harvested from sperm whales and were a major ingredient of industrial lubricants and transmission fluids. Banning of whale harvests led to searches for alternative biological sources of such structures. Jojoba, a desert shrub found in the American southwest, is the only plant species known to accumulate waxes (up to 60% dry weight) rather than TAG as a seed storage reserve. These waxes are mostly derived from C20–C24 monounsaturated fatty acids and alcohols and are synthesized by the elongation of oleate followed by reduction to alcohols by a fatty acid reductase. The wax storage lipid is formed by a fatty acyl-CoA: fatty alcohol acyltransferase, also referred to as wax synthase. The reductase and acyltransferase were purified from jojoba and the cognate cDNAs cloned. Coordinated expression of three genes, a *Lunaria annua* long-chain acyl-CoA elongase, and the jojoba reductase and acyltransferase in *Arabidopsis*, resulted in wax production at up to 70% of the oil present in mature seeds (Lardizabal, 2000). The high levels of accumulation indicated that all the genes necessary for this trait were identified. If this trait can be successfully transferred to commercial crops this would represent a large potential source of waxes used for a variety of applications including cosmetics and industrial lubricants.

### 12.2.3. Other industrial oils

Many of the unusual fatty acids produced by plants would have substantial value as industrial feedstocks if they were available in sufficient quantity at low prices. Examples in this category include fatty acids with hydroxy, epoxy, cyclopropane or branched chains. These specialty fatty acids are often produced in wild species which have not been optimized for high agronomic and oil yields, and therefore such specialty oils are expensive to produce. An alternative to the long-term effort required for domestication of such plants is the introduction of genes relevant to unusual fatty acid production into existing high-yielding oil crops. A step in this direction was made by the isolation of a cDNA for a fatty acid hydroxylase from the castor oil plant. When this gene was introduced into transgenic plants, approximately 20% hydroxy fatty acids were produced (Broun, 1997). Genes for specialty fatty acid production need not be isolated only from plants. As mentioned above, the stearoyl-CoA desaturases from animals and yeast are active in plants. Furthermore, the cyclopropane synthase of *E. coli* (Schmid, 1995) and desaturases from cyanobacteria and fungi have been successfully expressed in transgenic plants. Thus, in principle there are no fundamental barriers to producing a wide range of oil compositions using genes borrowed from diverse organisms. Furthermore, protein engineering offers even more possibilities to tailor the substrates and products of plant enzymes to produce 'designer oil crops' for specific end uses [6].

## 13. Future prospects

It is now possible to say that the biosynthetic pathways have been determined for all major plant lipids and most of the genes identified for enzymes in these pathways have been cloned. Clearly there has been great progress, although the biosynthetic pathways of sphingolipids, *trans*-$\Delta^3$-16:1 and some details of several pathways remain elusive. In addition, the enzymes involved in the production of many unusual fatty acids found in seed oils are largely unexplored.

One area of expanding interest is the production of lipid hormones and signal molecules. Several lipids including phosphatidylinositol phosphates, diacylglycerol and *N*-acylphosphatidylethanolamine have been implicated in signal transduction in plants. Another intriguing similarity between plants and animals is their use of oxygenated fatty acids in response to wounding. As mentioned in Section 10, jasmonate, a plant growth regulator derived from 18:3, is able at femtomolar concentrations to induce proteinase inhibitors and other plant defense genes. Like leukotriene synthesis in animals, jasmonate biosynthesis begins with the generation of a hydroperoxide by lipoxygenase. Jasmonate itself contains a cyclopentane ring comparable to those of prostaglandins. The common roles and origins of oxygenated fatty acids in plants and animals suggest a very early common ancestor for these pathways.

Application of molecular genetics and genomics to problems in lipid biochemistry should continue to expand. Particularly stimulating has been the complete sequencing of the *Arabidopsis* genome and the availability of over 1 million expressed sequence tag (EST) sequences from a variety of plants. A recent survey of *Arabidopsis* genes involved in plant acyl lipid metabolism identified over 450 genes

Table 6

Internet resources related to plant lipid metabolism

| Organization | Web site content | URL |
|---|---|---|
| National Plant Lipid Consortium (NPLC) | Directory of scientists involved in plant lipid research. E-mail newsgroup for information on plant lipids. Abstracts of NPLC meetings | http://www.msu.edu/user/ohlrogge/ |
| Michigan State University | Survey and catalog of genes for plant lipid metabolism. | http://www.canr.msu.edu/lgc/index.html http://www.plantbiology.msu.edu/ gene_survey/front_page.htm |
| | Gene expression profiles based on plant lipid ESTs and microarrays | http://www.bpp.msu.edu/Seed/SeedArray.htm |
| Kathy Schmid, Butler University | Links to many oilseed research laboratories and web sites | http://blue.butler.edu/~kschmid/lipids.html |
| Benning and Ohlrogge Laboratories | Database and analysis of >10,000 ESTs from developing *Arabidopsis* seeds | http://benningnt.bch.msu.edu |
| USDA Oilseed Database | Data on fatty acid composition of seeds of thousands of species | www.ncaur.usda.gov/nc/ncdb/search.html-ssi |

(http://www.plantbiology.msu.edu/gene_survey/front_page). Although most of these could be assigned a tentative function based on sequence similarity to previously characterized genes, only a handful have been examined individually at an experimental level, and therefore the precise function of hundreds of genes awaits further work. For example, there are 8 genes which are similar to acyl-CoA desaturases whose function has not yet been identified and there are 19 genes for lipid transfer proteins and 51 additional lipid transfer protein-like genes. The physiological reasons underlying the existence of large gene families for lipid transfer proteins, plastid ACPs, and stearoyl desaturases, but only one gene for keto acyl-ACP synthase III and for most fatty acid synthase and ACC subunits, etc. remain to be elucidated. The availability of large collections of T-DNA insertion mutants means that the impact of gene disruptions can be tested. However, because 60% of *Arabidopsis* genes are present as duplicates, such gene disruptions must be supplemented by other strategies of functional genomics. Some websites related to these efforts in the plant lipid field are presented in Table 6.

Much of past lipid research has focused on a reductionist approach in which cells are taken apart and their pieces analyzed. The overall success of this approach and the wealth of new clones and sequence information have given us an unprecedented knowledge of the pieces of the puzzle which represent lipid metabolism. However, as in any puzzle, it is not just complete knowledge of the pieces, but an understanding of how (and when) they fit together that defines the challenge. Microarrays that permit simultaneous monitoring of expression of many genes have begun to provide a more global overview of how genes work together to control seed metabolism (Girke, 2000). Together with the ability to rapidly over- and under-express genes in transgenic plants and the strengths of classical biochemistry, such recent advances in analytical techniques should allow us to enter a new stage of lipid research emphasizing the interplay between metabolic compartments and the control of lipid synthesis during the plant life cycle.

## Abbreviations

| | |
|---|---|
| ACC | acetyl-CoA carboxylase |
| ACP | acyl carrier protein |
| EST | expressed sequence tag |
| CoA | coenzyme A |
| TAG | triacylglycerol |

## References

1. Pyke, K.A. (1999) Plastid division and development. Plant Cell 11, 549–556.
2. Gueguen, V., Macherel, D., Jaquinod, M., Douce, R. and Bourguignon, J. (2000) Fatty acid and lipoic acid biosynthesis in higher plant mitochondria. J. Biol. Chem. 275, 5016–5025.
3. Gerhardt, B. (1993) Catabolism of fatty acids. In: T.S. Moore (Ed.), Lipid Metabolism in Plants. CRC Press, Boca Raton, FL, pp. 527–565.
4. Overath, P. and Stumpf, P. (1964) Fat metabolism in higher plants XXIII. Properties of a soluble fatty acid synthetase from avocado mesocarp. J. Biol. Chem. 239, 4103–4110.
5. Harwood, J.L. (1996) Recent advances in the biosynthesis of plant fatty acids. Biochim. Biophys. Acta 1301, 7–56.
6. Shanklin, J. and Cahoon, E.B. (1998) Desaturation and related modifications of fatty acids. Annu. Rev. Plant Physiol. Plant Mol. Biol. 49, 611–641.
7. Jones, A., Davies, H.M. and Voelker, T.A. (1995) Palmitoyl-acyl carrier protein (ACP) thioesterase and the evolutionary origin of plant acyl-ACP thioesterases. Plant Cell 7, 359–371.
8. Konishi, T., Shinohara, K., Yamada, K. and Sasaki, Y. (1996) Acetyl-CoA carboxylase in higher plants: most plants other than Gramineae have both the prokaryotic and the eukaryotic forms of this enzyme. Plant Cell Physiol. 37, 117–122.
9. Ohlrogge, J.B. and Jaworski, J.G. (1997) Regulation of fatty acid synthesis. Annu. Rev. Plant Physiol. Plant Mol. Biol. 48, 109–136.
10. Kunst, L., Browse, J. and Somerville, C. (1988) Altered regulation of lipid biosynthesis in a mutant of Arabidopsis deficient in chloroplast glycerol phosphate acyltransferase activity. Proc. Natl. Acad. Sci. USA 85, 4143–4147.
11. Kinney, A.J. (1993) Phospholipid headgroups. In: T.S. Moore Jr. (Ed.), Lipid Metabolism in Plants. CRC Press, Boca Raton, FL, pp. 259–284.
12. Froehlich, J.E., Benning, C. and Dormann, P. (2001) The digalactosyldiacylglycerol (DGDG) synthase DGD1 is inserted into the outer envelope membrane of chloroplasts in a manner independent of the general import pathway and does not depend on direct interaction with monogalactosyldiacylglycerol synthase for DGDG biosynthesis. J. Biol. Chem. 276, 31806–31812.
13. Frandsen, G.I., Mundy, J. and Tzen, J.T.C. (2001) Oil bodies and their associated proteins, oleosin and caleosin. Physiol. Plant. 112, 301–307.

14. Feussner, I., Kühn, H. and Wasternack, C. (2001) Lipoxygenase-dependent degradation of storage lipids. Trends Plant Sci. 6, 268–273.
15. van de Loo, F.J., Fox, B.G. and Somerville, C. (1993) Unusual fatty acids. In: T.S. Moore Jr. (Ed.), Lipid Metabolism in Plants. CRC Press, Boca Raton, FL, pp. 167–194.
16. Broun, P., Shanklin, J., Whittle, E. and Somerville, C. (1998) Catalytic plasticity of fatty acid modification enzymes underlying chemical diversity of plant lipids. Science 282, 1315–1317.
17. Voelker, T. and Kinney, A.J. (2001) Variations in the biosynthesis of seed-storage lipids. Annu. Rev. Plant Physiol. 52, 335–361.
18. Dahlqvist, A., Stahl, U., Lenman, M., Banas, A., Lee, M., Sandager, L., Ronne, H. and Stymne, S. (2000) Phospholipid:diacylglycerol acyltransferase: an enzyme that catalyzes the acyl-CoA-independent formation of triacylglycerol in yeast and plants. Proc. Natl. Acad. Sci. USA 97, 6487–6492.
19. Kolattukudy, P.E. (2001) Polyesters in higher plants. Adv. Biochem. Eng./Biotechnol. 71, 1–49.
20. Lichtenthaler, H.K. (2000) Non-mevalonate isoprenoid biosynthesis: enzymes, genes and inhibitors. Biochem. Soc. Trans. 28, 785–789.
21. Li, J., Biswas, M.G., Chao, A., Russell, D.W. and Chory, J. (1997) Conservation of function between mammalian and plant steroid 5α-reductases. Proc. Natl. Acad. Sci. USA 94, 3554–3559.
22. Sperling, P., Ternes, P., Moll, H., Franke, S., Zahringer, U. and Heinz, E. (2001) Functional characterization of sphingolipid C4-hydroxylase genes from Arabidopsis thaliana. FEBS Lett. 494, 90–94.
23. Lynch, D.V., Criss, A.K., Lehoczky, J.L. and Bui, V.T. (1997) Ceramide glucosylation in bean hypocotyl microsomes: evidence that steryl glucoside serves as glucose donor. Arch. Biochem. Biophys. 340, 311–316.
24. Schaller, F. (2001) Enzymes of the biosynthesis of octadecanoid-derived signalling molecules. J. Exp. Bot. 52, 11–23.
25. Okuley, J., Lightner, J., Feldmann, K., Yadav, N., Lark, E. and Browse, J. (1994) The Arabidopsis FAD2 gene encodes the enzyme that is essential for polyunsaturated lipid synthesis. Plant Cell 6, 147–158.
26. Lightner, J., Wu, J. and Browse, J. (1994) A mutant of Arabidopsis with increased levels of stearic acid. Plant Physiol. 106, 1443–1451.
27. McConn, M. and Browse, J. (1998) Polyunsaturated membranes are required for photosynthetic competence in a mutant of Arabidopsis. Plant J. 15, 521–530.
28. Arondel, V., Lemieux, B., Hwang, I., Gibson, S., Goodman, H.M. and Somerville, C.R. (1992) Map-based cloning of a gene controlling omega-3 fatty acid desaturation in Arabidopsis. Science 258, 1353–1355.
29. James, D.W., Lim, E., Keller, J., Plooy, I., Ralston, E. and Dooner, H.K. (1995) Directed tagging of the Arabidopsis fatty acid elongation 1 (FAE1) gene with the maize transposon activator. Plant Cell 7, 309–319.
30. Broun, P., Gettner, S. and Somerville, C. (1999) Genetic engineering of plant lipids. Annu. Rev. Nutr. 19, 197–216.
31. Dehesh, K., Edwards, P., Fillatti, J., Slabaugh, M. and Byrne, J. (1998) KAS IV: a 3-ketoacyl-ACP synthase from Cuphea sp. is a medium chain specific condensing enzyme. Plant J. 15, 383–390.
32. Knutzon, D.S., Hayes, T.R., Wyrick, A., Xiong, H., Davies, H. and Voelker, T.A. (1999) Lysophosphatidic acid acyltransferase from coconut endosperm mediates the insertion of laurate at the sn-2 position of triacylglycerols in lauric rapeseed oil and can increase total laurate levels. Plant Physiol. 120, 739–746.

D.E. Vance and J.E. Vance (Eds.) *Biochemistry of Lipids, Lipoproteins and Membranes (4th Edn.)*

# CHAPTER 5

# Oxidation of fatty acids in eukaryotes

Horst Schulz

*City College of CUNY, Department of Chemistry, Convent Avenue at 138th Street, New York, NY 10031,
USA, Tel.: +1 (212) 650-8323; Fax: +1 (212) 650-8322; E-mail: hoschu@sci.ccny.cuny.edu*

## 1. The pathway of β-oxidation: a historical account

Fatty acids are a major source of energy in animals. The study of their biological degradation began in 1904 when Knoop [1] performed the classical experiments that led him to formulate the theory of β-oxidation. In his experiments, Knoop used fatty acids with phenyl residues in place of the terminal methyl groups. The phenyl residue served as a *reporter group* because it was not metabolized, but instead was excreted in the urine. When Knoop fed phenyl substituted fatty acids with an odd number of carbon atoms, like phenylpropionic acid ($C_6H_5-CH_2-CH_2-COOH$) or phenylvaleric acid ($C_6H_5-CH_2-CH_2-CH_2-CH_2-COOH$), to dogs, he isolated from their urine hippuric acid ($C_6H_5-CO-NH-CH_2-COOH$), the conjugate of benzoic acid and glycine. In contrast, phenyl-substituted fatty acids with an even number of carbon atoms, such as phenylbutyric acid ($C_6H_5-CH_2-CH_2-CH_2-COOH$), were degraded to phenylacetic acid ($C_6H_5-CH_2-COOH$) and excreted as phenylaceturic acid ($C_6H_5-CH_2-CO-NH-CH_2-COOH$). These observations led Knoop to propose that the oxidation of fatty acids begins at carbon atom 3, the β-carbon, and that the resulting β-keto acids are cleaved between the α-carbon and β-carbon to yield fatty acids shortened by two carbon atoms. Knoop's experiments prompted the idea that fatty acids are degraded in a stepwise manner by successive β-oxidation. In the years following Knoop's initial study, Dakin [2] performed similar experiments with phenylpropionic acid. Besides hippuric acid he isolated the glycine conjugates of the following β-oxidation intermediates: phenylacrylic acid ($C_6H_5-CH=CH-COOH$), β-phenyl-β-hydroxypropionic acid ($C_6H_5-CHOH-CH_2-COOH$), and benzoylacetic acid ($C_6H_5-CO-CH_2-COOH$). At the same time, Embden and coworkers demonstrated that in perfused livers unsubstituted fatty acids are degraded by β-oxidation and converted to ketone bodies. Thus, by 1910 the basic information necessary for formulating the pathway of β-oxidation was available.

After a 30-year period of little progress, Munoz and Leloir in 1943, and Lehninger in 1944, demonstrated the oxidation of fatty acids in cell-free preparations from liver. Their work set the stage for the complete elucidation of β-oxidation. Detailed investigations with cell-free systems, especially the studies of Lehninger, demonstrated the need for energy to 'spark' the oxidation of fatty acids. ATP was shown to meet this requirement and to be essential for the activation of fatty acids. Activated fatty acids were shown by Wakil and Mahler, as well as by Kornberg and Pricer, to be thioesters formed from fatty acids and coenzyme A. This advance was made possible by earlier studies of Lipmann and coworkers who isolated and characterized coenzyme A, and Lynen [3] and coworkers who proved the structure of 'active acetate' to be acetyl-CoA. Acetyl-CoA was found to be identical with the two-carbon fragment removed from fatty acids during their degradation. The subcellular location of the β-oxidation system was finally established by Kennedy and Lehninger, who demonstrated that mitochondria were the cellular components most active in fatty acid oxidation. The mitochondrial location of this pathway agreed with the observed coupling of fatty acid oxidation to the citric acid cycle and to oxidative phosphorylation. The most direct evidence for the proposed β-oxidation cycle emerged from enzyme studies carried out in the fifties primarily in the laboratories of Green in Wisconsin, Lynen in Munich, and Ochoa in New York. Their studies were greatly facilitated by newly developed methods of protein purification and by the use of spectrophotometric enzyme assays with chemically synthesized intermediates of β-oxidation as substrates.

## 2. Uptake and activation of fatty acids in animal cells

Fatty acids are transported between organs either as unesterified fatty acids complexed to serum albumin or in the form of triacylglycerols associated with lipoproteins. Triacylglycerols are hydrolyzed outside of cells by lipoprotein lipase to yield free fatty acids. The mechanism by which free fatty acids enter cells remains poorly understood despite a number of studies performed with isolated cells from heart, liver, and adipose tissue [4]. Kinetic evidence has been obtained for both a saturable and non-saturable uptake of fatty acids. The saturable uptake, which predominates at nanomolar concentrations of free fatty acids, is presumed to be carrier-mediated, whereas the non-saturable uptake, which is significant only at higher concentrations of free fatty acids, has been attributed to non-specific diffusion of fatty acids across the membrane. Several suspected fatty acid transport proteins have been identified [5]. However, their specific function(s) in fatty acid uptake and their molecular mechanisms remain to be elucidated.

Once long-chain fatty acids have crossed the plasma membrane, they either diffuse or are transported to mitochondria, peroxisomes, and the endoplasmic reticulum where they are activated by conversion to their CoA thioesters. Whether this transfer of fatty acids between membranes is a facilitated process or occurs by simple diffusion is an unresolved issue. The identification of low-molecular-weight (14–15 kDa) fatty acid binding proteins (FABPs) in the cytosol of various animal tissues prompted the suggestion that these proteins may function as carriers of fatty acids in the cytosolic compartment [6]. FABPs may also be involved in the cellular uptake of fatty acids, their intracellular storage, or the delivery of fatty acids to sites of their utilization. The importance of FABPs in fatty acid metabolism is supported by the observation that the uptake and utilization of long-chain fatty acids are reduced in knock-out mice lacking heart FABP. These animals exhibit exercise intolerance and, at old age, develop cardiac hypertrophy. Nonetheless, the molecular mechanism of FABP function remains to be elucidated.

The metabolism of fatty acids requires their prior activation by conversion to fatty acyl-CoA thioesters. The activating enzymes are ATP-dependent acyl-CoA synthetases, which catalyze the formation of acyl-CoA by the following two-step mechanism in which E represents the enzyme:

$$E + R-COOH + ATP \xrightarrow{\text{Mg}^{2+}} (E:R-CO-AMP) + PP_i$$

$$(E:R-CO-AMP) + CoASH \longrightarrow R-CO-SCoA + AMP + E$$

The evidence for this mechanism was primarily derived from a study of acetyl-CoA synthetase. Although the postulated intermediate, acetyl-AMP, does not accumulate in solution, and therefore only exists bound to the enzyme, the indirect evidence for this intermediate is very compelling. Other fatty acids are assumed to be activated by a similar mechanism, even though less evidence in support of this hypothesis has been obtained. The activation of fatty acids is catalyzed by a group of acyl-CoA synthetases that differ with respect to their subcellular locations and their specificities for fatty acids of different chain lengths [7]. Their chain-length specificities are the basis for classifying these enzymes as short-chain, medium-chain, long-chain and very-long-chain acyl-CoA synthetases.

A short-chain-specific acetyl-CoA synthetase that is present in mammalian mitochondria has been purified and its cDNA has been cloned. This 71-kDa enzyme, which is most active with acetate as a substrate but exhibits some activity towards propionate, has been detected in mitochondria of heart, skeletal muscle, kidney, adipose tissue and intestine, but not in those of liver. A cytosolic 78-kDa acetyl-CoA synthetase has been identified in liver, intestine, adipose tissue and mammary gland, all of which have high lipogenic activities. Expression studies support the hypothesis that the cytosolic enzyme synthesizes acetyl-CoA for lipogenesis, whereas the mitochondrial acetyl-CoA synthetase activates acetate headed for oxidation.

Medium-chain acyl-CoA synthetases are present in mitochondria of various mammalian tissues. The partially purified enzyme from beef heart mitochondria acts on fatty acids with 3–7 carbon atoms, but is most active with butyrate. In contrast, the 66-kDa enzyme purified from bovine liver mitochondria activates fatty acids with 3–12 carbon atoms with hexanoate being the best substrate. This enzyme also activates aromatic carboxylic acids like benzoic acid and its substituted derivatives. Overall, liver mitochondria, in contrast to heart mitochondria, contain medium-chain acyl-CoA synthetase activities with much broader substrate specificities toward fatty acids of varying chain lengths and structures.

Long-chain acyl-CoA synthetase is a membrane-bound enzyme that is associated with the endoplasmic reticulum, peroxisomes, and the outer mitochondrial membrane. The enzyme acts efficiently on saturated fatty acids with 10–20 carbon atoms and on common unsaturated fatty acids having 16–20 carbon atoms. Molecular cloning and expression studies of long-chain acyl-CoA synthetase have revealed the existence of five different isozymes (ACS 1–5) in the rat. A detailed investigation of the hepatic acyl-CoA synthetases ACS 1, 4 and 5 suggests that ACS 1 and 4, which are present in the endoplasmic reticulum and related subcellular structures, function in lipid biosynthesis while ACS 5 with a mitochondrial localization may catalyze the activation of fatty acids for β-oxidation. Such functional commitment of isozymes has previously been recognized. For example, long-chain acyl-CoA synthetases ACS I and ACS II of

*Candida lipolytica* are thought to activate long-chain fatty acids for complex lipid synthesis and peroxisomal β-oxidation, respectively.

Very long-chain acyl-CoA synthetase activates fatty acids with 22 or more carbon atoms and also acts on long-chain and branched-chain fatty acids. This membrane-bound enzyme is strongly expressed in liver where it is associated with the endoplasmic reticulum and peroxisomes but not with mitochondria. Purification of the 70-kDa very long-chain acyl-CoA synthetase enabled the cloning of the rat and human cDNAs coding for this enzyme. Sequence homologies with other cDNAs resulted in the identification of choloyl-CoA synthetase and in the demonstration that fatty acid transport protein 1 (FATP 1), a suggested transporter of fatty acids across the plasma membrane, exhibits very long-chain acyl-CoA synthetase activity.

In addition to ATP-dependent acyl-CoA synthetases, several GTP-dependent acyl-CoA synthetases have been described. The best known of these enzymes is succinyl-CoA synthetase, which cleaves GTP to GDP plus phosphate and functions in the tricarboxylic acid cycle. Although a mitochondrial GTP-dependent acyl-CoA synthetase activity was described, the existence of a distinct enzyme with such activity has been questioned.

## 3. Fatty acid oxidation in mitochondria

In animal cells, fatty acids are degraded in both mitochondria and peroxisomes, whereas in lower eukaryotes β-oxidation is confined to peroxisomes. Mitochondrial β-oxidation provides energy for oxidative phosphorylation and generates acetyl-CoA for ketogenesis in liver. The oxidation of fatty acids with odd numbers of carbon atoms proceeds by β-oxidation and yields, in addition to acetyl-CoA, 1 mole of propionyl-CoA per mole of fatty acid. Propionyl-CoA is further metabolized to succinate.

### 3.1. Mitochondrial uptake of fatty acids

Fatty acyl-CoA thioesters that are formed at the outer mitochondrial membrane cannot directly enter the mitochondrial matrix, where the enzymes of β-oxidation are located, because the inner mitochondrial membrane is impermeable to CoA and its derivatives. Instead, carnitine carries the acyl residues of acyl-CoA thioesters across the inner mitochondrial membrane. The carnitine-dependent translocation of fatty acids across the inner mitochondrial membrane is schematically shown in Fig. 1 [8]. The reversible transfer of fatty acyl residues from CoA to carnitine is catalyzed by carnitine palmitoyltransferase I (CPT I), which is an enzyme of the outer mitochondrial membrane. The resultant acylcarnitines cross the inner mitochondrial membrane via the carnitine : acylcarnitine translocase. This carrier protein catalyzes a rapid mole for mole exchange of acylcarnitine for carnitine, carnitine for carnitine, and acylcarnitine for acylcarnitine. This exchange, especially of acylcarnitine for carnitine, is essential for the translocation of long chain fatty acids from the cytosol into mitochondria. In addition, the translocase facilitates a slow unidirectional flux of carnitine across the inner mitochondrial membrane. This unidirectional flux of carnitine may be important

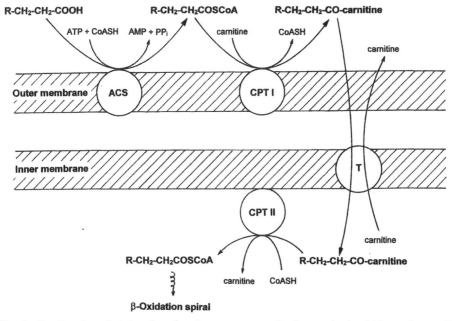

Fig. 1. Carnitine-dependent transfer of acyl groups across the inner mitochondrial membrane. ACS, acyl-CoA synthetase; CPT I and CPT II, carnitine palmitoyltransferase I and II, respectively; T, carnitine : acylcarnitine translocase.

for mitochondria of organs other than liver to acquire carnitine, which is synthesized in the liver. The rat liver translocase, which has a subunit molecular mass of 32.5 kDa, has been purified and its cDNA has been cloned. The protein has also been functionally reconstituted into proteoliposomes. In the mitochondrial matrix, carnitine palmitoyltransferase II (CPT II), an enzyme that is associated with the inner mitochondrial membrane, catalyzes the transfer of acyl residues from carnitine to CoA to form acyl-CoA thioesters that then enter the β-oxidation spiral. CPT II has been purified from mitochondria of bovine heart and rat liver. The purified enzyme has a subunit molecular mass of approximately 70 kDa and catalyzes the reversible transfer of acyl residues with 10–18 carbon atoms between CoA and carnitine. The cDNAs of rat and human CPT II have been cloned and sequenced. The predicted amino acid sequences of the corresponding proteins show a greater than 80% homology with each other. CPT I, in contrast to CPT II, is reversibly inhibited by malonyl-CoA, its natural regulator, and is covalently modified and inactivated by CoA derivatives of certain alkyl glycidic acids [8]. The latter property was utilized to label this protein for generating sequence information that permitted the molecular cloning of CPT I. Human and rat cDNAs code for 88-kDa proteins that are highly homologous (88%) to each other and also are very similar (50%) to CPT II. An isoform of liver CPT I (L-CPT I) is present in skeletal muscle (M-CPT I) while both isoforms are expressed in heart mitochondria. L-CPT I and M-CPT I are products of different genes and have different kinetic properties. M-CPT I compared to L-CPT I is much more sensitive toward malonyl-CoA ($K_i \approx 20$

nM vs. 2 µM) but has a lower affinity for carnitine ($K_m \approx 500$ µM vs. 30 µM). CPT I is anchored in the outer mitochondrial membrane via two transmembrane segments so that the 46 N-terminal residues and the large C-terminal catalytic domain remain on the cytosolic side of the membrane. CPT I together with CPT II and acyl-CoA synthetase seem to be concentrated at contact sites between the inner and outer mitochondrial membrane. The effective expression of active CPT I in the yeast *Pichia pastoris*, which is devoid of this enzyme, made it possible to study the structure–function relationship of this enzyme. The general conclusion of these studies is that the cytosolic N-terminal region of CPT I harbors positive and negative regulatory elements that determine the sensitivity of L-CPT I toward malonyl-CoA and the affinity of M-CPT I for carnitine [9].

In addition to CPT I and CPT II, mitochondria contain a carnitine acetyltransferase, which has been purified and its cDNA has been cloned. The enzyme from bovine heart has an estimated molecular mass of 60 kDa and is composed of a single polypeptide chain. It catalyzes the transfer of acyl groups with 2–10 carbon atoms between CoA and carnitine. The function of this enzyme has not been established conclusively. Perhaps, the enzyme regenerates free CoA in the mitochondrial matrix by transferring acetyl groups and other short-chain or medium-chain acyl residues from CoA to carnitine. The resultant acylcarnitines can leave mitochondria via the carnitine : acylcarnitine translocase and can be metabolized by the same or other tissues, or can be excreted in urine. In addition, carnitine acetyltransferase together with CPT II may convert acylcarnitines that were formed by the partial β-oxidation of fatty acids in peroxisomes to acyl-CoAs for further oxidation in mitochondria.

Short-chain and medium-chain fatty acids with less than 10 carbon atoms can enter mitochondria as free acids independent of carnitine. They are activated by short-chain and medium-chain acyl-CoA synthetases that are present in the mitochondrial matrix.

## 3.2. Enzymes of β-oxidation in mitochondria

The enzymes of β-oxidation either are associated with the inner mitochondrial membrane or are located in the mitochondrial matrix. The reactions catalyzed by these enzymes are shown schematically in Fig. 2, which also provides a hypothetical view of the physical and functional organization of these enzymes.

In the first of four reactions that constitute one cycle of the β-oxidation spiral acyl-CoA is dehydrogenated to 2-*trans*-enoyl-CoA according to the following equation.

$$R-CH_2-CH_2-CO-SCoA + FAD \longrightarrow R-CH=CH-CO-SCoA + FADH_2$$

Four acyl-CoA dehydrogenases with different but overlapping chain length specificities cooperate to assure the complete degradation of all fatty acids that can be metabolized by mitochondrial β-oxidation. The names of the four dehydrogenases, short-chain, medium-chain, long-chain, and very-long-chain acyl-CoA dehydrogenases, reflect their chain-length specificities. Purification of these enzymes has permitted detailed studies of their molecular and mechanistic properties [10,11]. The first three dehydrogenases are soluble matrix enzymes with similar molecular masses between 170 and 190 kDa. They are composed of four identical subunits, each of which carries a tightly, but non-covalently bound, flavin adenine dinucleotide (FAD). Their cDNAs

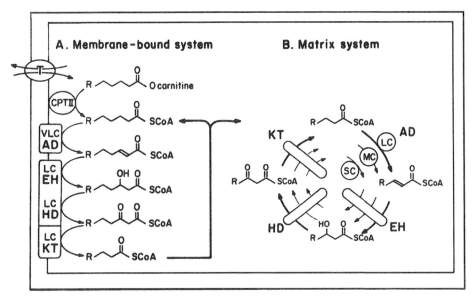

Fig. 2. Model of the functional and physical organization of β-oxidation enzymes in mitochondria. (A) β-Oxidation system active with long-chain (LC) acyl-CoAs; (B) β-Oxidation system active with medium-chain (MC) and short-chain (SC) acyl-CoAs. Abbreviations: T, carnitine : acylcarnitine translocase; CPT II, carnitine palmitoyltransferase II; AD, acyl-CoA dehydrogenase; EH, enoyl-CoA hydratase; HD, L-3-hydroxyacyl-CoA dehydrogenase; KT, 3-ketoacyl-CoA thiolase; VLC, very long-chain.

have been cloned and sequenced. High degrees of homology (close to 90%) have been observed for the same type of enzyme from man and rat and significant homologies (30–35%) are apparent when different enzymes from one source are compared. The crystal structure of medium-chain acyl-CoA dehydrogenases at 2.4 Å resolution confirmed the homotetrameric structure of the enzyme with one FAD bound per subunit in an extended conformation [12]. Very-long-chain acyl-CoA dehydrogenase, in contrast to the three other dehydrogenases, is a protein of the inner mitochondrial membrane. Purification of this enzyme and its molecular cloning established that it is a 133-kDa homodimer with one FAD bound per subunit. The four dehydrogenases differ with respect to their specificities for substrates of various chain lengths. Short-chain acyl-CoA dehydrogenase only acts on short-chain substrates like butyryl-CoA and hexanoyl-CoA. Medium-chain acyl-CoA dehydrogenase is most active with substrates from hexanoyl-CoA to dodecanoyl-CoA, whereas long-chain acyl-CoA dehydrogenase preferentially acts on octanoyl-CoA and longer-chain substrates. Very-long-chain acyl-CoA dehydrogenase extends the activity spectrum to longer-chain substrates, including those having acyl chains with 22 and 24 carbon atoms. However, long-chain acyl-CoA dehydrogenase may have a specific function in the β-oxidation of unsaturated fatty acids because this enzyme, in contrast to very long-chain acyl-CoA dehydrogenase, effectively dehydrogenates 4-enoyl-CoAs and 5-enoyl-CoAs, which are β-oxidation intermediates of unsaturated fatty acids. This hypothesis is supported by the observed accumulation of intermediates of unsaturated but not of saturated fatty acids in knock-out mice lacking long-chain acyl-CoA dehydrogenase. The dehydrogenation of acyl-CoA thioesters involves the removal of a proton from the $\alpha$-carbon of the substrate and hydride transfer from the $\beta$-carbon to the FAD cofactor of the enzyme to yield 2-*trans*-enoyl-CoA and enzyme-bound $FADH_2$ [12]. Studies based on X-ray crystallography, chemical modifications, and site-specific mutagenesis established that glutamate 376 is the base responsible for the $\alpha$-proton abstraction in medium-chain acyl-CoA dehydrogenase. Other acyl-CoA dehydrogenases follow a similar mechanism. Reoxidation of $FADH_2$ occurs by two successive single-electron transfers from the dehydrogenase to the FAD prosthetic group of a second flavoprotein named electron-transferring flavoprotein (ETF), which donates electrons to an iron–sulfur flavoprotein named ETF : ubiquinone oxidoreductase. The latter enzyme, a component of the inner mitochondrial membrane, feeds electrons into the mitochondrial electron transport chain via ubiquinone. The flow of electrons from acyl-CoA to oxygen is schematically shown below.

$$R-CH_2-CH_2-CO-SCoA \rightarrow FAD(acyl-CoA\ dehydrogenase) \rightarrow$$
$$FAD(ETF) \rightarrow FAD/[4Fe4S](ETF : ubiquinone\ oxidoreductase) \rightarrow$$
$$ubiquinone \rightarrow \rightarrow \rightarrow \rightarrow oxygen$$

ETF is a soluble matrix protein with a molecular mass of close to 60 kDa. It is composed of two non-identical subunits of similar molecular masses with one FAD per protein dimer. The crystal structure of ETF revealed the location of FAD in a cleft between the two subunits.

In addition to the four acyl-CoA dehydrogenases involved in fatty acid oxidation, two acyl-CoA dehydrogenases specific for metabolites of branched-chain amino acids have been isolated and purified. They are isovaleryl-CoA dehydrogenase and 2-methyl-branched chain acyl-CoA dehydrogenase.

In the second step of $\beta$-oxidation 2-*trans*-enoyl-CoA is reversibly hydrated by enoyl-CoA hydratase to L-3-hydroxyacyl-CoA as shown below.

$$R-CH=CH-CO-SCoA + H_2O \longrightarrow R-CH(OH)-CH_2-CO-SCoA$$

Two enoyl-CoA hydratases have been identified in mitochondria [4]. The better characterized of the two enzymes is enoyl-CoA hydratase or crotonase, which is a 161-kDa homohexamer. The best substrate of crotonase is crotonyl-CoA ($CH_3-CH=CH-CO-SCoA$), which is hydrated to form L($S$)-3-hydroxybutyryl-CoA. The activity of the enzyme decreases with increasing chain length of the substrate so that the activity with 2-*trans*-hexadecenoyl-CoA is only $1-2\%$ of the activity achieved with crotonyl-CoA. Crotonase also hydrates 2-*cis*-enoyl-CoA to D-3-hydroxyacyl-CoA and exhibits very low $\Delta^3,\Delta^2$-enoyl-CoA isomerase activity. The crystal structure of crotonase revealed that it is a member of the hydratase/isomerase superfamily with two active site glutamate residues that function as general acid and general base in the syn addition of water to crotonyl-CoA. The second enoyl-CoA hydratase, referred to as long-chain enoyl-CoA hydratase, is virtually inactive with crotonyl-CoA, but effectively hydrates medium-chain and long-chain substrates. The activities of crotonase and long-chain enoyl-CoA hydratase complement each other thereby assuring high rates of hydration of all enoyl-CoA intermediates. Long-chain enoyl-CoA hydratase is a component enzyme of the trifunctional β-oxidation complex, which additionally exhibits long-chain activities of L-3-hydroxyacyl-CoA dehydrogenase and 3-ketoacyl-CoA thiolase [13]. This β-oxidation complex is a protein of the inner mitochondrial membrane. It consists of equimolar amounts of a large α-subunit with a molecular mass of close to 80 kDa and of a small β-subunit with a molecular mass of approximately 48 kDa. Cloning and sequencing of the cDNAs that code for this complex revealed significant homologies of the amino-terminal and central regions of the large subunit with enoyl-CoA hydratase and L-3-hydroxyacyl-CoA dehydrogenase, respectively, and of the small subunit with 3-ketoacyl-CoA thiolase. These homologies are indicative of the locations of the component enzymes on the complex.

The third reaction in the β-oxidation cycle is the reversible dehydrogenation of L-3-hydroxyacyl-CoA to 3-ketoacyl-CoA catalyzed by L-3-hydroxyacyl-CoA dehydrogenase as shown in the following equation.

$$R-CH(OH)-CH_2-CO-SCoA + NAD^+$$
$$\longrightarrow R-CO-CH_2-CO-SCoA + NADH + H^+$$

Four L-3-hydroxyacyl-CoA dehydrogenases have been identified in mitochondria.

L-3-Hydroxyacyl-CoA dehydrogenase is a soluble matrix enzyme, which has a molecular mass of approximately 65 kDa and is composed of two identical subunits [4]. The enzyme and its cDNA have been sequenced. The crystal structure of the pig heart enzyme revealed a bilobal structure with the $NAD^+$ binding site in the N-terminal region and the substrate binding site in the cleft between the C-terminal and N-terminal domains. The enzyme is specific for $NAD^+$ as a coenzyme. It acts on L-3-hydroxyacyl-CoAs of various chain lengths but is most active with medium-chain and short-chain substrates. A second L-3-hydroxyacyl-CoA dehydrogenase was recently isolated from bovine liver. This enzyme is a homotetramer with a subunit molecular mass of 27 kDa. Its substrate specificity resembles that of the L-3-hydroxyacyl-CoA dehydrogenase described above. However, the enzyme's primary function may be in androgen metabolism and not in β-oxidation because it exhibits significant 17β-hydroxysteroid dehydrogenase activity and is absent or almost absent from some tissues with high β-oxidation activity. Long-chain L-3-hydroxyacyl-CoA dehydrogenase is a component enzyme of the trifunctional β-oxidation complex or trifunctional protein [13]. This dehydrogenase is active with medium- and long-chain substrates, but not with 3-hydroxybutyryl-CoA, and hence complements the soluble dehydrogenase to assure high rates of dehydrogenation over the whole spectrum of β-oxidation intermediates. A soluble short-chain L-3-hydroxy-2-methylacyl-CoA dehydrogenase is also present in the mitochondrial matrix. This enzyme, which acts on short-chain substrates with or without 2-methyl substituents, is believed to function only in isoleucine metabolism.

In the last reaction of the β-oxidation cycle 3-ketoacyl-CoA is cleaved by thiolase as shown below.

$$R-CO-CH_2-CO-SCoA + CoASH \longrightarrow R-CO-SCoA + CH_3-CO-SCoA$$

The products of the reaction are acetyl-CoA and an acyl-CoA shortened by two carbon atoms. The equilibrium of the reaction is far to the side of the thiolytic cleavage products thereby driving β-oxidation to completion. All thiolases that have been studied in detail contain an essential sulfhydryl group, which participates directly in the carbon–carbon bond cleavage as outlined in the following equations where E–SH represents thiolase.

$$E–SH + R–CO–CH_2–CO–SCoA \longrightarrow R–CO–S–E + CH_3–CO–SCoA$$
$$R–CO–S–E + CoASH \longrightarrow R–CO–SCoA + E–SH$$

According to this mechanism, 3-ketoacyl-CoA binds to the enzyme and is cleaved between its α and β carbon atoms. An acyl residue, which is two carbons shorter than the substrate, is transiently bound to the enzyme via a thioester bond, while acetyl-CoA is released from the enzyme. Finally, the acyl residue is transferred from the sulfhydryl group of the enzyme to CoA to yield acyl-CoA.

Several types of thiolases have been identified, some of which exist in multiple forms [4]. Mitochondria contain three classes of thiolases: (1) acetoacetyl-CoA thiolase or acetyl-CoA acetyltransferase, which is specific for acetoacetyl-CoA ($C_4$) as a substrate; (2) 3-ketoacyl-CoA thiolase or acetyl-CoA acyltransferase, which acts on 3-ketoacyl-CoA thioesters of various chain lengths ($C_4$–$C_{16}$) and (3) long-chain 3-ketoacyl-CoA thiolase, which acts on medium-chain and long-chain 3-ketoacyl-CoA thioesters, but not on acetoacetyl-CoA. The latter two enzymes are essential for fatty acid β-oxidation, whereas acetoacetyl-CoA thiolase most likely functions only in ketone body and isoleucine metabolism. Long-chain 3-ketoacyl-CoA thiolase is a component enzyme of the membrane-bound trifunctional β-oxidation complex [13], whereas the other two thiolases are soluble matrix enzymes. All mitochondrial thiolases have been purified and their cDNAs have been cloned and sequenced. A comparison of amino acid sequences proved all mitochondrial thiolases to be different, but homologous, enzymes. 3-Ketoacyl-CoA thiolase is composed of four identical subunits with a molecular mass of close to 42 kDa. This enzyme acts equally well on all substrates tested except for acetoacetyl-CoA, which is cleaved at half the maximal rate observed with longer chain substrates.

The absence or near absence of intermediates of β-oxidation from mitochondria prompted the idea of intermediate channeling due to the existence of a multienzyme complex of β-oxidation enzymes in intact mitochondria. The identification and characterization of at least two isozymes for each of the four reactions of the β-oxidation spiral led to the presentation of a model for their physical and functional organization as shown in Fig. 2 [14]. By this model, the membrane-bound, long-chain specific β-oxidation system, consisting of very-long-chain acyl-CoA dehydrogenase and the trifunctional β-oxidation complex, converts long-chain to medium-chain fatty acyl-CoAs, which are completely degraded by the matrix system of soluble enzymes that have a preference for medium-chain and short-chain substrates. An assumption underlying this model is that all enzymes thought to function in fatty acid β-oxidation are essential for this process. So far this assumption has proven to be correct. The characterization of inherited disorders of fat metabolism in humans has revealed that each of the many β-oxidation enzymes found to be deficient in a patient is essential for the normal degradation of fatty acids (for more detail see Section 5).

## 3.3. β-Oxidation of unsaturated fatty acids

Unsaturated fatty acids, which usually contain *cis* double bonds, also are degraded by β-oxidation. However, additional (auxiliary) enzymes are required to act on the pre-existing double bonds once they are close to the thioester group as a result of chain-shortening [15]. All double bonds present in unsaturated and polyunsaturated fatty acids can be classified either as odd-numbered double bonds, like the 9-*cis* double bond of oleic acid and linoleic acid or as even-numbered double bonds like the 12-*cis* double bond of linoleic acid. Since both classes of double bonds are present in linoleic acid, its degradation illustrates the breakdown of all unsaturated fatty acids. A summary of the β-oxidation of linoleic acid is presented in Fig. 3. Linoleic acid, after conversion to its CoA thioester(I), undergoes three cycles of β-oxidation to yield 3-*cis*,6-*cis*-dodecadienoyl-CoA(II) which is isomerized to 2-*trans*,6-*cis*-dodecadienoyl-CoA(III) by $\Delta^3,\Delta^2$-*trans*-enoyl-CoA isomerase, an auxiliary enzyme of β-oxidation. 2-*trans*,6-*cis*-Dodecadienoyl-CoA(III) is a substrate of β-oxidation and can complete one cycle to yield 4-*cis*-decenoyl-CoA(IV), which is dehydrogenated to 2-*trans*,4-*cis*-decadienoyl-CoA(V) by medium-chain acyl-CoA dehydrogenase. 2-*trans*,4-*cis*-Decadienoyl-CoA(V) cannot continue on its course through the β-oxidation spiral, but instead is reduced by NADPH in a reaction catalyzed by 2,4-dienoyl-CoA reductase. The product of this reduction, 3-*trans*-decenoyl-CoA(VI), is isomerized by $\Delta^3,\Delta^2$-enoyl-CoA isomerase to 2-*trans*-decenoyl-CoA(VII), which can be completely degraded by completing four cycles of β-oxidation. Altogether, the degradation of unsaturated fatty acids in mitochondria involves at least $\Delta^3,\Delta^2$-enoyl-CoA isomerase and 2,4-dienoyl-CoA reductase as auxiliary enzymes in addition to the enzymes of the β-oxidation spiral.

More recent is the demonstration that odd-numbered double bonds can be reduced at the stage of 5-enoyl-CoA intermediates formed during the β-oxidation of unsaturated fatty acids. Shown in Fig. 4 is the sequence of reactions that explains the NADPH-dependent reduction of 5-*cis*-enoyl-CoA(I) [16]. After introduction of a 2-*trans* double bond by acyl-CoA dehydrogenase, the resultant 2,5-dienoyl-CoA(II) is converted to 3,5-dienoyl-CoA(III) by $\Delta^3,\Delta^2$-enoyl-CoA isomerase. A novel enzyme, $\Delta^{3,5}\Delta^{2,4}$-dienoyl-CoA isomerase, converts 3,5-dienoyl-CoA(III) to 2-*trans*,4-*trans*-dienoyl-CoA(IV) by a concerted shift of both double bonds. Finally, 2,4-dienoyl-CoA reductase catalyzes the NADPH-dependent reduction of one double bond to produce 3-*trans*-enoyl-CoA(V), which, after isomerization to 2-*trans*-enoyl-CoA(VI) by $\Delta^3,\Delta^2$-enoyl-CoA isomerase, can reenter the β-oxidation spiral. Although the significance of this new pathway has not been fully explored, it seems likely that it provides an avenue for the metabolism of 3,5-dienoyl-CoAs that may be formed fortuitously by $\Delta^3,\Delta^2$-enoyl-CoA isomerase acting on 2,5-dienoyl-CoA intermediates.

Two $\Delta^3,\Delta^2$-enoyl-CoA isomerases exist in rat mitochondria. One is mitochondrial $\Delta^3,\Delta^2$-enoyl-CoA isomerase that has been purified and its cDNA has been cloned [17]. This enzyme is a multimer of one type of subunit with a molecular mass of 30 kDa. In addition to converting the CoA derivatives of 3-*cis*-enoic acids and 3-*trans*-enoic acids with 6–16 carbon atoms to the corresponding 2-*trans*-enoyl-CoAs, the enzyme catalyzes the conversion of 2,5-dienoyl-CoA to 3,5-dienoyl-CoA and of

138

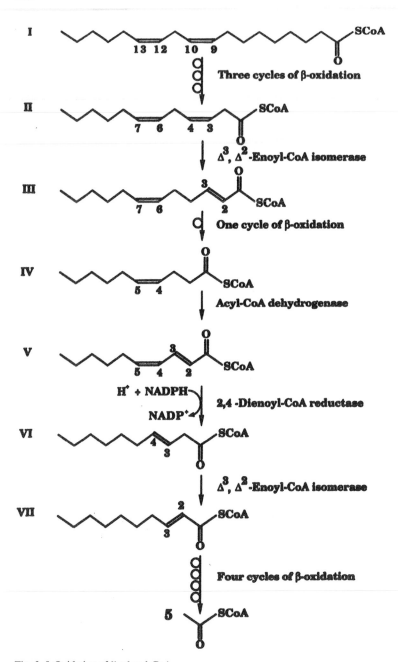

Fig. 3. β-Oxidation of linoleoyl-CoA.

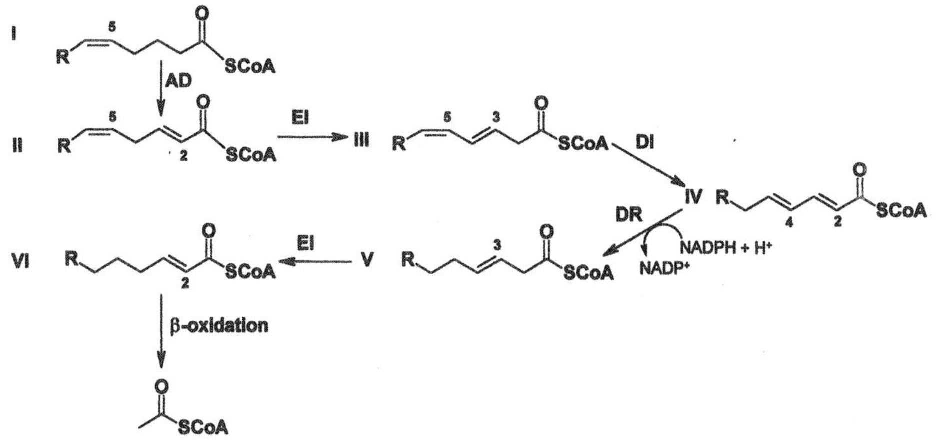

Fig. 4. β-Oxidation of 5-*cis*-enoyl-CoA. AD, acyl-CoA dehydrogenase; EI, $\Delta^3,\Delta^2$-enoyl-CoA isomerase; DI, $\Delta^{3,5},\Delta^{2,4}$-dienoyl-CoA isomerase; DR, 2,4-dienoyl-CoA reductase.

3-ynoyl-CoA to 2,3-dienoyl-CoA. A second $\Delta^3,\Delta^2$-enoyl-CoA isomerase has been identified in mitochondria. This isomerase is identical with peroxisomal $\Delta^3,\Delta^2$-enoyl-CoA isomerase and therefore has a dual subcellular localization. It is more active with long-chain than medium-chain substrates and has a preference for 3-*trans*-enoyl-CoA as compared to 3-*cis*-enoyl-CoAs.

2,4-Dienoyl-CoA reductase has been purified and its cDNA has been cloned. This enzyme is a homotetramer with a native molecular mass of 124 kDa. The reductase has a specific requirement for NADPH. A second mitochondrial 2,4-dienoyl-CoA reductase has been observed but has not yet been characterized.

$\Delta^{3,5}\Delta^{2,4}$-Dienoyl-CoA isomerase is a member of the hydratase/isomerase superfamily with a 32-kDa subunit in a homohexameric arrangement. It is present in both mitochondria and peroxisomes because of mitochondrial and peroxisomal targeting signals at its N-terminus and C-terminus, respectively. The crystal structure of this isomerase revealed the presence of one active site glutamate and aspartate residue each, which catalyze simultaneous proton transfers that facilitate the $3,5 \rightarrow 2,4$ double bond isomerization with substrates having acyl chains with 8–20 carbon atoms. It also catalyzes the isomerization of 3,5,7-trienoyl-CoA to 2,4,6-trienoyl-CoA. This reaction is a likely step in the β-oxidation of conjugated linoleic acid like 9-*cis*,11-*trans*-octadecadienoic acid.

### 3.4. Regulation of fatty acid oxidation in mitochondria

The rate of fatty acid oxidation is a function of the plasma concentration of unesterified fatty acids. Unesterified or free fatty acids are generated by lipoprotein lipase or are released from adipose tissue into the circulatory system, which carries them to other tissues or organs. Hormones like glucagon and insulin regulate the lipolysis of triacylglycerols in adipose tissue (see Chapter 10). The utilization of fatty acids for either oxidation or lipid synthesis depends on the nutritional state of the animal, more specifically on the availability of carbohydrates. Because of the close relationship among lipid metabolism, carbohydrate metabolism, and ketogenesis, the regulation of fatty acid oxidation in liver differs from that in tissues like heart and skeletal muscle, which have an overwhelming catabolic function. For this reason, the regulation of fatty acid oxidation in liver and heart will be discussed separately.

The direction of fatty acid metabolism in liver depends on the nutritional state of the animal. In the fed animal, the liver converts carbohydrates to fatty acids, while in fasted animals fatty acid oxidation, ketogenesis, and gluconeogenesis are the more active processes. Clearly, there exists a reciprocal relationship between fatty acid synthesis and fatty acid oxidation. Although it is well established that lipid and carbohydrate metabolism are under hormonal control, it has been more difficult to identify the mechanism that regulates fatty acid synthesis and oxidation. McGarry and Foster [18] have proposed that the concentration of malonyl-CoA, the first committed intermediate in fatty acid biosynthesis, determines the rate of fatty acid oxidation. The essential features of their hypothesis are presented in Fig. 5. In the fed animal, where glucose is actively converted to fatty acids, the concentration of malonyl-CoA is elevated. Malonyl-CoA at micromolar concentrations inhibits hepatic CPT I thereby decreasing the transfer of fatty acyl residues from CoA to carnitine and their translocation into mitochondria. Consequently, $\beta$-oxidation is depressed. When the animal changes from the fed to the fasted state, hepatic metabolism shifts from glucose breakdown to gluconeogenesis with a resultant decrease in fatty acid synthesis. The concentration of malonyl-CoA decreases, and the inhibition of CPT I is relieved. Furthermore, starvation causes an increase in the total CPT I activity and a decrease in the sensitivity of CPT I toward malonyl-CoA. Altogether, during starvation acylcarnitines are more rapidly formed and translocated into mitochondria thereby stimulating $\beta$-oxidation and ketogenesis.

It appears that the cellular concentration of malonyl-CoA is directly related to the activity of acetyl-CoA carboxylase, which is hormonally regulated. The short-term regulation of acetyl-CoA carboxylase involves the phosphorylation and dephosphorylation of the enzyme (see Chapter 6). In the fasting animal, a high [glucagon]/[insulin] ratio causes the phosphorylation and inactivation of acetyl-CoA carboxylase. As a consequence, the concentration of malonyl-CoA and the rate of fatty acid synthesis decrease, while the rate of $\beta$-oxidation increases. A decrease of the [glucagon]/[insulin] ratio reverses these effects. Thus, both fatty acid synthesis and fatty acid oxidation are regulated by the ratio of [glucagon]/[insulin].

It has been suggested that malonyl-CoA also regulates fatty acid oxidation in non-hepatic tissues like heart and skeletal muscle [19,20]. The formation of malonyl-CoA in these tissues is catalyzed by a 280-kDa isoform (ACC 2) of the 265-kDa acetyl-CoA carboxylase (ACC 1) that is the predominant form of lipogenic tissues. The disposal of malonyl-CoA is thought to be catalyzed by cytosolic malonyl-CoA decarboxylase. If so, the tissue concentration of malonyl-CoA is determined by the activities of both the carboxylase and decarboxylase. Both enzymes seem to be regulated. ACC 2 is phosphorylated and inactivated by AMP-dependent kinase in response to stress caused by ischemia/hypoxia and exercise and is activated allosterically by citrate. A concern about this model for the regulation of fatty acid oxidation in heart and skeletal muscle is the discrepancy between the micromolar tissue concentration of malonyl-CoA and

# FATTY ACID SYNTHESIS    FATTY ACID OXIDATION

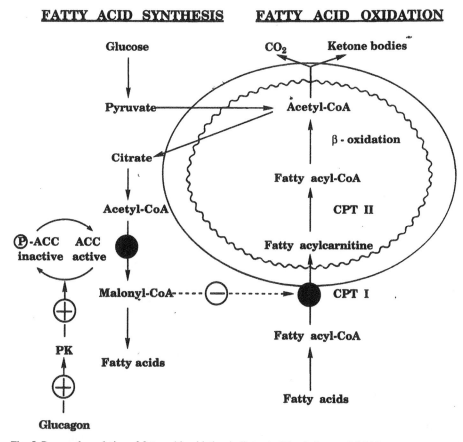

Fig. 5. Proposed regulation of fatty acid oxidation in liver. ⊕, Stimulation; ⊖, inhibition; •, enzymes subject to regulation. ACC, acetyl-CoA carboxylase; CPT, carnitine palmitoyltransferase; PK, protein kinase.

the nanomolar $K_i$ of muscle CPT I (M-CPT I) for malonyl-CoA. This enzyme should be completely inhibited at all times unless the effective malonyl-CoA concentration is lower due to binding to other proteins or due to intracellular compartmentation.

In heart, and possibly in other tissues, the rate of fatty acid oxidation is tuned to the cellular energy demand in addition to being dependent on the concentration of plasma free fatty acids [21]. At sufficiently high concentrations (>0.6 mM) of free fatty acids the rate of fatty acid oxidation is only a function of the cellular energy demand. Studies with perfused hearts and isolated heart mitochondria have shown that a decrease in the energy demand results in elevated concentrations of acetyl-CoA and NADH and in lower concentrations of CoA and $NAD^+$. The resultant increases in the ratios of [acetyl-CoA]/[CoA] and [NADH]/[$NAD^+$] in the mitochondrial matrix may be the cause for the reduced rate of β-oxidation. Experiments with isolated heart mitochondria have provided support for this view. Moreover, these experiments support the conclusion that the ratio of [acetyl-CoA]/[CoA], and not of [NADH]/[$NAD^+$], controls the rate of β-oxidation. Although the site of this regulation has not been identified unequivocally, it is possible that the [acetyl-CoA]/[CoA] ratio regulates the activity of 3-ketoacyl-CoA thiolase and thereby controls the flux of fatty acids through the β-oxidation spiral.

## 4. Fatty acid oxidation in peroxisomes

Peroxisomes and glyoxysomes, collectively referred to as microbodies, are subcellular organelles capable of respiration. They do not have an energy-coupled electron transport system like mitochondria, but instead contain flavin oxidases, which catalyze the substrate-dependent reduction of oxygen to $H_2O_2$. Because catalase is present in these organelles, $H_2O_2$ is rapidly reduced to water. Thus, peroxisomes and glyoxysomes are organelles with a primitive respiratory chain where energy released during the reduction of oxygen is lost as heat. Glyoxysomes are peroxisomes that contain the enzymes of the glyoxylate pathway in addition to flavin oxidases and catalase. Peroxisomes or glyoxysomes are found in all major groups of eukaryotic organisms including yeasts, fungi, protozoa, plants and animals.

An extramitochondrial system capable of oxidizing fatty acids was first detected in glyoxysomes of germinating seeds. When rat liver cells were shown to contain a β-oxidation system in peroxisomes [22], the interest in the peroxisomal pathway was greatly stimulated and one set of β-oxidation enzymes was soon identified and characterized [23,24]. It should be noted that peroxisomal β-oxidation occurs in all eukaryotic organisms, whereas mitochondrial β-oxidation seems to be restricted to animals. Studies of peroxisomal β-oxidation were aided by the use of certain drugs, e.g. clofibrate and di(-2-ethylhexyl)phthalate, which induce the expression of the enzymes of peroxisomal β-oxidation and in addition cause the proliferation of peroxisomes in rodents. The induction of peroxisomal β-oxidation by xenobiotic proliferators or fatty acids involves the peroxisomal proliferator-activated receptors (PPARs), which are members of the nuclear hormone receptor family and which recognize peroxisomal proliferator response elements upstream of the affected structural genes [4]. Although rat liver peroxisomes are capable of chain-shortening regular long-chain fatty acids, their main function seems to be the partial β-oxidation of very-long-chain fatty acids, methyl-branched carboxylic acids like pristanic acid, prostaglandins, dicarboxylic acids, xenobiotic compounds like phenyl fatty acids, and hydroxylated 5-β-cholestanoic acids, formed during the conversion of cholesterol to cholic acid.

### 4.1. Fatty acid uptake by peroxisomes

The mechanism of fatty acid uptake by peroxisomes is poorly understood. Although small molecules like substrates and cofactors can freely cross the membrane of isolated peroxisomes from animals, it seems that in vivo the peroxisomal membrane constitutes a permeability barrier that would require transporters to facilitate the uptake of substrates and cofactors [25]. In animal and yeast cells, long-chain fatty acids are activated outside of the peroxisomal membrane. Long-chain acyl-CoAs are thought to enter peroxisomes in a facilitated process involving half ABC transporters like ALDP, ALDRP, PMP70, and PMP69 in animal cells and Pxa1p and Pxa2p in yeast. In contrast, very long-chain fatty acids in animal cells and medium-chain fatty acids in yeast can be activated in the peroxisomal matrix. The β-oxidation of medium-chain fatty acids in yeast requires two membrane proteins that are assumed to facilitate the uptake of medium-chain fatty acids and ATP, respectively.

## 4.2. Pathways and enzymology of peroxisomal α-oxidation and β-oxidation

The first step in peroxisomal β-oxidation (see Fig. 6) is the dehydrogenation of acyl-CoA to 2-*trans*-enoyl-CoA catalyzed by acyl-CoA oxidase. This enzyme, in contrast to the mitochondrial dehydrogenases, transfers two hydrogens from the substrate to its FAD cofactor and then to $O_2$, which is reduced to $H_2O_2$. Rat liver contains three acyl-CoA oxidases with different substrate specificities. Their names, palmitoyl-CoA oxidase, pristanoyl-CoA oxidase, and trihydroxycoprostanoyl-CoA oxidase are indicative of their preferred substrates [26]. Interestingly, human liver contains only one branched-chain acyl-CoA oxidase besides palmitoyl-CoA oxidase. Cloning and sequencing of the gene and cDNAs coding for rat acyl-CoA oxidase revealed two

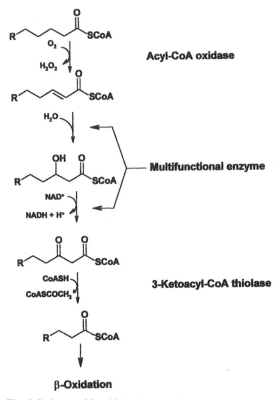

Fig. 6. Pathway of β-oxidation in peroxisomes.

isoforms of this enzyme as the result of the alternative use of two exons [24]. The rat liver palmitoyl-CoA oxidase is a homodimer with a molecular mass of close to 150 kDa. Ligands of PPARs induce the expression of this enzyme but not of the other two acyl-CoA oxidases. Palmitoyl-CoA oxidase is inactive with butyryl-CoA and hexanoyl-CoA as substrates, but dehydrogenates all longer chain substrates with similar maximal velocities. Acyl-CoA oxidases from organisms other than mammals are either active with substrates of all chain lengths or the organisms express more than one acyl-CoA oxidase with complementing chain length specificities. Consequently, fatty acids can be completely degraded in yeasts, plants, and other lower eukaryotic organisms, but not in mammals.

The next two reactions of β-oxidation, the hydration of 2-enoyl-CoA to 3-hydroxyacyl-CoA and the $NAD^+$-dependent dehydrogenation of 3-hydroxyacyl-CoA to 3-ketoacyl-CoA, are catalyzed in rat liver peroxisomes by multifunctional enzymes 1 (MFE 1) and 2 (MFE 2) [25,27]. Both multifunctional enzymes harbor enoyl-CoA hydratase and 3-hydroxyacyl-CoA dehydrogenase activities. MFE 1 additionally exhibits $\Delta^3,\Delta^2$-enoyl-CoA isomerase activity. 3-Hydroxyacyl-CoA intermediates formed and acted on by MFE 1 have the L-configuration, whereas the hydroxy intermediates produced by MFE 2 have the D-configuration. Both MFE 1 and MFE 2 convert medium-chain and long-chain 2-*trans*-enoyl-CoAs to 3-ketoacyl-CoAs but show little or no activity with short-chain substrates. MFE 2 is slightly more active than MFE 1 with longer-chain substrates. However, MFE 2 alone acts on substrates with 2-methyl branches like those formed during the β-oxidation of pristanic acid and hydroxylated 5β-cholestanoic acid. MFE 1 consists of a single 80-kDa polypeptide while MFE 2 is a homodimer with a molecular mass of approximately 150 kDa. Yeast and fungi contain only one multifunctional enzyme each with D-specific enoyl-CoA hydratase and 3-hydroxyacyl-CoA dehydrogenase activities.

The last reaction of β-oxidation, the CoA-dependent cleavage of 3-ketoacyl-CoA, is catalyzed by 3-ketoacyl-CoA thiolase. Two rat 3-ketoacyl-CoA thiolases coded for by different genes have been detected. One is constitutively expressed, whereas the expression of the other is highly induced in response to peroxisomal proliferators [25,27]. Both enzymes are homodimers with molecular masses of close to 80 kDa. Both exhibit little activity toward acetoacetyl-CoA, but are active with all longer chain substrates except for 3-keto-2-methylacyl-CoA intermediates formed during the β-oxidation of pristanic acid and hydroxylated 5β-cholestanoic acids. However, these intermediates are acted upon by another 3-ketoacyl-CoA thiolase ($SCP_x$-thiolase) that was identified in peroxisomes during a study of the 58-kDa precursor of sterol carrier protein-2. The C-terminal segment of this 58-kDa protein is identical with sterol carrier protein-2, whereas the N-terminal domain harbors the thiolase, which is most active with medium-chain substrates. The crystal structure of the peroxisomal 3-ketoacyl-CoA thiolase from *Saccharomyces cerevisiae* at 2.8 Å resolution shows two cysteine residues in close proximity at the presumed active site.

Because peroxisomes contain at least two enzymes for each step of β-oxidation, specific functions for these enzymes were inferred from their substrate specificities. Most of the predictions were verified by analyzing fatty acids that accumulate in patients and/or knock-out mice deficient for individual enzymes. Together, these data support the proposal that branched-chain acyl-CoA oxidase, MFE 2, and SCP$_x$-thiolase are essential for the degradation of pristanic acid and hydroxylated 5-β-cholestanoic acid. The β-oxidation of very-long-chain fatty acids requires the involvement of acyl-CoA oxidase and MFE 2. 3-Ketoacyl-CoA thiolase is believed, but has not been proven to function in the breakdown of straight-chain fatty acids. Surprisingly the knock-out mouse for MFE 1 does not exhibit an obvious phenotype, thus leaving the function of this enzyme in doubt.

Unsaturated fatty acids are degraded in peroxisomes by the pathways outlined in Figs. 3 and 4. Two $\Delta^3,\Delta^2$-enoyl-CoA isomerases are present in rat liver peroxisomes. One is the monofunctional $\Delta^3,\Delta^2$-enoyl-CoA isomerase that is also present in mitochondria, the other is the $\Delta^3,\Delta^2$-enoyl-CoA isomerase activity of multifunctional enzyme 1. The monofunctional $\Delta^3,\Delta^2$-enoyl-CoA isomerase has a preference for long-chain substrates and may play the major role in the partial β-oxidation of long-chain unsaturated fatty acids. So far this isomerase has only been obtained by cloning and heterologous expression based on its homology to the sole $\Delta^3,\Delta^2$-enoyl-CoA isomerase of yeast. The crystal structure of yeast $\Delta^3,\Delta^2$-enoyl-CoA isomerase revealed the presence of a single glutamate residue at the active site, which catalyzes a 1,3-proton transfer that results in the shift of the double bond. A 2,4-dienoyl-CoA reductase distinct from the mitochondrial reductase but homologous with the yeast 2,4-dienoyl-CoA reductase has been identified in mammalian peroxisomes by a cloning and expression approach. This enzyme acts on a wide spectrum of 2,4-dienoyl-CoAs but is most active with medium-chain substrates. $\Delta^{3,5}\Delta^{2,4}$-Dienoyl-CoA isomerase of mammalian peroxisomes is the same enzyme that is also present in mitochondria (see Section 3.3).

The products of peroxisomal β-oxidation in animals are chain-shortened acyl-CoAs, acetyl-CoA, and NADH. The β-oxidation of chain-shortened acyl-CoAs is completed in mitochondria. For this purpose, acyl-CoAs, including acetyl-CoA, are thought to leave peroxisomes as acylcarnitines, which can be formed by peroxisomal carnitine octanoyltransferase and/or carnitine acetyltransferase [25]. These reactions, as well as the observed hydrolysis of acetyl-CoA to acetate, would regenerate CoA in peroxisomes. The transporters that facilitate the exit of the β-oxidation products from peroxisomes have not yet been identified.

Phytanic acid (3,7,11,15-tetramethylhexadecanoic acid), a component of the human diet that is derived from phytol, a constituent of chlorophyll, is not degraded by β-oxidation because its 3-methyl group interferes with this process. Instead it is chain shortened by α-oxidation in peroxisomes as outlined in Fig. 7 [25,27]. Activation of phytanic acid(I) to phytanoyl-CoA(II) by long-chain or very long-chain acyl-CoA synthetase converts it to a substrate of hydroxyphytanoyl-CoA hydroxylase. Cleavage of the resultant 2-hydroxyphytanoyl-CoA(III) by 2-hydroxyphytanoyl-CoA lyase yields pristanal(IV) and formyl-CoA that are oxidized to pristanic acid(V) and $CO_2$, respectively. Pristanic acid after activation to pristanoyl-CoA is a substrate of β-oxidation because the 2-methyl group does not interfere with the process as long as the 2-methyl group has the S configuration. A 2-methylacyl-CoA racemase that is present in both peroxisomes and mitochondria epimerizes (2R)-pristanoyl-CoA and (25R)-trihydroxycholestanoyl-CoA to their S isomers that are substrates of peroxisomal β-oxidation.

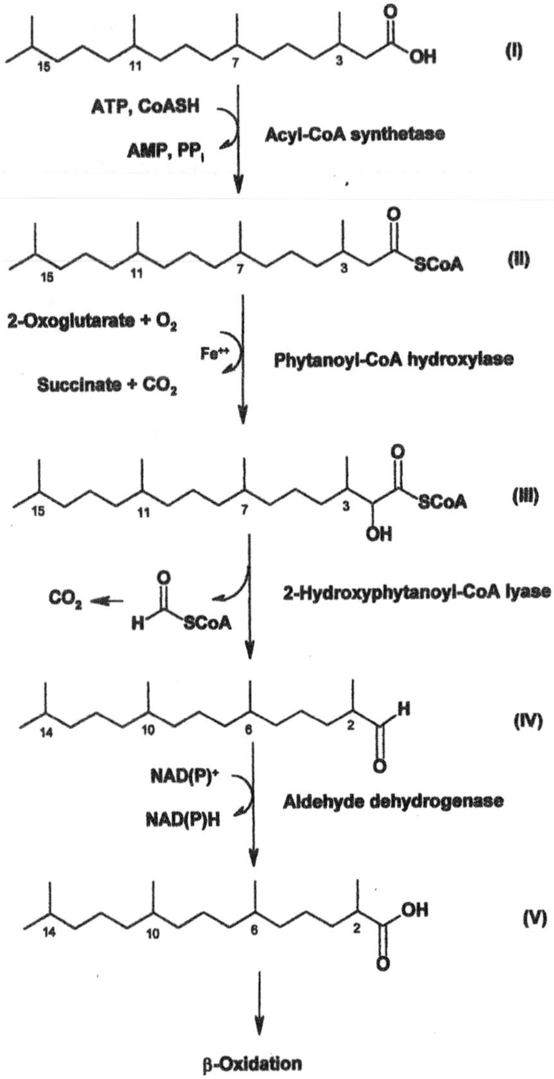

Fig. 7. α-Oxidation of phytanic acid.

## 5. Inherited diseases of fatty acid oxidation

Disorders of fatty acid oxidation were first described in 1973 when deficiencies of carnitine and carnitine palmitoyltransferase (CPT II) were identified as causes of muscle weakness [28]. Patients with low levels (5–45% of normal) of CPT II have recurrent episodes of muscle weakness and myoglobinuria, often precipitated by prolonged exercise and/or fasting. Almost a decade later, a deficiency of medium-chain dehydrogenase was identified in patients with a disorder of fasting adaptation [28]. This relatively common disorder is characterized by episodes of non-ketotic hypoglycemia provoked by fasting during the first 2 years of life. Between episodes, patients with medium-chain acyl-CoA dehydrogenase deficiency appear normal. Therapy is aimed at preventing fasting, if necessary by the intravenous administration of glucose, and includes carnitine supplementation. The molecular basis of medium-chain acyl-CoA dehydrogenase deficiency is an A → G base transition in 90% of the disease causing alleles. This mutation results in the replacement of lysine-329 by a glutamate residue, which impairs the assembly of subunits into the functional tetrameric enzyme. In the years following the identification of medium-chain acyl-CoA dehydrogenase deficiency, fatty acid oxidation disorders due to the following enzymes deficiencies have been described: short-chain acyl-CoA dehydrogenase, very-long-chain acyl-CoA dehydrogenase, electron-transferring flavoprotein (ETF), ETF:ubiquinone oxidoreductase, 3-hydroxyacyl-CoA dehydrogenase, long-chain 3-hydroxyacyl-CoA dehydrogenase, trifunctional β-oxidation complex, 2,4-dienoyl-CoA reductase, carnitine palmitoyltransferase I, and carnitine:acylcarnitine translocase [28,29]. A deficiency of mitochondrial acetoacetyl-CoA thiolase impairs isoleucine and ketone body metabolism, but not fatty acid oxidation.

Many of these disorders are associated with the urinary excretion of acylcarnitines, acyl conjugates of glycine, and dicarboxylic acids that are characteristic of the metabolic block. A general conclusion derived from studies of these disorders is that an impairment of β-oxidation makes fatty acids available for microsomal ω-oxidation by which fatty acids are oxidized at their terminal (ω) methyl group or at their penultimate (ω − 1) carbon atom. Molecular oxygen is required for this oxidation and the hydroxylated fatty acids are further oxidized to dicarboxylic acids. Long-chain dicarboxylic acids can be chain-shortened by peroxisomal β-oxidation to medium-chain dicarboxylic acids, which are excreted in urine.

Several disorders associated with an impairment of peroxisomal β-oxidation have been described [30]. Of these, Zellweger syndrome and neonatal adrenoleukodystrophy are characterized by the absence, or low levels, of peroxisomes due to the defective biogenesis of this organelle. As a result of this deficiency, compounds that are normally metabolized in peroxisomes, for example very long-chain fatty acids, dicarboxylic acids, hydroxylated 5-β-cholestanoic acids, and also phytanic acid, accumulate in plasma [25,30]. Infants with Zellweger syndrome rarely survive longer than a few months due to hypotonia, seizures and frequently cardiac defects. In addition to disorders of peroxisome biogenesis, defects of each of the three enzymes of the peroxisomal β-oxidation spiral and of the peroxisomal very long-chain acyl-CoA synthetase (X-linked adrenoleukodystrophy) have been reported [25,30]. Most of these patients were hypotonic, developed seizures, failed to make psychomotor gains, and died in early childhood. The importance of α-oxidation in humans has been established as a result of studying Refsum's disease, a rare and inherited neurological disorder. Patients afflicted with this disease accumulate large amounts of phytanic acid due to a deficiency of phytanoyl-CoA hydroxylase [25,30].

## 6. Future directions

Fatty acid oxidation has been studied for almost a century with the result that a fairly detailed view of this process has emerged. The molecular characterization of most β-oxidation enzymes has yielded a wealth of structural information while the dynamics of this pathway remain less well understood. This is in part due to an absence of information about the organization of the β-oxidation enzymes and the impact such organization has on the control of the process. Also a number of questions about the regulation of this process remain unanswered, especially about its regulation in extrahepatic tissues. Even the extensively studied regulation of hepatic fatty acid oxidation by malonyl-CoA continues to be further investigated to provide an understanding of the regulatory mechanism at the molecular level. In spite of impressive progress in the area of peroxisomal β-oxidation, aspects of this process remain unresolved. For example, it is unclear how fatty acids enter peroxisomes and how products exit from this organelle. Also, the transcriptional regulation of this process has not been fully explored. Moreover, the cooperation between peroxisomes and mitochondria in fatty acid oxidation remains to be studied. Not all of the reactions of the β-oxidation spiral have been verified experimentally and hence some may not take place as envisioned. Finally, the complete characterization of known disorders of β-oxidation in humans and the identification of new disorders will raise questions about some accepted features of this process and will prompt re-investigations of issues thought to be resolved.

## Abbreviations

ACC     acetyl-CoA carboxylase
ACS     acyl-CoA synthetase
CPT     carnitine palmitoyltransferase
ETF     electron-transferring flavoprotein
FABP    fatty acid binding protein
MFE    multifunctional enzyme
PPAR   peroxisomal proliferator-activated receptor
SCP     sterol carrier protein

## References

1. Knoop, F. (1904) Der Abbau aromatischer Fettsäuren im Tierkörper. Ernst Kuttruff, Freiburg, Germany.
2. Dakin, H. (1909) The mode of oxidation in the animal organism of phenyl derivatives of fatty acids. Part IV. Further studies on the fate of phenylpropionic acid and some of its derivatives. J. Biol. Chem. 6, 203–219.
3. Lynen, F. (1952–1953) Acetyl coenzyme A and the fatty acid cycle. Harvey Lect. Ser. 48, 210–244.
4. Kunau, W.-H., Dommes, V. and Schulz, H. (1995) Beta oxidation of fatty acids in mitochondria, peroxisomes, and bacteria. Prog. Lipid Res. 34, 267–341.
5. Abumrad, N., Coburn, C. and Ibrahimi, A. (1999) Membrane proteins implicated in long-chain fatty

acid uptake by mammalian cells: CD36, FATP and FABPm. Biochim. Biophys. Acta 1441, 4–13.

6. Coe, N.R. and Bernlohr, D.A. (1998) Physiological properties and functions of intracellular fatty acid-binding proteins. Biochim. Biophys. Acta 1391, 287–306.

7. Watkins, P.A. (1997) Fatty acid activation. Prog. Lipid Res. 36, 55–83.

8. McGarry, J.D. (2001) Travels with carnitine palmitoyltransferase I: from liver to germ cell with stops in between. Biochem. Soc. Trans. 29, 241–245.

9. Zammit, V.A., Price, N.T., Fraser, F. and Jackson, V.N. (2001) Structure–function relationship of the liver and muscle isoforms of carnitine palmitoyltransferase I. Biochem. Soc. Trans. 29, 287–292.

10. Ikeda, Y. and Tanaka, K. (1990) Purification and characterization of five acyl-CoA dehydrogenases from rat liver mitochondria. In: K. Tanaka and P.M. Coates (Eds.), Fatty Acid Oxidation. Clinical, Biochemical and Molecular Aspects. Alan R. Liss, New York, pp. 37–54.

11. Izai, K., Uchida, Y., Orii, T., Yamamoto, S. and Hashimoto, T. (1992) Novel fatty acid β-oxidation enzymes in rat liver mitochondria. I. Purification and properties of very-long-chain acyl coenzyme A dehydrogenase. J. Biol. Chem. 267, 1027–1033.

12. Thorpe, C. and Kim, J.-J. (1995) Structure and mechanism of action of the acyl-CoA dehydrogenases. FASEB J. 9, 718–725.

13. Uchida, Y., Izai, K., Orii, T. and Hashimoto, T. (1992) Novel fatty acid β-oxidation enzymes in rat liver mitochondria. II. Purification and properties of enoyl coenzyme A (CoA) hydratase/3-hydroxyacyl-CoA dehydrogenase/3-ketoacyl-CoA thiolase trifunctional protein. J. Biol. Chem. 267, 1034–1041.

14. Liang, X., Le, W., Zhang, D. and Schulz, H. (2001) Impact of the intramitochondrial enzyme organization on fatty acid oxidation. Biochem. Soc. Trans. 29, 279–282.

15. Schulz, H. and Kunau, W.-H. (1987) Beta-oxidation of unsaturated fatty acids: a revised pathway. Trends Biochem. Sci. 12, 403–406.

16. Smeland, T.E., Nada, M., Cuebas, D. and Schulz, H. (1992) NADPH-dependent β-oxidation of unsaturated fatty acids with double bonds extending from odd-numbered carbon atoms. Proc. Natl. Acad. Sci. USA 89, 6673–6677.

17. Hiltunen, J.K. and Qin, Y.-M. (2000) β-Oxidation — strategies for the metabolism of a wide variety of acyl-CoA esters. Biochim. Biophys. Acta 1484, 117–128.

18. McGarry, J.D. and Foster, D.W. (1980) Regulation of hepatic fatty acid oxidation and ketone body production. Annu. Rev. Biochem. 49, 395–420.

19. Lopaschuk, G.D., Belke, D.D., Gamble, J., Itoi, T. and Schönekess, B.O. (1994) Regulation of fatty acid oxidation in the mammalian heart in health and disease. Biochim. Biophys. Acta 1213, 263–276.

20. Ruderman, N.B., Saha, A.K., Vavas, D. and Witters, L.A. (1999) Malonyl-CoA, fuel sensing, and insulin resistance. Am. J. Physiol. 276, E1–E18.

21. Schulz, H. (1994) Regulation of fatty acid oxidation in heart. J. Nutr. 124, 165–171.

22. Lazarow, P.B. and de Duve, C. (1976) A fatty acyl-CoA oxidizing system in rat liver peroxisomes; enhancement by clofibrate, a hypolipidemic drug. Proc. Natl. Acad. Sci. USA 73, 2043–2046.

23. Hashimoto, T. (1990) Purification, properties, and biosynthesis of peroxisomal β-oxidation enzymes. In: K. Tanaka and P.M. Coates (Eds.), Fatty Acid Oxidation. Clinical, Biochemical and Molecular Aspects. Alan R. Liss, New York, pp. 137–152.

24. Osumi, T. (1990) Molecular cloning and sequencing of the peroxisomal β-oxidation enzymes. In: K. Tanaka and P.M. Coates (Eds.), Fatty Acid Oxidation. Clinical, Biochemical and Molecular Aspects. Alan R. Liss, New York, pp. 681–696.

25. Wanders, R.J.A., Vreken, P., Ferdinandusse, S., Jansen, G.A., Waterham, H.R., van Roermund, C.W.T. and van Grunsven, E.G. (2001) Peroxisomal fatty acid α- and β-oxidation in humans: enzymology, peroxisomal metabolite transporters and peroxisomal diseases. Biochem. Soc. Trans. 29, 250–267.

26. Van Veldhoven, P.P., Vanhove, G., Asselberghs, S., Eyssen, H.J. and Mannaerts, G.P. (1992) Substrate specificities of rat liver peroxisomal acyl-CoA oxidases: palmitoyl-CoA oxidase (inducible acyl-CoA oxidase), pristanoyl-CoA oxidase (non-inducible acyl-CoA oxidase), and trihydroxycoprostanoyl-CoA oxidase. J. Biol. Chem. 267, 20065–20074.

27. Van Veldhoven, P.P., Casteels, M., Mannaerts, G.P. and Baes, M. (2001) Further insights into peroxisomal lipid breakdown via α- and β-oxidation. Biochem. Soc. Trans. 29, 292–298.

28. Roe, C.R. and Coates, P.M. (1995) Mitochondrial fatty acid oxidation disorders. In: C.R. Scriver, A.L.

Beaudet, W.S. Sly and D. Valle (Eds.), The Metabolic and Molecular Basis of Inherited Disease. 7th edn., Vol. I, McGraw-Hill, New York, pp. 1501–1533.

29. Bennett, M.J., Rinaldo, P. and Strauss, A.W. (2001) Inborn errors of mitochondrial fatty acid oxidation. Crit. Rev. Clin. Lab. Sci. 37, 1–44.

30. Lazarow, P.B. and Moser, H.W. (1995) Disorders of peroxisome biogenesis. In: C.R. Scriver, A.L. Beaudet, W.S. Sly and D. Valle (Eds.), The Metabolic and Molecular Basis of Inherited Disease. 7th edn., Vol. II, McGraw-Hill, New York, pp. 2287–2324.

D.E. Vance and J.E. Vance (Eds.) *Biochemistry of Lipids, Lipoproteins and Membranes (4th Edn.)*

# CHAPTER 6

# Fatty acid synthesis in eukaryotes

Vangipuram S. Rangan[*] and Stuart Smith

*Children's Hospital Oakland Research Institute, 5700 Martin Luther King Jr. Way,
Oakland, CA 94611, USA, Tel.: +1 (510) 450-7675; Fax: +1 (510) 450-7910;
E-mail: ssmith@chori.org; vrangan@medarex.com*

## 1. Introduction

Fatty acids fulfill several crucial roles in animals. They represent a major storage form of energy, they provide an essential structural component of membranes, through direct covalent linkage they are used to modify and regulate the properties of many proteins and, as components of certain lipid signaling molecules, they perform important roles in metabolic regulation. This chapter will focus primarily on the structure, mechanism of action and regulation of the enzymes responsible for the biosynthesis of long-chain saturated fatty acids, de novo. All of the carbon atoms of fatty acids are derived from the two-carbon precursor, acetyl-CoA and until the demonstration in 1958 that $CO_2$ was required for the biosynthesis of fatty acids de novo, it had been assumed that the pathway utilized the enzymes of β-oxidation operating in the reverse direction (S.J. Wakil, 1958). Studies during the late 1950s and early 1960s, primarily in the laboratories of S.J. Wakil, P.R. Vagelos and F. Lynen, established clearly that the thermodynamic barrier posed by condensation of two acetyl-CoA molecules is circumvented by the introduction of an energy-dependent carboxylation step that generates malonyl-CoA as a co-substrate. Much of the early progress in identifying the individual enzymes involved uniquely in the biosynthetic route was made using the *Escherichia coli* system and it was not until the mid-1970s that it became clear that in eukaryotes, the enzymes are covalently linked in 'multifunctional' polypeptides. In prokaryotes, exemplified by *E. coli*, more than 10 individual proteins are involved in the biosynthesis of long-chain saturated fatty acids from acetyl-CoA (Chapter 3). This system, in which each enzyme is present on a single polypeptide, became known as the 'type II' FAS system and the multifunctional polypeptide system, exemplified by yeast and animals, became known as the 'type I' FAS system. Remarkably, in animals, the catalytic components required for the entire fatty acid biosynthetic pathway are integrated into two multifunctional polypeptides, acetyl-CoA carboxylase (ACC) and fatty acid synthase (FAS) [1]. Both proteins are posttranslationally modified by the covalent attachment of vitamin derivatives that play essential roles as 'swinging arms' in the translocation of intermediates between different catalytic sites. In ACC, a biotin moiety, attached to the ε-amino group of a lysine residue, serves as a carboxyl carrier between the carboxylase and transcarboxylase catalytic domains and in the FAS, a phosphopantetheine moiety, attached to a serine hydroxyl, serves to transport substrates and the growing acyl chain through the various catalytic centers of the complex. Expression of both enzymes is regulated at the transcriptional level, in a tissue-specific manner, in response to various developmental, nutritional, and hormonal signals [2]. In addition, ACC is subject to short-term regulation by allosteric and phosphorylation

[*] Present address: Medarex Inc., 521 Cottonwood Drive, Milpitas, CA 95035, USA.

mechanisms and is commonly regarded as the pace setting enzyme for fatty acid synthesis [3]. The only free intermediate in the entire biosynthetic pathway is malonyl-CoA; all other intermediates exist only as covalently bound acyl-enzyme complexes [1]. Recently, this metabolite has been recognized as playing a critical role in a fuel sensing and signaling mechanism that regulates food intake and energy metabolism [4,5].

## 2. Acetyl-CoA carboxylase

### 2.1. The reaction sequence

The formation of malonyl-CoA from acetyl-CoA is a two-step reaction involving first, the ATP-dependent carboxylation of the biotinyl moiety, followed by transfer of the carboxyl to an acetyl-CoA acceptor.

$$\text{Enzyme-biotin} + HCO_3^- + ATP \xleftrightarrow[Mg^{2+}]{biotin\ carboxylase} \text{Enzyme-biotin-}CO_2^- + ADP + Pi$$

$$\text{Enzyme-biotin-}CO_2^- + \text{acetyl-CoA} \xrightarrow{transcarboxylase} \text{Enzyme-biotin} + \text{malonyl-CoA}$$

### 2.2. Domain organization

The amino acid sequence and domain structure of ACC is highly conserved in eukaryotes, such that antibodies raised against animal and fungal forms cross-react with each other. The two catalytic domains, biotin carboxylase and transcarboxylase, are located in the amino- and carboxyterminal halves of the polypeptide, respectively (Fig. 1). Between these domains lies the conserved biotin-binding site motif Met–Lys–Met within the biotin

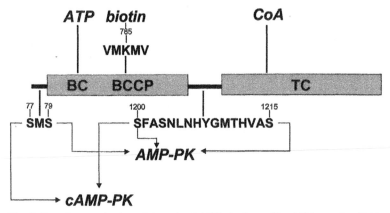

Fig. 1. Domain organization of the animal ACCα-isoform. The ACC contains 2346 residues. The precise locations of boundaries between the biotin carboxylase (BC), biotin carboxyl carrier protein (BCCP) and transcarboxylase (TC) domains has not been established; none of the domains has been expressed as individual proteins. The locations of ATP, biotin and CoA-binding sites are shown above the map and the location of phosphorylation sites shown to influence activity are shown below the map. Adapted from Corton and Hardie [6] and Kim [7].

carboxyl carrier protein domain. The presumed ATP and $HCO_3^-$ binding sites within the biotin carboxylase domain and the acetyl-CoA-binding site, within the transcarboxylase domain, have been identified by sequence analysis but have yet to be confirmed by mutagenesis. The major differences between animal and yeast ACC are in the first 100 or so aminoterminal residues and near the center of the polypeptides, both regions that are implicated in regulation of activity of the animal form by phosphorylation.

The minimal functional unit (protomer) is a homodimer. In the presence of citrate, the inactive protomer rapidly assumes a catalytically active conformation and undergoes a slower, reversible polymerization to a long filamentous polymeric structure, $\sim$100 Å wide and up to 5000 Å long, containing as many as 20 protomers [8]. The polymeric structure may stabilize the enzyme protomers in an active conformation. Depolymerization and inactivation of the enzyme is promoted by long-chain acyl-CoA thioesters, which are direct competitors with citrate, and malonyl-CoA.

## 2.3. Isoforms

Two major isoforms of ACC, $\alpha$ and $\beta$, have been described in animals (K.-H. Kim, 1988, 1996; S.J. Wakil, 1995) that, paradoxically, have distinctly different roles in metabolism [4]. The 265 kDa $\alpha$-isoform is the predominant form expressed in the soluble cytoplasm of lipogenic tissues such as adipose, liver and mammary gland, where it plays an essential role in providing malonyl-CoA as a carbon source for fatty acid synthesis. The 280 kDa $\beta$-isoform, on the other hand, is expressed mainly in heart and skeletal muscle, tissues with very low lipogenic capacity, but also to some extent in liver, and is associated with the outer mitochondrial membrane. In these tissues, malonyl-CoA produced by the $\beta$-isoform functions primarily as a negative regulator of carnitine palmitoyltransferase I and, in so doing, controls the flux of fatty acids into the mitochondria for $\beta$-oxidation (Chapter 5). The two isoforms exhibit extensive sequence similarity, the major difference being the presence of $\sim$120 additional residues at the aminoterminus of the $\beta$-form that serve to anchor the enzyme in the outer mitochondrial membrane [7]. The presence of a third isoform, containing eight additional amino acids, four residues upstream from Ser1200, has been inferred by detection of a mRNA species containing an additional 24 nucleotides that may result from alternative splicing of the gene for the $\alpha$-isoform [9]. The mRNA encoding the longer version of the $\alpha$-isoform dominates in liver and adipose, whereas that encoding the shorter one dominates in lactating mammary gland. Activity of the two $\alpha$-isoforms appears to be regulated by different phosphorylation mechanisms (Section 4.2).

## 3. Fatty acid synthase

### 3.1. The reaction sequence

The overall reaction catalyzed by the animal FAS can be summarized by the equation:

$$\text{Acetyl-CoA} + 7\text{Malonyl-CoA} + 14\text{NADPH} + 14\text{H}^+ \longrightarrow$$
$$\text{Palmitic Acid} + 7\text{CO}_2 + 8\text{CoA} + 14\text{NADP}^+ + 6\text{H}_2\text{O}$$

154

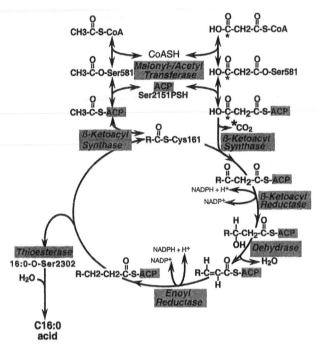

Fig. 2. Reaction sequence catalyzed by the fatty acid synthase. The location of all covalent acyl-enzyme intermediates is shown. PSH, phosphopantetheine moiety. The condensation reaction involves the stereochemical inversion of the C2 of the malonyl moiety. Reduction of the 3-ketoacyl moiety to a (3R)-hydroxyacyl moiety involves transfer of the prochiral 4S hydrogen of NADPH. Dehydration to the *trans*-enoyl moiety occurs through the syn-elimination of the prochiral 2S hydrogen and the 3R hydroxyl group. Reduction of the *trans*-enoyl moiety proceeds by the transfer of the prochiral 4R hydrogen of NADPH to the prochiral 2S position [11]. Recent studies indicate that the malonyl C-3 is released as $HCO_3^-$, rather than $CO_2$ in the decarboxylation reaction (S. Smith, 2002).

The pathway can be visualized as a cyclic process in which the acetyl primer undergoes a series of Claisen condensation reactions with seven malonyl extender molecules and, following each condensation, the β-carbon of the β-ketoacyl moiety formed is completely reduced by a three-step ketoreduction–dehydration–enoylreduction process [10]. The saturated acyl chain product of one cycle becomes the primer substrate for the following cycle, so that two saturated carbon atoms are added to the primer with each turn of the cycle (Fig. 2). The final product is released as a free fatty acid by the animal FAS and as a CoA ester by the yeast complex.

*3.2. The catalytic components*

Several catalytic elements are required for the biosynthetic process: acyltransferases that load the primer and extender substrates onto the FAS complex; a posttranslationally phosphopantetheinylated acyl carrier protein, which translocates the various intermediates between catalytic sites; a β-ketoacyl synthase, which performs the condensation reaction; the β-ketoacylreductase, dehydrase and enoylreductase enzymes responsible

Animal FAS, $\alpha_2$

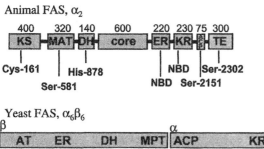

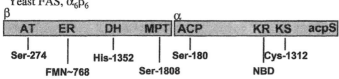

Fig. 3. Domain maps of the animal and yeast fatty acid synthases. The rat fatty acid synthase contains 2505 residues. KS, β-ketoacyl synthase; MAT, malonyl/acetyltransferase; DH, dehydrase; ER, enoylreductase; KR, ketoreductase; ACP, acyl carrier protein; TE, thioesterase; NBD, nucleotide-binding domain. The approximate number of residues in each domain is indicated above the map. The MAT, ACP and thioesterase domains have been isolated, by limited proteolysis, and expressed as independent proteins. The central core region has no known catalytic function, but is thought to contain a dimerization domain that promotes interaction between the subunits. The location of essential active-site residues, and the nucleotide-binding region of the two reductases are shown below the map. The yeast β subunit contains 1887 residues and the α subunit 1845. AT, acetyltransferase; MPT, malonyl/palmitoyltransferase; acpS, holo-ACP synthetase. The boundaries between component domains of the yeast protein are less clearly defined; none of these domains has been isolated or expressed as independent proteins. Adapted from Smith [13] and Fichtlscherer et al. [14].

for the β-carbon processing reactions and a chain terminating enzyme responsible for product release. Although the animal and yeast FASs employ essentially the same overall reaction scheme, they exhibit minor differences in the enzymatic details and major differences in their macromolecular architecture. (1) Whereas the animal FAS utilizes the same acyltransferase for loading of both the acetyl and malonyl substrates (S. Smith, 1996, 1997), the yeast FAS uses two different enzymes, one specific for acetyl moieties, the other able to load acetyl or malonyl [12]. (2) The animal FAS releases the product as a free fatty acid, through the action of a thioesterase, which severs the bond between fatty acid and phosphopantetheine thiol (S. Smith, 1978) whereas the yeast FAS uses the same broad specificity acyltransferase responsible for substrate loading to transfer the product to a CoA acceptor, so that the product is an acyl-CoA thioester [12]. (3) The enoylreductase of the animal FAS utilizes only NADPH, but the yeast enoylreductase requires both NADPH and FMN as cofactors [1].

*3.3. Domain organization*

In eukaryotes, the enzymes required for fatty acid synthesis de novo are integrated into large multifunctional polypeptides that are located in the cytosol. However, the domain organization and overall molecular architecture of the animal and yeast complexes are quite different (Fig. 3). Thus, the animal FAS is a dimer of identical, 272 kDa polypeptides [10,11,13], whereas the yeast FAS contains six copies each of two different subunits, α and β, of molecular masses 208 kDa and 220 kDa, respectively (E. Schweizer, 1973).

Remarkably, the ordering of the catalytic domains in the yeast multifunctional polypeptides is quite dissimilar to that of the animal and the two multifunctional proteins appear to have evolved along quite different gene fusion pathways. A unique feature of the yeast FAS is that phosphopantetheinylation is catalyzed by a discrete domain located at the carboxyterminus of the α subunit (E. Schweizer, 2000). This endogenous phosphopantetheinyl transferase cooperates with the ACP domain of the second α subunit within the $\alpha_2\beta_2$ protomer. Phosphopantetheinylation of the animal, prokaryotic and plant ACPs is catalyzed by separate, discrete enzymes. The phosphopantetheinyl transferase acting on the animal FAS has not yet been characterized.

Both eukaryotic forms of FAS require the oligomeric structure for activity: the animal $\alpha_2$ complex has two equivalent sites for fatty acid synthesis (S.J. Wakil, 1984; S. Smith, 1995) and the yeast $\alpha_6\beta_6$ complex has six (S.J. Wakil, 1980). Thus far the complexes have been refractory to crystallographic analysis and structural analysis has been limited to lower-resolution methods. Electron microscopy and small-angle neutron scattering studies (S.J. Wakil, 1987; A. Ikai, 1988) reveal the animal FAS as an ellipsoid structure, $\sim 216 \times 144 \times 72$ Å, containing two cavities, one at each end of the molecule. The two cavities may represent the two sites for fatty acid synthesis in the dimeric structure. Electron micrographic images of the complex labeled with Fab fragments derived from antibodies raised against the thioesterase domain indicate that the two carboxylterminal domains are located one at each pole of the ellipsoid. Imaging of complexes labeled by Fab fragments from antibodies raised against the region between the dehydrase and enoylreductase domains, indicate that this non-catalytic core is located near the center of the ellipsoid structure. This core region contains dimerization domains that may serve to stabilize interactions between the two subunits [13].

A more detailed 25 Å resolution structure of the yeast FAS has been computed from galleries of electron microscopic images (J. Stoops, 1996). The complex appears as a barrel-shaped structure, $\sim 245 \times 220$ Å, formed by three zig-zag-shaped, α subunit pairs, each with a pair of over- and under-lying arch-shaped β subunits that cap the ends of the barrel. Thus the protomeric unit is an $\alpha_2\beta_2$ structure. The barrel contains six cavities that likely constitute the six equivalent sites for fatty acid synthesis. Each cavity has two funnel-shaped openings of $\sim 20$ Å that may allow diffusion of substrates into, and products out of, the catalytic centers.

## 3.4. Chain initiation

Since the animal FASs utilize a common loading site for both the primer and chain-extender substrates, acetyl- and malonyl-CoA are mutually competitive inhibitors of each other. These FASs cannot order the sequential loading of one acetyl moiety and seven malonyl moieties but instead rely on a self-editing process, in which both substrates are rapidly exchanged between CoA ester and enzyme-bound forms [11,13]. Only when the appropriate pair of substrates is bound, that is an acetyl, or longer saturated acyl moiety, at the cysteine active site and a malonyl moiety at the phosphopantetheine site, does condensation take place. Nonproductive binding, for example when the malonyl moiety binds before the acetyl moiety, or when two acetyl moieties are bound sequentially, results in rapid translocation of the substrates back to their CoA ester form. Paradoxically then, for the system to function efficiently, free CoA, which appears on the product side of the equation, must be available at all times (S. Smith, 1982).

The yeast FAS employs two different acyltransferases for loading of acetyl and malonyl substrates and does not rely on a self-editing mechanism to facilitate translocation of the appropriate substrates to the condensation sites [12]. Nevertheless, since these complexes terminate acyl-chain growth by transfer of the acyl moiety to a CoA acceptor, free CoA is also required for efficient operation of the pathway. Both the animal $\alpha_2$ and yeast $\alpha_6\beta_6$ FASs exhibit the expected stoichiometry for acyl chain assembly when all substrates and cofactors are present (i.e. two fatty acyl chains per $\alpha_2$, six fatty acyl chains per $\alpha_6\beta_6$). However, neither the animal (S. Smith, 1985) nor the yeast [12] FAS can be fully saturated with substrate, when only malonyl-CoA or acetyl-CoA is present. This substoichiometric binding of substrates by the yeast FAS has been referred to by Schweizer and colleagues as a negative cooperativity that ensures that the complex is not overloaded with substrates that would otherwise compete for sites required for processing the various acyl-enzyme intermediates present on the enzyme [12].

## 3.5. Chain termination and product specificity

The eukaryotic FASs synthesize predominantly the 16-carbon saturated product with smaller amounts of 14- and 18-carbon products. The enzymatic basis of product specificity of the animal FAS has been studied in some detail in Gordon Hammes', and in our, laboratory. The $\beta$-ketoacyl synthase has a relatively broad chain-length specificity and is able to transfer efficiently saturated acyl moieties with 2 to 14 carbon atoms from the phosphopantetheine thiol to the active-site cysteine thiol (Fig. 4). However, longer chain-length acyl moieties are transferred between thiols with increasing difficulty

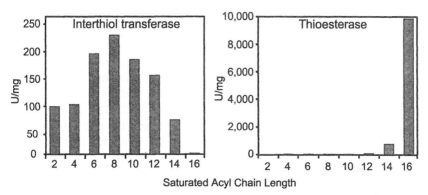

Fig. 4. Chain-length specificities of the interthiol transferase activity associated with the 3-ketoacyl synthase and the chain-terminating thioesterase. The interthiol transferase activity was assayed on a recombinant form of FAS in which functionality of the malonyl/acetyl transferase, ACP and thioesterase domains had been inactivated by mutation; acyl-CoAs were used as model acyl donors and pantetheine as a model acceptor [15]. Thioesterase activity was assayed on a recombinant form of the isolated thioesterase domain and acyl-CoAs were used as model substrates (S. Smith, 1978).

[11]. In contrast, the chain-terminating thioesterase has very limited ability to remove acyl moieties with less than 16 carbon atoms from the phosphopantetheine thiol (S. Smith, 1978, 1991). Thus the specificities of the chain-elongating and chain-terminating enzymes complement each other perfectly, ensuring that the 16-carbon fatty acid is the major product (Fig. 4). The β-ketoacyl synthase cannot transfer incompletely reduced intermediates (β-keto-, β-hydroxy or enoyl) from the phosphopantetheine thiol to the active-site cysteine, so only the saturated intermediates are elongated (S. Smith, 1997).

Some specialized animal tissues, such as the lactating mammary gland of mammals and the preen glands of birds are able to use the FAS to produce the shorter chain-length fatty acids characteristic of milk fat and the oily secretion used for waterproofing feathers, respectively. In the mammary glands of nonruminants (J. Knudsen, 1981; S. Smith, 1981) and the preen glands of birds (P.E. Kolattukudy, 1981), formation of these unusual products is attributable to the expression of a separate, discrete 29 kDa thioesterase that is able to access and release saturated acyl chain intermediates directly from the phosphopantetheine of the FAS complex. Ruminants, on the other hand, possess a FAS with unusual properties. In these species, the acyltransferase responsible for substrate loading on the FAS has a very broad acyl chain-length specificity so that saturated C4, C6, C8 and C10 intermediates can equilibrate between enzyme-bound and CoA ester forms (J. Knudsen, 1981). In the lactating mammary gland, this pool of short chain-length acyl-CoAs can be utilized for triglyceride synthesis and milk fat production. The FASs of some specialized sebaceous glands, such as the harderian, meibomian and preen glands, produce significant amounts of odd carbon-number and methyl-branched fatty acids, primarily as a result of the availability of significant amounts of propionyl-CoA and methylmalonyl-CoA as primer and chain-extender substrates (P.E. Kolattukudy, 1978, 1987).

*3.6. Interdomain communication*

The 20-Å-long, phosphopantetheine 'swinging arm' of the animal and fungal FASs has long been accepted as a key factor in allowing each of the six catalytic centers access to the ACP domains. However, distances between some of the catalytic sites of the animal FAS have been estimated, by fluorescence energy transfer, as being considerably greater than 20 Å; for example, 48 Å between the phosphopantetheine thiol and the thioesterase active site and 40 Å between the phosphopantetheine thiol and the two nucleotide-binding domains (P.E. Kolattukudy, 1985; G.G. Hammes, 1986). These observations suggest that cooperation between the acyl carrier domain and the catalytic domains during fatty acid biosynthesis may involve extensive conformational changes within the complex. Such conformational changes may be mediated in part through the presence of flexible linker regions between the catalytic domains.

Although the dimeric form of the animal FAS contains two sites for fatty acid synthesis, the monomeric form is incapable of transferring substrates from the CoA ester form to the phosphopantetheine thiol and cannot catalyze the condensation reaction. In 1991, the demonstration by J. Stoops and S.J. Wakil that the cysteine active-site thiol could be cross-linked to the phosphopantetheine thiol of the opposite subunit by dibromopropanone inspired the proposal that the two subunits of the animal

FAS are positioned in a fully extended antiparallel orientation, such that two centers for fatty acid synthesis are formed at the subunit interface. The arrangement of domains in the multifunctional polypeptide is consistent with a model in which each center for fatty acid synthesis requires cooperation of the three aminoterminal domains of one subunit (β-ketoacyl synthase, malonyl/acetyl transferase and dehydrase) with the four carboxyterminal domains of the other subunit (enoylreductase, β-ketoacylreductase, acyl carrier protein and thioesterase) [10,11,13]. The development, in our laboratory, of procedures for the production of recombinant animal FAS dimers containing different mutations on each subunit has afforded a unique opportunity to test this model (S. Smith, 1998). Several of our findings cannot be explained adequately by the prevailing model [16]. First, in vitro mutant complementation analyses have revealed that the β-ketoacyl synthase and malonyl/acetyltransferase can cooperate functionally with the acyl carrier protein domains of *both* subunits, although the intrasubunit functional interaction is less efficient than is the intersubunit interaction. Second, mutant complementation analysis also indicates that the β-hydroxy dehydration reaction is catalyzed exclusively by cooperation between the dehydrase and acyl carrier protein domains associated with the *same* subunit.

Third, the results of a thorough reinvestigation of the specificity of dibromopropanone interaction with the animal FAS are inconsistent with the original interpretation that cross-linking of active-site cysteine and phosphopantetheine thiols by this reagent occurs exclusively intersubunit. Thus, treatment of animal FAS dimers with dibromopropanone generates three new molecular species with decreased electrophoretic mobilities; none of these species is formed by fatty acid synthase mutant dimers lacking either the active-site cysteine of the β-ketoacyl synthase domain (Cys161Ala FAS mutant) or the phosphopantetheine thiol of the acyl carrier protein domain (Ser2151Ala FAS mutant). When dimers carrying one or both mutations on one or both subunits were treated with dibromopropanone and analyzed by a combination of sodium dodecyl sulfate/polyacrylamide gel electrophoresis, Western blotting, gel filtration and matrix-assisted laser desorption mass spectrometry, the two slowest moving of the three cross-linked species were identified as doubly and singly cross-linked dimers, respectively, whereas the fastest moving species was identified as originating from internally cross-linked subunits. This internally cross-linked species accounted for as much as 35% of the total cross-linked species [17].

On the other hand, certain features of the original model are supported by results of the mutant complementation studies. Thus, the substrate loading and condensation reactions are catalyzed most efficiently by cooperation of the β-ketoacyl synthase and malonyl/acetyltransferase domains with the ACP domain of the opposite subunit and the β-carbon processing and chain-terminating reactions are catalyzed by cooperation of the phosphopantetheine moiety with catalytic domains associated with the same subunit [16].

Perhaps the most important implication of these recent findings is that the structural organization of the animal FAS must permit head-to-tail interactions between domains located on opposite subunits and must allow for functional interactions between domains located distantly on the same subunit. These requirements could be met by a modified head-to-tail model in which the two ACP domains, at each of the subunit 'tails', have

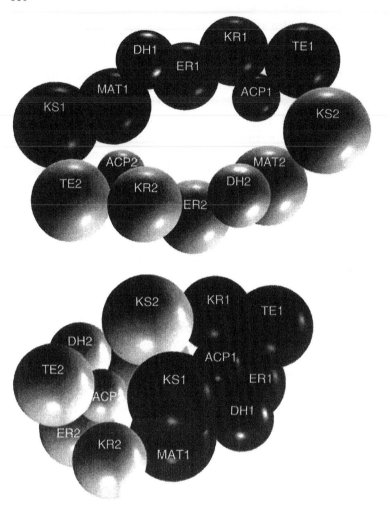

Fig. 5. Alternative functional models for the animal fatty acid synthase. (Top) The classical 'head-to-tail' model [10,11,13,16]. (Bottom) The alternative model [16]. The two subunits are distinguished by dark and light shading and by the numbers 1 or 2 on each domain. KS, β-ketoacyl synthase; MAT, malonyl/acetyl-transferase; DH, dehydrase, ER, enoyl reductase; KR, β-ketoacyl reductase; ACP, acyl carrier protein, TE, thioesterase. For simplicity, only those domains for which catalytic functions have been established are shown in the model; thus the non-catalytic core region has been omitted.

access to the β-ketoacyl synthase and malonyl/acetyltransferase domains at the 'heads' of either subunit (Fig. 5, top). Although such a mechanism would require a remarkable degree of flexibility within the FAS polypeptides, to ensure productive interactions both inter- and intrasubunit, it cannot be formally excluded at this time.

An alternative model has been envisaged that requires much less conformational flexibility in the subunits (Fig. 5, bottom). In this model, the β-ketoacyl synthase and malonyl/acetyltransferase domains of both subunits are located near the center of the dimer, where they can freely access the ACP domains of both subunits, and the domains responsible for the β-carbon processing reactions are located at opposite ends of the dimer and therefore have access to only one ACP domain. Thus the model retains the original concept of head-to-tail oriented subunits, but also allows for head-to-tail interactions between domains of the same subunit.

## 4. Short-term regulation of fatty acid synthesis

Fatty acids, esterified in triglycerides, represent the major form of stored energy in animals. A relatively small proportion of excess caloric intake in the form of carbohydrate is stored as glycogen, most is converted to fat via the de novo lipogenic pathway. Thus the flux of substrate through the pathway is high in animals that have ingested high amounts of carbohydrate and is low during periods of fasting. In most animals, the liver and adipose tissue are the major sites of fatty acid biosynthesis, although the pathway is also of vital importance in other tissues during certain stages of development. In animals having access to a regular daily supply of food, fatty acid synthesis in the liver varies dramatically during the diurnal cycle. In the laboratory rat fed ad libitum, the peak in hepatic lipogenesis occurs during the feeding period, in the dark phase of the daily cycle, and is paralleled by changes in the activity of ACC (N. Iritani, 1985, 1990). Untreated diabetic rats do not exhibit this diurnal rhythm in lipogenic activity and ACC activity remains low throughout each cycle; administration of insulin restores the normal diurnal rhythm. However, most investigations into the role of hormones and nutrients in the regulation of lipogenesis have not studied these diurnal changes, but instead have focused on models in which the animals are first subjected to the stress of a prolonged fast of one or two days, prior to reintroduction of a food supply.

### 4.1. Regulation of substrate supply for fatty acid synthesis

Following a period of fasting, the initial stimulus for activation of the lipogenic pathway appears to be the influx of dietary carbohydrate. Glucose stimulates secretion of insulin from pancreatic β-cells and, as the ratio of insulin to glucagon increases in the blood, the activities of several enzymes in the glycolytic and lipogenic pathways are elevated, while the activities of key gluconeogenic enzymes are decreased (Fig. 6). These early changes in activity are brought about primarily by changes in the catalytic efficiency of key enzymes, whereas the longer-term responses can also involve changes in enzyme concentration (Section 5.4).

The resulting increased flux of pyruvate into the mitochondria and the subsequent activation of pyruvate dehydrogenase causes a rapid increase in the production and export of citrate via the tricarboxylate anion carrier. The elevated concentration of cytosolic citrate can simultaneously provide acetyl-CoA, the carbon source for fatty acid synthesis, through the action of ATP : citrate lyase, and stimulate the conversion of acetyl-CoA to malonyl-CoA, by activation of ACC. The reducing equivalents required for fatty acid synthesis are provided largely through the action of malic enzyme and

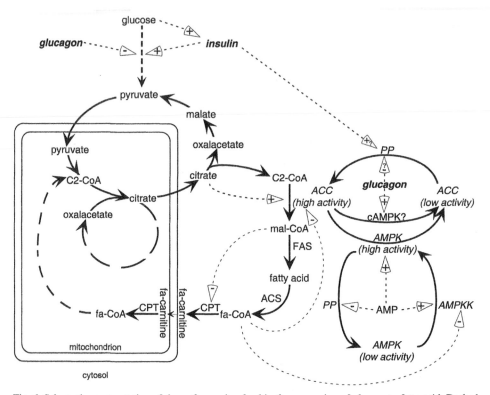

Fig. 6. Schematic representation of the pathways involved in the conversion of glucose to fatty acid. Dashed arrows with + or − signs, indicate points of positive or negative regulation. ACS, acyl-CoA synthetase; PP, protein phosphatase; AMPKK, AMPK kinase.

the two pentose phosphate cycle enzymes, glucose-6-phosphate dehydrogenase and 6-phosphogluconate dehydrogenase. The supply of NADPH by these enzymes appears to be largely dependent on its rate of utilization and is unlikely to be limiting in the lipogenic pathway.

The two pancreatic hormones, insulin and glucagon exert essentially opposite effects on energy metabolism. Thus, during the intervals between feeding and during prolonged periods of fasting, when the ratio of insulin to glucagon decreases in the blood, the activities of key enzymes in the glycolytic and lipogenic pathways are decreased, while the activities of key gluconeogenic enzymes are increased resulting in a reversal of the process described above.

## 4.2. Regulation of acetyl-CoA carboxylase α activity by reversible phosphorylation

ACC preparations purified from lipogenic tissues contain up to eight phosphate residues per subunit. Six of the phosphorylation sites are located within the first 100 aminoterminal residues, the other two near the center of the polypeptide (Fig. 2). Mild, limited proteolysis of ACC results in cleavage of the highly phosphorylated aminoterminal domain and full activation of the enzyme in the absence of citrate. Unless special precautions are observed to prevent activation of protein kinases during tissue disruption, ACC preparations typically are highly phosphorylated and require addition of supraphysiological concentrations of citrate (1–5 mM) for full activity, regardless of the nutritional status of the animal (N.B. Madsen, 1987; S.J. Wakil, 1988 and D.G. Hardie, 1992). However, when livers of fed animals are freeze-clamped prior to isolation of the enzyme, preparations are obtained that have low phosphate content and can be fully activated by physiological concentrations of citrate ($\sim$0.2 mM). On the other hand, the enzyme purified from freeze-clamped livers of fasted animals has a high phosphate content and requires high added citrate concentrations for full activity. A plethora of kinases has been implicated in the phosphorylation of the enzyme (Fig. 2) but, despite the flurry of activity that occurred in this field in the 1980s and early 1990s, there is still no clear consensus as to which of them is physiologically the most important in regulating ACC activity. Two kinases, one dependent on cAMP for activity (cAMPK) the other dependent on 5′-AMP (AMPK) can inactivate ACC by phosphorylation in vitro and have been considered the strongest candidates [6,7]. The cAMPK phosphorylates ACC at Ser77 and Ser1200 and causes primarily an increase in the $K_a$ for citrate and a slight reduction in $V_{max}$, whereas the AMPK phosphorylates mainly at Ser79, Ser1200 and, to a lesser extent, at Ser1215 and produces a large decrease in $V_{max}$. The AMPK has been considered as a particularly attractive candidate, since it is also able to phosphorylate and inactivate 3-hydroxy-3-methylglutaryl coenzyme-A reductase, and so potentially is in a position to coordinately regulate both the fatty acid and sterol biosynthetic pathways [6]. Treatment in vitro with cAMPK and AMPK of recombinant ACCs, mutated in one or more of the potential phosphorylation sites, indicated that when only Ser79 is available as a phosphorylation target, AMPK is able to inactivate ACC and when only Ser1200 is available, cAMPK, but not AMPK is able to inactivate ACC [7]. Thus, the critical target for cAMPK appears to be Ser1200 and for AMPK, Ser79. Strong arguments have been made supporting roles for both cAMPK and AMPK as regulators of ACC activity. During fasting, glucagon secretion raises the intracellular concentration of cAMP, which potentially can result in phosphorylation and inactivation of ACC by cAMPK, and indeed glucagon treatment of hepatocytes increases phosphorylation at Ser1200 (D.G. Hardie, 1988). Similarly, during fasting, as ATP production falls into deficit and AMP concentrations rise, AMPK could be activated, resulting in phosphorylation and inactivation of ACC. Indeed not only is AMPK activated directly by AMP, but so is an AMPK kinase that activates AMPK by phosphorylation, whereas a AMPK phosphatase that inactivates AMPK is inhibited by AMP (Fig. 6). Nevertheless, some puzzling observations remain unresolved. First, a report from Ki-Han Kim's laboratory in 1990 concluded that the $\alpha$-isoform of ACC containing eight additional residues near Ser1200 is not phosphorylated at this site by the cAMPK. However, the same report documented that it is precisely this putative cAMPK-resistant, longer isoform that dominates in liver, the major site of lipogenesis and a tissue that is responsive to glucagon; this finding casts serious doubt on the role of the cAMPK in regulation of ACC activity. Second, although the diurnal rhythm in the activation state of ACC correlates nicely with variations in the phosphate content at position Ser79, AMPK activity remains at the same level throughout the light and dark cycles [6]. Furthermore, no reduction in AMPK activity is observed when either hepatocytes or pancreatic β-cell lines are exposed to high levels of glucose. The possibility that it is in fact a phosphatase with specificity for Ser79 of ACC that is the regulated enzyme has yet to be resolved.

Collectively, these observations indicate that the catalytic activity of ACC is determined by a complex interplay between citrate-induced conformational changes and phosphorylation events. However, the question as to exactly how the covalent modification and conformational changes in ACC bring about changes in activity has not yet been addressed at the structural level.

## 4.3. Malonyl-CoA, fuel sensing and appetite control

Within the last several years there has grown an increasing awareness of the important role played by malonyl-CoA as a fuel sensing and signaling molecule [4]. Of 28 intermediates formed between acetyl-CoA and palmitic acid, malonyl-CoA is the only true free intermediate and, as a regulator of CPTI, is uniquely positioned to control both the rate of biosynthesis and degradation of fatty acids.

Malonyl-CoA formed in liver is utilized primarily as a substrate for fatty acid synthesis, but at the same time, as an inhibitor of carnitine palmitoyltransferase I, it controls entry of fatty acids into the mitochondria for oxidation (J.D. McGarry, 1980) (Fig. 6). Thus, following feeding, when excess dietary carbohydrate is being converted to fat and tissue malonyl-CoA levels are elevated, activity of carnitine palmitoyltransferase I is inhibited and flux of fatty acids into the mitochondria is prevented [4]. This control mechanism ensures that the fatty acid biosynthetic and oxidation pathways are not simultaneously activated in the same tissue, thus avoiding the establishment of a futile cycle. Conversely, during periods of fasting hepatic malonyl-CoA levels are lowered, flux of substrate through the fatty acid biosynthetic pathway is halted and the block in entry of fatty acids into the mitochondria is removed.

Recent studies have revealed a previously unsuspected correlation between intracellular concentration of malonyl-CoA and appetite control [5]. As anticipated, mice treated with cerulenin, an inhibitor of the β-ketoacyl synthase activity of the FAS (Chapter 3), exhibit decreased rates of hepatic lipogenesis and elevated intracellular malonyl-CoA. However, these animals also cease feeding, maintain normal activity, exhibit a dramatic loss of adipose tissue mass and suffer a major reduction in body weight. Treatment with 5-(tetradecyloxy)-2-furoic acid, an inhibitor of ACC, also results in decreased fatty acid synthesis but has no effect on feeding behavior. Indeed, administration of 5-(tetradecyloxy)-2-furoic acid restores normal feeding behavior in mice treated with a FAS inhibitor, indicating that it is the elevated intracellular concentration of malonyl-CoA that triggers loss of appetite. The attenuation effect is observed when the compounds are administered intracerebroventricularly, implicating the central nervous system as the site of action of the inhibitors. This conclusion is further supported by the observation that expression of the hypothalamic neuropeptide Y, an appetite stimulant, is dramatically reduced in mice treated with an inhibitor of FAS. Since both ACC and FAS are expressed in hypothalamic neurons, it appears likely that the malonyl-CoA concentration sensing mechanism may be directly 'hard-wired' to the appetite control center in the hypothalamus. Under normal conditions, the rise in tissue malonyl-CoA concentration following feeding may provide a 'satiety signal' indicating that excess caloric intake is now being channeled into fat stores, so that neuropeptide Y production is turned off and feeding ceases. Thus the condition induced by treatment with FAS inhibitors likely mimics the normal fed state so that the animal is deceived into a prolonged period of fasting for the duration of the treatment.

Perhaps it would appear paradoxical that in mice treated with FAS inhibitors, where food intake is dramatically reduced and intracellular malonyl-CoA levels rise in the lipogenic tissues, fatty acid oxidation provides the major source of energy and adipose mass is reduced. The answer to this puzzle likely is attributable to unique metabolic control mechanisms operative in muscle, the main site of fatty acid oxidation during physical activity [4]. The activity of the ACCβ-form and the concentration of malonyl-CoA in skeletal muscle decreases within seconds of the initiation of exercise. The mechanism by which activity of the β-form is regulated in muscle appears to involve both allosteric regulation by citrate and reversible phosphorylation, since activity of AMPK is elevated during muscle contraction. The mechanism of disposal of malonyl-CoA produced in muscle by ACCβ is not well understood. Most likely malonyl-CoA is utilized by a cytosolic malonyl-CoA decarboxylase, since genetic defects in this enzyme produce a phenotype characteristic of mitochondrial fatty acid oxidation disorders (S.J. Gould, 1999). Almost certainly, in view of the scarcity of FAS in muscle, malonyl-CoA is not utilized for fatty acid synthesis, so that administration of the FAS inhibitors is unlikely to impact significantly the concentration of malonyl-CoA in muscle. The regulation of malonyl-CoA levels by different mechanisms in muscle and lipogenic tissues allows for use of fatty acids as metabolic fuel in muscle, even when malonyl-CoA levels may be high in liver, as in the case of animals treated with FAS inhibitors.

## 5. Regulation of the intracellular concentration of lipogenic enzymes

### 5.1. Strategies and methodology

The intracellular concentrations of lipogenic enzymes potentially can be altered through changes in their rate of synthesis or degradation. Thus changes in the rates of either transcription or translation, as well as changes in the stability of specific mRNAs and their encoded proteins need to be examined when evaluating factors that may regulate changes in enzyme concentration. In reality, the decrease in concentration of ACC and FAS that occurs during relatively long-term fasting appears to be due to both a cessation in transcription of their genes and a decrease in the half-life of the proteins themselves. On refeeding of animals that have been subjected to long-term fasting, transcription of ACC and FAS is activated and the half-life of the proteins is lengthened until a new steady state concentration of enzymes is established. During differentiation of preadipocytes into mature adipocytes, in vitro, the increase in the intracellular concentration of lipogenic enzymes can be attributed to both stabilization of their mRNAs and increased rate of transcription of their genes. In recent years investigators have focused primarily on factors that influence the rate of gene transcription, so that relatively few details are known as to how changes in the stability of the mRNAs and proteins are brought about.

Cell culture systems are widely used to study the regulation of gene transcription, as these cells can usually be induced to take up and express heterologous reporter gene constructs and can be manipulated to mimic changes in the hormonal and nutritional environment that occur in intact animals. The most commonly used cell culture systems are primary hepatocytes and hepatoma cell lines, as models for studying the effects of hormones and nutrients on expression of lipogenic enzymes in the liver, and 3T3-L1 and 30A5 cells, as adipocyte models. These adipocyte lines are derived from fibroblasts which, at confluence, can be modulated to differentiate and accumulate fat droplets. Upon differentiation of these preadipocytes, lipogenic enzymes accumulate to levels similar to that characteristic of normal adipose tissue, so that these cell lines are commonly used to study gene transcription in both differentiating and mature adipocytes. However, the differentiating preadipocyte culture system does not mimic all of the characteristics of intact adipose tissue and the responsiveness of hepatoma cells to hormones and nutrients is typically muted, compared to that of hepatocytes or the liver of whole animals. Therefore, caution needs to be exercised in interpreting results obtained with these model systems, particularly when corroborating evidence from in vivo experiments is lacking. The use of transgenic animals affords a powerful model system for evaluating the roles of putative cis-regulatory and trans-acting proteins in mediating tissue-specific gene expression in response to nutritional status.

Regulation of gene transcription involves interaction of DNA regulatory sequences, termed cis-acting elements, that are usually found in the 5′-flanking regions of genes, with nuclear proteins, termed trans-acting factors. Understanding the mechanism of transcriptional regulation of a gene, therefore, requires the identification of both the cis-acting elements in the gene and the trans-acting factors in the nuclei that bind to them as well as elucidation of the signaling events that lead to changes in the interaction between the cis elements and trans factors.

## 5.2. The acetyl-CoA carboxylase promoter

The two major isoforms of ACC, α and β, which exhibit about 80% amino acid sequence similarity, are encoded by separate genes that map to chromosomes 17q21and 12q23.1, respectively. The α-isoform is expressed predominantly in tissues that exhibit high rates of fatty acid synthesis, such as liver and adipose tissue, whereas the β-isoform is expressed mainly in heart and skeletal muscle, and to a lesser extent in liver. Transcription of the α gene is under the control of two promoters, designated as PI and PII that are separated by 12.3 kbp. Five different species of mRNA, all of which contain the same base sequence in the coding region but differ in the 5′-untranslated region, are generated by the alternative splicing of the first four exons [18].

The usage of α gene promoters is tissue-specific (Fig. 7). Thus, only promoter PI is active in adipose tissue and promoter PII in mammary gland, whereas both promoters are active in liver. PII, which is active in all tissues at least at a low level, is sometimes described as a 'housekeeping promoter', although it contains a strong enhancer element and several regulatory cis-elements that confer responsiveness to external stimuli. A third ACCα promoter, PIII, has been described in sheep that gives rise to an N-terminal variant of ACCα that exhibits tissue-restricted mode of expression (M.C. Barber, 2001).

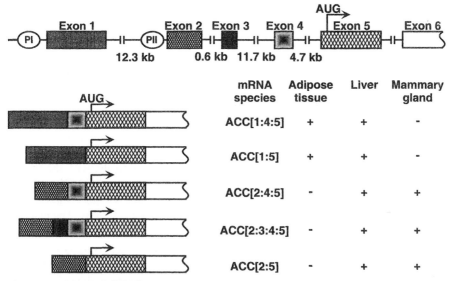

Fig. 7. Structure of the 5′-flanking region of the rat ACCα gene and tissue-specific distribution of the various mRNA species transcribed. The upper section of the figure shows a genomic map of the first six exons of the gene and includes the location and sizes of the intronic sequences. The circles labeled PI and PII indicate the positions of the two ACC gene promoters. The AUG codon, indicated by a right-angled arrow in exon 5, represents the translation initiation site. The lower section of the figure depicts the structural organization of the 5′-flanking region of the various mRNA transcripts. The mRNAs are named based on their exon content. The presence (+) or absence (−) of particular mRNA species in three different rat tissues is indicated by + and − signs, respectively. This figure is adapted from Kim [7].

The choice of gene promoter usage also seems to vary with animal species and developmental status. Thus, when animals are fasted and refed a high-carbohydrate, low-fat diet, transcripts from both PI and PII promoters are elevated in the liver of 12 days old chicks, whereas mainly the transcript from the PI promoter-driven transcript is induced in adult rats; similarly, changes in the thyroid hormone level affect both PI and PII promoter activity in chickens but only PI promoter activity in adult rats. The ACCβ gene is also transcribed by two promoters, PIβ and PIIβ, that are located immediately upstream of the first and second exons, respectively. However, the promoter regions of the α and β genes exhibit little sequence similarity as the ACCβ promoter region contains *cis*-acting elements that bind muscle-specific transcription factors (K.-S. Kim, 2001).

### 5.3. The fatty acid synthase promoter

FAS genomic clones containing the 5′-flanking DNA have been isolated and partially characterized from rat, goose, chicken and human. The FAS gene reportedly is transcribed from two promoters in humans [19], but from only one in other species. Promoter I, which contains recognizable TATA and CAAT boxes, controls transcription from an initiation site near the beginning of exon 1, whereas promoter II, which contains neither a TATA nor CAAT box, controls transcription from two sites, one in intron 1, the other 49 nt upstream of the ATG start codon in exon 2. Transient transfection experiments with FAS-reporter gene chimeras indicate that promoter I is about 15 times stronger than promoter II and that the presence of promoter II attenuates the transcriptional activity of promoter I. Based on these results, a model has been proposed (S.J. Wakil, 1996) in which intronic promoter II exerts a roadblock for RNA polymerase II that has initiated transcription from promoter I. Promoter II of the human FAS gene, which has not yet been well characterized, may play a role in low-level constitutive expression of FAS, whereas promoter I may be preferred for rapid transcription, under demand for high lipogenic rates.

The rat and goose genes exhibit 90 and 61% identity, respectively, with the human FAS gene in the promoter region containing the TATA box, an inverted CCAAT box and a functional Sp1-binding site. In common with the human FAS gene, the rat gene contains a long first intron that exerts a negative effect on promoter activity. The negative regulatory element has been mapped to a region between +405 to +1083 in the rat gene. Sequence elements within this region, which are capable of binding a variety of nuclear proteins, also confer a negative effect on transcription via a heterologous promoter, in an orientation-dependent manner. However, the results of run-on assays with nuclei derived from tissues that express FAS at either high or low levels indicate that the different rates of transcription can be accounted for by differences in the extent of initiation, rather than by operation of a transcriptional pausing mechanism, as proposed for the human FAS gene (S. Smith, 1997). The sequence of the first intronic region is poorly conserved between rat and human genes and the physiological significance, if any, of the use of an intronic promoter in the human, but not in the rat, FAS gene remains to be demonstrated.

Table 1 summarizes various cis-acting elements that have been identified in the FAS and ACC promoters and the trans-acting factors that bind to them.

## 5.4. Transcriptional regulation of lipogenic gene expression in response to dietary carbohydrate

The adaptive response of gene expression to nutrient availability is vital to the survival of all organisms. In animals, the response to availability of dietary carbohydrate is mediated both by effects of glucose metabolism itself and by secondary effects on hormone secretion. Thus, glucose entering the blood stream provides the initial stimulus for the synthesis and secretion of insulin, which in turn promotes the entry of glucose into cells and its subsequent metabolism. Insulin release at the onset of feeding also stimulates the conversion of thyroxine into the more potent hormone 3,5,3'-triiodothyronine (T3) and may increase the level of T3 receptors in the liver. Thus the feeding stimulus activates both insulin and T3-mediated signaling pathways, both of which have the effect of increasing lipogenesis [20].

The initial, rapid hepatic response to carbohydrate intake involves activation of key glycolytic and lipogenic enzyme activities (Fig. 6). The long-term response is accompanied by an increased enzyme production, primarily as a result of increased gene transcription. Conversely, glucagon production is elevated during fasting, resulting in down-regulation of glycolytic and lipogenic enzymes. The involvement of insulin,

Table 1

Cis-elements and trans-factors involved in the regulation of FAS and ACC genes

| Consensus cis-acting motif | Trans-acting factor | Effect on transcription | Location in rat FAS promoter | Location in ACCα promoter |
|---|---|---|---|---|
| TATA box (TTTAAAT) | TBP | Initiates transcription | −33/−26 | Not present |
| E-Box (CANNTG) | USF and other bHLH-LZ family members | Mediates the insulin response in FAS and ACCα PI promoter | −60/−65, −326/−332 | −931/−936 (Rat, PII); −109/−114 (Sheep, PI) |
| Inverted CAAT box (ATTGGCC) | NF-Y | Activates the basal transcription; required for conferring sterol response to FAS promoter; mediates cAMP response in FAS promoter | −98/−92, −516/−498 | Not identified |
| GC-Box (G/TG/AAGGCG/TG/AG/AG/T) | Sp1 family members | Activates transcription; participates in conferring sterol response to ACC and FAS promoters; mediates glucose response in ACCα PII promoter | −91/−83, −168/−160, −226/−218, −242/−284, −482/−474 and −557/−549 | −254/−242, −329/−317 (Rat, PII) |
| Sterol response element (TCACNCCAC) | SREBP family members | Mediates the sterol response; mediates the nutritional response | −73/−43, −150/−141 | −273/−256, −977/−967 (Rat, PII) |
| Thyroid hormone response element (A/GGGT/AC/AAN₄/GGGT/AC/AA) | TR/RXR and/or LXR/RXR | Mediates the thyroid hormone response | −120/−125, −669/−650 | −108/−82 (Chicken, PII) |
| Carbohydrate response element (CATGN₇CGTG) | ChoRF/USF | Mediates the carbohydrate response in FAS and ACCα PI promoter | −7214/−7190 | −126/−102 (Rat, PI) |

TBP, TATA box binding protein; USF and bHLH-LZ, upstream stimulatory factor and basic helix–loop–helix leucine zipper; NF-Y, nuclear factor Y; Sp1, stimulatory protein 1; SREBP, sterol-response element binding protein; TR/RXR, a heterodimer consisting of thyroid hormone receptor and retinoid receptor; LXR/RXR, a heterodimer consisting of liver receptor X and retinoid receptor; ChoRF, carbohydrate-response element binding factor. Consensus cis-motifs are given in 5′ to 3′ orientation and N represents any nucleotide. It should be noted that many variations of the consensus motif are functional in various genes.

glucagon and T3 in the regulation of lipogenic gene expression has been appreciated for some time, but only recently have carbohydrates themselves been implicated in this process. The identity of the intracellular signaling molecule produced by the metabolism of glucose has not yet been determined. One candidate is glucose-6-phosphate, since intracellular concentration of this metabolite varies in parallel with ACC and FAS mRNA concentrations in liver and adipose tissue, in response to dietary carbohydrate (P. Ferre, 1997). Another is xylulose-5-phosphate, an intermediate in the pentose phosphate cycle, since this metabolite can mimic the effect of hyperglycemia on glucose-regulated genes, without causing an elevation of glucose-6-phosphate [21].

### 5.4.1. The role of SREBPs

The effects of glucose, insulin and glucagon on transcription of lipogenic enzymes involves recruitment of a number of transcriptional activators, including SREBP, USF, NF-Y and Sp1. Pioneering work in the Goldstein and Brown laboratories established that SREBPs belong to the basic helix–loop–helix leucine zipper family of transcription factors and are involved in the regulation of the biosynthesis of both cholesterol and fatty acids [22] (Chapter 15). These DNA-binding proteins form dimers that can recognize both the direct repeat sterol regulatory element 5′-TCACNCCAC-3′ and the inverted repeat E-box, 5′-CANNTG-3′ [23].

The SREBPs are synthesized as ~1150 amino acid precursor proteins, bound to the endoplasmic reticulum and nuclear membrane, that undergo a sequential two-step proteolytic cleavage under low cellular sterol concentrations. This proteolytic cleavage results in the release of the aminoterminal section of SREBP which enters the nucleus and activates the transcription of genes involved in cholesterol and fatty acid synthesis, by binding to regulatory elements within their promoter regions. There are three major SREBP isoforms encoded by two different genes. The single SREBP-2 isoform produced by the SREBP-2 gene is believed to function primarily in the maintenance of cholesterol homeostasis [23]. The SREBP-1 gene can be transcribed from two different promoters, generating two isoforms, SREBP-1c and SREBP-1a, the latter having 29 additional amino acids at the aminoterminus. The additional aminoterminal residues of SREBP-1a carry a binding site for NF-Y (also known as CBP) and imparts a strong promoter enhancing effect to this isoform; SREBP-1c, on the other hand, is a relatively weaker transcriptional activator, unless additional sequences are present in the promoter region that can recruit co-activators such as NF-Y and Sp1. The relative levels of SREBP-1a and 1c mRNAs vary significantly in different tissues. In tissues active in lipogenesis, such as liver and adipose tissue, the SREBP-1c mRNA is more abundant, whereas in spleen and most cultured cell lines, the SREBP-1a mRNA predominates.

The results of various in vivo and in vitro studies indicate that SREBPs play a major role in the regulation of lipogenic gene expression. For example, transcription of SREBP-1c, ACC and FAS genes is shut off during fasting of normal animals and turned on following refeeding a high-carbohydrate diet. However, transgenic animals carrying a disrupted SREBP-1 gene fail to modulate transcription of the lipogenic genes in response to these changes in nutritional status (H. Shimano, 1999). Over-expression of SREBP-1a or 1c in cultured preadipocytes or in transgenic animal liver activates transcription of lipogenic genes [24]. Transgenic animals over-expressing SREBP-1c in

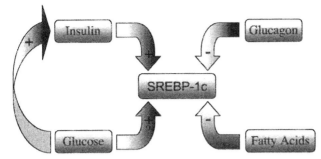

Fig. 8. Central role of SREBP-1c in the regulation of lipogenesis. Influx of glucose leads to increased secretion of insulin by pancreatic β-cells. Insulin and glucose induce lipogenic gene expression by increasing synthesis of SREBP-1c, whereas glucagon and fatty acids exert negative effect by decreasing SREBP-1c synthesis. Arrows carrying + and − symbol indicate positive and negative effect, respectively, on expression of SREBP-1c by corresponding external stimuli. The question mark indicates that direct activation of SREBP-1c expression by glucose has not been demonstrated definitively.

adipose cells develop insulin-resistant hyperglycemia, accumulate triglycerides in the liver and serum, and fail to down-regulate hepatic lipogenic gene expression in response to fasting. SREBP-1c also has been shown to be involved in the insulin-induced expression of the glucokinase gene [25,26]. The observation that insulin activates the expression of the SREBP-1c gene in primary hepatocytes whereas glucagon and cAMP have the opposite effect suggests that SREBP could mediate some of the effects of insulin on lipogenic gene expression in the liver, indirectly, by increasing glucose flux through glucokinase, and/or directly, by activation of lipogenic gene expression (Fig. 8).

### 5.4.2. The role of SREBP co-activators

As relatively weak transcriptional activators, SREBPs typically operate synergistically with other general transcriptional activators such as Sp1, NF-Y and cAMP-response element binding protein (T.F. Osborne, 2000). However, between different promoters, there are significant differences in the identity of these co-regulators and in the position of their binding sites relative to the position and number of SREBP-binding sites. These differences may be attributable, at least in part, to the differential recruitment of SREBP isoforms having different co-regulatory factor requirements. Sp1 may function synergistically with SREBP in mediating the glucose/insulin activation of both the FAS and ACCα genes. Increased binding of Sp1 to GC-rich regions between nucleotides −329/−317 and −254/−242 of the ACCα II promoter has been observed in 30A5 adipocytes exposed to glucose. Disruption of either Sp1-binding site by mutation eliminates glucose responsiveness of the ACCα II promoter. The increased binding of Sp1 has been attributed to a glucose-mediated induction of protein phosphatase I in the nucleus, which dephosphorylates Sp1, generating a form that binds more effectively to the promoter (K.-H. Kim, 1996).

NF-Y, a heterotrimeric transcription factor that binds to the inverted CCAAT-box motif of the FAS promoter, also appears to play a role as an activator of basal transcription of this gene. The cAMP-mediated down-regulation of FAS transcription that occurs on fasting appears to involve the interaction of NF-Y with the inverted CCAAT-box, although the mechanistic details have not yet been worked out [27].

### 5.4.3. The role of USFs

The ubiquitous DNA-binding proteins USF1 and USF2 were initially proposed as the key mediators of the glucose/insulin-response in lipogenic genes. USF1 and USF2, members of a basic helix–loop–helix leucine zipper family of transcription factors, bind to the E-box sequence motif, CANNTG. Both transient transfection experiments in cell culture and the use of transgenic animal models have implicated two USF-binding sites in the glucose/insulin responsiveness of the FAS gene, one at −65, the other at −332 [28]. The USF-binding site at position −65 in the rat FAS promoter is overlapped by an SREBP-binding site. Although this region of the gene can bind USF and SREBP independently, studies with transgenic mice suggest that it is the binding of USF at this site that is more important in mediating the fasting/refeeding response. Thus, it appears that it is SREBP binding at the second site, −150 to −141, that is essential in modulating transcription of the FAS gene. The nucleotide sequence in this region is a motif that is highly conserved in rat, human and goose FAS promoters. An E-box located at −114 of the ACCα I promoter that binds USF1 and 2, but not SREBP-1, has also been implicated in the insulin responsiveness of the ovine gene [29]. As yet, it is unclear as to whether, or how, SREBPs and USFs interact in modulating transcription of the FAS and ACC genes in response to fasting and refeeding. Quite possibly, the USF proteins act synergistically with SREBP, as do other co-activators such as Sp1 and NF-Y.

### 5.4.4. The role of carbohydrate-response elements

Additional sequence elements have been identified within the ACC and FAS genes that play a role in modulation of transcription in response to glucose, independent of insulin. These sequences, termed glucose-response elements, or carbohydrate-response elements, consist of two E-box half-sites related to the sequence 5′-CACG, in either direct or inverted orientation and separated by 7 or 9 nucleotides, and are similar to those identified as carbohydrate-response elements in the pyruvate kinase and S14 genes (H.C. Towle, 2000). The carbohydrate-response elements, which are located at −126 to −102 in the rat ACCα I promoter, and at −7240 to −7190 in the rat FAS gene, bind both USF and an additional, as yet unidentified *trans*-acting factor [30,31]. This factor, termed a carbohydrate-response factor, presumably is linked to the metabolism of glucose by an as yet unidentified signaling pathway.

### 5.4.5. Signaling pathways

Although it is well established that genes involved in glucose catabolism and lipid biosynthesis are up-regulated by high glucose and insulin concentrations in both liver and adipose tissues, the signaling pathway from glucose/insulin to the transcriptional machinery is not clearly understood. The binding of insulin to its receptor activates the associated tyrosine kinase and results in autophosphorylation of the receptor. The signal is then propagated by two main routes: the insulin receptor substrates/phosphatidylinositol 3-kinase pathway and the Ras/mitogen-activated protein kinase pathway. It is the phosphatidylinositol 3-kinase signaling pathway that has been implicated in the regulation of FAS transcription by insulin, although the identity of the participants remains to be established. The AMPK which is activated under various stress conditions due to a rise in the cellular AMP to ATP ratio, plays a significant role in the regulation of glucose-activated genes including FAS and ACC. Glucose-mediated transcriptional activation of FAS and ACC genes is inhibited by activated AMPK possibly due to inhibition of a yet to be identified activator or activation of an unidentified repressor. However, AMPK plays no role in the induction of glucose-activated gene expression since basal AMPK activity is not reduced by high levels of glucose in the cell [32]. A similar situation may be operative in the regulation of ACC activity by reversible phosphorylation. Here too, clear evidence implicates AMPK in the down-regulation (phosphorylation), but not in the up-regulation (dephosphorylation) of ACC activity.

*5.4.6. The role of thyroid hormone*
Thyroid hormone plays a significant role in regulation of expression of lipogenic enzymes in response to changes in nutritional status. T3 stimulates transcription of ACC and FAS in a tissue-specific manner, exerting its strongest effect in liver. In chick embryo hepatocytes, the rate of lipogenesis and the levels of lipogenic enzymes are low until the newly hatched chicks are fed and transcription of the lipogenic enzymes is activated. The induction of lipogenic enzymes can be simulated in vitro by treatment of chick embryo hepatocytes with T3. The mechanism of thyroid hormone-induced gene expression involves activation of the thyroid hormone receptor to form a heterodimer with the retinoid X receptor (RXR) or liver X receptor (LXR), which in turn binds to the consensus thyroid hormone-response element (Table 1) and activates the transcription. Treatment of chick embryo hepatocytes with T3 results in the stimulation of the promoter II-driven transcript of the ACCα gene, whereas in Hep G2 cells, T3 increases transcription from the human FAS promoter I, but has no effect on promoter II (S.J. Wakil, 1998; F.B. Hillgartner, 2001). The T3 response is mediated by the consensus thyroid hormone-response element half-site sequence, arranged as direct repeats on the noncoding strand of ACCα promoter II and the coding strand of FAS promoter I. The FAS promoter I also harbors an additional thyroid hormone-response element motif having only one half-site of consensus sequence. The amount of T3-receptor/RXR heterodimers that binds to these thyroid hormone-response elements is increased in the presence of T3. As yet, the details of the mechanism by which T3 regulates ACCα and FAS transcription remain unclear. F.B. Hillgartner and colleagues (2001) have proposed that T3 regulates ACCα transcription by a novel mechanism involving changes in the composition of nuclear receptor complexes bound to thyroid hormone-response element. Heterodimeric complexes containing LXR/RXR ensure basal level of ACCα expression whereas the complexes containing LXR/RXR and T3-receptor/RXR mediate T3-induced ACCα expression. Insulin also contributes to the T3-mediated activation of lipogenic gene expression by facilitating the conversion of thyroxin to T3 and increasing the level of nuclear T3 receptor.

*5.4.7. Down-regulation during fasting*
The down-regulation of transcription of the ACC and FAS genes that accompanies fasting is most likely initiated by a change in the relative amounts of glucagon and

insulin produced by the pancreas that results in decreased transcription of SREBP-1, phosphorylation of Sp1, and withdrawal of the carbohydrate-response element-mediated stimulatory effect and down-regulation of T3 production. The elevation of intracellular cAMP concentration that results from increased glucagon production also plays a role in the down-regulation process. However, the *cis*-acting elements and *trans*-acting factors involved appear to be different for ACC and FAS. NF-Y binding at an inverted CCAAT-box has been implicated in the case of the FAS promoter and AP-2 binding at variants of well-characterized cAMP-response element in the ACCα PII promoter. Phosphorylation of AP-2 by cAMPK facilitates the binding of AP-2 to the cAMP-response element [33]. Further investigation is required to determine the precise mechanism whereby cAMP exerts its effects on these *trans*-acting factors. The *cis*-elements and the *trans*-factors that confer cAMP responsiveness to the ACCα PI promoter still remain to be identified.

*5.4.8. Summary*

In summary, a tentative model for the transcriptional regulation of the ACC and FAS genes, in response to fasting and refeeding a carbohydrate-rich diet, can be envisaged in which the carbohydrate-response factor, together with SREBP, play central roles and Sp1, NF-Y and USF function as co-activators of SREBP. However, experimental validation of this model is incomplete in that it remains unclear as to exactly how functionality of the putative co-activators NF-Y and USF is modulated. For example, binding of NF-Y to the inverted CCAAT-box of the FAS gene is not altered either in cultured hepatoma cells treated with cAMP, or in the livers of fasted animals, although FAS transcription is down-regulated in both situations. Similarly, binding of USF to the E-box at $-68/-52$ of the rat FAS gene is unaltered in the livers of refed animals as well as in cultured hepatoma or adipose cells treated with insulin/glucose, although this interaction appears essential for mediating the fasting/refeeding response. Possibly, covalent modification of these co-activators of SREBP may alter their functionality within the transcriptional initiation complex, but experimental evidence is lacking at present.

*5.5. Transcriptional regulation of lipogenic gene expression in response to dietary polyunsaturated fatty acids*

Dietary fish oils containing n-3 fatty acids, such as eicosapentaenoic acid and docosa-hexaenoic acid, decrease blood triglyceride concentrations in hypertriglycemic patients and are considered to have protective effects against cardiovascular diseases. Moreover, polyunsaturated fatty acids, when fed to rodents inhibit hepatic de novo lipogenesis, triglyceride secretion, and increase activities of enzymes involved in fatty acid oxidation. These effects on lipid metabolism are due to activation or suppression of transcription of key metabolic enzymes. Thus, polyunsaturated fatty acids inhibit the transcription of the FAS and ACC genes and activate the transcription of genes encoding acyl-CoA oxidase and cytochrome *P*450 4A2, enzymes that are involved in peroxisomal and mitochondrial fatty acid oxidation, respectively (S.D. Clarke, 1994). The transcriptional activation of genes encoding fatty acid oxidation enzymes is mediated by a nuclear peroxisome proliferator-activated receptor (Chapter 5). However, these receptors are not involved in transcriptional regulation of the genes encoding lipogenic enzymes. The mechanism of down-regulation of lipogenic enzyme expression appears to be mediated, at least in part, through SREBP-1, since feeding of dietary fish oil to mice results in a decrease of the hepatic mRNA for SREBP-1, but not in that for SREBP-2 [34,35].

### 5.6. Transcriptional regulation during development

Fatty acids are important components of membrane phospholipids and are essential for phosphatidylcholine biosynthesis in proliferating cells. Nevertheless, transcription of ACC and FAS genes is not regulated by cell-cycling, but is regulated by certain growth factors such as colony stimulating factor 1 (S. Jackowski, 2000). The transcription of ACC and FAS genes is also regulated during the differentiation and development of several tissues. For example, in mammary glands, expression of lipogenic enzymes is increased during late pregnancy and early lactation, in preparation for the production of milk fat. In brain, FAS expression is elevated during the myelination phase of development, but is low in mature animals (P.R. Vagelos, 1972, 1973). In the lung, FAS activity increases during late gestation, in preparation for the synthesis of the palmitoyl component of lung surfactant that is required for air breathing (L.W. Gonzales, 1999). Our knowledge of the tissue-specific mechanisms regulating most of these processes is, as yet, scant. Perhaps the most extensively studied differentiation model is that of the preadipocyte, in which the maturation process is characterized by the coordinate induction of the lipogenic enzymes and the accumulation of intracellular triglycerides (Chapter 10). Paradoxically, cAMP, which is implicated in the down-regulation of lipogenic enzyme gene expression during fasting, is required as a positive effector in the transcriptional activation of lipogenic enzymes in both the developing lung and the differentiating preadipocyte. In culture, preadipocytes must first be sensitized by exposure to cAMP before they will respond to insulin and turn on the expression of lipogenic enzymes (K.-H. Kim, 1991). In the case of ACC, this effect appears to be mediated primarily through the PII promoter. However, it still remains to be established whether this synergistic interaction between insulin and cAMP occurs in vivo or is specific to differentiation of 30A5 cells in culture.

## 6. Future directions

A major obstacle in understanding how the enzymes of the multifunctional polypeptide system cooperate to effect the biosynthesis of fatty acids is the absence of a detailed three-dimensional structure for either ACC or FAS. Crystal structures of many of the individual type II FAS enzymes have been obtained in recent years and have provided valuable information about the various catalytic mechanisms that likely operate in their type I counterparts. However, the most interesting aspects of the type I FASs clearly concern the overall architecture of the complex. The intriguing question as to how an ACP domain of the animal FAS is able to make functional contacts with eight catalytic domains, six on the same subunit and two on the companion subunit, can only be answered by generating high-quality images of the entire 540 kDa dimer, either by X-ray crystallography or high-resolution electron microscopy. Similarly, a detailed structural analysis of the animal ACC is required in order to understand how reversible phosphorylation and the conformational changes induced by citrate modulate the activity of this enzyme.

Substantial progress has been made during the last 5 years in identifying the various *trans*-acting factors that regulate transcription of the ACC and FAS genes in response to nutritional stimuli and we are likely to see many of the remaining details of these processes elaborated in the near future.

Arguably, the most exciting recent discovery in the field of fatty acid biosynthesis is that both ACC and fatty acid synthase may constitute legitimate targets for the development of therapeutic agents for the treatment of obesity and cancer. The potential usefulness of inhibitors of the two enzyme complexes rests largely on their ability to modulate intracellular malonyl-CoA concentrations. Cerulenin (2,3-epoxy, 4-oxo, 7,10-dienoyl, dodecylamide), the fatty acid synthase inhibitor that, when administered to mice, causes loss of appetite and reduction in adipose mass, is too unstable for therapeutic use, but a chemically stable inhibitor has been synthesized (C75; 3-carboxy-4-octyl-2-methylenebutyrolactone) that is equally effective in suppressing appetite and inducing weight loss [36]. Both cerulenin and C75 are also potent inhibitors of DNA replication in many types of tumors that characteristically overexpress fatty acid synthase. The toxicity of these compounds appears to be due to their ability to inhibit fatty acid synthesis and cause an elevation in intracellular malonyl-CoA concentration that ultimately results in apoptosis of the tumor cell. Tumor cells exposed to these inhibitors are unable to synthesize or oxidize fatty acids, since the elevated malonyl-CoA concentration inhibits CPT1 and blocks entry of fatty acids into the mitochondria. The effects of cerulenin can be mimicked by the combined administration of 5-(tetradecyloxy)-2-furoic acid and etomoxir, inhibitors of ACC and carnitine palmitoyltransferase I, respectively, supporting the theory that it is the inhibition of both fatty acid synthesis and fatty acid oxidation that triggers the apoptotic process (F.P. Kuhajda, 2001). The exact mechanism by which apoptosis is initiated by the disruption in fatty acid metabolism is not yet completely understood. One possibility is that fatty acids are diverted away from mitochondrial oxidation into ceramide synthesis, a situation that is known to induce apoptosis. Alternatively, the process could involve interaction of carnitine palmitoyltransferase I with Bcl-2, a mitochondrial membrane protein that regulates programmed cell death.

In another ground-breaking study, ACCβ-isoform knockout mice were generated and found to have a higher-than-normal rate of fatty acid oxidation and lower amounts of adipose fat (S.J. Wakil, 2001). The absence of malonyl-CoA in muscle tissue appears to free the carnitine palmitoyltransferase from its normal down-regulation control mechanism and lead to efficient oxidation of fatty acids that have been mobilized from adipose tissue. Most surprisingly, the knockout mice accumulate 50% less adipose fat, despite consuming 20% more food than wild-type mice!

Although further study is needed to carefully evaluate the long-term consequences of administration of inhibitors of ACC and fatty acid synthase, undoubtedly these findings will stimulate searches for new inhibitors that could offer novel therapies for the treatment of both obesity, its associated disorders, and of cancer.

## Abbreviations

| | |
|---|---|
| ACC | acetyl-CoA carboxylase |
| FAS | fatty acid synthase |
| ACP | acyl carrier protein |
| cAMPK | cAMP-dependent protein kinase |
| AMPK | AMP-dependent protein kinase |
| T3 | $3,5,3'$-triiodothyronine |
| SREBP | sterol-response element binding protein |
| USF | upstream stimulatory factor |
| NF-Y | nuclear factor Y |
| Sp1 | stimulatory protein 1 |
| RXR | retinoid X receptor |
| LXR | liver X receptor |

## References

1.  Wakil, S.J., Stoops, J.K. and Joshi, V.C. (1983) Fatty acid synthesis and its regulation. Annu. Rev. Biochem. 52, 537–579.
2.  Iritani, N. (1992) Nutritional and hormonal regulation of lipogenic-enzyme gene expression in rat liver. Eur. J. Biochem. 205, 433–442.
3.  Numa, S. and Tanabe, T. (1984) Acetyl-CoA carboxylase and its regulation. In: S. Numa (Ed.), Fatty Acid Metabolism and its Regulation. Elsevier, Amsterdam, pp. 1–27.
4.  Ruderman, N.B., Saha, A.K., Vavvas, D. and Witters, L.A. (1999) Malonyl-CoA, fuel sensing, and insulin resistance. Am. J. Physiol. 276, E1–E18.
5.  Loftus, T.M., Jaworsky, D.E., Frehywot, G.L., Townsend, C.A., Ronnett, G.V., Lane, M.D. and Kuhajda, F.P. (2000) Reduced food intake and body weight in mice treated with fatty acid synthase inhibitors. Science 288, 2379–2381.
6.  Corton, J.M. and Hardie, D.G. (1996) Regulation of lipid biosynthesis by the AMP-activated protein kinase and its role in the hepatocellular response to stress. Prog. Liver Dis. 14, 69–99.
7.  Kim, K.H. (1997) Regulation of mammalian acetyl-coenzyme A carboxylase. Annu. Rev. Nutr. 17, 77–99.
8.  Lane, M.D., Moss, J. and Polakis, S.E. (1974) Acetyl coenzyme A carboxylase. Curr. Top. Cell. Regul. 8, 139–195.
9.  Kong, I.-S., Lopez-Casillas, F. and Kim, K.-H. (1990) Acetyl-CoA carboxylase mRNA species with or without inhibitory coding sequence for Ser-1200 phosphorylation. J. Biol. Chem. 265, 13695–13701.
10. Wakil, S.J. (1989) Fatty acid synthase, a proficient multifunctional enzyme. Biochemistry 28, 4523–4530.
11. Chang, S.I. and Hammes, G.G. (1990) Structure and mechanism of action of a multifunctional enzyme complex: fatty acid synthase. Accid. Chem. Res. 23, 363–369.
12. Schuster, H., Rautenstrauss, B., Mittag, M., Stratmann, D. and Schweizer, E. (1995) Substrate and product binding sites of yeast fatty acid synthase: stoichiometry and binding kinetics of wild-type and in vitro mutated enzymes. Eur. J. Biochem. 228, 417–424.
13. Smith, S. (1994) The animal fatty acid synthase: one gene, one polypeptide, seven enzymes. FASEB J. 8, 1248–1259.
14. Fichtlscherer, F., Wellein, C., Mittag, M. and Schweizer, E. (2000) A novel function of yeast fatty acid synthase. Subunit alpha is capable of self-pantetheinylation. Eur. J. Biochem. 267, 2666–2671.
15. Witkowski, A., Joshi, K.A. and Smith, S. (1997) Characterization of the interthiol acyltransferase reaction catalyzed by the β-ketoacyl synthase domain of the animal fatty acid synthase. Biochemistry 36, 16338–16344.

16. Rangan, V.S., Joshi, A.K. and Smith, S. (2001) Mapping the functional topology of the animal fatty acid synthase by mutant complementation in vitro. Biochemistry 40, 10792–10799.

17. Witkowski, A., Joshi, A.K., Rangan, V.S., Falick, A.M., Witkowska, H.E. and Smith, S. (1999) Dibromopropanone cross-linking of the phosphopantetheine and active-site cysteine thiols of the animal fatty acid synthase can occur both inter- and intra-subunit: Re-evaluation of the side-by-side, antiparallel subunit model. J. Biol. Chem. 274, 11557–11563.

18. Luo, X., Park, K., Lopez-Casillas, F. and Kim, K.-H. (1989) Structural features of the acetyl-CoA carboxylase gene: Mechanisms for the generation of mRNAs with 5' end heterogeneity. Proc. Natl. Acad. Sci. USA 86, 4042–4046.

19. Hsu, M.H., Chirala, S.S. and Wakil, S.J. (1996) Human fatty acid synthase gene: evidence for the presence of two promoters and their functional interaction. J. Biol. Chem. 271, 13584–13592.

20. Hillgartner, F.B., Salati, L.M. and Goodridge, A.G. (1995) Physiological and molecular mechanisms involved in nutritional regulation of fatty acid synthesis. Physiol. Rev. 75, 47–76.

21. Massillon, D., Chen, W., Barzilai, N., Prus-Wertheimer, D., Hawkins, M., Liu, R., Taub, R. and Rossetti, L. (1998) Carbon flux via the pentose phosphate pathway regulates the hepatic expression of the glucose-6-phosphatase and phosphoenolpyruvate carboxykinase genes in conscious rats. J. Biol. Chem. 273, 228–234.

22. Brown, M.S. and Goldstein, J.L. (1997) The SREBP pathway: regulation of cholesterol metabolism by proteolysis of a membrane-bound transcription factor. Cell 89, 331–340.

23. Osborne, T.F. (2000) Sterol regulatory element-binding proteins (SREBPs): key regulators of nutritional homeostasis and insulin action. J. Biol. Chem. 275, 32379–32382.

24. Horton, J.D., Bashmakov, Y., Shimomura, I. and Shimano, H. (1998) Regulation of sterol regulatory element binding proteins in livers of fasted and refed mice. Proc. Natl. Acad. Sci. USA 95, 5987–5992.

25. Foretz, M., Guichard, C., Ferre, P. and Foufelle, F. (1999) Sterol regulatory element binding protein-1c is a major mediator of insulin action on the hepatic expression of glucokinase and lipogenesis-related genes. Proc. Natl. Acad. Sci. USA 96, 12737–12742.

26. Foretz, M., Pacot, C., Dugail, I., Lemarchand, P., Guichard, C., Le Liepvre, X., Berthelier-Lubrano, C., Spiegelman, B., Kim, J.B., Ferre, P. and Foufelle, F. (1999) ADD1/SREBP-1c is required in the activation of hepatic lipogenic gene expression by glucose. Mol. Cell. Biol. 19, 3760–3768.

27. Rangan, V.S., Oskouian, B. and Smith, S. (1996) Identification of an inverted CCAAT box motif in the fatty acid synthase gene as an essential element for mediation of transcriptional regulation by cAMP. J. Biol. Chem. 271, 2307–2312.

28. Sul, H.S., Latasa, M.J., Moon, Y. and Kim, K.H. (2000) Regulation of the fatty acid synthase promoter by insulin. J. Nutr. 130, 315S–320S.

29. Travers, M.T., Vallance, A.J., Gourlay, H.T., Gill, C.A., Klein, I., Bottema, C.B. and Barber, M.C. (2001) Promoter I of the ovine acetyl-CoA carboxylase-alpha gene: an E-box motif at −114 in the proximal promoter binds upstream stimulatory factor (USF)-1 and USF-2 and acts as an insulin-response sequence in differentiating adipocytes. Biochem. J. 359, 273–284.

30. O'Callaghan, B.L., Koo, S.H., Wu, Y., Freake, H.C. and Towle, H.C. (2001) Glucose regulation of the acetyl-CoA carboxylase promoter PI in rat hepatocytes. J. Biol. Chem. 276, 16033–16039.

31. Rufo, C., Teran-Garcia, M., Nakamura, M.T., Koo, S.H., Towle, H.C. and Clarke, S.D. (2001) Involvement of a unique carbohydrate-responsive factor in the glucose regulation of rat liver fatty-acid synthase gene transcription. J. Biol. Chem. 276, 21969–21975.

32. Woods, A., Azzout-Marniche, D., Foretz, M., Stein, S.C., Lemarchand, P., Ferre, P., Foufelle, F. and Carling, D. (2000) Characterization of the role of AMP-activated protein kinase in the regulation of glucose-activated gene expression using constitutively active and dominant negative forms of the kinase. Mol. Cell. Biol. 20, 6704–6711.

33. Park, K. and Kim, K.-H. (1993) The site of cAMP action in the insulin induction of gene expression of acetyl-CoA carboxylase is AP-2. J. Biol. Chem. 268, 17811–17819.

34. Kim, H.-J., Takahashi, M. and Ezaki, O. (1999) Fish oil feeding decreases mature sterol regulatory element-binding protein 1 (SREBP-1) by down-regulation of SREBP-1c mRNA in mouse liver. J. Biol. Chem. 274, 25892–25898.

35. Worgall, T.S., Sturley, S.L., Seo, T., Osborne, T.F. and Deckelbaum, R.J. (1998) Polyunsaturated fatty

acids decrease expression of promoters with sterol regulatory elements by decreasing levels of mature sterol regulatory element-binding protein. J. Biol. Chem. 273, 25537–25540.

36. Kuhajda, F.P., Pizer, E.S., Li, J.N., Mani, N.S., Frehywot, G.L. and Townsend, C.A. (2000) Synthesis and antitumor activity of an inhibitor of fatty acid synthase. Proc. Natl. Acad. Sci. USA 97, 3450–3454.

D.E. Vance and J.E. Vance (Eds.) *Biochemistry of Lipids, Lipoproteins and Membranes (4th Edn.)*

# CHAPTER 7

# Fatty acid desaturation and chain elongation in eukaryotes

Harold W. Cook and Christopher R. McMaster

*Atlantic Research Centre, Departments of Pediatrics and Biochemistry and Molecular Biology, Dalhousie University, Halifax, NS B3H 4H7, Canada, Tel.: +1 (902) 494-7066; Fax: +1 (902) 494-1394; E-mail: h.cook@dal.ca; cmcmaste@is.dal.ca*

## 1. Introduction

Although the contribution of lipid molecules to the hydrophobic character of membranes was recognized late in the nineteenth century, the nutritional importance of specific lipid molecules was first revealed through the pioneering work of Burr and Burr in 1929. They fed rats a fat-free diet and observed that retarded growth, scaly skin, tail necrosis and eventual death were reversed by feeding specific fats. Linoleic acid was recognized as the active agent and the term 'essential fatty acid' was coined.

During the 1950s, the advent of chromatographic techniques (gas–liquid and thin-layer chromatography) and greater availability of appropriate substrates and precursors labelled with radioisotopes contributed in a major way to current understanding of fatty acid metabolism. In the 1960s, studies focused on in vitro assays of specific enzymatic steps [1]. $\Delta 9$ desaturase was measured in yeast, rat liver microsomes and plants. It was determined that polyunsaturated fatty acid (PUFA) formation in animal tissues involved $\Delta 6$ and $\Delta 5$ desaturation. Chain elongation of long chain fatty acids was found in the endoplasmic reticulum (ER) and mitochondria [2,3]. Relationships between essential fatty acids and prostaglandins (Chapter 13) were elucidated. Subsequent work led to an understanding of primary, alternate and competitive pathways of desaturation and chain elongation and of factors that regulate fatty acid metabolism.

The 1990s and 21st century have seen the advent of molecular and genetic systems that support in-depth analysis of the regulation of fatty acid elongation and desaturation pathways [4–9]. More complete understanding of the roles of particular unsaturated fatty acids in the regulation of cell biology and human health and disease continues to evolve.

### 1.1. Physical consequences of fatty acyl chain desaturation and elongation

Among the forces holding lipid molecules within membranes, London–Van der Waals interactions have a major impact along the fatty acyl chains. These relatively weak forces are additive, proportional to the number of overlapping methylene groups, and inversely proportional to the distance between them. Thus, for long fatty acyl chains, total London–Van der Waals bonding strength is greater than electrostatic and hydrogen bonding of polar head groups. Accordingly, length of a fatty acyl chain is a crucial

| FATTY ACID | ABBREVIATION | MELTING POINT | SPATIAL WIDTH(nm) | POSSIBLE CONFIGURATION |
|---|---|---|---|---|
| Stearic Acid | 18:0 | 70° | 0.25 |  |
| Oleic Acid | c-18:1n-9 | 16° | 0.72 | 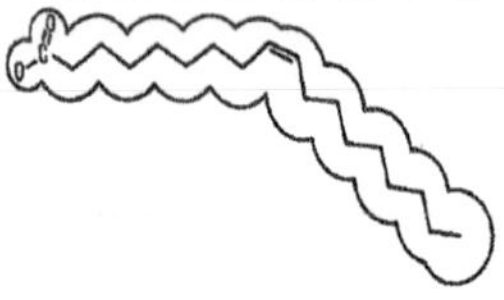 |
| Elaidic Acid | t-18:1n-9 | 43° | 0.31 |  |
| Linoleic Acid | c,c-18:2n-6 | -5° | 1.13 | 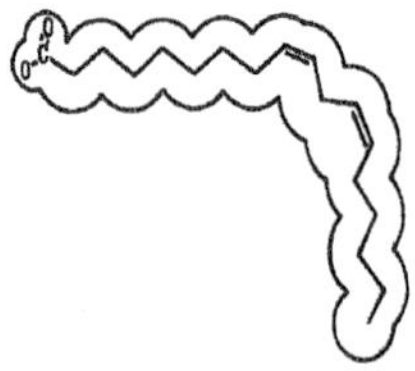 |

Fig. 1. Physical characteristics of fatty acids. See Table 1 for nomenclature of the fatty acids. Single bonds have a length of 0.154 nm and an angle of approximately 111°; double bonds have a length of 0.133 nm and an angle of approximately 123°. *c*, *cis*; *t*, *trans*.

parameter in its contribution to membrane structure and stability. Restricted solubility in an aqueous environment is governed by the tendency of acyl chains to remain associated with one another (oil does not dissolve well in water). Similarly, response to temperature is modulated by the extent to which thermal influences dissociate acyl chains. Thus, as acyl chain length increases, solubility decreases and melting point increases. Where more loosely packed membrane structures are advantageous, rigidity of lengthy saturated acyl chains can be countered by double bonds. A double bond of *cis* geometric configuration results in a 30° angle change from the linearity of the saturated chain (Fig. 1). Double bonds also are non-rotating, restrict chain movement and increases polarity through charge concentration.

Double bonds cause bends in fatty acyl chains within biological membranes (potentially, but probably not in a membrane milieu, an acyl chain with six double bonds could assume nearly circular shape) and decrease rigidity. Accordingly, within membranes, acyl chain length and the number and position of double bonds markedly influence fluidity, permeability and stability of biological membranes.

*1.2. Fatty acyl chains and biology*

Many (but not all) required fatty acids also can be biosynthesized by mammalian cells. Altering fatty acyl composition of membrane lipids influences physical properties and thus changes the function of integral proteins. Release of specific fatty acids from membrane stores regulates cellular signal transduction and gene transcription with specificity determined by differences in the unsaturation and chain length [10–13].

Fatty acids with several double bonds are known as PUFA. Some PUFA can only be synthesized from dietary fats as animal cells do not have the ability to make double bonds at some positions along the fatty acyl chain. Required dietary fatty acids are known as 'essential fatty acids'. PUFA serve as precursors of biologically active prostaglandins and leukotrienes (Chapter 13) that are required to mount immune and inflammatory responses to infection and pain. In the nucleus, PUFA also control gene transcription through separate classes of receptors called peroxisomal proliferator activated receptors (PPARs) [14] and sterol response elements binding proteins (SREBPs) [15] (also see Chapters 10 and 15). PUFA released from membrane phospholipids by agonist-induced stimulation of phospholipases (Chapter 11) can be involved in signal transduction through modulation of cellular proteins including isoforms of protein kinase C or cyclases, and in translocation of specific enzymes to membranes [12]. Many of these diverse functions are modulated by a specific type of fatty acid. Thus, a large variety of fatty acyl chains are required in the lipids of biological membranes and storage depots. Types obtained through dietary intake and de novo synthesis are insufficient to meet the varied demands of cells, so there is substantial metabolism and rearrangement as development, growth, and aging proceed. Knowledge of how the array of fatty acyl chains is derived and modified, and what regulates metabolism of fatty acyl chains by desaturation and chain elongation are described in sections that follow.

## 2. Chain elongation of long chain fatty acids

De novo synthesis of fatty acids (Chapter 6) produces mainly 16-carbon palmitic acid, with minor amounts of 18-carbon stearic acid (Table 1). Quantitatively, these chain lengths are major components of many membrane lipids; qualitatively, they are related to optimal width of the membrane lipid bilayer. Many major chains are longer than 16 or 18 carbons, constituting more than half of the total acyl chains of many tissues. For example, in myelin surrounding axonal processes of neuronal cells, fatty acyl chains of 18 carbons or greater are more than 60% of the total, and in sphingolipids, 24-carbon acyl chains are prominent. Chain lengths of 28–36 carbons have been reported in phospholipids of retinal photoreceptors.

Many eukaryotic cells have capacity for 2-carbon chain elongation of both endogenously synthesized and dietary fatty acids (Fig. 2). In liver, brain and other tissues, two primary systems — one in the endoplasmic reticulum (ER) and the other in mitochondria — provide chain elongation. The ER system predominates quantitatively. Mitochondrial elongation is distinct from reversal of fatty acid catabolism through β-oxidation (Chapter 5). Peroxisomes also contain an acetyl-CoA dependent elongation system that is enhanced markedly after treatment with specific fatty acids and drug mimics known collectively as peroxisomal proliferators (Chapter 10).

Table 1

Nomenclature and bond positions of major long chain fatty acids

| Common name | Systematic name [a] | Abbreviations | Δ Bond positions |
|---|---|---|---|
| Palmitic acid | hexadecanoic acid | 16:0 | |
| Palmitoleic acid | 9-hexadecenoic acid | 16:1 n-7 | 9 |
| | 6-hexadecenoic acid | 16:1 n-10 | 6 |
| Stearic acid | octadecanoic acid | 18:0 | |
| Oleic acid | 9-octadecenoic acid | 18:1 n-9 | 9 |
| Vaccenic acid | 11-octadecenoic acid | 18:1 n-7 | 11 |
| Petroselenic acid | 6-octadecenoic acid | 18:1 n-12 | 6 |
| Elaidic acid | t-9-octadecenoic acid | t-18:1 n-9 | 9 |
| Linoleic acid | 9,12-octadecadienoic acid | 18:2 n-6 | 9,12 |
| Linoelaidic acid | t,t-9-12-octadecadienoic acid | t,t-18:2 n-6 | 9,12 |
| α-Linolenic acid | 9,12,15-octadecatrienoic acid | 18:3 n-3 | 9,12,15 |
| γ-Linolenic acid | 6,9,12-octadecatrienoic acid | 18:3 n-6 | 6,9,12 |
| Stearidonic acid | 6,9,12,15-octadecatetraenoic acid | 18:4 n-3 | 6,9,12,15 |
| Arachidic acid | eicosanoic acid | 20:0 | |
| Gadoleic acid | 9-eicosenoic acid | 20:1 n-11 | 9 |
| Gondoic acid | 11-eicosenoic acid | 20:1 n-9 | 11 |
| Dihomo-γ-linolenic acid | 8,11,14-eicostrienoic acid | 20:3 n-6 | 8,11,14 |
| Mead acid | 5,8,11-eicosatrienoic acid | 20:3 n-9 | 5,8,11 |
| Arachidonic acid | 5,8,11,14-eicosatetraenoic acid | 20:4 n-6 | 5,8,11,14 |
| Timnodonic acid | 5,8,11,14,17-eicosapentaenoic acid | 20:5 n-3 | 5,8,11,14,17 |
| Behenic acid | docosanoic acid | 22:0 | |
| Cetoleic acid | 11-docosenoic acid | 22:1 n-11 | 11 |
| Erucic acid | 13-docosenoic acid | 22:1 n-9 | 13 |
| Adrenic acid | 7,10,13,16-docosatetraenoic acid | 22:4 n-6 | 7,10,13,16 |
| Docosapentaenoic acid | 4,7,10,13,16-docosapentaenoic acid | 22:5 n-6 | 4,7,10,13,16 |
| Docosapentaenoic acid | 7,10,13,16,19-docosapentaenoic acid | 22:5 n-3 | 7,10,13,16,19 |
| Clupanodonic acid | 4,7,10,13,16,19-docosahexaenoic acid | 22:6 n-3 | 4,7,10,13,16,19 |
| Lignoceric acid | tetracosanoic acid | 24:0 | |
| Nervonic acid | 15-tetracosenoic acid | 24:1 n-9 | 15 |
| Cerotic acid | hexacosanoic acid | 26:0 | |
| Ximenic acid | 17-hexacosenoic acid | 26:1 n-9 | 17 |

[a] All double bonds are of cis geometric configuration except where t indicates a trans double bond.

## 2.1. Endoplasmic reticulum elongation system

Fatty acyl chain elongation associated with the ER is highly active. Fatty acids must be activated to CoA derivatives; the CoA moiety provides energy for attachment of the donor group to a growing fatty acyl chain. Several fatty acyl-CoA synthetases with specificities for chain length and degree of unsaturation have been purified and their cDNAs cloned. Describing their precise subcellular locations is a needed step in determining links to the synthesis of particular lipids.

Distinctive to the ER elongation system, the 2-carbon elongation unit is donated from 3-carbon malonyl-CoA; interestingly, malonyl-CoA itself is not synthesized in the ER. Microsomal chain elongation is active with both saturated and unsaturated fatty acids, with γ-linolenate, 18:3 n-6, being the most effective substrate. In a manner analogous

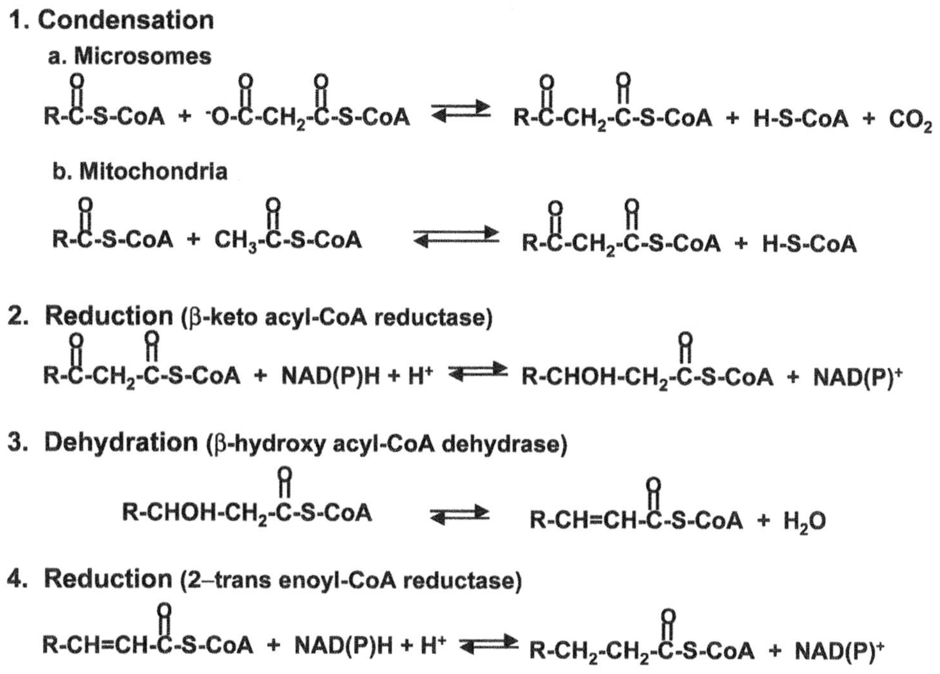

## 1. Condensation

### a. Microsomes

$$R\text{-}\overset{O}{\overset{\|}{C}}\text{-S-CoA} + {}^{-}O\text{-}\overset{O}{\overset{\|}{C}}\text{-CH}_2\text{-}\overset{O}{\overset{\|}{C}}\text{-S-CoA} \rightleftharpoons R\text{-}\overset{O}{\overset{\|}{C}}\text{-CH}_2\text{-}\overset{O}{\overset{\|}{C}}\text{-S-CoA} + H\text{-S-CoA} + CO_2$$

### b. Mitochondria

$$R\text{-}\overset{O}{\overset{\|}{C}}\text{-S-CoA} + CH_3\text{-}\overset{O}{\overset{\|}{C}}\text{-S-CoA} \rightleftharpoons R\text{-}\overset{O}{\overset{\|}{C}}\text{-CH}_2\text{-}\overset{O}{\overset{\|}{C}}\text{-S-CoA} + H\text{-S-CoA}$$

## 2. Reduction (β-keto acyl-CoA reductase)

$$R\text{-}\overset{O}{\overset{\|}{C}}\text{-CH}_2\text{-}\overset{O}{\overset{\|}{C}}\text{-S-CoA} + NAD(P)H + H^+ \rightleftharpoons R\text{-CHOH-CH}_2\text{-}\overset{O}{\overset{\|}{C}}\text{-S-CoA} + NAD(P)^+$$

## 3. Dehydration (β-hydroxy acyl-CoA dehydrase)

$$R\text{-CHOH-CH}_2\text{-}\overset{O}{\overset{\|}{C}}\text{-S-CoA} \rightleftharpoons R\text{-CH=CH-}\overset{O}{\overset{\|}{C}}\text{-S-CoA} + H_2O$$

## 4. Reduction (2–trans enoyl-CoA reductase)

$$R\text{-CH=CH-}\overset{O}{\overset{\|}{C}}\text{-S-CoA} + NAD(P)H + H^+ \rightleftharpoons R\text{-CH}_2\text{-CH}_2\text{-}\overset{O}{\overset{\|}{C}}\text{-S-CoA} + NAD(P)^+$$

Fig. 2. Reactions in 2-carbon chain elongation of long chain fatty acids.

to de novo synthesis of 16-carbon acyl chains, a four-component reaction occurs in the 2-carbon elongation process of long chain fatty acyl-CoA (Fig. 2). Condensation of fatty acyl-CoA and malonyl-CoA to β-ketoacyl-CoA is the initial step. It is rate-limiting, determines fatty acyl specificity, and results in addition of the 2-carbon moiety. The second reaction is catalyzed by a reductase that utilizes NADPH (in preference to NADH) to form β-hydroxy acyl-CoA. It is not known if the ability to use both electron donors is due to two separate reductases. Flow of reducing equivalents from NADPH or NADH to β-keto reductase appears to involve cytochrome $b5$ and cytochrome $P$-450 reductase. The third reaction in chain elongation involves dehydration to enoyl-CoA, and the final reaction is a second reductase step catalyzed by 2-*trans*-enoyl-CoA reductase that requires NADPH. With possible exception of the dehydrase, active sites of all components of the ER elongation system have a cytosolic orientation. Partial purification of components of the elongation system suggest that the enzymes are discrete entities (in contrast to a single polypeptide for de novo fatty acid synthase of animal tissues); acyl-CoA derivatives can be isolated at each intermediate step. Despite such evidence, some investigations suggest covalent linkage of acyl-CoA to a multi-functional complex of the enzymes after condensation.

Regulation of the ER chain elongation system is not well understood. The rate-limiting condensation enzyme, but not the reductases or dehydrase, can be influenced by diet. Fasting depresses elongation and refeeding a carbohydrate diet increases overall chain elongation, whereas refeeding a high protein diet does not. Dietary effects seem similar for saturated and PUFA substrates. In rat liver, elongation activity peaks at two weeks after birth and declines before another increase after weaning; old adult rats have lower elongation activities than young adult rats. Elongation of 16:0 by rat brain microsomes declines with age but elongation of PUFA increases and remains high for several months. Little is known about developmental changes of elongation activities in human tissues. Inhibition of elongation by induction of a diabetic state and reversal by insulin is greater for 16:0 than for 18:3 n-3 suggesting divergent regulation of elongation of acyl chain classes. Xenobiotics, including plasticizers or hypolipidemic agents such as fibrates, markedly alter chain elongation, primarily at the condensation step. Studies with brain indicate different elongation enzymes reacting with dissimilar acyl chain lengths. In Quaking mice (a genetic mutant with defective myelination), chain elongation of 20:0 to 22:0 and 24:0 was reduced by 70%, whereas elongation of 16:0 and 18:0 was unaltered relative to control mice. Thus, more than one elongation system with specificity based on length and unsaturation of acyl chains appears to be operative in the ER.

## 2.2. Mitochondrial elongation system

Although less active than the microsomal system, mitochondrial chain elongation has been extensively investigated, particularly in liver and brain. The 2-carbon condensing donor in mitochondria is acetyl-CoA (Fig. 2; reaction 1b). Generally, monounsaturated fatty acyl-CoA is more active than saturated-CoA and both support higher activity than PUFA, particularly in brain. Maximal mitochondrial elongation in liver, brain, kidney, and adipose tissue seems to require both NADPH and NADH, whereas heart, aorta and muscle require only NADH.

During the 1970s, mechanisms of mitochondrial chain elongation were elucidated [3]. Although β-oxidation (Chapter 5) and mitochondrial chain elongation have the same organelle location, reversal of β-oxidation is not feasible; the FAD-dependent acyl-CoA dehydrogenase of β-oxidation is substituted by a more thermodynamically favorable enzyme, enoyl-CoA reductase (Fig. 2; reaction 4), to produce overall negative free energy for the sequence. Enoyl-CoA reductase from liver mitochondria is distinct from the ER reductase, based on pH optima and specificities for saturated and unsaturated acyl-CoAs. Kinetic studies suggest that enoyl-CoA reductase is rate-limiting in mitochondrial elongation.

## 2.3. Functions of elongation systems

Chain elongation in the ER appears to be the most important source of acyl chains greater than 16 carbons for membrane phospholipids during growth and maturation, when long chain acids may not be supplied adequately in the diet. 18- to 24-carbon saturates and monoenes and 20- and 22-carbon polyunsaturates are required for neural growth and myelination, regardless of dietary fluctuations.

The function of the mitochondrial elongation system is less clear but it may participate in biogenesis of mitochondrial membranes or in transfer of reducing equivalent between carbohydrates and lipids. However, in view of the relatively low activity toward

16- and 18-carbon acyl chains, a primary role in the formation of long acyl chains for membrane synthesis is questionable.

Specific roles for elongation in peroxisomes have not been defined but this organelle may produce the very long chain saturated and polyenoic fatty acids of 24–36 carbons. Import of very long chain fatty acids into peroxisomes is defective in patients with the inherited disease, X-linked adrenoleukodystrophy, and this results in increased cellular and serum levels of very long chain fatty acids. The gene whose mutation leads to the degenerative disease apparently codes for a transporter required for importing very long chain fatty acids into peroxisomes for subsequent activation by a peroxisomal very long chain fatty acyl-CoA synthase. The fate of long chain acyl-CoAs formed in peroxisomes is not yet understood.

Studies using the yeast *Saccharomyces cerevisiae* resulted in the isolation of three genes required for elongation of fatty acyl-CoAs — *ELO1*, *ELO2*, and *ELO3* [7]. *ELO1* encoded protein prefers to elongate 14-carbon fatty acyl-CoAs, the *ELO2* encoded protein elongates 20- and 22-carbon chains, and the *ELO3* gene product elongates a broad range of fatty acyl CoA-chains to 26-carbon products. Inactivation of the *ELO2* or *ELO3* gene results in defective sphingolipid synthesis. In both yeast and mammals, sphingolipids store fatty acids greater than 24 carbons indicating that *ELO2* and *ELO3* elongation enzymes are coupled directly to synthesis of sphingolipids. Inactivation of *ELO2* or *ELO3* genes gives yeast cells the ability to survive in the absence of v-SNARES required for targeting of membrane vesicles from the Golgi to the plasma membrane. This suggests a role for long chain fatty acids or sphingolipids, mediated by *ELO2* and *ELO3* encoded proteins, in regulated fusion of biological membranes.

# 3. Formation of monounsaturated fatty acids by oxidative desaturation

The spectrum of fatty acyl chains needed to meet requirements of lipid storage, membrane synthesis and maintenance, and lipid regulation of cellular processes cannot be provided by diet or de novo fatty acid synthesis and chain elongation alone. Unsaturated fatty acids also must be synthesized in cells, supplemented by essential fatty acids in the diet.

## 3.1. Nomenclature to describe double bonds

Before discussing desaturation enzymes, abbreviations to describe the number and position of double bonds in acyl chains (Table 1) will be outlined using linoleic acid as an example.

(1) To indicate that linoleic acid is an 18-carbon fatty acid with two double bonds, the shorthand 18:2 is used. The number before the colon denotes the number of carbon atoms and the number following refers to the number of double bonds.

(2) To assign the position of an individual double bond or specificity of an enzyme inserting it, the delta ($\Delta$) nomenclature is used. This describes a bond position relative to the carboxyl (number one) carbon of the acyl chain. For linoleic acid, the double bonds are in the $\Delta 9$ and $\Delta 12$ positions, between carbons 9–10 and

12–13, and are introduced into the 18-carbon chain by Δ9 and Δ12 desaturase enzymes.

(3) To designate an individual fatty acid within a 'family' of structurally related acids, the n- nomenclature is used. Here, the position of the first double bond from the methyl end is described. Thus, 18:2 n-6 indicates that the double bond closest to the methyl end is 6 carbons from the methyl group and in the Δ12 position. This convention is particularly useful in designating groups of fatty acids derived from the same parent compound as metabolic reactions do not occur on the methyl side of an existing double bond.Among other conventions, the ω6 designation is still widely used to describe the position of a double bond from the methyl end (ω-carbon). It is similar to the n- nomenclature. Linoleic acid is an ω6 fatty acid.

(4) To indicate the geometric configuration of a double bond, the designation is preceded by a c- for cis or t- for trans. Thus, c,c-18:2 n-6 distinguishes linoleic acid from a trans-containing isomer, such as a conjugated linoleic acid (CLA; e.g., 9c,11t-18:2). Generally, double bond configuration is cis.

### 3.2. Characteristics of monounsaturated fatty acid-forming desaturation enzymes

Monounsaturated fatty acids are formed in mammalian systems by direct oxidative desaturation (removal of two hydrogens resulting in the introduction of a double bond) of a preformed long chain saturated fatty acid. The oxygenase type of enzyme is associated with the ER of liver, mammary gland, brain, testes and adipose tissue. The Δ9 desaturase is the predominant, if not exclusive, desaturation enzyme for saturated acids in these tissues and is rate-limiting in the formation of 18:1 n-9. Δ9 desaturase acts on fatty acyl-CoA. For most tissues, 14- to 18-carbon saturated fatty acyl chains are good substrates, with stearoyl-CoA being most active. Reduced pyridine nucleotide is required and generally NADH is more active than NADPH. The Δ9 desaturase has an absolute requirement for molecular oxygen which acts as an electron acceptor for two pairs of hydrogens — one from NADH and the other from the fatty acyl-CoA.

The Δ9 desaturation system consists of three major protein components: (1) NADH-cytochrome $b5$ reductase, (2) cytochrome $b5$, and (3) a desaturase domain that is cyanide-sensitive and rate-limiting in double bond formation (Fig. 3). In some systems, such as those in insects, cytochrome $P$-450 may substitute for cytochrome $b5$. Integral association of Δ9 desaturase with the ER has retarded characterization of the complex. Loss of activity during solubilization was largely overcome through use of controlled ratios of detergents to protein and combinations of mild extraction solvents. The purified desaturase component has one atom of non-heme iron per molecule as the prosthetic group and 62% non-polar amino acid residues. Thus, the desaturase is largely within the microsomal membrane, with the active centre exposed to the cytosol. Interaction with specific reagents suggests that arginyl residues play a role at the binding site for the negatively charged CoA moiety of the substrate and tyrosine residues are involved in chelation of the iron prosthetic group. Isolation, characterization and site-directed mutagenesis of the cDNA coding for mammalian Δ9 desaturase has confirmed the essential catalytic role of histidine residues.

NADH-cytochrome $b5$ reductase and cytochrome $b5$ are more readily solubilized

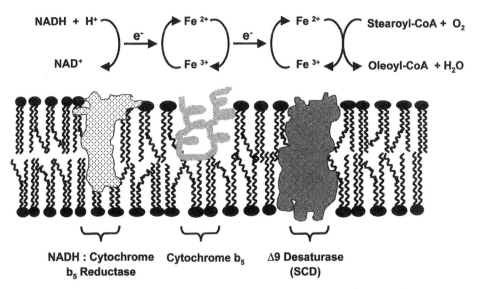

Fig. 3. Representation of the Δ9 (stearoyl-CoA) desaturase (SCD) complex, including electron transport proteins.

than the desaturase. Cytochrome $b5$ has a hydrophilic region of 85 residues (including the NH2-terminal) and the protein terminates in a hydrophobic COOH-terminal tail of 40 amino acids that attaches the protein to the membrane. It is uncertain whether one or two cytochrome $b5$ molecules per complex are required to transfer two electrons from NADH to oxygen.

Stereochemical studies have shown that the D-hydrogens at positions 9 and 10 are removed concertedly to give a *cis* double bond without involving an oxygen-containing intermediate. Attempts to demonstrate a hydroxyacyl intermediate have been negative and hydroxyacyl-CoAs are not readily desaturated. In insect systems oxygen free radical is involved.

Δ9 desaturase is translated on soluble cytoplasmic polysomes with post-translational binding of iron and insertion into the ER. Plasmid expression vectors have been used to synthesize desaturase peptides that can be reconstituted with cytochrome $b5$ and cytochome b5 reductase to give Δ9 desaturase activity. Dietary induction of Δ9 desaturase facilitated isolation of poly(A+) RNA and preparation of cDNA for stearoyl-CoA desaturase (SCD). mRNA increased 40- to 60-fold following refeeding of fasted animals and during induced differentiation of preadipocytes. Two genes (*SCD-1* and *SCD-2*) encoding isoforms of Δ9 desaturase with 87% homology were identified. Whereas liver exclusively expresses SCD-1, brain, spleen, heart and lymphocytes express only SCD-2; both SCD-1 and SCD-2 are expressed in adipose tissue, lung and kidney.

A structural gene for Δ9 desaturase from yeast (*OLE1*) has been isolated and characterized [16]. The *OLE1* gene encodes the only fatty acid desaturase in yeast and *OLE1* protein is found in the ER. The predicted protein amino acid sequence has a high degree of similarity to mammalian SCD. Indeed, expression of rat SCD in yeast can substitute for yeast *OLE1*. If the *OLE1* gene is genetically inactivated, yeast are no longer viable, although life can be rescued in medium containing 14:1, 16:1, 18:1, or 18:2 fatty acids. A temperature-sensitive allele of the *OLE1* gene results in intact protein at 25°C but inactive protein at 37°C; at the latter temperature, there is an increase in saturated, and decrease in unsaturated, fatty acids. At the non-permissive temperature, dividing cells are unable to transfer mitochondria into newly made daughter cells, but the defect can be rescued by addition of 18:1 n-9. The molecular link between fatty acid desaturation and mitochondrial inheritance has yet to be identified.

### 3.3. Modification of Δ9 desaturase activities in vitro

Cyanide completely inhibits Δ9 desaturation in rat liver by acting on the terminal desaturase component (cyanide-sensitive factor), apparently related to accessibility to the non-heme iron. Some Δ9 desaturases (e.g., in insects and yeast) are not inhibited by cyanide.

Cyclopropenoid fatty acids found in stercula and cotton seeds are potent inhibitors of Δ9 desaturase. Sterculoyl- and malvaloyl-CoA (18- and 16-carbon derivatives with cyclopropene rings in the Δ9 position) specifically inhibit Δ9 desaturation. Hens fed meal containing cyclopropene fatty acids have decreased 18:1/18:0 ratios in their egg yolks. Cyclopropene acids have been used in vitro to differentially alter Δ9 and Δ6 desaturase activities, distinguishing relative contributions of these two enzymes in perinatal brain development.

### 3.4. Age-related, dietary and hormonal regulation of Δ9 desaturase

Extreme response to dietary alterations is a remarkable feature of Δ9 desaturase. Levels of *SCD-1* and *SCD-2* encoded Δ9 desaturase are highly regulated at the levels of transcription and of transcript and protein stability [6,8,9,15]. When rats are not fed for 12–72 h, liver Δ9 desaturase (SCD-1) activity declines to less than 5% of control values (Fig. 4). After refeeding, Δ9 desaturase activity increases dramatically to more than 2-fold above normal. The restoration has been termed 'super-induction' as levels of enzyme activity can be more than 100-fold above the fasted state, particularly when the rats are refed a fat-free diet enriched in carbohydrate or protein. Protein synthesis inhibitors and immunological techniques have been used to show that synthesis of the desaturase component is altered quantitatively. Liver SCD-1 mRNA is dramatically altered by dietary changes; for example, a nearly 100-fold increase occurs in rat pups nursed by mothers on an essential fatty acid deficient diet compared to controls, and a 45-fold increase occurs in liver upon refeeding fasted mice with a fat-free, high-carbohydrate diet. Responses of liver enzyme to dietary intake probably explain the 'circadian changes' in Δ9 desaturase, where liver activities can fluctuate 4-fold over a 24-h period; highest activity (around midnight) corresponds to maximal food intake in the nocturnal rat.

In contrast to the liver enzyme, brain Δ9 desaturase (SCD-2) is little altered by dietary restrictions. This ensures continuing activity during crucial stages of brain

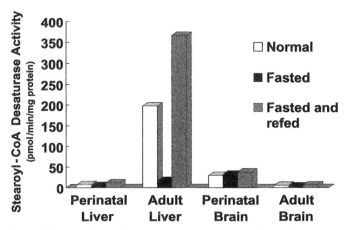

Fig. 4. Effects of fasting for 48 h and subsequent refeeding with a normal diet for 24 h on in vitro Δ9 desaturase activities of brain and liver from 10-day-old and adult rats. Adapted from Cook and Spence [29].

development. Brain Δ9 desaturase activity is greatest during the perinatal and suckling period in rats and is generally higher than in liver. However, when rats are weaned, brain Δ9 desaturase activity slowly declines. In contrast to the extreme changes in liver SCD-1 noted above, brain SCD-2 mRNA increases only 2-fold in neonates that are suckling mothers on an essential fatty acid deficient diet.

Transcription of *SCD-1* and *SCD-2* is regulated through specific elements in the promoter of each gene [8]. Although there are differences in the promoter region of *SCD-1* and *SCD-2* that lead to tissue-specific expression, shared regions appear to be responsible for decreased transcription in response to cholesterol (or conversely, increased transcription in response to sterol depletion). The regulation of SCD-1 and SCD-2 in response to cholesterol is through the binding of the transcription factor SREBP-1a and nuclear factor-Y (that heterodimerize) to three conserved regions within the *SCD-1* and *SCD-2* promoters. Addition of PUFA, particularly 18:2 n-6 or 20:4 n-6 [17], also decreases the level of SCD-1 and SCD-2 mRNA; whether this is due to decreased transcription or increased mRNA degradation, or both, is still unclear. SCD proteins also turn over rapidly in cells, possibly to ensure that transcriptional responses quickly alter SCD protein levels. The N-terminal 33-amino acids of SCD are required for turnover. The protease system responsible for degradation of SCD protein is not known.

Hormonal regulation of Δ9 desaturase is complex and not understood fully. Rats with genetic diabetes, or made diabetic by destruction of pancreatic β-cells, have depressed Δ9 desaturase activity in liver, mammary gland and adipose tissue. Insulin restores activity in vivo but is without effect in an in vitro assay. Insulin appears essential for basal transcription of the *SCD-1* gene and markedly induces transcription through a process requiring synthesis of a protein regulator. Significant changes in cytochrome *b*5 and the reductase are not elicited by insulin. Other hormones and effectors, such as glucagon and cyclic AMP, do not alter Δ9 desaturase activity, whereas epinephrine and thyroxine do.

Studies of SCD mutations in mice support an integration of monoene formation with neutral lipid and lipoprotein metabolism as well as unpredicted effects on the skin and eyes [17–19]. Mice containing a natural mutation that inactivates SCD-1 display a dramatic decrease in synthesis of liver triacylglycerols and cholesterol esters, leading to lower levels of these lipids in very low and low density lipoproteins. Specific genetic inactivation of the mouse SCD-1 gene produces similar effects on triacylglycerols. In addition, these mice develop skin abnormalities and atrophied sebaceous glands (that secret oil into the skin) and meibomian glands (that produce lipid secretions for the eye), resulting in an inability of these mice to blink.

## 4. Formation of polyunsaturated fatty acids

### 4.1. Characteristics in animal systems

All eukaryotic organisms contain polyenoic fatty acyl chains in their membrane lipids, and most mammalian tissues modify acyl chain composition by introducing more than one double bond.

(1) The first double bond introduced into a saturated acyl chain is generally in the $\Delta9$ position so that substrates for further desaturation contain either a $\Delta9$ double bond or one derived from the $\Delta9$ position by chain elongation. Relatively large amounts of 16:1 n-10 and 18:1 n-10 in neonatal rat brain are exceptions but the qualitative significance during brain development is unclear.

(2) Like $\Delta9$ desaturation that inserts the first double bond, further desaturation is an oxidative process requiring molecular oxygen, reduced pyridine nucleotide and electron transfer involving a cytochrome and related reductase.

(3) Animal systems cannot introduce double bonds beyond the $\Delta9$ position. Thus, second and subsequent double bonds are always inserted between an existing bond and the carboxyl end of the acyl chain, never on the methyl side (Fig. 5). Plants, on the other hand, introduce second and third double bonds between the existing double bond and the terminal methyl group (Chapter 4). Diatoms, *Euglena*, insects, snails and slugs can desaturate on either side of an existing bond; at least 15 insect species form 18:2 n-6 by de novo synthesis and some produce 20:4 n-6. Consequently, double bonds are found at the $\Delta9$, $\Delta6$, $\Delta5$ and $\Delta4$ positions as a result of desaturation in animals, at the $\Delta9$, $\Delta12$ and $\Delta15$ positions in plants, and at the $\Delta5$, $\Delta6$, $\Delta9$, $\Delta12$ and $\Delta15$ positions in insects and other invertebrates. Well-established evidence confirms distinct $\Delta9$, $\Delta6$ and $\Delta5$ desaturases in many animal tissues. In contrast, the longstanding assumption that a $\Delta4$ desaturase exists in the classical pathways of PUFA metabolism (Fig. 6) has not been confirmed by direct enzyme characterization. The abundance of long chain polyenoic acids containing $\Delta4$ double bonds in tissues such as brain, retina and testes, and their formation in vivo and in vitro from more saturated precursors, indicate that mammalian tissues can produce PUFA with a $\Delta4$ double bond. A new dimension to the classical pathway, erroneously thought to include direct $\Delta4$ desaturation, is now recognized. In rat liver, fish and human monocytes (and probably other

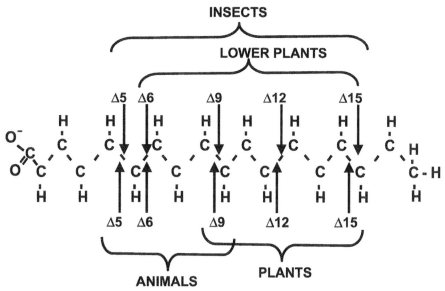

Fig. 5. Positions of fatty acyl chain desaturation by enzymes of animals, plants, insects and lower plants.

human tissues), formation of 22:6 n-3 from 22:5 n-3 or of 22:5 n-6 from 22:4 n-6 involves direct chain elongation of these respective precursors and then Δ6 desaturation followed by a 2-carbon chain shortening, the latter by β-oxidation in peroxisomes. This alternative to direct Δ4 desaturation has become known as the Sprecher pathway [20]. Comparative studies with tissue explants, primary cultures and neoplastic cells suggest that introduction of the Δ4 bond into 22:6 n-3 is characteristic of differentiated tissue and may be restricted in undifferentiated cells.

(4) In most organisms, and certainly in higher animals, methylene interruption between double bonds is usually maintained with conjugated double bonds being rare.

(5) All bonds introduced by oxidative desaturation in animals are in the *cis* geometric configuration.

### 4.2. Essential fatty acids — a contribution of plant systems

Requirements for PUFA cannot be met by de novo metabolic processes in mammalian tissues. Animals are absolutely dependent on plants (or insects) for providing double bonds in the Δ12 and Δ15 positions of the two major precursors of the n-6 and n-3 fatty acids, linoleic and linolenic acids. In animal tissues these acyl chains are converted to fatty acids containing 3–6 double bonds. Even very long chain PUFA of 28–36 carbons apparently do not have more than 6 double bonds.

### 4.3. Families of fatty acids and their metabolism

Relationships among fatty acids in the metabolic pathways can be evaluated by considering groups or families of fatty acids based on the parent unsaturated acid in the

194

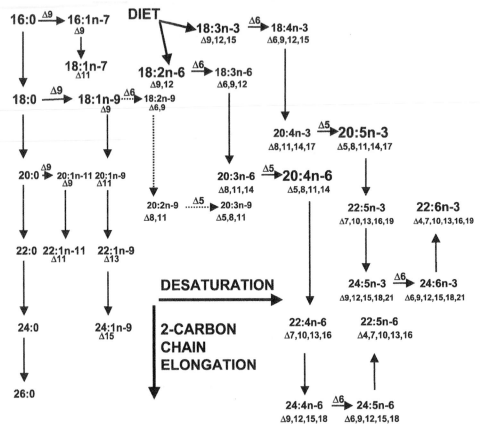

Fig. 6. Major pathways of fatty acid desaturation and chain elongation in animal tissues. Note the alternating sequence of desaturation in the horizontal direction and chain elongation in the vertical direction in the formation of polyunsaturated fatty acids from dietary essential fatty acids.

sequence. The predominant fatty acid families (Fig. 6) are:
  n-6 acids derived from 18:2 n-6
  n-3 acids derived from 18:3 n-3
  n-9 acids derived from 18:1 n-9, and
  n-7 acids derived from 16:1 n-7.

### 4.3.1. The n-6 family

Arachidonate, 20:4 n-6, is an abundant polyenoic acyl chain found in most animal tissues. 20:4 n-6 can be derived from 18:2 n-6 by the alternating sequence of $\Delta6$ desaturation, chain elongation of the 18:3 n-6 intermediate, and $\Delta5$ desaturation of 20:3 n-6 (Fig. 6). $\Delta6$ desaturation is considered rate-limiting in most situations. 20:4 n-6 is a component of phospholipids contributing to the structural integrity of membranes and is the primary precursor of important biological signalling molecules (e.g., prostaglandins and leukotrienes) with a variety of biological activities (Chapter 13). Frequently, 20:4

n-6 is referred to as an essential fatty acid as there is an absolute requirement for it; however, 18:2 n-6 can be converted to adequate 20:4 n-6 by most mammalian cell types. Neutrophils are a major exception, requiring 20:4 (-6) for leukotriene production but being unable to synthesize it from 18:2 n-6.

In liver and most other tissues of animals in a normal, balanced state, only 18:2 n-6 and 20:4 n-6 accumulate in relatively large quantities; much lower levels of the intermediates, 18:3 n-6 and 20:3 n-6, are detected. Such observations support a rate-limiting role for Δ6 desaturase in the sequence. When Δ6 desaturase activity is limited, this can be by-passed by providing 18:3 n-6 from enriched sources (such as evening primrose oil) for subsequent elongation and Δ5 desaturation to 20:4 n-6.

The genes and cDNAs coding for mammalian Δ6 and Δ5 desaturase enzymes have been isolated and characterized [4,5]. In humans, both desaturases are on chromosome 11. Mouse and human Δ6 desaturases are comprised of 444 amino acids with 87% homology between species. A surprising feature in analysis of the amino acid sequences is that both Δ6 and Δ5 desaturases have a cytochrome $b5$ domain at the N-terminus of the polypeptide chain with a C-terminal catalytic domain very similar to that found in Δ9 desaturase, including the positioning of specific catalytic histidine residues. Yeast Δ9 desaturase also has an N-terminus cytochrome $b5$ domain; genetic inactivation of the yeast cytochrome $b5$ gene still allows for functional Δ9 desaturation, while deletion of the *OLE1* encoded Δ9 desaturase cytochrome $b5$ domain results in inactive Δ9 desaturase. Thus, cellular cytochrome $b5$ can not substitute for the covalently attached N-terminal cytochrome $b5$ domain. By analogy, the N-terminal cytochrome $b5$ domain also is likely functional and essential for mammalian Δ6 and Δ5 desaturases.

mRNA transcript levels for Δ6 and Δ5 desaturases, determined for various mammalian tissues, correlate reasonably with detectable enzyme activities. Highest levels of transcript and activity for both Δ6 and Δ5 desaturases are found in liver, followed by lower amounts in heart, brain, placenta and lung. Liver Δ6 and Δ5 desaturase mRNA levels and enzyme activities are affected by fasting and refeeding with the level of PUFA in the diet being highly influential; high PUFA diets result in decreased liver Δ6 and Δ5 desaturase mRNA and enzyme activity levels and diets lacking PUFA result in high Δ6 and Δ5 desaturase mRNA and enzyme activities. Enhancement of desaturase activity observed upon refeeding after fasting can be suppressed by glucagon or dibutyryl cAMP for Δ6 desaturase activity but not for Δ5 desaturase. In rat liver, Δ6 desaturase activity is low during fetal and neonatal development, increases dramatically around weaning and remains high throughout adulthood. Brain Δ6 desaturase activity is relatively high in the fetus and neonates and markedly declines by weaning, remaining low throughout adulthood. Δ5 desaturase follows similar trends. Liver microsomes from human neonates have Δ6 and Δ5 desaturases at lower levels than for adult humans or rodents.

Relationships between PUFA metabolism and insulin are complex: PUFA formation is impaired in diabetes, insulin increases the activities of Δ6 and Δ5 desaturases, essential fatty acid deficiency prevents induced diabetes and supplementation with fish-oil (containing n-3 fatty acids) worsens hyperglycemia in diabetic patients. Epinephrine suppresses Δ6 and Δ5 desaturases through β-receptors. Triiodothyronine inhibits both Δ6 and Δ5 desaturases. Generally, Δ6 and Δ5 desaturases respond similarly to glucocorticoids, other steroids and adrenocorticotropic hormone. Potential roles of protein kinases as mediators of hormone action in control of PUFA biosynthesis remain to be defined. Vitamin E supplementation increases brain $\Delta6$ desaturase and vitamin B6 deficiency markedly decreases liver $\Delta6$ desaturation. Some cultured cell lines have lost $\Delta6$ desaturase but retain $\Delta5$ desaturase activity. The seed extract sesamin inhibits $\Delta5$, but not $\Delta6$, desaturase and the hypocholesterolemic drug simvastatin specifically increases $\Delta5$ desaturation. These differences in activity are likely mediated through alterations in mRNA levels, although whether this is due to changes in the level of transcription or mRNA stability has yet to be determined.

In some tissues, 22:4 n-6 and 22:5 n-6 are quantitatively significant. Although numerous earlier reports concluded that there is $\Delta4$ desaturation, the Sprecher pathway appears to be responsible primarily for production of 22:5 n-6 [20]. Exogenous 22:4 n-6 also is a substrate for 'retroconversion'. This process of partial degradation involves loss of either a 2-carbon fragment or a double bond and 2 or 4 carbons by $\beta$-oxidation. In general, retroconversion utilizes fatty acids of 20 carbons or greater. Since only double bonds in the $\Delta4$ position are lost, this process could provide fatty acids with the first double bond in the $\Delta5$ position. The quantitative significance of this process is not established. Deficiency of retroconversion of 22:6 n-3 to 20:5 n-3 in fibroblasts from a Zellweger patient with defective assembly of peroxisomes supports a role for peroxisomal $\beta$-oxidation (Chapter 5) in retroconversion.

A physical relationship of desaturases with chain elongation enzymes in ER membranes remains to be demonstrated. Indirect evidence supports a sequence of reactions (including esterification of products into phospholipids) that proceeds in a concerted manner without release of free fatty acyl intermediates [21].

### 4.3.2. The n-3 family

The most abundant n-3 acyl chains are 20:5 n-3 and 22:6 n-3. These fatty acids are esterified to phospholipids in cerebral cortex, retina, testes, muscle and liver. In retinal rod outer segments, the fatty acids in phospholipids are 40–60% 22:6 n-3. Many animal tissues convert dietary 18:3 n-3 to 20:5 n-3 and 22:6 n-3 by desaturation and elongation combined with peroxisomal chain shortening (Fig. 6). Whether the $\Delta6$ desaturase in the Sprecher pathway converting 22:5 n-3 to 22:6 n-3 (or 22:4 n-6 to 22:5 n-6) is the same as the $\Delta6$ desaturase acting directly on 18:3 n-3 (or 18:2 n-6) is unresolved with evidence for and against a common enzyme. Alternative sequences for initial steps of 18:3 n-3 (and 18:2 n-6) metabolism involving chain elongation followed by sequential $\Delta5$ and $\Delta8$ desaturation have been detected but their significance is unclear.

Potential benefits of increased consumption of fish and fish oil products abundant in 20:5 n-3 and 22:6 n-3 have been investigated extensively in animals and humans [22]. Enzymes in phytoplankton consumed by fish, or in fish themselves, produce these fatty acids from 18:3 n-3. Consumption of n-3 fatty acids leads to altered acyl chain composition of phospholipids in plasma, platelets, neutrophils and red cells of mammals, with further enrichment of n-3 fatty acids in peripheral body tissues taking longer. Generally, increases in n-3 content are at the expense of n-6 acids. Some populations, such as Greenland Inuit, routinely consume high levels of n-3 fatty acids and have a lower incidence of ischemic heart disease and longer bleeding times related to a reduction in platelet aggregation. Clinical trials support positive effects of n-3 fatty acids in decreasing platelet aggregation, lowering blood pressure, reducing circulating triacylglycerols and producing modest beneficial changes in cholesterol and lipoproteins.

Some studies indicate that n-3 PUFA reduce severity of arthritis or impaired kidney functions involving abnormal immune and inflammatory responses. Other areas of potential n-3 fatty acid involvement include enhanced insulin sensitivity in the diabetic state, countering n-6 fatty acids in chemically induced, transplanted or metastatic tumors, and altered visual acuity and response to learning tests. Each fatty acid family has a role in overall nutritional balance believed largely to be manifested at the level of eicosanoid production (Chapter 13). The high concentration of n-3 fatty acids in some body tissues with specialized functions (e.g., brain and retina) and tenacious retention of n-3 acids during dietary deprivation suggest important structural and physical roles.

Suggestions that 0.2–0.3% of energy as n-3 fatty acids is adequate for adults and 0.5% during pregnancy, lactation and infancy must be matched with the need for n-3 to n-6 ratios of 1:4 to 1:8 to promote normal growth and development. Reported deficiencies in humans on long-term intravenous or gastric tube feeding have been corrected by supplementation with n-3 fatty acids. Interest in potential positive benefits of concentrated fish oils in the diet should be balanced by attention to potential negative or toxological side effects.

Competition among fatty acids of the n-3 and n-6 families for desaturation and chain elongation enzymes is extensively documented from a variety of in vivo and in vitro experiments. 18:3 n-3 is a better substrate for $\Delta 6$ desaturase than is 18:2 n-6; accordingly, abundance of 18:3 n-3 can effectively decrease formation of 20:4 n-6. In humans, conversion of 18:3 n-3 to 20:5 n-3 and 22:6 n-3 is much greater than conversion of 18:2 n-6 to 20:4 n-6.

### 4.3.3. The n-9 family

The prominent acyl chain in this family is 18:1 n-9. Generally, competition from 18:2 n-6 and 18:3 n-3 for $\Delta 6$ desaturase prevents formation and accumulation of more unsaturated n-9 acids. However, in animals on a diet deficient in essential fatty acids, competition by 18:2 n-6 is removed and 18:1 n-9 is utilized as a substrate for the rate-limiting $\Delta 6$ desaturase (Fig. 6). Further chain elongation of 18:2 n-9 to 20:2 n-9 and $\Delta 5$ desaturation results in accumulation of 20:3 n-9. While 20:3 n-9 may partially substitute for some physical functions within membranes, it is not a precursor of prostaglandins and cannot alleviate the signs of essential fatty acid deficiency.

Since deficiency of essential fatty acids markedly reduces 20:4 n-6 while increasing 20:3 n-9, a ratio of triene to tetraene (20:3 n-9/20:4 n-6) of less than 0.2 in tissues and serum usually indicates essential fatty acid deficiency. However, use of this ratio has limitations as inhibition of $\Delta 6$ desaturase will reduce formation of both 20:3 n-9 and 20:4 n-6 and result in a deficiency state without altering the ratio. Total amount of n-6 acids may be a better reflection of essential fatty acid deficiency.

### 4.3.4. The n-7 family

The primary n-7 acid in membranes and circulating lipids is 16:1 n-7. As most analyses do not distinguish specific 18:1 isomers, the contribution of 18:1 n-7 to the 18:1 fraction is seldom appreciated even though in developing brain 18:1 n-7 comprises 25% of total 18-carbon monoene. Potentially, 20:4 n-7, with only a single carbon shift of the double bonds compared to 20:4 n-6, could be formed from 16:1 n-7; however, high levels of PUFA derived from 16:1 n-7 are not detected even on a fat-free diet, although increased levels of 16:1 n-7 frequently accompany deficiency of essential fatty acids.

# 5. Unsaturated fatty acids with trans double bonds

## 5.1. General properties

*Trans*-unsaturated fatty acids, the geometric isomers of naturally occurring *cis*-acids, are not produced by mammalian enzymes but are formed by microorganisms in the gastrointestinal tract of ruminant mammals and chemically during commercial partial hydrogenation of fats and oils [23,24]. *Trans*-acids have been inaccurately labelled as unnatural, foreign or non-physiological. From diets containing beef fat, milk fat, margarines and partially hydrogenated vegetable oils, *trans*-acids are ingested, incorporated and modified in animal tissues.

Early studies suggested selective exclusion of *trans*-acids from metabolic processes and incorporation into membrane lipids, particularly in the central nervous system. Collectively, further studies have indicated little selectivity for absorption, esterification or β-oxidation compared to *cis*-isomers. Discrimination against specific positional isomers of *trans*-acids (e.g., $\Delta$13 *trans*) may occur. In general, *trans*-monoenoic acids are recognized as a distinct class with properties intermediate between saturated and *cis*-monounsaturated acids, particularly in specificity for esterification to phospholipids. Accumulation of *trans*-acids in tissues generally is proportional to dietary levels. Lack of preferential accumulation over the long term suggests that *cis* and *trans* isomers turn over similarly. Nonetheless, potential negative influences of *trans*-unsaturated fatty acids in biological systems continue to be controversial.

Specific interactions of *trans*-acids with desaturation and chain elongation enzymes of animal tissues have been reported. Positional isomers of *t*-18:1 (except for the $\Delta$9 isomer) are substrates for $\Delta$9 desaturase of rat liver microsomes, resulting in *cis,trans*-dienoic isomers that, in some cases, are desaturated again by $\Delta$6 desaturase to unusual polyunsaturated structures.

## 5.2. Trans-polyenoic fatty acids

Isomers of *trans,trans*-dienoic fatty acids, including *t,t*-18:2 n-6, are substrates for $\Delta$6 desaturation in liver and brain, albeit at a lower rate than for *c,c*-18:2 n-6. Dienoic *trans*-isomers of 18:2 clearly lack properties of essential fatty acids and interfere with normal conversion of 18:2 n-6 to 20:4 n-6, primarily at the $\Delta$6 desaturase. Diets containing *trans*-acids intensify signs of mild essential fatty acid deficiency; conversion of available 18:2 n-6 to 20:4 n-6 is reduced as is 18:1 n-9 to 20:3 n-9. Although complex interactions are possible, *trans*-dienes are minor components (usually less than 1%) of hydrogenated vegetable oils used in human food products.

Many investigators conclude that *trans*-acids are undesirable dietary components. The actual influence of increased *trans* fatty acid intake during early development, on genetic disorders or on cardiovascular disease remains controversial and unresolved. It is reasonable to attempt to reduce their human consumption but continued investigations of *trans*-acids as a distinct and significant class of fatty acids are necessary.

*5.3. Fatty acids with conjugated double bonds*

While research interests in *trans* unsaturated fatty acids have oscillated somewhat over the last few decades, much recent attention has been directed to a specific class of dietary conjugated fatty acids, some with *trans* bonds, known as conjugated linoleic acid isomers (CLA; e.g., 9c,11t; 9t,11t; 10t,12c; 10t,12t; 10c,12c isoforms — compared to 9c,12c-linoleic acid) [24,25]. Intriguing, but poorly understood, health benefits have been implicated for CLAs including suppression of cancer cell growth, weight gain, diabetes and atherosclerosis. CLA isoforms can be metabolized to isoforms of 20:4 n-6 and some interfere with $\Delta 9$ desaturation or 20:4 n-6 and prostaglandin production. However, information on their relative metabolism, interference with normal *cis* isomers such as 18:1 n-9 and 18:2 n-6 or alterations by deficiency and disease states is limited.

# 6. Abnormal patterns of distribution and metabolism of long chain saturated and unsaturated fatty acids

Despite the diversity of enzymes involved in unsaturated fatty acyl chain formation, documented clinical defects specific to unsaturated acyl chain metabolism are few. Most alterations of acyl chain patterns reflect dietary deficiency states or impaired activities but not absence of specific fatty acid metabolizing enzymes. This indicates that desaturation and elongation enzymes are essential in supporting life-sustaining cellular processes. However, defects or deficiencies resulting in abnormal patterns of unsaturated fatty acid distribution have been documented.

*6.1. Deficiency of essential fatty acids and related nutrients*

In experimental animals, severe effects are observed in the absence of dietary essential fatty acids. These include a dramatic decrease in weight, dermatosis and increased permeability to water, enlarged kidneys and reduced adrenal and thyroid glands, cholesterol accumulation, impaired reproduction, and ultimate death. Substantial changes in fatty acid composition of tissue and circulating lipids occur. Brain is exceptionally resistant to loss of essential fatty acids, but modification of acyl patterns can be achieved if a deficient diet is started at an early age and continued long enough. The four n-6 acids in the sequence from 18:2 n-6 to 20:4 n-6 individually have similar potency in reversing effects of deficiency, whereas the capacity of 18:3 n-3 is much lower. Collective assessment of nutritional studies indicates that 2–4% 18:2 n-6 and 0.2–0.5% 18:3 n-3 are adequate, but such requirements depend on the demands of specific tissues and the stage of growth, development and metabolism.

Inadequate supply of essential fatty acids resulting in deficiency signs described for rats is rare in humans. Normal diets contain enough 18:2 n-6 and 18:3 n-3, or their metabolic products, to meet tissue demands. Adipose stores provide a protective buffer against temporary limited intake. However, severe deficiency states have been observed in humans (especially premature infants with restricted adipose stores) on prolonged intravenous feeding or artificial milk formulations without adequate lipid supplements. Marked alterations of serum fatty acid patterns characterized by depletion of n-6 acids and a major increase in the 20:3 n-9 to 20:4 n-6 ratio are accompanied by severe skin rash, loss of hair and irritability. These signs are reversed rapidly by supplementation with lipid emulsions containing 18:2 n-6.

Functions for 18:2 n-6, in addition to a role as precursor of 20:4 n-6, seem likely. Some fatty acids that cannot serve as prostaglandin precursors prevent signs of essential fatty acid deficiency. Cats apparently require both 18:2 n-6 and 20:4 n-6 in their diets but may have relatively low, rather than absence of, $\Delta6$ desaturase activity.

Gross signs of zinc deficiency are similar to those observed in essential fatty acid deficiency. Possible relationships between zinc, PUFA and eicosanoids have been proposed, but direct connections at the metabolic level have not been shown.

In several human diseases, abnormal patterns of PUFA, attributable to insufficient dietary 18:2 n-6 or to abnormal metabolism, have been described [26]. Some, including acrodermatitis enteropathica, cystic fibrosis, Crohn disease, peripheral neuropathy, and congenital liver disease have diminished capabilities for desaturation or chain elongation of PUFA or conversion to eicosanoids. Alcoholism, cirrhosis, Reye's syndrome, and chronic malnutrition are accompanied by significantly abnormal patterns of essential fatty acids in serum phospholipids. As high intake of n-6 may be proinflammatory, a countering interaction of n-6 PUFA could protect in situations such as rheumatoid arthritis, autoimmune disease (e.g., AIDS) or malignant tumor progression. Balance between PUFA families may be particularly important during stress of recovery from surgery or burn injury.

## 6.2. Relationships to plasma cholesterol

Considerable evidence supports a correlation between high intake of dietary saturated fats, relative to PUFA, and the occurrence of atherosclerosis and coronary disease. Risk of coronary disease is proportional to serum cholesterol levels and total serum cholesterol can be decreased following dietary intake of lipids enriched in PUFA [22]. Factors such as platelet aggregation, blood pressure and vascular obstruction may be influenced through some of the oxygenated derivatives of PUFA.

# 7. Regulation through sensors and receptors

## 7.1. Membrane sensing factors and response elements

Cells closely regulate membrane content of unsaturated fatty acids, although precisely how the level is sensed and a response is initiated is not understood. Genetic studies in yeast resulted in discovery of an unsaturated fatty acids sensing system. A transcription factor encoded by the yeast *STP23* gene is required for efficient transcription of the Δ9 desaturase gene, *OLE1*. Addition of exogenous unsaturated fatty acids results in decreased *OLE1* transcription and synthesis of unsaturated fatty acids. The regulatory *SPT23* protein contains a C-terminal, membrane-spanning domain that anchors the protein to the ER. Upon addition of exogenous unsaturated (but not saturated) fatty acids, *SPT23* protein is ubiquinated, resulting in targeted cleavage of the C-terminus. The N-terminal *SPT23* transcription factor moves to the nucleus and shuts off transcription of the *OLE1* gene [27,28]. Thus, this ER-bound protein is a precursor of a nuclear transcription factor and is similar to SREBP that is derived from an ER precursor protein (Chapter 15).

Based on similarities of the yeast system to mammalian SREBP processing in response to altered sterol levels (Chapter 15), a system for sensing of unsaturated fatty acids levels in mammalian cells also may exist. Both SCD1 and SCD2 are regulated by sterols through SREBP. In transgenic mice, truncated forms of SREBP give a 3- to 4-fold increase in both SCD1 and 2. Both SREBP and PUFA regulation of SCDs through promoter reporter regions are dependent on sterol- and adjacent-response elements.

## 7.2. Peroxisomal proliferator-activated receptors (PPARs)

Fatty acids and their oxygenated derivatives are among the known activators of members of the PPAR family of nuclear receptors [14]. PPARγ is a transcription factor and critical modulator of fat cell differentiation and function, thus providing a direct link between fatty acid concentrations and regulation of gene transcription in adipocytes. Drugs that act as agonists or antagonists of PPARγ can be used to alter lipid metabolism, insulin resistance and diabetes and to influence related diseases such as obesity, atherosclerosis and hypertension. Although paradoxes remain in the complex interactions mediated by PPARs, studies with gene knockouts or specifically designed modulator molecules are shedding new light on PPARs in relation to uptake, storage and oxidation of fatty acids.

# 8. Future directions

Progress in understanding desaturation and chain elongation of fatty acyl chains and the influence of fatty acids on metabolic functions has been exciting over the last half century. At the same time, expansion of knowledge of mechanisms, regulation and functions of these processes, particularly the extent to which information applies to humans, is needed. Appropriate levels and balance of n-6 and n-3 fatty acids and the modulating role of n-3 acids, especially relating to fetal and neonatal development, to stresses such as surgery, accidental injury or incompetent immunity, and to inherited or acquired diseases require continuing study. Surprisingly, potential deleterious effects of *trans* unsaturated fatty acids still has not been resolved fully. Are monenoic fatty acids a logical substitution if other dietary fatty acids are reduced? In establishing nutritional standards for classes of fatty acids, both individual and species variability and the complex interplay of a variety of physiological functions must be considered.

Application of cloning, molecular probes and genetic technologies to desaturation and chain elongation enzymes has substantially progressed recently but the full power of molecular probes and techniques is yet to be realized in this realm. Specific molecular probes for Δ6 desaturase will be beneficial for determining whether the enzyme(s) acting on 18:2 n-6, 18:3 n-3, 24:4 n-6 and 24:5 n-3 are identical. Modulation of Δ5 desaturase relative to preceding desaturation and elongation steps could be enlightening. Studies based on mutant cell lines or transgenic and gene disrupted mice, deficient in or overexpressing one or more components of the desaturation–chain elongation sequences, should provide exciting insights into regulation. To what extent are the components coordinately regulated and is the sequence of activities physically associated to provide channelled conversion to the major end products? Is the location of desaturation and elongation confined to the ER or might there be significant activities within peroxisomes or nuclear and plasma membranes?

Intracellular signalling roles for PUFA, in addition to being precursors of eicosanoids, are recognized but require more definition particularly in relation to PPAR (or other orphan receptors) classes and functions. How do PUFA functions relate to diseases and the type and amount of dietary fatty acids we ingest? Details of alternatives to classical pathways of n-3 and n-6 fatty acid metabolism must be pursued.

## Abbreviations

| ER | endoplasmic reticulum |
|---|---|
| PPAR | peroxisomal proliferator activated receptor |
| PUFA | polyunsaturated fatty acid |
| SCD | stearoyl-CoA desaturase |
| SREBP | sterol response element binding protein |

## References

1. Mead, J.F. (1981) The essential fatty acids: past, present and future. Prog. Lipid Res. 20, 1–6.
2. Cinti, D.L., Cook, L., Nagi, M.N. and Suneja, S.K. (1992) The fatty acid chain elongation system of mammalian endoplasmic reticulum. Prog. Lipid Res. 31, 1–51.
3. Seubert, W. and Podack, E.R. (1973) Mechanisms and physiological roles of fatty acid chain elongation in microsomes and mitochondria. Mol. Cell. Biochem. 1, 29–40.
4. Cho, H.P., Nakamura, M. and Clarke, S.D. (1999) Cloning, expression, and fatty acid regulation of the human Delta-5 desaturase. J. Biol. Chem. 274, 37335–37339.
5. Cho, H.P., Nakamura, M.T. and Clarke, S.D. (1999) Cloning, expression, and nutritional regulation of the mammalian Delta-6 desaturase. J. Biol. Chem. 274, 471–477.

6. Los, D.A. and Murata, N. (1998) Structure and expression of fatty acid desaturases. Biochim. Biophys. Acta Lipids Lipid Metab. 1394, 3–15.

7. Oh, C.S., Toke, D.A., Mandala, S. and Martin, C.E. (1997) *ELO2* and *ELO3*, homologues of the *Saccharomyces cerevisiae ELO1* gene, function in fatty acid elongation and are required for sphingolipid formation. J. Biol. Chem. 272, 17376–17384.

8. Tabor, D.E., Kim, J.B., Spiegelman, B.N. and Edwards, P.A. (1999) Identification of conserved *cis*-elements and transcription factors required for sterol-regulated transcription of stearoyl-CoA desaturase 1 and 2. J. Biol. Chem. 274, 20603–20610.

9. Tocher, D.R., Leaver, M.J. and Hodgson, P.A. (1998) Recent advances in the biochemistry and molecular biology of fatty acyl desaturases. Prog. Lipid Res. 37, 73–117.

10. Clarke, S.D. and Jump, D.B. (1994) Dietary polyunsaturated fatty acid regulation of gene transcription. Annu. Rev. Nutr. 14, 83–98.

11. Duplus, E., Glorian, M. and Forest, C. (2000) Fatty acid regulation of gene transcription. J. Biol. Chem. 275, 30749–30752.

12. Graber, R., Sumida, C. and Nunez, E.A. (1994) Fatty acids and cell signal transduction. J. Lipid Mediat. 9, 91–116.

13. Stubbs, C.D. and Smith, A.D. (1990) Essential fatty acids in membrane: physical properties and function. Biochem. Soc. Trans. 18, 779–781.

14. Willson, T.M., Lambert, M.H. and Kliewer, S.A. (2001) Peroxisome proliferator-activated receptor γ and metabolic disease. Annu. Rev. Biochem. 70, 341–367.

15. Hannah, V.C., Ou, J.F., Luong, A., Goldstein, J.L. and Brown, M.S. (2001) Unsaturated fatty acids down-regulate SREBP isoforms 1a and 1c by two mechanisms in HEK-293 cells. J. Biol. Chem. 276, 4365–4372.

16. Gonzalez, C.I. and Martin, C.E. (1996) Fatty acid-responsive control of mRNA stability — unsaturated fatty acid-induced degradation of the *Saccharomyces OLE1* transcript. J. Biol. Chem. 271, 25801–25809.

17. Miyazaki, M., Kim, Y.C., Gray-Keller, M.P., Attie, A.D. and Ntambi, J.M. (2000) The biosynthesis of hepatic cholesterol esters and triglycerides is impaired in mice with a disruption of the gene for stearoyl-CoA desaturase 1. J. Biol. Chem. 275, 30132–30138.

18. Miyazaki, M., Man, W.C. and Ntambi, J.M. (2001) Targeted disruption of stearoyl-CoA desaturase1 gene in mice causes atrophy of sebaceous and meibomian glands and depletion of wax esters in the eyelid. J. Nutr. 131, 2260–2268.

19. Miyazaki, M., Kim, Y.C. and Ntambi, J.M. (2001) A lipogenic diet in mice with a disruption of the stearoyl-CoA desaturase 1 gene reveals a stringent requirement of endogenous monounsaturated fatty acids for triglyceride synthesis. J. Lipid Res. 42, 1018–1024.

20. Sprecher, H. (2000) Metabolism of highly unsaturated n-3 and n-6 fatty acids. Biochim. Biophys. Acta Mol. Cell Biol. Lipids 1486, 219–231.

21. Cook, H.W., Clarke, J.T.R. and Spence, M.W. (1983) Concerted stimulation and inhibition of desaturation, chain elongation, and esterification of essential fatty acids by cultured neuroblastoma cells. J. Biol. Chem. 258, 7586–7591.

22. Fernandes, G. and Venkatraman, J.T. (1993) Role of omega-3 fatty acids in health and disease. Nutr. Res. 13, S19–S45.

23. Carlson, S.E., Clandinin, M.T., Cook, H.W., Emken, E.A. and Filer Jr., L.J. (1997) *Trans* fatty acids: infant and fetal development. Am. J. Clin. Nutr. 66, 717S–736S.

24. Ip, C. (1997) Review of the effects of *trans* fatty acids, oleic acid, n-3 polyunsaturated fatty acids, and conjugated linoleic acid on mammary carcinogenesis in animals. Am. J. Clin. Nutr. 66, 1523S–1529S.

25. West, D.B., Delany, J.P., Camet, P.M., Blohm, F., Truett, A.A. and Scimeca, J. (1998) Effects of conjugated linoleic acid on body fat and energy metabolism in the mouse. Am. J. Physiol. Regul. Integr. Comp. Physiol. 275, R667–R672.

26. Holman, R.T. and Johnson, S. (1981) Changes in essential fatty acid profiles of serum phospholipids in human disease. Prog. Lipid Res. 20, 67–73.

27. Hoppe, T., Matuschewski, K., Rape, M., Schlenker, S., Ulrich, H.D. and Jentsch, S. (2000) Activation of a membrane-bound transcription factor by regulated ubiquitin/proteasome-dependent processing. Cell 102, 577–586.

28. Zhang, S.R., Skalsky, Y. and Garfinkel, D.J. (1999) *MGA2* or *SPT23* is required for transcription of the Delta9 fatty acid desaturase gene, *OLE1*, and nuclear membrane integrity in *Saccharomyces cerevisiae*. Genetics 151, 473–483.
29. Cook, H.W. and Spence, M.W. (1973) J. Biol. Chem. 248, 1793–1796.

D.E. Vance and J.E. Vance (Eds.) *Biochemistry of Lipids, Lipoproteins and Membranes (4th Edn.)*

# CHAPTER 8

# Phospholipid biosynthesis in eukaryotes

## Dennis E. Vance

*CIHR Group on Molecular and Cell Biology of Lipids and Department of Biochemistry, University of Alberta, Edmonton, AB T6G 2S2, Canada, Tel.: +1 (780) 492-8286; Fax: +1 (780) 492-3383; E-mail: dennis.vance@ualberta.ca*

## 1. Introduction

The objective of this chapter is to provide an overview of eukaryotic phospholipid biosynthesis at an advanced level. Phospholipids make up the essential milieu of cellular membranes and act as a barrier for entry of compounds into cells. Phospholipids also function as precursors of second messengers such as diacylglycerol (DG) and inositol-1,4,5-$P_3$ which is covered in Chapter 12. A third, and usually overlooked function of phospholipids, is storage of energy in the form of fatty acyl components. This function is probably quantitatively important only under extreme conditions such as starvation.

## 2. Phosphatidic acid biosynthesis and conversion to diacylglycerol

Phosphatidic acid (PA) is an intermediate that occurs at a branch point in glycerolipid biosynthesis as shown in Fig. 1. Significant developments in elucidation of the biosynthetic pathway occurred in the 1950s when Kornberg and Pricer demonstrated that fatty acids are activated to acyl-CoA prior to reaction with glycerol-3-P. Subsequent studies from the laboratories of Kennedy, Shapiro, Hübscher and others delineated the biosynthetic pathway for PA.

An important step in PA biosynthesis is the activation of fatty acids by acyl-CoA synthetases to yield acyl-CoA (Fig. 2). Five different forms of rat acyl-CoA synthetase have been cloned, each encoded by a separate gene (R.A. Coleman, 2001). Different forms of the enzyme have been found on endoplasmic reticulum (ER), mitochondria and mitochondrial associated membrane (MAM), a subfraction of the ER (J.E. Vance, 1990). Hence, these synthetases may function differently in providing substrate for phospholipid and triacylglycerol (TG) biosynthesis.

### 2.1. Glycerol-3-P acyltransferase

This enzyme catalyzes the first committed reaction in the biosynthesis of PA. The relative importance of this acyltransferase in regulation of phospholipid biosynthesis has not been clearly established. In mammals, two glycerol-3-P acyltransferases have been identified, one associated with mitochondria and the other on the ER (Fig. 2). The ER acyltransferase is inhibited by *N*-ethylmaleimide whereas the mitochondrial

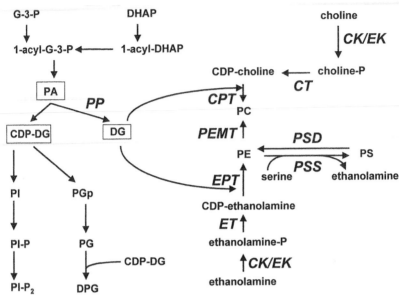

Fig. 1. Phospholipid biosynthetic pathways in animal cells. The abbreviations are: DHAP, dihydroxyace-tone phosphate; G-3-P, glycerol-3-phosphate; PA, phosphatidic acid; DG, diacylglycerol; CDP-DG, cytidine diphosphodiacylglycerol; PI, phosphatidylinositol; PG, phosphatidylglycerol; PGp, phosphatidylglycerol phosphate; DPG, diphosphatidylglycerol; PP, phosphatidic acid phosphatase; PE, phosphatidylethanolamine; PC, phosphatidylcholine; PEMT, phosphatidylethanolamine *N*-methyltransferase; CT, CTP: phosphocholine cytidylyltransferase; PS, phosphatidylserine; CK/EK, choline/ethanolamine kinase; CPT, CDP-choline: 1,2-diacylglycerol cholinephosphotransferase; EPT, CDP-ethanolamine: 1,2-diacylglycerol ethanolaminephos-photransferase; ET, CTP: phosphoethanolamine cytidylyltransferase; PSD, phosphatidylserine decarboxy-lase; PSS, phosphatidylserine synthase.

enzyme is not. The mitochondrial acyltransferase prefers palmitoyl-CoA as an acyl donor compared to oleoyl-CoA, whereas the ER enzyme does not show a preference for saturated versus unsaturated acyl-CoAs. For this and other reasons the mitochondrial enzyme is thought to be primarily responsible for the abundance of saturated fatty acids in the SN-1 position of glycerophospholipids (D. Halder, 1994). Mitochondrial glycerol-3-P acyltransferase has been localized to the outer mitochondrial membrane with the active site facing the cytosol (R.A. Coleman, 2001). Mitochondrial glycerol-3-P acyltransferase has been purified and the cDNA cloned.

Transcription of the mitochondrial acyltransferase gene is decreased by starvation and glucagon, and increased by a high carbohydrate diet [1]. These responses make physiological sense since animals do not need to make triacylglycerols and to a lesser extent phospholipids when energy supply is limited. The microsomal activity is not significantly altered by these treatments. The 5′ flanking region of the murine gene for the mitochondrial acyltransferase was linked to a luciferase reporter plasmid and expressed in 3T3-L1 pre-adipocytes. Deletion analysis on the promoter indicated that sequences between −86 and −55 bp were important for the expression of luciferase activity. Subsequent studies identified −78 to −55 bp as sites that bind sterol response

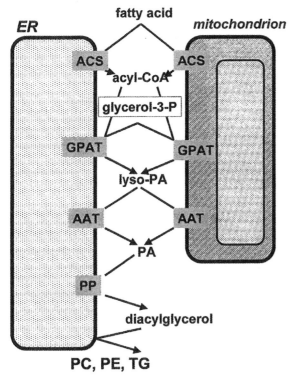

Fig. 2. Biosynthesis of phosphatidic acid (PA) can occur on both the endoplasmic reticulum and the outer membrane of mitochondria. The abbreviations are: ACS, acyl-CoA synthetase; GPAT, glycerol-3-P acyltransferase; AAT, 1-acylglycerol-3-P acyltransferase; PP, phosphatidic acid phosphatase; PC, phosphatidylcholine; PE, phosphatidylethanolamine; TG, triacylglycerol.

element binding protein (SREBP) and NF-Y transcription factors (P.A. Edwards, 1997). As indicated in Chapters 6, 7 and 15, SREBP is a key regulator of cholesterol and fatty acid synthesis, and fatty acid desaturation. Thus, it makes physiological sense that SREBP would also enhance glycerolipid, particularly TG, biosynthesis by increasing the expression of glycerol-3-P acyltransferase. Disruption of the mitochondrial glycerol-3-P acyltransferase gene in mice caused a decrease in the TG and cholesterol content in liver and striking changes in glycerolipid fatty acid composition (R.A. Coleman, 2002).

## 2.2. 1-Acylglycerol-3-P acyltransferase

Much less is known about the second step in the PA biosynthetic pathway (Fig. 2). The activity of this acyltransferase is much lower in mitochondria than in ER. It is presumed that much of the lyso-PA formed in mitochondria is transferred to ER for the second acylation. In vitro studies indicate that a carrier protein is not required (A.K. Hajra, 1992). The esterification at position 2 is specific for unsaturated fatty acids. However, the types of fatty acyl-CoAs available also influence the acyl-CoA selected for transfer to lyso-PA. Two human isoforms of 1-acylglycerol-3-P acyltransferase have been cloned and expressed [2].

## 2.3. Dihydroxyacetone-P acyltransferase

This enzyme is an integral membrane protein exclusively localized to the luminal side of peroxisomes (A. Poulos, 1993). Reports on the localization to other organelles are likely a result of peroxisomal contamination. Once 1-acyldihydroxyacetone-P is formed it can be used as a substrate for 1-alkyldihydroxyacetone-P synthesis (Chapter 9) or reduced to lyso-PA by a peroxisomal acyldihydroxyacetone-P reductase (Fig. 1) which also utilizes 1-alkyldihydroxyacetone-P as a substrate.

## 2.4. Phosphatidic acid phosphatase

This enzyme hydrolyses PA to DG which can be converted to TG, phosphatidylcholine (PC) or phosphatidylethanolamine (PE) (Figs. 1 and 2). There are two forms of the phosphatase. The cytosolic ER form is dependent on $Mg^{2+}$ and inhibited by thiol reagents such as $N$-ethylmaleimide. The activity of this enzyme can be regulated by reversible translocation between cytosol and ER. The cytosolic form of the enzyme is inactive and is translocated to the ER membrane in the presence of fatty acids, fatty acyl-CoAs and PA. Since the substrate, PA, is found on the ER it is logical to expect that the ER is where the enzyme functions in the cell.

The second PA phosphatase is neither inhibited by $N$-ethylmaleimide nor stimulated by $Mg^{2+}$. The cDNA for this phosphatase was cloned and expressed [3]. It appears to be a glycosylated protein on the plasma membrane with 6 putative transmembrane regions. This lipid phosphatase also hydrolyzes ceramide-1-phosphate, sphingosine-1-phosphate as well as lyso-PA. The active site is predicted to face outside the cell and has activity on lyso-PA but low activity on PA (H. Kanoh, 2000). Thus, the plasma membrane phosphatase probably has no role in intracellular phospholipid biosynthesis. Consistent with a signaling role, the plasma membrane phosphatase shares 34% sequence identity with Wunen protein (also has phosphatase activity) that functions in germ cell migration in *Drosophila* embryos. Furthermore, both of these proteins are very similar to Dri42, a protein that is upregulated in expression during epithelial differentiation in rat intestinal mucosa.

Yeast also has two PA phosphatases, a 104-kDa, microsomal form and a 45-kDa, mitochondrial form [4]. Addition of inositol induces the expression of the 45-kDa enzyme but not the 104-kDa enzyme (see Section 10 for more on regulation of phospholipid biosynthesis in yeast).

A novel DG pyrophosphate phosphatase has also been identified in yeast [4]. The enzyme has PA phosphatase activity but prefers DG pyrophosphate as substrate. This phosphatase has broad specificity but only DG pyrophosphate and PA have been shown to be substrates in vivo. A yeast mutant defective in the pyrophosphatase gene was viable and accumulated DG pyrophosphate (G.M. Carman, 1998). The enzyme expression is induced by inositol and zinc deprivation (G.M. Carman, 2001). The enzyme is similar to the mammalian plasma membrane PA phosphatase which also hydrolyzes DG pyrophosphate. The function of DG pyrophosphate is unknown. Since it occurs at a very low level in yeast (0.18 mol% of total phospholipids), DG pyrophosphatase may have a signaling function [4].

## 3. Phosphatidylcholine biosynthesis

### 3.1. Historical background

PC was first described by Gobley in 1847 as a component of egg yolk and named 'lecithin' after the Greek equivalent for egg yolk (*lekithos*). In the 1860s Diakonow and Strecker demonstrated that lecithin contained two fatty acids linked to glycerol and that choline was attached to the third hydroxyl by a phosphodiester linkage. The first significant advance in understanding PC biosynthesis occurred in 1932 with the discovery by Charles Best that animals had a dietary requirement for choline. In the 1950s the CDP-choline pathway for PC biosynthesis (Fig. 3) was described by Eugene Kennedy and coworkers. A key observation was that CTP, rather than ATP, was the activating nucleotide for PC biosynthesis [5]. CTP is required not only for PC biosynthesis but also for the de novo synthesis of all phospholipids (prokaryotic and

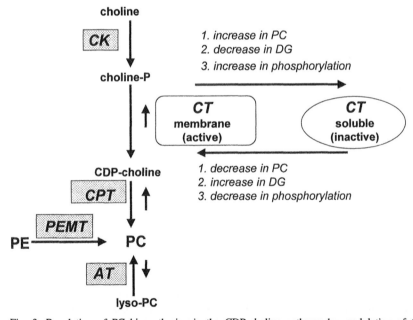

Fig. 3. Regulation of PC biosynthesis via the CDP-choline pathway by modulation of the binding of CTP:phosphocholine cytidylyltransferase (CT) to membranes. Three different modes of regulation of CT activity are indicated. The abbreviations are: CK, choline kinase; CPT, CDP-choline:1,2-diacylglycerol cholinephosphotransferase; PEMT, phosphatidylethanolamine *N*-methyltransferase; AT, lyso-PC acyltransferase; PC, phosphatidylcholine; PE, phosphatidylethanolamine; DG, diacylglycerol.

eukaryotic, excluding PA which can be considered to be an intermediate in glycerolipid biosynthesis).

An alternative pathway for PC biosynthesis, of quantitative significance only in liver, is the conversion of PE to PC via PE methylation (Fig. 3). The first observation of this pathway was in 1941 when Stetten fed [$^{15}$N]ethanolamine to rats and isolated [$^{15}$N]choline. Two decades later Bremer and Greenberg detected a microsomal enzyme that converted PE to PC via transfer of methyl groups from $S$-adenosylmethionine.

### 3.2. Choline transport and oxidation

Choline is not made de novo in animal cells except by methylation of PE to PC and subsequent hydrolysis of the choline moiety. Therefore, choline must be imported from extracellular sources. There are two distinct transport mechanisms for choline [6]; a high affinity ($K_m$ or $K_t < 5$ μM), Na-dependent transporter and a lower affinity ($K_t > 30$ μM), Na-independent transporter. Several cDNAs encoding proteins that show high affinity transport of choline have been reported. A human cDNA is predicted to have 13 transmembrane spanning domains (R.D. Blakely, 2000).

Once choline is inside the cell, its normal fate is rapid phosphorylation by choline kinase (Fig. 3). In neurons choline is also converted to the neurotransmitter, acetylcholine. Choline is also oxidized to betaine [$^-$OOC–CH$_2$–N$^+$(CH$_3$)$_3$] in the liver and kidney. In liver betaine is an important donor of methyl groups for methionine biosynthesis and the one carbon pool. Betaine is produced in mitochondria into which choline is transported by a specific transporter on the inner membrane. Next, choline is oxidized to betaine aldehyde by choline dehydrogenase on the inner leaflet of the inner mitochondrial membrane. The conversion to betaine is catalyzed by betaine-aldehyde dehydrogenase located in the mitochondrial matrix.

Betaine can be transported into kidney medulla by a betaine transporter. In renal medulla, eubacteria, halotolerant plants, marine invertebrates and cartilaginous fish, betaine accumulates as an osmolyte (a small organic solute that accumulates in response to hypertonicity without adverse affect to the cell or organism) (J.S. Handler, 1992). Hypertonicity of the renal medulla is important for the kidney's ability to concentrate urine.

### 3.3. Choline kinase

The enzyme was first demonstrated in yeast extracts by J. Wittenberg and A. Kornberg (more famous for his contributions to DNA replication) in 1953. The enzyme was purified by K. Ishidate (1984) from rat kidney and shown also to phosphorylate ethanolamine [6]. This kinase is now referred to as choline/ethanolamine kinase β. The cDNA for a rat liver choline/ethanolamine kinase encoded an enzyme that is now referred to as choline/ethanolamine kinase α1. Northern analyses indicate that the mRNA for choline/ethanolamine kinase α1 is most abundant in testis. Choline/ethanolamine kinase α2 appears to be a splice variant of choline/ethanolamine kinase α1. The choline/ethanolamine kinase α and β genes have been characterized. The length of the gene was 40 kb for the choline/ethanolamine kinase α gene whereas the β gene was only 3.5 kb in length (K. Ishidate, 2000).

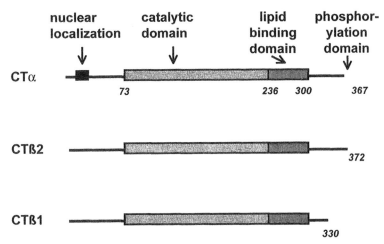

Fig. 4. Domain structures of CTP:phosphocholine cytidylyltransferase (CT) α, β1 and β2. CTα contains a nuclear localization signal, a N-terminal catalytic domain, an amphipathic helical (lipid binding) domain and a C-terminal phosphorylation domain. The CTβ forms lack the nuclear localization signal but contain catalytic and amphipathic helical domains. CTβ1 is missing the phosphorylation domain whereas CTβ2 has a phosphorylation domain that is different from that of CTα.

Choline is not only required in the diet of animals but also in the medium of animal cells in culture (H. Eagle, 1955). Choline is essential because of the cell's requirement for PC to grow and divide.

There is evidence that the activity of choline kinase might be regulatory for cell division in some cases [6].

*3.4. CTP:phosphocholine cytidylyltransferase*

This enzyme activity was first described by Kennedy and Weiss in 1955 [5]. Over three decades later CT was finally purified to homogeneity (P.A. Weinhold, 1987). The CT gene was cloned from S. cerevisiae (S. Yamashita, 1987) by complementation of a yeast mutant defective in CT activity. The cDNA of rat liver CT was subsequently cloned (R.B. Cornell, 1990). CT is a homodimer in soluble extracts of rat liver and is also found on membranes. In most cells CT is thought to exist in an inactive reservoir in its soluble form and to be active when associated with membranes (Fig. 3).

Two genes encode different forms of CT, α and β. The CTα gene spans approximately 26 kb. Exon 1 is untranslated, exon 2 encodes the translation start site and a nuclear localization signal, exons 4–7 encode the catalytic domain, exon 8 codes for the alpha helical membrane binding domain and exon 9 encodes a C-terminal phosphorylation domain (I. Tabas, 1997) (Fig. 4).

The CTβ gene is located on the X chromosome and encodes two isoforms, CTβ1 and CTβ2 (Fig. 4), presumably derived by mRNA splicing. Both isoforms differ from CTα at the amino terminus, lack the nuclear localization signal and are found in the cytoplasm of animal cells [7]. The primary sequences of CTβ1 and CTβ2 are identical except at the carboxyl terminal. CTβ1 lacks most of the phosphorylation domain that is present in CTβ2. There are significant differences between the sequences of the phosphorylation domains of CTα and CTβ.

CT has classically been considered to be a cytoplasmic enzyme since its activity is found in the cytosol and on microsomal membranes in cellular homogenates. However, Kent and coworkers demonstrated that CT was found in the nuclear matrix and associated with the nuclear membrane [8]. The role of the nuclear localization signal in CTα was explored by mutagenesis. Mutated CTα was expressed in a CHO mutant (MT-58) that was temperature-sensitive for CT activity (C. Raetz, 1980). In MT-58 cells, CT activity is present at low levels and the cells grow at 33°C. At the restrictive temperature of 40°C, there was no CT activity and the cells died via apoptosis (F. Tercé, 1996). Expression of CTα in which residues 8–28 (the nuclear localization signal) was deleted resulted in expression of CT largely, but not exclusively, in the cytoplasm [8]. These cells were able to survive at the restrictive temperature. Since some CTα was expressed in the nucleus, the experiment does not yet prove that cells can grow and divide when CTα is present only in the cytoplasm. There is intriguing evidence that CTα migrates into the cytoplasm during the G1 phase of the cell cycle, a time when PC biosynthesis is activated (R.B. Cornell, 1999). Thus, the role the nuclear localization signal of CT plays in cellular PC biosynthesis remains an intriguing question.

The lipid binding domain and the phosphorylated domains are involved in the regulation of CT activity. These domains of CTα have been deleted by either proteolysis with chymotrypsin or by construction of CTα truncation mutants [9]. CTα cDNAs that were truncated in the region of residue 314 (Fig. 4) lacked the phosphorylation segment, and CT truncated at residues 236, 231 or 228 lacked both the phosphorylation and lipid binding domains. When the lipid binding and phosphorylation domains were deleted, CT was a soluble, active enzyme that did not bind to membranes. Thus, the lipid binding domain is regulatory for the binding to membranes and the activation of CT. The binding of phospholipids to CT appears to activate the enzyme by decreasing the apparent $K_m$ value for CTP (S.L. Pelech, 1982; S. Jackowski, 1995).

CT activity is modulated by phosphorylation. Experiments with CT truncation mutants have demonstrated that the phosphorylation domain is not required for lipid binding or CT activity. In vitro, CT is phosphorylated by casein kinase II, cdc2 kinase, cAMP kinase, protein kinase C and glycogen synthase kinase-3 but not by MAP kinase. However, the stoichiometry of phosphorylation is less than 0.2 mol P/mol CT with any of the kinases and in vitro phosphorylation does not affect enzyme activity. Exactly what role phosphorylation of CT plays in a physiologically relevant system remains to be demonstrated (see Section 4.4).

### 3.5. CDP-choline : 1,2-diacylglycerol cholinephosphotransferase

This enzyme was also discovered by Kennedy and coworkers [5] and is considered to be located on the ER but is also found on the Golgi, MAM and nuclear membranes [10]. Even though the enzyme has been known for more than four decades and despite intense efforts in many laboratories, the cholinephosphotransferase has never been purified. The difficulty is that the enzyme is an intrinsic membrane-bound protein that

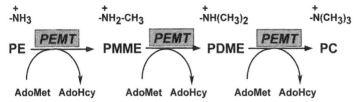

Fig. 5. Reactions catalyzed by phosphatidylethanolamine N-methyltransferase (PEMT). AdoMet, S-adenosylmethionine; AdoHcy, S-adenosylhomocysteine; PMME, phosphatidylmonomethylethanolamine; PDME, phosphatidyldimethylethanolamine; PC, phosphatidylcholine.

requires detergents for solubilization. Moreover, the detergents complicate purification procedures commonly used such as gel filtration because the protein binds to micelles that are hard to separate on the basis of molecular size. The purification of membrane-bound enzymes has been described as 'masochistic enzymology' (D.E. Vance, 1990).

Yeast genetics and molecular biology have, however, allowed for the cholinephosphotransferase to be cloned. Two genes, *CPT1* and *EPT1*, each account for 50% of the cholinephosphotransferase activity in yeast extracts [10]. By the use of null mutations in these two genes, it has been established that *CPT1* is responsible for 95% of the PC made and *EPT1* gene product accounts for 5%. The *EPT1* gene product utilizes both CDP-choline and CDP-ethanolamine whereas *CPT1* catalyzes only reactions with CDP-choline.

More recently a human choline/ethanolaminephosphotransferase cDNA (hCEPT1) was cloned and expressed (C.R. McMaster, 1999). The open reading frame predicts a protein with 7 membrane-spanning domains. Subsequently, the same lab cloned a human cDNA that encoded for a CDP-choline-specific enzyme (hCPT1) with 60% sequence identity to hCEPT1. hCEPT1 mRNA was detected in all tissues tested whereas the expression of hCPT1 was highest in heart, testis, intestine and colon.

Cholinephosphotransferase acts at a branch point in the metabolism of DG that can also be converted to PE, TG or PA (Fig. 1). Most studies indicate that there is an excess of cholinephosphotransferase in cells, hence, the amount of active enzyme does not limit PC biosynthesis. However, it is clear that the in vivo activity of cholinephosphotransferase is regulated by substrate supply. The supply of CDP-choline is regulated by the activity of CT (Section 3.4). The supply of DG in liver seems to be controlled by the supply of fatty acids. Excess DG not utilized for PC or PE biosynthesis is stored in liver as TG.

### 3.6. Phosphatidylethanolamine N-methyltransferase

All nucleated cells contain PC and the CDP-choline pathway. Thus, it was not obvious why the pathway for PE methylation (Fig. 5) survived during evolution. Nor was it obvious why PE methyltransferase (PEMT) activity is mostly found in liver whereas 2% or less of the hepatic PEMT activity is found in other tissues of the body.

PEMT was purified from rat liver microsomes although it is an intrinsic membrane protein (N.R. Ridgway, 1987). Sequence of the amino terminal enabled the cloning of

214

the cDNA for PEMT (Z. Cui, 1993). Preparation of an antibody to the deduced sequence of the carboxyl terminal peptide permitted subcellular localization of the enzyme. The major activity for PEMT is found on the ER but the antibody only recognized a protein that was exclusively localized to MAM (J.E. Vance, 1990). This isoenzyme of PEMT is referred to as PEMT2 and the activity on the ER is called PEMT1. Both PEMTs catalyze all three transmethylation reactions that convert PE to PC (Fig. 5).

A mouse was generated in which the *Pemt* gene was disrupted and there was no PEMT activity [11]. The $Pemt^{-/-}$ mice lived and bred normally and there was a 50% increase in CT activity in their livers. Since the mice retained the CDP-choline pathway, the lack of an obvious phenotype was not surprising. However, when the mice were fed a choline-deficient diet for 3 days, which attenuates PC synthesis via the CDP-choline pathway, the $Pemt^{-/-}$ mice exhibited severe liver failure [12]. $Pemt^{+/+}$ mice fed a choline-deficient diet were normal with no obvious liver pathology. Thus, it seems that the PEMT pathway has survived in evolution to provide PC at times when the CDP-choline pathway is less active such as might occur during starvation. Moreover, pregnant rats and suckling mothers can also have choline reserves depleted (S.H. Zeisel, 2000), hence, the PEMT pathway might provide an evolutionary advantage in this respect. The structurally related compound, dimethylethanolamine [$HOCH_2-CH_2-N^+(CH_3)_2$] would not substitute for choline in the $Pemt^{-/-}$ mice even though it was converted to phosphatidyldimethylethanolamine (K.A. Waite, 2002). Thus, it seems that the third methyl group on the phospholipid has a critical function in mice.

Further studies on $Pemt^{-/-}$ mice showed that male, but not female, $Pemt^{-/-}$ mice fed a high fat/high cholesterol diet have a defect in secretion of very low density lipoproteins that contain apolipoprotein B100 (A. Noga, 2002) (Chapter 19). The mechanism(s) for this sexual dimorphism is not clear.

The human gene encoding PEMT has been cloned and characterized. Whereas only one mRNA transcript has been identified in mice, human liver has three separate mRNAs that differ only at the 5' end, in a non-coding region of the transcript (D.J. Shields, 2001). Thus, the three transcripts encode the same protein. The function of separate PEMT mRNAs is going to be difficult to study in humans.

Yeast also has both the PE methylation pathway and the CDP-choline pathway. In yeast two enzymes are used for the conversion of PE to PC [13]. The methylation of PE to phosphatidylmonomethylethanolamine is catalyzed by the *PEM1/CHO2* gene product whereas the subsequent two methylations are catalyzed by the *PEM2/OPI3* gene product. Deletion of both *PEM1* and *PEM2* genes is lethal unless the yeast is supplied with choline. Yeast normally grows in the absence of choline and depend on the PEMT pathway. Thus, the CDP-choline pathway and the PE methylation pathway can compensate for each other in yeast.

Bacteria generally do not contain PC but *Rhodobacter sphaeroides* make PC by methylation of PE. Interestingly, this enzyme is soluble and has virtually no homology to PEMT or the yeast enzymes (V. Arondel, 1993). Also in one bacterium, a novel choline-dependent pathway was recently discovered in *Sinorhizobium meliloti* in which choline reacts with CDP-DC to form PC (O. Geiger, 2000).

## 4. Regulation of phosphatidylcholine biosynthesis

### 4.1. The rate-limiting reaction

The CT reaction usually limits the rate of PC biosynthesis. The first evidence in favor of this conclusion was measurement of pool sizes of the aqueous precursors (in rat liver, choline = 0.23 mM, phosphocholine = 1.3 mM, CDP-choline = 0.03 mM). These values assume that 1 g wet tissue is 1 ml and there is no compartmentation of the pools. The second assumption may not be valid as there is evidence for compartmentation of PC precursors (M. Spence, 1989). Nevertheless, the relative amounts of these compounds might be correct in the biosynthetic compartment(s). The concentration of phosphocholine is 40-fold higher than CDP-choline which is consistent with a 'bottleneck' in the pathway at the reaction catalyzed by CT.

Pulse-chase experiments demonstrate this bottleneck more vividly. After a 0.5 h pulse of hepatocytes with [*methyl*-$^3$H]choline, more than 95% of the radioactivity in the precursors of PC was in phosphocholine, the remainder in choline and CDP-choline. When the cells were chased with unlabeled choline in the medium, labeled phosphocholine was quantitatively converted to PC (Fig. 6). The radioactivity in CDP-choline remained low during the chase and CDP-choline was rapidly converted to PC. There was minimal radioactivity in choline which suggests that choline is immediately phosphorylated after it enters the cell.

One additional point should be made. If a cell or tissue is in a steady state, pool sizes and reaction rates are not changing. Thus, although the rate of PC synthesis is determined by the CT reaction, the rates of the choline kinase and cholinephosphotransferase reactions will be the same as that catalyzed by CT. Otherwise, changes in the pool sizes of precursors would occur. For example, if the choline kinase reaction were faster than

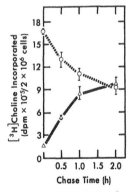

Fig. 6. Incorporation of [$^3$H-*methyl*]choline into phosphocholine and PC as a function of time. Hepatocytes from rat liver were incubated with labeled choline for 30 min. Subsequently, the cells were washed and incubated (chased) for various times with unlabeled choline. The disappearance of radioactivity from phosphocholine (dashed line) and its appearance in PC (solid line) are shown. Adapted from fig. 1 of Pelech et al. (1983), J. Biol. Chem. 258, 6783, with permission.

the CT reaction, there would be an increase in the amount of phosphocholine. Thus, CT sets the pace, but the other reactions proceed at the same rate.

## 4.2. The translocation hypothesis

CT is recovered from cells and tissues in both cytosol and microsomal fractions. However, in the early 1980s evidence from several laboratories suggested a close correlation between CT activity on the microsomal membranes and the rate of PC biosynthesis. The hypothesis was that the active form of the enzyme was on cellular membranes and CT in the cytosol acted as a reservoir (Fig. 3). In agreement with this proposal, cytosolic fractions contain essentially no phospholipid and CT requires phospholipids for activity. Thus, cells have a facile mechanism for altering the rate of PC biosynthesis by a reversible translocation of CT between a soluble, inactive reservoir and cellular membranes. This mechanism for activation applies to CTα and both CTβ1 and CTβ2. This hypothesis remains basically valid except much of the 'cytosolic' CT may originate from the nucleus and activated CT may be associated with the nuclear membrane as well as the endoplasmic reticulum.

Binding of CT to membranes begins by electrostatic adsorption followed by hydrophobic interactions that involve intercalation of the protein into the nonpolar core of the membrane [9] (Fig. 7). When insertion of CT into the membrane lipids is blocked by using viscous gel phase lipids, CT binds electrostatically to the membrane but is not activated. Four properties of membranes promote CT insertion [9]: (1) interfacial packing defects as might occur when lipids with small head groups such as DG are in the membrane; (2) low lateral surface pressure (loose packing) as observed in highly curved compared to planar bilayers; (3) acyl chain disorder that can be caused by oxidation of the fatty acyl chains; (4) curvature strain that would occur when membranes are enriched in hexagonal phase-preferring lipids such as PE and DG. Synthesis of PC would reverse these properties of membranes and form a more stable bilayer.

## 4.3. Regulation of phosphatidylcholine biosynthesis by lipids

As indicated in Fig. 3, the association of CT with membranes and CT activation can be modulated by lipids. Both feed-forward and feed-back mechanisms for regulation of CT activity have been identified. DG may alter the rate of PC biosynthesis both as a substrate and as a modulator of CT binding to membranes. In vitro an increase in the content of DG in membranes enhanced the binding of CT.

Feedback regulation of CT and PC biosynthesis by PC has also been described (H. Jamil, 1990). Regulation of a metabolic pathway by product inhibition is commonly observed. In hepatocytes derived from choline-deficient rats, the rate of PC biosynthesis was inhibited by approximately 70% compared to choline-supplemented rats, the amount of PC declined and there was a corresponding increased binding of CT to membranes. CT appeared to sense a need for increased PC biosynthesis and was poised on the membrane prepared for catalysis. However, in choline-deficient cells there is less substrate, phosphocholine, so increased PC biosynthesis could not occur. When choline-deficient hepatocytes were supplied with choline, there was a positive

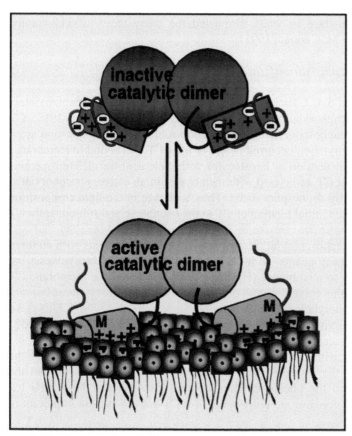

Fig. 7. Translocation of CTP:phosphocholine cytidylyltransferase (CT) from an inactive soluble form to a membrane-associated activated form. The reversible interaction of CT with membranes involves the amphipathic helical region lying on the surface of the membrane with the hydrophilic side interacting with the negatively charged lipid head-groups and the hydrophobic side intercalating into the membrane core. From Cornell and Northwood [9] with permission.

correlation between the increased level of PC and the release of soluble CT. Similar correlations were observed when the level of PC was increased, either by providing methionine for enhanced conversion of PE to PC, or by providing lyso-PC which is imported into hepatocytes and acylated to PC.

An elegant feedback regulation of CT has been shown in the yeast *Saccharomyces cerevisiae* (V.A. Bankaitis, 1995). SEC14p is a phospholipid transfer protein that when assayed in vitro prefers phosphatidylinositol (PI) and PC and is an essential gene product (Chapter 17). SEC14p inhibited the CDP-choline pathway when PC was bound to SEC14p. In contrast, when PI was bound to SEC14p, there was minimal inhibition of CT. Thus, in yeast under conditions where PC is abundant, there is a feedback inhibition of CT and the CDP-choline pathway.

CTP, has also been implicated as regulatory in animal systems and yeast. Overexpression of CTP synthetase in yeast stimulated the biosynthesis of PC via the CDP-choline pathway (G.M. Carman, 1995).

### 4.4. Phosphorylation of cytidylyltransferase

As mentioned in Section 3.4, CTα has a domain that is extensively phosphorylated. Moreover, the state of phosphorylation can affect CT activity [8] (S.L. Pelech, 1982). CT bound to membranes is dephosphorylated compared to soluble CT. The question arose whether or not dephosphorylation occurred before or after CT was bound to membranes (M. Houweling, 1994). Incubation of hepatocytes with oleic acid for different periods of time demonstrated that CT associated with membranes in an active, phosphorylated form and was subsequently dephosphorylated. Thus, a change in the lipid composition of membranes mediated the initial binding of CT to the membrane and subsequently CT was dephosphorylated.

Activation of CT by dephosphorylation was implicated in experiments with cultured macrophages. Loading macrophages with cholesterol stimulated PC biosynthesis but did not alter CT binding to membranes (I. Tabas, 1995). However, the membrane CT increased in activity and this coincided with dephosphorylation of CT.

Deletion of the phosphorylation domain does not impair the ability of CT to make enough PC for cells to survive. This was demonstrated in a line of CHO cells (MT-58) that have a temperature-sensitive defect in the CT reaction (C. Kent, 1995). Stable transfection of the cells with CTα lacking the phosphorylation domain allowed the MT-58 cells to grow at the restricted temperature. Interestingly, CT that lacked the lipid binding domain and the phosphorylation domain also rescued these cells (C. Kent, 1999). Thus, these domains are not essential for CT activity but are important in regulating CT activity.

### 4.5. Transcriptional and post-transcriptional regulation of CTα

Most studies on CT activity and PC biosynthesis have not indicated regulation at the level of gene expression. The ability of a cell to activate the soluble form of CT would normally satisfy the cell's requirement for PC. Nevertheless, some control over the expression of the genes encoding CT must occur. The proximal promoter of the CTα gene has numerous potential regulatory elements (I. Tabas, 1997). Subsequent studies showed that Sp1, the first mammalian transcription factor purified and cloned (R. Tjian, 1986, 1987), had an important role in regulating the expression of the CTα gene [14]. The related nuclear factor, Sp3, could also activate CTα transcription (M. Bakovic, 2000). DNase protection assays indicated several elements in the proximal promoter bound unidentified nuclear factors. The yeast one hybrid system was utilized to clone the cDNA for one of these factors and transcription enhancer factor 4 (TEF4) was identified as a regulator of CTα transcription (H. Sugimoto, 2001). These initial studies were done in experiments involving transfections with various cDNA constructs. The first experiments to search for transcriptional regulation in a cell physiology-relevant system were on CTα expression in a murine fibroblast cell line as a function of the cell cycle (L. Golfman, 2001). During the G0 to G1 phase of the cell cycle there is an increase in PC biosynthesis (S. Jackowski, 1996; L. Golfman, 2001) but there was no enhanced transcription of the CTα gene. Instead, increased transcription occurred during the S phase of the cell cycle, possibly to increase the amount of CTα in preparation for mitosis. Obviously, there is much to be done to elucidate the factors and DNA elements involved in transcriptional regulation of the CTα gene. There are no reports on transcriptional regulation of the CTβ gene.

SREBPs play a critical role in regulating the expression of genes involved in fatty acid (Chapters 6 and 7) and cholesterol (Chapter 15) metabolism. Thus, several labs explored whether or not SREBPs might alter the expression of the CTα gene. Interestingly, one report indicated there was no direct modulation of CTα transcription (N. Ridgway, 2000) whereas research from another lab implicated sterols and SREBPs in the regulation of CTα transcription (P.A. Edwards, 2001). Further work is required on the relationship between SREBPs and PC biosynthesis.

The level of CTα mRNA can also be regulated by alterations in mRNA stability. When a macrophage cell line was depleted of colony stimulating factor and then repleted there was a 4-fold induction of mRNA for CTα (S. Jackowski, 1991). The stability of CTα mRNA increased after the addition of colony stimulating factor. An increase in CTα mRNA in fetal lung type II cells has also been ascribed to enhanced mRNA stability (M. Post, 1996).

Finally, enhanced turnover of CTα via the ubiquitin–proteosome pathway appears to be the mechanism by which tumor necrosis factor decreases the level of CTα in alveolar type II cells (R. Mallampalli, 2000).

### 4.6. Transgenic and gene-disrupted murine models of CTα

To determine if enhanced PC biosynthesis would protect macrophages from excess cholesterol-induced toxicity, genetically modified mice have been generated. A truncated version of CTα lacking the phosphorylation domain was expressed specifically in macrophages of mice under control of the scavenger receptor (Chapter 21) promoter (I. Tabas, 1999). These cells were protected from cholesterol-induced toxicity. In another approach, CTα expression was eliminated in macrophages using the Cre-lox method for selective disruption of a gene in specific cells (I. Tabas, 2000). The lack of the CTα gene and hence decreased PC biosynthesis caused enhanced sensitivity to cholesterol loading. In the absence of cholesterol loading, the macrophages without CTα appeared normal, possibly due to increased expression of CTβ2.

# 5. Phosphatidylethanolamine biosynthesis

### 5.1. Historical background and biosynthetic pathways

PE was first alluded to in a book published by Thudichum in 1884. He described 'kephalin' as a nitrogen- and phosphorus-containing lipid that was different from lecithin. In 1913, Renall and Baumann independently isolated ethanolamine from kephalin. In 1930, Rudy and Page isolated the first pure preparation of PE. The structure of PE was established in 1952 by Baer and colleagues.

The biosynthesis of PE in eukaryotes can occur via four pathways (Fig. 8). The route via CDP-ethanolamine constitutes de novo synthesis of PE. The other pathways arise as a result of the modification of a pre-existing phospholipid. The CDP-ethanolamine pathway was first described by Kennedy and Weiss in 1956. The decarboxylation of phosphatidylserine (PS) to yield PE (Fig. 8) was shown in 1960 to occur in animal cells. PS decarboxylation is the only route for PE biosynthesis in *E. coli* (Chapter 3). The PE generated by this pathway can react with serine to generate PS and ethanolamine (Fig. 8). This appears to be one mechanism by which ethanolamine is made in cells. The other involves degradation of sphingosine (Chapter 14). The ethanolamine generated by either pathway can be utilized for PE biosynthesis via the CDP-ethanolamine pathway. No one has ever been able to show the decarboxylation of serine to ethanolamine in animal cells. Such a reaction was shown to occur in a plant, *Arabidopsis thaliana* (A.D. Hanson, 2001). PE can also be formed by reacylation of lyso-PE or reaction of ethanolamine with PS (Fig. 8).

*5.2. Enzymes of the CDP-ethanolamine pathway*

As mentioned in Section 3.3, the phosphorylation of ethanolamine in liver can be catalyzed by choline/ethanolamine kinase (Figs. 1 and 8). The cDNA encoding an ethanolamine kinase was cloned from *Drosophila* (P. Pavlidis, 1994). These scientists did not plan on cloning this cDNA since their approach was to determine the gene responsible for the easily shocked (*eas*) phenotype in this insect. These mutant flies display transient paralysis following a brief mechanical shock. In the *eas* mutant, a 2 base pair deletion caused formation of a premature stop codon in the ethanolamine kinase gene. Analysis of the phospholipids showed a decrease in PE from 59% of the total phospholipid in wild type to 56% in *eas*. How this minor change mediates the paralysis is not known. The difference may reflect a major change in PE content in a particular tissue or subcellular membrane. More recently, the gene that encodes a yeast ethanolamine kinase (G.M. Carman, 1999) and a human cDNA for ethanolamine-specific kinase (S. Jackowski, 2001) were cloned and expressed.

The second step in the CDP-ethanolamine pathway is catalyzed by CTP:phospho-ethanolamine cytidylyltransferase [15]. The enzyme is distinct from CT and is not activated by lipids. Although the phosphoethanolamine cytidylyltransferase is recovered in cytosol from cell extracts, much of the enzyme has been localized to rough ER of rat liver by immunoelectronmicroscopy. Unlike CTα, there is no report of the phosphoethanolamine cytidylyltransferase in the nucleus.

CDP-ethanolamine:1,2-diacylglycerol ethanolaminephosphotransferase is an integral membrane protein found on the ER, Golgi and MAM. The enzyme shows a distinct preference for DG species that contain 1-palmitoyl-2-docosahexaenoyl (22:6) fatty acids. In hepatocytes in culture, nearly 50% of PE made via the ethanolaminephosphotransferase reaction is this species. The purpose of this extraordinary selectivity is unknown. The bovine hepatic enzyme was purified and exhibited both ethanolamine- and choline-phosphotransferase activity (L. Binaglia, 1999).

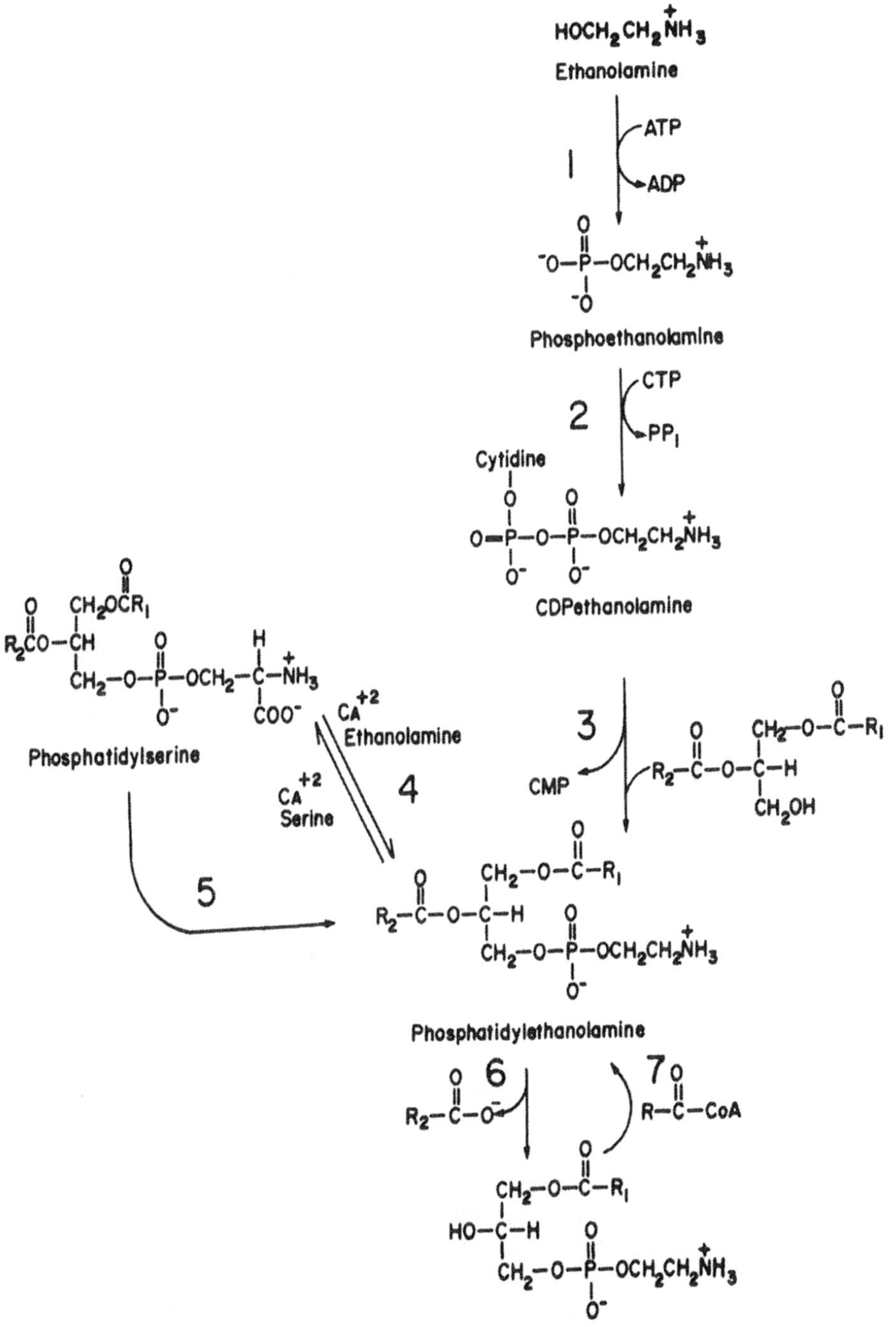

Fig. 8. Pathways for the biosynthesis of PE and PS. The numbers indicate the enzymes involved. 1, ethanolamine(choline) kinase; 2, CTP:phosphoethanolamine cytidylyltransferase; 3, CDP-ethanolamine:1,2-diacylglycerol ethanolaminephosphotransferase; 4, PS synthase; 5, PS decarboxylase; 6, phospholipase A2; 7, acyl-CoA:lyso-PE acyltransferase.

A yeast gene (*EPT1*) that encodes an ethanolaminephosphotransferase and a human cDNA that encodes a choline/ethanolaminephosphotransferase have been cloned as discussed in Section 3.5.

## 5.3. Regulation of the CDP-ethanolamine pathway

Unlike CT, there is minimal literature on the mechanisms that control the activity of phosphoethanolamine cytidylyltransferase. Åkesson and Sundler in the 1970s found that phosphoethanolamine cytidylyltransferase was rate-limiting for PE biosynthesis. However, the supply of DG as a substrate can also limit the rate of PE biosynthesis (L.B.M. Tijburg, 1989). Thus, both the supply of CDP-ethanolamine from the cytidylyltransferase reaction and the supply of DG can regulate PE biosynthesis. Two studies have implicated channeling of intermediates in the biosynthesis of PE in mammalian cells [15].

## 5.4. Phosphatidylserine decarboxylase

PS decarboxylase is found in both prokaryotes (Chapter 3) and the mitochondria of eukaryotes. The enzyme activity was first described by Kanfer and Kennedy in 1964. The enzyme has not been purified from a eukaryotic source but the gene has been cloned and expressed from CHO cells (M. Nishijima, 1991) and yeast [16]. The yeast gene (*PSD1*) encodes a protein that is localized to mitochondria. However, when *PSD1* was disrupted in yeast, 5% PS decarboxylase activity remained and the yeast continued to grow. Subsequently, a second gene, *PSD2*, was isolated. When both *PSD1* and *PSD2* were disrupted, the yeast became ethanolamine auxotrophs. The PSD2 protein has been localized to the vacuolar and Golgi compartments. The function of *PSD2* is not known other than it can supply enough PS decarboxylase to allow growth of yeast in the absence of *PSD1*. The rate of PS decarboxylation is determined by the rate of PS transport into mitochondria (Chapter 17).

# 6. Phosphatidylserine biosynthesis

## 6.1. Historical developments and biosynthesis

PS accounts for 5–15% of the phospholipids in eukaryotic cells. The lower concentration of PS compared to PC and PE is probably the reason PS was not discovered as a separate component of 'kephalin' (originally identified to be only PE in 1930) until 1941 by Folch. The correct structure was proposed by Folch in 1948 and confirmed by chemical synthesis in 1955 by Baer and Maurukas. PS is a required cofactor for protein kinase C and is required for initiation of the blood clotting cascade. In the plasma membrane of cells PS is normally located on the inner monolayer. During apoptosis, exposure of PS on the cell surface (outer monolayer) leads to recognition and removal of these cells by macrophages (V. Fadok, 1992).

PS is made in prokaryotes (Chapter 3), in some plants and yeast (S. Yamashita, 1997) via the CDP-diacylglycerol pathway. This route does not exist in animals. Instead, PS is made by a base-exchange reaction catalyzed by PS synthase first described by Hübscher in 1959 (reaction 4 in Fig. 8) in which the head group of a PC or PE is exchanged for serine.

## 6.2. Chinese hamster ovary cell mutants and regulation

CHO mutants were generated that were auxotrophic for PS and demonstrated that these cells have two PS synthases [17]. PS synthase 1 utilizes PC and serine as substrates whereas PS synthase 2 utilizes only serine and PE. The two PS synthases, when coupled with PS decarboxylase, yield PS at the expense of PC and generate both choline and ethanolamine which could be recycled into the biosynthesis of PC and PE. As a result, PS and PE can both be generated without a decline in the amount of PC.

$$PC + serine \xrightarrow{\text{PS synthase 1}} PS + choline$$

$$PS \xrightarrow{\text{PS decarboxylase}} PE + CO_2$$

$$PE + serine \xrightarrow{\text{PS synthase 2}} PS + ethanolamine$$

The sum of the reactions is:

$$PC + two \ serines \longrightarrow PS + choline + ethanolamine + CO_2$$

A CHO mutant defective in PS synthase 1 was used to clone by complementation the cDNA for this enzyme [17]. The deduced amino acid sequence for murine PS synthase 1 was >90% identical to the CHO enzyme (S. Stone, 1998). The cDNA for PS synthase 2 from CHO cells was cloned and shown to be 32% identical in amino acid sequence to PS synthase 1 [17]. Immunoblot analysis indicated that both of the murine PS synthases are mainly localized to MAM (S. Stone, 2000). The source of the substantial PS synthase activity in the rough and smooth ER remains unknown, possibly a third PS synthase activity. The mRNAs encoding PS synthases 1 and 2 were found in all murine tissues examined but PS synthase 2 was enriched in testis and kidney (J.E. Vance, 2001).

Our understanding of regulation of PS biosynthesis is in its infancy. Addition of exogenous PS to the medium of CHO cells feedback inhibited the biosynthesis of PS [17]. CHO mutants in which Arg-95 of PS synthase 1, or Arg-97 in PS synthase 2, were replaced by lysine were no longer sensitive to inhibition by PS (M. Nishijima, 1998, 1999). There also appears to be 'cross-talk' between PE biosynthesis via the CDP-ethanolamine pathway and PS synthase 1/PS decarboxylation pathway since over-expression of PS synthase 1 increased production of PE from decarboxylation of PS and decreased PE biosynthesis via the CDP-ethanolamine pathway (S. Stone, 1999). Interestingly, over-expression of PS synthase 2 did not alter the activity of the CDP-ethanolamine pathway.

The murine gene for PS synthase 1 has been cloned and characterized (J.E. Vance, 2001). This is an important step toward the generation of mice with a disrupted gene for PS synthase 1. The gene for PS synthase 2 has been disrupted and the mice are viable (S. Young, 2002).

# 7. Inositol phospholipids

## 7.1. Historical developments

A major fate of PA is conversion to DG that is metabolized to PC, PE and TG (Fig. 1). Alternatively, PA can react with CTP to form CDP-DG that is utilized for the biosynthesis of the inositol phospholipids, phosphatidylglycerol (PG) and diphosphatidylglycerol (DPG) (Fig. 1).

Inositol is a cyclohexane derivative in which all 6 carbons are substituted with hydroxyl groups. The most common isoform is *myo*-inositol but other less abundant inositols with different structures also occur. The first report of an inositol-containing lipid was in 1930 from *Mycobacteria* [18] which is ironic since inositol lipids are rarely found in bacteria. Brain is the richest source of these lipids, as first discovered by Folch and Wooley in 1942. In 1949, Folch described PI phosphate (PI-P) which was later found to include PI and PI bisphosphate (PI-P$_2$). The chemical structures of PI, PI-P and PI-P$_2$ were determined by Ballou and coworkers between 1959 and 1961. PI (1.7 $\mu$mol/g liver) constitutes around 10% of the phospholipids in a cell or tissue. PI-P and PI-P$_2$ are present at much lower concentrations (1–3% of PI). Agranoff et al. published the first experiments in 1958 on the incorporation of [$^3$H]inositol into PI. Subsequently, Paulus and Kennedy showed that CTP was the preferred nucleotide donor.

## 7.2. CDP-diacylglycerol synthase

Regulation of the conversion of PA to CDP-DG is not well understood. The enzyme, CDP-DG synthase, is largely microsomal but is also found in the mitochondrial inner membrane.

A cDNA encoding CDP-DG synthase 1 was cloned from *Drosophila* (C.S. Zuker, 1995). This isoform is specifically located in photoreceptor cells of *Drosophila*. Mutations in this isoform lead to a defect in PI-P$_2$ biosynthesis. As a result mutant photoreceptor cells show severe defects in their phospholipase C-mediated signal transduction that can be rescued by re-introduction of the CDP-DG synthase cDNA.

CDNAs encoding human and murine CDP-DG synthases 1 and 2 were more recently cloned (S. Jackowski, 1997; B. Franco, 1999). CDP-DG synthase 2 is expressed during embryogenesis in the central nervous system whereas CDP-DG synthase 1 had a high level of expression in adult retina.

Curiously, in *Saccharomyces cerevisiae*, CDP-DG synthase activity is found in microsomes and the mitochondrial inner membrane even though only one gene encodes this activity [19]. Since only a single mRNA species was found, there may not be alternative splicing of the yeast gene. The yeast CDP-DG synthase gene is essential for cell viability as well as germination of spores.

## 7.3. Phosphatidylinositol synthase

Three potential sources for cellular inositol are: diet, de novo biosynthesis and recycling of inositol. Biosynthesis of inositol from glucose occurs in the brain and testes, and other tissues to a lesser extent. The rate-limiting step appears to be the synthesis of inositol-3-phosphate from glucose-6-phosphate [20]. Inositol-3-phosphate is hydrolyzed to inositol by a phosphatase.

PI synthase was purified from human placenta [21]. When the cDNAs encoding either CDP-DG synthase 1 or phosphatidylinositol synthase, or both, were over-expressed in COS 7 cells, there was no change in the rate of PI biosynthesis indicating that the level of these enzymes was not limiting for PI biosynthesis (S. Jackowski, 1997).

Disruption of the PI synthase gene in yeast is lethal indicating that PI is essential [22]. Further information on the inositol phospholipids and their functions is covered in Chapter 12.

## 8. Polyglycerophospholipids

### 8.1. Historical developments and biosynthetic pathways

Diphosphatidylglycerol (DPG), commonly known as cardiolipin, was discovered in 1942 in beef heart by Pangborn. The correct structure (Fig. 9) was proposed in 1956–1957 and confirmed by chemical synthesis in 1965–1966 by de Haas and van Deenen. Phosphatidylglycerol (PG) was first isolated in 1958 from algae by Benson and Mauro. The structure was confirmed by Haverkate and van Deenen in 1964–1965. The third lipid in this class, bis(monoacylglycerol)phosphate was recovered from pig lung by Body and Gray in 1967. The stereochemistry differs from PG and DPG since bis(monoacylglycerol)phosphate contains $sn$-(monoacyl)glycerol-1-phospho-$sn$-1′-(monoacyl)-glycerol rather than a $sn$-glycerol-3-phospho linkage.

These three lipids (Fig. 9) are widely distributed in animals, plants, and microorganisms. In animals, DPG is found in highest concentration in cardiac muscle (9–15% of phospholipid), hence the name cardiolipin, and is exclusively found in the mitochondria. PG is generally present at a concentration of less than 1% of total cellular phospholipids, except in lung, where it comprises 2–5% of the phospholipid. In pulmonary surfactant and alveolar type II cells, PG is 7–11% of the total lipid phosphorous. Bis(monoacylglycerol)phosphate comprises less than 1% of total phospholipids in animal tissues, except in alveolar (lung) macrophages where it is 14–18% of total phospholipid.

The biosynthesis of PG was elucidated by Kennedy and coworkers in 1963 (Fig. 1). For DPG biosynthesis PA is transferred from CDP-DG to PG to yield DPG. DPG synthesis in $E.$ $coli$ differs and involves the condensation of two molecules of PG (Chapter 3).

Understanding the biosynthesis of bis(monoacylglycerol)phosphate has been a particular challenge because the carbon linked to the phosphate residue is the $sn$-1 rather than $sn$-3 configuration. The likely biosynthetic pathway is depicted in Fig. 10 [23].

An intermediate in the biosynthesis of bis(monoacylglycerol)phosphate is 1-acyl-lyso-PG (Fig. 10), also known as lysobis-PA. Recent studies have shown that the inner membranes of late endosomes are enriched in lysobis-PA and that these membranes play an important role in the sorting of insulin growth factor receptor 2 and the mannose-

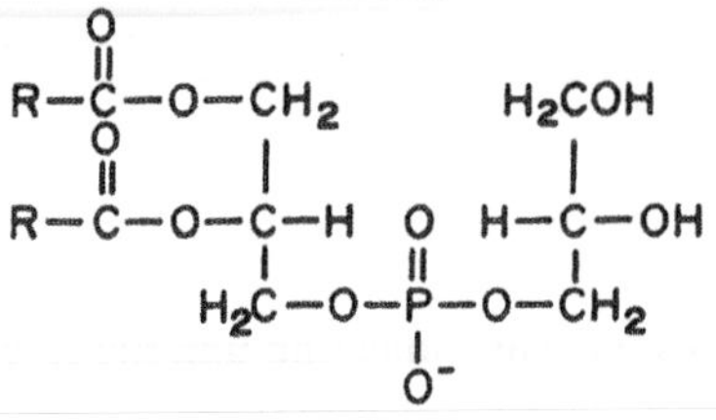

**Phosphatidylglycerol**

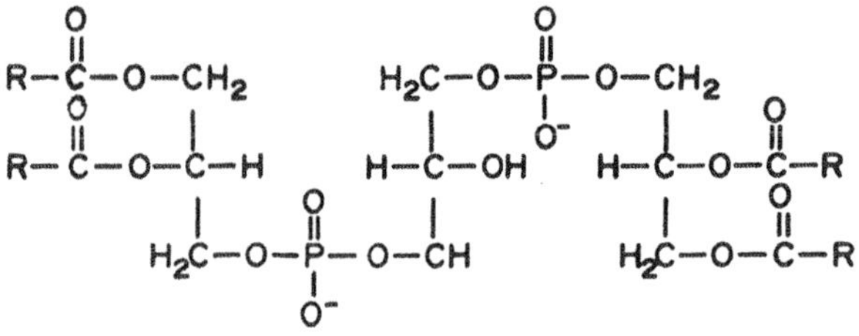

**Diphosphatidylglycerol**

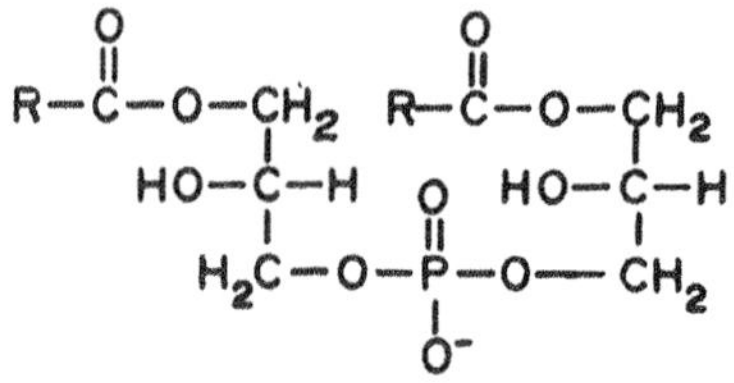

**Bis(monoacylglycero)phosphate**

Fig. 9. Structures of polyglycerophospholipids.

6-phosphate receptor [24]. Moreover, lysobis-PA cross-reacts with antibodies produced in patients with antiphospholipid syndrome. Possibly, some of the pathological defects in this disease could arise from disruption of endosomal traffic. Moreover, the defect in cholesterol trafficking in Niemann-Pick C disease (Chapter 17) may also involve lysobis-PA (J. Gruenberg, 1999).

## 8.2. Enzymes and subcellular location

PG can be made in mitochondria and microsomes from various animal cells and, except for lung, appears to be primarily converted to DPG. DPG is biosynthesized exclusively on the matrix side of the mitochondrial inner membrane and is found only in this organelle. DPG synthase requires $Co^{2+}$ for activity (K.Y. Hostetler, 1991). There is evidence that the rate-limiting step in DPG biosynthesis is the conversion of PA into CDP-DG (G.M. Hatch, 1994). Consistent with this idea, the levels of CTP have been shown to regulate DPG biosynthesis in cardiac myoblasts (G.M. Hatch, 1996).

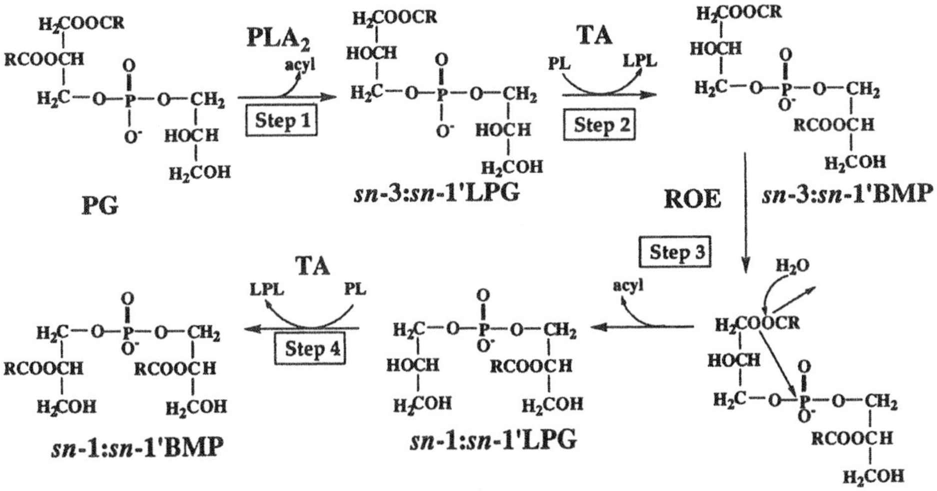

Fig. 10. Proposed pathway for the biosynthesis of bis(monoacylglycero)phosphate. Phospholipase A₂ (PLA₂) hydrolyzes PG to 1-acyl-lyso-PG (LPG). LPG is then acylated by a transacylase (TA), using a phospholipid (PL) as the acyl donor, to form bis(monoacylglycero)phosphate (BMP) that still retains the *sn*-3 : *sn*-1′ stereoconfiguration of the original PG and a lysophospholipid (LPL). The glycerol backbone of the *sn*-3 : *sn*-1′-BMP is reoriented by an enzymatic activity (ROE) to yield *sn*-1 : *sn*-1′-LPG (step 3). The final product, *sn*-1 : *sn*-1′-BMP, is formed upon acylation of *sn*-1 : *sn*-1′-LPG (step 4). The assignment of the acyl residues to the *sn*-2 positions of both glycerol moieties is based on their being primarily unsaturated and from degradation studies. It is believed that spontaneous rearrangement can occur so that the acyl residues end up on the *sn*-3 carbons as shown in Fig. 9. Figure from Amidon et al. [23] with permission.

Using techniques developed by Raetz and coworkers [25] M. Nishijima (1993) and coworkers isolated a temperature-sensitive mutant in PG-P synthase of CHO cells. The mutant had 1% of wild type CHO PG-P synthase activity at 40°C and a temperature-sensitive defect in PG and DPG biosynthesis. This mutant was used to show that DPG is required for the NADH-ubiquinone reductase (complex I) activity of the respiratory chain.

In yeast DPG synthesis has been genetically interrupted [26]. The yeast grows at temperatures between 16 and 30°C without DPG but fails to grow at 37°C on fermentable carbon sources such as glucose even though intact mitochondria are, therefore, not required for ATP synthesis. Thus, mitochondria must have some necessary function in yeast survival other than generating energy [26].

The fatty acyl content of phospholipids can also impact on mitochondrial function. Incubation of cardiomyocytes with palmitic acid increased the palmitic acid content of PA and PG and decreased DPG levels in mitochondria with a concomitant release of cytochrome *c* leading to apoptosis (W. Dowhan, 2001).

## 9. Remodeling of the acyl substituents of phospholipids

Phospholipids are made de novo with the fatty acid compositions present in the precursors DG and CDP-DG. Once the phospholipid is made, the fatty acid substituents

228

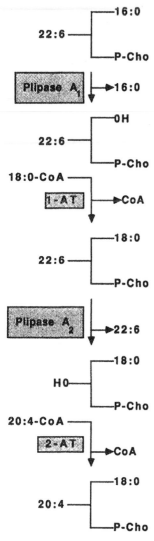

Fig. 11. Fatty acids at both the *sn*-1 and *sn*-2 positions of PC can be deacylated by phospholipases and reacylated by acyltransferases. Palmitic acid (16:0) can be removed from the *sn*-1 position and replaced with stearic acid (18:0). The fatty acid at the *sn*-2 position is depicted as docosahexaenoic acid (22:6) which can be replaced with 20:4 or 18:2. If the fatty acid at the *sn*-2 position were oleic acid, it could also be deacylated and reacylated. Alternatively, deacylation/reacylation could occur initially at the *sn*-2 position. Plipase, phospholipase; 1-AT, acyl-CoA:lyso-PC 1-acyltransferase; 2-AT, acyl-CoA:lyso-PC 2-acyltransferase; cho, choline.

can be remodeled via deacylation–reacylation reactions (Fig. 11). Remodeling can occur on either the *sn*-1 or *sn*-2 positions of the glycerolipid. For example, a major molecular species formed from the conversion of PE to PC is 16:0–22:6-PC (R.W. Samborski, 1990). This species of PC has a half-life of less than 6 h and appears not to be significantly degraded but rather converted to other molecular species, particularly those with 18:0 on the *sn*-1 position and 20:4, 18:2 or 22:6 on the *sn*-2 position. Other studies have suggested that the main products of de novo PC and PE biosynthesis are 16:0–18:2, 16:0–18:1, 16:0–22:6 and 18:1–18:2. The major remodeled product is 18:0–20:4 for both PC and PE (H.H.O. Schmid, 1995). Why 18:0–20:4-PC and -PE are made by this circuitous route, rather than directly, is not known.

## 10. Regulation of gene expression in yeast

The pathways for the biosynthesis of phospholipids in yeast were largely elucidated by Lester and coworkers in the late 1960s (Fig. 12). These pathways are similar to those found in other eukaryotes except PS in yeast is made via a pathway similar to that found in *E. coli* where CDP-DG reacts with serine to yield PS and CMP.

Considerable interest in yeast as a model system has developed over the past two

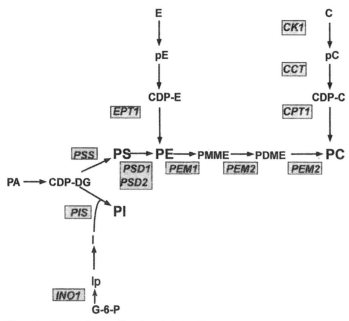

Fig. 12. The pathway for phospholipid biosynthesis in yeast and designation of the genes (italics in boxes) encoding the enzymes that catalyze the reactions. The abbreviations are: E, ethanolamine; pE, phosphoethanolamine; CDP-E, CDP-ethanolamine; C, choline; pC, phosphocholine; CDP-C, CDP-choline; PE, phosphatidylethanolamine; PMME, phosphatidylmonomethylethanolamine; PDME, phosphatidyldimethylethanolamine; PC, phosphatidylcholine; PS, phosphatidylserine; PA, phosphatidic acid; CDP-DG, CDP-diacylglycerol; PI, phosphatidylinositol; I, inositol; Ip, inositol phosphate; G-6-P, glucose-6-phosphate. The genes encode the following enzymes: *INO1*, I-1-P synthase; *PIS*, PI synthase; *PSS* (also known as *CHO1*), PS synthase; *EPT1*, CDP-E:1,2-diacylglycerol ethanolaminephosphotransferase; *PEM1* (*CHO2*), PE methyltransferase; *PEM2* (*OPI3*), phospholipid methyltransferase; *CK1*, choline kinase; *CCT*, CTP:phosphocholine cytidylyltransferase (abbreviated as CT elsewhere in this chapter); *CPT1*, CDP-C:1,2-diacylglycerol cholinephosphotransferase; *PSD1* and *PSD2*, PS decarboxylase.

decades. Reasons for choosing *Saccharomyces cerevisiae* include a large knowledge base in classical genetics, the ease of making mutant strains and the ability to grow large amounts of yeast. Whereas understanding the regulation of expression of phospholipid biosynthetic enzymes in animal cells is still in its infancy, considerable progress has been made in the yeast system [26–28]. When yeast cells are grown in the presence of choline and inositol, the expression of the enzymes involved in the conversion of PA and glucose-6-P to PI, PC and PE is depressed (Fig. 12).

Both positive and negative regulatory factors are involved in the regulation of expression of phospholipid biosynthetic enzymes in yeast. The *INO2* and *INO4* genes encode transcription factors that are required for the expression of inositol-1-P synthase (*INO1*). In vitro transcribed and translated proteins derived from *INO2* and *INO4* form a heterodimer that binds a specific DNA fragment of the *INO1* gene referred to as UAS$_{INO}$ (S.A. Henry, 1994). Ino4p (the protein encoded by *INO4*) and Ino2p exhibit basic helix–loop–helix domains. The Ino2p–Ino4p heterodimer binds to UAS$_{INO}$ of the *INO1* promoter that contains two copies of a binding site (CANNTG) for basic helix–loop–helix-containing proteins.

The *OPI1* gene encodes a protein that is a negative regulatory factor for phospholipid biosynthesis [27]. Opi1p contains a leucine zipper, a motif implicated in protein–DNA interactions and transcriptional control. *Opi1* mutants exhibit a two-fold increase in the constitutive expression of inositol-1-P synthase and other enzymes involved in PI, PC and PE biosynthesis. The mechanism by which Opi1p mediates its negative regulatory role is unknown. Opi1p does not interact directly with UAS$_{INO}$ or with Ino2p or Ino4p. Phosphorylation of Opi1 by protein kinase C may be involved (G.M. Carman, 2001).

Recent experiments have identified other proteins that interact with Ino4p (J.M. Lopez, 2000) indicating that there is still much to learn about transcriptional regulation of phospholipid biosynthetic genes in yeast. How the regulatory genes (*INO2*, *INO4*, *OPI1*) are themselves regulated is just beginning to be studied (J.M. Lopes, 2001).

## 11. Future directions

Since the first edition of this book was published in 1985 there have been astonishing developments in phospholipid metabolism. Some of these advances have dictated that a separate chapter be devoted to the role of glycerophospholipids in signal transduction (Chapter 12). The purification of some enzymes involved and the use of genetic screens has allowed molecular biological techniques to be used to clone and express cDNAs and genes for eukaryotic phospholipid biosynthetic enzymes. In addition, genetically modified mouse models are being developed.

(1) We can expect that crystal structures of some of the soluble proteins will be reported.
(2) More genes that encode phospholipid biosynthetic enzymes will be cloned and characterized. Elements of the genes involved in regulation of transcription will be mapped and positive and negative transcription factors should be identified.
(3) We can expect that more transgenic mice that over-express some of these enzymes,

as well as mice in which phospholipid biosynthetic genes have been disrupted, will be produced. Such studies should provide valuable insight into the role of these enzymes in whole animal physiology.

(4) The yeast system will continue to be exploited for studies on gene function and expression as well as regulation of phospholipid biosynthesis.

(5) There should be progress in understanding the regulation of PE, PI, PS and DPG biosynthesis.

(6) In the process of testing hypotheses and asking fundamental questions about phospholipid biosynthesis, we can continue to expect the unexpected.

## Abbreviations

| | |
|---|---|
| CDP-DG | CDP-diacylglycerol |
| CHO | Chinese hamster ovary |
| CT | CTP : phosphocholine cytidylyltransferase |
| DPG | diphosphatidylglycerol (cardiolipin) |
| DG | diacylglycerol |
| ER | endoplasmic reticulum |
| MAM | mitochondria associated membrane |
| PA | phosphatidic acid |
| PC | phosphatidylcholine |
| PE | phosphatidylethanolamine |
| PEMT | phosphatidylethanolamine N-methyltransferase |
| PG | phosphatidylglycerol |
| PI | phosphatidylinositol |
| PS | phosphatidylserine |
| SREBP | sterol response element binding protein |
| TG | triacylglycerol |

## References

1. Dircks, L.K. and Sul, H.S. (1997) Mammalian mitochondrial glycerol-3-phosphate acyltransferase. Biochim. Biophys. Acta 1348, 10–16.
2. Coleman, R.A., Lewin, T.M. and Muoio, D.M. (2000) Physiological and nutritional regulation of enzymes of triacylglycerol synthesis. Annu. Rev. Nutr. 20, 77–103.
3. Kanoh, H., Kai, M. and Wada, I. (1997) Phosphatidic acid phosphatase from mammalian tissues: discovery of channel-like proteins with unexpected functions. Biochim. Biophys. Acta 1348, 56–62.
4. Carman, G.M. (1997) Phosphatidate phosphatases and diacylglycerol pyrophosphate phosphatases in *Saccharomyces cerevisiae* and *Escherichia coli*. Biochim. Biophys. Acta 1348, 45–55.
5. Kennedy, E.P. (1989) Discovery of the pathways for the biosynthesis of phosphatidylcholine. In: D.E. Vance (Ed.), Phosphatidylcholine Metabolism. CRC Press, Boca Raton, FL, pp. 1–9.
6. Ishidate, K. (1997) Choline/ethanolamine kinase from mammalian tissues. Biochim. Biophys. Acta 1348, 70–78.
7. Lykidis, A., Baburina, I. and Jackowski, S. (1999) Distribution of CTP : phosphocholine cytidylyl-

transferase (CCT) isoforms: identification of a new CCTβ splice variant. J. Biol. Chem. 274, 26992–27001.

8. Kent, C. (1997) CTP:phosphocholine cytidylyltransferase. Biochim. Biophys. Acta 1348, 79–90.

9. Cornell, R.B. and Northwood, I.C. (2000) Regulation of CTP:phosphocholine cytidylyltransferase by amphitropism and relocalization. Trends Biochem. Sci. 25, 441–447.

10. McMaster, C.R. and Bell, R.M. (1997) CDP–choline: 1,2-diacylglycerol cholinephosphotransferase. Biochim. Biophys. Acta 1348, 100–110.

11. Walkey, C.J., Donohue, R., Agellon, L.B. and Vance, D.E. (1997) Disruption of the murine gene encoding phosphatidylethanolamine N-methyltransferase. Proc. Natl. Acad. Sci. USA 94, 12880–12885.

12. Walkey, C.J., Yu, L., Agellon, L.B. and Vance, D.E. (1998) Biochemical and evolutionary significance of phospholipid methylation. J. Biol. Chem. 273, 27043–27046.

13. Kanipes, M.I. and Henry, S.A. (1997) The phospholipid methyltransferases in yeast. Biochim. Biophys. Acta 1348, 134–141.

14. Bakovic, M., Waite, K., Tang, W., Tabas, I. and Vance, D.E. (1999) Transcriptional activation of the murine CTP:phosphocholine cytidylyltransferase gene *(Ctpct)*: combined action of upstream stimulatory and inhibitory *cis*-acting elements. Biochim. Biophys. Acta 1438, 147–165.

15. Bladergroen, B.A. and van Golde, L.M.G. (1997) CTP:phosphoethanolamine cytidylyltransferase. Biochim. Biophys. Acta 1348, 91–99.

16. Voelker, D.R. (1997) Phosphatidylserine decarboxylase. Biochim. Biophys. Acta 1348, 236–244.

17. Kuge, O. and Nishijima, M. (1997) Phosphatidylserine synthases I and II of mammalian cells. Biochim. Biophys. Acta 1348, 151–156.

18. Hawthorne, J.N. (1982) Inositol phospholipids. In: J.N. Hawthorne and G.B. Ansell (Eds.), Phospholipids. Elsevier, Amsterdam, pp. 263–278.

19. Dowhan, W. (1997) CDP–diacylglycerol synthase of microorganisms. Biochim. Biophys. Acta 1348, 157–165.

20. Downes, C.P. and MacPhee, C.H. (1990) Myo-inositol metabolites as cellular signals. Eur. J. Biochem. 193, 1–18.

21. Antonsson, B. (1997) Phosphatidylinositol synthases from mammalian tissues. Biochim. Biophys. Acta 1348, 179–186.

22. Nikawa, J.-I. and Yamashita, S. (1997) Phosphatidylinositol synthase from yeast. Biochim. Biophys. Acta 1348, 173–178.

23. Amidon, B., Schmitt, J.D., Thuren, T., King, L. and Waite, M. (1995) Biosynthetic conversion of phosphatidylglycerol to *sn*-1:*sn*-1′ bis(monoacylglycerol)phosphate in a macrophage-like cell line. Biochemistry 34, 5554–5560.

24. Kobayashi, T., Stang, E., Fang, K.S., de Moerloose, P., Parton, R.G. and Gruenberg, J. (1998) A lipid associated with the antiphospholipid syndrome regulates endosome structure and function. Nature 392, 193–197.

25. Zoeller, R.A. and Raetz, C.R.H. (1992) Strategies for isolating somatic cell mutants defective in lipid biosynthesis. Methods Enzymol. 209, 34–51.

26. Schlame, M., Rua, D. and Greenberg, M.L. (2000) The biosynthesis and functional role of cardiolipin. Prog. Lipid Res. 39, 257–288.

27. Nikoloff, D.M. and Henry, S.A. (1991) Genetic analysis of yeast phospholipid biosynthesis. Annu. Rev. Genet. 25, 559–583.

28. Swede, M.J., Hudak, K.A., Lopes, J.M. and Henry, S.A. (1992) Strategies for generating phospholipid synthesis mutants in yeast. Methods Enzymol. 209, 21–34.

D.E. Vance and J.E. Vance (Eds.) *Biochemistry of Lipids, Lipoproteins and Membranes (4th Edn.)*

# CHAPTER 9

# Ether-linked lipids and their bioactive species

Fred Snyder[1], Ten-ching Lee[1] and Robert L. Wykle[2]

[1] *Oak Ridge Associated Universities (retired), Oak Ridge, TN 37831, USA*
[2] *Department of Biochemistry, Wake Forest University Medical Center, Winston-Salem, NC 27517, USA*

## 1. Introduction

Naturally occurring ether lipids contain either $O$-alkyl or $O$-alk-1-enyl groupings. Those possessing the $O$-alk-1-enyl moiety with a *cis* double bond adjacent to the ether bridge are referred to as plasmalogens, as well as vinyl ethers. Both the $O$-alkyl and $O$-alk-1-enyl substituents are generally located at the *sn*-1 position of the glycerol moiety although di- and tetra-$O$-alkylglycerolipids have been described in some cells. Unlike the diverse types of acyl moieties present in glycerolipids, the predominant $O$-alkyl and $O$-alk-1-enyl ether-linked chains generally consist mainly of 16:0, 18:0, and 18:1 aliphatic groupings, but other types of chain lengths, degrees of unsaturation, and occasional branched-chains do exist as minor components. Except for intermediary metabolites and certain bioactive lipids, ether linkages in phospholipids of mammalian cells exist almost exclusively in the choline and ethanolamine glycerolipid classes. The majority of the $O$-alkyl moieties normally occur as plasmanylcholines[1], whereas the $O$-alk-1-enyl grouping is mainly associated with the plasmenylethanolamines with the exception of heart where plasmenylcholines predominate. Some neutral lipids such as alkyldiacylglycerols (glyceryl ether diesters) and alkylacylglycerols, analogs of triacylglycerols and diacylglycerols, respectively, are also found in cells. Fig. 1 illustrates the chemical structures of the most common ether lipids found in mammals.

A number of books [1–6] and review articles [7–14] on ether lipids, some specifically emphasizing platelet-activating factor (PAF), are recommended as reading material. These sources provide a comprehensive listing of published papers.

## 2. Historical highlights

The early literature concerning ether-linked lipids has also been covered in detail [1,2,9,15]. Perhaps the first evidence, albeit circumstantial, to suggest the existence of $O$-alkyl lipids in nature was the isolation of an unsaponifiable fraction of lipids from starfish that was referred to as 'astrol', which was subsequently shown to have similar

---

[1] Plasmanyl designates the radical '1-alkyl-2-acyl-*sn*-glycero-3-phospho-', whereas plasmenyl represents the radical '1-alk-1-enyl-2-acyl-*sn*-glycero-3-phospho-'; the prefix phosphatidyl is used only to denote the radical '1,2-diacyl-*sn*-glycero-3-phospho-'. 'Radyl' is used as a prefix in glycerolipid nomenclature when the aliphatic substituents are unknown at the *sn*-positions of the glycerol moiety or when either acyl, alkyl, or alk-1-enyl moieties would be of equal importance.

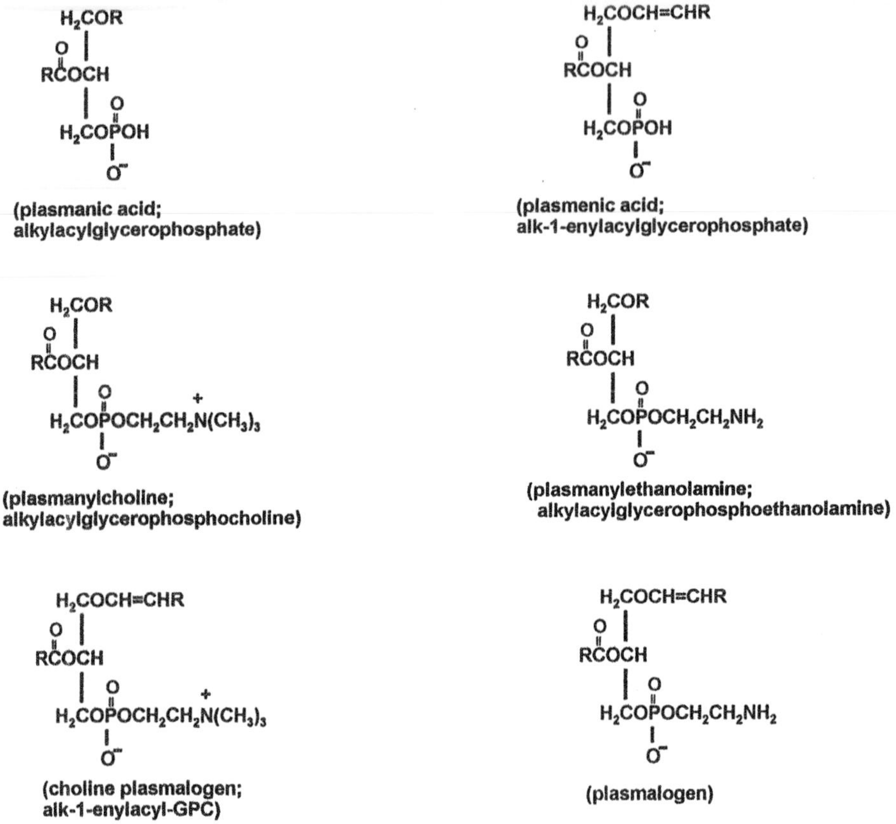

$$H_2COR$$
$$RCOCH \ (O)$$
$$H_2COPOH \ (O) \ O^-$$

(plasmanic acid;
alkylacylglycerophosphate)

$$H_2COCH=CHR$$
$$RCOCH \ (O)$$
$$H_2COPOH \ (O) \ O^-$$

(plasmenic acid;
alk-1-enylacylglycerophosphate)

$$H_2COR$$
$$RCOCH \ (O)$$
$$H_2COPOCH_2CH_2N(CH_3)_3^+ \ (O) \ O^-$$

(plasmanylcholine;
alkylacylglycerophosphocholine)

$$H_2COR$$
$$RCOCH \ (O)$$
$$H_2COPOCH_2CH_2NH_2 \ (O) \ O^-$$

(plasmanylethanolamine;
alkylacylglycerophosphoethanolamine)

$$H_2COCH=CHR$$
$$RCOCH \ (O)$$
$$H_2COPOCH_2CH_2N(CH_3)_3^+ \ (O) \ O^-$$

(choline plasmalogen;
alk-1-enylacyl-GPC)

$$H_2COCH=CHR$$
$$RCOCH \ (O)$$
$$H_2COPOCH_2CH_2NH_2 \ (O) \ O^-$$

(plasmalogen)

Fig. 1. Chemical structures of biologically significant types of ether-linked lipids found in mammalian cells.

properties to batyl alcohol, an alkylglycerol possessing an 18-carbon aliphatic chain at the sn-1 position of the glycerol moiety. During the same period the presence of alkyl ether lipids in liver oils of various saltwater fish was described by the Japanese scientists M. Tsujimoto and Y. Toyama (1922). The common names of the alkylglycerols, chimyl [16:0 alkyl], batyl [18:0 alkyl], and selachyl [18:1 alk-9-enyl] alcohols, are based on the fish species from which they were originally isolated. Complete proof of the precise chemical nature of the alkyl linkage at the sn-1 position in these glycerolipids was provided by W.H. Davies, I.M. Heilbron, and W.E. Jones (1933) from England.

The German scientists, R. Feulgen and K. Voit (1924) originally described plasmalogens in a variety of fresh tissue slices preserved in a HgCl₂ solution after being erroneously treated with a fuchsin–sulfurous acid reagent without the normal fixation and related histological processing with organic solvents. Only the cytoplasm of cells, but not the nuclei, was stained a red–violet color, which led to the conclusion that an aldehyde was present in the cell plasma. This substance was called 'plasmal'. If the histological preparations were treated with a lipid-extracting solvent before exposure to the dye, no colored stain appeared in the cytoplasm. This unknown precursor of the cytosolic aldehyde that reacted with the dye was called plasmalogen, a name still retained as the generic term for all alk-1-enyl-containing glycerolipid classes.

It was not until the 1950s that the precise chemical structure of the alk-1-enyl linkage in ethanolamine plasmalogens was proven, primarily through the combined efforts of M.M. Rapport and G.V. Marinetti in the United States, G.M. Gray in England, E. Klenk and H. Debuch in Germany, and their various co-workers. The first cell-free systems to synthesize the alkyl ether bond were described independently in 1969 by F. Snyder, R.L. Wykle, and B. Malone and by A. Hajra. Shortly thereafter, studies by R.L. Wykle, M.L. Blank, B. Malone, and F. Snyder and by F. Paltauf and A. Holasek demonstrated that the *O*-alkyl moiety of an intact phospholipid could be enzymatically desaturated to the alk-1-enyl grouping (see Section 6.2.5). One of the most significant developments in the ether-lipid field occurred in 1979 when one of the most potent bioactive molecules known, an acetylated form of a choline-containing alkylglycerolipid called platelet-activating factor or PAF, was identified independently by three separate groups.

## 3. Natural occurrence

Chemical, chromatographic, and mass spectral methods for analyzing ether-linked glycerolipids have been reviewed [16,17]. Ether-linked phospholipids generally are isolated as a mixture with their ester-linked counterparts.

Ether-linked lipids occur throughout the animal kingdom and are even found as minor components in several higher plants. Some mammalian tissues, and avian, marine, molluscan, protozoan, and bacterial lipid extracts contain significant proportions of ether-linked lipids. Highest levels of ether lipids in mammals occur in nervous tissue, heart muscle, testes, kidney, preputial glands, tumor cells, erythrocytes, bone marrow, spleen, skeletal tissue, neutrophils, eosinophils, macrophages, platelets, and lipoproteins. The large quantities of ethanolamine plasmalogens associated with various lipoproteins from rat serum and human plasma (36% and 50%, respectively, of the total ethanolamine phosphatides) is of particular interest since the liver contains relatively low amounts of ether lipids and the plasmalogens are not acquired in the lipoproteins after their secretion. Although the dietary consumption of ether lipids by humans has largely been ignored by nutritionists, it is clear that certain meats and seafoods can contain relatively high amounts of these lipids.

Analogs of triacylglycerols have also been described. 1-Alkyl-2,3-diacyl-*sn*-glycerols are characteristically elevated in tumor lipids and 1-alk-1-enyl-2,3-diacyl-*sn*-glycerols (neutral plasmalogens) have also been detected in tumors and adipose tissue of mammals and in fish liver oil. In fact, even alkylacetylacylglycerols have been shown to be formed by human leukemic cells.

1-Alkyl-2-acyl-*sn*-glycero-3-phosphocholine (Fig. 1), a significant component of platelets, neutrophils, macrophages, eosinophils, basophils, monocytes, and endothelial, mast, and HL-60 cells (a human promyelocytic leukemic cell line), is a precursor of platelet-activating factor (PAF, 1-alkyl-2-acetyl-*sn*-glycero-3-phosphocholine; see Fig. 2). Thus, this precursor appears to be a constituent of all cells known to produce

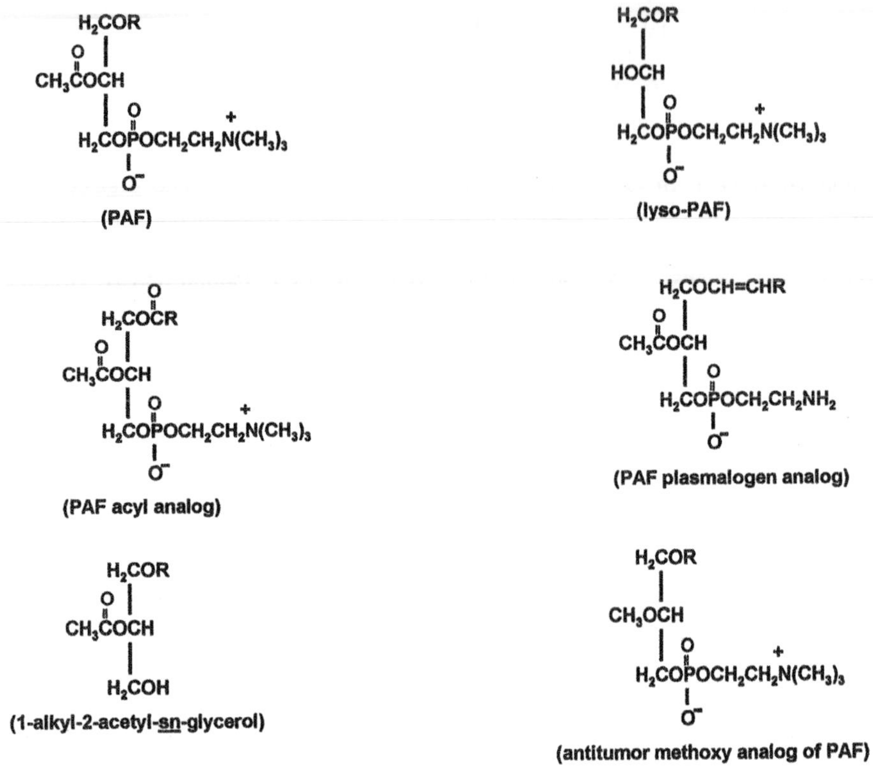

Fig. 2. Chemical structures of PAF and structurally related ether-linked glycerolipids possessing biological activities.

PAF by the remodeling pathway; in human neutrophils and eosinophils the alkyl subclass comprises 45 and 70 mol% of the choline-linked phosphoglycerides, respectively, while the ethanolamine-linked class contains 60–65 mol% plasmalogen. PAF is also found in saliva, urine, and amniotic fluid, which indicates that other cell types could be the source of PAF in these fluids.

Dialkylglycerophosphocholines have been reported as minor constituents of bovine heart and spermatozoa. Moreover, heart tissue is unique with respect to its plasmalogen content, since in some animal species, this is the only mammalian tissue known to contain significant amounts of choline plasmalogens instead of the usually encountered ethanolamine plasmalogens.

Halophilic bacteria contain an unusual dialkyl type of glycerolipid (a diphytanyl ether analog of phosphatidylglycerophosphate) that has an opposite stereochemical configuration from all other known ether-linked lipids, i.e., the ether linkages are located at the sn-2 and sn-3 positions. The biosynthetic pathway for the formation of the ether bond in halophiles is still unknown. Acidophilic thermophiles contain tetraalkyl glycerolipids with their two glycerol moieties linked across their membranes, which prevents them from being freeze-fractured.

Many anaerobic bacteria are highly enriched in plasmalogens. For example, *Clostridium butyricum* contains significant amounts of ethanolamine plasmalogens and *Megasphaera elsdenii* has been reported to contain very large quantities of plasmenyl ethanolamine and plasmenylserine. However, despite the large pool of plasmalogens in such anaerobes, no information has emerged about how they synthesize the alk-1-enyl ether bond.

## 4. Physical properties

Replacement of ester linkages in glycerolipids with ether bonds mainly affects hydrophobic–hydrophilic interactions. Nevertheless, the closer linear packing arrangement attainable with ether-linked moieties also is capable of influencing the polar head group region of phospholipids. The novel placement of the Δ1 double bond in plasmalogens can also exert effects on stereochemical relationships and therefore, the presence of an ether-linkage in phospholipids can modify both the configuration and functional properties of membranes.

In model membranes, ether-linked lipids have been shown to decrease ion permeability, surface potential, and lower the phase temperature of membrane bilayers when compared to their diacyl counterparts. *Clostridium butyricum* appears to be able to regulate the stability of the bilayer arrangement of membranes by altering the ratio of ether versus acyl type of ethanolamine phospholipids in response to changes in the degree of lipid unsaturation of the membranes. The experiments with bacteria indicate that the substitution of plasmenylethanolamine for phosphatidylethanolamine in biomembranes would have only small effects on lipid melting transitions, whereas the tendency to form non-lamellar lipid structures would be significantly increased.

## 5. Biologically active ether lipids

### 5.1. Platelet-activating factor

In 1979, the chemical structure of PAF was identified as 1-alkyl-2-acetyl-*sn*-glycero-3-phosphocholine (Fig. 2). The semisynthetic preparation tested in these initial experiments caused aggregation of platelets at concentrations as low as $10^{-11}$ M and induced an antihypertensive response when as little as 60 ng were administered intravenously to hypertensive or normotensive rats. It is now known that PAF, a phospholipid secreted by numerous cells, exerts many different types of biological responses (Table 1) and it has been implicated as a contributing factor in the pathogenesis of such diverse disease processes as asthma, hypertension, allergies, inflammation, and anaphylaxis, to name only a few.

PAF has been isolated and very well characterized from a number of cellular sources. Basophils, neutrophils, platelets, macrophages, monocytes and mast, endothelial, and HL-60 cells are among the highest producers of PAF when stimulated by agonists such as chemotactic peptides, zymosan, thrombin, calcium ionophores, antigens, bradykinin,

Table 1

Biological activities associated with platelet-activating factor

I. In vivo responses:
1. Bronchoconstriction ↑
2. Systemic blood pressure ↓
3. Pulmonary resistance ↑
4. Dynamic lung compliance ↓
5. Pulmonary hypertension and edema ↑
6. Heart rate ↑
7. Hypersensitivity responses ↑
8. Vascular permeability ↑

II. Cellular responses:
1. Aggregation of neutrophils and platelets ↑
2. Degranulation of platelets, neutrophils, and mast cells ↑
3. Shape changes in platelets, neutrophils, and endothelial cells ↑
4. Chemotaxis and chemokinesis in neutrophils ↑

III. Biochemical responses:
1. $Ca^{2+}$ uptake ↑
2. Respiratory burst and superoxide production ↑
3. Protein phosphorylation ↑
4. Arachidonate turnover ↑
5. Phosphoinositide turnover ↑
6. Protein kinase ↑
   – protein kinase C
   – mitogen-activated protein kinase
   – G-protein receptor kinases
   – protein tyrosine kinase
7. Glycogenolysis ↑
8. Tumor necrosis factor production ↑
9. Interleukin 2 production ↓
10. Activation of immediate-early genes, e.g., c-*fos* and c-*jun*, zif/268 ↑

ATP, $C_{5a}$, collagen, and disease states. The amount of PAF produced by various stimuli is dependent on the cell type and the specific agonist used. Most animal tissues also have the capacity to produce PAF by de novo synthesis (see Section 6.3.2).

Other acetylated glycerolipids that are structurally related to PAF include 1-alkyl-2-acetyl-*sn*-glycerols and the plasmalogen and acyl analogs of PAF that possess choline or ethanolamine moieties (Fig. 2). Both the alkylacetylglycerols (perhaps via phosphorylation) and the choline plasmalogen analog of PAF can mimic the actions of PAF, perhaps through their interactions with the PAF receptor. Biological potencies of the PAF analogs range from 5- to 4000-fold weaker than PAF [18].

An unnatural chemically synthesized analog of PAF, 1-alkyl-2-methoxy-*sn*-glycero-3-phosphocholine (Fig. 2) and related derivatives, possesses unique highly selective antitumor activity [19]. Clinical studies in Europe have shown a promising therapeutic potential for the methoxy analog in treating certain types of human cancers. Although its mode of action has been difficult to ascertain, the primary site of action of these PAF analogs is the plasma membrane rather than the cell nucleus. The cytotoxic activity of this antineoplastic phospholipid is apparently due to its ability to prevent the formation of membranes by blocking phosphatidylcholine synthesis (Chapter 8) via the inhibition of CTP:phosphocholine cytidylyltransferase, the rate-limiting enzyme in phosphatidylcholine biosynthesis [20].

## 5.2. Other ether-linked mediators

In addition to PAF and eicosanoids, 1-alkyl-2-acyl-*sn*-glycero-3-phosphocholine yields 1-alkyl-2-acyl-*sn*-glycero-3-phosphate when acted upon by phospholipase D [21]. Both the alkylacyl- and diacylglycerols share in their ability to increase responses (priming) of neutrophils to other stimuli of arachidonic acid release and the oxidative burst. However, only the diacylglycerol primes for the formation of lipoxygenase products.

## 5.3. Oxidized phospholipids

Oxidation of the phospholipids of plasma lipoproteins generates bioactive phospholipid species with PAF-like activity [13]. A complex mixture of oxidation products is formed but the species that bind and act through the PAF receptor are alkyl ether-linked and contain short-chain oxidized residues in the *sn*-2 position derived from polyunsaturated acyl chains. Normally, PAF analogs containing *sn*-2 chains longer than four carbons have little activity but the introduction of an oxidized group at the end of the chain yields longer-chain active analogs. The PAF acetylhydrolase (Section 7.3.1) can remove the oxidized chains to inactivate these products. These oxidized species may play an important role in inflammation and development of atherosclerotic plaques and other cardiovascular disorders.

## 5.4. Receptors, overexpression, and knockout mice

The cDNA encoding the cell surface PAF receptor has been cloned and the primary structure sequenced from a number of cells/tissues including guinea pig lung, human neutrophils, HL-60 cells (granulocytic form), and human heart [13,14]. Human and guinea pig receptors consist of 342 amino acids with a C-terminal cytoplasmic tail possessing serine and threonine residues which could be potential sites for regulation via phosphorylation. The PAF receptor is typical of other members of the family of G-protein-coupled receptors, which span the membrane seven times (e.g., rhodopsin, $\beta_1$ and $\beta_2$ adrenergic, $D_2$-dopamine, and M1–M5 muscarinic). Based on modeling and site-directed mutagenesis of the receptor it is proposed that the central portion of the receptor and the histidine residues 188, 248, and 249 form the PAF binding pocket. It is also deemed likely that there is a disulfide bond between the cysteine residues at positions 90 and 163. A mutation in the third transmembrane domain resulted in a constitutively active receptor. It has been shown in other studies that the third intracellular loop is necessary for initiating phosphatidylinositol turnover. The fate of the receptor-bound PAF is unknown.

The role of PAF in vivo has been examined by overexpressing the guinea pig receptor in mice [8,14]. These mice had an increased death rate in response to endotoxin and surprisingly developed melanocytic tumors of the skin. Ishii and Shimizu [14] have also generated PAF receptor knockout mice. These mice had less severe anaphylactic responses including diminished cardiovascular instability, alveolar edema, and airway constriction than did wild-type mice. Unexpectedly, the receptor-deficient animals reproduced normally and remained susceptible to endotoxin. The existence of a second PAF receptor might explain these results. Further studies are required to resolve the somewhat contradictory observations. In humans lacking PAF acetylhydrolase, an enzyme that degrades PAF, an increased severity of asthma, coronary artery disease and stroke was observed. Several review articles [7,8,13,14] discuss the role of PAF receptors in signal transduction, cloning, sequencing, and related studies.

*5.5. Receptor antagonists*

A number of PAF antagonists are available that block binding of PAF to its receptor. Some of these antagonists are derived from plants, such as *Ginkgo biloba*, while others are structural analogs of PAF, and yet others are chemically synthesized compounds found through screening. Although some of the inhibitors effectively block PAF responses in certain systems, they have not proven highly effective as anti-inflammatory drugs. It is possible that the drugs do not gain access to all PAF receptors in vivo, or that the network of inflammatory mediators synergizes to overcome suppression of the PAF receptor.

# 6. Enzymes involved in ether lipid synthesis

In view of the vast literature about the enzymes involved in the metabolism of ether-linked lipids and PAF, the reader should consult the various reviews on this subject [5,10–13].

*6.1. Ether lipid precursors*

*6.1.1. Acyl Co-A reductase*
Fatty alcohol precursors of ether lipids are derived from acyl-CoAs via a fatty aldehyde intermediate in a reaction sequence catalyzed by a membrane-associated acyl-CoA reductase (Fig. 3A). A cytosolic form of the reductase from bovine heart has also been described.

Acyl-CoA reductases associated with membrane systems use acyl-CoA substrates, and in mammalian cells, they exhibit a specific requirement for NADPH. The fatty alcohols produced by the reductase can be oxidized back to the fatty acids by microsomes in the presence of NAD.

Although only traces of fatty aldehydes can normally be detected as an intermediate in these reactions, the use of trapping agents such as semicarbazide has documented that aldehydes are indeed formed as intermediates. Acyl-CoA reductase prefers saturated substrates over acyl-CoAs that are unsaturated; in fact, the enzyme in brain microsomes is not able to convert polyunsaturated moieties to fatty alcohols. Some evidence

A)

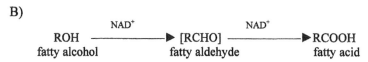

B)

Fig. 3. Enzymatic synthesis (A) and oxidation (B) of long-chain fatty alcohols by (I) acyl-CoA reductase and (II) fatty alcohol oxidoreductase, respectively.

indicates that, at least in brain, acyl-CoA reductase is localized in microperoxisomes. Topographical studies of microperoxisomal particles have revealed the acyl-CoA reductase activity is located at the cytosolic surface of these membranes. In rabbit Harderian glands and *Euglena gracilis*, the NADPH-dependent reductase appears to be closely coupled with fatty acid synthase and it has been suggested that the fatty acid bound to acyl carrier protein, rather than acyl-CoA, is the substrate for this reductase.

### 6.1.2. Dihydroxyacetone-P acyltransferase

Presumably, dihydroxyacetone-P acyltransferase is present in all mammalian cells that synthesize alkylglycerolipids, since the acylation of dihydroxyacetone-P is an obligatory step in the biosynthesis of the ether bond in glycerolipids. On the other hand, the quantitative importance of the pathways utilizing dihydroxyacetone-P versus *sn*-glycerol-3-P in the biosynthesis of glycerolipid esters (Chapter 8) has never been firmly established in intact cells.

Current evidence indicates that dihydroxyacetone-P acyltransferase, as well as alkyl-dihydroxyacetone-P synthase, is localized in microperoxisomes. Nevertheless, many studies of these enzymes have been done with microsomal and/or mitochondrial preparations; however, it is well known that microperoxisomes sediment with microsomes and large peroxisomes sediment with mitochondria under the usual preparation conditions for these subcellular fractions. Investigations of the topographical orientation of dihydroxyacetone-P acyltransferase in membrane preparations from rabbit Harderian glands and rat brains indicate that unlike most other enzymes in glycerolipid metabolism, dihydroxyacetone-P acyltransferase appears to be located on the internal side of microsomal vesicles.

### 6.2. Ether lipids

### 6.2.1. O-Alkyl bond: mechanism of formation

Formation of the alkyl ether bond in glycerolipids is catalyzed by alkyldihydroxyacetone-P synthase (Fig. 4). This reaction, which forms alkyldihydroxyacetone-P as the first detectable intermediate in the biosynthetic pathway for ether-linked glycerolipids, is

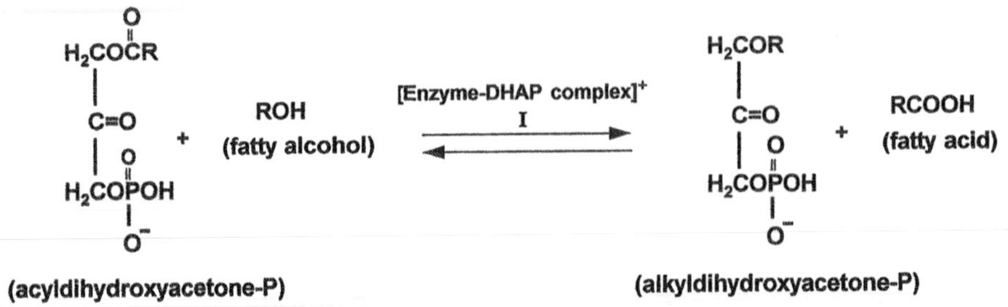

$$H_2COCR \quad (O)$$

(acyldihydroxyacetone-P)  (alkyldihydroxyacetone-P)

Fig. 4. The reaction that forms the *O*-alkyl bond is catalyzed by (I) alkyldihydroxy acetone-P synthase and is thought to proceed via a ping-pong mechanism. The abbreviation DHAP in this illustration designates dihydroxyacetone-P. Upon binding of acyl-DHAP to the enzyme, alkyl-DHAP synthase, the pro-*R* hydrogen at carbon atom 1 is exchanged by an enolization of the ketone, followed by release of the acyl moiety to form an activated enzyme–DHAP complex. The carbon atom at the 1 position of DHAP in the enzyme complex is thought to carry a positive charge that may be stabilized by an essential sulfhydryl group of the enzyme; thus, the incoming alkoxide ion reacts with the carbon 1 atom to form the ether bond of alkyl-DHAP. It has been proposed that a nucleophilic cofactor at the active site covalently binds the DHAP portion of the substrate.

unique in mammals since it is the only one known where a fatty alcohol can be directly substituted for a covalently linked acyl moiety. Alkyldihydroxyacetone-P synthase has been primarily investigated in microsomal preparations; however, as with dihydroxyacetone-P acyltransferase, there is general agreement that the synthase activity is peroxisomal. The alkyl synthase cDNAs from guinea pig and human liver reveal that both proteins contain a peroxisomal targeting signal 2 [22]. The importance of peroxisomes in ether lipid synthesis has been highlighted by the finding that patients with Zellweger syndrome (lacking peroxisomes) and related peroxisomal-deficient diseases have extremely low levels of plasmalogens and ether lipids.

Kinetic experiments with a partially purified enzyme from Ehrlich ascites cells have suggested the reaction catalyzed by alkyldihydroxyacetone-P synthase involves a ping-pong rather than sequential-type mechanism, with an activated enzyme–dihydroxyacetone-P intermediary complex playing a central role. The existence of this intermediate would explain the reversibility of the reaction, since the enzyme–dihydroxyacetone-P complex can react with either fatty alcohols (forward reaction) or fatty acids (back reaction). This unusual enzymatic mechanism is also consistent with other known properties of alkyldihydroxyacetone-P synthase. Acyldihydroxyacetone-P acylhydrolase does not appear to participate in this mechanism since its activity is not present in the purified synthase preparation.

A number of novel features characterize the reaction that forms alkyldihydroxyacetone-P. The pro-*R* hydrogen at C-1 of the dihydroxyacetone-P moiety of acyldihydroxyacetone-P exchanges with water, without any change in the configuration of the C-1 carbon. Cleavage of the acyl group of acyldihydroxyacetone-P occurs before the addition of the fatty alcohol, and either fatty acids or fatty alcohols can bind to the activated enzyme–dihydroxyacetone-P complex to produce acyldihydroxyacetone-P or alkyldihydroxyacetone-P, respectively. There is no evidence for a Schiff's base being formed. Nevertheless, a ketone function is an essential feature of the substrate, acyldihydroxyacetone-P. In addition, mass spectrometric analyses have clearly shown that the oxygen of the ether bond is donated by the fatty alcohol and both oxygens of the acyl linkage of acyldihydroxyacetone-P are recovered in the fatty acid released.

The cDNAs for human and guinea pig alkyldihydroxyacetone-P synthase have been cloned and expressed. The apparent molecular mass of the enzyme from guinea pig is 65 kDa on polyacrylamide gel electrophoresis. The mature enzymes of both human and guinea pig have a predicted mass of 67 kDa. The enzyme is synthesized as a 658 amino acid precursor containing an N-terminal presequence of 58 amino acids encoding the peroxisomal targeting signal 2 motif, which is removed in the mature protein. In studies of the structure and mechanism of action of the enzyme, de Vet et al. [22] made the surprising finding that the enzyme contains a FAD binding domain. They demonstrated the presence of FAD in the enzyme and found that the FAD cofactor is required for activity of the enzyme. The FAD participates directly in catalysis and becomes reduced upon incubation with acyldihydroxyacetone-P. This finding suggests that the dihydroxyacetone-P moiety has been oxidized as the acyl chain is removed. Evidence indicated that the unidentified oxidized intermediate is not covalently linked to the enzyme but can be washed off the enzyme. Addition of fatty alcohol and synthesis of alkyldihydroxyacetone-P results in reoxidation of the $FADH_2$. Normally, acylhydrolase reactions proceed by acyl oxygen fission in which only one of the oxygens of the ester bond remains with the acyl chain; alkyl oxygen fission, where both oxygens of the ester bond remain with the released acyl chain, as catalyzed by the alkyl synthase, is very unusual. The proposed oxidized dihydroxyacetone-P intermediate is yet to be identified. These new findings and available systems may soon reveal the exact mechanism by which this exciting enzyme is able to synthesize an ether bond.

Alkyldihydroxyacetone-P synthase exhibits a very broad specificity for fatty alcohols of different carbon chain lengths. On the other hand, the specificity of the synthase for acyldihydroxyacetone-P possessing different acyl chains is less well understood, primarily because of their lack of availability.

### 6.2.2. O-Alkyl analog of phosphatidic acid and alkylacylglycerols

Once alkyldihydroxyacetone-P is synthesized, it can be readily converted to the O-alkyl analog of phosphatidic acid (Fig. 5) in a two-step reaction sequence. The NADPH-dependent oxidoreductase, located on the cytosolic side of peroxisomal membranes, is capable of reducing the ketone group of both the alkyl and acyl analogs of dihydroxyacetone-P. Dietary ether lipids can also enter this pathway, since alkylglycerols formed via the catabolism of dietary ether-linked lipids during absorption are known to be phosphorylated by an ATP:alkylglycerol phosphotransferase to form 1-alkyl-2-lyso-sn-glycerol-3-P (Fig. 5, Reaction IV), which can then be acylated by an acyl-CoA acyltransferase to produce plasmanic acid, the O-alkyl analog of phosphatidic acid. The latter can be dephosphorylated to alkylacylglycerols which occupy an important branch point in the ether lipid pathway in a manner analogous to the diacylglycerols. Reaction steps beginning with 1-alkyl-2-acyl-sn-glycerol in the routes leading to the more complex ether-linked neutral lipids and phospholipids (Fig. 5, Reactions V, VI, and VII) are thought to be catalyzed by the same enzymes as those involved in the

244

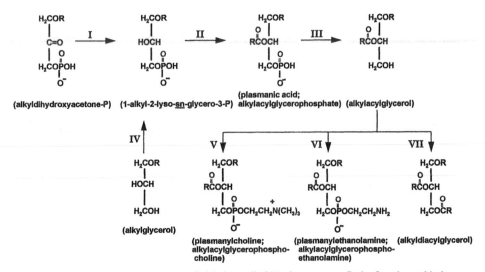

Fig. 5. Biosynthesis of membrane phospholipids from alkyldihydroxyacetone-P, the first detectable intermediate formed in the biosynthetic pathway for ether-linked glycerolipids. Enzymes responsible for catalyzing the reactions shown are (I) NADPH:alkyldihydroxyacetone-P oxidoreductase, (II) acyl-CoA:1-alkyl-2-lyso-*sn*-glycero-3-P acyltransferase, (III) 1-alkyl-2-acyl-*sn*-glycero-3-P phosphohydrolase, (IV) ATP:1-alkyl-*sn*-glycerol phosphotransferase, (V) CDP-choline:1-alkyl-2-acyl-*sn*-glycerol cholinephosphotransferase, (VI) CDP-ethanolamine:1-alkyl-2-acyl-*sn*-glycerol ethanolaminephosphotransferase, and (VII) acyl-CoA:1-alkyl-2-acyl-*sn*-glycerol acyltransferase.

pathways originally established by Kennedy and co-workers in the late 1950s for the diacylglycerolipids (Chapter 8).

*6.2.3. Neutral ether-linked glycerolipid*

Alkyldiacylglycerols, the *O*-alkyl analog of triacylglycerols, are produced by acylation of 1-alkyl-2-acyl-*sn*-glycerols in a reaction catalyzed by an acyl-CoA acyltransferase (Fig. 5, Reaction VII). The acyltransferase can also acylate 1-alk-1-enyl-2-acyl-*sn*-glycerols to form the 'neutral plasmalogen' analog of triacylglycerols. In addition, an acetylated *O*-alkyl analog of triacylglycerols has been shown to be synthesized from 1-alkyl-2-acetyl-*sn*-glycerols in HL-60 cells. The biological function of these ether-linked neutral lipids is unknown at the present time.

*6.2.4. O-Alkyl choline- and ethanolamine-containing phospholipids*

1-Alkyl-2-acyl-*sn*-glycerols, derived from the alkyl analog of phosphatidic acid by the action of a phosphohydrolase, also are utilized as substrates by cholinephosphotransferase (Fig. 5, Reaction V) and ethanolaminephosphotransferase (Fig. 5, Reaction VI) to form plasmanylcholines and plasmanylethanolamines, the alkyl analogs of phosphatidylcholine and phosphatidylethanolamine. Plasmanylcholine is the membrane source of lyso-PAF, the ether lipid precursor of the potent biologically active phospholipid, PAF, whereas plasmanylethanolamine is the direct precursor of ethanolamine plasmalogens.

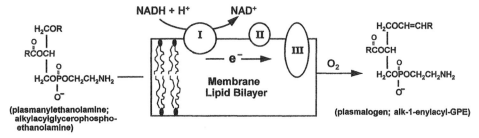

Fig. 6. Biosynthesis of ethanolamine plasmalogens by Δ1-alkyl desaturase. Components of the enzyme complex responsible for this unusual type of desaturation between carbons 1 and 2 of the O-alkyl chain are (I) NADH cytochrome $b_5$ reductase, (II) cytochrome $b_5$, and (III) Δ1-alkyl desaturase. GPE, glycerophosphoethanolamine.

### 6.2.5. Ethanolamine plasmalogens

The Δ1-alkyl desaturase, a microsomal mixed-function oxidase system, is responsible for the biosynthesis of ethanolamine plasmalogens from alkyl lipids (Fig. 6). The alkyl desaturase, which produces the alk-1-enyl grouping, is a unique enzyme, since it can specifically and stereospecifically abstract hydrogen atoms from C-1 and C-2 of the O-alkyl chain of an intact phospholipid molecule, 1-alkyl-2-acyl-*sn*-glycero-3-phospho-ethanolamine, to form the *cis* double bond of the O-alk-1-enyl moiety. Only intact 1-alkyl-2-acyl-*sn*-glycero-3-phosphoethanolamine is known to serve as a substrate for the alkyl desaturase.

Like the acyl-CoA desaturases (Chapter 7), the Δ1-alkyl desaturase exhibits the typical requirements of a microsomal mixed-function oxidase: molecular oxygen, a reduced pyridine nucleotide, cytochrome $b_5$, cytochrome $b_5$ reductase, and a terminal desaturase protein that is sensitive to cyanide. The precise reaction mechanism responsible for the biosynthesis of the ethanolamine plasmalogens is unknown. However, it is clear from an investigation with a tritiated fatty alcohol, that only the 1S and 2S (*erythro*) labeled hydrogens are lost during the formation of the alk-1-enyl moiety of ethanolamine plasmalogens.

### 6.2.6. Choline plasmalogens

Δ1-Alkyl desaturase does not utilize 1-alkyl-2-acyl-*sn*-glycero-3-phosphocholine as a substrate. In fact, biosynthesis of the significant quantities of choline plasmalogens that occurs in some heart tissues remains an enigma, although most available data strongly imply that they are derived from the ethanolamine plasmalogens. Considerable evidence has accumulated to indicate that a combination of phospholipase $A_2$, lysophospholipase D, acyltransferase, phosphohydrolase, and cholinephosphotransferase activities participate in the conversion of plasmenylethanolamine to plasmenylcholine. Direct base exchange, coupled phospholipase C/cholinephosphotransferase reactions, and the methylation of the ethanolamine moiety could also contribute to the synthesis of plasmenylcholine [23,24]. Available evidence indicates that direct polar head group remodeling mechanisms (Fig. 7) or a combined enzymatic modification of the *sn*-2 and *sn*-3 positions of ethanolamine plasmalogens (Fig. 8) best explain how choline plasmalogens are formed.

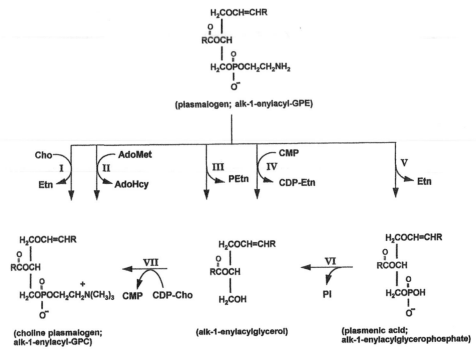

Fig. 7. Possible pathways for biosynthesis of choline plasmalogens via the modification of the *sn*-3 polar head group of ethanolamine plasmalogens are catalyzed directly by (I) a base exchange enzyme or (II) *N*-methyltransferase. A combination of other enzymatic reactions can also result in the replacement of the ethanolamine moiety of plasmenylethanolamine to produce plasmenylcholines; the enzymes responsible include (III) phospholipase C, (IV) the reverse reaction of ethanolamine phosphotransferase, (V) phospholipase D, (VI) a phosphohydrolase, and (VII) cholinephosphotransferase. Abbreviations: AdoMet, *S*-adenosyl-L-methionine; AdoHcy, *S*-adenosyl-L-homocysteine; Etn, ethanolamine; GPE, glycerophospho-ethanolamine.

### 6.3. PAF and related bioactive species

#### 6.3.1. Remodeling route

The remodeling pathway of PAF synthesis (Fig. 9) is thought to be the primary contributor to hypersensitivity reactions and for this reason this route has been implicated in most pathological responses involving PAF. Biosynthesis of PAF during inflammatory cellular responses or following various agonist stimulation occurs via the enzymatic remodeling of membrane-bound alkylacylglycerophosphocholines by replacing an acyl moiety with an acetate group. The enzymes responsible for catalyzing the hydrolytic deacylation step appear to be highly specific for the molecular species of alkylacylglycerophosphocholines possessing an arachidonoyl moiety at the *sn*-2 position. The initial reaction that produces lyso-PAF requires either the combined actions of a membrane-associated CoA-*independent* transacylase/phospholipase A$_2$ (Fig. 9, Reaction II; indirect route) or can be catalyzed in a single direct hydrolytic step by a phospholipase A$_2$ (Fig. 9, Reaction I). A CoA-*dependent* transacylase (reversal of an acyl-CoA acyltransferase

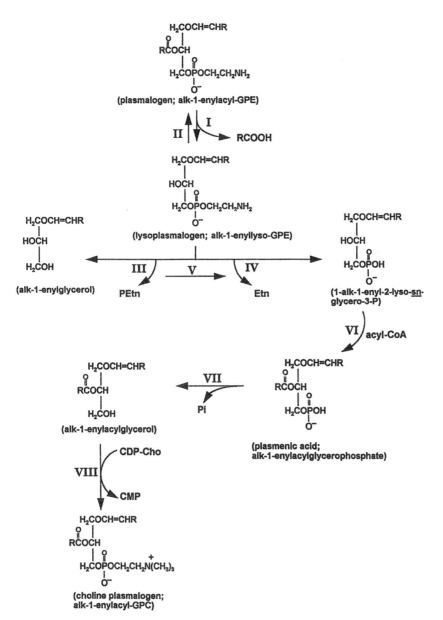

Fig. 8. Biosynthesis of plasmenylcholine via the modification of both the *sn*-2 acyl and *sn*-3 polar head group moieties of plasmenylethanolamine. (I) phospholipase $A_2$, (II) CoA-independent transacylase, (III) lysophospholipase C, (IV) lysophospholipase D, (V) a phosphotransferase, (VI) acyl-CoA acyltransferase, (VII) phosphohydrolase, and/or (VIII) cholinephosphotransferase. Abbreviations: Etn, ethanolamine; Cho, choline; GPE, *sn*-glycero-3-phosphoethanolamine; GPC, *sn*-glycero-3-phosphocholine.

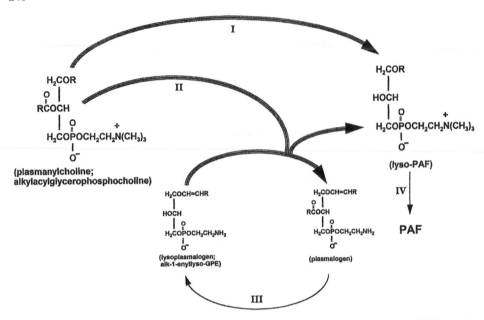

Fig. 9. Biosynthesis of PAF via the remodeling pathway. Lyso-PAF, the immediate precursor of PAF, can be formed from 1-alkyl-2-acyl-*sn*-3-glycerophosphocholine through the direct action of (I) a phospholipase $A_2$ or (II) a CoA-*independent* transacylase. The lysoplasmenylethanolamine (or other potential ethanolamine- or choline-containing lysoglycerophospholipids) is thought to be generated by (III) a phospholipase $A_2$ that exhibits a high degree of selectivity for substrates having an arachidonoyl moiety at the *sn*-2 position. The transacylase (II) appears to possess both acyl transfer and phospholipase $A_2$ hydrolytic activities, which could exist as a single protein or as a tightly associated complex of two distinctly different proteins. The lyso-PAF produced by either the transacylation (II) or direct phospholipase $A_2$ (I) reactions can then be acetylated to form PAF by (IV) an acetyl-CoA acetyltransferase.

reaction) is also capable of generating lyso-PAF (Fig. 10, Reaction I). Two reviews have focused on the different types of transacylase reactions involved in the remodeling of phospholipids [25,26].

A number of studies indicate that the 85 kDa cytosolic phospholipase $A_2$ is likely the phospholipase responsible for release of arachidonic acid and PAF synthesis in stimulated cells. It is highly selective for arachidonate as is the CoA-independent transacylase. One of the most convincing findings showing that this enzyme is responsible for initiating the remodeling pathway is that macrophages from cytosolic phospholipase $A_2$ knockout mice almost completely lose their ability to synthesize both PAF and eicosanoids [13,14]. Since the cytosolic phospholipase $A_2$ does not distinguish between the ester and ether linkage in the *sn*-1 position, acetylated products recovered from cells reflect the composition of the choline-containing phosphoglycerides. The activity is regulated by phosphorylation of the enzyme and by translocation from the cytosol to membranes, which requires concentrations of $\mu M$ $Ca^{2+}$; $Ca^{2+}$ is not required for the hydrolytic mechanism of cytosolic phospholipase $A_2$.

The acetyltransferase responsible for the final step in the synthesis of PAF (Fig. 9,

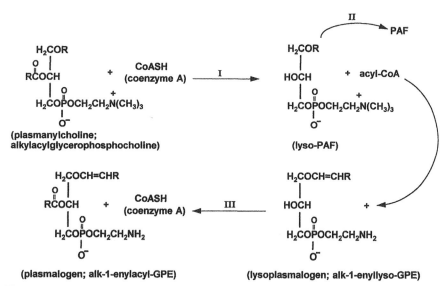

Fig. 10. Involvement of a CoA-*dependent* transacylase in the production of lyso-PAF for the synthesis of PAF and the remodeling of the $sn-2$ acyl group of membrane phospholipids. The enzymes responsible for catalyzing these reactions are (I) the CoA-*dependent* transacylase (with CoA as the acyl acceptor), (II) acetyl CoA:lyso-PAF acetyltransferase, and (III) an acyl-CoA:lysophospholipid acyltransferase. The reaction depicted for the CoA-dependent transacylase represents the reversal of the reaction catalyzed by acyl-CoA:lysophospholipid acyltransferase. GPE designates *sn*-3-glycerophosphoethanolamine.

Reaction IV) is membrane-bound and, like the CoA-independent transacylase, has neither been purified nor its cDNA cloned. This membrane-bound enzyme can also acetylate both the alk-1-enyl and acyl analogs of lyso-PAF and utilizes short-chain acyl-CoAs ranging from $C_2$ to $C_6$ as substrates. Several studies indicate that the enzyme is activated by phosphorylation even though the unphosphorylated enzyme appears to have a basal activity. In studies of human neutrophils, Nixon et al. [27] have concluded from studies with MAP kinase inhibitors and recombinant cytosolic phospholipase $A_2$ and MAP kinases that the acetyl-CoA:lyso-PAF acetyltransferase is specifically activated by the p38 stress-activated MAP kinase but not by p42 and p44 ERKs. In contrast, cytosolic phospholipase $A_2$ is activated in the cells by both the ERKs and p38 kinase. Related findings suggest that the production of phosphatidic acid by phospholipase D specifically activates the p38 kinase cascade but not the ERKs. Identification of the specific protein kinases responsible for the direct activation of cytosolic phospholipase $A_2$ and the acetyltransferase in intact cells is complicated by cross-talk between the kinases, including protein kinase C.

Regulation of the CoA-independent transacylase activity (Fig. 9, Reaction II) is poorly understood. It has been demonstrated that production of lyso-PAF via the transacylation step can occur in either a CoA-*independent* (Fig. 9) or CoA-*dependent* (Fig. 10) manner [20,21]. With the CoA-*independent* transacylase, ethanolamine lyso-plasmalogens as well as other ethanolamine- or choline-containing lysoglycerophosphatides serve as the acyl acceptor for the selective transfer of an arachidonoyl moiety from alkylacylglycerophosphocholine. CoA itself, instead of a lysophospholipid, is the acyl acceptor in the reaction catalyzed by the CoA-dependent transacylase (Fig. 10). This type of transacylation is thought to represent the reverse reaction of that catalyzed by an acyl-CoA:lyso-PAF acyltransferase but is not selective for arachidonate. In addition to participating in the formation of lyso-PAF, the transacylases also serve an important role in the remodeling of acyl moieties located at the $sn$-2 position of the choline- and ethanolamine-containing phospholipids.

The lysoplasmalogen or other lysophospholipid acceptors that are substrates for the transacylases appear to be formed by the direct action of a phospholipase $A_2$ on the appropriate membrane-associated phospholipid which simultaneously releases arachidonic acid for its subsequent metabolism to bioactive eicosanoid products (Chapter 13). In both the direct and indirect routes of lyso-PAF production, the action of a phospholipase $A_2$ is required; it is plausible that both routes participate in PAF synthesis to varying degrees depending on conditions. Since both eicosanoid and PAF mediators can be formed via the remodeling pathway and these mediators can act synergistically, the assessment of biological responses following cell activation can often be difficult to interpret.

### 6.3.2. De novo route

PAF biosynthesis via the de novo pathway [10,11] is thought to be the primary source of the physiological levels of PAF in cells and blood (Fig. 11). Both fatty acids and neurotransmitters can stimulate the de novo synthesis of PAF. The sequence of enzymatic reactions (Fig. 11) involved in the de novo route include (a) acetylation of 1-alkyl-2-lyso-$sn$-glycero-3-P by an acetyl-CoA-dependent acetyltransferase (Reaction I), (b) dephosphorylation of 1-alkyl-2-acetyl-$sn$-glycero-3-P (Reaction II), and (c) the transfer of phosphocholine from CDP-choline to 1-alkyl-2-acetyl-$sn$-glycerol by a dithiothreitol-insensitive cholinephosphotransferase (Reaction III) to form PAF. The acetyltransferases associated with the remodeling (Fig. 9) and de novo routes (Fig. 11) possess distinctly different properties and substrate specificities. Also, the dithiothreitol-insensitivity of this cholinephosphotransferase contrasts with the inhibitory effect of dithiothreitol on the cholinephosphotransferase that synthesizes phosphatidylcholine and plasmanylcholine from diacylglycerols and alkylacylglycerols, respectively. In addition, the two dissimilar cholinephosphotransferase activities that synthesize PAF and phosphatidylcholine exhibit different pH optima and respond differently to detergents, ethanol, temperature, and substrates. Although the enzymes in the de novo pathway exhibit a relatively high degree of substrate specificity, the $sn$-1 acyl analogs of the corresponding $O$-alkyl equivalents can also be utilized as substrates by the acetyltransferase, phosphohydrolase, and the dithiothreitol-insensitive cholinephosphotransferase.

### 6.3.3. PAF transacetylase

Two novel CoA-independent transacetylases that use PAF as the donor molecule (Fig. 12) are PAF:lysophospholipid transacetylase and PAF:sphingosine transacetylase [28]. Both transacetylases have no requirement for $Ca^{2+}$, $Mg^{2+}$, or CoA. The PAF:lysophospholipid transacetylase transfers the acetyl group from PAF to a variety

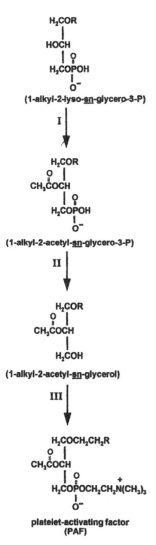

Fig. 11. Biosynthesis of PAF via the de novo pathway. The three-step reaction sequence in this route, beginning with 1-alkyl-2-lyso-*sn*-glycero-3-P as the precursor, is catalyzed by (I) acetyl-CoA:alkyllysoglycero-P acetyltransferase, (II) alkylacetylglycero-P phosphohydrolase, and (III) CDP-choline:alkylacetylglycerol cholinephosphotransferase.

of lysophospholipids to form a series of PAF analogs. Among all the lysophospholipids tested, acyllysoglycerophosphocholine is the most active acetyl group acceptor. In addition, *cis*-9-octadecen-1-ol can also serve as acetate acceptor, whereas alkylglycerol, acylglycerol, or cholesterol are inactive. Biochemical studies suggest that the CoA-independent transacetylase differs from the CoA-independent transacylase that transfers long-chain acyl moieties.

252

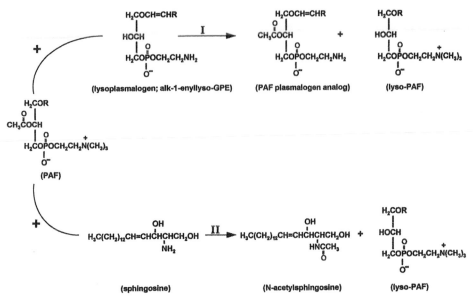

Fig. 12. PAF transacetylase transfers the acetate moiety of PAF to other selective substrates to produce a plasmalogen analog (I) and acetylsphingosine (II). GPE, glycerophosphoethanolamine.

The PAF : sphingosine transacetylase transfers the acetate group from PAF to sphingosine forming $N$-acetylsphingosine ($C_2$-ceramide). The enzyme has a narrow substrate specificity and strict stereochemical configuration requirement. Ceramide, sphingosylphosphocholine, stearylamine, sphingosine-1-phosphate, or sphingomyelin are not substrates, whereas sphinganine has a limited capacity to accept the acetate from PAF. Only the naturally synthesized D-*erythro* isomer but not the synthetic L-*erythro*-, D-*threo*-, or L-*threo*-isomer of sphingosine can serve as a substrate. Both PAF : lysophospholipid transacetylase and PAF : sphingosine transacetylase have similar tissue distributions. The PAF : sphingosine transacetylase is located in mitochondria, microsomes, and cytosol with mitochondria having the highest specific activity. Physiological levels of $C_2$-ceramide (in μM range) have been detected in both undifferentiated and differentiated HL-60 cells.

Rat kidney transacetylases from mitochondria/microsomes and cytosols have an apparent molecular mass of 40 kDa. Both purified enzymes from membranes and cytosols contain three catalytic activities; PAF : lysophospholipid transacetylase, PAF : sphingosine transacetylase, and PAF acetylhydrolase (PAF-AH). A search using a protein sequence data bank indicates that these sequences have homology with the sequences present in bovine PAF-AH II (Section 7.3.1).

The substrate specificity, kinetic parameters, and inhibitor effects suggest that the three individual catalytic activities of the transacetylase have different dependencies on the thiol-containing residue(s) of the enzyme, i.e., cysteine. Furthermore, the nonresponsiveness of the purified cytosolic transacetylase to phosphatidylserine activation indicates that membrane and cytosolic transacetylase may be posttranslationally distinct.

Analysis of a series of site-directed mutant PAF-AH II proteins in CHO-K1 cells shows that lysophospholipid transacetylase is decreased, whereas PAF-AH activity is not affected in C120S and G2A mutants. Thus, Cys$^{120}$ and Gly$^2$ are implicated in the catalysis of the lysophospholipid transacetylase reaction in this enzyme. It appears that N-myristoylation is not required for PAF-AH activity.

Several lines of evidence indicate that transacetylase activity has a physiological role in vivo. With intact differentiated HL-60 cells, [$^3$H]acetate from [$^3$H]PAF can be incorporated into alk-1-enyl[$^3$H]acetylglycerophosphoethanolamine in the presence of ionophore A23187, but not in its absence. In endothelial cells stimulated by ATP, bradykinin, and ionophore A23187, acylacetylglycerophosphocholine is the predominant product and the radiolabelled acetate group of PAF is incorporated into acylacetylglycerophosphocholine in a time-dependent fashion.

In ATP-activated cells, PAF : acyllysoglycerophosphocholine transacetylase and formation of acylacetylglycerophosphocholine are concurrently and transiently induced, while PAF : sphingosine transacetylase and PAF-AH activities remain unchanged. Evidence indicates that tyrosine protein kinase and protein kinase C are directly or indirectly involved in the activation of the transacetylase activity through protein phosphorylation. In addition, ATP induces the translocation of acyllysoglycerophosphocholine transacetylase from cytosol to membranes and also increases the specific enzyme activity on the membrane. Collectively, the three catalytic activities of the transacetylase are regulated in agonist-activated cells through posttranslational modifications (such as reversible phosphorylation/dephosphorylation, myristoylation, etc.) and translocation of the enzyme from cytosol to membranes.

# 7. Catabolic enzymes

## 7.1. Ether lipid precursors

### 7.1.1. Fatty alcohols
Fatty alcohols are oxidized to fatty acids via a fatty alcohol:NAD$^+$ oxidoreductase, a microsomal enzyme found in most mammalian cells (Fig. 3B). The high activity of this enzyme probably accounts for the extremely low levels of unesterified fatty alcohols generally found in tissues or blood. Detection of fatty aldehydes, by trapping them as semicarbazide derivatives during oxidation of the alcohol, suggests that the fatty alcohol oxidoreductase catalyzes a two-step reaction that involves an aldehyde intermediate.

### 7.1.2. Dihydroxyacetone-P and acyldihydroxyacetone-P
Dihydroxyacetone-P can be diverted from its precursor role in ether lipid synthesis when it is converted to sn-glycerol-3-P by glycerol-3-P dehydrogenase. Another bypass that prevents the formation of alkyldihydroxyacetone-P occurs if the ketone function of acyldihydroxyacetone-P is first reduced by an NADPH-dependent oxidoreductase, since the product, 1-acyl-2-lyso-sn-glycerol-3-P, can then be converted to different diacyl types of glycerolipids. Obviously, the metabolic removal and/or formation of fatty

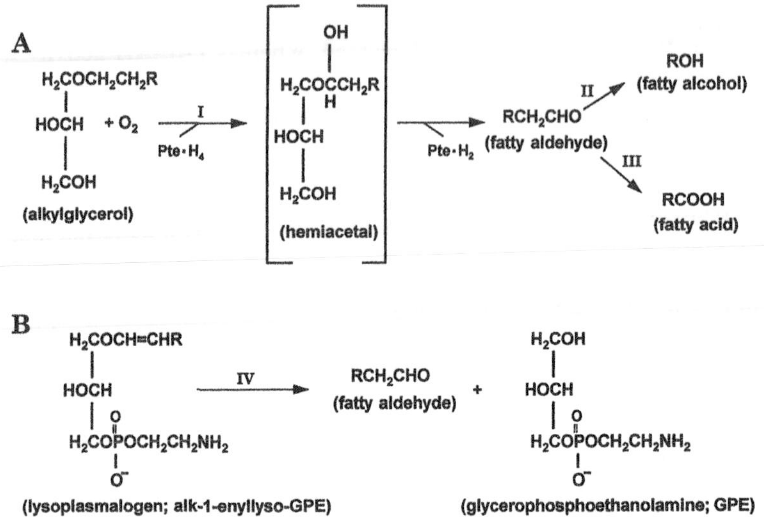

**A**

$H_2COCH_2CH_2R$
|
$HOCH \quad + O_2$  $\xrightarrow[Pte \cdot H_4]{I}$
|
$H_2COH$

(alkylglycerol)

$$\begin{bmatrix} & OH \\ & | \\ H_2COCCH_2R \\ & | \\ & H \\ HOCH \\ & | \\ H_2COH \\ \\ (hemiacetal) \end{bmatrix}$$

$\xrightarrow[Pte \cdot H_2]{}$ $RCH_2CHO$ (fatty aldehyde)

$\xrightarrow{II}$ ROH (fatty alcohol)

$\xrightarrow{III}$ RCOOH (fatty acid)

**B**

$H_2COCH=CHR$
|
$HOCH$
|  O
|  ‖
$H_2COPOCH_2CH_2NH_2$
|
$O^-$

(lysoplasmalogen; alk-1-enyllyso-GPE)

$\xrightarrow{IV}$ $RCH_2CHO$ (fatty aldehyde) +

$H_2COH$
|
$HOCH$
|  O
|  ‖
$H_2COPOCH_2CH_2NH_2$
|
$O^-$

(glycerophosphoethanolamine; GPE)

Fig. 13. Cleavage of the *O*-alkyl linkage in glycerolipids (A) is catalyzed by (I) tetrahydropteridine (Pte·H$_4$)-dependent alkyl monooxygenase. The fatty aldehyde product can be either reduced to a long-chain fatty alcohol by (II) a reductase or oxidized to a fatty acid by (III) an oxidoreductase. Removal of the *O*-alk-1-enyl moiety from plasmalogens (B) is catalyzed by a plasmalogenase. As with the *O*-alkyl monooxygenase, the fatty aldehyde can be converted either to the corresponding fatty alcohol or fatty acid. GPE, glycerophosphoethanolamine.

alcohols, dihydroxyacetone-P, or acyldihydroxyacetone-P from the ether lipid precursor pool represent important control points for regulating the ether lipid pathway.

### 7.2. Ether-linked lipids

#### 7.2.1. O-Alkyl cleavage enzyme

Oxidative cleavage of the *O*-alkyl linkage in glycerolipids is catalyzed by a microsomal tetrahydropteridine (Pte·H$_4$)-dependent alkyl monooxygenase (Fig. 13A). The required cofactor, Pte·H$_4$, is regenerated from the Pte·H$_2$ by an NADPH-linked pteridine reductase, a cytosolic enzyme. Oxidative attack on the ether-linked grouping in lipids is similar to the enzymatic mechanism described for the hydroxylation of phenylalanine. Fatty aldehydes produced via the cleavage reaction can be either oxidized to the corresponding acid or reduced to the alcohol by appropriate enzymes.

Alkyl cleavage enzyme activities are highest in liver and intestinal tissue, whereas most other cells/tissues possess very low activities. Tumors and other tissues that contain significant quantities of alkyl lipids generally have very low alkyl cleavage enzyme activities, which is consistent with the overall premise that the level of ether-linked glycerolipids is inversely proportional to the activity of the alkyl cleavage enzyme.

Structural features of glycerolipid substrates utilized by the alkyl cleavage enzyme are (a) an *O*-alkyl moiety at the *sn*-1 position, (b) a free hydroxyl group at the *sn*-2 position, and (c) a free hydroxyl or phosphobase group at the *sn*-3 position. If the hydroxyl group at the sn-2 position is replaced by a ketone or acyl grouping, or when a free phosphate is at the sn-3 position, the O-alkyl moiety at the sn-1 position is not cleaved by the Pte·H$_4$-dependent monooxygenase. Thus, 1-alkyl-2-lysophospholipids (e.g., lyso-PAF) are substrates for the cleavage enzyme, but they are attacked at much slower rates than are alkylglycerols.

### 7.2.2. Plasmalogenases

Plasmalogenases (Fig. 13B) are capable of hydrolyzing the O-alk-1-enyl grouping of plasmalogens or lysoplasmalogens. The products of this reaction are a fatty aldehyde and either 1-lyso-2-acyl-sn-glycero-3-phosphoethanolamine (or choline) or sn-glycero-3-phosphoethanolamine (or choline), depending on the chemical structure of the parent substrate. Plasmalogenase activities have been described in microsomal preparations from liver and brain of rats, cattle, and dogs, but their biological significance is poorly understood.

### 7.2.3. Phospholipases and lipases

In general, the sn-2 and sn-3 ester groupings associated with either the alkyl or alk-1-enyl glycerolipids are hydrolyzed by lipolytic enzymes with the same degree of substrate specificity as their acyl counterparts (Fig. 14). However, the ether linkages themselves are not hydrolyzed by lipases or phospholipases and the presence of an ether linkage at the sn-1 position of the glycerol moiety generally reduces the overall reaction rate to the extent that certain lipases have been successfully used to remove diacyl contaminants in the purification of some ether-linked phospholipids. The only lipolytic enzyme (other than those that cleave the ether linkages) known to exhibit an absolute specificity for ether-linked lipids is lysophospholipase D. The uniqueness of lysophospholipase D is that it exclusively recognizes only 1-alkyl-2-lyso-sn-glycero-3-phosphobases or 1-alk-1-enyl-2-lyso-sn-glycero-3-phosphobases as substrates; thus, lyso-PAF is a substrate for this novel enzyme (Fig. 14, Reaction II).

### 7.3. PAF and related bioactive species

### 7.3.1. Acetylhydrolase

PAF acetylhydrolase (AH) enzymes (Fig. 14, Reaction I) are a specific group of Ca$^{2+}$-independent phospholipases A$_2$ that remove the acetyl moiety at the sn-2 position of PAF [29–31]. Mammalian PAF-AHs can be classified into intracellular and extracellular types. Intracellular PAF-AHs consist of at least three groups of enzymes, namely, PAF-AH I, PAF-AH II, and erythrocyte-type PAF-AH. Extracellular PAF-AH occurs as plasma AH.

The erythrocyte enzyme appears to be a homodimer comprised of the 25-kDa polypeptide and is different from PAF-AH II. PAF-AH is a serine esterase and requires reducing agents for maximal activity. The most likely role of the erythrocyte PAF-AH in vivo is to hydrolyze the products of oxidative fragmentation of membrane phospholipids.

PAF-AH I, rich in brain and exclusively located in cytosols, is an unusual G-protein-like ($\alpha_1/\alpha_2$)$\beta$ heterotrimer complex composed of 45($\beta$)-, 30($\alpha_2$)-, and 29($\alpha_1$)-kDa

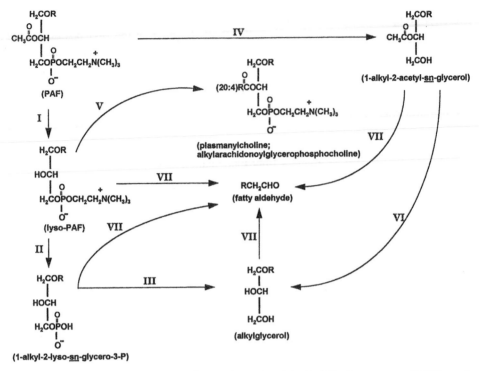

Fig. 14. Catabolism of PAF and its metabolites can be catalyzed by the following enzymes: (I) PAF acetyl-hydrolase, (II) lysophospholipase D, (III) phosphohydrolase, (IV) phospholipase C, (V) CoA-*independent* or CoA-*dependent* transacylase, and/or (VI) alkylacetylglycerol acetylhydrolase. The *O*-alkyl linkage in those products that contain free hydroxyl groups can be cleaved by (VII) the *O*-alkyl Pte·H₄-dependent monooxygenase.

subunits. The tertiary fold of $\alpha_1$ subunit is similar to that found in p21$^{ras}$ and other GTPases. The active site is made up of a trypsin-like triad of Ser 47, His 195, and Asp 193. A sequence of ~30 amino acids adjacent to the active serine residue exhibits significant similarity to the first transmembrane region of the PAF receptor. The catalytic 30($\alpha_2$)-kDa subunit is highly homologous (63.2% identity) to that of the 29($\alpha_1$)-kDa subunit, especially (86%) in the catalytic and PAF receptor homologous domains.

The 45($\beta$) kDa subunit, which is not essential for catalytic activity, exhibits striking homology (99%) with a protein encoded by the causal gene (*LIS-1*) for the onset of Miller–Dieker lissencephaly, a human brain malformation manifested by a smooth cerebral surface and impaired neural migration. In addition, the 45-kDa subunit contains a 7-tandem WD-40 repeat in its primary structure. This repeat is thought to be important for interactions with other protein components, especially with pleckstrin-homology (PH) domains. Therefore, the hydrolysis of PAF may induce conformational changes in the heterotrimeric PAF-AH I complex that affect the ability of the 45-kDa subunit to interact with cytoskeletal proteins.

Furthermore, the $\alpha_1$ subunit appears to be expressed specifically in neurons of fetal and neonatal brain of rats. Significant levels of the $\alpha_1$ subunit are not expressed in any adult rat tissues. In contrast, $\alpha_2$ and $\beta$ transcripts and proteins are almost constantly expressed from fetal stages through adulthood. The catalytic subunits switch from the $\alpha_1/\alpha_2$ heterodimer to the $\alpha_2/\alpha_2$ homodimer and along with the above-mentioned data suggest that PAF-AH I in brain is involved in brain development through regulation of neuronal migration.

PAF-AH type II (PAF-AH II), expressed most abundantly in liver and kidney, is a monomeric 40-kDa protein and a member of the serine esterase family. This enzyme exhibits broader substrate specificity than PAF-AH I. PAF-AH II hydrolyzes oxidized phospholipids as effectively as PAF, whereas PAF-AH I is more specific for PAF. Furthermore, unlike PAF-AH I, PAF-AH II is distributed in both the membrane and soluble fractions.

PAF-AH II is a $N$-myristoylated enzyme, the first reported among lipases and phospholipases. It translocates between cytosol and membranes in response to the redox state. When overexpressed in cells, it protects against oxidative stress-induced cell death. PAF-AH II may function as an antioxidant phospholipase and promote the hydrolysis of oxidized phospholipids produced during reactive oxygen species-induced apoptosis.

Plasma PAF-AH is unique since it is mainly associated with the high density and low density lipoproteins. Plasma PAF-AH also hydrolyzes PAF and structurally related oxidized phospholipids with up to 9 carbon $sn$-2 acyl chains. The cDNA for this enzyme encodes a 44-kDa secretory protein that contains a typical signal sequence and a serine esterase/neutral lipase consensus motif GXSXG. Serine 273 (of the GXSXG motif), Asp 296, and His 351 are required for catalysis. Plasma PAF-AH displays $\sim$40% homology with intracellular PAF-AH II, but little homology/similarity with PAF-AH I over the whole amino acid sequence.

Pretreatment of animals with recombinant plasma PAF-AH blocks PAF-induced inflammation. Furthermore, deficiency of plasma PAF-AH is an autosomal recessive syndrome that is associated with severe asthma in Japanese children. A point mutation of exon 9 of the plasma PAF-AH gene results in the production of an inactive protein. In addition, an increase in enzyme activity has been reported in humans or rats with hypertension, and in the plasma of patients with atherosclerosis. The level of plasma PAF-AH decreased markedly near the end of pregnancy in rabbits. It has been proposed that this is a component of the mechanism for initiating the onset of labor. Since PAF stimulates the contraction of uterine muscle, a decrease in PAF-AH activity will allow the accumulation of PAF

## 8. Metabolic regulation

Regulatory mechanisms that control the metabolism of ether-linked lipids are still poorly understood. In fact, most progress in this area has concerned PAF metabolism, primarily because of the high degree of interest in this potent mediator. Nevertheless, a variety of factors are known to influence the overall rates of ether lipid metabolism, but such studies have mainly been of the descriptive type and none have addressed the molecular enzymatic mechanisms involved. Regulatory controls that must be considered in the metabolism of ether-linked lipids are those that influence (a) the enzymes responsible for catalyzing the biosynthesis and catabolism of the ether lipid precursors (fatty alcohols and dihydroxyacetone-P), (b) alkyldihydroxyacetone-P synthase which is responsible for the synthesis of alkyldihydroxyacetone-P, and (c) branch point enzymes, e.g., those steps that utilize diradylglycerols.

Glycolysis plays an important role in controlling the levels of ether lipids at the precursor level. For example, the high glycolytic rate of tumors generates significant quantities of dihydroxyacetone-P, which could explain the relatively high levels of ether lipids found in cancerous cells.

Factors responsible for the regulation of biosynthetic and catabolic enzyme activities that catalyze specific reaction steps in the metabolic pathways for ether-linked lipids appear to be very complex. Although the rate-limiting steps are poorly understood, two important intermediary branch points in the biosynthesis of ether-linked lipids involve 1-alkyl-2-lyso-$sn$-glycero-3-P and 1-alkyl-2-acyl-$sn$-glycerols. The 1-alkyl-2-lyso-$sn$-glycero-3-P can be ultimately converted to either PAF via de novo route or to 1-alkyl-2-acyl-$sn$-glycerols following an acylation and dephosphorylation step. The alkylacylglycerols represent a branch point since they are the direct precursors of plasmanylcholines, plasmanylethanolamines, and alkyldiacylglycerols. Conditions that influence either branch point would have profound effects on the proportion of the different types of ether-linked lipids formed. Since the choline- and ethanolamine-phosphotransferases appear to be able to utilize both diacyl- and alkylacyl-glycerols, it is apparent that the availability of specific diradylglycerols is crucial in controlling the diacyl and alkylacyl species composition of membranes. Most of the catabolic enzymes of ether lipid metabolism have received far less attention than those associated with the biosynthetic pathways.

Studies of the regulation of PAF metabolism are still in the early stages of development. Rate-limiting steps in the de novo pathway of PAF biosynthesis are the acetyl-CoA:1-alkyl-2-lyso-$sn$-glycero-3-P acetyltransferase and the cytidylyltransferase that forms CDP-choline for the cholinephosphotransferase catalyzed step (fig. 10, Chapter 8). Any factor that stimulates these rate-limiting reactions (e.g., activation of cytidylyltransferase by fatty acids) is also known to enhance the de novo synthesis of PAF.

In the remodeling pathway, it is clear that arachidonic acid can influence the formation of PAF at the substrate level since cells depleted of alkylarachidonoylglycerophosphocholines lose their ability to synthesize PAF. Therefore, the transacylase/phospholipase $A_2$ step (Fig. 9, Reaction II) as well as a specific phospholipase $A_2$ (Fig. 9, Reactions I or III) can be rate-limiting. Regulation of the acetyl-CoA:lyso-PAF acetyltransferase in the remodeling pathway (Fig. 9, Reaction IV) appears to be controlled by a phosphorylation/phosphohydrolase system.

Acetylhydrolase and other catabolic enzymes in PAF metabolism also have an important regulatory role in controlling PAF levels since it is known that the activity of acetylhydrolase can drastically change during various diseases, pregnancy, and macrophage differentiation.

# 9. Functions

## 9.1. Membrane components

Cellular functions of ether-linked glycerolipids are especially poorly understood, but their ability to serve as both membrane components and as cellular mediators is now well established. Both the alkyl and alk-1-enyl phospholipids that contain long-chain acyl groups at the sn-2 position appear to be essential structural components of many membrane systems. Some species of the ether lipids associated with membranes act as storage reservoirs for polyunsaturated fatty acids. The apparent protective nature of ether-linked groups against lipolytic actions is reflected by their ability to slow the rate of hydrolysis of acyl moieties at the sn-2 position by phospholipase $A_2$. The preferential sequestering of polyunsaturated fatty acids in ether-linked phospholipids has been observed even in essential fatty acid deficiency. The plasmalogens of inflammatory cells are highly enriched in arachidonate; in human neutrophils 80% of the cellular arachidonate is found in the plasmenylethanolamine fraction. There is one report with photosensitized cells exposed to long wavelength ultraviolet radiation that has suggested that plasmalogens might have a role in protecting membranes against certain forms of oxidative stress. However, the true function(s) of plasmalogens remains a puzzle.

## 9.2. Cell mediators

The multifaceted responses generated by PAF in vivo and in target cells and the ubiquitous distribution of PAF-related enzymes in mammalian cells has emphasized the important role of bioactive ether-linked lipids as diverse regulators of metabolic and cellular processes. Also, identification of the PAF receptor as a member of the general family of G-protein-coupled receptors further strengthens the potential importance of PAF as a mediator in cell signalling pathways. Activation of phosphatidylinositol-specific phospholipase C by the binding of PAF to its receptor elevates the intracellular levels of $Ca^{2+}$ and diacylglycerols, events that activate protein kinase C (Chapter 12). The latter catalyzes the phosphorylation of specific proteins and it is clearly documented that 20- and 40-kDa proteins are phosphorylated in PAF-treated rabbit platelets. However, an array of kinase cascades and cellular alterations need to be considered in any proposed biochemical mechanism for explaining the diverse actions of PAF.

# 10. Future directions

There are many unanswered questions about the dual role that ether lipids serve as membrane components and as cellular signaling molecules. Although it is clear that arachidonic acid is closely associated and tenaciously retained by ether lipids in membranes, even in essential fatty acid deficiency, much remains to be elucidated about the enzymatic systems and regulatory controls that affect the release of this sequestered pool of arachidonic acid for its subsequent conversion to bioactive eicosanoid metabolites.

260

The mechanisms that account for the synergistic actions of PAF and eicosanoids are still poorly understood. In addition, the significance of ether lipids as a dietary nutrient has received little attention even though they occur in a variety of foods and it is known that ether lipid supplements are readily incorporated into cellular lipids.

It is noteworthy that organisms living in harsh environments of high temperatures or high salt and low pH contain only ether lipids suggesting that they serve as the Teflon of lipids. Invasive cells such as neutrophils and eosinophils also have high levels of ether lipids. Based on these observations it is tempting to speculate that the high levels of ether lipids found in most tumor cells may contribute to their invasiveness.

Now that the alkyl synthase has been cloned and sequenced and shown to contain a flavin, further elucidation of the mechanism of ether bond synthesis is anticipated. Despite the advances made in cloning and sequencing of the PAF receptor, the exact mechanisms for explaining how PAF participates in signal transduction, the generation of second messengers, and gene expression remain poorly understood. Moreover, even the binding site of PAF to its receptor has not yet been rigorously identified and the presence of intracellular PAF binding sites still are unresolved issues. Certainly, the significance of PAF in physiological and disease processes needs to be more firmly established.

A major enigma is the function of plasmalogens. Despite the large quantities of ethanolamine plasmalogens found in nervous tissue and other cells, their cellular role or the molecular mechanism and regulatory controls for the alkyl desaturase responsible for their formation are still unknown. The alkyl desaturase has yet to be purified, cloned and carefully compared to fatty acid desaturases. Likewise, the biosynthesis of choline plasmalogens is still not fully understood, although compelling evidence exists to indicate that they originate from ethanolamine plasmalogens via remodeling mechanisms.

## References

1. Snyder, F. (Ed.) (1972) Ether Lipids: Chemistry and Biology. Academic Press, New York, NY, 433 pp.
2. Mangold, H.K. and Paltauf, F. (Eds.) (1981) Ether Lipids: Biochemical and Biomedical Aspects. Academic Press, New York, NY, 439 pp.
3. Snyder, F. (Ed.) (1987) Platelet-Activating Factor and Related Lipid Mediators. Plenum Press, New York, NY, pp.
4. Barnes, P.J., Page, C.P. and Henson, P.M. (Eds.) (1989) Platelet Activating Factor and Human Disease. Frontiers in Pharmacology and Therapeutics, Blackwell, Oxford, 334 pp.
5. Snyder, F., Lee, T.-c. and Wykle, R.L. (1985) Ether-linked glycerolipids and their bioactive species; enzymes and metabolic regulation. In: A.N. Martonosi (Ed.), The Enzymes of Biological Membranes. Vol. 2. Plenum Press, New York, NY, pp. 1–58.
6. Braquet, P., Touqui, L., Shen, T.Y. and Vargaftig, B.B. (1987) Perspectives in platelet activating factor research. Pharmacol. Rev. 39, 97–145.
7. Chao, W. and Olson, M.S. (1993) Receptors and signal transduction. Biochem. J. 292, 617–629.
8. Izumi, T. and Shimizu, T. (1995) Platelet-activating factor receptor: gene expression and signal transduction. Biochim. Biophys. Acta 1259, 317–333.
9. Snyder, F. (1999) The ether lipid trail: a historical perspective. Biochim. Biophys. Biochim. Acta 1436, 265–278.

10. Snyder, F. (1995) Platelet-activating factor: the biosynthetic and catabolic enzymes. Biochem. J. 305, 689–705.
11. Snyder, F. (1995) PAF and its analogs: metabolic pathways and related intracellular processes. Biochim. Biophys. Acta 254, 231–249.
12. Lee, T.-c. (1998) Biosynthesis and possible biological functions of plasmalogens. Biochim. Biophys. Acta 1394, 129–145.
13. Prescott, S.M., Zimmerman, G.A., Stafforini, D.M. and McIntyre, T.M. (2000) Platelet-activating factor and related lipid mediators. Annu. Rev. Biochem. 69, 419–445.
14. Ishii, S. and Shimizu, T. (2000) Platelet-activating factor (PAF) receptor and genetically engineered PAF receptor mutant mice. Prog. Lipid Res. 39, 41–82.
15. Debuch, H. and Seng, P. (1972) The history of ether-linked lipids through 1960. In: F. Snyder (Ed.), Ether Lipids: Chemistry and Biology. Academic Press, New York, NY, pp. 1–24.
16. Blank, M.L. and Snyder, F. (1994) Chromatographic analysis of ether-linked glycerolipids, including platelet-activating factor and related cell mediators. In: T. Shibamoto (Ed.), Lipid Chromatographic Analysis. Marcel Dekker, New York, NY, pp. 291–316.
17. Murphy, R.C., (1993) Mass spectrometry of lipids. In: F. Snyder (Ed.), Handbook of Lipid Research. Plenum Press, New York, NY, 290 pp.
18. O'Flaherty, J.T., Tessner, T., Greene, D., Redman, J.R. and Wykle, R.L. (1994) Comparison of 1-*O*-alkyl-, 1-*O*-alk-1-enyl-, and 1-*O*-acyl-2-acetyl-*sn*-glycero-3-phosphoethanolamines and -3-phosphocholines as agonists of the platelet-activating factor family. Biochim. Biophys. Acta 1210, 209–216.
19. Lohmeyer, M. and Bittman, R. (1994) Antitumor ether lipids and alkylphosphocholines. Drugs Future 19, 1021–1037.
20. Boggs, K.P., Rock, C.O. and Jackowski, S. (1995) Lysophosphatidylcholine and 1-*O*-Octadecyl-2-*O*-methyl-*rac*-glycero-3-phosphocholine inhibit the CDP–choline pathway of phosphatidylcholine synthesis at the CTP:phosphocholine cytidylyltransferase step. J. Biol. Chem. 270, 7757–7764.
21. Daniel, L.W., Huang, C., Strum, J.C., Smitherman, P.K., Greene, D. and Wykle, R.L. (1993) Phospholipase D hydrolysis of choline phosphoglycerides is selective for the alkyl-linked subclass of Madin–Darby canine kidney cells. J. Biol. Chem. 268, 21519–21526.
22. de Vet, E.C.J.M., Hilkes, Y.H.A., Fraaije, M.W. and van den Bosch, H. (2000) Alkyl-dihydroxyacetonephosphate synthase: Presence and role of flavin adenine dinucleotide. J. Biol. Chem. 275, 6276–6283.
23. Blank, M.L., Fitzgerald, V., Lee, T.-c. and Snyder, F. (1993) Evidence for biosynthesis of plasmenylcholine from plasmenylethanolamine in HL-60 cells. Biochim. Biochim. Acta 1166, 309–312.
24. Strum, J.C., Emilsson, A., Wykle, R.L. and Daniel, L.W. (1992) Conversion of 1-*O*-alkyl-2-acyl-*sn*-glycero-3-phosphocholine to 1-*O*-alk-1′-enyl-2-acyl-*sn*-glycero-3-phosphoethanolamine. A novel pathway for the metabolism of ether-linked phosphoglycerides. J. Biol. Chem. 267, 1576–1583.
25. MacDonald, J.I.S. and Sprecher, H. (1991) Phospholipid fatty acid remodeling in mammalian cells. Biochim. Biophys. Acta 1084, 105–121.
26. Snyder, F., Lee, T.-c. and Blank, M.L. (1992) The role of transacylases in the metabolism of arachidonate in platelet-activating factor. In: R.T. Holman, H. Sprecher and J.L. Harwood (Eds.), Progress in Lipid Research. Vol. 31, Pergamon Press, New York, NY, pp. 65–86.
27. Nixon, A.B., O'Flaherty, J.T., Salyer, J.K. and Wykle, R.L. (1999) Acetyl-CoA:1-O-alkyl-2-lyso-*sn*-glycero-3-phosphocholine acetyltransferase is directly activated by p38 kinase. J. Biol. Chem. 274, 5469–5473.
28. Bae, K.-a., Longobardi, L., Karasawa, K., Malone, B., Inoue, T., Aoki, J., Arai, H., Inoue, K. and Lee, T.-c. (2000) Platelet-activating factor (PAF)-dependent transacetylase and its relationship with PAF acetylhydrolases. J. Biol. Chem. 275, 26704.
29. Tjoelker, L.W., Wilder, C., Eberhardt, C., Stafforini, D.M., Dietsch, G., Schimpf, B., Hooper, S., Trong, H.L., Cousens, L.S., Zimmerman, G.A., Yamada, Y., McIntyre, T.M., Prescott, S.M. and Gray, P.W. (1995) Anti-inflammatory properties of a platelet-activating factor acetylhydrolase. Nature 374, 549–552.
30. Stafforini, D.M., McIntyre, T.M., Zimmerman, G.A. and Prescott, S.M. (1997) Platelet-activating factor acetylhydrolase. J. Biol. Chem. 272, 17895–17898.

31. Manya, H., Aoki, J., Watanabe, M., Adachi, T., Asou, H., Inoue, Y., Arai, H. and Inoue, K. (1998) Switching of platelet-activating factor acetylhydrolase catalytic subunits in developing rat brain. J. Biol. Chem. 273, 18567–18572.

D.E. Vance and J.E. Vance (Eds.) *Biochemistry of Lipids, Lipoproteins and Membranes (4th Edn.)*
© 2002 Elsevier Science B.V. All rights reserved

# CHAPTER 10

# Adipose tissue and lipid metabolism

David A. Bernlohr, Anne E. Jenkins and Assumpta A. Bennaars

*Department of Biochemistry, Molecular Biology and Biophysics, University of Minnesota,
321 Church St. SE, Minneapolis, MN 55455, USA, Tel.: +1 (612) 624-2712;
Fax: +1 (612) 625-2163; E-mail: bernl001@umn.edu*

## 1. Introduction

The development of adipose tissue and the biochemistry of the adipocyte are research areas that have intrigued investigators for decades. Originally considered as simply a storage organ for triacylglycerol, interest in the biology of adipose tissue has increased substantially within the last decade, coming to the forefront in areas such as molecular genetics, endocrinology and neurobiology. Recent advances have demonstrated that the adipocyte is not a passive lipid storage depot but a dynamic cell that plays a fundamental role in energy balance and overall body homeostasis. Moreover, the fat cell functions as a sensor of lipid levels, transmitting information to a neural circuit affecting hunger, satiety and sleep.

This chapter will focus on the biochemistry of the adipocyte. Adipocytes make up approximately one-half of the cells in adipose tissue, the remainder being blood and endothelial cells, adipose precursor cells of varying degrees of differentiation, and fibroblasts. The reader is referred to excellent reviews by G. Ailhaud (1992) and P. Cornelius (1994) that focus exclusively on the differentiation process. While touching on adipose cell biology, this chapter will focus on the biochemistry of triacylglycerol metabolism.

## 2. Adipose development

### 2.1. Development of white and brown adipose tissue in vivo

The study of white adipose tissue (WAT) development in mammals has been facilitated by the use of experimental animal models. Rodents, guinea pigs, rabbits, pigs, as well as humans, have all been evaluated for the development of white adipose tissue. In general, WAT is not detected at all in mice or rats during embryogenesis, but in pigs and humans it is evident during the last third of gestation. In humans, small clusters of adipocytes are present that increase in size during gestation. Larger clusters of fat cells are associated with tissue vascularization and a general increase in cluster size is positively correlated with larger blood vessels. Paracrine/autocrine factors play a significant role in both capillary growth and adipose conversion.

After birth, sex- and site-dependent differences in fat deposits are well known in humans and several animal species. The diet plays a critical role in the degree of lipid filling within an adipocyte. However, controversy surrounds the question of new fat cell development following a long-term fast. In general, starvation conditions lead to a loss of adiposity and some apparent diminution in the number of fat cells. Refeeding restores lipid levels and the apparent number of adipocytes. Consequently, fasting/refeeding typically has little effect on the number of adipocytes in the body. It is generally accepted that adipose precursor cells are present throughout life and that removal of adipocytes, either by diet or surgical methods, will ultimately result in a restoration of adipose levels.

In contrast to white adipose tissue, brown adipose tissue (BAT) develops during fetal life and is morphologically and biochemically identifiable at birth. Using the uncoupling protein 1 as a BAT marker (specific for brown adipose tissue mitochondria, see Section 3.6), brown fat development has been shown to occur maximally during the last third of gestation. Two conditions have been shown to enhance the development of brown fat hyperplasia in rodents: cold-acclimation and hyperphagia. Both conditions result in a metabolic demand for high-energy expenditure, either in the form of increased heat need or increased metabolism. Brown adipose tissue is common in rodents, camels and hibernating animals such as bears and marmots. The oxidation of triacylglycerol stores in brown adipose tissue depots during hibernation or fasting provides certain animals with a source of water and energy during nutrient deprivation. In humans, although still somewhat controversial, it is generally accepted that brown fat is not present to any significant extent and that white adipose tissue carries out the body's energy storage functions.

In general it is assumed that WAT and BAT develop from different immediate precursor cells. However, a common precursor for both cannot be ruled out and the possible transformation of BAT into WAT has been considered. The conversion from BAT to WAT would be correlated with a decrease in BAT-specific gene products such as the uncoupling protein. However, the reverse does not appear to take place. That is, cold-adapted rodents do not lose WAT and redevelop BAT in response to a low-temperature challenge.

## 2.2. In situ models of adipose conversion

Ailhaud and colleagues have described the adipoblast to adipocyte conversion as a multistep process initiating with the determination of pluripotent proliferative cells to the adipocyte pathway (Table 1). To better characterize the differentiation process and examine the molecular basis of adipose development, a number of murine, hamster, and rat model cell lines (3T3-L1, 3T3-F442A, 10T½, Ob1771) have been established. In general, committed preadipocytes express few markers associated with mature fat cells and are still capable of DNA replication and cell division. Hormonal stimulation by IGF-1 (or high concentrations of insulin), the addition glucocorticoids (e.g., dexamethasone), as well as a phosphodiesterase inhibitor (isobutylmethylxanthine), are frequently used to induce terminal differentiation in culture [1]. Following differentiation the mature adipocytes exhibit a massive triacylglycerol accumulation and are responsive to hormonal stimulation.

For decades, brown fat metabolism has been studied with tissue explants. While

Table 1
Stages of adipose conversion

| Stages | Cell type | Characteristics |
|--------|-----------|-----------------|
| Stage 1 | Mesenchymal/pluripotent | Multipotential — ability to differentiate into muscle, cartilage or fat |
| | *Determination* | |
| Stage 2 | Adipoblasts | Unipotential — can only differentiate into adipocytes |
| | *Commitment* | |
| Stage 3 | Preadipocytes | No lipid accumulation, early transcription factors, e.g. C/EBPβ, and early markers of differentiation expressed, e.g. lipoprotein lipase |
| | *Terminal differentiation* | |
| Stage 4 | Adipocytes | Lipid accumulation and expression of late transcription factors, e.g. PPARγ and C/EBPα, and late markers of differentiation, e.g. PEPCK, aP2, FATP |

white adipose tissue is important for the storage of energy in the form of triacylglycerol, brown fat functions to dissipate energy in the form of heat through the action of a specific mitochondrial proton transporter, the uncoupling protein 1. While the 3T3-L1 or 3T3-F442A cell lines provided a convenient method to study white adipose metabolism, similar brown fat models were, until recently, lacking. However, by expressing the SV40 early genes under control of the strong fat-cell specific adipocyte fatty acid-binding protein (aP2) promoter in transgenic animals, brown fat tumors developed due to t-antigen-induced oncogenesis [2]. Such tumors were used to derive hibernoma cell lines (rapidly growing brown fat cells) exhibiting the properties of brown fat. The brown fat hibernomas express the mRNA for the uncoupling protein upon stimulation with cAMP, cAMP analogs, or a variety of $\beta_2$ and $\beta_3$-receptor agonists. Such cell lines have been extremely useful for the study of brown fat gene expression and metabolism.

## 2.3. Transcriptional control during development

To characterize the molecular basis for differential gene expression, a number of laboratories have identified transcription factors regulating genes expressed in adipocytes. Of those genes most actively studied, the adipocyte fatty acid-binding protein (aP2) gene and the insulin-stimulatable glucose transporter gene (GLUT4) have proven particularly useful. The aP2 gene is expressed in an adipose-specific manner and is up-regulated at least 50-fold as a consequence of adipose conversion [3]. The aP2 gene is regulated by glucocorticoids, insulin, and polyunsaturated fatty acids while the insulin-stimulatable glucose transporter is regulated primarily by insulin, cAMP and fatty acids [4,5]. Using these and other adipose genes as templates, three different transcription factor families have been identified as critical components of the adipocyte differentiation program.

### 2.3.1. C/EBP family of transcription factors
The CCAAT/enhancer-binding proteins (C/EBP) are a family of transcription factors strongly implicated in the control of genes involved in intermediary metabolism. Origi-

nally cloned by McKnight and colleagues, the C/EBPs are leucine-zipper transcription factors, a family of proteins whose sequences are characterized by the presence of a basic region followed by a leucine-rich motif. Leucine-zipper proteins are capable of forming coil–coil interactions with other similar types of factors. As such, the C/EBPs form homo- and heterodimers with other family members, thereby allowing for their binding to cis-regulatory elements within the promoter/enhancers of genes regulated by C/EBPs.

A number of genes involved in adipose lipid metabolism are regulated by the C/EBP family of transcription factors. Binding sites for the C/EBP proteins reside within the promoters of the aP2, stearoyl-CoA desaturase and insulin-stimulatable glucose transporter genes [4]. Transient transfection studies have revealed that the C/EBP sites within the promoter of the glucose transporter gene are functional and that these transcription factors play the central role in regulating the expression of the gene in differentiated adipocytes. Expression of antisense C/EBPα RNA in 3T3-L1 preadipocytes blocked the expression of C/EBPα and concomitant expression of several adipocyte genes including the insulin-stimulatable glucose transporter and adipocyte fatty acid-binding protein. Moreover, in such antisense C/EBPα-expressing cells, the accumulation of cytoplasmic triacylglycerol was blocked, suggesting that a global inhibition of genes expressing proteins of adipose lipid metabolism was occurring. Consistent with a central role for C/EBPα in lipid metabolism, mice bearing a targeted disruption in the C/EBPα allele fail to accumulate triacylglycerol in both adipose and liver.

Three members of the C/EBP family of transcription factors are expressed in adipocytes: α, β and δ. The temporal expression of the three isoforms during 3T3-L1 differentiation suggests that the C/EBP genes are subject to exquisite regulatory controls. For example, within the C/EBPα promoter resides a C/EBP binding site that suggests that C/EBPβ and/or C/EBPδ may be responsible for the activation of expression of the C/EBPα gene. In addition, insulin regulates the transcription of the C/EBPα, β, and δ genes in fully differentiated 3T3-L1 adipocytes. Insulin addition to 3T3-L1 adipocytes represses the expression of C/EBPα while inducing the expression of C/EBPβ and C/EBPδ. Furthermore, glucocorticoids reciprocally regulate expression of the C/EBPα and δ genes in 3T3-L1 adipocytes and white adipose tissue [6]. This observation may provide a mechanistic connection between the accumulation of central adipose tissue and hypercortisolemia associated with Cushing's syndrome.

### 2.3.2. PPAR/RXR family of transcription factors

While the C/EBP family of transcription factors has been implicated as central to the control of gene expression in the differentiated adipocyte, a different family of DNA-binding proteins is apparently instrumental in regulating the differentiation of preadipocytes into mature fat cells. With the adipocyte fatty acid-binding protein gene as a template, Spiegelman and colleagues employed transgenic animal technology to map the region of the adipocyte fatty acid-binding protein gene necessary and sufficient to direct the expression of a chloramphenicol acetyl transferase transgene in a fat-cell-specific manner. Surprisingly, they found that while the region of DNA necessary for C/EBP action was essential for regulation in the mature adipocyte, a distinct 518 bp enhancer region some 5.4 kb upstream of the start of transcription was required for fat-cell-specific expression. By using DNA gel mobility shift analysis, an adipose-specific factor (ARF6) was identified which bound to a DNA element (ARE6) within the upstream enhancer. Importantly, the ARE6 element exhibited sequence similarity to the consensus nuclear hormone response elements. The ARE6 DNA element was similar to that which bound a heterodimer between the nuclear retinoid X receptor (specific for 9-*cis* retinoic acid ligands) and the peroxisome proliferator activated receptor (PPAR).

PPARs belong to a family of nuclear transcription factors that function in a ligand-dependent manner. Once bound by a ligand, the receptors heterodimerize with retinoid X receptors (RXR). They can activate transcription in target genes by recognizing and binding to specific DNA recognition elements termed peroxisome proliferator activated receptor response elements, PPRE (direct repeat of AGGTCA spaced by one or two nucleotides). This PPRE element is identical to the ARE6 DNA element originally identified by Spielgelman and his colleagues. To date, three different PPAR isoforms α, δ/β and γ, and splice variants have been identified that are encoded by separate genes [7]. The tissue-specific expression pattern of these transcription factors is indicative of their function in those tissues [8].

PPARα target genes involved in fatty acid catabolism (β and ω oxidation pathways) and is most abundant in liver although also found in kidney, heart and brown adipose tissue [8]. The δ isoform is most widely distributed, and is found in a variety of tissues including heart, kidney, brain, intestine, muscle, spleen, lung and adrenal. The γ isoform is the most highly restricted in its expression pattern with primary sources being adipose, macrophage and mammary tissue. Alternate promoter usage coupled with differential mRNA splicing result in two closely related PPARγ isoforms that differ by only 30 amino terminal amino acid residues [9]. The γ1 isoform is found in adipose tissue and to a lesser extent in liver, kidney and heart. The γ2 isoform is found almost exclusively in white adipose tissue. During the adipocyte differentiation program, the level of expression of the three different PPAR isoforms is temporally regulated. Low expression levels of PPARα in white adipose tissue as compared to brown adipose tissue suggests that its role in differentiation is a minor one.

In cultured cell lines, PPARγ, which is expressed during the late stages of adipocyte differentiation, has been shown to be the most adipogenic of the three isoforms. A balance between the expression of PPARγ and retinoic acid receptor forms (RAR and RXR) controls heterodimer formation during adipogenesis [7]. Some of the genes regulated by PPARγ in adipose tissue include adipocyte fatty acid-binding protein (aP2), lipoprotein lipase, fatty acid transport protein (FATP1), acyl-CoA synthetase, stearoyl-CoA synthetase, and phospho-enol pyruvate carboxykinase [10].

A variety of hydrophobic ligands including polyunsaturated and oxidized fatty acids have been suggested as natural ligands for PPARs. Moreover, the antidiabetic glitazone drugs are targeted towards PPARs [11]. Some eicosanoids, in particular, 15-deoxy $\Delta^{12,14}$-prostaglandin J2, 15 HETE and 9- and 13-hydroxyoctadecadienoic acids derived from linoleic acid (Chapter 13), have been identified as potent natural ligands for PPARγ [2]. Other arachidonic acid metabolites derived from the lipoxygenase pathway, namely 8-*S*-hydroxyeicosatetraeinoic acid (8SHETE) and leukotriene B4 have also been suggested as ligands for PPARα [8]. The expression of PPARγ in adipose tissue is upregulated by hormones like insulin and glucocorticoids while cytokines, tumor necrosis factor α in particular, have been shown to decrease the expression of PPARγ and C/EBPα [8].

The PPARγ activity is regulated in concert with a nuclear coactivator termed PGC-1. While PGC-1 was originally believed to function only in white and brown adipocytes, PGC-1 is also expressed in hepatocytes where it plays a primary role in controlling the expression of genes involved in gluconeogenesis.

### 2.3.3. SREBP family of factors

Sterol regulatory element-binding proteins (SREBPs) belong to a family of transcription factors that regulate genes involved in cholesterol and fatty acid metabolism [10]. SREBPs consist of three major isoforms: SREBP1a, 1c (ADD1) and 2. The SREBP1 gene has two alternate promoters which result in the synthesis of two proteins (SREBP1a and 1c) of different lengths. SREBP1a is a more potent activator of transcription because of its longer acidic amino terminal transactivation domain and its capacity to induce a wider range of target genes. SREBP2 is encoded by a separate gene.

SREBPs are basic helix–loop–helix leucine zipper (bHLH-LZ) DNA-binding proteins. These proteins have dual DNA specificity. They can recognize and bind to inverted repeat sequences known as E box motif (5′-CANNTG-3′) as well as direct repeat sterol regulatory elements. SREBPs contain a tyrosine residue within their basic domain which confers a conformation change within the protein allowing for the recognition of sterol regulatory elements or related sites [12].

SREBPs are unique transcription factors because they are initially synthesized as precursor membrane-bound proteins present in the endoplasmic reticulum and the nuclear envelope. Both their amino and carboxyl terminals are cytosolic. Cleavage at two specific sites within the amino terminal by proteases releases the transcription activation and DNA-binding domain which represents the mature SREBP [12]. The cleavage process is regulated by the cellular concentration of sterol which directly regulates the activity of the proteases and by a chaperone protein, SREBP-cleavage activating protein (SCAP), which is involved in trafficking the precursor SREBPs to the Golgi network for cleavage [13,14] (see Chapter 15).

All cultured cell lines express SREBP1a and 2 [12]. Liver and adipose tissue express predominantly SREBP 2 and 1c. Overexpression of SREBP1a in cultured cells and animal livers results in an increase in the expression of cholesterol and fatty acid metabolism genes. An increase in the levels of cholesterol and triacylglycerols was observed in the liver; however, there was no change in the serum levels of these metabolites. A decrease in the mass of triacylglycerols present in white adipose tissue was also noted [14].

Overexpression of SREBP1c in adipose cells led to the development of hyperglycemia, fatty liver and an increase in the levels of serum triacylglycerols while overexpression of SREBP2 led to a significant increase in the accumulation of primarily cholesterol. In mature adipocytes, insulin has been shown to upregulate the expression of SREBP1c which can then induce the expression of genes involved in lipogenesis, e.g. lipoprotein lipase and fatty acid synthetase [15] and possibly enzymes necessary for synthesis of an endogenous PPARγ ligand.

Targeted disruption of the SREBP1 gene (lack of SREBP1a and c) has resulted in negligible effects on white adipose tissue mass and function possibly due to the upregulation of SREBP2. No difference in the expression of genes involved in fatty acid catabolism or triacylglycerol synthesis was observed [10]. SREBP2 knockout mice are embryonic lethal.

## 3. Biochemical aspects of lipid metabolism

### 3.1. Lipid delivery to adipose tissue

The primary function of adipose tissue is to serve as a storage site for the excess energy derived from food consumption. This energy can then be utilized by the organism to fulfill subsequent metabolic requirements during times of little or no consumption. In the case of white adipose tissue, these requirements entail efficient storage of large amounts of energy in a form that can be mobilized readily to supply the needs of organs and tissues elsewhere in the body. Lipids, particularly fatty acids, are an exceptionally efficient fuel storage species. The highly reduced hydrocarbon tail can be readily oxidized to produce large quantities of reduced coenzymes and subsequently ATP. At the same time, the very hydrophobic nature of the fatty tail precludes concomitant storage of excess water that would increase the mass and spatial requirements of the organism considerably. Also, the relatively straight, chain-like structure of the fatty acid permits dense packing of many molecules into each cell, maximizing the use of storage space available. Brown adipose also stores energy in lipid form, but more frequently produces heat by oxidizing fatty acids within the adipocyte, rather than supplying free fatty acids for use by other cell types.

### 3.2. Fatty acid uptake and trafficking

At the adipose tissue beds, fatty acids are liberated from triacylglycerol-rich lipoproteins through the action of lipoprotein lipase (Chapter 20). Released fatty acids are bound by albumin and are the donors of lipid for fatty acid uptake. Two schools of thought have dominated hypotheses dealing with fatty acid uptake. First, local protonation of fatty acids due to the relative acidity at the plasma membrane, coupled with the low aqueous solubility of fatty acids at neutral pH and high permeability of fatty acids into the hydrophobic environment of the plasma membrane, creates a sufficient driving force for diffusion across the outer and inner leaflets of the membrane. A second, and drastically different viewpoint, is that there are protein cofactors called lipid transporters that facilitate the transfer process.

Several studies by Hamilton and others have demonstrated that fatty acids can diffuse through membranes at rates, and with properties, consistent with biological transport. Adsorption, flip-flop, and desorption of fatty acids at the plasma membrane functions to transfer protons across the membrane. Acidification of micelles, blood neutrophils, and hepatocytes accompanies fatty acid uptake, consistent with a diffusional mechanism and indicates that protein-mediated transfer is not necessary.

In contrast, a number of plasma membrane fatty acid-binding proteins have been identified and argued to facilitate transbilayer flux. Abumrad and colleagues have identified a murine homologue of human cell surface antigen CD36 as a lipid-binding protein involved in lipid transport and termed the protein fatty acid translocase. Fatty acid translocase is also found in muscle and cardiac myocytes and has been genetically linked to hyperlipidemia and hypertension. Lodish and co-workers have identified the fatty acid transporter family of plasma membrane proteins and used expression cloning to demonstrate fatty acid transport activity. FATP1 and FATP4, exhibit very long-chain acyl-CoA synthetase activity suggesting that esterification of very long-chain fatty acids at the plasma membrane may be coupled to uptake.

Once inside the cell, free fatty acids are minimally soluble in the aqueous cytoplasm. The charged carboxylate group provides enough electrostatic hindrance to prevent association with the neutral triacylglycerols, whereas the hydrocarbon tail reduces solubility in water. At high enough concentrations fatty acids exert a detergent-like effect that would disrupt membranes and/or they could cluster together in micelles in the crowded cytoplasm. To alleviate this problem, the adipocyte and other lipid-metabolizing cell types have evolved intracellular fatty acid-binding proteins, a family of small, soluble, highly abundant proteins that bind and sequester free fatty acids [16]. Adipocytes express two fatty acid-binding proteins, the products of the FABP4 and FABP5 genes.

The adipocyte fatty acid-binding protein (aP2) has become a paradigm for in vitro studies of protein–lipid interactions. Ligands for this protein are long-chain ($>14$ carbon) fatty acids and/or retinoic acid. Its small size ($\approx15$ kDa), high solubility and stability have facilitated purification and characterization of many features of the protein. Crystal structures of the adipocyte fatty acid-binding protein have been solved at high resolution for wild type and site-directed mutant forms, both in the absence and presence of bound ligands. Despite a widely varying degree of primary sequence similarity, the intracellular lipid-binding proteins as a family share a virtually superimposable tertiary structure consisting of ten antiparallel β-strands arranged in a flattened barrel [16] (Fig. 1). A single lipid ligand is bound inside the barrel within a large interior water-filled cavity, and held in place by the concerted effect of general surface contacts and specific electrostatic interactions between highly conserved cavity residues and the ligand's polar head group.

### 3.3. Glucose transport and the generation of the triacylglycerol backbone

The immediate backbone precursor for acylglycerol formation is primarily glycerol 3-phosphate, derived from glycolysis or glycerolgenesis within adipocytes. Fat cells express specific glucose transporters on the plasma membrane to ensure a ready supply of glycolytic intermediates for triacylglycerol synthesis. There are two types of glucose transport proteins in adipose: GLUT1 and GLUT4 [17]. Both are structurally similar with 12 membrane-spanning α-helices with intracellular amino and carboxyl termini. Both proteins are expected to have a large, hydrophilic intracellular loop separating transmembrane domains six and seven, as well as an extracellular loop containing N-glycosylation site(s) demarcated by transmembrane domains one and two. The

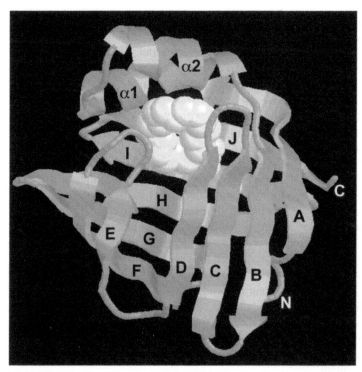

Fig. 1. Diagram of arachidonate buried within the cavity of crystalline adipocyte lipid-binding protein. The protein is depicted as a ribbon drawing with the 10 β-strands labeled A through J and the two α-helices indicated by α1 and α2. The bound arachidonate is illustrated by the yellow space-filling model. N- and C-termini are labelled as N and C, respectively. Note that the carboxyl function of arachidonate is found buried within the cavity, oriented away from the surface of the protein. The entire contact surface of the fatty acid is found within the binding cavity, sequestered from the surrounding milieu.

majority of GLUT1 has been shown by Cushman and co-workers to be present in the plasma membrane of cells unstimulated by insulin, constitutively facilitating transport of glucose down a concentration gradient. However, the bulk of insulin-stimulated glucose transport results from the activity of GLUT4. In the basal state, GLUT4 is largely found in small, intracellular vesicles, but rapidly translocates to the plasma membrane following insulin stimulation. In addition, insulin promotes a change in the rate of intracellular GLUT4 recycling and trafficking which results in a net 10- to 15-fold stimulation of hexose transport in response to insulin [11]. Once inside the cell, facilitative transport of glucose by GLUT1 and GLUT4 is rendered unidirectional by the action of hexokinase. Glucose 6-phosphate can only proceed to the glycolytic pathway because adipocytes do not express significant levels of glucose 6-phosphatase.

### 3.4. Fatty acid and triacylglycerol biosynthesis

Adipocytes readily convert the products of glycolysis into fatty acids via the de novo biosynthetic pathway (Chapter 6). Briefly, surplus citrate is transported from the mitochondrion and cleaved to produce cytoplasmic acetyl-CoA. Cytoplasmic acetyl-CoA is acted upon by acetyl-CoA carboxylase that produces malonyl-CoA. The next steps of the fatty acid biosynthetic pathway are carried out by the multifunctional fatty acid synthase enzyme that utilizes NADPH to catalyze multiple condensations of malonyl-CoA with acetyl-CoA or the elongating lipid, eventually generating palmitate.

In adipocytes, two pathways exist for the production of phosphatidic acid. In one, glycerolphosphate is sequentially esterified with two acyl-CoAs to produce 1-acylglycerolphosphate and 1,2-diacylglycerolphosphate. In the second pathway, dihydroxyacetone phosphate is esterified with acyl-CoA to produce acyl dihydroxyacetone phosphate. An acyl dihydroxyacetone phosphate reductase subsequently produces 1-acylglycerolphosphate that leads to the production of 1,2-diacylglycerolphosphate. The resultant phosphatidic acid is dephosphorylated generating 1,2-diacylglycerol and triacylglycerol formed through the activity of diacylglycerol acyltransferase (DGAT).

The DGAT catalyzed reaction is crucial for it represents a branch point for hydrocarbon flow towards either the triacylglycerol or phospholipid pathways. However, this view has been modified through the production of DGAT null mice. DGAT-deficient mice are viable, synthesize triglycerols normally and are resistant to diet-induced obesity. This implies that either alternate biosynthetic pathways are being utilized or additional DGAT isoforms are present in adipose cells. To address this, Farese and colleagues have identified a second DGAT (DGAT2) with kinetic properties distinct from the original DGAT (now termed DGAT1). Moreover, additional DGAT-like sequences are present in the murine and human genome suggesting an unappreciated complexity in diacylglycerol metabolism.

## 3.5. Triacylglycerol mobilization

Lipolysis refers to the process by which triacylglycerol molecules are hydrolyzed to free fatty acids and glycerol. During times of metabolic stress (i.e. during fasting or prolonged strenuous exercise when the body's energy needs exceed the circulating nutrient levels), the adipocyte's triacylglycerol droplet is degraded to provide free fatty acids to be used as an energy source by other tissues. Numerous stimuli are capable of eliciting the lipolytic response in adipocytes. However, ultimately the same pair of enzymes, hormone-sensitive lipase and monoacylglycerol lipase, is responsible for catalyzing the hydrolysis of the triacylglycerol ester bonds.

Complete hydrolysis of triacylglycerol involves the breakage of three ester bonds to liberate three fatty acids and a glycerol moiety (Fig. 2). The same enzyme, hormone-sensitive lipase, is responsible for facilitating hydrolysis of the esters at positions 1 and 3 of the triacylglycerol. A second enzyme, 2-monoacylglycerol lipase, catalyzes hydrolysis of the remaining ester to yield a third free fatty acid and glycerol. Glycerol must be shuttled back to the liver for use in oxidation or gluconeogenesis. Glycerol has no alternative fate in the adipocyte; adipocytes do not express a glycerol kinase and so are unable to reuse glycerol. Glycerol is effluxed out of adipocytes via an aquaporin-type of transport molecule. Mono- and diacylglycerols can be re-esterified by the endoplasmic reticulum acyltransferases. During a lipolytic stimulus, re-esterification is thought to be minimized so that the net direction of these reactions is toward lipolysis.

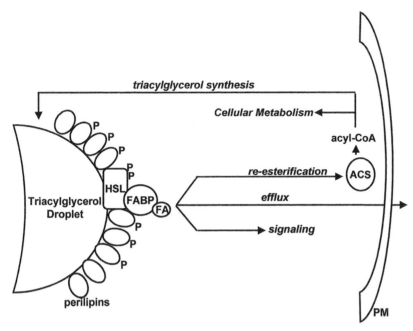

Fig. 2. Schematic representation of the key steps in lipolysis. Triacylglycerol and diacylglycerol depots, surrounded by lipid-associated proteins such as perilipin, are hydrolyzed by hormone-sensitive lipase (HSL) releasing fatty acids (FA) and generating monoacylglycerol. Monoacylglycerol lipase catalyzes the last step in lipolysis formation of fatty acid and glycerol. Glycerol is released from the adipocyte while fatty acids can be bound by intracellular lipid-binding proteins (FABP) and shuttled out of the cell, to signaling depots, or to sites of re-esterification by acyl-CoA synthetase (ACS) and subsequent metabolism or regeneration of triacylglycerol. PM, plasma membrane; P, phosphorylation.

However, under maximal lipolytic conditions, substantial recycling of fatty acids occurs such that on average about two fatty acid molecules are released per glycerol molecule. Outside the adipocyte, fatty acids are immediately bound to serum albumin and carried in the bloodstream to the liver, muscle and other tissues for oxidation.

To avoid futile cycling of fatty acids (and concomitant loss of large amounts of energy), and to maintain proper energy balance between storage and expenditure, triacylglycerol synthesis and hydrolysis are carefully regulated. This regulation is present on several levels, including hormonal secretions from the endocrine system, neurotransmitter secretions from the sympathetic nervous system, intracellular G-protein-mediated signal cascades, gene expression, post-translational modification and product inhibition (Fig. 3). The proximal target of regulatory action is the enzyme responsible for initiating fatty acid mobilization: hormone-sensitive lipase (HSL).

HSL catalyzes the rate-limiting step in lipolysis and is regulated through the stimulation of β-adrenergic receptors by catecholamines or, more recently discovered, by natriuretic peptides [18] and the activation of cAMP-dependent protein kinase. HSL is an 84 kDa protein modeled to be organized into N-terminal (1–300) and C-terminal (300–767) domains. The regulatory and catalytic activity of the enzyme lies within

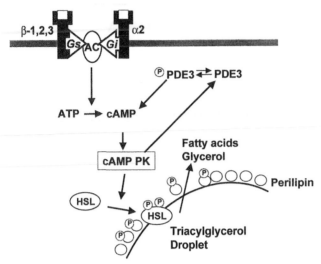

Fig. 3. Activation of lipolysis via adrenoreceptor-coupled systems. Binding of lipolytic agonists to β-adrenoreceptors (β₁, β₂ and β₃) couples to the G-protein which in turn activates adenylyl cyclase (AC) thereby producing cAMP. cAMP activation of protein kinase A (cAMP PK) results in phosphorylation and activation of hormone-sensitive lipase (HSL) and perilipin and subsequent translocation of HSL to and movement of perilipin around the lipid droplet. cAMP PK also phosphorylates and activates the cGMP-inhibited cAMP phosphodiesterase (PDE), providing a feedback system to lower intracellular cAMP. α₂-adrenoreceptor activation results in coupling with Gᵢ and a decrease in adenylyl cyclase activity. Dynamic interplay between β and α₂ adrenoreceptors regulates the activity of adenylyl cyclase and, therefore, cAMP PK.

the predicted α/β hydrolase fold of the C-terminal domain. Three residues (Asp703, His733 and Ser423) form the catalytic triad. Phosphorylation of the C-terminal domain results in a modest increase in specific activity of the enzyme and is correlated with a translocation of the protein from the cytoplasm to the surface of the lipid droplet.

Multiple phosphorylation sites on HSL have been identified, and its regulation by phosphorylation/dephosphorylation makes it unique among lipases. Two phosphorylation sites originally thought to be basal and regulatory sites [19], have been extensively examined. Ser565 is a basal phosphorylation site, while Ser563, once thought to be the site of activation, is now of unknown function. More recently two additional serine phosphorylation sites have been mapped and linked to translocation. Ser659 and Ser660 have been determined to be regulatory sites essential for the activation of HSL and mutation of these sites inhibits the translocation of HSL to the lipid droplet [20].

Phosphorylation and translocation of HSL to the droplet surface is thought to be functionally linked to association with one or more droplet-associated proteins. Present at the lipid droplet are a number of proteins, including perilipins, ADRP, lipotransin, and adipophilin. These proteins are thought to form a functional complex surrounding the lipid droplet, both protecting it from uncontrolled hydrolysis and assisting in amassing lipid. Phosphorylation of perilipin is correlated with hormone-stimulated lipolysis suggesting that phosphorylated perilipin has reduced affinity for droplets and dissociates from the surface concomitant with HSL association. Consistent with this model, perilipin null mice are found to have constitutively active HSL, are hyperphagic, but maintain normal body weights, suggesting that perilipin acts as a barrier to HSL at the droplet surface. Perilipin null mice also reverse the obese phenotype when bred into db/db mice.

The N-terminal domain of HSL has recently been shown to be a docking domain for interaction with the adipocyte fatty acid-binding protein (aP2). Using a combination of yeast two hybrid analysis co-immunoprecipitation, and deletion analysis, it has been shown that the adipocyte fatty acid-binding protein forms a complex with the N-terminal domain. Moreover, the FABP stimulates the activity of HSL and relieves product inhibition by fatty acids. In experiments using aP2 null mice, lipolysis is markedly reduced while fatty acids accumulate intracellularly suggesting that the fatty acid-binding protein docks onto HSL, binds a product fatty acid and facilitates the intracellular trafficking of lipolyzed fatty acids in an efflux shuttle.

Multiple lines of HSL null mice have been developed that exhibit somewhat different characteristics. However, there are some generalizations that can be made. HSL-deficient mice exhibit reduced, but not eliminated hormone-stimulated lipolysis. This suggests that there may be additional lipases linked to hormone stimulation although compensatory mechanisms not normally associated with adipose tissue in vivo have not been ruled out. The HSL null mice are sterile due to the lack of neutral cholesterol ester hydrolase activity inherent to HSL and essential for spermatozoa production.

### 3.5.1. Catecholamines and adrenoreceptors in adipocytes

The catecholamines epinephrine and norepinephrine (adrenaline and noradrenaline) originate in the inner medullar region of the adrenal glands. Stimulation of the adrenal by the sympathetic nervous system leads to secretion of catecholamines into the bloodstream. In addition, adipose tissue is itself directly innervated by the sympathetic nervous system. Various types of metabolic stress trigger the sympathetic nervous system to release its neurotransmitter, norepinephrine, directly into adipose, where its effects on the adipocyte are mediated by specific plasma membrane adrenoreceptors. Rapid reflex responses are primarily stimulated by the sympathetic nervous system, whereas more long-term (i.e. on the scale of hours, days, and weeks) and/or basal effects are subject to regulation by catecholamine secretion.

The effects of catecholamines and the mechanisms that mediate them have been extensively studied in adipocytes. Adipocytes express a combination of five different adrenoreceptor isoforms: $\alpha_1$, $\alpha_2$, $\beta_1$, $\beta_2$, and $\beta_3$ [21]. Lipolysis is signaled by $\beta$-adrenergics. An anti-lipolytic signal is transduced by the $\alpha_2$-adrenergics, and the $\alpha_1$-adrenergics are involved in a separate pathway. In short, although lipolysis is the observed outcome of catecholamine stimulation, it is the steady state result of competition between two opposing pathways triggered by the same signal.

The mechanisms of signal transduction are reasonably well known. Binding of catecholamines to the $\beta$-adrenoreceptors activates adenylyl cyclase via a stimulatory G-protein ($G_s$) (Fig. 3). Adenylyl cyclase catalyzes the conversion of ATP to cAMP. cAMP binds the regulatory subunit of protein kinase A, releasing the active catalytic subunit. Active protein kinase A in turn phosphorylates the HSL and perilipins which increase translocation to and from the triacylglycerol droplet, respectively. The same signal bound to the $\alpha_2$-adrenoreceptor affects an inhibitory G-protein ($G_i$), which inhibits the activity of adenylyl cyclase. Disappearance of cAMP eventually causes cAMP to dissociate from the regulatory subunit of protein kinase A, which then inactivates the catalytic subunit by reassociation. In the absence of continued phosphorylation, dephosphorylation mediates reverse translocation of HSL and perilipins to the basal state.

With simultaneous activation of opposing pathways, the relative contribution of each receptor type becomes very important. Small mammals, such as rats and hamsters, express mainly $\beta_1$ and $\beta_3$ while rats express very little of the $\alpha_2$ isotype. Large mammals (e.g. humans and monkeys) express almost exclusively $\beta_1$ and $\beta_2$ and a significant amount of $\alpha_2$ receptor. It appears that the $\beta_3$ receptor is expressed to a greater extent in brown adipocytes than in white adipocytes.

A second observed pattern of receptor regulation was demonstrated by the use of agonists and antagonists for each receptor isotype. At very low agonist concentrations, only $\alpha_2$-receptor activity is observed (i.e. anti-lipolysis). As the agonist concentration is increased, $\beta_1$ becomes active and initiates lipolysis. Only under much more stimulatory agonist conditions do $\beta_3$ receptors become active. $\beta_2$, in animals that express it, seems to be active under conditions more similar to $\beta_1$. Affinity for ligands and level of expression of receptors are two methods utilized by adipocytes to regulate catecholamine effects. The interplay between the various isotypes is responsible for the adrenergic balance of lipolysis and anti-lipolysis. In general, $\alpha_2$-mediated anti-lipolysis modulates resting adipocyte activity, whereas during stress-induced norepinephrine release, increased binding to the $\beta$-adrenergics overcomes the $\alpha_2$ inhibitory effect and $\beta$-mediated lipolysis prevails.

### 3.5.2. Glucagon

Although catecholamines are perhaps the strongest physiological lipolytic stimulus, other hormones also play an important role in mediating energy balance. One such hormone is glucagon that is one of three polypeptide hormones secreted by endocrine tissues located within the pancreas. Glucagon is secreted into the circulation in response to low blood glucose levels and the result of its action is mobilization of stored energy.

Stimulation by glucagon takes place by a virtually identical pathway to stimulation by catecholamines. Glucagon binds extracellularly to a specific seven-transmembrane-domain receptor, activating adenylyl cyclase via a stimulatory G-protein. Protein kinase A is subsequently activated and phosphorylates HSL, which begins to hydrolyze triacylglycerol stores. Protein kinase A also phosphorylates (and activates) enzymes in the glycogen degradation pathway, and inhibits de novo fatty acid synthesis by phosphorylation of acetyl-CoA carboxylase in concert with AMP-activated protein kinase (Chapter 6).

Because the same regulatory pathway is activated, the same feedback mechanisms used to modulate chronic catecholamine effects are equally significant for prolonged glucagon stimulation. At some level of cAMP production, the cAMP response element-binding protein transcription factors become phosphorylated by protein kinase A and upregulate cAMP-responsive gene expression leading to increased receptor expression.

However, protein kinase A phosphorylation of the cell surface receptors leads to uncoupled G-protein activity and heterologous desensitization to both the glucagon and catecholamine signals. It is also relevant that protein kinase A can phosphorylate and thereby activate cGMP-inhibited phosphodiesterase [22], which cleaves cAMP and probably helps modulate its effects to minimize desensitization.

### 3.5.3. Steroid and thyroid hormone

In addition to the major metabolic regulators in adipocytes (catecholamines, glucagon, and insulin), many diverse types of hormones have effects on adipocyte metabolism. The most notable results are effected by glucocorticoids, sex steroids, and thyroid hormones.

Glucocorticoids are steroid hormones secreted by the adrenal cortex in response to stress or starvation. Glucocorticoids display a permissive effect on lipolysis stimulated by catecholamines. Glucocorticoid response elements have been observed in the upstream regions of $\beta_1$ and $\beta_2$ adrenoreceptor genes and, in fact, an increase in numbers of expressed $\beta$-receptors in response to glucocorticoids has been reported. Glucocorticoid response elements have also been identified in the upstream regions of the C/EBP family of transcription factors. Additionally, the activity of the stimulatory G-protein is enhanced by glucocorticoids. These effects are consistent with the finding that adrenalectomy reduces $G_s$ protein and mRNA levels, and that subsequent administration of a glucocorticoid such as dexamethasone can restore those levels. Glucocorticoids, therefore, probably ensure maintenance of catecholamine-induced lipolysis by enhancing transcription of the genes involved in that signal cascade.

Sex steroids (primarily estrogen in females, which is synthesized by the ovaries, and testosterone in males, synthesized by the testes) like glucocorticoids, also affect gene transcription by binding to nuclear Zn-finger transcription factors that recognize steroid response elements. In female rats, ovariectomy was shown to diminish lipolysis by decreasing the effectiveness of the adenylyl cyclase catalytic activity. Lipolysis was restored to normal levels in these animals by administration of estrogen but not progesterone. Castrated male rats exhibited decreased lipolysis that appeared to be caused both by defective adenylyl cyclase catalysis and a decreased number of $\beta$-adrenergic receptors, again implying desensitization to catecholamines. Normal lipolytic levels could be restored by administration of testosterone.

The circulating thyroid hormones thyroxine and its more potent derivative triiodothyronine are secreted from the thyroid gland in response to hypothalamus/pituitary stimuli. The effect of elevated thyroid hormone is increased lipolysis, which appears to be mediated by an increase in $\beta_1/\beta_2$-adrenergic receptor expression and a decrease in inhibitory G-protein expression. These alterations effectively sensitize the adipocyte to catecholamine stimulation.

### 3.5.4. Insulin and anti-lipolysis

Pancreatic $\beta$ cells secrete the polypeptide hormone insulin in response to elevated blood glucose levels (hyperglycemia). Insulin is the most important physiological stimulus for energy storage. Its effect directly counteracts the effects of glucagon and the catecholamines. Insulin receptors are found in many diverse cell and tissue types, not the least significant of which is adipose.

The insulin receptor is an integral membrane protein that functions as a tetramer composed of two α and two β subunits. The β subunits each span the plasma membrane once, and the α subunits are covalently attached to the β subunit extracellularly by disulfide bonds. The insulin binding site is external. The intracellular domains contain many tyrosine phosphorylation sites and the receptor is itself a tyrosine kinase. Ligand-binding induces autophosphorylation of several intracellular domains, activating the kinase activity of each β subunit. A complex series of interactions follows in which the insulin receptor phosphorylates some of its substrates directly (insulin receptor substrate-1, IRS-2) or recruits various adaptor proteins such as Shc and Grb2 that transmit the insulin signal [23].

Insulin-binding to the adipocyte insulin receptor simultaneously stimulates lipogenesis and inhibits lipolysis. Insulin action effectively clears fatty acids and glucose from the blood both by increasing uptake and storage, and by decreasing mobilization of stored energy. The mechanisms by which these effects are accomplished are highly complex and have not been entirely elucidated although some aspects of the process are clear. The insulin receptor tyrosine kinase is capable of inducing phosphorylation and activation of the cGMP-inhibited phosphodiesterase and several protein serine-phosphatases (most likely protein phosphatases 1, 2A and 2C). Phosphatidyl inositol 3-kinase has been demonstrated as an essential component in the insulin-stimulated activation of the cGMP-inhibited phosphodiesterase but, as previously mentioned, protein kinase A can also fulfill this role in the absence of insulin. Thus, insulin inhibits the cAMP cascade (including activation of hormone-sensitive lipase) through cleavage of cAMP and direct dephosphorylation of protein kinase A-activated substrates. Dephosphorylation also activates acetyl-CoA carboxylase, the enzyme that catalyzes the first committed step in de novo fatty acid synthesis, and fatty acyl-CoA synthetase, the first enzyme in the triacylglycerol synthetic pathway. Glucose transport is stimulated via GLUT4 translocation to the plasma membrane, and lipoprotein lipase secretion (and, therefore, fatty acid uptake) is enhanced. In addition, insulin reduces dramatically the number of cell surface β-adrenergic receptors, which further desensitizes the adipocyte to lipolytic stimuli.

The concerted insulin-induced actions of fatty acid/glucose uptake and triacylglycerol synthesis reduce blood glucose. Eventually the diminished glucose levels signal the pancreas to stop secreting insulin and initiate secretion of glucagon. Intermediate stress such as fright or strenuous exercise is capable of stimulating lipolysis via sympathetic nervous system secretion of norepinephrine. In well fed, resting adipocytes, insulin effects are supported and strengthened by the anti-lipolytic action of $\alpha_2$-adrenergics. Additional anti-lipolytic influence is exerted by adenosine and certain prostaglandins (notably prostaglandin $E_2$ in mature adipocytes and prostacyclin in preadipocytes). Adenosine effects are modulated through the A1 adenosine receptor, a member of the family of purinergic receptors identified in various tissues. Prostaglandins also bind specific cell surface receptors (Chapter 13). Both types of receptors are known to act through an inhibitory G-protein, and so transduce a signal similar to insulin. Thus, the delicate balance between energy storage and mobilization is ensured by complex (and differential) interplay of many regulatory systems and factors.

Table 2
Comparison of major features of white and brown adipose tissue

| Major feature | White adipose | Brown adipose |
|---|---|---|
| Vascularization | Some, limited | Extensive |
| Distribution | Extensive, many sites | Restricted |
| Sympathetic innervation | Some, limited | Extensive |
| Fatty acid role(s) | Synthesis, storage, signaling | Storage, oxidation, signaling |
| Uncoupling protein | Present | Highly expressed |
| Thermogenesis | Present | Highly developed |
| Insulin effects | Extensive | Extensive |
| Adrenoreceptors | Primarily $\alpha_2$, $\beta_1$, $\beta_2$, $\beta_3$ | Primarily $\alpha_1$, $\beta_1$, $\beta_3$ |
| Droplet size | Large, single | Small, multiple |
| Mitochondria | Few | Many, densely packed |

*3.6. Brown fat lipid metabolism*

White and brown adipocytes are obviously labeled as a result of their difference in color. The variation in appearance between these tissues is a direct reflection of the very different role each performs in the organism and results from specific morphological differences on a cellular level. Whereas the purpose of white fat is to store and release energy in the form of free fatty acids, the essential function executed by brown fat is the expenditure of fatty acid-derived energy for maintenance of the organism's thermal stability (Table 2).

Brown fat derives its color from extensive vascularization and the presence of many densely packed mitochondria (due to the heme cofactors in the mitochondrial enzyme cytochrome oxidase). Brown fat is traversed by many more blood vessels than is white fat. These blood vessels assist in delivering fuel for storage and oxidation, and in dispersing heat generated by the numerous mitochondria to other parts of the body. Brown adipocytes differ in appearance from white adipocytes by the presence of many small triacylglycerol droplets, as opposed to a single large droplet (i.e. multilocular, rather than unilocular). Regulation of brown fat activity is accomplished primarily through the action of the sympathetic nervous system. The blood vessels and each individual brown adipocyte are directly innervated by sympathetic nervous system nerve endings that exert control by release of norepinephrine. Stimulation by the sympathetic nervous system in response to external temperature decrease is essential to the maintenance of brown fat function, and atrophy occurs when regular sympathetic nervous system activity declines [24].

Brown adipocytes also differ from white adipocytes at the molecular level. The major adrenergic receptor subtype expressed by brown adipocytes is the $\beta_3$, but $\alpha_1$ and $\beta_1$ are also found. The most notable difference between brown and white adipocytes is the production of uncoupling protein by the former and its relative absence in the latter [21,24]. Brown adipocytes also express a type II 5'-deiodinase enzyme, which converts the thyroid hormone thyroxine to its more potent form, triiodothyronine, and are capable of secreting triiodothyronine into circulation.

Uncoupling proteins confer to the brown adipocyte the ability to catabolize fatty acids

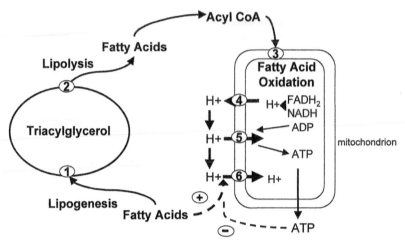

Fig. 4. Thermogenesis in the brown fat mitochondrion. The major fuel pathways of brown fat are represented. Triacylglycerol levels are balanced by the processes of lipogenesis (1) and lipolysis (2). When excess fuel is present, or when heat is needed, fatty acids produced by lipolysis are activated with CoA and transferred to the mitochondrion. Via β-oxidation (3), the long-chain fatty acids are degraded to acetyl-CoA and reduced coenzymes (NADH and FADH$_2$). The coenzymes transfer their reducing equivalents across the mitochondrial inner membrane (4) against the concentration gradient. Typically the proton gradient is dissipated by the action of the proton-ATPase (5) which uses the energy to drive ATP synthesis. However, brown fat mitochondria possess the uncoupling protein (6), which allows for proton transport across the membrane, down the concentration gradient, with the change in free energy lost as heat. The uncoupling protein is positively regulated by fatty acids and inhibited by purines.

inefficiently (that is, without the usual concomitant ATP production) and dissipate the heat generated by this excessive catabolic activity to other tissues via the bloodstream (Fig. 4). This process is known as thermogenesis and is characterized on two levels. Obligatory thermogenesis occurs in all cell types as the result of ubiquitous nominal inefficiencies in metabolism. Facultative thermogenesis occurs specifically in response to stimuli such as cold (non-shivering thermogenesis in adipose, shivering thermogenesis in muscle) or overfeeding (diet-induced thermogenesis). Facultative thermogenesis, particularly non-shivering thermogenesis, is the specific role of brown adipose. It should be noted that uncoupling protein isoforms are expressed ubiquitously throughout the body suggesting that thermogenesis and/or proton dissipation are common regulatory themes.

### 3.6.1. Triacylglycerol synthesis and storage
Brown adipose tissue, in its cold/epinephrine-activated state (as opposed to an atrophied or quiescent state), demonstrates increases in blood flow, lipoprotein lipase activity, triacylglycerol synthesis, 5'-deiodinase activity, and triiodothyronine-enhanced uncoupling protein gene expression. The processes of fatty acid uptake and triacylglycerol synthesis are essentially the same in both brown and white fat. However, norepinephrine release by the sympathetic nervous system in acute cold exposure stimulates brown adipose tissue to enhance expression and secretion of lipoprotein lipase to its sites in the vascular

epithelium. Lipoprotein lipase releases fatty acids from passing chylomicrons and very low-density lipoprotein (Chapter 20), causing an influx of fatty acids into the brown adipocytes.

Free fatty acids and norepinephrine inhibit acetyl-CoA carboxylase. Increased fatty acid uptake, lipolysis and esterification occur simultaneously in brown adipose tissue. BAT fatty acyl-CoA synthetase and acyltransferases, associated primarily with the endoplasmic reticulum, catalyze triacylglycerol formation as in white adipose. Triacylglycerol synthesis is decreased during fasting, and increases sharply in an insulin- and norepinephrine-dependent fashion, upon refeeding. Probably the increased triacylglycerol synthesis is required by active brown adipocytes to accommodate the enhanced fuel influx, which is in turn required for thermogenesis. Concomitant synthesis of triacylglycerol and degradation of fatty acids probably constitutes a futile process that is itself thermogenic.

### 3.6.2. Fatty acid oxidation, bioenergetics and thermogenesis

Fatty acids utilized by BAT for thermogenesis are derived from several sources including dietary triacylglycerol (via chylomicrons), very low-density lipoprotein triacylglycerol from the liver, free fatty acids from white adipose bound to circulating albumin, hydrolysis of internal acyl-CoA molecules, and hydrolysis of internal triacylglycerol stores by HSL. In fact, the capacity of BAT for lipolysis actually exceeds its capacity for thermogenesis, such that it becomes an exporter of fatty acids at very high norepinephrine concentrations. Norepinephrine stimulates HSL via $\beta_1$ and $\beta_3$ adrenoreceptors as described for white adipose (Section 3.5.1). Increased synthesis of thyroxine 5'-deiodinase, responsible for increased levels of triiodothyronine and uncoupling protein, is mediated by $\alpha_1$ adrenoreceptors [21,24].

Non-shivering thermogenesis is induced by heat loss when the temperature of the environment is significantly below the temperature of the organism. Thermogenesis can be suppressed by fever, exercise, and environmental temperatures similar to body temperature. Heat generation occurs through the dissipative relaxation of the proton gradient independent of ATP production (Fig. 4). The uncoupling protein facilitates proton movement down a concentration gradient across the mitochondrial membrane without simultaneous production of ATP by the proton-dependent ATP synthetase. Since the rate of ATP synthesis is usually the limiting factor of respiration and is dependent on utilization of energy from proton movement along the gradient, dissipation of the gradient by uncoupling protein uncouples oxidation from its rate limitations. Unlimited oxidation produces the large amounts of heat that are distributed by brown adipose during thermal distress.

The original uncoupling protein (termed UCP1) was discovered nearly 25 years ago, and is now recognized to be one member of a burgeoning UCP multigene family. The proteins are found in several tissues and across species. UCP1 is expressed predominantly in brown fat and is a ~306 amino acid (~33 kDa) protein predicted to span the inner mitochondrial membrane several times, projecting its C-terminus into the intermembrane space [25]. Its activity is regulated by free fatty acids, which interact with the protein in the membrane and probably lower the membrane potential for proton translocation, facilitating gradient dissipation. Uncoupling protein also has a highly pH-dependent C-terminal purine nucleotide binding site that may serve as a regulator of the protein's activity as well. Small changes in pH drastically affect ADP- and ATP-binding to this site, and ADP/ATP-binding has been shown to inhibit proton translocation in reconstituted phospholipid vesicles.

The oxidative fuel for thermogenesis is exclusively fatty acids even if glucose is available. This is interesting because insulin facilitates uptake of large amounts of glucose during thermogenesis — much more than the cell requires for synthesis of glycerol backbones. During thermogenesis, norepinephrine activates key regulated glycolytic enzymes such as phosphofructokinase and pyruvate dehydrogenase, thus upregulating glycolysis as well as fatty acid oxidation. It has been postulated that upregulation of glycolysis may be essential for ATP production by substrate-level phosphorylation. Since ATP synthesis is uncoupled from oxidation, the cell's ATP requirements must be met another way. In addition, the cell continues to utilize large amounts of reduced cofactors to produce heat. These too can be replenished by an elevated glycolytic rate.

The transcription of uncoupling protein mRNA is upregulated by norepinephrine activation of adrenoreceptors and increases in cAMP. This upregulation can be enhanced by the presence of triiodothyronine and abolished if the deiodinase activity of the cell is inhibited. BAT has a nuclear receptor for triiodothyronine that functions as a transcription factor, and probably binds upstream of the uncoupling protein gene to activate transcription. The presence of both thyroid response elements and cAMP response elements is likely to be required.

## 4. Molecular cell biology of adipose tissue

### 4.1. Energy balance and basal metabolic rate

The balance of food intake and energy expenditure is critical for survival. Each organism represents a unique energy equation based upon its feeding habits, exercise patterns, body composition, and environmental conditions. The net result of the organism's solution to this equation determines its basal metabolic rate (BMR), which is defined in the laboratory setting as the output of a metabolite per unit time, measured at rest after an overnight fast.

Intake and storage of fatty acids must be counterbalanced by an equivalent expenditure of stored energy to maintain constant body mass. The capacity of an organism for expenditure is indexed by its BMR. So, what factors determine individual BMR? Prolonged exposure to harsh environmental conditions, such as extreme cold, leads to an elevated BMR in rodents via thermogenesis and dissipation of heat by brown adipose. Starvation or semi-starvation lowers BMR while regular strenuous and/or prolonged exercise enhances BMR. Resting muscle metabolizes primarily fatty acids, so lean body mass enhances BMR in the fed state. Obesity also elevates BMR due to the grossly increased lipolytic rates observed for an enlarged, excessively proliferated adipose tissue. However, the increased lipolysis is nonetheless overbalanced by consumption in this syndrome. Also, adipocytes of obese individuals proliferate more readily in culture than adipocytes derived from lean individuals, suggesting fat tissue may form more easily in obesity. Increased body mass in the absence of a concomitant increase in fat-free (i.e. metabolically active) mass does not enhance BMR, since futile cycling of fatty acids between the triacylglycerol-esterified and non-esterified states is energetically not very costly. The distribution of fat, however, is a relevant factor for BMR. Sex hormones, for example, tend to direct proliferation of abdominal (visceral) adipocytes in males, but preferentially direct deposition of adipose to the lower body (gluteal–femoral region) of females. Visceral adipose displays inherently reduced sensitivity to the anti-lipolytic effects of insulin and, therefore, elevated lipolysis that contributes to higher BMR, whereas lower body adipose tends to the opposite. In addition, female adipocytes produce and secrete estrogen, which stimulates further production of preadipocytes, amplifying the estrogen effect on adipose deposition. The net effect, as in obesity, is an increased body cell mass independent of fat-free mass and a decreased BMR in females relative to males of comparable mass.

## 4.2. The hypothalamus-adipocyte circuit and the ob gene

In lean individuals, energy balance is maintained by equilibrium between consumption and expenditure. The central control for this complex mechanism is localized in the brain, and specifically to the hypothalamus. The idea that fat cell-derived secreted proteins were responsible for regulating energy intake was first proposed in the 1950s. Studies of mouse models for obesity led to the hypothesis that the adipocyte was capable of sensing when its stores were replete and transmitting a signal to that effect. The theory was expanded as several secreted products have been identified that link the biology of the fat cell to energy homeostasis (Fig. 5). The most notable of these products is leptin, the product of the *ob* gene.

There are five mouse models of obesity identified as resulting from single gene mutations (Table 3). Of these five, *ob*, and its protein product leptin, have received the most attention. Administration of leptin to obese *ob/ob* mice halts hyperphagia and reduces adiposity without dramatically adverse effects on overall body homeostasis [27]. This adipocyte-derived hormone circulates at levels that are proportional to the body fat content and has been extensively investigated for its role in regulation of body fat stores, food intake, satiety and energy expenditure. As part of its mechanism of action, leptin traverses the blood–brain barrier into cerebrospinal fluid and binds to

Table 3

Genetic models for obesity

| Gene name | Allele | Protein | Physiological function |
|---|---|---|---|
| Lethal yellow | Ay | Agouti | Antagonist of melanocortin receptors |
| Obese | ob | Leptin | Hormone — regulates food intake |
| Diabetes | db | Leptin receptor | Expressed in brain and peripheral tissues — binds leptin |
| Fat | fat | Carboxypeptidase E | Processing and sorting neuropeptides and prohormones |
| Tubby | tub | Tubby protein | Possible transcription factor in hypothalamus |

284

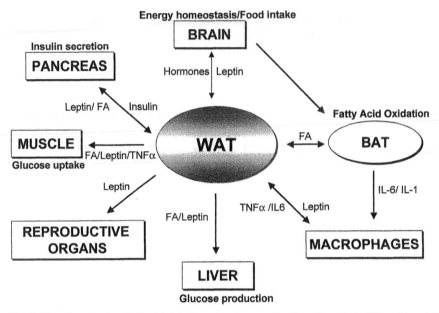

Fig. 5. The adipocyte circuit. Depicted is the network connecting the adipocyte to different target tissues by the factors it secretes. The expression and secretion of leptin by the adipocyte in response to energy balance and food intake directly affect hypothalamic function, liver (glucose production), muscle (glucose uptake) and pancreas (insulin secretion) function, macrophages and reproductive organs. Numerous endocrine factors (catecholamines, glucagon, insulin, and cytokines) secreted by these tissues also affect brown and white fat metabolism. FA, fatty acids; TNFα, tumor necrosis factor; IL-6, interleukin-6; IL-1, interleukin-1.

receptors (products of the *db* gene) on the hypothalamus, thereby controlling feeding behavior. Although the brain expresses the highest number of leptin receptors there are also leptin receptors ubiquitously expressed in peripheral tissues like adipose tissue [28] mammarygland, ovarian follicles, placenta and bone [29]. Therefore, fluctuations in leptin levels not only affect metabolism as a whole but also have a wider impact on reproduction, bone formation and the immune system. Leptin receptors are members of the class 1 cytokine receptor family and are linked to JAK-STAT signaling systems [28].

The short-term regulation of leptin expression is influenced by a number of factors that regulate C/EBPα and PPARγ activity on the leptin promoter. Consumption of a high fat diet, insulin, glucose and glucocorticoids induce ob expression [29], while fasting, cold exposure, β3 agonists, catecholamine stimulation and thiazolidinediones all decrease leptin expression [28,29]. Leptin works to suppress appetite by affecting the synthesis and/or action of a complex system of neuropeptides including neuropeptide Y, MCH, POMC, alpha-MSH, all of which directly or indirectly influence food intake.

Insulin is an additional factor that has been implicated in energy homeostasis. Insulin, although an anabolic hormone when acting directly upon liver, muscle, and adipose, mediates a completely opposite effect via the brain. Ventromedial hypothalamic lesions in rodents are associated with obesity and hyperphagia [26]. Administration of insulin directly to the ventromedial hypothalamus of rats has been shown to cease feeding and initiate weight loss, effects that disappeared when insulin was removed. In addition, feeding was stimulated by direct administration of anti-insulin antibodies to this region. Moreover, the presence of insulin receptors in the hypothalamus and correlation of plasma insulin levels to body mass provide additional support for the putative involvement of insulin in regulation of food intake and body fat stores. One possible vehicle for this proposed insulin-signaling mechanism is the hypothalamic neurotransmitter neuropeptide Y. Administration of neuropeptide Y directly to the hypothalamus strongly stimulates feeding, an effect opposite from that of insulin. Synthesis and release of neuropeptide Y are known to be stimulated by caloric deprivation and there is evidence that insulin may reduce neuropeptide Y mRNA levels. However, it is evident that insulin, whose effects are widespread in various tissues, is not specific enough to solely exert appetite control on the time scale in which satiety and cessation of feeding occur.

*4.3. Cytokine control of adipose lipid metabolism*

Adipocytes have profound sensitivity to tumor necrosis factor-$\alpha$ (TNF-$\alpha$), interferons $\alpha$, $\beta$ and $\gamma$, and interleukins 1, 6 and 11 [30] (Fig. 5). In general, such cytokines inhibit lipogenesis and triacylglycerol storage by adipocytes, activate lipolysis and antagonize insulin. Additionally, many cytokines interfere with proliferation of preadipocytes in 3T3-L1 or 3T3-F442A cell lines and/or diminish adipogenesis in vivo.

TNF-$\alpha$ is expressed at high levels in adipose tissue derived from several genetically defined rodent models of obesity and insulin resistance. In *db/db* obese mice, TNF-$\alpha$ expression is elevated but not the expression of other cytokines such as interleukin-1 or interleukin-6. Measurements of insulin receptor tyrosine kinase activity in obese *fa/fa* rats shows a reduced function, which could be restored by administering soluble TNF-$\alpha$ receptor to sequester the secreted TNF-$\alpha$ protein. TNF-$\alpha$ interferes with insulin signaling by affecting the phosphorylation of the insulin receptor, thus contributing to the development of insulin resistance in these animals. In cultured 3T3L1 cells, TNF-$\alpha$ inhibits the expression of C/EBP$\alpha$ and PPAR$\gamma$, thus leading to reduced expression of genes that are involved in triacylglycerol accumulation and metabolism. Interleukin-6 is secreted by white adipose tissue under basal conditions and its expression is increased by TNF-$\alpha$ and by other conditions linked to wasting disorders (cancer, cachexia, HIV).

TNF-$\alpha$ and the interferons decrease lipogenesis, down-regulating the mRNA levels of the key lipogenic enzymes acetyl-CoA carboxylase and fatty acid synthase; the interferons diminish mRNA levels of fatty acid synthase but not of acetyl-CoA carboxylase. TNF-$\alpha$, interleukin-1, and some interferons also increase lipolysis, although the mechanism is probably post-transcriptional, since Northern blot analysis shows that TNF-$\alpha$ and the interferons decrease the level of HSL mRNA [30].

The adipocyte is involved in expressing and secreting a diverse range of factors (Table 4) which have a direct impact on adipocyte metabolism. These biomolecules have varied physiological functions in other target tissues thus placing the adipocyte as a central player in regulating energy homeostasis overall.

Table 4

Summary of several adipocyte secreted factors

| Adipose secreted factor | Expression in the obese state | Function in vivo |
| --- | --- | --- |
| Acylation stimulating protein (ASP) | Decreased | Stimulates triacylglycerol synthesis |
| Adiponectin/AdipoQ/Acrp 30 | Decreased | Associated with insulin resistance |
| Adipsin | Decreased | Activation of alternative complement pathway |
| Angiotensinogen | Increased | Regulates blood pressure |
| Insulin-like growth factor 1 | Increased | Proliferation/mediates effects of growth hormone |
| Interleukin-6 | Increased | Immune response/glucose and lipid metabolism |
| Leptin | Increased | Energy expenditure, reproduction satiety factor |
| Plasminogen activator inhibitor | Increased | Cardiovascular function/wound healing |
| Prostaglandin $E_2$ | Increased | Antilipolytic, suppresses cAMP production |
| Prostaglandin $F_2$ | Increased | Inhibits adipogenesis |
| Prostaglandin I | Increased | Adipogenic in preadipocytes |
| Transforming growth factor β | Increased | Involved in proliferation, differentiation and apoptosis |
| Tumor necrosis factor | Increased | Contributes to insulin resistance and type 2 diabetes |

## 5. Future directions

Obesity is an increasing worldwide public health concern. In the United States, six out of ten adults are overweight according to a recent National Health and Nutrition survey and approximately, 300,000 deaths annually are a direct result of obesity or complications that result from the disease. While obesity may seem on the surface to simply be the excess consumption of food compared to energy expenditure, it is a complex metabolic disorder centered on adipose lipid metabolism and cellular signaling systems linked to it. Adipocytes are now generally accepted to be a complex cell type involved in generating a number of signals which include cytokines, hormones and growth factors that not only affect itself and the neighboring cells but also impact target tissues involved in energy metabolism. When Albert Lehninger authored the second edition of his famous text *Biochemistry* in 1975, the topic of adipose tissue was given one and a half thin pages while discussions of liver metabolism occupied four such pages. That was not too bad considering that discussions of the metabolic functions of skeletal muscle or brain were given less than one page. At the time, Lehninger's text was the most common entry point for students interested in the topic of adipose metabolism. In the last two decades our appreciation and awareness of adipocytes as a dynamic cell type with connections to both the endocrine and nervous systems has increased dramatically. As described within, adipocytes play the central role in maintenance of the energy balance. As such, pathophysiologic conditions such as obesity and non-insulin-dependent diabetes mellitus, two of the most common disease states of the Western Hemisphere, bring adipose tissue to center stage.

Over the next decade, more transgenic and knockout mouse models will be developed to test hypotheses concerning regulation of adipose triacylglycerol metabolism (L. Chan, 2000; C. Londos, 2001) and its relationship to the overall energy balance. In addition, known genetic alleles linked to control of lipid synthesis will be explored in detail providing additional insights into control points affecting lipid storage and oxidation (K. Reue, 1999). The combination of natural variants and laboratory model systems provides a wealth of experimental systems designed to explore lipid metabolism. Adipocytes will continue to be a favorite experimental system for the analysis of hormone action. The major effects of insulin on fat-cell metabolism have made adipocytes the system of choice for probing the mechanistic basis of insulin-stimulated glucose transport, adrenergic receptor activation of lipolysis and the connections of cytokine biology to wasting diseases. Genomic and proteomic technology platforms now support the analysis of fat cells under a variety of hormonal, metabolic and genetic backgrounds. The ability to integrate complex gene expression patterns with protein analysis into a comprehensive view of lipid metabolism and its control is within reach. As such, it remains to be seen in 20 to 25 years how many pages fat cell metabolism will warrant in the modern textbooks.

## *Abbreviations*

| | |
|---|---|
| ACS | acyl-CoA synthetase |
| ADRP | adipose differentiation-related protein |
| alpha-MSH | alpha-melanocyte-stimulating hormone |
| aP2 | adipocyte fatty acid-binding protein |
| ARE | adipocyte response element |
| ARF | adipocyte regulatory factor |
| BAT | brown adipose tissue |
| bHLH-LZ | basic helix–loop–helix leucine zipper |
| BMR | basal metabolic rate |
| C/EBP | CCAAT/enhancer-binding protein |
| DGAT | diacylglycerol acyltransferase |
| FABP | fatty acid-binding protein |
| FATP | fatty acid transport protein |
| GLUT | glucose transporter |
| HETE | hydroxyeicosatetraeinoic acid |
| HSL | hormone-sensitive lipase |
| IGF-1 | insulin-like growth factor 1 |
| IRS-1 | insulin receptor substrate-1 |
| MCH | melanin-concentrating hormone |
| NUC1 | nuclear transcription factor 1 |
| PDE | phosphodiesterase |
| PGC-1 | nuclear receptor coactivator 1 |
| PKA | cyclic AMP-dependent protein kinase A |
| POMC | pro-opiomelanocortin |
| PPAR | peroxisome proliferator-activated receptor |
| PPRE | peroxisome proliferator-activated receptor response elements |
| RXR | retinoid X receptor |
| SCAP | SREBP-cleavage activating protein |

| SRE | sterol regulatory elements |
| SREBP | sterol regulatory element-binding proteins |
| SV40 | simian virus 40 |
| TNFα | tumor necrosis factor |
| UCP | uncoupling protein |
| WAT | white adipose tissue |

## References

1. Rangwala, S.M. and Lazar, M.A. (2000) Transcriptional control of adipogenesis. Annu. Rev. Nutr. 20, 535–559.
2. Ross, S.R., Choy, L., Graves, R.A., Fox, N. and Solevjeva V. et al. (1992) Hibernoma formation in transgenic mice and isolation of a brown adipocyte cell line expressing the uncoupling protein gene. Proc. Natl. Acad. Sci. USA 89, 7561–7565.
3. Bernlohr, D.A., Doering, T.L., Kelly, T.J. and Lane, M.D. (1985) Tissue specific expression of p422 protein, a putative lipid carrier, in mouse adipocytes. Biochem. Biophys. Res. Commun. 132, 850–855.
4. Ezaki, O., Flores-Riveros, J.R., Kaestner, K.H., Gearhart, J. and Lane, M.D. (1993) Regulated expression of an insulin-responsive glucose transporter (GLUT4) minigene in 3T3-L1 adipocytes and transgenic mice. Proc. Natl. Acad. Sci. USA 90, 3348–3352.
5. Grimaldi, P.A., Knobel, S.M., Whitesell, R.R. and Abumrad, N.A. (1992) Induction of aP2 gene expression by nonmetabolized long-chain fatty acids. Proc. Natl. Acad. Sci. USA 89, 10930–10934.
6. MacDougald, O.A., Cornelius, P., Lin, F.T., Chen, S.S. and Lane, M.D. (1994) Glucocorticoids reciprocally regulate expression of the CCAAT/enhancer-binding protein alpha and delta genes in 3T3-L1 adipocytes and white adipose tissue. J. Biol. Chem. 269, 19041–19047.
7. Debril, M.B., Renaud, J.P., Fajas, L., Auwerx, J. and Chinetti G. et al. (2001) The pleiotropic functions of peroxisome proliferator-activated receptor gamma. J. Mol. Med. 79, 30–47.
8. Chinetti, G., Fruchart, J.C. and Staels, B. (2000) Peroxisome proliferator-activated receptors (PPARs): nuclear receptors at the crossroads between lipid metabolism and inflammation. Inflamm. Res. 49, 497–505.
9. Zhang, Y., Proenca, R., Maffei, M., Barone, M. and Leopold L. et al. (1994) Positional cloning of the mouse obese gene and its human homologue. Nature 372, 425–432.
10. Kersten, S., Chinetti, G., Fruchart, J.C. and Staels, B. (2001) Mechanisms of nutritional and hormonal regulation of lipogenesis. EMBO Rep. 2, 282–286.
11. Holman, G.D., Lo Leggio, L. and Cushman, S.W. (1994) Insulin-stimulated GLUT4 glucose transporter recycling. A problem in membrane protein subcellular trafficking through multiple pools. J. Biol. Chem. 269, 17516–17524.
12. Osborne, T.F., Chinetti, G., Fruchart, J.C. and Staels, B. (2000) Sterol regulatory element-binding proteins (SREBPs): key regulators of nutritional homeostasis and insulin action. J. Biol. Chem. 275, 32379–32382.
13. Thewke, D., Kramer, M. and Sinensky, M.S. (2000) Transcriptional homeostatic control of membrane lipid composition. Biochem. Biophys. Res. Commun. 273, 1–4.
14. Shimomura, I., Hammer, R.E., Richardson, J.A., Ikemoto, S. and Bashmakov Y. et al. (1998) Insulin resistance and diabetes mellitus in transgenic mice expressing nuclear SREBP-1c in adipose tissue: model for congenital generalized lipodystrophy. Genes Dev. 12, 3182–3194.
15. Rosen, E.D., Walkey, C.J., Puigserver, P., Spiegelman, B.M. and Shimomura I. et al. (2000) Transcriptional regulation of adipogenesis. Genes Dev. 14, 1293–1307.
16. LaLonde, J.M., Bernlohr, D.A. and Banaszak, L.J. (1994) The up-and-down beta-barrel proteins. FASEB J. 8, 1240–1247.
17. Pessin, J.E., Bell, G.I., LaLonde, J.M., Bernlohr, D.A. and Banaszak, L.J. (1992) Mammalian facilitative glucose transporter family: structure and molecular regulation. Annu. Rev. Physiol. 54, 911–930.

18. Sengenes, C., Berlan, M., De Glisezinski, I., Lafontan, M. and Galitzky, J. (2000) Natriuretic peptides: a new lipolytic pathway in human adipocytes. FASEB J. 14, 1345–1351.

19. Shen, W.J., Patel, S., Natu, V. and Kraemer, F.B. (1998) Mutational analysis of structural features of rat hormone-sensitive lipase. Biochemistry 37, 8973–8979.

20. Anthonsen, M.W., Ronnstrand, L., Wernstedt, C., Degerman, E. and Holm, C. (1998) Identification of novel phosphorylation sites in hormone-sensitive lipase that are phosphorylated in response to isoproterenol and govern activation properties *in vitro*. J. Biol. Chem. 273, 215–221.

21. Lafontan, M. and Berlan, M. (1993) Fat cell adrenergic receptors and the control of white and brown fat cell function. J. Lipid Res. 34, 1057–1091.

22. Rahn, T., Ridderstrale, M., Tornqvist, H., Manganiello, V. and Fredrikson G. et al. (1994) Essential role of phosphatidylinositol 3-kinase in insulin-induced activation and phosphorylation of the cGMP-inhibited cAMP phosphodiesterase in rat adipocytes. Studies using the selective inhibitor wortmannin. FEBS Lett. 350, 314–318.

23. White, M.F., Kahn, C.R., Rahn, T., Ridderstrale, M. and Tornqvist H. et al. (1994) The insulin signaling system. J. Biol. Chem. 269, 1–4.

24. Himms-Hagen, J., Rahn, T., Ridderstrale, M., Tornqvist, H. and Manganiello V. et al. (1989) Brown adipose tissue thermogenesis and obesity. Prog. Lipid Res. 28, 67–115.

25. Jezek, P., Orosz, D.E., Modriansky, M. and Garlid, K.D. (1994) Transport of anions and protons by the mitochondrial uncoupling protein and its regulation by nucleotides and fatty acids. A new look at old hypotheses. J. Biol. Chem. 269, 26184–26190.

26. Schwartz, M.W., Figlewicz, D.P., Woods, S.C., Porte Jr, D. and Baskin D.G. et al. (1993) Insulin, neuropeptide Y, and food intake. Ann. N.Y. Acad. Sci. 692, 60–71.

27. Halaas, J.L., Gajiwala, K.S., Maffei, M., Cohen, S.L. and Chait B.T. et al. (1995) Weight-reducing effects of the plasma protein encoded by the obese gene. Science 269, 543–546.

28. Kim, S. and Moustaid-Moussa, N. (2000) Secretory, endocrine and autocrine/paracrine function of the adipocyte. J. Nutr. 130, 3110S–3115S.

29. Rayner, D.V. and Trayhurn, P. (2001) Regulation of leptin production: sympathetic nervous system interactions. J. Mol. Med. 79, 8–20.

30. Doerrler, W., Feingold, K.R. and Grunfeld, C. (1994) Cytokines induce catabolic effects in cultured adipocytes by multiple mechanisms. Cytokine 6, 478–484.

D.E. Vance and J.E. Vance (Eds.) *Biochemistry of Lipids, Lipoproteins and Membranes (4th Edn.)*

# CHAPTER 11

# Phospholipases

## David C. Wilton[1] and Moseley Waite[2]

[1] *Division of Biochemistry and Molecular Biology, School of Biological Sciences,
University of Southampton, Bassett Crescent East, Southampton SO16 7PX, UK, Tel.: +44 (2380) 594308;
Fax: +44 (2380) 594459; E-mail: dcw@soton.ac.uk*
[2] *Department of Biochemistry, Wake Forest University School of Medicine, Winston-Salem, NC 27157, USA*

## 1. Overview

### 1.1. Definition of phospholipases

Phospholipases (PLs) are a ubiquitous group of enzymes that share the property of hydrolyzing a common substrate, phospholipid. Nearly all share another property; they are more active on aggregated substrate above the phospholipid's critical micellar concentration (cmc). As shown in Fig. 1, phospholipases have low activity on monomeric substrate but become activated when the substrate concentration exceeds the cmc. The properties of phospholipids that define the aggregation state (micelle, bilayer vesicle, hexagonal array, etc.) are described in Chapter 1.

The phospholipases are diverse in the site of action on the phospholipid molecule, their function and mode of action, and their regulation. The diversity of function suggests that phospholipases are critical to life since the continual remodeling of

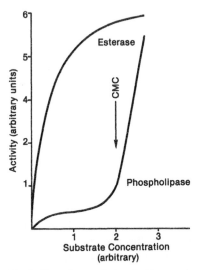

Fig. 1. Dependence of phospholipase and nonspecific esterase activity on substrate concentration. Esterase exhibits Michaelis–Menten kinetics on soluble substrates, whereas a phospholipase becomes fully active above the cmc of the substrate.

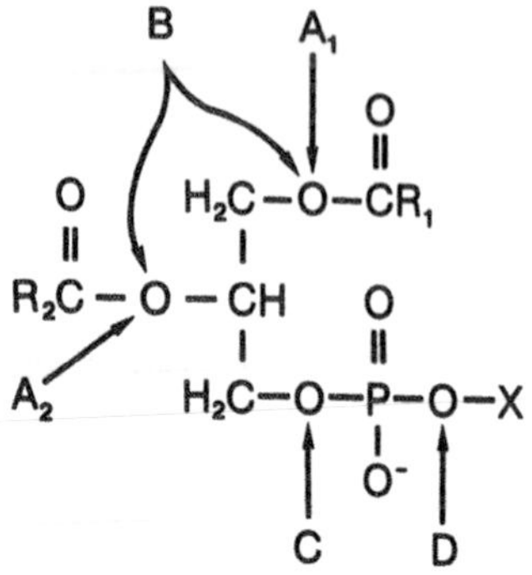

```
      B       A₁
       \     |    O
        \    ↓    ‖
   O    (  H₂C-O-CR₁
   ‖    {      |
 R₂C - O - CH      O
   /         |     ‖
 A₂      H₂C-O-P-O-X
              ↑ |↑ |
                O⁻
              C   D
```

Fig. 2. Sites of hydrolysis by phospholipases.

cellular membranes requires the action of one or more phospholipase. Their functions go beyond their role in membrane homeostasis; they also function in such diverse roles from the digestion of nutrients to the formation of bioactive molecules involved in cell regulation. There are indications that a few phospholipases may carry out a biological function independent of their catalytic activity by binding to a regulatory membrane receptor. Phospholipase-like proteins with toxic properties, yet which lack a functional catalytic site, are found in venoms. It is of interest that most, but not all, phospholipases studied in detail thus far are soluble proteins. The soluble nature of many phospholipases suggests that their interaction with cellular membranes is one of the regulatory mechanisms that exist to prevent membrane degradation or to precisely control the formation of phospholipid-derived signalling molecules.

The classification of the phospholipases, based on their site of attack, is given in Fig. 2. The phospholipases A (PLAs) are acyl hydrolases classified according to their hydrolysis of the 1-acyl ester (PLA₁) or the 2-acyl ester (PLA₂). Some phospholipases will hydrolyze both acyl groups and are called phospholipase B. In addition, lysophospholipases remove the remaining acyl groups from monoacyl(lyso)phospholipids. Cleavage of the glycerophosphate bond is catalyzed by phospholipase C (PLC) while the removal of the base group is catalyzed by phospholipase D (PLD). The phospholipases C and D are therefore phosphodiesterases.

## 1.2. Assay of phospholipases

The subject of phospholipase assays has been concisely reviewed [1–4]. The simplest procedure is to measure protons released during hydrolysis and is usually made by continuous titration with the aid of a pH Stat. However, this technique lacks the sensitivity and specificity of other types of assays. A second commonly used approach employs synthetic substrates for spectrophotometric or fluorometric assays [3,4]. These substrates permit a continual assay well suited for kinetic studies and are of reasonable sensitivity. The major drawback is that the substrates are not natural substrates and as such, their use should be considered as a model that may or may not reflect the enzyme's kinetic properties in biological systems. With fluorescent substrates the product is physically removed from the substrate aggregate, resulting in a spectral change in order to achieve kinetic analysis. This separation can be accomplished by the addition of albumin to absorb the product or by partitioning the product of a short-chain substrate into the aqueous phase. The thioacylester analogs of phospholipids provide a sensitive spectrophotometric assay for some $PLA_1$ or $PLA_2$ assays based on the reaction of released thiol with Ellmann's reagent.

An alternative fluorescence approach that does not involve synthetic substrates is to use a fluorescent displacement assay where the normal product of hydrolysis such as a long-chain fatty acid displaces a fluorescent probe from a protein that binds long-chain fatty acids. The change in fluorescence is monitored and the assay has the advantage that any source of natural phospholipid substrate can be used including cell membranes and lipoproteins (D.C. Wilton 1990, 1991; G.V. Richieri, 1995).

A third approach employs phospholipids with radioisotopes incorporated into specific positions in the molecule. The products are separated from the substrate by partitioning or chromatographic procedures. By the appropriate choice of labeling, the specificity of the enzymes can readily be established and as little as a few picomoles of product can be detected. The use of isotopes has been helpful in the measurement of phospholipase activity using the membranes of whole cells or isolated subcellular fractions previously labeled with radioactive phospholipid precursors.

A fourth and elegant method of phospholipase assay employs a monomolecular film of phospholipid [5]. With this technique the interfacial properties of the lipid substrate are carefully controlled and zero-order kinetics are obtained. This technique provides important data on the effect of lipid surface pressure and enzyme penetration on catalysis.

## 1.3. Interaction of phospholipases with interfaces

The catalytic turnover of phospholipases at the interface distinguishes them from the general class of esterases (Fig. 1). Therefore, the study of phospholipases must include an understanding of their interaction with the lipid interface. The increased enzyme activity seen when phospholipids are present above their cmc (Fig. 1) implies that phospholipases have an interfacial binding surface and that interfacial binding will precede normal catalysis. Interfacial binding is crucial because the cmc of normal cellular phospholipids is very low ($\ll 10^{-9}$ M) and hence the concentration of monomeric substrate in the aqueous phase is sub-nanomolar. Moreover, the half-time for desorption of a long-chain phospholipid from the bilayer interface is of the order of hours. What this means in practice is that the phospholipase must first bind to the interface and that the interfacial binding step has a profound effect on overall catalysis. This concept is illustrated in Fig. 3 and interfacial binding ($E \rightarrow E^*$) often provides the basis for the physiological regulation of these types of enzyme. Unless the phospholipase can bind productively to the phospholipid surface, it cannot access substrate and express activity.

The nature of interfacial binding and the rate enhancements that are achieved are controversial areas. However, it is clear that both polar and non-polar interactions are involved and the precise contribution of each must depend on the nature of the

294

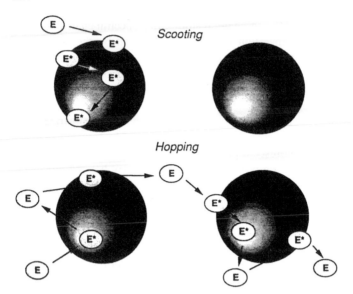

Fig. 3. Schematic illustration of the two modes of interfacial catalysis with vesicles. In the scooting mode (top) the enzyme bound to the interface does not dissociate and exhibits first-order-type kinetics on that vesicle. The excess vesicles which do not contain enzyme are not hydrolyzed. In the hopping mode (bottom) the enzyme desorbs from the interface after each or a few turnover cycles. All vesicles, therefore, are accessible for hydrolysis. E, enzyme in solution; $E^*$ enzyme bound to the vesicle. (Adapted from Berg et al. [7].)

phospholipid interface and the interfacial binding surface of the phospholipase. A number of factors can be considered that could make a major contribution to the enhanced hydrolysis at interfaces and these relate primarily either to the *substrate* or to the *enzyme*.

### 1.3.1. Substrate effects

Phospholipids in aqueous solution form aggregates the nature of which is determined by the structure of the phospholipid. Long-chain phospholipids normally form structures containing bilayers whereas shorter-chain phospholipids form micelles. Once the phospholipid is present as an aggregate it is the concentration of substrate in the surface of the aggregate that is critical and is defined as mole fraction, being unity with a pure phospholipid. Therefore, provided the enzyme can bind to the interface, the effective concentration of substrate will rise dramatically producing a corresponding rate enhancement.

Several assay systems have been developed to describe kinetic events at the interface. Each system has certain advantages yet have limitations that need recognition. An elegant kinetic analysis of secreted $PLA_2$s has been developed by Jain and coworkers (reviewed in Refs. [6,7]. This group of $PLA_2$s in general have a high affinity for anionic phospholipids such as the model system, phosphatidylmethanol. The high dissociation constant for the phosphatidylmethanol interface, $\ll 10^{-10}$ M, allowed the identification of the scooting mode of action of the porcine pancreatic phospholipase (Fig. 3). Under those conditions the enzyme remains attached to the vesicle interface thus promoting hydrolysis of the outer phospholipid monolayer without loss of vesicle integrity or release of enzyme from the vesicle. Care must be taken, however, to demonstrate that the integrity of the vesicle is maintained (low $Ca^{2+}$ concentrations) and that the number of enzyme molecules is substantially less than the number of vesicles present. Under these 'scooting' conditions, hydrolysis can be measured in the absence of interfacial effects and has provided direct evidence that interfacial binding is separate from catalysis.

Scooting conditions are the preferred assay method the assessment of potential enzyme inhibitors [6,7]. This is because the inhibitor or other factors will not affect interfacial binding of the enzyme and only the effect of the inhibitor on classical catalysis ($E^* + S \rightarrow E^*S \rightarrow E^* + P$) will be measured. By contrast, in the 'hopping' model (Fig. 3) the enzyme is able to leave the vesicle during successive rounds of catalysis and will rebind to different vesicles thus producing a continuum of substrate composition and structure as hydrolysis proceeds, making kinetic analysis more difficult [6,7].

Another useful assay system is the Triton mixed-micelle substrate. This model developed by Reynolds et al. [3] can employ a wide range of substrates and is an effective way of measuring enzyme activity. Since in this system there is an exchange of substrate molecules between the micelles that is more rapid than the catalytic rate (M.J. Thomas, 1999) a large excess of substrate is desirable. As a result, linear kinetics can be obtained until 40–50% of total substrate is degraded. One limitation that needs consideration is the affinity of the enzyme for Triton. This, however, can be determined (R.A. Burns, 1982).

Important factors when considering the enhanced hydrolysis at interfaces are the substrate environment in the monolayer and the need to transfer a substrate molecule from this monolayer to the active site. Interfacial disorder may provide an important parameter that facilitates such transfer of substrate to the active site. Phospholipase activity is enhanced under conditions that affect phospholipid fluidity, packing density of the phospholipids and polymorphism of the aggregate. A highly ordered structure seen with phosphatidylcholine phospholipids either above or below the transition temperature tends to give low rates of hydrolysis. Discontinuities in such ordered structures such as assays at temperatures close to the transition temperatures and the presence of other lipids such as anionic lipids or non-bilayer forming phospholipids, promote catalysis by perturbing the interface. In the case of anionic lipids it may be difficult to distinguish the contribution to catalysis due to enhanced interfacial binding from effects on the structure of the interface that promote phospholipid transfer to the active site.

## 1.3.2. Enzyme effects

An important question is if binding of the phospholipase to the interface promotes a conformational change in the structure of the enzyme that facilitates catalysis compared with the enzyme structure in free solution. This is clearly the case with many lipases where lid opening is seen on binding to the aggregate. In the case of the porcine pancreatic secreted $PLA_2$, NMR analysis of the enzyme when bound to anionic micelles compared to the enzyme in solution has revealed significant differences compared with the X-ray structure. In particular, the N-terminal region of the enzyme is disordered in free solution but in crystal structures is condensed as an α-helix when bound to an anionic micelle (B. van den Berg, 1995).

The effect of specific phospholipid molecules on overall protein conformation and hence activity may be considered as an example of allostericity. Such allosteric behavior has been discussed in terms of a second phospholipid-binding site or as discrete interactions between the head groups of phospholipids and specific residues on the interfacial binding surface of the enzyme. The emerging evidence for secreted PLA$_2$s would support a model involving multiple interactions between enzyme and interfacial lipid [7].

## 2. The phospholipases

Many types of phospholipases have now been purified and characterized including full crystal structures. Further, the cellular function of many phospholipases has been examined using various techniques of molecular biology including gene transfection, gene knockouts and antisense strategies. Since it is impossible to cover all phospholipases that have been characterized, only examples will be discussed in detail where significant information relating structure to function is available. The roles of phospholipases in signal transduction are dealt with in Chapter 12.

### 2.1. Phospholipase A$_1$

The phospholipases A$_1$ comprise a large group of 1-acyl hydrolases, some of which also degrade neutral lipids (lipases) or remove the acyl group at position 2 in addition to that at position 1 (phospholipase B) and thus must have lysophospholipase activity. Where the enzyme appears to show low selectivity for the *sn*-1 or *sn*-2 positions, the term phospholipase A is used. The term phospholipase B should be restricted to those enzymes where the mechanism involves minimal accumulation of lysophospholipid product. In this section we also consider various enzymes of the PLA type that do not fit a more precise definition in terms of acyl chain selectivity.

#### 2.1.1. Escherichia coli *phospholipases A*

Two phospholipases A have been purified from *Escherichia coli* based on their differential sensitivity to treatment with detergents [2]. A detergent-insensitive enzyme is localized in the outer membrane, whereas a detergent-sensitive enzyme is found on the cytoplasmic membrane and in soluble fractions. The outer membrane enzyme, known as outer membrane phospholipase A has broad substrate specificity, and demonstrates PLA$_1$, PLA$_2$, lysophospholipase A$_1$ and lysophospholipase A$_2$ activity as well as hydrolyzing mono- and diacylglycerols. The recent crystal structure allows a more detailed discussion of what is an integral membrane phospholipase [8].

The protein is a 12-stranded anti-parallel β-barrel with amphipathic β-strands travers-

**Ca²⁺**

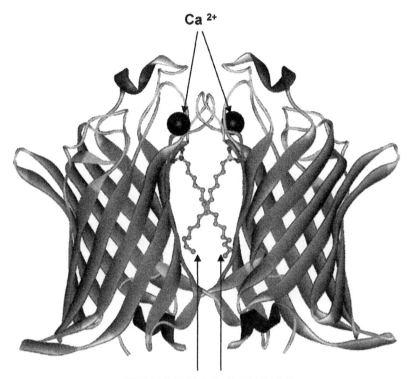

**HEXADECYLSULPHONYL**

Fig. 4. Crystal structure of the outer membrane phospholipase A dimer from *E. coli* shown in the plane of the membrane. The top half of the molecule is located in the lipopolysaccharide monolayer facing the exterior of the cell. The phospholipid monolayer of the outer cell membrane would be located around the bottom half of the protein. Two calcium ions are shown at the active sites while Ser-144 of each active site is covalently modified with a hexadecylsulphonyl moiety represented in a ball and stick format. Structure is adapted from Snijder and Dijkstra [8] using MSI Weblab Viewer.

ing the membrane (Fig. 4). The active site catalytic residues are similar to a classical serine hydrolase triad except that in addition to the serine (Ser-144) and histidine (His-142) there is an asparagine (Asn-156) in place of the expected aspartic acid. Calcium at the active site is predicted to be involved in polarization of the substrate ester carbonyl group and stabilization of the negatively charged reaction intermediate.

The active site is located at the exterior of the β-barrel at the outer leaflet side of the membrane (Fig. 4). The lipopolysaccharide outer leaflet does not normally contain phospholipid and this structural isolation of active site from substrate present in the inner leaflet is believed to be important in preventing uncontrolled phospholipase activity. It is proposed that appropriate bacterial stimulation results in phospholipid moving to the outer monolayer and enzyme activation, including dimerization (Fig. 4), allows phospholipid hydrolysis. This hydrolysis enhances the permeability of the outer cell membrane producing the appropriate adaptations such as the release of bacteriocins. The fact that *E. coli* mutants deficient in this phospholipase A have normal growth characteristics and phospholipid turnover is consistent with the specialized role of this enzyme in bacterial adaptation.

### 2.1.2. Lipases with phospholipase $A_1$ activity

Lipoprotein lipase and hepatic lipase are two lipases that degrade triacylglycerols in lipoproteins but also demonstrate significant phospholipase $A_1$ activity. The enzymes have 50% sequence identity and are members of a superfamily of lipases and phospholipases that share the G–X–S–X–G motif at the active site and an Asp–His–Ser triad that is required for catalysis. Physiologically, lipoprotein lipase is primarily responsible for the degradation of the neutral lipids in triacylglycerol-rich chylomicrons and very low density lipoproteins whereas hepatic lipase prefers smaller denser particles such as high density lipoproteins (Chapter 20). Hepatic lipase is about 2–3 fold more efficient at hydrolyzing phospholipids than lipoprotein lipase and has lysophospholipase activity. Another lipase, intestinal lipase, has lysophospholipase activity and is also referred to as a phospholipase B. The cDNA encoding this enzyme has now been cloned and essential catalytic residues have been identified (T. Lu, 2001).

### 2.2. Phospholipase B and lysophospholipases

The distinction between phospholipase B and lysophospholipases is not clear [2] since both diacyl- and monoacyl-phospholipids are substrates. As discussed above (Section 2.1) a working definition of these enzymes can be based on the extent to which the lysophospholipid accumulates upon hydrolysis of the diacyl substrate. Those phospholipases B for which there is significant structural and functional information are discussed. In addition mammalian lysophospholipases are reviewed.

### 2.2.1. Phospholipase B from microorganisms

The amino acid sequence of the phospholipase B from *Penicillium notatum* was deduced from its cDNA (N. Masuda, 1991). The details of the overall catalytic mechanism are unknown, but intriguing, as substrate specificity is affected by the presence or absence of detergent as well as the glycosylation state of the enzyme. Phospholipase B activity has subsequently been identified in other fungi including *Cryptococcus neoformans*, *Saccharomyces cerevisiae* and *Candida albicans*. In the case of *C. albicans*, gene disruption has identified this phospholipase B as a major factor in host cell penetration and hence pathogenicity (S.D. Leidich, 1998). It is likely that this secreted enzyme has a similar role in other fungi. Phospholipase B activity has also been isolated from *Mycobacterium phlei* and *Mycobacterium lepraemurium* where product analysis indicates that the initial activity is that of a PLA$_1$ (S. Maeda, 1996).

### 2.2.2. Mammalian lysophospholipases

Lysophospholipids [9] are generally found in very low concentrations (0.5–6% of total lipid membrane weight) in biological membranes. High concentrations of lysophospholipids affect membrane properties and membrane enzymes even leading to cell lysis. Increased lysophospholipid levels are associated with atherosclerosis while lysophosphatidylcholine is highly abundant in atherogenic lipoproteins such as oxidatively modified low density lipoprotein where it constitutes up to 40% of the total lipid. Other diseases where there are links with lysophospholipid levels include inflammation, hyperlipidemia and lethal dysrhythmias in myocardial ischemia. At low concentrations lysophospholipids such as lysophosphatidylcholine and lysophosphatidic acid have the properties of second messengers. Against this background the role of lysophospholipases in mammalian tissues assumes a considerable significance.

There are two small mammalian lysophospholipases (I and II) which despite their similar size appear to be the products of separate genes rather than splice variants or post-translational modifications [9]. The enzymes, which are calcium-independent, have been purified from a variety of tissues and lack $PLA_1$, $PLA_2$ or acyltransferase activity. Their mechanism of action appears to involve the catalytic triad characteristic of serine proteases and lipases. Using an NMR based assay, lysophospholipase I hydrolyzed 1-palmitoyl-lysophosphatidylcholine and 2-palmitoyl-lysophosphatidylcholine at similar rates [9].

In addition to the small lysophospholipases, there are a few high molecular weight enzymes that display lysophospholipase activity. Two such enzymes, the group IV cytosolic $PLA_2$ and the 85 kDa calcium-independent $PLA_2$ are discussed in Sections 2.3.2.1 and 2.3.2.2. Similarly, a number of $PLA_1$s also display significant lysophospholipase activity including enzymes that are specific for phosphatidylserine (Y. Nagai, 1999) and phosphatidic acid (M.H. Han, 2001).

### 2.3. Phospholipase $A_2$

The phospholipases $A_2$ were the first of the phospholipases to be recognized. Over a century ago, Bokay (1877–1878) observed that phosphatidylcholine was degraded by some component in pancreatic fluid that is now known to be the pancreatic phospholipase $A_2$. At the turn of the century, cobra venom was shown to have hemolytic activity directed toward the membranes of erythrocytes (P. Keyes, 1902). The lytic compound produced by the venom phospholipase was identified a decade later and termed lysocithin (later, lysolecithin). These studies spurred further investigation of this intriguing class of enzymes and their mechanism of attack on water-insoluble substrates.

An increasing number of $PLA_2$s have now been identified within this expanding super-family and this has required a re-evaluation of classification criteria. Recently, Six and Dennis have categorized these enzymes into two types based on catalytic mechanism [10]. In one group (Table 1) are those enzymes that utilize a catalytic histidine as the primary catalytic residue whereas the other group involves a catalytic serine (Table 2) and normally an acyl-serine intermediate. Within these two categories individual enzymes retain their historic grouping.

### 2.3.1. The 14 kDa secreted phospholipases $A_2$

These enzymes now include secreted $PLA_2$s from such diverse sources as venoms, mammalian and plant tissues (Table 1). They are characterized by a requiring mM concentration of calcium for activity and involving an active site histidine and aspartate pair. They are typically extra-cellular enzymes with a large number of disulfide bonds. The enzymes in groups I–III have provided the foundation for our understanding of

Table 1

Characteristics of secreted PLA$_2$s using a histidine residue for catalysis [a]

| Group | | Sources | Size (kDa) | Di-S (no.) | Unique Di-S | C-term extension (no. of residues) | Molecular characteristics |
|---|---|---|---|---|---|---|---|
| I | A | Cobra/krait venom | 13–15 | 7 | 11-77 | None | |
| | B | Mammalian pancreas | 13–15 | 7 | 11-77 | None | 5 residue pancreatic/elapid loop, propeptide |
| II | A | Human synovial fluid/platelets, rattlesnake venom | 13–15 | 7 | 50-137 | 7 | |
| | B | Gaboon viper venom | 13–15 | 6 | 50-137 | 6 | Lacks disulfide at C61–C94 |
| | C | Rat/mouse testes | 15 | 8 | 50-137, 86-92 | 7 | |
| | D | Human/mouse spleen/pancreas | 14–15 | 7 | 50-137 | 7 | |
| | E | Human/mouse brain/heart/ uterus | 14–15 | 7 | 50-137 | 7 | |
| | F | Mouse testes/embryo | 16–17 | 7 | 50-137 | 30 | Extra cysteine in C-terminal extension |
| III | | Bee, lizard, scorpion, human | 15–18 | 5 | N/A | N/A | Human form (55 kDa) has novel C- and N-terminal domains |
| V | | Mammalian heart/lung/ macrophage | 14 | 6 | None | None | |
| IX | | Marine snail venom | 14 | 6 | N/A | N/A | |
| X | | Human spleen/thymus/ leukocyte | 14 | 8 | 11-77, 50-137 | 8 | |
| XI | A | Green rice shoots | 12.4 | 6 | N/A | N/A | |
| | B | Green rice shoots | 12.9 | 6 | N/A | N/A | |

[a] Adapted from Six and Dennis [10]. N/A, not available.

these secreted enzymes and sufficient quantities of natural and mutant enzymes from these groups have been obtained for X-ray crystallographic analysis. Also, the sequences of a very large number of these phospholipases are now known and have been used to demonstrate their structural, functional, and evolutionary relatedness [11].

The conserved active site residues, His-48 and Asp-99 (pancreatic enzyme numbering), provide the catalytic dyad and, with the availability of the crystal structure of the pancreatic enzyme, produced the proton-relay mechanism (Fig. 5) in 1980 [12]. In this mechanism a water molecule directly replaces the serine found in the classical protease/lipase catalytic triad mechanisms. More recently, an alternative mechanism has been proposed (J. Rogers, 1996) that involves two water molecules (W5 and W6) seen at the active site of the crystal structure (Fig. 6). In this mechanism, proposed by Jain and referred to as a 'calcium coordinated oxyanion' mechanism, the attacking nucleophile (W5) is coordinated to the calcium thus enhancing its nucleophilicity. This water is connected to His-48 by a second water molecule (W6). Thus the major formal difference between the two mechanisms is that the latter mechanism involves a second water molecule while the first water molecule is activated by coordination to

Table 2
Characteristics of PLA$_2$s using a serine residue for catalysis[a]

| Group | | Alternative name | Sources | Size (kDa) | Ca$^{2+}$ require-ment/role | Molecular characteristics |
|---|---|---|---|---|---|---|
| IV | A | cPLA$_2$ α | Human U937 cells/ platelets, rat kidney, RAW 264,7 | 85 | < μM Membrane translocation | C2 domain, α/β-hydrolase, phosphorylation |
| | B | cPLA$_2$ β | Human, liver/pancreas/ brain/heart | 114 | < μM Membrane translocation | C2 domain, α/β-hydrolase |
| | C | cPLA$_2$ γ | Human heart/skeletal muscle | 64 | None | Prenylated, α/β-hydrolase |
| VI | A-1 | iPLA$_2$-A | P388D$_1$ macrophages, CHO cells | 84–85 | N/A | 8 ankyrin repeats |
| | A-2 | iPLA$_2$-B | Human B-lymphocytes | 88–90 | N/A | 7 ankyrin repeats |
| | B | iPLA$_2$-γ/2 | Human heart/skeletal muscle | 88 | N/A | Membrane-bound |
| VII | A | PAF-AH | Mammalian plasma | 45 | N/A | Secreted, α/β-hydrolase, Ser/His/Asp triad |
| | B | PAF-AH (II) | Human/bovine liver/kidney | 40 | N/A | Myristoylated, Ser/His/ Asp triad |
| VIII | A | PAF-AH Ib α$_1$ (subunit of trimer) | Human brain | 26 | N/A | G-protein fold, Ser/His/ Asp triad, dimeric |
| | B | PAF-AH Ib α$_2$ (subunit of trimer) | Human brain | 26 | N/A | G-protein fold, Ser/His/ Asp triad, dimeric as hetero- or homodimer |

[a] Adapted from Six and Dennis [10]. N/A, Not available; c, cytosolic; i, (calcium) independent; PAF-AH, platelet-activating factor acetylhydrolase.

the calcium. It is argued that the rate limiting step lies during the decomposition of the tetrahedral intermediate whereas in the originally proposed mechanism the formation of this tetrahedral intermediate is rate limiting [7].

*2.3.1.1. Group I secreted PLA$_2$s.* This group historically contains group IA enzymes such as those from cobra and krait venom while the IB enzymes are the mammalian pancreatic enzymes. Both the cobra and pancreatic enzymes were early models for structure–function analysis [14] including crystal structures of both apo-enzyme and enzyme with bound phospholipid inhibitors. However, at this time there are no crystal structures of enzymes bound to a lipid interface or with a phospholipid substrate at the active site.

A space-fitting model of substrate bound to the active site of the cobra venom enzyme (Fig. 7) gives insight into how the enzyme functions even though bulk interactions with the lipid interface are missing. The most obvious feature is existence of the active-site tunnel into which the substrate enters. However, the enzyme interacts loosely with the first 9–10 carbons of the acyl group at position 2 of the glycerol that may account for the enzyme's lack of acyl specificity. This model also suggests that the substrate molecule is not completely withdrawn from the bilayer and significant

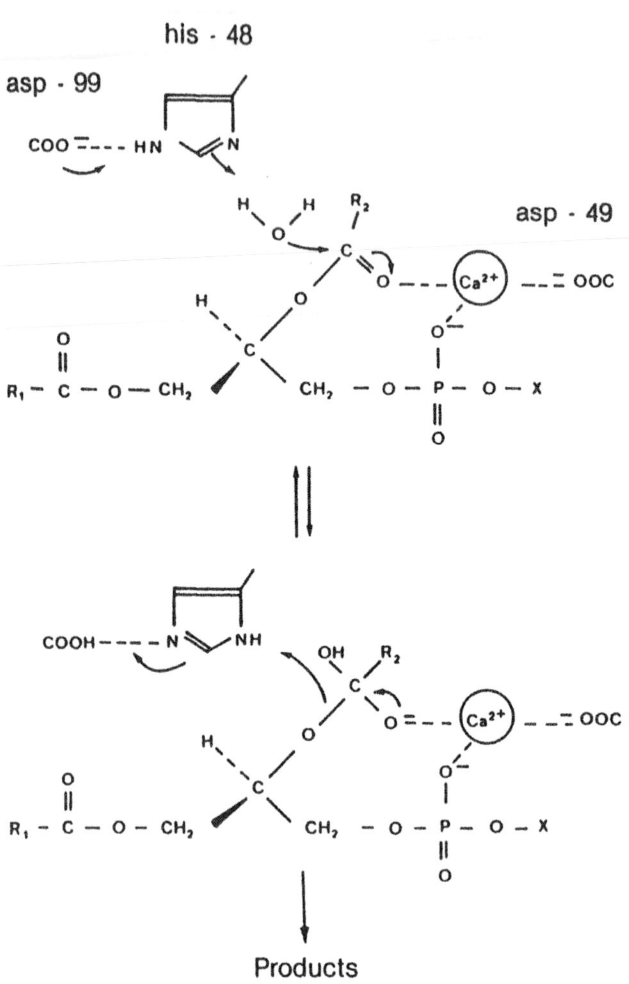

Fig. 5. Proton-relay mechanism of hydrolysis proposed by Verheij et al. [12] for secreted phospholipases A$_2$.

hydrophobic interactions of the molecule undergoing hydrolysis and the interface are maintained. A particular feature is the ability of the cobra venom enzyme to hydrolyze phosphatidylcholine in vesicles and cell membranes, a feature that may in part reflect the presence of tryptophans and other aromatic residues on the interfacial surface. Such residues, particularly tryptophan, are able to partition into the interfacial region of a phosphatidylcholine interface (W.M. Yau, 1998) promoting interfacial binding and catalysis [15]. The presence of such residues in the presumptive interface region can be clearly seen in Fig. 7.

The mammalian pancreatic enzymes (group IB) have primarily a digestive role and are secreted as the pro-enzyme (zymogen) that requires subsequent proteolytic cleavage to remove a hexapeptide at the N-terminus. The pro-enzyme is unable to bind to the phospholipid interface. Unlike the cobra venom enzyme, the pancreatic

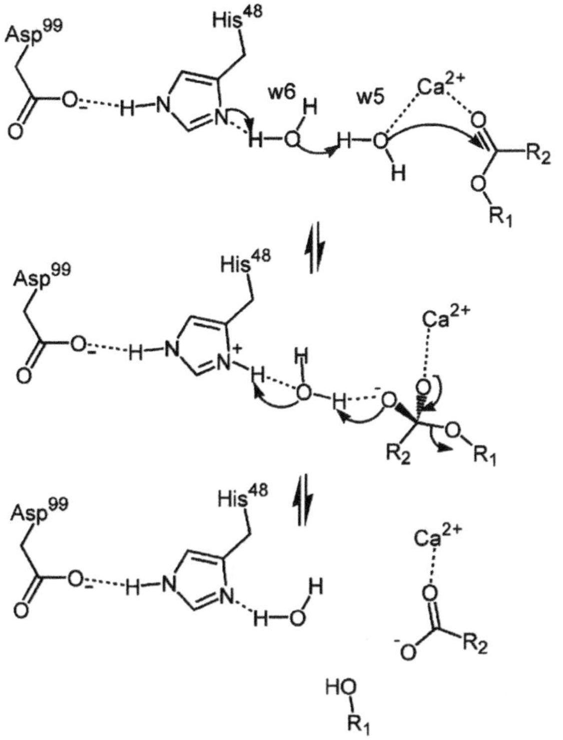

Fig. 6. Calcium coordinated oxyanion mechanism of hydrolysis proposed by Jain for secreted phospholipases A$_2$ (J. Rogers, 1996). The two water molecules at the active site that are implicated in catalysis are shown as W5 and W6.

enzyme expresses low activity with a zwitterionic interface such as that provided by phosphatidylcholine. The enzyme has a considerable preference for anionic interface that can be provided by anionic phospholipid per se or by the inclusion of other anionic lipids in the phosphatidylcholine interface. Presumably the anionic bile salts provide such negative charge in the mixed micelles produced in the intestine to allow lipid digestion. The expression of the pancreatic enzyme in other tissues and the presence of cell surface receptors for this enzyme (as well as other secreted phospholipases A$_2$) suggest additional physiological roles [11,16].

*2.3.1.2. Group IIA–F secreted PLA$_2$s.* Historically, the group IIA secreted PLA$_2$s included venom enzymes from rattlesnakes and vipers. They are characterized by a C-terminal extension while they lack a surface loop region (elapid loop) present in group I enzymes. The most interesting member of this group IIA is the mammalian enzyme that was isolated from the synovial fluid of patients with rheumatoid arthritis and from platelets. This was the first mammalian non-pancreatic enzyme to be identified and was implicated as a key enzyme in arachidonic acid release from phospholipids, the first step in the production of the inflammatory eicosanoids. However, it has subsequently been

304

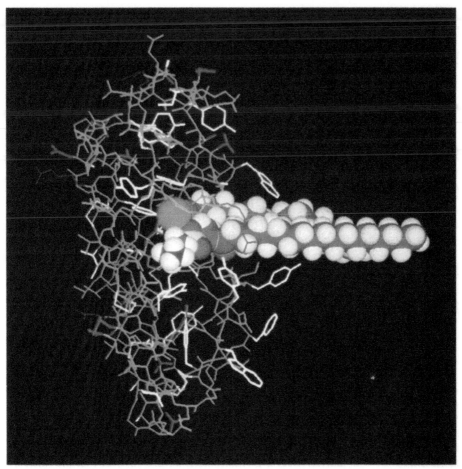

Fig. 7. X-ray crystal structure of cobra venom (*Naja naja naja*) phospholipase A$_2$ with bound Ca$^{2+}$ showing a space-filling model of dimyristoyl phosphatidylethanolamine bound in the catalytic site. The ends of the fatty acid chains stick out of the enzyme and are presumably associated with the micelle or membrane [13].

established that it is the intracellular group IV enzyme that plays the dominant role in eicosanoid production (see Section 2.3.2.1) and hence the precise physiological role of this IIA enzyme in the inflammatory response remains to be clarified.

It is clear that the human IIA enzyme behaves as an acute phase protein (R.M. Crowl, 1991) and extra-cellular levels increase dramatically in acute inflammatory conditions such as septicemia where blood levels can rise over 100-fold. The detailed structural properties of the enzyme are unusual and appear to reflect at least some of its physiological roles. The enzyme is highly cationic (p*I* > 10) and has a net positive charge of +19 due mostly to cationic residues distributed across the surface of the protein, a feature that is linked to the high affinity of the protein for heparin. An unusual characteristic of the enzyme is its marked preference for anionic phospholipid interfaces and very low affinity for a zwitterionic interface as provided by phosphatidylcholine.

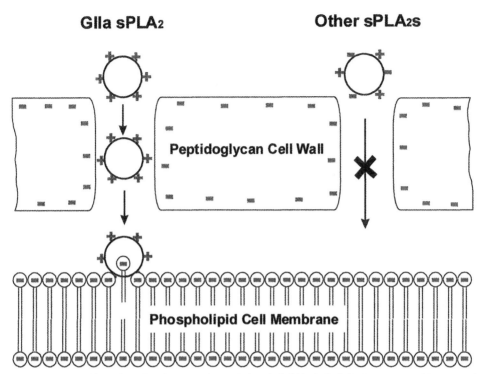

**GIIa sPLA2**              **Other sPLA2s**

Peptidoglycan Cell Wall

Phospholipid Cell Membrane

Fig. 8. A model of bacterial cell wall penetration and phospholipid hydrolysis by human group IIA secreted PLA$_2$ (gIIA sPLA$_2$). The extreme and global nature of the positive charge of this secreted PLA$_2$ has been proposed to allow the enzyme to move through the negatively charged cell wall mediated by the continual making and breaking of electrostatic bonds. For simplicity bacterial phospholipids are shown as being negatively charged, although the bacterial membrane will contain some zwitterionic phospholipids, normally phosphatidylethanolamine.

This preference is expressed at the stage of interfacial binding while active site substrate preference is modest and favors phospholipids such as phosphatidylglycerol over phosphatidylcholine. These surface properties are consistent with one particular physiological role for the enzyme, namely as an antibacterial protein that shows specificity towards gram-positive bacteria [17]. The highly cationic nature of the protein allows the enzyme (but not other more neutral secreted PLA$_2$s) to penetrate the highly anionic bacterial cell wall. Moreover, the anionic bacterial cell membrane that is rich in phosphatidylglycerol provides an optimum substrate for this enzyme (Fig. 8). In contrast, this enzyme is essentially inactive against the host cell membranes that are normally zwitterionic being rich in phosphatidylcholine and sphingomyelin. This lack of activity against host membranes (which is partly due to lack of an interfacial tryptophan residue (S.F. Baker, 1998)) is important as the serum levels of the enzyme can rise to above 1 µg/ml.

The enzyme is released from cells with known antibacterial activity (macrophages and Paneth cells) and is present at very high concentrations (~10 µg/ml) in human tears

(X.D. Qu, 1998) along with lysozyme. The increased sensitivity of mice lacking the group IIA enzyme to gram-positive bacteria provides further support for the antibacterial role of the enzyme (V.J.O. Laine, 1999). The preference of this enzyme for anionic interfaces also means that this enzyme may be active in helping to destroy apoptotic and damaged cells [16].

Notwithstanding an antibacterial role for the enzyme, the involvement of this enzyme in the inflammatory response, where it is linked to the delayed release of prostaglandins, has been demonstrated and requires a molecular explanation. The ability of the enzyme, after secretion, to bind to cell surface heparan sulfate proteoglycans via cationic residues on the protein surface has been the focus of attention. It is proposed that the enzyme binds to one particular type of proteoglycan, namely the glypicans and that a continual translocation of glypicans from the cell surface to the nucleus occurs during cell activation [16]. This is proposed to be the result of a caveolae-mediated endocytotic event called potocytosis which transfers the group IIA secreted $PLA_2$ within calcium-rich vesicles to the nuclear region of the cell in proximity to the cyclooxygenase (Chapter 13) [16].

*2.3.1.3. Group IIB–F secreted $PLA_2$s.* Group IIB includes the gaboon viper enzyme that is missing one of the highly conserved disulfides. Groups IIC–F are recently discovered mammalian enzymes of which the IIC is present only as a pseudo-gene in humans. These human genes, along with the IIA and V, map to the same chromosome locus; however, the function of these enzymes remains to be elucidated (M. Murakami, 2001).

*2.3.1.4. Group III secreted $PLA_2$.* The bee venom enzyme was historically placed in group III and is the primary allergen in bee venom. The enzyme is significantly larger than other secreted $PLA_2$s with less structural homology. The availability of a crystal structure has allowed extensive mutagenesis. In particular, the charge reversal mutagenesis of 5 of the 6 cationic residues on the interfacial binding surface produced a mutant with minimal effects on binding to anionic interfaces. This result highlighted the importance of non-electrostatic interaction in interfacial binding (F. Ghomashchi, 1998). The incorporation of spin labels on the protein surface following cysteine mutagenesis provided a method for defining the topological relationship between the protein and the membrane interface (Y. Lin, 1998). The results demonstrated the interaction of hydrophobic residues but not cationic residues with the interfacial region of the membrane.

A group III human enzyme has recently been identified that possesses long N- and C-terminal extensions but as yet no function has been ascribed to this enzyme.

*2.3.1.5. Group V secreted $PLA_2$.* This enzyme is the most studied human secreted $PLA_2$ after the group IIA enzyme. The enzyme has a tryptophan on the interfacial binding surface, Trp-31, which is partly responsible for its ability to hydrolyze cell membranes. Exogenously added group V enzyme can catalyze phosphatidylcholine hydrolysis in the cell surface with subsequent activation of group IV cytosolic $PLA_2$ and enhanced leukotriene biosynthesis. Removal of Trp-31 by mutagenesis greatly reduces the effectiveness of added enzyme [18]. Like the IIA enzyme, the group V enzyme binds to heparan sulfate proteoglycans on the cell surface and is internalized. However, the dynamics of this process and also mutagenic studies (K.P. Kim, 2001) indicate that internalization is linked to degradation of the enzyme and not to eicosanoid production in human neutrophils and mast cells. In contrast, internalization results in eicosanoid production in human embryonic kidney 293 cells [16,21] (Y.J. Kim, 2002). Thus, the apparent differences between the cellular fate of the group IIA and V enzymes may reflect the types of cells being used for these studies.

*2.3.1.6. Group X secreted PLA$_2$.* This enzyme is effective in the hydrolysis of phosphatidylcholine vesicles and the plasma membrane when it is added exogenously to adherent mammalian cells. Hydrolysis is accompanied by enhanced prostaglandin E2 production. The enzyme does not bind to heparin so heparan sulfate proteoglycan-associated internalization and degradation seen with the group IIA and V enzymes, respectively, cannot occur with this enzyme (S. Bezzine, 2000). The crystal structure of the group X enzyme is now available (Y.H. Pan, 2002).

*2.3.1.7. Group IX, group XI and other secreted PLA$_2$s.* At this time these groups are represented by proteins from a snail venom (group IX) (conodipine-M) and green rice shoots (group XI) respectively (Table 1) while an increasing number of secreted PLA$_2$s are being identified from analysis of genomic data bases in both the plant and animal kingdoms [11]. The cloning and expression of these enzymes and related proteins will be the first step in understanding their physiological functions.

*2.3.2. Phospholipases A$_2$ that involve a catalytic serine residue*
*2.3.2.1. Group IV cytosolic PLA$_2$.* Over the past decade or more the group IV cytosolic PLA$_2$ has taken center stage as the phospholipase that is primarily involved in the regulation of prostaglandin and leukotriene biosynthesis (Chapter 13) as part of the inflammatory response [19]. Even though this enzyme translocates to a membranous fraction, it is recovered from the cytosolic fraction of the cell and hence it is termed cytosolic. This enzyme is distinct from the mammalian secreted phospholipases A$_2$ in its size (85 versus 14 kDa), stimulation by Ca$^{2+}$ (at micromolar versus millimolar concentration), specificity at the *sn*-2 position of the substrate (arachidonate versus no specificity) and catalytic mechanism. This enzyme shows minimal head-group specificity and phosphatidylcholine is the normal substrate. The discovery of the cytosolic phospholipase A$_2$ has provided new insights into the cell signalling events that initiate the 'arachidonate cascade' described in Chapter 13.

The sequence of the cytosolic PLA$_2$ was deduced from the cDNA isolated from a number of species and is now referred to as cytosolic PLA$_2$ α as the result of the more recent discovery of further isozymes (β and γ). The human enzyme, mapped to chromosome 1, is highly conserved amongst mammalian species but the sequence can differ up to 20–30% between mammals and non-mammalian vertebrates. Distinct regions of the enzyme have been identified including two domains, an N-terminal C2 domain and a larger C-terminal catalytic domain (Fig. 9). The γ-isozyme lacks the C2 domain and is prenylated at the C-terminus. The complete crystal structure of human cytosolic PLA$_2$ α (A. Dessen, 1999) has been reviewed [20] and provides a logical basis for discussion of the function and regulation of this important enzyme.

*The catalytic domain.* The enzyme shows lysophospholipase and transacylase activities that are consistent with the formation of an acyl-serine intermediate characteristic of lipases, and Ser-228 has been identified as the involved residue. In addition Asp-549 has been demonstrated to be the second member of the predicted catalytic triad. However, histidine has not been identified as the third member and none of the 19 histidine residues in the protein has been shown to play any catalytic role. At present a dyad mechanism must be invoked while Arg-200 is in a position to stabilize the oxyanion intermediate. Interestingly, 1-palmitoyl lysophosphatidylcholine is degraded at a rate comparable to 2-arachidonoyl phosphatidylcholine which raises the possibility that the enzyme serves multiple functions in the cell.

*The C2 domain.* The enzyme has a Ca$^{2+}$-dependent phospholipid binding domain (CaLB or C2 domain) common to many proteins that translocate to membranes from the cytosol in the presence of Ca$^{2+}$. The activity of the cytosolic PLA$_2$ increases several fold as the Ca$^{2+}$ concentration is increased to concentrations found in activated cells (300 nM). Since Ca$^{2+}$ is not involved in the catalytic event, Ca$^{2+}$ promotion of enzyme–membrane interaction probably accounts for Ca$^{2+}$'s stimulatory effect. The molecular mechanism by which Ca$^{2+}$ binding promotes membrane interactions has been the subject of intensive investigations. The C2 domain consists of an anti-parallel β-sandwich composed of two four-stranded sheets (Fig. 9) while the structure is capped

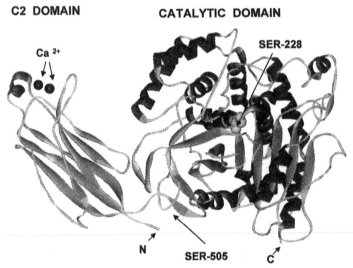

Fig. 9. Crystal structure of group IV cytosolic PLA$_2$ showing the C2 and catalytic domain. The two calcium ions bond to the C2 domain and the active site Ser-228 are highlighted. The phospholipid interface would be parallel to the top surface of the molecule as shown. A flexible region between residues 499 and 538 is not seen in the crystal structure but contains Ser-505, the approximate position of which is indicated by the arrow. The N- and C-terminals of the protein are indicated. Adapted from A. Dessen (1999) and Dessen [20] using MSI Weblab Viewer.

by three $Ca^{2+}$ binding loops known as the calcium-binding regions. It would appear that the binding of $Ca^{2+}$, allows penetration of two of the loops into the membrane providing a stable membrane–protein interaction (O. Perisic, 1999; L. Bittova, 1999). The net result is that the enzyme translocates to the membrane surface to allow catalysis. Immunofluorescence studies have revealed that the enzyme preferentially translocates to the nuclear envelope region, a location that also contains the cyclooxygenase and 5-lipooxygenase, presumably allowing facile transfer of the released arachidonic acid to these first enzymes in the prostaglandin and leukotriene pathways. The mechanism by which cytosolic PLA$_2$ selectively translocates to this specific nuclear region is unclear in view of the lack of obvious phospholipid head-group selectivity of the enzyme. It has been proposed that the 60 kDa protein vimentin acts as an adaptor for membrane targeting and interacts directly with the C2 domain in a $Ca^{2+}$-dependent manner [21].

*Phosphorylation of cytosolic PLA$_2$ α.* There are multiple sites for phosphorylation on the cPLA$_2$ and a number of protein kinases appear to use the enzyme as substrate. A critical site is Ser-505, the site phosphorylated by mitogen-activated protein kinase. Phosphorylation at this site increases the activity of the enzyme both in vitro and in vivo. However, other phosphorylation sites may be involved; in particular, Ser-727 may also have an important role in some cell systems (M.A. Gijon, 2000). The Ser-505 is located in a flexible loop that connects the C2 and catalytic domains (Fig. 9). It is possible that phosphorylation of this residue produces the optimum orientation of the two domains with respect to the membrane interface or affects the interaction of these two domains with another membrane protein such as vimentin [20].

*Gene knockouts.* The very large number of enzymes that have now been discovered with PLA$_2$ activity increases the difficulty of being able to unambiguously define the role of a regulator of the inflammatory response involving arachidonic acid to one particular enzyme. Gene knockout studies in mice involving cytosolic PLA$_2$ α have been particularly successful in both defining the primary importance of cytosolic PLA$_2$ in inflammation and identifying specific physiological roles [22]. Thus cytosolic PLA$_2$ knockout mice have revealed important roles in normal fertility, generation of eicosanoids from inflammatory cells, brain injuries and allergenic responses. Other forms of PLA$_2$ cannot replace these functions and hence the enzyme becomes a prime pharmacological target.

*2.3.2.2. Group VI phospholipase A$_2$.* This group consists of the intracellular calcium-independent PLA$_2$s and at this time it is the group VIA enzyme for which most information is available [23]. An additional calcium-independent PLA$_2$ has recently been described (D.J. Mancuso, 2000) and has been categorized as group VIB.

The group VIA enzyme is 85–88 kDa and consists of multiple splice variants that contain 7 or 8 ankyrin repeats. The enzymes are widely distributed and have both a cytosolic and membrane location within the cell. They exhibit lysophospholipase and transacylase activity while their ability to be inactivated by hydrophobic serine-reactive inhibitors is indicative of a catalytic serine and acyl-serine intermediates. A variety of evidence indicates that these enzymes are involved in basal phospholipid fatty acid remodeling as a general housekeeping function within most cell types. It is possible that these enzymes also have roles in signal transduction and other physiological functions [23].

*2.3.2.3. Group VII and group VIII PLA₂s.* The group VII and VIII enzymes (Table 2) are better known as platelet-activating factor acetylhydrolases [24]. These enzymes catalyze hydrolysis of the *sn*-2 ester bond of platelet-activating factor (see Chapter 9) and related pro-inflammatory phospholipids, and thus attenuate their bioactivity. They exist as both secreted (plasma) and intracellular forms and the plasma form has attracted most attention. The plasma form not only hydrolyzes platelet-activating factor but also a range of oxidatively damaged phosphatidylcholines from cell membranes and lipoproteins that originally contained arachidonic acid at the *sn*-2 position. Such oxidative damage can be significant in reperfusion injury, cigarette smoking and inflammation and as such their formation is essentially uncontrolled. Therefore, the sole mechanism for regulating the biological impact of these compounds lies in their degradation by the acetylhydrolases.

The clinical importance of this enzyme is highlighted by the lack of platelet-activating factor acetylhydrolase activity in a significant proportion of the Japanese population and in the majority of cases this is due to a mutation, V279F. This mutation is associated with asthma, stroke, myocardial infarction, brain hemorrhage and non-familial cardiomyopathy. It is possible that treatment with recombinant platelet-activating factor acetylhydrolase may be beneficial [24].

Intracellular forms of the enzyme all hydrolyze platelet-activating factor acetyl hydrolase but some are more selective and do not hydrolyze oxidized phospholipids [24].

## 2.4. Phospholipase C

### 2.4.1. Bacterial phospholipases C
Phospholipases C have been known to be associated with bacteria since the classic demonstration by Macfarlane and Knight (1941) that α-toxin in *Clostridium perfringens* was a phospholipase C (reviewed in Waite [2]). The most extensively studied phospholipases C are those from *Bacillus cereus* and provided the first crystal structure for a PLC (E. Hough, 1989). Another PLC that is specific for phosphatidylinositol is also secreted in large amounts by *B. cereus* and has provided a crystal structure (D.W. Heinz, 1995).

### 2.4.2. Mammalian phospholipases C
These phospholipases are primarily involved in signal transduction and are reviewed in detail in Chapter 12. The structure and mechanism of action of phosphatidylinositol-specific PLCs from both mammalian and bacterial sources has been presented [25,26]. The structure of mammalian phosphatidylinositol-PLC δ is shown in Fig. 10 and highlights the domain structure. This protein was crystallized in the absence of its PH domain that is at the N-terminal. A 'tether and fix' model of membrane association is suggested (L.-O. Essen, 1996) whereby initial specific interaction is via the PH domain that will target membrane phosphoinositides followed by the binding of the C2 and catalytic domains. A two-step catalytic mechanism is proposed for both mammalian and bacterial enzymes [26] involving a cyclic phosphodiester intermediate which in the case of the bacterial enzyme can be the major product of the reaction.

The major pathway of sphingomyelin degradation involves a special phospholipase

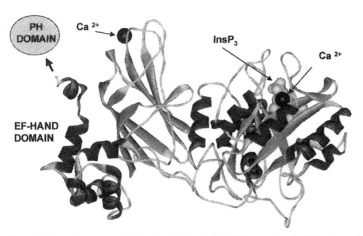

C2 DOMAIN       CATALYTIC DOMAIN

Fig. 10. Crystal structure of phosphatidylinositol-PLC δ showing the C2 and catalytic domains. The position of the PH domain that would be attached to the EF-hand domain is indicated. The membrane surface would be parallel to the top surface of the molecule as shown. Calcium ions are shown bound to the active site and to the C2 domain. The position of inositol trisphosphate in space-filling format at the active site is indicated. Adapted from L.-O. Essen (1996) and Heinz et al. [26] using MSI Weblab Viewer.

C, a sphingomyelinase. The enzyme is secreted by *B. cereus* while several mammalian sphingomyelinases are achieving a prominent role in signal transduction in mammalian systems. The description of sphingolipids and their metabolism is covered in Chapter 14.

### 2.5. Phospholipase D

Classically, plants and bacteria have been the major sources for the purification of phospholipases D. The function of phospholipase D in plants is not known although it may be involved in cell turnover and energy utilization during different cycles in plant life. Bacterial phospholipases D in some cases are toxins and can lead to severe cellular damage either alone or in combination with other proteins secreted from bacteria. These bacterial enzymes may also serve to help provide nutrients for the cell such as inorganic phosphate, as do the bacterial phospholipases C [2].

All phospholipases D characterized thus far act by a phosphatidate exchange reaction that has a covalent phosphatidyl-enzyme as an intermediate [27]. For this reason, the enzyme can catalyze a 'base-exchange' reaction in which alcohols can substitute for water as the phosphatidate acceptor. In fact, alcohols are better than water as phosphatidate acceptors; about 1% of an alcohol (e.g., ethanol) in water yields phosphatidylethanol almost exclusively and this principle is the basis of some PLD assays.

In the case of the member of the PLD family, Nuc, when the enzyme was incubated with $^{32}$P-labeled inorganic phosphate, a phospho-enzyme intermediate was produced and identified as phospho-histidine (E.B. Gottlin, 1998). Recently, the first crystal structure of a PLD from *Streptomyces* sp. has been published (I. Leiros, 2000) and

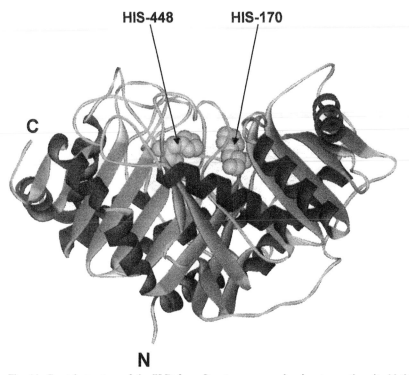

**HIS-448**     **HIS-170**

**C**

**N**

Fig. 11. Crystal structure of the PLD from *Streptomyces* sp. showing two active site histidines in space-filling format together with the N- and C-termini of the protein. Adapted from I. Leiros (2000) using MSA Weblab Viewer.

is shown in Fig. 11. The positions of two active site histidines, His-170 and His-448, are indicated and it is proposed that His-170 is the more likely residue to act as the nucleophile and thus be covalently modified during catalysis.

Mammalian PLDs are now the subject of intense interest as they appear to be intimately involved in signal transduction. Two mammalian PLDs have been identified and these contain several domains often associated with signal transduction proteins including a PH domain and a separate phosphatidylinositol-*bis*-phosphate binding domain. These proteins appear to be controlled by a variety of regulatory molecules [28]. The potential roles of the enzymes in cell signaling [29] and membrane traffic [30] are discussed in detail in Chapter 12.

Phospholipases C and D that act on glycosylphosphatidylinositol-anchored proteins on the cell surface are of growing importance and are discussed in Chapter 2.

## 3. Future directions

Since the last edition of 'Biochemistry of Lipids, Lipoproteins and Membranes' there has been remarkable progress in our knowledge of phospholipases. This progress has included the crystal structures of many more phospholipases and the use of the techniques of molecular genetics in order to define the precise physiological function of individual enzymes. However, at this time the rate of elucidation of the structure and function of known enzymes is being matched by the discovery of new phospholipases in a very diverse array of organisms. This is particularly the case with the secreted $PLA_2$s and a major challenge is to understand the role or roles of many of these enzymes in their respective organisms. A major factor that has evolved is the important role of binding to the phospholipid interface in determining and regulating physiological function. Our knowledge of the molecular interactions at membrane interfaces is limited and will require advances in both X-ray crystallography and NMR to elucidate such details. There are exciting times ahead!

## Abbreviations

| | |
|---|---|
| cmc | critical micellar concentration |
| PLs | phospholipases |
| $PLA_1$ | phospholipase $A_1$ |
| $PLA_2$ | phospholipase $A_2$ |
| PLC | phospholipase C |
| PLD | phospholipase D |

## References

1. Van den Bosch, H. and Aarsman, A.J. (1979) A review on methods of phospholipase A determination. Agents Actions 9, 382–389.
2. Waite, M. (1987) The Phospholipases. In: D.J. Hanahan (Ed.), Handbook of Lipid Research. Vol. 5, Plenum, New York, NY, p. 332.
3. Reynolds, L.J., Washburn, W.N., Deems, R.A. and Dennis, E.A. (1991) Assay strategies and methods for phospholipases. Methods Enzymol. 197, 3–23.
4. Hendrickson, H.S. (1994) Fluorescence-based assays for lipases, phospholipases and other lipolytic enzymes. Anal. Biochem. 219, 1–8.
5. Verger, R. (1980) Enzyme kinetics of lipolysis. Methods Enzymol. 64B, 340–392.
6. Gelb, M.H., Jain, M.K., Hanel, A.M. and Berg, O.G. (1995) Interfacial enzymology of glycerolipid hydrolases: Lessons from secreted phospholipases $A_2$. Annu. Rev. Biochem. 64, 653–688.
7. Berg, O.G., Gelb, M.H., Tsai, M.-D. and Jain, M.K. (2001) Interfacial enzymology: The secreted phospholipase $A_2$-paradigm. Chem. Rev. 101, 2613–2654.
8. Snijder, H.J. and Dijkstra, B.W. (2000) Bacterial phospholipase A: structure and function of an integral membrane phospholipase. Biochim. Biophys. Acta 1488, 91–101.
9. Wang, A.J. and Dennis, E.A. (1999) Mammalian lysophospholipases. Biochim. Biophys. Acta 1439, 1–16.
10. Six, D.A. and Dennis, E.A. (2000) The expanding superfamily of phospholipase $A_2$ enzymes: classification and characterisation. Biochim. Biophys. Acta 1488, 1–19.
11. Valentin, E. and Lambeau, G. (2000) Increasing molecular diversity of secreted phospholipases $A_2$ and their receptors and binding proteins. Biochim. Biophys. Acta 1488, 59–70.
12. Verheij, H.M., Slotboom, A.J. and DeHaas, G.H. (1981) Structure and function of phospholipase $A_2$. Rev. Physiol. Biochem. Pharmacol. 91, 91–203.

13. Dennis, E.A. (1994) Diversity of group types, regulation, and function of phospholipase $A_2$. J. Biol. Chem. 269, 13057–13060.

14. Yuan, C. and Tsa, M.-D. (1999) Pancreatic phospholipase $A_2$: new views on old issues. Biochim. Biophys. Acta 1441, 215–222.

15. Gelb, M.H., Cho, W. and Wilton, D.C. (1999) Interfacial binding of secreted phospholipase $A_2$: more than electrostatics and a major role for tryptophan. Curr. Opin. Struct. Biol. 9, 428–432.

16. Murakami, M. and Kudo, I. (2001) Diversity and regulatory functions of mammalian secretory phospholipase $A_2$. Adv. Immunol. 77, 163–194.

17. Buckland, A.G. and Wilton, D.C. (2000) The antibacterial properties of secreted phospholipases $A_2$. Biochim. Biophys. Acta 1488, 71–82.

18. Cho, W. (2000) Structure, function and regulation of group V phospholipase $A_2$. Biochim. Biophys. Acta 1488, 48–58.

19. Leslie, C.C. (1997) Properties and regulation of cytosolic phospholipase $A_2$. J. Biol. Chem. 272, 16709–16712.

20. Dessen, A. (2000) Structure and mechanism of human cytosolic phospholipase $A_2$. Biochim. Biophys. Acta 1488, 40–47.

21. Murakami, M., Nakatani, Y., Kuwata, H. and Kudo, I. (2000) Cellular components that functionally interact with signaling phospholipase $A_2$s. Biochim. Biophys. Acta 1488, 159–166.

22. Sapirstein, A. and Bonventre, J.V. (2000) Specific physiological roles cytosolic phospholipase $A_2$ as defined by gene knockouts. Biochim. Biophys. Acta 1488, 139–148.

23. Winstead, M.V., Balsinde, J. and Dennis, E.A. (2000) Calcium-independent phospholipase $A_2$: structure and function. Biochim. Biophys. Acta 1488, 28–39.

24. Tjoelker, L.W. and Stafforini, D.M. (2000) Platelet-activating factor acetylhydrolases in health and disease. Biochim. Biophys. Acta 1488, 102–123.

25. Katan, M. (1998) Families of phosphoinositide-specific phospholipase C: structure and function. Biochim. Biophys. Acta 1436, 5–17.

26. Heinz, D.W., Essen, L.O. and Williams, R.L. (1998) Structural and mechanistic comparison of prokaryotic and eukaryotic phosphoinositide-specific phospholipase C. J. Mol. Biol. 275, 635–650.

27. Waite, M. (1999) The PLD superfamily: insights into catalysis. Biochim. Biophys. Acta 1439, 187–197.

28. Frohman, M.A., Sung, T.C. and Morris, A.J. (1999) Mammalian phospholipase D structure and regulation. Biochim. Biophys. Acta 1439, 175–186.

29. Exton, J.H. (1999) Regulation of phospholipase D. Biochim. Biophys. Acta 1439, 121–133.

30. Jones, D., Morgan, C. and Cockroft, S. (1999) Phospholipase D and membrane traffic. Biochim. Biophys. Acta 1439, 229–244.

D.E. Vance and J.E. Vance (Eds.) *Biochemistry of Lipids, Lipoproteins and Membranes (4th Edn.)*
© 2002 Elsevier Science B.V. All rights reserved

# CHAPTER 12

# Glycerolipids in signal transduction

## Linda C. McPhail

*Department of Biochemistry, Wake Forest University School of Medicine, Winston-Salem, NC 27157, USA,*
*Tel.: +1 (336) 716-2621; Fax: +1 (336) 716-7671; E-mail: lmcphail@wfubmc.edu*

## 1. Introduction

Glycerolipids have emerged as essential molecules for the regulation of cell functions by hormones, neurotransmitters, growth factors, and inflammatory cytokines. Membrane phospholipids are acted upon by a host of phospholipases, lipid kinases and phosphatases to generate signaling lipids. Fig. 1 illustrates the major enzymes involved in the production of phospholipid-derived second messengers. In general, formation of signaling lipids is initiated by ligand binding to a specific cell-surface receptor, which leads to the activation of phospholipases and/or lipid kinases. The lipid-derived second messengers generated then act on specific target proteins to influence cell function. Signaling lipids and inositol phosphates, like most second messengers, are quickly metabolized to limit the response or to generate another signaling molecule.

The first glycerolipid pathway known to be associated with cell signaling was the turnover of phosphatidylinositol (PI), which leads to changes in the levels of

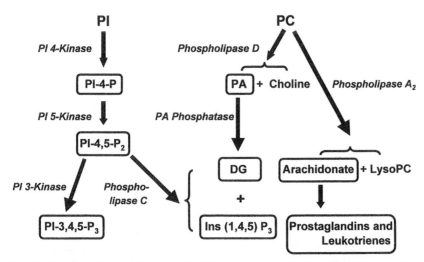

Fig. 1. Overview of phospholipase and lipid kinases that participate in signal transduction mechanisms. Abbreviations: PI, phosphatidylinositol; PI-4-P, PI-4-phosphate; PI-4,5-$P_2$, PI-4,5-bisphosphate; PI-3,4,5-$P_3$, PI-3,4,5-trisphosphate; PC, phosphatidylcholine; PA, phosphatidic acid; DG, diacylglycerol; Ins(1,4,5)$P_3$, inositol trisphosphate.

various phosphoinositides and the generation of diacylglycerol (DG) and inositol-1,4,5-trisphosphate (Ins-1,4,5-P$_3$). Both of these products are important second messengers. The turnover of phosphatidylcholine (PC) also generates second messengers: phosphatidic acid (PA), by the action of phospholipase D (PLD), and arachidonic acid and lyso-PC (LPC), by the action of phospholipase A$_2$. PA has an emerging role as a second messenger and also can be converted to lyso-PA (LPA) by certain phospholipases A. Arachidonic acid also has direct intracellular targets as a second messenger, but primarily is metabolized to prostaglandins and leukotrienes (Chapter 13). Alkyl species of lyso-PC are converted to platelet-activating factor (Chapter 9). PA and DG are interconvertible by the action of the enzymes PA phosphatase and DG kinase. DG, PA, Ins-1,4,5-P$_3$ and the phosphoinositides primarily have intracellular sites of action. In contrast, prostaglandins, leukotrienes, platelet-activating factor, LPC and LPA exit the cell and act as autocrine, paracrine, or circulating hormones by binding to specific cell-surface heptahelical receptors on the same, neighboring, or distant target cells.

This chapter describes the mechanisms by which lipid-derived second messengers regulate various cell functions. Current information about the pathways that regulate the formation and metabolism of signaling lipids and inositol phosphates, as well as the progress in identifying protein targets for the second messengers, will be discussed. Excluded is coverage of arachidonic acid-derived signaling lipids and platelet-activating factor, as they receive extensive discussion in other chapters (Chapters 9 and 13).

## 2. Inositol phosphates

### 2.1. Mechanisms of generation and metabolism

#### 2.1.1. Phosphoinositide-hydrolyzing phospholipase C

Mabel and Lowell Hokin were the first to observe (in 1953) that the turnover of PI increased greatly in response to cholinergic stimulation of pigeon pancreatic slices [1]. This began the study and elucidation of the PI cycle, illustrated in Fig. 2. A series of two lipid kinases (PI 4-kinase and PI 5-kinase) sequentially add phosphates to positions 4 and 5 of the inositol ring to form PI-4,5-bisphosphate (PI-4,5-P$_2$). PI-4,5-P$_2$ is the major substrate for the phosphoinositide-hydrolyzing phospholipase C (PLC) family of enzymes, which are activated in response to hormone or other agonist stimulation. Hydrolysis of PI-4,5-P$_2$ by PLC produces the second messengers DG and Ins-1,4,5-P$_3$. The function of these messengers is discussed below (Sections 2.2, 3.2, 3.3). Both products are metabolized further and used for the resynthesis of PI, as shown in Fig. 2. The situation is more complex than shown, since the action of additional kinases and phosphatases results in the formation of inositol phosphate species phosphorylated on various combinations of up to all six hydroxyls of the inositol ring. Many of the over 60 potential species have not been isolated from biological sources, but several inositol tetrakisphosphates, Ins-1,3,4,5,6-P$_5$ and Ins-1,2,3,4,5,6-P$_6$, are present in animal cells [2]. Known and potential functions of these molecules are described in Section 2.2.

The mammalian PLC family consists of 11 isozymes divided into four subgroups (β, γ, δ, ε), all of which catalyze the hydrolysis of PI-4,5-P$_2$ to Ins-1,4,5-P$_3$ and DG in

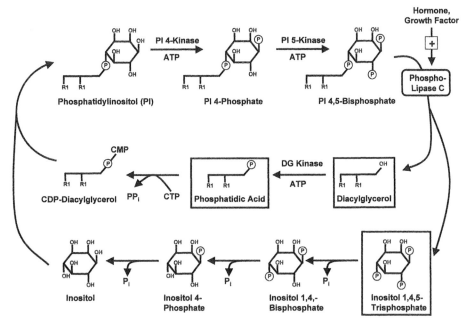

Fig. 2. Phosphatidylinositol cycle. See text for details. Shown in bold boxes are the signaling molecules derived from the PI cycle. DG kinase, diacylglycerol kinase; PI, phosphatidylinositol.

response to agonist binding to over 100 cell-surface receptors [3]. The four subgroups all contain similar catalytic X and Y domains, but differ in size and in their regulatory regions (Fig. 3). The divergent regulatory regions are responsible for differences in the types of receptors and the mechanisms by which members in each subgroup are activated. Three of the subgroups ($\delta$, $\beta$, $\gamma$) contain three regulatory regions in common:

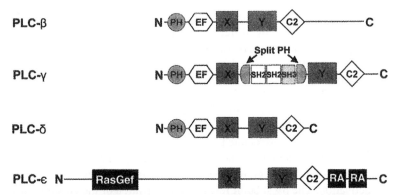

Fig. 3. Domain structures of the four isoforms of phosphatidylinositol-specific phospholipase C (PLC). The X and Y catalytic regions are conserved among all the isoforms. PH, pleckstrin homology domain; EF, EF hands; SH2 and SH3, Src homology domains 2 and 3; C2, C2 domain; RasGEF, Ras guanine nucleotide exchange factor-like domain; RA, Ras-binding domain. See text for discussion.

318

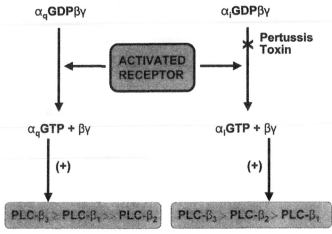

Fig. 4. Scheme for activation of phospholipase C-β (PLC-β) isoforms by G protein α and βγ subunits. The left side of the figure shows the pertussis toxin-insensitive activation of PLC-β isoforms by $G_q\alpha$ subunits. The right side of the figure shows the pertussis toxin-sensitive activation of PLC-β isoforms by βγ subunits derived from a $G_i$ type of G protein.

an N-terminal pleckstrin homology (PH) domain, an EF hand domain and a C-terminal C2 domain. PH domains bind certain phosphoinositides and/or proteins (see Section 5.2.1), EF hand domains usually bind calcium [4], and C2 domains usually bind anionic phospholipids in the presence of calcium (see Section 3.2) [5]. The PLC-γ subgroup contains a second PH domain that is split by two Src homology 2 (SH2) domains and a Src homology 3 (SH3) domain. SH2 regions usually interact with phosphorylated tyrosine residues in other proteins and SH3 domains bind specific polyproline motifs in target proteins. The PLC-ε subgroup contains a C2 domain, as well as an N-terminal Ras guanine nucleotide exchange factor-like domain and one or two C-terminal Ras binding domains. The crystal structure of PLC-δ (minus the PH domain) has been solved (Chapter 11), which clearly illustrates the domain structure of the enzyme.

Each PLC subgroup is regulated by different mechanisms, triggered by agonist binding to specific receptors and involving recruitment of the soluble PLC to the plasma membrane. The PLC-β subgroup is regulated by interaction with α and βγ subunits of certain heterotrimeric G proteins [3] (Fig. 4). Agonist binding to heptahelical receptors triggers the exchange of GTP for GDP on the α subunit of heterotrimeric G proteins, causing the α and βγ subunits to dissociate. Both free subunits then interact with effector enzymes to regulate intracellular signaling. G protein α subunits hydrolyze the bound GTP back to GDP and the subunits reassociate to terminate the signaling process. There are four subfamilies of G protein α subunits, and one of these ($G\alpha_q$) interacts with the C2 domain and the extended C-terminal region of PLC-β to aid in membrane recruitment and induce enzyme activation. The βγ subunit released by activation of the $G_i$ subfamily also can recruit and activate PLC-β by interaction with the PH domain. Treatment of cells with pertussis toxin can determine which G protein family is responsible for PLC activation. Pertussis toxin inhibits the activation of the $G_i$ family of G proteins, but has no effect on the $G_q$ family. Different isozymes in the PLC-$\beta$ subfamily demonstrate differing abilities to respond to G protein $\alpha$ and $\beta\gamma$ subunits. For example, PLC-$\beta 1$ is quite sensitive to $G\alpha_q$ and poorly responsive to $G\beta\gamma$. Finally, PI-3-P, generated by a PI 3-kinase, binds the PH domain of PLC-$\beta 1$ and aids in recruitment of the enzyme to the membrane.

PLC-$\gamma$ is activated by a large group of cell-surface receptors, including receptor protein tyrosine kinases, antigen and immunoglobulin receptors, and heptahelical receptors [3]. The basic mechanism used by all receptors involves tyrosine phosphorylation of PLC-$\gamma$ and targeting of the enzyme to the plasma membrane. Receptors with intrinsic tyrosine kinase activity (e.g. growth factor receptors) first recruit PLC-$\gamma$ to phosphorylated tyrosine residues on the receptor by interaction with the N-terminal SH2 domain. This allows the receptor tyrosine kinase to phosphorylate PLC-$\gamma$, inducing enzyme activation. Recruitment of PLC-$\gamma$ to the membrane also involves the interactions of PI-3,4,5-$P_3$ with the C-terminal SH2 domain and the PH domain. Regulatory roles for the EF hand, SH3 and C2 domains of PLC-$\gamma$ are not known. Activation of PLC-$\gamma$ by antigen involves intracellular non-receptor protein tyrosine kinases that are either recruited to the receptor or activated by other pathways. These tyrosine kinases phosphorylate PLC-$\gamma$ and, along with PI-3,4,5-$P_3$, induce enzyme activation. With heptahelical receptors, a non-receptor tyrosine kinase phosphorylates and activates a growth factor receptor, which then recruits and phosphorylates PLC-$\gamma$, as described above.

The pathways used by membrane receptors to activate the PLC-$\delta$ isozymes are unclear [3]. PLC-$\delta$ is more sensitive to $Ca^{2+}$ than the other subgroups, so an increase in intracellular $Ca^{2+}$ levels induced by receptor-mediated activation of one of the other PLC subtypes may be responsible for PLC-$\delta$ activation. Also, a newly discovered type of GTP-binding protein, $G_h$ (high-molecular-weight G protein) may regulate PLC-$\delta$ after stimulation of cells through $\alpha_1$-adrenergic, oxytocin, or the $\alpha$ subtype of thromboxane $A_2$ receptors.

The newly discovered PLC-$\varepsilon$ appears to be regulated by two mechanisms [3]. Growth factor receptors lead to activation of the small GTPase Ras, which binds to PLC-$\varepsilon$ via the C-terminal Ras binding domain and activates the enzyme. The N-terminal Ras guanine nucleotide exchange factor-like domain may prolong PLC-$\varepsilon$ activation by reactivating exchange of GTP for GDP on Ras. PLC-$\varepsilon$ is also activated by the $\alpha$ subunit of the heterotrimeric G protein $G_{12}$, which couples to LPA receptors, which are members of the large heptahelical (G protein-coupled) receptor family.

### 2.1.2. Inositol phosphate kinases and phosphatases

The initial product of PLC, Ins-1,4,5-$P_3$, acts as a second messenger for the release of $Ca^{2+}$ from intracellular stores (Section 2.2.1). However, Ins-1,4,5-$P_3$ is rapidly metabolized by sequential phosphatase action to free inositol or by Ins-1,4,5-$P_3$ 3-kinase to inositol-1,3,4,5-tetrakisphosphate (Ins-1,3,4,5-$P_4$). Inositol polyphosphate phosphatases comprise a family of enzymes, which use both inositol phosphates and the phosphoinositides as substrates to varying degrees [6]. Group I inositol polyphosphate 5-phosphatases act on Ins-1,4,5-$P_3$ and Ins-1,3,4,5-$P_4$ and appear to be key enzymes for dampening $Ca^{2+}$ signaling. Inositol polyphosphate 1-phosphatase hydrolyzes Ins-1,4-$P_2$ and Ins-1,4,5-$P_3$ and may be the target of lithium inhibition during therapy for manic depression.

Ins-1,4,5-P$_3$ 3-kinases, which also remove Ins-1,4,5-P$_3$, are activated by the Ca$^{2+}$-binding protein calmodulin as a feedback mechanism when Ins-1,4,5-P$_3$ causes a rise in Ca$^{2+}$ levels [2]. Other members of the inositol phosphate kinase multi-gene family include inositol phosphate multikinases, which can phosphorylate Ins-1,4,5-P$_3$ to form Ins-1,3,4,5,6-P$_5$, and InsP$_6$ kinases, which can form Ins-1,2,3,4,5,6-P$_6$ from multiple inositol phosphate substrates [2].

## 2.2. Cellular targets

### 2.2.1. Control of intracellular calcium levels by inositol-1,4,5-trisphosphate
Ca$^{2+}$ is an important intracellular signaling molecule, which regulates a host of cell functions [7]. These include contraction, proliferation, fertilization, secretion, vesicle trafficking, apoptosis, and intermediary metabolism. In keeping with its signaling role, the cytosolic level of calcium is very low (~100 nM), but can rise to ~1000 nM when cells are activated by agonist–receptor interaction. Receptor activation triggers pathways leading to the opening of Ca$^{2+}$ channels in the endoplasmic reticulum and the plasma membrane, causing the marked elevation in intracellular Ca$^{2+}$ concentration. The increase is transient because the second messengers acting to open Ca$^{2+}$ channels are rapidly metabolized and Ca$^{2+}$ pumps and exchangers remove Ca$^{2+}$ from the cytosol to restore the resting state. A major pathway for increasing Ca$^{2+}$ levels is the formation of Ins-1,4,5-P$_3$ by activation of PLC (Section 2.1.1). Ins-1,4,5-P$_3$ binds to its receptor, which is a Ca$^{2+}$ channel in the endoplasmic reticulum. The role of Ins-1,4,5-P$_3$ may be to increase the sensitivity of the receptor to stimulation by cytosolic Ca$^{2+}$, rather than directly causing the channel to open.

### 2.2.2. Targets for other inositol phosphates
More highly phosphorylated inositol phosphates also may regulate cell function [2]. Ins-1,3,4,5-P$_4$ has a higher affinity than Ins-1,4,5-P$_3$ for hydrolysis by inositol polyphosphate 5-phosphatase and can prolong the lifetime of Ins-1,4,5-P$_3$, thus enhancing increases in Ca$^{2+}$ concentration. Ins-1,3,4,5-P$_4$ directly activates Ca$^{2+}$ channels in the plasma membrane in some cells, also leading to enhanced Ca$^{2+}$ levels. Ins-3,4,5,6-P$_4$ appears to inhibit plasma membrane Ca$^{2+}$-regulated chloride channels. Inositol hexakisphosphate (InsP$_6$), also known as phytic acid, has several proposed functional targets. One is a protein kinase that phosphorylates pacsin/syndapin I, a protein involved in vesicle trafficking. A second is a DNA-dependent protein kinase that regulates DNA end-joining. InsP$_6$ also may regulate mRNA transport out of the nucleus.

# 3. Diacylglycerols

## 3.1. Mechanisms of generation and metabolism

### 3.1.1. Hydrolysis of phospholipids
The initial step for DG generation in response to agonist–receptor engagement is the hydrolysis of PI-4,5-P$_2$ by PLC (Section 2.1.1). A PC-hydrolyzing PLC activity has also

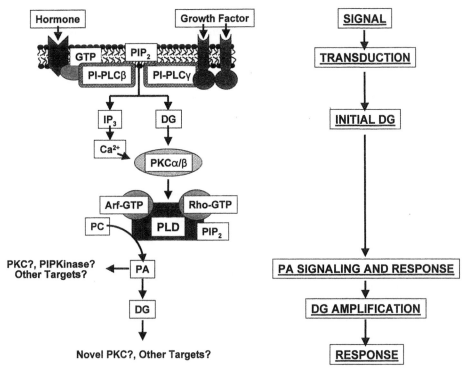

Fig. 5. Phospholipase cascade model for activation of phospholipase D (PLD) and amplification of diacylglycerol generation. Abbreviations: PI-PLC, phosphatidylinositol-specific phospholipase C; DG, diacylglycerol; PC, phosphatidylcholine; PA, phosphatidic acid; PKC, protein kinase C; $PIP_2$, PI-4,5-bisphosphate; PIPkinase, PI phosphate kinase.

been described (e.g. [8]), but is not yet identified at the molecular level. In many cell types (e.g. human neutrophils [9]), two 'waves' of DG production take place caused by a phospholipase cascade (Fig. 5). PI-4,5-$P_2$-hydrolyzing PLC is responsible for the initial rapid wave of DG production. The second, more sustained wave of DG production is from the activation of PC-hydrolyzing PLD to generate PA, followed by conversion of the PA to DG by a PA phosphatase [10,11]. PA has its own roles as a second messenger (Section 4.3). Depending on the molecular species of the PC pools acted upon by PLD, a mixture of diradylglycerols (diacylglycerols, alkylacylglycerols, alkenylacylglycerols) may be formed. Clearly, the fatty acid composition of the DG species derived from PI-4,5-$P_2$ and PC differs [11], suggesting their functional roles are not the same. Indeed, the DG and other diradylglycerol species derived from PC may not activate protein kinase C (PKC) — the major target of DG in cells [11]. However, DG binds to proteins other than PKC (Section 3.3), raising the possibility that PC-derived diradylglycerols have other targets. Alternatively, the diradylglycerols derived from PA phosphatase may enter lipid biosynthetic pathways (Fig. 2).

The PA phosphatases that convert PA to DG comprise a family of enzymes, which function in both general lipid metabolism and glycerolipid signaling [10]. Type I PA phosphatase (EC 3.1.3.2) is a cytosolic enzyme absolutely dependent on $Mg^{2+}$ and is likely involved in lipid biosynthetic reactions. It has not been characterized at the molecular level. Type II PA phosphatase does not require divalent cations and is an integral membrane protein. It is more properly termed a lipid phosphate phosphatase, since it hydrolyzes a variety of signaling lipid phosphates. Substrates include PA, LPA, ceramide 1-phosphate, sphingosine 1-phosphate, and diacylglycerol pyrophosphate. The cDNAs encoding several Type II lipid phosphate phosphatases have been cloned and these enzymes belong to a phosphatase superfamily. Superfamily members include bacterial non-specific acid phosphatases, yeast lipid phosphate phosphatases, fungal haloperoxidases, mammalian glucose-6-phosphatase, and several others. All family members share three highly conserved domains that likely comprise the active site. The mechanisms that regulate the lipid phosphate phosphatases are not clear, but several of the enzymes contain putative phosphorylation sites. Additional studies are needed to determine the functional roles of the lipid phosphate phosphatases, but they clearly can degrade a number of lipid second messengers. Thus, their primary function may be to attenuate cellular responses to agonist–receptor interaction.

### 3.1.2. Diacylglycerol kinases

DG is metabolized by DG kinases, which phosphorylate DG to form PA [12]. This action likely serves to attenuate DG-mediated responses, but the generated PA may be another second messenger (Section 4.3). Mammalian DG kinases comprise a family of nine isoenzymes ($\alpha$, $\beta$, $\gamma$, $\delta$, $\eta$, $\varepsilon$, $\zeta$, $\iota$, $\theta$), differing in their primary structure (Fig. 6), substrate specificity and tissue distribution. All contain a conserved catalytic domain and two cysteine-rich (C1) domains. C1 domains in PKC and other proteins bind DG (Sections 3.2 and 3.3); however, the role of C1 domains in DG kinase isoforms is not known. The presence of other protein : protein and protein : lipid interaction domains (e.g. PH, EF hand, proline-rich region, ankyrin repeats) divides the isoenzymes into five classes

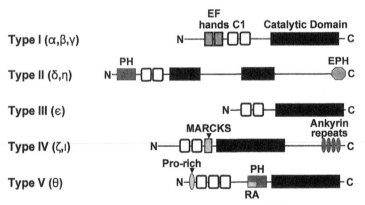

Fig. 6. Domain structures of mammalian diacylglycerol (DG) kinases. Domains are labeled from top of figure. PH, pleckstrin homology; EPH, ephrin C-terminal tail homology; MARCKS, sequence homologous to the myristoylated alanine-rich C-kinase substrate (MARCKS) phosphorylation site domain; Pro-rich, proline-rich domain; RA, Ras-associating domain.

(Fig. 6), suggesting that each class may participate in different signaling pathways. Many of the isoenzymes also have putative phosphorylation sites. Thus, regulation of DG kinases involves phosphorylation reactions and specific protein : protein and protein : lipid interactions, resulting in alterations in localization and activity of the enzymes. The diversity and complexity of the DG kinase family suggests that, like PKC (Section 3.2) and PLC (Section 2.1.1), these enzymes are crucial participants in lipid-mediated signaling processes. However, specific roles for DG kinases in the regulation of cell functions await elucidation.

## 3.2. Protein kinase C is an important target for diacylglycerol

PKC was the first lipid-regulated protein kinase to be described and its discovery by Nishizuka and colleagues [13,14] initiated a new era of lipid-mediated signal transduction. DG and $Ca^{2+}$ were quickly identified as the physiological activators of PKC, placing PKC as a primary downstream participant in the hormonally activated PI pathway described by the Hokins [1] (Section 2.1.1). PKC regulates almost all cell functions, including cell differentiation, proliferation, metabolism, and apoptosis. The discovery that PKC is the target of the tumor-promoting phorbol diesters illustrated its importance for the regulation of carcinogenesis [15].

PKC is now known to be a family of twelve mammalian isoforms, divided into four classes based on structural differences and cofactor dependence [15,16] (Fig. 7). All PKCs are a single polypeptide chain, consisting of an N-terminal regulatory region and a C-terminal catalytic domain connected by a protease-sensitive hinge region. The conserved catalytic domain consists of the C3 ATP-binding region and the C4 substrate-binding domain. The conventional isoforms ($\alpha$, $\beta$I, $\beta$II, $\gamma$) are activated by

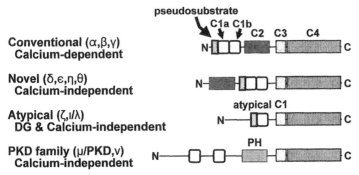

Fig. 7. Domain structure of protein kinase C (PKC) isoforms. The N-terminal half of the protein contains regulatory regions (C1, C2), while the C-terminal half contains the catalytic regions (C3, C4). The conventional isoforms contain tandem C1 domains and one C2 domain, which bind diacylglycerol (C1), $Ca^{2+}$ (C2) and anionic phospholipids (C2). Adjacent to the C1 domains is a pseudosubstrate sequence involved in maintaining the auto-inhibitory conformation of the PKCs. The novel isoforms contain an altered C2 domain, which no longer binds $Ca^{2+}$. The atypical isoforms lack a C2 domain and have one atypical C1 domain, accounting for their lack of activation by diacylglycerol (DG) and $Ca^{2+}$. The PKD family is larger in size, has two non-adjacent C1 domains and a pleckstrin homology (PH) domain instead of a C2 domain. It is $Ca^{2+}$-independent, but is activated by DG and anionic phospholipids.

$Ca^{2+}$, DG, and phosphatidylserine. The C2 region binds phosphatidylserine and $Ca^{2+}$, while DG binds to the two C1 domains. A pseudosubstrate region, containing sequences resembling phosphorylation sites in PKC substrates but without a phosphorylatable residue, is at the N-terminal end of the first C1 domain. In the inactive state, PKC is folded such that the pseudosubstrate region interacts with the C4 substrate binding domain, blocking access of exogenous substrates to the active site. Binding of DG, phosphatidylserine and $Ca^{2+}$ to the regulatory domain disrupts this interaction, leading to PKC activation.

The novel PKC isoforms ($\delta$, $\varepsilon$, $\theta$, $\eta$) have two DG-binding C1 domains and an atypical C2 region near the N-terminus, which binds phosphatidylserine, but not $Ca^{2+}$. These isoforms are $Ca^{2+}$-independent and are activated by phosphatidylserine and DG. The atypical PKCs ($\zeta$, $\iota/\lambda$) lack the C2 region and have an atypical C1 region that does not bind DG. Thus, these isoforms are $Ca^{2+}$- and DG-independent, but can still be activated by phosphatidylserine. Other mechanisms of activating the atypical isoforms likely exist. Members of a fourth subclass ($\mu$, $\nu$) are larger than the other isoforms and have two separated C1 regions and a PH domain in the extended regulatory region [17]. These isoforms are activated by phosphatidylserine and DG, but may have additional mechanisms of regulation.

It has become clear that regulation of PKC involves more than binding of activators to induce conformational changes in the enzyme. The lipid activators of PKC also target the enzyme to membranes, which regulates full activation of the enzyme, access of the enzyme to substrates, and proteolytic degradation of the enzyme [16]. Membrane targeting involves the cooperation between lipid binding to the C1 and C2 domains and probable interactions of other regions with membrane lipids. Protein:protein interactions also regulate PKC localization and function [15]. PKC isoforms interact with a wide variety of proteins, including RACKs (receptors for activated C kinase), STICKs (substrates that interact with C kinase), cytoskeletal proteins, and scaffolding proteins. These interactions function either to target specific PKC isoforms to sites of activator generation and/or substrates or to integrate PKC into other signaling pathways. Finally, all PKC isoforms require phosphorylation during maturation of the enzyme for catalytic competence. Phosphorylation within the activation loop by phosphoinositide-dependent protein kinase-1 or a related kinase is required for enzyme activity. Also, two sites in the C-terminal tail of PKC must be autophosphorylated for proper localization and catalytic competence of the enzyme. The multiplicity of PKC isoforms and the complexity of their regulation affirm that this family of enzymes plays critical roles in various signaling pathways.

### 3.3. Non-PKC targets of diacylglycerol

Most of the PKC isoforms are targets of DG and mediate much of the signaling by this lipid second messenger. However, other proteins contain C1 domains that may bind DG. Some of these appear to be atypical C1 domains, similar to the C1 domain in PKC-$\zeta$ [16]. These include DG kinase isoforms, the protein kinase Raf, and the Rac guanine nucleotide exchange factor Vav. In contrast, DG and phorbol diesters have been found to bind to several non-kinase proteins, including the chimaerins, which are Rac GTPase-activating proteins, Unc-13 and related proteins, which may be scaffolding proteins, and RasGRP, a guanine nucleotide release protein for Ras [15]. The main function of DG may be to recruit these proteins to membrane sites where they regulate other enzyme activities (e.g. RasGRP activates Ras). Thus, DG may have other functional roles beyond activation of PKC.

## 4. Phospholipase D and the generation of phosphatidic acid

### 4.1. Discovery and molecular nature of phospholipase D

PLD (EC 3.1.4.4) was discovered in plants over 50 years ago as an enzyme that catalyzed the hydrolysis of PC to form PA and choline [18]. However, PLD was not of interest in the field of signal transduction until mammalian PC-hydrolyzing PLD was shown to be activated by extracellular stimuli in the late 1980's (reviewed in [18]). Mammalian PLD is activated by a wide variety of hormones, neurotransmitters, cytokines, and growth factors and plays essential roles in the regulation of numerous cellular functions [18,19]. A number of other eukaryotic PLD activities have been described, including oleate-activated PLD, phosphoinositide-specific PLD, phosphatidylserine/phosphatidylethanolamine-hydrolyzing PLDs, N-acyl-phosphatidylethanolamine PLD, lyso-PLD, and glycosylphosphatidylinositol-PLD. With the exception of the glycosylphosphatidylinositol-PLD, which is a serum enzyme and has no signaling role, none of these other enzymes have been identified at the molecular level. Thus, this section will focus on the molecularly characterized PC-hydrolyzing PLDs that clearly function in glycerolipid signaling.

PC-PLD enzymes are members of a diverse gene superfamily that includes cardiolipin and phospholipid synthases, endonucleases and certain viral proteins, which share a common 'HKD' motif involved in catalysis (E.V. Koonin, 1996; C.P. Ponting, 1996). PC-PLDs are present in most organisms, including bacteria, plants, flies, yeast, worms, and mammals [18]. Some organisms express only one PC-PLD gene (e.g. yeast), while others express multiple genes (plants, mammals). The domain structure of representative PC-PLDs from various organisms is shown in Fig. 8. All have four conserved domains (I–IV), which comprise the catalytic core of the enzyme. The exact functions of Domains I and III are unknown. Domains II and IV contain the short conserved sequence $HxKx_4Dx_6GG/S/TxN$ termed the 'HKD motif' or the phosphatidyltransferase motif, present in all members of the PLD superfamily. PLD activity requires both HKD motifs and Domains I–IV may form a bilobed, 'dimer' structure, such that both HKD motifs are present in the active site (I. Leiros, 2000). The histidines are thought to attach the phosphatidyl moiety to the enzyme, forming a phosphohistidine intermediate. This is a hallmark of the 'transphosphatidylation' reaction catalyzed by PC-PLD and other members of the PLD superfamily. The phosphatidyl moiety is transferred to a $H_2O$ molecule to form PA, or it can be transferred to an alcohol acceptor to form a phosphatidylalcohol. This property is the basis for a commonly used PLD assay, in which cells are incubated in the presence of a short-chain primary alcohol (e.g. 1-butanol, ethanol) and the formation of the phosphatidylalcohol (which is poorly metabolized) is measured.

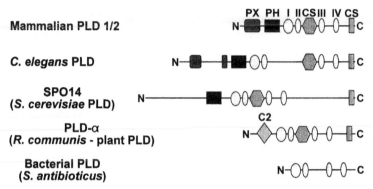

Fig. 8. Comparison of domain structures of phospholipase D (PLD) isoforms from various organisms. Domains are labeled at the top of the figure, except for the C2 domain found only in plant PLD isoforms. All isoforms contain four conserved regions involved in catalysis (I, II, III, IV). All isoforms except the bacterial enzymes contain two conserved regions (CS) of unknown function. PX, phox homology domain; PH, pleckstrin homology domain. The PX domain in *C. elegans* PLD is split.

Incubation of cells with primary alcohols can be used to reduce the production of PA by PLD, thus interfering with PLD-dependent regulation of cellular functional responses.

Most of the PC-PLDs from higher organisms are inactive in unstimulated cells and are regulated by the interaction of intracellular signaling proteins and lipids with domains present in the various proteins [18] (Fig. 8). The active site of PLD is likely blocked by the folded structure of the protein and requires a conformational change to release the autoinhibition. Mammalian and *Caenorhabditis elegans* PLDs each contain an atypical PH domain and a PX (phox homology) domain in the N-terminal half of the molecule. The *Saccharomyces cerevisiae* enzyme contains only the PH domain. The PH domain in human PLD1 binds PI-4,5-$P_2$ (M.N. Hodgkin, 2000), but it is controversial whether or not this binding is responsible for the PI-4,5-$P_2$ dependence of PLD activity. PX domains in other proteins were recently shown to target to membranes by binding selected phosphoinositides [20]. The function of the PX domain in PLDs is not known. The plant PLDs lack PH and PX domains, but have a C2 domain in the N-terminal region (Fig. 8). The C2 domain binds $Ca^{2+}$ and anionic phospholipids and is likely responsible for the dependence of the plant enzymes on $Ca^{2+}$ for activity. All eukaryotic PC-PLDs also contain conserved regions between catalytic Domains II and III and at the C-terminus. The functions of these regions are unknown.

### 4.2. Localization and regulation of phospholipase D

Numerous receptor agonists induce activation of mammalian PC-PLD in vivo [19]. Activation does not involve direct interaction of occupied receptors with PLD, but instead uses intracellular signaling intermediates. Evidence is strong that PLD is regulated in vivo by PKC, the ARF, Rho, and Ras/RalA families of small GTPases, and the intracellular levels of PI-4,5-$P_2$. The cDNAs for two mammalian PLD isoforms (PLD1, PLD2) have been cloned and each gene can be alternatively spliced to yield two products. In vitro assays of the recombinant PLD enzymes show that PLD1a and PLD1b require PI-4,5-P$_2$ for activity and are synergistically activated by PKC-$\alpha$ or -$\beta$, ARF1, and RhoA (see Fig. 5). It is not known if such synergism between multiple activators takes place in intact cells. In contrast, PLD2a and PLD2b require PI-4,5-P$_2$, are only slightly activated by ARF1 and are insensitive to PKC and RhoA. For the most part, the functional binding sites on the PLD molecule for these regulators are unknown. The PI-4,5-P$_2$-binding site may be the PH domain or the conserved region between Domains II and III (Fig. 8). The ARF-binding site has not been identified, despite intensive effort. The PKC-$\alpha$/$\beta$ binding site appears to be in the N-terminal third of PLD1 and the RhoA-binding site is in the C-terminal third, but the specific motifs involved have not been defined.

PLD1 and PLD2 usually are differentially localized in cell types that have been examined [18]. PLD1 is associated with intracellular membrane compartments, most likely endosomal/lysosomal membranes. PLD2 is associated with the plasma membrane and has been found in caveolae. Given the in vitro differences in the regulation of PLD1 and PLD2 and their differential localization, it is likely that each has unique functions in cellular regulation.

### 4.3. Functions regulated by phospholipase D and phosphatidic acid

Studies that selectively manipulate PLD1 and PLD2 levels and activity in intact cells, in order to identify PLD isoform-dependent downstream signaling elements and cellular functions, are still in the early stages. Similarly, while PA regulates a number of proteins in vitro, it is not clear whether PA has the same role in vivo for most of the proteins. However, pharmacological approaches and PLD localization studies have strongly implicated PLD and PA as important regulators of several cell functions.

#### 4.3.1. Membrane trafficking, secretion and cell proliferation
Both PLD1 and PLD2 have been implicated in membrane trafficking events, including budding of vesicles from the Golgi apparatus and the *trans*-Golgi network and ARF-dependent clathrin coat assembly in the endosomal/lysosomal system [18]. The production of PA by PLD may promote the recruitment of adapter and coat proteins to sites of vesicle budding. PA also can induce activation of Type I PI 4-phosphate kinases, leading to increased levels of PI-4,5-P$_2$ and additional protein recruitment and PLD activation. Scission of vesicles is mediated by the GTPase dynamin or its analogues, which are also regulated by anionic phospholipids. Thus, the increased levels of PA and PI-4,5-P$_2$ in the budding membrane are thought to mediate recruitment of the proteins necessary to complete the budding process.

Evidence also implicates PLD in the regulation of stimulated secretory and degranulation responses, such as the insulin-stimulated translocation of GLUT4 to the plasma membrane, release of granule contents from mast cells and neutrophils, exocytosis from chromaffin cells, and the release of matrix metalloproteases from cancer cells [18]. The mechanisms by which the PLD product PA mediates secretion are not known, but also may involve recruitment of particular proteins to secretory vesicle and granule membranes.

PA is thought to have a similar recruitment role in the regulation of Raf1, a key protein kinase that regulates cell proliferation (M.A. Rizzo, 2000). Raf1 is a target of the small GTPase Ras and triggers cell proliferation through activation of the mitogen-activated protein kinase cascade. PLD activation is required for the activation of Raf1 in intact cells and a PA-binding site in Raf1 has been identified. However, PA has no direct effect on Raf1 activity. Mutation of the PA-binding site prevents recruitment of Raf1 to the membrane and inhibits activation of the downstream mitogenic kinase cascade. This suggests that the role of PA is to recruit Raf1 to the membrane where Ras is located.

### 4.3.2. NADPH oxidase

Pharmacological data strongly indicate that PLD activation is needed for receptor-mediated activation of the phagocyte multi-component enzyme NADPH oxidase [21]. This enzyme functions in host defense against infection and in tissue damage during inflammatory diseases. Study of the processes regulating this enzyme is providing insight into the complex mechanisms by which signaling lipids mediate functional responses. Activation of NADPH oxidase requires recruitment of four cytosolic proteins, including the small GTPase Rac, to assemble with heterodimeric flavocytochrome $b_{558}$ in the membrane. The flavocytochrome then undergoes a presumed conformational change, allowing it to transfer electrons from NADPH to $O_2$ to form superoxide anion. The cytosolic proteins exist in inactive complexes and must be dissociated or undergo conformational changes to move to the membrane. The entire process involves phosphorylation of components (by lipid-activated protein kinases, such as PKC), lipid- and phosphorylation-regulated SH3 region-mediated protein : protein interactions, and direct interaction of signaling lipids (PA, DG, arachidonic acid) with various oxidase proteins. PA and arachidonic acid each cause conformational changes in a cytosolic oxidase component, p47$^{phox}$, and the membrane-bound flavocytochrome. PA (plus DG) or arachidonic acid (alone) induce phosphorylation-independent oxidase activation in purified cell-free systems containing only oxidase proteins, presumably by their direct interactions with p47$^{phox}$ and flavocytochrome $b_{558}$. Two of the cytosolic oxidase components, p47$^{phox}$ and p40$^{phox}$, have N-terminal PX domains, which bind phosphoinositides and likely participate in the recruitment process [20]. New data indicate that PA binds to the p47$^{phox}$ PX domain with greater affinity and efficacy than phosphoinositides (L.C. McPhail, unpublished results). Thus, PA may be the physiological ligand for this particular PX domain and may regulate NADPH oxidase activation partially through its PX domain interaction. Studies are ongoing, but it is clear that PA and other signaling lipids induce activating conformational changes in NADPH oxidase components and regulate membrane localization of oxidase proteins by both direct binding and protein kinase-dependent mechanisms.

## 5. Phosphoinositides

The phosphoinositides comprise a large group of inositol-containing phospholipids, with phosphates attached to positions D3, D4, or D5 (or various combinations of these positions) of the inositol ring (Figs. 9 and 10). They are present at low concentrations in cells: PI, the precursor phospholipid, is present at less than 10% of total membrane

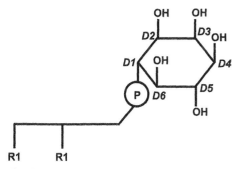

Fig. 9. Numbering of the inositol ring. Phosphates can be attached at positions D1 through D6 on the inositol ring. Shown is phosphatidylinositol, in which inositol is attached to the glycerol backbone by a phosphate in position D1.

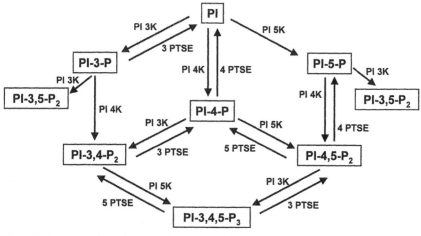

Fig. 10. Interconversion of phosphoinositides. The pathways by which phosphoinositide kinases and phosphatases synthesize and degrade various phosphoinositides are shown. Abbreviations: PI, phosphatidylinositol; PI-4-P, PI-4-phosphate; PI-4,5-$P_2$, PI-4,5-bisphosphate; PI-3,4,5-$P_3$, PI-3,4,5-trisphosphate; PI 3K, PI 3-kinase; 3 PTSE, PI 3-phosphatase.

lipid; PI-4,5-$P_2$, the substrate for PLC, is about 5%; phosphoinositides phosphorylated at the 3 position are less than 0.25% of the total inositol-containing lipids [22]. The phosphoinositides regulate a large number of signaling pathways and cellular functions by binding to specific sites on proteins to alter their localization or activity (Section 5.2). A complex set of PI kinases and phosphatases is involved in the interconversion (Fig. 10) (Section 5.1), and considerable progress in identifying these enzymes at the molecular level has been made. Phosphoinositide-metabolizing enzymes are regulated by receptor-triggered signaling processes, which results in highly localized changes in the levels of specific phosphoinositides to coordinate temporal and spatial regulation of cellular functions.

*5.1. Phosphoinositide kinases and phosphatases*

*5.1.1. Phosphatidylinositol 4- and 5-kinases*
The enzymes that phosphorylate positions D4 and D5 of the inositol ring fall into two major categories, i.e. those that classically act on PI-4-P and those that utilize PI [23]. The PI 4-kinases convert PI to PI-4-P and are divided into two types (II and III), based on biochemical differences (sensitivity to detergents and adenosine). The cDNA for a Type II enzyme has recently been cloned and is suggested to establish a novel family of PI 4-kinases because it lacks the phosphoinositide kinase (PIK) domain (S. Minogue, 2001). The PIK domain is found in the Type III PI kinases and PI 3-kinases (Section 5.1.2) and has an unknown function. No homology to other known proteins was found in the Type II enzyme. Several Type III enzymes have been identified molecularly in yeast and mammals and are categorized as $\alpha$ and $\beta$ isoforms, based on sequence homology. All contain the PIK domain and a conserved catalytic region. The mammalian $\alpha$ isoform additionally contains ankyrin-like repeats, a SH3 domain, and a polyproline motif. Although it is likely that the PI kinases are regulated by extracellular signals, the mechanisms involved are not known.

The enzymes that act on monophosphorylated forms of PI (PIP) are divided into two families, based on their preferred site of phosphorylation [23,24]. The PIP 5-kinases (also known as Type I) phosphorylate position 5 and PIP 4-kinases (Type II) phosphorylate position 4 of the inositol ring. The PIP 5-kinases are more promiscuous in substrate specificity and phosphorylate both PI-4-P (to form PI-4,5-$P_2$) and PI-3-P (to form PI-3,5-$P_2$ or PI-3,4-$P_2$). In addition, they can add two phosphates to PI-3-P in vitro to form PI-3,4,5-$P_3$. The PIP 4-kinases prefer PI-5-P as substrate and form PI-4,5-$P_2$. Multiple isoforms of each have been found in plants and animals. The two groups share homology in their kinase domain, which resembles only slightly the kinase domains in other phosphoinositide and protein kinases. The Type II enzymes contain polyproline sequences, indicating possible interaction with SH3 domains. The two groups do not possess other known domains or motifs. A crystal structure of a PIP 4-kinase shows the enzyme exists as a dimer, forming an extended basic flat face that interacts with membranes containing anionic phospholipids (V.D. Rao, 1998).

Regulation of the PIP 5- and 4-kinases by receptor-mediated signals is under intensive study, but physiologically relevant mechanisms are unclear [23,24]. Some of the Type II enzymes associate with cell-surface growth factor and cytokine receptors. The Type I enzymes are markedly stimulated by PA in vitro and this also may occur in intact cells (D.R. Jones, 2000). The Type I enzymes may be regulated by GTP-binding proteins, since the non-hydrolyzable GTP analog GTPγS increases PIP 5-kinase activity in cells.

*5.1.2. Phosphoinositide 3-kinases*
Phosphoinositide 3-kinases catalyze the addition of phosphate to the D3 position of the inositol ring of phosphoinositides. Phosphoinositide 3-kinases comprise a large family of enzymes divided into three classes (I, II, and III), with each class showing differences in structure (Fig. 11), substrate specificity, and regulation [23]. The class I enzymes are heterodimeric proteins, consisting of a 110 kDa catalytic subunit and a 50–101 kDa adaptor/regulatory subunit. The adaptor/regulatory subunit inhibits catalytic activity

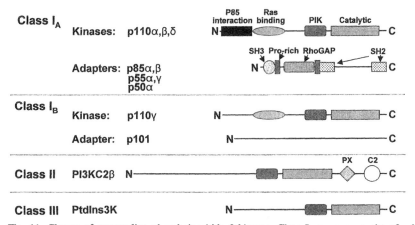

Fig. 11. Classes of mammalian phosphoinositide 3-kinases. Class I enzymes consist of a kinase subunit and an adapter (regulatory) subunit. Functional domains are labeled at the top of the figure or above each isoform. Domains include: p85 interaction domain; Ras binding domain; phosphoinositide kinase (PIK) domain; catalytic domain; Src homology 2 (SH2) and 3 (SH3) domains; proline-rich region (Pro-rich); Rho-GTPase-activating protein (RhoGAP) domain; phox homology (PX) domain; and C2 domain.

and recruits the enzyme to specific subcellular sites by regulated interaction with other proteins. In vitro, the class I enzymes use PI, PI-4-P, and PI-4,5-$P_2$ as substrates. However, the major products in vivo are PI-3,4-$P_2$ and PI-3,4,5-$P_3$, so intracellular regulatory mechanisms must limit substrate specificity.

The class I enzymes are further subdivided into two classes (IA and IB), based on differences in sequence and receptor regulation (Fig. 11). The class IA enzymes are regulated by receptor protein tyrosine kinases and consist of three 110 kDa catalytic subunits ($\alpha$, $\beta$, $\delta$) from different genes and at least five regulatory subunits (p85$\alpha$, p85$\beta$, p55$\alpha$, p55$\gamma$, p50$\alpha$), which can interact interchangeably. The p55$\alpha$ and the p50$\alpha$ proteins are alternatively spliced products of the p85$\alpha$ gene. By convention, all of these subunits are referred to as p85. The class IB enzyme is activated by G protein-coupled receptors and is made up of one p110 catalytic subunit ($\gamma$) and a 101 kDa regulatory subunit, with no homology to the class IA adaptor/regulatory subunits. The p110$\gamma$ subunit cannot interact with the class IA regulatory subunits, because it lacks an N-terminal interaction region present in p110$\alpha$, $\beta$, and $\delta$ (Fig. 11). The p110 subunits share a number of other functional domains, including the C-terminal catalytic domain, a PIK domain shared with the Type III PI 4-kinases, and a Ras-binding domain. When activated by receptor agonists, the small GTPase Ras can bind to p110 and increase catalytic activity, indicating that the class I enzymes are Ras effectors.

The class IA p85 subunits also share a number of protein-interaction domains (Fig. 11), whose function is to recruit the holoenzyme to specific cellular locations by interaction between the domains and specific target proteins. All p85 proteins contain two SH2 domains, which bind to specific motifs containing phosphorylated tyrosine residues. Tyrosine phosphorylation occurs in many receptor complexes (e.g. receptor tyrosine kinases such as the epidermal growth factor receptor) in response to extracellular stimuli. Binding of the SH2 domains of p85 to phosphotyrosine motifs in the receptor complex mediates the recruitment of p110 to the membrane, where it can access substrates and undergo additional activation by Ras. The longer forms of p85 also contain a SH3 domain in the N-terminal region, which binds to polyproline motifs in target proteins, and either two (p85α) or three (p85β) polyproline (Pro-rich) motifs. The shorter p85 isoforms contain only one polyproline motif. Only a few binding partners (e.g. the GTPase dynamin for the SH3 region, the Src family of protein tyrosine kinases for the polyproline motifs) for the N-terminal region are known. The longer forms of p85 contain a Rho-GTPase-activating protein homology domain, which mediates interaction of the Rho family of GTPases (e.g. Rac, Cdc42) with p85. Whether these different protein-interaction motifs cooperate to localize and activate the enzyme, antagonize each other under certain conditions, or allow the enzyme to participate in different signaling pathways is not clear.

The class IB catalytic subunit (p110γ) is activated by interaction with βγ subunits of heterotrimeric G proteins. It shares the catalytic, PIK, and Ras-binding domains with the class 1A p110 subunits (Fig. 11). The function of the p101 adaptor/regulatory subunit is not clear, since it has no homology to the p85 subunits. However, p101 appears to be necessary for the activation of p110γ by βγ subunits.

The class II phosphoinositide kinases (α, β) are single polypeptides, sharing the catalytic and PIK domains with the class I p110 subunits. These kinases also contain an atypical C2 domain, which may mediate the binding of the protein to anionic phospholipids in membranes. The C2 domain lacks residues that are important for $Ca^{2+}$ binding, so $Ca^{2+}$ most likely does not play a role in phospholipid binding. The two mammalian isoforms both have a C-terminal PX domain, which may bind PI-4,5-$P_2$ (X. Song, 2001). Both the C2 and the PX domains are likely to regulate membrane recruitment and activation of the class II enzymes. In vitro, the class II enzymes prefer PI and PI-4-P as substrates, thus forming PI-3-P and PI-3,4-$P_2$. Their specificity in vivo is not known.

The class III phosphoinositide 3-kinases share only the catalytic and PIC domains with the enzymes in the other two classes. The first class III enzyme was identified in yeast, where it is essential for transport of proteins. These enzymes phosphorylate only PI and are thought to be the major contributors to the synthesis of PI-3-P. A putative adaptor/regulator for the class III enzymes has been identified, which is a protein serine/threonine kinase. It was thought that the class III enzymes are not subject to receptor-mediated regulation, but recent work suggests that they regulate the process of phagosome maturation in professional phagocytic cells (O.V. Vieira, 2001). Since several protein domains have been shown to bind PI-3-P (Section 5.2.2), it is likely that the class III enzymes have important signaling roles.

### 5.1.3. PTEN and other phosphoinositide phosphatases

Inositol phosphatases remove phosphates from the inositol ring and many of these enzymes can use both soluble inositol phosphates and lipid phosphoinositides as substrates. The focus in this section is on the lipid phosphatases that hydrolyze phosphate from the D3, D4, or D5 position of the phosphoinositides. The overall function of these phosphatases is to down-regulate phosphoinositide-dependent signaling reactions.

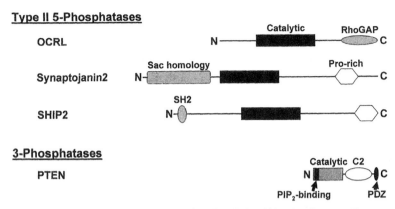

Fig. 12. Domain structures of representative phosphoinositide phosphatases. Shown are selected members of the Type II 5-phosphatases and the 3-phosphatases. Domains include: catalytic (different for the 5- and the 3-phosphatases); Rho-GTPase-activating protein (RhoGAP) domain; Sac homology domain, a region with lipid phosphatase activity; polyproline (Pro-rich) region; Src homology 2 (SH2) domain; C2 domain; a phosphatidylinositol-4,5-bisphosphate binding motif (PIP$_2$-binding) and a PDZ motif.

Mutations in genes encoding some of these enzymes are responsible for certain dystrophies and myopathies and also are common in various tumors. These phosphatases are classified into three families, based on selectivity for the different positions on the inositol ring: the 5-phosphatases, the 3-phosphatases, and the 4-phosphatases [25]. Domain structures of representatives from the 5- and 3-phosphatase families are shown in Fig. 12.

The largest family is the 5-phosphatases, which is divided into two major groups, based on substrate preference and structural differences [6,25]. They all share a homologous phosphatase domain, defined by two signature motifs involved in catalysis and divalent metal ion binding (Y. Tsujishita, 2001). Type I enzymes hydrolyze only water-soluble substrates, as discussed in Section 2.1.2. Type II enzymes prefer PI-4,5-P$_2$ and/or PI-3,4,5-P$_3$ as substrates. Although they share the 5-phosphatase catalytic domain, Type II members differ with respect to other regions. One subclass is typified by OCRL-1 that, when mutated, causes Lowe syndrome or oculocerebrorenal dystrophy. Characteristics of the disease include cataracts, mental retardation and the development of renal failure. OCRL-1 contains a C-terminal Rho-GTPase-activating protein domain of unknown function, which could be involved in localization of the enzyme.

A second Type II subclass consists of the synaptojanins, which participate in synaptic vesicle trafficking. These proteins have a C-terminal polyproline region, presumably mediating interactions with specific SH3 domain-containing proteins. Synaptojanins also contain an N-terminal Sac homology domain, which possesses a separate lipid phosphatase activity (W.E. Hughes, 2000). The Sac domain shows selectivity for monophosphoinositides or phosphoinositides without adjacent phosphates (i.e. PI-3,5-P$_2$). Thus, synaptojanins have two phosphatase domains, which allows hydrolysis of PI-4,5-P$_2$ to PI.

A third Type II sub-class consists of SHIP1 and SHIP2, which show specificity for 3-phosphate-containing substrates (e.g. PI-3,4,5-P$_3$). These enzymes contain a SH2 domain at the N-terminus, which is responsible for their acronym (*SH*2-containing *inositol phosphatase*). SHIP1 is restricted to hematopoietic cells, while SHIP2 is more widely distributed. The SHIPs also can be phosphorylated on C-terminal region Tyr residues, which allows interaction with SH2 domain-containing proteins (e.g. Shc). SHIPs additionally contain one or more polyproline motifs in the C-terminal half of the protein, which mediates binding to SH3 domain-containing proteins. The SH2 domains and the other motifs mediate assembly of SHIPs into growth factor and cytokine receptor signaling complexes, where the enzymes can hydrolyze newly formed $PI-3,4,5-P_3$. This limits the duration of $PI-3,4,5-P_3$-mediated signaling.

The fourth Type II subclass of 5-phosphatases (not shown in Fig. 12), like the SHIPs, shows selectivity for $PI-3,4,5-P_3$. These enzymes lack SH2 domains, but contain numerous N-terminal polyproline motifs, as well as an immunoreceptor tyrosine-based activation motif, usually found only in receptor molecules (M.V. Kisseleva, 2000). Binding partners for these motifs, which presumably regulate localization and function of the enzyme, are not identified.

The 3-phosphatases are a novel family of proteins, which contain a protein tyrosine phosphatase active site motif [26]. PTEN/MMAC1/TEP1 (acronyms for *p*hosphatase and *ten*sin homolog, *m*utated in *m*ultiple *a*dvanced *c*ancers, and *T*GFβ-regulated and *e*pithelial cell-enriched *p*hosphatase) was the first member identified and it was originally thought to be a dual specificity protein tyrosine kinase involved in tumor suppression. Mutations in the PTEN gene cause tumorigenesis and contribute to specific syndromes with increased risk for breast and thyroid cancers. The known functions of PTEN are to switch off cell proliferation and to permit programmed cell death (apoptosis), which clearly are tumor suppressive activities. Maehama and Dixon (Maehama, 1998) first reported that PTEN had lipid phosphatase activity, hydrolyzing the D3 phosphate from $PI-3,4,5-P_3$ with high efficiency. Hydrolysis of $PI-3,4-5-P_3$ seems to be the primary function of the enzyme in vivo, based on studies using mutations that eliminate lipid, but not protein, phosphatase activity. Regulation of PTEN function is most likely by membrane translocation, mediated by a N-terminal $PI-4,5-P_2$ binding domain, a C2 domain in the C-terminal region, and a C-terminal PDZ-binding site. Both the $PI-4,5-P_2$ binding domain and the C2 domain mediate binding to membrane phospholipids. Interaction with PDZ proteins, which are scaffolding proteins, may aid in assembly of PTEN with signaling complexes.

A second group of enzymes, the myotubularins, also hydrolyze the D3 phosphate with preference for PI-3-P [26]. Mutations in two myotubularin family members, MTM1 and MTMR2, cause two human diseases: X-linked myotubular myopathy and Type 4B Charcot–Marie–Tooth syndrome, respectively. The mutations usually affect catalytic activity, suggesting the requirement for phosphatase activity in the normal functioning of the proteins. However, the physiological functions controlled by the myotubularins are not understood. These proteins contain several regulatory domains, including a PDZ-binding motif. Some family members also contain lipid-binding domains (i.e. PH or FYVE domains), which may help to localize the enzyme near its substrate.

The 4-phosphatases are also involved in regulation of 3-phosphoinositide signaling. These enzymes hydrolyze the D4 phosphate of $PI-3,4-P_2$ to yield PI-3-P [6]. This action will limit the duration of receptor-mediated increases in the levels of $PI-3,4-P_2$ (see

Section 5.2.1). The 4-phosphatases are $Mg^{2+}$-independent and may be down-regulated by receptor-triggered proteolysis. Little is known of their physiological functions.

## 5.2. Cellular targets

Phosphoinositides exert their effects by binding to specific domains in target proteins. Lipid binding may cause a conformational change in the target protein, thereby altering enzyme activity, or it may affect the localization of the target protein and access to enzymatic substrates. Multiple domains and motifs have been shown to bind phosphoinositides in vitro [25,27]. PH domains generally show specificity for PI-4,5-$P_2$, PI-3,4-$P_2$ or PI-3,4,5-$P_3$. PI-4,5-$P_2$ also binds to the ENTH (acronym for epsin N-terminal homology) domain and to certain Lys/Arg-rich motifs in proteins. Some SH2 and PTB domains bind PI-3,4,5-$P_3$ in vitro, but the relevance of this interaction in vivo is unclear. FYVE domains [named after the first four proteins shown to contain it: *Fab1p*, *Y*OTB, *Vac1p*, and *e*arly endosome antigen 1 (EEA1)] and certain PX domains selectively bind PI-3-P. Other PX domains show greater specificity for other lipids, as discussed in Section 5.2.2.

## 5.2.1. Domains that bind polyphosphoinositides

PH domains contain about 120 amino acids and were first identified as targets for PI-4,5-$P_2$. These domains are found in over 100 proteins, including protein kinases, phospholipases, and regulatory proteins for small GTPases (exchange factors and GTPase-activating proteins). PH domains are divided into four groups, based on their selectivity for binding various phosphoinositides [27]. Group I, typified by PH domains in Bruton's tyrosine kinase, exchange factors for the small GTPase Arf (ARNO, GRP1), and GTPase-activating proteins for Arfs (centaurin α) and Ras (GAP1), show high affinity binding for PI-3,4,5-$P_3$. Group II PH domains selectively bind PI-4,5-$P_2$ and are found in PLCδ, oxysterol-binding protein, β-adrenergic receptor kinase, and β-spectrin. Group III PH domains are found in the protein kinases Akt/protein kinase B (PKB) and phosphoinositide-dependent kinase 1 (PDK1) and bind PI-3,4-$P_2$ and PI-3,4,5-$P_3$ with equal affinities. Group IV PH domains bind phosphoinositides with relatively low affinity, typified by the PH domain in the endocytic GTPase dynamin.

The Group III PH domain-containing proteins are important cellular targets for PI-3,4,5-$P_3$ and PI-3,4-$P_2$, because of their central roles in the regulation of cell survival and proliferation. Akt/PKB is a 57 kDa Ser/Thr protein kinase identified as the cellular homolog of the viral oncoprotein v-Akt. It was also shown to have high homology to protein kinases A and C (hence the name PKB). Akt/PKB is activated by the binding of PI-3,4,5-$P_3$ or PI-3,4-$P_2$, which localizes the enzyme to the membrane and induces conformational changes allowing phosphorylation of a critical Thr residue by PDK1. PDK1 is a 63 kDa Ser/Thr kinase, which binds to membranes containing polyphosphoinositides, placing the enzyme in close proximity to Akt/PKB. After interaction with another protein kinase (PKC-related kinase 1), PDK1 is then able to phosphorylate a C-terminal Ser residue in Akt/PKB, rendering the enzyme fully active. Active Akt/PKB phosphorylates a variety of substrates involved in the regulation of apoptosis, cell metabolism, and gene expression. The overall effect of activated Akt/PKB is to promote cell survival, increase cell size, and enhance cell proliferation.

PI-4,5-$P_2$ has additional binding sites in proteins, i.e. stretches of Lys/Arg-rich sequences, found primarily in cytoskeletal proteins and in PTEN, and the ENTH domain, a region of about 140 residues found in several proteins involved with endocytosis [27]. Cytoskeletal protein targets include the ezrin/radixin/moesin family of proteins and vinculin, which help to cross-link actin to the plasma membrane, and proteins that regulate actin polymerization, such as gelsolin, cofilin, and profilin (M.J. Bottomley, 1998). ENTH domain-containing proteins include epsin and other proteins involved in assembly of the clathrin lattice. PI-4,5-$P_2$ also accumulates in the early phagosome, where it recruits PLC. Thus, PI-4,5-$P_2$ plays major roles in the regulation of the cytoskeleton, endocytosis, and phagocytosis (T. Nebl, 2000; A. Simonsen, 2001).

### 5.2.2. Domains that bind phosphatidylinositol-3-phosphate

A role for phosphoinositides in endosomal trafficking was first discovered in yeast, where PI-3-P was shown to be the major phosphoinositide involved. Recently, a specific binding domain for PI-3-P, the FYVE domain, was discovered (H. Stenmark, 1996). This domain consists of about 60–80 residues and coordinates two $Zn^{2+}$ atoms [27]. Binding of PI-3-P involves initial contact between a hydrophobic loop of the domain with the membrane, followed by interaction with the headgroup of PI-3-P. Use of the isolated domain coupled to green fluorescent protein demonstrated that PI-3-P is selectively enriched in early endosomes. Many FYVE domain-containing proteins target to endosomes, where they regulate endocytosis, endosomal fusion, and vacuolar transport. About 60 proteins are known to contain FYVE domains, including early endosomal antigen 1, which is essential for endosomal fusion, and PI-3-P-metabolizing enzymes (some of the myotubularin-related 3-phosphatases and the PI 5-kinase PIKfyve).

Another binding target for PI-3-P is the PX domain, first identified as a conserved motif of about 120 residues in the N-terminal regions of the NADPH oxidase proteins p47$^{phox}$ and p40$^{phox}$. It is present in over 70 human and yeast proteins, and PX domain-containing proteins are also found in plants, flies, and worms [27]. The proteins have diverse functions, although a large group includes the sorting nexins, which are involved in targeting of proteins to lysosomes. Recent studies established that several PX domains bind selectively to PI-3-P, including those in several sorting nexins, the yeast SNARE protein Vam7p, and the NADPH oxidase component p40$^{phox}$. However, PX domains from several other proteins (p47$^{phox}$, class II PI 3-kinase, cytokine-independent survival kinase) preferentially bind polyphosphoinositides (PI-4,5-$P_2$, PI-3,4-$P_2$, PI-3,4,5-$P_3$). The PX domain of p47$^{phox}$ also binds PA with higher affinity than the polyphosphoinositides (L.C. McPhail, unpublished results). The lipid-binding specificity for numerous other PX domains is not known. The diversity in lipid-binding and in the types of proteins containing PX domains suggest that PX domains have multiple functions in lipid-signaling pathways.

# 6. Lysophosphatidic acid and other lysophospholipids

For many years it was believed that lysophospholipids were simply intermediates in the metabolism of glycerolipids. However, it has now been clearly established that LPA, LPC and the sphingolysolipids sphingosylphosphorylcholine and sphingosine-1-phosphate act as extracellular signaling molecules [28]. These lipids bind to a family of related cell-surface heptahelical receptors and are implicated in tumorigenesis, angiogenesis, immunity, atherosclerosis, and neuronal survival. Because sphingolipids are discussed in a separate chapter, the focus here will be on LPA and lysophosphatidylcholine.

## 6.1. Sources of lysophospholipids

A number of cell types have been shown to produce lysophospholipids, including platelets, macrophages, other leukocytes, some epithelial cells, and some tumor cells [29]. The regulation of the synthesis and metabolism of extracellular lysophospholipids is not well understood. Phosphatidylcholine is likely to be the major source for both LPC and LPA and several pathways are possible. The action of a phospholipase A directly yields LPC. LPA can be formed by the sequential actions of PLD and a phospholipase A or of a phospholipase A and a lyso-PLD. The nature of the phospholipases $A_2$ or $A_1$ involved may vary, depending on cellular source and stimulus, and is not clearly delineated. Although lyso-PLD activity has been detected in various cell types, the cDNA encoding the enzyme has not been cloned. It is not known how lysophospholipids exit the cell, once they are synthesized. Albumin is probably the main LPA-binding protein in plasma, although plasma gelsolin also binds LPA with higher affinity than albumin. LPC is present in oxidized low density lipoproteins. Metabolism of lysophospholipids is by reacylation and/or, for LPA, dephosphorylation. The enzyme LPA acyltransferase can convert LPA to PA. Lipid phosphate phosphatase-1 is positioned in the plasma membrane of cells such that it can dephosphorylate extracellular LPA to form monoacylglycerol.

## 6.2. Lysophospholipids are extracellular signaling molecules

LPA induces a number of cellular responses, including proliferation, decreased apoptosis, platelet aggregation, smooth muscle contraction, chemotaxis, and tumor cell invasion. LPC has inflammatory effects in the body, including up-regulation of endothelial cell adhesion molecules and growth factors, and activation of macrophages. Both LPA and LPC may be elevated in certain pathophysiological conditions, such as ovarian cancer and artherosclerosis. LPA and LPC are ligands for members of a family of heptahelical receptors, formerly known as the EDG receptors and now termed the LP receptors [28]. Family members number 12 so far and can be divided into three distinct groups. The first group consists of three receptors (termed $LPA_{1-3}$) showing high selectivity for LPA. The second group (5 receptors, termed $S1P_{1-5}$) preferentially binds sphingosine-1-phosphate. The third group ($SPC_1$, $LPC_1$, PSY, GPR4) is more divergent and shows greater specificity for LPC and sphingosylphosphorylcholine. As a group, the LP receptors couple to several G proteins, including $G_i$, $G_q$, and $G_{12/13}$. Thus, the lysophospholipids exert their effects on cell functions by triggering a variety of G protein-linked second messenger pathways, including activation of PLC, PLD, phosphoinositide 3-kinases, PKC, MAP kinases, and small GTPases. $LPA_1$ receptor (also called EDG-2) null mice have a complex phenotype, with 50% neonatal lethality and neuronal and other abnormalities (J.J. Contos, 2000). This suggests a critical role for $LPA_1$ in multiple aspects of development. $LPC_1$ null mice show T cell abnormalities and develop a late-onset autoimmune disease, similar to systemic lupus erythematosus (L.Q. Li, 2001). Creation of null mice deficient in other receptors in the LP family is needed to help elucidate the spectrum of biological functions of these lysophospholipids.

## 7. Future directions

Much remains to be learned about glycerolipid signaling. It is becoming clear that the signaling intermediates and protein targets involved in the regulation of specific cell functions assemble and disassemble in dynamic complexes at specific intracellular locations. Movement of proteins in and out of these complexes is likely orchestrated by the transient presence of specific signaling lipids, along with the cytoskeleton. It will be a challenge to design tools to visualize these processes at high resolution within living cells without disrupting the signaling pathways or the functions being regulated. The engineering of fluorescently tagged protein domains that recognize specific lipids is providing some insight [20]. With the dawn of the genomic age, it should soon be possible to identify all of the enzymes involved in the synthesis and metabolism of the signaling lipids, based on homologies with known proteins. Similarly, additional targets for the lipids will be found. However, discerning the biological functions of the numerous enzymes and target proteins involved in lipid signaling will continue to require extensive research using biochemical, cell biological, and molecular tools. Such a global and integrated approach has been established for understanding G protein-dependent cell signaling (The Alliance for Cellular Signaling, http://cellularsignaling.org/), which, of course, encompasses glycerolipid signaling.

## Abbreviations

| | |
|---|---|
| DG | Diacylglycerol |
| Ins-1,4,5-$P_3$ | Inositol-1,4,5-trisphosphate |
| $InsP_6$ | Inositol hexakisphosphate |
| LPA | Lysophosphatidic acid |
| LPC | Lysophosphatidylcholine |
| PA | Phosphatidic acid |
| PC | Phosphatidylcholine |
| PH | Pleckstrin homology |
| PKB | Protein kinase B |
| PKC | protein kinase C |

| PLC | Phospholipase C |
| PLD | Phospholipase D |
| PI | Phosphatidylinositol |
| PIP | Phosphatidylinositol monophosphate |
| PI-4,5-P$_2$ | Phosphatidylinositol-4,5-bisphosphate |
| PX | Phox homology |
| SH2 | Src homology 2 |
| SH3 | Src homology 3 |

## References

1. Hokin, L.E. (1985) Receptors and phosphoinositide-generated second messengers. Annu. Rev. Biochem. 54, 205–235.
2. Irvine, R.F. and Schell, M.J. (2001) Back in the water: the return of the inositol phosphates. Nat. Rev. Mol. Cell Biol. 2, 327–338.
3. Rhee, S.G. (2001) Regulation of phosphoinositide-specific phospholipase C. Annu. Rev. Biochem. 70, 281–312.
4. Lewit-Bentley, A. and Rety, S. (2000) EF-hand calcium-binding proteins. Curr. Opin. Cell Biol. 10, 637–643.
5. Hurley, J.H. and Misra, S. (2000) Signaling and subcellular targeting by membrane-binding domains. Annu. Rev. Biophys. Biomol. Struct. 29, 49–79.
6. Majerus, P.W., Kisseleva, M.V. and Norris, F.A. (1999) The role of phosphatases in inositol signaling reactions. J. Biol. Chem. 274, 10669–10672.
7. Berridge, M.J., Lipp, P. and Bootman, M.D. (2000) The versatility and universality of calcium signalling. Nat. Rev. Mol. Cell Biol. 1, 11–21.
8. Ramoni, C., Spadaro, F., Menegon, M. and Podo, F. (2001) Cellular localization and functional role of phosphatidylcholine-specific phospholipase C in NK cells. J. Immunol. 167, 2642–2650.
9. Dougherty, R.W., Dubay, G.R. and Niedel, J.E. (1989) Dynamics of the diradylglycerol responses of stimulated phagocytes. J. Biol. Chem. 264, 11263–11269.
10. Waggoner, D.W., Xu, J., Singh, I., Jasinska, R., Zhang, Q.X. and Bradley, D.N. (1999) Structural organization of mammalian lipid phosphate phosphatases: implications for signal transduction. Biochim. Biophys. Acta 1439, 299–316.
11. Wakelam, M.J.O. (1998) Diacylglycerol — when is it an intracellular messenger? Biochim. Biophys. Acta 1436, 117–126.
12. Van Blitterswijk, W.J. and Houssa, B. (2000) Properties and functions of diacylglycerol kinases. Cell Signal. 12, 595–605.
13. Takai, Y., Kishimoto, A., Iwasa, Y., Kawahara, Y., Mori, T. and Nishizuka, Y. (1979) Calcium-dependent activation of a multifunctional protein kinase by membrane phospholipids. J. Biol. Chem. 254, 3692–3695.
14. Takai, Y., Kishimoto, A., Mori, T. and Nishizuka, Y. (1979) Unsaturated diacylglycerol as a possible messenger for the activation of calcium-activated, phospholipid-dependent protein kinase system. Biochem. Biophys. Res. Commun. 91, 1218–1224.
15. Ron, D. and Kazanietz, M.G. (1999) New insights into the regulation of protein kinase C and novel ester receptors. FASEB J. 13, 1658–1676.
16. Newton, A.C. and Johnson, J.E. (1998) Protein kinase C: a paradigm for regulation of protein function by two membrane-targeting modules. Biochim. Biophys. Acta 1376, 155–172.
17. Hayashi, A., Seki, N., Hattori, A., Kozuma, S. and Saito, T. (1999) PKCν, a new member of the protein kinase C family, composes a fourth subfamily with PKCμ. Biochim. Biophys. Acta 1450, 99–106.
18. Liscovitch, M., Czarny, M., Fiucci, G. and Tang, X. (2000) Phospholipase D: molecular and cell biology of a novel gene family. Biochem. J. 345, 401–415.

19. Exton, J.H. (1999) Regulation of phospholipase D. Biochim. Biophys. Acta 1439, 121–133.
20. Sato, T.K., Overduin, M. and Emr, S.D. (2001) Location, location, location: membrane targeting directed by PX domains. Science 294, 1881–1885.
21. McPhail, L.C., Waite, K.A., Regier, D.S., Nixon, J.B., Qualliotine-Mann, D., Zhang, W.X., Wallin, R. and Sergeant, S. (1999) A novel protein kinase target for the lipid second messenger phosphatidic acid. Biochim. Biophys. Acta 1439, 277–290.
22. Rameh, L.E. and Cantley, L.C. (1999) The role of phosphoinositide 3-kinase lipid products in cell function. J. Biol. Chem. 274, 8347–8350.
23. Fruman, D.A., Meyers, R.E. and Cantley, L.C. (1998) Phosphoinositide kinases. Annu. Rev. Biochem. 67, 481–507.
24. Anderson, R.A., Boronenkov, I.V., Doughman, S.D., Kunz, J. and Loijens, J.C. (1999) Phosphatidylinositol phosphate kinases, a multifaceted family of signaling enzymes. J. Biol. Chem. 274, 9907–9910.
25. Vanhaesebroeck, B., Leevers, S.J., Ahmadi, K., Timms, J., Katso, R., Driscoll, P.C., Woscholski, R., Parker, P.J. and Waterfield, M.D. (2001) Synthesis and function of 3-phosphorylated inositol lipids. Annu. Rev. Biochem. 70, 535–602.
26. Maehama, T., Taylor, G.S. and Dixon, J.E. (2001) PTEN and myotubularin: novel phosphoinositide phosphatases. Annu. Rev. Biochem. 70, 247–279.
27. Xu, Y., Seet, L.F., Hanson, B. and Hong, W. (2001) The Phox homology (PX) domain, a new player in phosphoinositide signalling. Biochem. J. 360, 513–530.
28. Hla, T., Lee, M.J., Ancellin, N., Paik, J.H. and Kluk, M.J. (2001) Lysophospholipids — receptor revelations. Science 294, 1875–1878.
29. Goetzl, E.J. (2001) Pleiotypic mechanisms of cellular responses to biologically active lysophospholipids. Prostaglandins Other Lipid Mediat. 64, 11–20.

D.E. Vance and J.E. Vance (Eds.) *Biochemistry of Lipids, Lipoproteins and Membranes (4th Edn.)*

# CHAPTER 13

# The eicosanoids: cyclooxygenase, lipoxygenase, and epoxygenase pathways

## William L. Smith[1] and Robert C. Murphy[2]

*[1] Department of Biochemistry and Molecular Biology, Michigan State University, East Lansing, MI 48824-1319, USA, Tel.: +1 (517) 355-1604; Fax: +1 (517) 353-9334; E-mail: smithww@msu.edu*
*[2] Department of Pediatrics, Division of Cell Biology, National Jewish Medical and Research Center, 1400 Jackson Street, Room K929, Denver, CO 80206-2762, USA, Tel.: +1 (303) 398-1849; Fax: +1 (303) 398-1694; E-mail: murphyr@njc.org*

## 1. Introduction

### 1.1. Terminology, structures, and nomenclature

The term 'eicosanoids' is used to denote a group of oxygenated, twenty carbon fatty acids (Fig. 1) [1]. The major precursor of these compounds is arachidonic acid (all *cis* 5,8,11,14-eicosatetraenoic acid), and the pathways leading to the eicosanoids are known collectively as the 'arachidonate cascade'. There are three major pathways within the cascade, including the cyclooxygenase, lipoxygenase, and epoxygenase pathways. In each case, these pathways are named after the enzyme(s) that catalyzes the first committed step. The prostanoids, which include the prostaglandins and thromboxanes, are formed via the cyclooxygenase pathway. The first part of our discussion will focus on the prostanoids. Later in this chapter, we will describe the lipoxygenase and epoxygenase pathways.

The structures and biosynthetic interrelationships of the most important prostanoids are shown in Fig. 2 [1]. PG is the abbreviation for prostaglandin, and TX is the abbreviation for thromboxane. Naturally occurring prostaglandins contain a cyclopentane ring, a *trans* double bond between C-13 and C-14, and an hydroxyl group at C-15. The letters following the abbreviation PG indicate the nature and location of the oxygen-containing substituents present in the cyclopentane ring. Letters are also used to label thromboxane derivatives (e.g., TXA and TXB). The numerical subscripts indicate the number of carbon–carbon double bonds in the side chains emanating from the cyclopentane ring (e.g., $PGE_1$ vs. $PGE_2$). In general, those prostanoids with the '2' subscript are derived from arachidonate; the '1' series prostanoids are formed from 8,11,14-eicosatetraenoate, and the '3' series compounds are derived from 5,8,11,14,17-eicosapentaenoate. Greek subscripts are used to denote the orientation of ring hydroxyl groups (e.g., $PGF_\alpha$).

Prostanoids formed by the action of cyclooxygenases have their aliphatic side chains emanating from C-8 and C-12 of the cyclopentane ring in the orientations shown in Fig. 2. Prostanoids known as isoprostanes have their aliphatic groups in various other orientations [2]. Isoprostanes are formed from arachidonic acid by nonenzymatic autooxidation, and, somewhat surprisingly, isoprostanes and their metabolites are found in greater quantities in urine than metabolites of prostanoids formed enzymatically via

342

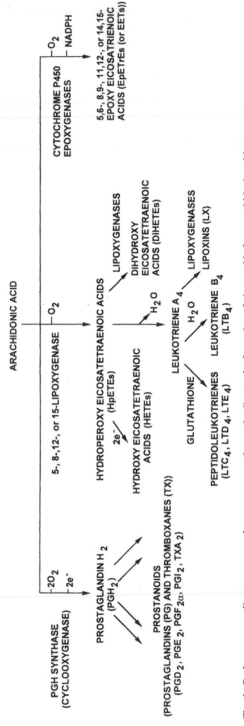

Fig. 1. Cyclooxygenase, lipoxygenase, and epoxygenase pathways leading to the formation of eicosanoids from arachidonic acid.

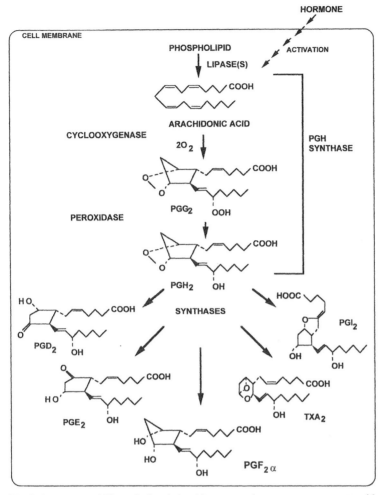

Fig. 2. Structures and biosynthetic relationships among the most common prostanoids.

the cyclooxygenase. Particularly in pathological conditions which support autooxidation (e.g., $CCl_4$ toxicity), isoprostanes are produced in abundance [2].

## 1.2. Prostanoid chemistry

Prostaglandins are soluble in lipid solvents below pH 3.0 and are typically extracted from acidified aqueous solutions with ether, chloroform/methanol, or ethyl acetate. PGE, PGF, and PGD derivatives are relatively stable in aqueous solution at pH 4–9; above pH 10, both PGE and PGD are subject to dehydration. $PGI_2$, which is also known as prostacyclin, contains a vinyl ether group that is very sensitive to acid-catalyzed hydrolysis; $PGI_2$ is unstable below pH 8.0. The stable hydrolysis product of PGI$_2$ is 6-keto-PGF$_{1\alpha}$. PGI$_2$ formation is usually monitored by measuring 6-keto-PGF$_{1\alpha}$ formation. TXA$_2$, which contains an oxane–oxetane grouping in place of the cyclopentane ring, is hydrolyzed rapidly ($t_{1/2} = 30$ s at 37°C in neutral aqueous solution) to TXB$_2$; TXA$_2$ formation is assayed by quantifying TXB$_2$. Prostaglandin derivatives are commonly quantified with immunoassays or by mass spectrometry using deuterium-labeled internal standards.

## 2. Prostanoid biosynthesis

Eicosanoids are not stored by cells, but rather are synthesized and released rapidly (5–60 s) in response to extracellular hormonal stimuli. The pathway for stimulus-induced prostanoid formation as it might occur in a model cell is illustrated in Fig. 2 [1]. Prostanoid formation occurs in three stages: (a) mobilization of free arachidonic acid (or 2-arachidonyl-glycerol (2-AG); see below) from membrane phospholipids; (b) conversion of arachidonate (or 2-AG) to the prostaglandin endoperoxide PGH$_2$ (or 2-PGH$_2$-glycerol); and (c) cell-specific conversion of PGH$_2$ (or 2-PGH$_2$-glycerol) to one of the major prostanoids.

### 2.1. Mobilization of arachidonate

Prostaglandin synthesis is initiated by the interaction of various hormones (e.g., bradykinin, angiotensin II, thrombin) with their cognate cell surface receptors (Figs. 2 and 3) which, in turn, causes the activation of one or more cellular lipases. Although in principle, there are a variety of lipases and phospholipases that could participate in this arachidonate mobilizing phase, the high molecular weight cytosolic phospholipase A$_2$ (PLA$_2$) and certain of the nonpancreatic, secretory PLA$_2$s appear to be the relevant lipases (Chapter 11). The current consensus regarding the roles of PLA$_2$s in prostanoid synthesis is that typically cytosolic PLA$_2$ is involved directly in mobilizing arachidonic acid for the constitutive, prostaglandin endoperoxide H synthase-1 (PGHS-1) whereas cytosolic PLA$_2$ is indirectly, and secretory PLA$_2$ is directly, involved in mobilizing arachidonate for the inducible PGHS-2 (Fig. 3); [3–5]. This is discussed in more detail below in describing the functions of the PGHS isoforms.

### 2.2. Cytosolic and secreted phospholipase A$_2$s

Cytosolic PLA$_2$ is found in the cytosol of resting cells, but as illustrated in Fig. 3, hormone-induced mobilization of intracellular Ca$^{2+}$ leads to the translocation of cytosolic PLA$_2$ to the ER and nuclear envelope. There, cytosolic PLA$_2$ cleaves arachidonate from the $sn2$ position of phospholipids on the cytosolic surface of the membranes. The arachidonate then traverses the membrane where it acts as a substrate for PGHSs which are located on the luminal surfaces of the ER and the associated inner and outer membranes of the nuclear envelope [1]. The activity of cytosolic PLA$_2$ is also augmented by phosphorylation by a variety of kinases [3]. The translocation of cytosolic PLA$_2$ involves the binding of Ca$^{2+}$ to an N-terminal CalB domain and then the binding of the

## A. cPLA₂

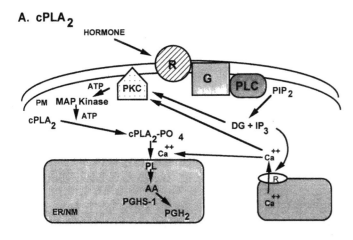

## B. sPLA₂

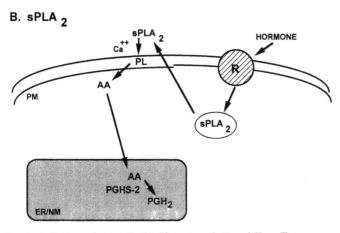

Fig. 3. Activation of cytosolic (c) PLA₂ (panel A) and Type II nonpancreatic secretory (s) PLA₂ (panel B) involved in mobilizing arachidonic acid from phospholipid. Abbreviations include: PL, phospholipid; ER/NM, endoplasmic reticulum/nuclear membrane; PM, plasma membrane; AA, arachidonic acid; PGHS, prostaglandin endoperoxide H synthase; R, receptor; G, G protein, PLC, phospholipase C; PIP₂, phosphatidylinositol-4,5-*bis*-phosphate; PKC, protein kinase C. The activation of secretory PLA₂ requires participation of cytosolic PLA₂ but the mechanism has not yet been determined.

$Ca^{2+}$/CalB domain to intracellular membranes [6]; $Ca^{2+}$ is involved in translocation of the enzyme but not in the catalytic mechanism of cytosolic PLA₂ (see Dessen et al., 1999).

Stimulus-dependent arachidonate mobilization by secretory PLA₂s depends on the ability of these enzymes to be released from cells and to rebind to the cell surface through heparin sulfate containing proteoglycans (Fig. 3). Once bound, these enzymes appear to act on phosphatidylcholine present on the extracellular face of the plasma membrane [7]. Secretory PLA₂ is shown in Fig. 3 in a vesicle that moves to the surface of the cell in response to an agonist, is secreted into the medium and then binds to a proteoglycan on the cell surface. Presumably, free arachidonate released through the actions of secretory $PLA_2$ can enter the cell and make its way to the ER, where it is acted upon by PGHSs. There are many different nonpancreatic secretory $PLA_2$s but to date, only secretory $PLA_2$s IIA and V have been shown to be involved in releasing arachidonate for prostanoid synthesis [7].

The crystal structure of secretory $PLA_2$ IIA from human synovial fluid has been determined. Secretory $PLA_2$s require high concentrations of $Ca^{2+}$ ($\sim 1$ mM) such as those found extracellularly for maximal activity. $Ca^{2+}$ is involved in phospholipid substrate binding and catalysis by secretory $PLA_2$s. Unlike cytosolic $PLA_2$, secretory $PLA_2$ shows no specificity toward either the phospholipid head group or the acyl group at the $sn2$ position. The levels of secretory $PLA_2$s are regulated transcriptionally in response to cell activation.

Cytosolic $PLA_2$ is directly involved in the immediate arachidonate release that occurs when a cell is challenged with a circulating hormone or protease. For example, thrombin acting through its cell surface receptor activates cytosolic $PLA_2$ in platelet cells to cause arachidonate release that results in $TXA_2$ formation. This entire process occurs in seconds. secretory $PLA_2$, on the other hand, plays a prominent role in 'late-phase' prostaglandin formation which occurs 2–3 h after cells have been exposed to a mediator of inflammation (e.g., endotoxin or interleukin-1) or a growth factor such as platelet-derived growth factor.

### 2.3. Mobilization of 2-arachidonyl-glycerol (2-AG)

In 2000, Kozak et al. made the insightful discovery that 2-AG is an efficient substrate for PGHS-2 but not PGHS-1 [8]. PGHS-2 converts 2-AG to 2-$PGH_2$-glycerol, and this intermediate is converted to the 2-prostanyl-glycerol derivatives but not 2-thromboxane-glycerol. At the time of this writing, it is not clear under what conditions or to what degree alternative substrates such as 2-AG are used in vivo to form products like the 2-prostanyl-glycerol derivatives. However, one can imagine that these products have a unique set of biological roles that distinguish them from the classical prostaglandins derived from arachidonic acid itself. Although the pathway leading to the formation of 2-AG itself has not been defined in the context of prostanoid metabolism, 2-AG could be formed from phosphatidylcholine through the sequential actions of phospholipase C and acylglycerol lipase [8].

### 2.4. Prostaglandin endoperoxide $H_2$ ($PGH_2$) formation

Once arachidonate is released, it can be acted upon by PGHS [9]. There are two PGHS isozymes, PGHS-1 and PGHS-2. PGHS-1 appears to use fatty acids such as arachidonate exclusively as substrates. In contrast, PGHS-2 utilizes both fatty acids and 2-AG about equally well [8]. The PGHSs exhibit two different but complementary enzymatic activities (Fig. 2): (a) a cyclooxygenase (*bis*-oxygenase) which catalyzes the formation of $PGG_2$ (or 2-$PGG_2$-glycerol) from arachidonate (or 2-AG) and two molecules of $O_2$; and (b) a peroxidase which facilitates the two-electron reduction of the 15-hydroperoxyl group of $PGG_2$ (or 2-$PGG_2$-glycerol) to $PGH_2$ (or 2-$PGH_2$-glycerol)

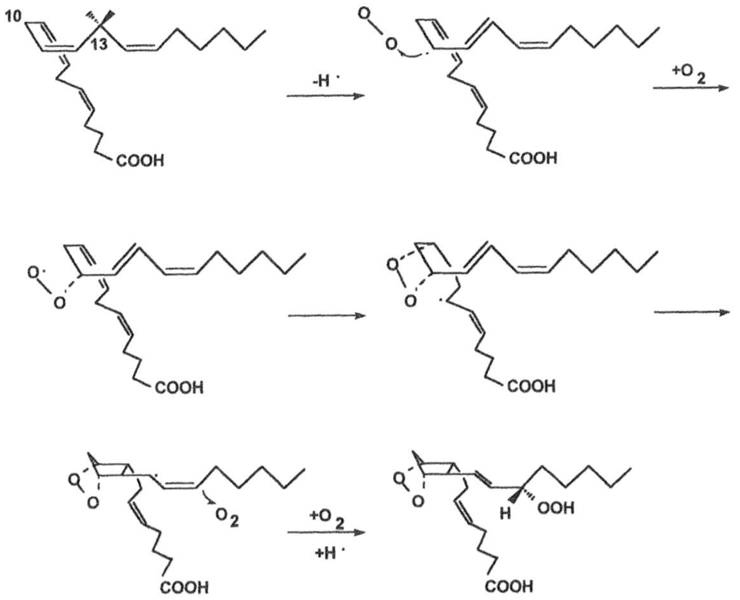

Fig. 4. Mechanism for the cyclooxygenase reaction showing the conversion of arachidonic acid and two molecules of oxygen to PGG$_2$.

(Fig. 2). The oxygenase and peroxidase activities occur at distinct but interactive sites within the protein.

The initial step in the cyclooxygenase reaction is the stereospecific removal of the 13-pro-S hydrogen from arachidonate. As depicted in Fig. 4 [9], an arachidonate molecule becomes oriented in the cyclooxygenase active site with a kink in the carbon chain at C-9. Abstraction of the 13-pro-S hydrogen and subsequent isomerization leads to a carbon-centered radical at C-11 and attack of molecular oxygen at C-11 from the side opposite that of hydrogen abstraction. The resulting 11-hydroperoxyl radical adds to the double bond at C-9, leading to intramolecular rearrangement and formation of another carbon-centered radical at C-15. This radical then reacts with another molecule of oxygen. The 15-hydroperoxyl group of PGG$_2$ can undergo a two-electron reduction to an alcohol yielding PGH$_2$ in a reaction catalyzed by the peroxidase activity of PGHSs.

## 2.5. PGHS active site

Depicted in Fig. 5 is a model of the cyclooxygenase and peroxidase active sites of ovine PGHS-1 [1]. The cyclooxygenase is an unusual activity that exhibits a requirement for hydroperoxide and undergoes a suicide inactivation [1]. The reason for the hydroperoxide activating requirement is that in order for the cyclooxygenase to function, a hydroperoxide must oxidize the heme prosthetic group located at the peroxidase active site to an oxo-ferryl heme radical cation. This oxidized heme intermediate abstracts an electron from Tyr385. Finally, the resulting Tyr385 tyrosyl

348

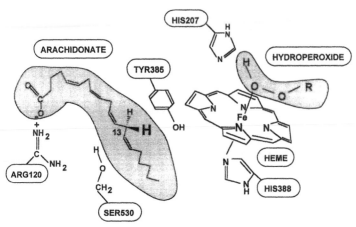

Fig. 5. Model of the cyclooxygenase and peroxidase active sites of the ovine PGHS-1.

radical abstracts the 13-pro-$S$ hydrogen from arachidonate, initiating the cyclooxygenase reaction. Once the cyclooxygenase reaction begins, newly formed $PGG_2$ can serve as the source of the activating hydroperoxide; prior to $PGG_2$ formation, ambient cellular hydroperoxides apparently serve to initiate heme oxidation and cyclooxygenase catalysis. Ser530, the site of acetylation of PGHS-1 by aspirin (Section 2.7), is shown within the cyclooxygenase active site in Fig. 5. Also shown is Arg120. The guanidino group of this residue serves as the counterion for the carboxylate group of arachidonate.

## 2.6. Physico-chemical properties of PGHSs

PGHS-1 was first purified from ovine vesicular gland, and most biochemical studies have been performed using this protein. PGHS-1 is associated with the luminal surfaces of the ER and the inner and outer membranes of the nuclear envelope. Detergent-solubilized ovine PGHS-1 is a dimer with a subunit molecular mass of 72 kDa. The reason for the existence of the dimer is unknown, but separation of the enzyme into monomers eliminates enzyme activity. The protein is $N$-glycosylated and is a hemoprotein containing one protoporphyrin IX per monomer [1]. The sequences of cDNA clones for PGHS-1 from many mammals indicate that initially the protein has a signal peptide of 24–26 amino acids, that is cleaved to yield a mature protein of 574 amino acids.

PGHS-2 was discovered in 1991 as an immediate early gene product in phorbol ester-activated murine 3T3 cells and in v-src-transformed chicken fibroblasts [1,10]. PGHS-1 and PGHS-2 from the same species have amino acid sequences that are 60% identical. The major sequence differences are in the signal peptides and the membrane binding domains (residues 70–120 of PGHS-1); in addition, PGHS-2 contains a unique 18 amino acid insert near its carboxyl terminus. The role of this 18 amino acid cassette in PGHS-2 is unknown.

The crystal structures of PGHS-1 and PGHS-2 have been determined. As noted earlier, two subunits of the enzymes form homodimers. Each monomer contains three

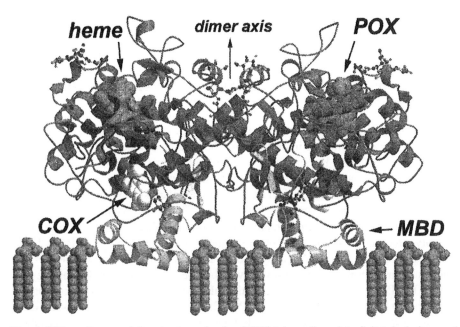

Fig. 6. Ribbon diagram of the structure of ovine PGHS-1 homodimer interdigitated via its membrane binding domain (MBD) into the luminal surface of the endoplasmic reticulum. The arrow denotes the dimer interface. Abbreviations include: POX, peroxidase; COX, cyclooxygenase.

sequential folding domains: an N-terminal epidermal growth factor-like domain of about 50 amino acids, an adjoining region containing about 70 amino acids that serves as the membrane binding domain, and a C-terminal globular catalytic domain (Fig. 6).

PGHS-1 and PGHS-2 are integral membrane proteins. However, their interactions with membranes do not involve typical transmembrane helices. Instead, analysis of the crystal structures and membrane domain labeling studies have established that PGHSs interact *monotopically* with only one surface of the membrane bilayer [1]. The interaction involves four short amphipathic α-helices present in the membrane binding domain noted above. The side chains of hydrophobic residues located on one surface of these helices interdigitate into and anchor PGHSs to the luminal surface of the ER and the inner and outer membranes of the nuclear envelope (Fig. 6).

## 2.7. PGHSs and nonsteroidal anti-inflammatory drugs

Prostaglandin synthesis can be inhibited by both nonsteroidal anti-inflammatory drugs (NSAIDs) and anti-inflammatory steroids. Both PGHS isozymes are pharmacological targets of common NSAIDs (e.g., aspirin, ibuprofen, naproxen). However, only prostaglandin synthesis mediated by PGHS-2 is inhibited by anti-inflammatory steroids, which block the synthesis of PGHS-2, at least in part, at the level of transcription [1]. Additionally, PGHS-2 is inhibited by 'COX-2 inhibitors' including rofecoxib and celecoxib. These drugs belong to a special class of NSAIDs specific for this isoform and are often referred to as COX-2 drugs. The selectivity of these latter drugs depends on subtle structural differences between the cyclooxygenase active sites of PGHS-1 and PGHS-2 [1]. These same differences account, at least in part, for the ability of PGHS-2 but not PGHS-1 to bind and oxygenate 2-arachidonyl-glycerol.

The best known NSAID is aspirin, acetylsalicylic acid. Aspirin binds to the cyclooxygenase active site, and, once bound, can acetylate Ser530 (Fig. 5) [1]. Acetylation of this active site serine causes irreversible cyclooxygenase inactivation. Curiously, the hydroxyl group of Ser530 is not essential for catalysis, but acetylated Ser530 protrudes into the cyclooxygenase site and interferes with the binding of arachidonic acid [1].

Acetylation of PGHSs by aspirin has important pharmacological consequences. Besides the analgesic, anti-pyretic, and anti-inflammatory actions of aspirin, low-dose aspirin treatment — one 'baby' aspirin daily or one regular aspirin every three days — is a useful anti-platelet cardiovascular therapy [11]. This low-dosage regimen leads to selective inhibition of platelet thromboxane formation (and platelet aggregation) without appreciably affecting the synthesis of other prostanoids in other cells. Circulating blood platelets lack nuclei and are unable to synthesize new protein. Exposure of the PGHS-1 of platelets to circulating aspirin causes irreversible inactivation of the platelet enzyme. Of course, PGHS-1 (and PGHS-2) inactivation also occurs in other cell types, but cell types other than platelets can resynthesize PGHSs relatively quickly. For new PGHS-1 activity to appear in platelets, new platelets must be formed. Because the replacement time for platelets is five to ten days, it takes time for the circulating platelet pool to regain its original complement of active PGHS-1.

There are many NSAIDs other than aspirin [1,12]. In fact, this is one of the largest niches in the pharmaceutical market, currently accounting for about five billion dollars in annual sales. Like aspirin, other NSAIDs act by inhibiting the cyclooxygenase activity of PGHS [1,11,12]. However, unlike aspirin, most of these drugs cause reversible enzyme inhibition simply by competing with arachidonate for binding. A well-known example of a reversible nonsteroidal anti-inflammatory drug is ibuprofen. All currently available NSAIDs inhibit both PGHS-1 and PGHS-2 [1,12]. However, inhibition of PGHS-2 appears to be primarily responsible for both the anti-inflammatory and analgesic actions of NSAIDs [11]. Dual inhibition of PGHS-1 and PGHS-2 with common NSAIDs causes unwanted ulcerogenic side-effects [1,5,12]. Indeed, the newly developed COX-2 drugs rofecoxib and celecoxib exhibit the anti-inflammatory and analgesic actions of classical NSAIDs but have lesser side-effects, particularly gastro-intestinal side-effects [11].

Recent attention has been focussed on COX-2 inhibitors as prophylactic agents in the prevention of colon cancer. About 85% of tumors of the colon express elevated levels of PGHS-2 and classical NSAIDs and COX-2 inhibitors reduce mortality due to colon cancer (Takuku et al., 1998; Dubois, 2001).

## 2.8. Regulation of PGHS-1 and PGHS-2 gene expression

PGHS-1 and PGHS-2 are encoded by separate genes [1]. Apart from the first two exons, the intron/exon arrangements are similar. However, the PGHS-2 gene (~8 kb) is considerably smaller than the PGHS-1 gene (~22 kb). The PGHS-1 gene is on human chromosome 9, while the PGHS-2 gene is located on human chromosome 1.

The expressions of the PGHS-1 and PGHS-2 genes are regulated in quite different ways. PGHS-1 is expressed more or less constitutively in almost all tissues, whereas PGHS-2 is absent from cells unless induced in response to cytokines, tumor promoters, or growth factors [1]. Cells use PGHS-1 to produce prostaglandins needed to regulate 'housekeeping activities' typically involving rapid responses to circulating hormones (Fig. 2). PGHS-2 produces prostanoids which function during specific stages of cell differentiation or replication; it is not yet clear what these latter functions are nor whether they involve products formed from arachidonate or 2-AG or both. There is some indirect evidence suggesting that at least some of the products formed via PGHS-2 operate at the level of the nucleus through peroxisomal proliferator-activated receptors to modulate transcription of specific genes (Chapters 10 and 16) [1,5].

Relatively little is known concerning the regulation of expression of PGHS-1, although the enzyme must be under developmental control. The regulation of expression of PGHS-2 continues to be an area of intensive investigation. Much of what is known about PGHS-2 comes from studies with cultured fibroblasts, endothelial cells and macrophages [1]. Typically, PGHS-2 is induced rapidly (1–3 h) and dramatically (20- to 80-fold). Platelet-derived growth factor, phorbol ester, and interleukin-1β induce PGHS-2 expression in fibroblasts and endothelial cells. Bacterial lipolysaccharide, interleukin-1β, and tumor necrosis factor a stimulate PGHS-2 in monocytes and macrophages. While only a limited number of tissues and cell types have been examined, it is likely that PGHS-2 can be induced in almost any cell or tissue with the appropriate stimuli. Importantly, as noted earlier, PGHS-2 expression, but not PGHS-1 expression, can be completely inhibited by anti-inflammatory glucocorticoids such as dexamethasone [1,12].

The promoters of the two PGHS genes are indicative of their mode of regulation. PGHS-1 has a TATA-less promoter, a feature common to housekeeping genes. Reporter plasmids constructed with the 5'-upstream region of the PGHS-1 gene have failed to show any significant inducible transcription from this promoter, supporting the concept that regulation of PGHS-1 occurs only developmentally. The PGHS-2 promoter, on the other hand, contains a TATA box, and experiments with reporter plasmids containing the PGHS-2 promoter and upstream 5'-flanking sequence have demonstrated that PGHS-2 is highly regulatable. Transcriptional activation of the PGHS-2 gene appears to be one important mechanism for increasing PGHS-2 expression. Transcription of PGHS-2 can be controlled by multiple signaling pathways including the cAMP pathway, the protein kinase C pathway (phorbol esters), viral transformation (*src*), and other pleiotropic pathways such as those activated by growth factors, endotoxin, and inflammatory cytokines. These latter agents (e.g., platelet-derived growth factor, lipopolysaccharide, interleukin-1β, tumor necrosis factor α) likely share convergent pathways involving nuclear factor κB (NFκB) and the CAAT enhancer binding protein (C/EBP), two transcription factors common to inflammatory responses, and one or more of the established mitogen activated protein kinase cascades: ERK1/2, JNK/SAPK, and p38/RK/Mpk2.

The nucleotide sequence of the PGHS-2 gene promoter is known but identification of *cis*-elements responsible for the regulation of this gene is incomplete. Regulatory elements in the 5'-flanking regions of the PGHS-2 gene that are known to regulate transcription are the overlapping E-box and CRE-1 sequence most proximal to the TATA box, a C/EBP-2 sequence, a NFκB binding site and a downstream CRE-2. In fibroblasts and endothelial cells, the most critical of the regulatory sequences is the CRE-1. This type of response element typically is activated by hetero- and homo-dimers of the c-Fos, c-Jun and ATF families of bZIP proteins (AP-1) and the cAMP regulatory binding protein (CREB). Activation of c-Jun is required for transcriptional activation of the PGHS-2 promoter in response to growth factors. It is not known if c-Jun forms homodimers, or interacts with other bZIP proteins.

In macrophages bacterial endotoxin stimulates PGHS-2 expression through cooperative activation via the CRE-1, C/EBP-2, NFκB and CRE-2 regulatory elements (Wingerd, 2002). Tumor necrosis factor α-stimulated expression is dependent on a NFκB site and the C/EBP-1 site. The C/EBPβ and C/EBPδ transcription factors are commonly involved in the regulation of inflammatory responses, and C/EBP regulatory elements are frequently found in promoters of so called acute phase genes. This family of transcription factors is activated by most of the inflammatory stimuli which induce PGHS-2 expression. C/EBPs bind to PGHS-2 promoters and function in conjunction with USF-1, NFκB, and c-Jun cis-regulatory proteins to activate transcription.

## 2.9. PGH₂ metabolism

Although all the major prostanoids are depicted in Fig. 2 as being formed by a single cell, prostanoid synthesis appears to be cell-specific [1]. For example, platelets form mainly $TXA_2$, endothelial cells form $PGI_2$ as their major prostanoid, and $PGE_2$ is the major prostanoid produced by renal collecting tubule cells. The syntheses of $PGE_2$, $PGD_2$, $PGF_{2\alpha}$, $PGI_2$, and $TXA_2$ from $PGH_2$ are catalyzed by PGE synthase, PGD synthase, $PGF_\alpha$ synthase, PGI synthase, and TXA synthase, respectively [1]. Formation of $PGF_{2\alpha}$ involves a two-electron reduction of $PGH_2$, and a $PGF_\alpha$ synthase utilizing NADPH can catalyze this reaction. All other prostanoids are formed via isomerization reactions involving no net change in oxidation state from $PGH_2$.

PGI synthase and TXA synthase are hemoproteins with molecular weights of 50–55,000. Both of these proteins are cytochrome P-450s. Both enzymes, like PGHSs, undergo suicide inactivation during catalysis. TXA synthase is found in abundance in platelets and lung. PGI synthase is localized to endothelial cells, as well as both vascular and nonvascular smooth muscle [13]. Both TXA and PGI synthases are found on the cytoplasmic face of the ER. $PGH_2$ formed in the lumen of the ER via PGHSs diffuses across the membrane and is converted to a prostanoid end product on the cytoplasmic side of the membrane.

At least four different proteins have been shown to have PGE synthase activity including a cytosolic PGE synthase, an inducible, membrane-associated PGE synthase and two cytosolic glutathione-S-transferase isozymes [13]. Cytosolic PGE synthase appears to be more tightly coupled to PGHS-1. Microsomal PGE synthase has been reported to be more tightly coupled to PGHS-2 and interestingly, its expression like that of PGHS-2 is inhibited by anti-inflammatory steroids such as dexamethasone. All PGE synthases require reduced glutathione as a cofactor. Glutathione facilitates cleavage of the endoperoxide group and formation of the 9-keto group [14]. $PGF_\alpha$ synthase activity has been partially purified from lung. Structurally, the enzyme is a member of the aldose reductase family of proteins. Glutathione-dependent and -independent PGD synthases have been isolated. The glutathione-dependent forms also exhibit glutathione-S-transferase activity. A glutathione-independent form of PGD synthase has been purified from brain [13].

## 3. Prostanoid catabolism and mechanisms of action

### 3.1. Prostanoid catabolism

Once a prostanoid is formed on the cytoplasmic surface of the ER, it diffuses to the cell membrane and exits the cell probably via carrier-mediated transport [14]. Prostanoids are local hormones that act very near their sites of synthesis. Unlike typical circulating hormones that are released from one major endocrine site, prostanoids are synthesized and released by virtually all organs. In addition, all prostanoids are inactivated rapidly in the circulation. The initial step of inactivation of $PGE_2$ is oxidation to a 15-keto compound in a reaction catalyzed by a family of 15-hydroxyprostaglandin dehydrogenases. Further catabolism involves reduction of the double bond between C-13 and C-14, ω-oxidation, and β-oxidation.

### 3.2. Physiological actions of prostanoids

Prostanoids act both in an autocrine fashion on the parent cell and in a paracrine fashion on neighboring cells [15,16]. Typically, the role of a prostanoid is to coordinate the responses of the parent cells and neighboring cells to the biosynthetic stimulus, a circulating hormone. The actions of prostanoids are mediated by G-protein-linked prostanoid receptors of the seven *trans* membrane domain receptor superfamily [16].

Those examples which have been studied in the most detail are the renal collecting tubule-thick limb interactions involving $PGE_2$ synthesized by the collecting tubule and the platelet–vessel wall interactions involving $PGI_2$ and $TXA_2$ [14]. For example, in the case of platelets $TXA_2$ is synthesized by platelets when they bind to subendothelial collagen that is exposed by microinjury to the vascular endothelium. Newly synthesized $TXA_2$ promotes subsequent adherence and aggregation of circulating platelets to the subendothelium. In addition, $TXA_2$ produced by platelets causes constriction of vascular smooth muscle. The net effect is to coordinate the actions of platelets and the vasculature in response to deendothelialization of arterial vessels. Thus, prostanoids can be viewed as local hormones which coordinate the effects of circulating hormones and other agents (e.g., collagen) that activate their synthesis.

### 3.3. Prostanoid receptors

The identification and characterization of prostanoid receptors has occurred during the past ten years [16]. These results coupled with studies of PGHS and prostanoid receptor knockout mice [5,15] have been critical in beginning to rationalize earlier results of studies on the physiological and pharmacological actions of prostanoids that had been somewhat confusing and difficult to interpret because prostaglandins were found to cause such a wide variety of seemingly paradoxical effects.

The seminal step in understanding the structures of prostanoid receptors and their coupling to second messenger systems resulted from the cloning of receptors for each of the prostanoids by Narumiya and others [16]. Prostanoid receptor cloning began with the TXA/PGH receptor known as the TP receptor. cDNA encoding this receptor was cloned using oligonucleotide probes designed from protein sequence data obtained from the TP receptor purified from platelets. The results confirmed biochemical predictions that the TP receptor was a seven-membrane spanning domain receptor of the rhodopsin family. Subsequent cloning of other receptors was performed by homology screening using receptor cDNA fragments as cross-hybridization probes. All of these prostanoid receptors are of the G-protein-linked receptor family.

It is now clear that there are pharmacologically distinct receptors for each of the known prostanoids. In the case of $PGE_2$, four different prostaglandin E (EP) receptors have been identified and designated as EP1, EP2, EP3, and EP4 receptors. Based on studies with selective agonists for each of the EP receptors and their effects on second messenger production, it appears that EP1 is coupled through $G_q$ to the activation of phospholipase C, EP2 and EP4 are coupled via $G_s$ to the stimulation of adenylate cyclase and EP3 receptors are coupled via $G_i$ to the inhibition of adenylate cyclase [16].

In order to determine the physiological roles of various prostanoid receptors a number of knockout mice have been developed [15,16]. These studies indicate that prostacyclin receptors are involved in at least some types of pain responses, EP3 receptors are involved in the development of fever, PGE receptors are important in allergy, EP2 and EP4 function in bone resorption and EP1 receptors are involved in chemically induced colon cancer. The availability of cloned prostanoid receptors provides a rationale and the appropriate technology to search for receptor agonists and antagonists that might provide some specificity beyond the currently available cyclooxygenase inhibitors which prevent broadly the synthesis of all prostaglandins.

The realization that the gene for PGHS-2 is an immediate early gene associated with cell replication and differentiation suggests that prostanoids synthesized via PGHS-2 may have nuclear effects. As noted above there have been several reports indicating that prostanoid derivatives can activate some isoforms of peroxisomal proliferator activated receptors (PPARs). There is evidence that $PGI_2$ can be involved in PPARδ-mediated responses such as decidualization and apoptosis [5] (S.K. Dey, 2000).

## 4. Leukotrienes and lipoxygenase products

### 4.1. Introduction and overview

Leukotrienes are produced by the action of 5-lipoxygenase (5-LO) which carries out the insertion of a diatomic oxygen at carbon atom-5 of arachidonic acid yielding 5(S)-hydroperoxy eicosatetraenoic acid (5-HpETE). A subsequent dehydration reaction catalyzed by the same enzyme, 5-LO, results in formation of leukotriene $A_4$ (LTA$_4$),

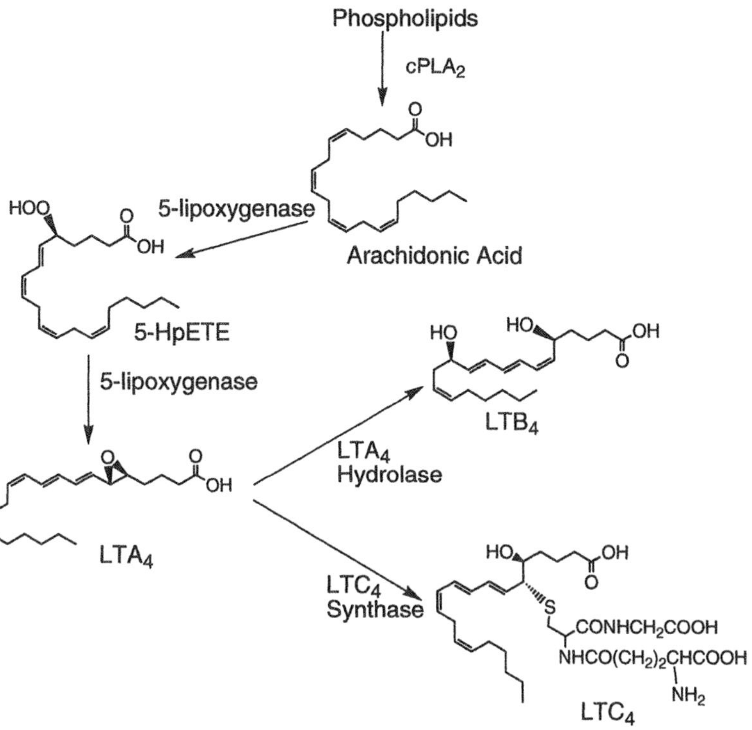

Fig. 7. Biochemical pathway of the metabolism of arachidonic acid into the biologically active leukotrienes. Arachidonic acid released from phospholipase by cytosolic (c) PLA₂ is metabolized by 5-lipoxygenase to 5-hydroperoxyeicosatetraenoic acid (5-HpETE) and leukotriene A₄ (LTA₄) which is then enzymatically converted into leukotriene B₄ (LTB₄) or conjugated by glutathione to yield leukotriene C₄ (LTC₄).

the chemically reactive precursor of biologically active leukotrienes. As was the case with prostanoid biosynthesis, leukotriene biosynthesis depends upon the availability of arachidonic acid as a free carboxylic acid as the 5-LO substrate, which typically requires the action of cytosolic phospholipase $A_2$ to release arachidonic acid from membrane phospholipids. Also, leukotrienes are not stored in cells, but synthesis and release from cells are rapid events following cellular activation. Interest in the leukotriene family of arachidonate metabolites arises from the potent biological activities of two products derived from $LTA_4$, that being leukotriene $B_4$ ($LTB_4$) and leukotriene $C_4$ ($LTC_4$) (Fig. 7). $LTB_4$ is a very potent chemotactic and chemokinetic agent for the human polymorphonuclear leukocyte, while $LTC_4$ powerfully constricts specific smooth muscle such as bronchial smooth muscle, and mediates leakage of vascular fluid in the process of edema [17]. The name leukotriene was conceived to capture two unique attributes of these molecules. The first attribute relates to those white blood cells derived from the bone marrow that have the capacity to synthesize this class of eicosanoid, for example the polymorphonuclear *leukocyte*. The last part of the name refers to the unique chemical structure, a conjugated *triene*, retained within these eicosanoids.

There are numerous other biochemical products of arachidonate metabolism formed by lipoxygenase enzymes other than 5-LO. Other monooxygenases expressed in mammalian cells include 12-lipoxygenase, 15-lipoxygenase and much less frequently 8-lipoxygenase. These enzymes are named in accordance with the carbon atom position of arachidonate initially oxygenated even though other polyunsaturated fatty acids can be substrates. In addition, arachidonic acid can be oxidized by specific isozymes of cytochrome P-450, leading to a family of epoxyeicosatrienoic acids (EETs). Methyl terminus oxidized arachidonate as well as lipoxygenase-like monohydroxy eicosatetraenoic acid (HETE) products are also formed. In general, much more is known about the biochemical role of prostaglandins and leukotrienes as mediators of biochemical events; little is known concerning the exact role played by the other lipoxygenase products or cytochrome P-450 products. In accord with the body of information available, various pharmacological tools are available to inhibit 5-LO as well as specific leukotriene receptors.

## 4.2. Leukotriene biosynthesis

The arachidonate 5-lipoxygenase (5-LO, EC 1.13.11.34) is a metalloenzyme with bound iron coordinated by four histidine residues. This 77,852 Da protein (human 5-LO) catalyzes the addition of molecular oxygen to the 1,4-*cis*-pentadienyl structural moiety closest to the carboxyl group of arachidonic acid to yield a conjugated diene hydroperoxide, typical for all lipoxygenase reactions. Some details of the mechanism of 5-LO are known in that 5-LO removes the pro-*S* hydrogen atom from carbon-7 of arachidonic acid, leading to reduction of Fe(III) to Fe(II) likely in a radical type mechanism [18]. Molecular oxygen then adds to carbon-5 to yield the hydroperoxy radical. The hydroperoxy radical abstracts a hydrogen atom to yield 5(*S*)-HpETE (5(*S*)-hydroperoxy-6,8,11,14-(*E,Z,Z,Z*)-eicosatetraenoic acid). A second enzymatic activity of 5-LO, sometimes termed LTA$_4$ synthase activity, catalyzes the stereospecific removal of the pro-*R* hydrogen atom at carbon-10 of the 5(*S*)-HpETE through a second redox cycle followed by a sigmatropic shift of the electrons to form the conjugated triene epoxide and loss of hydroxide (Fig. 8). One unique feature of the iron redox cycle is that these reactions involve one electron rather than two electron transfers, typical of most peroxidase enzymes, in what has been called a pseudoperoxidase reaction [19]. The product of this reaction, 5(*S*),6(*S*)-oxido-7,9,11,14-(*E,E,Z,Z*)-eicosatetraenoic acid,

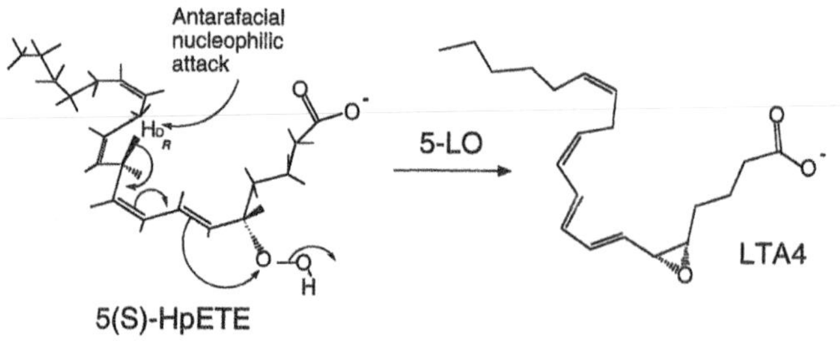

Fig. 8. Detailed mechanism of the 5-lipoxygenase reaction where the pro-*R* hydrogen from carbon-10 in 5-HpETE is removed by 5-lipoxygenase followed by sigmatropic rearrangement of electrons to form LTA$_4$.

is LTA$_4$, a conjugated triene epoxide that is highly unstable. LTA$_4$ undergoes rapid hydrolysis in water with a half-life of less than 10 s at pH 7.4 and also can react with proteins as well as DNA. Nonetheless within cells, LTA$_4$ is stabilized by binding to proteins that remove water from the immediate environment of the epoxide structure. The nonenzymatic hydrolysis products of LTA$_4$ include several biologically inactive and enantiomeric 5,12- and 5,6-diHETEs. However, the hydrolysis of LTA$_4$ catalyzed by LTA$_4$ hydrolase [20] produces the biologically active LTB$_4$, 5($S$),12($R$)-dihydroxy-6,8,10,14-($Z,E,E,Z$)-eicosatetraenoic acid. A second pathway for LTA$_4$ metabolism is prominent in cells expressing the enzyme LTC$_4$ synthase [21] which catalyzes the addition of glutathione to carbon-6 of the triene epoxide yielding 5($S$),6($R$)-$S$-glutathionyl-7,9,11,14-($E,E,Z,Z$)-eicosatetraenoic acid (LTC$_4$). LTC$_4$ synthase has been found to be localized on the nuclear membrane and is a unique glutathione ($S$) transferase. The formation of either LTC$_4$ or LTB$_4$ is controlled by the expression of either LTA$_4$ hydrolase or LTC$_4$ synthase by specific cell types. The human neutrophil, for example, expresses LTA$_4$ hydrolase and produces LTB$_4$ while the mast cell and eosinophil produce LTC$_4$, since they express LTC$_4$ synthase. Interestingly, cells have been found which do not have 5-LO, but do express either LTA$_4$ hydrolase (e.g., erythrocytes and lymphocytes) or LTC$_4$ synthase (e.g., platelets and endothelial cells). Studies have found that cells in fact cooperate in the production of biologically active leukotrienes through a process termed transcellular biosynthesis (J. Maclouf, 1989) where a cell such as the neutrophil or mast cell generates LTA$_4$ which is then released from the cell and is then taken up by either a platelet to make LTC$_4$ or red blood cell to make LTB$_4$. In spite of the chemical reactivity of LTA$_4$, this process is known to be highly efficient and approximately 60 to 70% of the LTA$_4$ produced by the activated neutrophil can be released to another cell for transcellular biosynthesis of leukotrienes (A. Sala, 1996).

### 4.3. Enzymes involved in leukotriene biosynthesis

### 4.3.1. 5-Lipoxygenase
The human 5-LO gene is unusually large and is present on chromosome 10. This gene covers more than 80 kb of DNA and has 14 exons that encode a 673 amino acid protein (without the initiator methionine residue) [22]. 5-Lipoxygenase has been purified from human, pig, rat, and guinea pig leukocytes, all with close to 90% homology. It is interesting to note that the purified or recombinant enzyme requires several cofactors for activity that include Ca$^{2+}$, ATP, fatty acid hydroperoxides, and phosphatidylcholine in addition to the arachidonic acid and molecular oxygen substrates [17,19]. Purified 5-LO was found to catalyze the initial oxidation of arachidonic acid to yield 5-HpETE as well as the second enzymatic reaction to convert 5-HpETE into LTA$_4$. Recombinant human 5-LO has been expressed in osteosarcoma cells, cos-M6 cells, baculovirus-infected SF9 insect cells, yeast, and *Escherichia coli* [19].

Low concentrations of calcium ion (1–2 μM) are required for maximal activity of purified 5-LO, but the major role of calcium appears to be that of increasing lipophilicity of 5-LO in order to promote membrane association. ATP has a stimulatory effect on 5-LO at 20 nM and lipid hydroperoxides are important to initiate the 5-LO catalytic cycle since they readily form Fe(III) within 5-LO by the pseudoperoxidase mechanism. Microsomal membranes as well as phosphatidylcholine vesicles can stimulate purified 5-LO activity since 5-LO performs the oxidation of arachidonic acid at the interface between the membrane and cytosol in a manner similar to that of cytosolic $PLA_2$. Calcium ions increase the association of 5-LO with phosphatidylcholine vesicles that likely recapitulates events within the cell where 5-LO becomes associated with the nuclear membrane [23] and to which arachidonic acid is presented by a second gene product which has been termed 5-LO activating protein (FLAP) [19]. In cells such as the neutrophil and mast cell where 5-LO is found in the cytosol, 5-LO is only catalytically active when bound to a membrane, typically the nuclear membrane. In fact, in some cells 5-LO is found to be constitutively associated with the nuclear membrane, likely a result of a process of cellular activation while 5-LO is found in the alveolar macrophage within the nucleus itself [23].

### 4.3.2. 5-Lipoxygenase activating protein (FLAP)

FLAP was found during the course of development of the drug MK-886 by workers at Merck-Frosst in Montreal [19]. They found this drug bound to a novel protein that was essential for the production of leukotrienes in stimulated, intact cells and hence, the name 'five lipoxygenase activating protein'. FLAP is a unique 161 amino acid containing protein (18,157 Da). While the role of FLAP is not entirely clear, experiments using a [125]I-labeled photoaffinity analog of arachidonic acid suggested that FLAP functions as a substrate transfer protein and in this manner stimulates 5-LO catalyzed formation of leukotrienes. Recently, it has been found that $LTC_4$ synthase has 31% amino acid identity to FLAP with a highly conserved region possibly involved in arachidonate binding for both proteins [19].

### 4.3.3. LTA₄ hydrolase

$LTA_4$ hydrolase catalyzes the stereochemical addition of water to form the neutrophil chemotactic factor $LTB_4$. $LTA_4$ hydrolase contains 610 amino acids (excluding the first methionine) with a molecular weight of 69,399 Da. $LTA_4$ hydrolase contains one zinc atom per enzyme molecule and this metal ion is essential for the catalytic activity [20]. $LTA_4$ hydrolase is also a member of a family of zinc metalloproteases and exhibits some protease activity. The finding of this activity led to the discovery of several drugs such as bestatin and captopril that are inhibitors of this enzyme. $LTA_4$ hydrolase is found in many cells including those which do not contain 5-LO and it is felt that these cells play an important role in transcellular biosynthesis of $LTB_4$ through cell–cell cooperation. $LTA_4$ is thought to be localized in the cytosol of the cell and is the only protein in the leukotriene biosynthetic cascade that is not found on the nuclear membrane following cellular activation. Therefore, in order to efficiently metabolize the chemically reactive $LTA_4$, either the $LTA_4$ hydrolase must come in close contact with the nuclear membrane during $LTA_4$ biosynthesis or a carrier protein must present $LTA_4$ to $LTA_4$ hydrolase in the cytosol. $LTA_4$ hydrolase is known to be efficiently suicide inactivated by $LTA_4$ when the electrophilic epoxide becomes covalently bound to the enzyme presumably within the active site (F.A. Fitzpatrick, 1990). A specific tyrosine residue has been found to be modified by $LTA_4$ and this residue has been implicated as the potential proton donor in the suggested mechanism of $LTA_4$ hydration. The 3-dimensional structure of $LTA_4$ hydrolase has now been determined by X-ray crystallography [20].

### 4.3.4. $LTC_4$ synthase

The conjugation of the tripeptide glutathione ($\gamma$-glutamyl-cysteinyl glycine) to the triene epoxide $LTA_4$ is carried out by $LTC_4$ synthase (EC 2.5.1.37). This enzyme is found localized on the nuclear envelope of cells and has little homology to soluble glutathione ($S$) transferases. $LTC_4$ synthase does have some primary amino acid sequence homology to the recently described microsomal glutathione ($S$) transferases and FLAP [21]. $LTC_4$ synthase has a restricted distribution and is found predominantly in mast cells, macrophages, eosinophils, and monocytes. However, human platelets and endothelial cells have been found to express $LTC_4$ synthase. Purified recombinant $LTC_4$ synthase will conjugate glutathione to both $LTA_4$ and $LTA_4$ methyl ester with a $K_m$ of approximately 2–4 $\mu$M for the free acid and 7–10 $\mu$M for the methyl ester. The $K_m$ for glutathione is approximately 2 mM. The drug MK-886 was found to inhibit $LTC_4$ synthase with an $IC_{50}$ of approximately 2–3 $\mu$M [19]. The gene for $LTC_4$ synthase is located on human chromosome 5, distal to that of cytokine, growth factor, and receptor genes relevant to the TH2 phenotype (T.D. Bigby, 2000).

### 4.4. Regulation of leukotriene biosynthesis

The biosynthesis of leukotrienes within cells is highly regulated and depends not only on the availability of arachidonic acid and molecular oxygen, but also on the subcellular location of 5-LO. Resting neutrophils synthesize little if any leukotrienes; however, following the elevation of intracellular calcium either through a physiological event such as phagocytosis or by pharmacological manipulation with the calcium ionophore A23187, the neutrophil produces a substantial amount of 5-LO products including leukotrienes from either endogenous or exogenous arachidonic acid. The need for an increase in intracellular calcium ion concentrations is a distinguishing feature of 5-LO that differentiates this enzyme from other lipoxygenases and cyclooxygenase.

A major determinant of 5-LO activity is the translocation of 5-LO to the nuclear membrane (Fig. 9). It has even been possible to demonstrate that the site of localization of 5-LO is at the inner nuclear membrane. The $Ca^{2+}$-dependent translocation event is thought to bring 5-LO to the same region where FLAP and cytosolic $PLA_2$ translocate. Regulation of leukotriene biosynthesis is thus a process of assembly of the leukotriene biosynthetic machine at a nuclear envelope site. It is this site where arachidonic acid is released from nuclear membrane phospholipids, then converted to $LTA_4$ and ultimately conjugated with glutathione by nuclear membrane $LTC_4$ synthase. The mechanism by which 5-LO is trafficked to the nuclear membrane is as yet undefined, but the discovery of this site of leukotriene biosynthesis was unexpected and suggests novel intracellular actions of leukotrienes or of 5-LO itself within the immediate nuclear environment [23].

The activity of 5-LO is also regulated by a suicide inactivation mechanism where $LTA_4$ rapidly inactivates 5-LO, likely through a covalent modification mechanism [18]. Continued synthesis of leukotrienes then requires synthesis of new 5-LO. The production of 5-LO is known to be regulated at the level of gene transcription as well

360

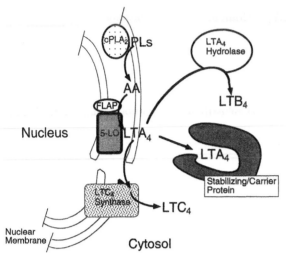

Fig. 9. Proposed model for the location of biosynthetic events occurring during leukotriene biosynthesis at the nuclear membrane of cells. Arachidonic acid is released from membrane glycerophospholipids (PLs) by translocated and nuclear membrane associated cytosolic (c) PLA$_2$ and is then presented to 5-lipoxygenase (5-LO) by way of FLAP. LTA$_4$ is either converted by nuclear membrane associated LTC$_4$ synthase into LTC$_4$ or carried to LTA$_4$ hydrolase possibly by a fatty acid binding protein which can stabilize the chemically reactive LTA$_4$ that is available for transcellular biosynthesis.

as mRNA translation in addition to the translocation mechanism [22,23]. The 5-LO gene promoter contains several consensus-binding sites for known transcription factors including Sp1 and EGR-1. In a region located 212 to 88 base pairs upstream from the translation start site of the human 5-LO gene, a highly rich G + C region is found that contains the consensus sequence or SP1 and EGR-1 transcription factors. Furthermore, fairly common variations have been found in human populations in which deletions of one or two of the Sp1 binding motifs were observed [22]. This genetic polymorphism could have substantial effects on the induction of 5-LO transcription.

Various biochemical mechanisms can also alter leukotriene biosynthesis within cells. Because of the complexity of 5-LO activation and the requirement of the Ca$^{2+}$-dependent translocation event, modification of the signal transduction pathways are known to alter leukotriene biosynthesis. For example, elevation of cellular levels of cyclic AMP have been known for some time to inhibit leukotriene biosynthesis even when synthesis is stimulated by the powerful calcium ionophore A23187. Adenosine and A2A receptor agonists are known to inhibit production of leukotrienes in human neutrophils, most likely through an enhanced production of cyclic AMP (P. Borgeat, 1999). The product in neutrophil leukotriene biosynthesis, namely LTB$_4$, can also inhibit the synthesis of leukotrienes when initiated during the course of phagocytosis. These effects can be observed at 1–3 nM, are mediated through the LTB$_4$ receptor, and likely represent a feedback-like inhibition of leukotriene biosynthesis through inactivation of 5-LO as well as cytosolic PLA$_2$ (J. Fiedler, 1998). Pharmacological agents have been developed to inhibit leukotriene production through direct action on 5-LO. The drug

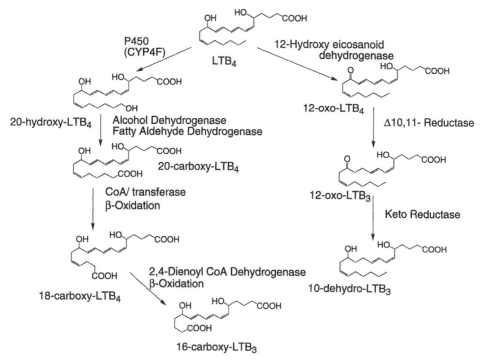

Fig. 10. Common metabolic transformations of LTB$_4$ either by the cytochrome P-450 (CYP4F-family members) and ω-oxidation followed by β-oxidation or by the 12-hydroxyeicosanoid dehydrogenase pathway which leads to reduction of the $\Delta^{10,11}$ double bond.

zileuton likely causes the reduction of activated 5-LO Fe(III) to the inactive 5-LO Fe(II) or prevents the oxidation of Fe(II) by lipid hydroperoxides, the pseudoperoxidase step, by serving as a competitive substrate [19]. Inhibitors of FLAP have been discussed above and include MK-886 and the related agent BAY x1005. Because of its close similarity to the FLAP protein, LTC$_4$ synthase can be inhibited by MK-886 albeit at higher concentrations.

*4.5. Metabolism of leukotrienes*

The conversion of leukotrienes into alternative structural entities is an important feature of inactivation of these potent biologically active eicosanoids. Metabolism of leukotrienes is rapid and the exact pathway depends upon whether the substrate is LTB$_4$ or LTC$_4$. LTB$_4$ is rapidly metabolized through both oxidative and reductive pathways (Fig. 10) [24]. The most prominent pathway present in the human neutrophil (CYP4F3) as well as hepatocyte (CYP4F2) involves specific and unique cytochrome P-450s of the CYP4F family. cDNAs encoding sixteen different proteins have now been cloned and expressed in several animal species [25] and each of these enzymes efficiently converts LTB$_4$ into 20-hydroxy-LTB$_4$. 20-Hydroxy-LTB$_4$ has some biological activity since it is a competitive agonist for the LTB$_4$ receptor. In the human neutrophil, 20-hydroxy-LTB$_4$

is further metabolized into 20-carboxy-LTB$_4$ by CYP4F3. In the hepatocyte and other tissues, 20-hydroxy-LTB$_4$ is further metabolized by alcohol dehydrogenase to form 20-oxo-LTB$_4$ then by fatty aldehyde dehydrogenase to form 20-carboxy-LTB$_4$ both of which reactions require NAD$^+$ (Fig. 10).

A unique reductive pathway has been observed to be highly expressed in cells such as keratinocytes, endothelial cells and kidney cells. This pathway arises from an initial oxidation of the 12-hydroxy group to a 12-oxo moiety followed by reduction of the conjugated dienone and double bond $\Delta^{10,11}$. The products of the 12-hydroxy eicosanoid dehydrogenase pathway have been found to be devoid of biological activity and in certain cells represent the major pathway of inactivation.

A secondary metabolic pathway is β-oxidation which was found to predominate from the 20-carboxy terminus of 20-carboxy-LTB$_4$ [14]. These events take place both within the peroxisome as well as the mitochondria of the hepatocyte. β-oxidation can also occur from the C-1 carboxyl moiety of LTB$_4$ which eventually results in the loss of the C-5 hydroxyl group. The importance of the each oxidation processes in LTB$_4$ metabolism is seen in human subjects with various genetic abnormalities. Deficiencies in peroxisomal metabolism (Zellweger disease) leads to a reduction in β-oxidation and in these individuals LTB$_4$ and 20-carboxy-LTB$_4$ can be measured as urinary excretion products (E. Mayatepek, 1999). Individuals with a deficiency in fatty aldehyde dehydrogenase termed Sjogren–Larsson syndrome were found to excrete measurable levels of LTB$_4$ and 20-hydroxy-LTB$_4$ (M.A. Willemsen, 2001). None of these compounds can be measured in the urine of normal individuals even when exogenous LTB$_4$ is administered.

The metabolism of LTC$_4$ results in activation as well as inactivation of biological activity of the sulfidopeptide leukotrienes. Initial peptide cleavage reactions including γ-glutamyl transpeptidase and various dipeptidases lead to the production of LTD$_4$ and LTE$_4$, both of which are biologically active metabolites (Fig. 11). Sulfidopeptide leukotrienes can also be metabolized specifically at the sulfur atom through oxidation reactions initiated by reactive oxygen species. More specific metabolic processing of the sulfidopeptide leukotrienes include ω-oxidation by cytochrome P-450 followed by β-oxidation from the ω-terminus resulting in a series of chain-shortened products [24]. Formation of 20-carboxy-LTE$_4$ results in complete inactivation of the biological activities of this molecule due to poor receptor recognition. Acetylation of the terminal amino group in LTE$_4$ and formation of N-acetyl-LTE$_4$ is an abundant metabolite in rodent tissue. In man, some LTE$_4$ is excreted in urine and has been used to reflect whole body production of sulfidopeptide leukotriene in vivo.

## 4.6. Biological activities of leukotrienes

LTB$_4$ is thought to play an important role in the inflammatory process by way of its chemotactic and chemokinetic effects on the human polymorphonuclear leukocyte. LTB$_4$ induces the adherence of neutrophils to vascular endothelial cells and enhances the migration of neutrophils (diapedesis) into extravascular tissues. The biological activity of LTB$_4$ is mediated through two specific G-protein-coupled receptors termed BLT$_1$ and BLT$_2$ [26]. BLT$_1$ (human receptor 37,591 Da) is almost exclusively expressed in human polymorphonuclear leukocytes and to a much lesser extent on macrophages

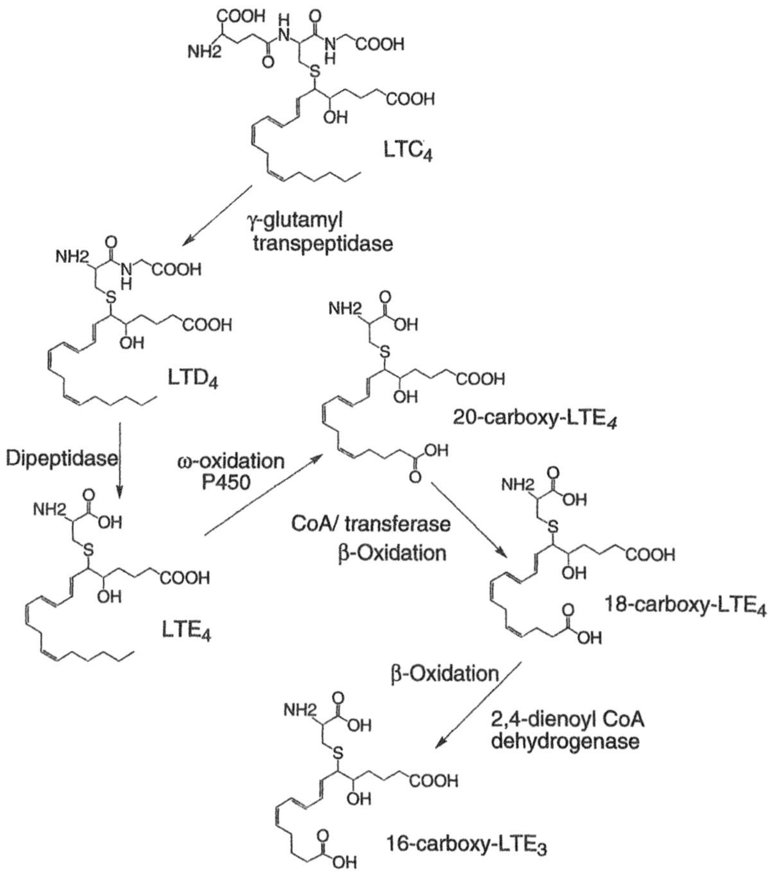

Fig. 11. Common metabolic transformations of LTC₄ to the biologically active sulfidopeptide leukotrienes, LTD₄ and LTE₄. Subsequent ω-oxidation of LTE₄ by cytochrome P-450 leads to the formation of 20-carboxy-LTE₄ which can undergo β-oxidation after formation of the CoA ester into a series of chain-shortened cysteinyl leukotriene metabolites.

and in tissues such as thymus and secretory PLeen. The human BLT gene is located on chromosome 14. The chemotactic effect of $LTB_4$ was shown to be mediated through the $BLT_1$ and $BLT_2$ receptors. Several specific agents have been developed by pharmaceutical companies to inhibit the $LTB_4$ receptor; however, none has been fully developed to be used in humans.

$LTC_4$ and the peptide cleavage products $LTD_4$ and $LTE_4$ have been identified as mediators causing bronchial smooth muscle contraction in asthma. These sulfidopeptide leukotrienes also increase vascular leakage leading to edema. The discovery of leukotrienes was, in fact, a result of the search for the chemical structure of the biologically active principle called 'slow reacting substance of anaphylaxis' (R.C. Murphy, 1979). Two receptors for the cysteinyl leukotriene have recently been characterized and termed $CysLT_1$ and $CysLT_2$. The $CysLT_1$ receptor was found to be a G-protein-coupled receptor with seven *trans* membrane regions [27]. The CysLT$_1$ receptor has limited distribution in tissues with the most prominent being in smooth muscle of the lung and small intestine and both LTD$_4$ and LTC$_4$ activate this receptor. Several drugs are now available for inhibition of the CysLT$_1$ receptor in human subjects. These are montelukast (2–5 nM), pranlukast (4–7 nM), and zafirlukast (2–3 nM). Interestingly, the gene encoding the human CysLT$_1$ receptor is located on the X chromosome [27].

### 4.7. Other lipoxygenase pathways

Numerous lipoxygenases occur within the plant and animal kingdoms and these enzymes have in common several aspects. First, iron is an essential component of the catalytic activity of these enzymes and is held in place through histidine residues rather than by heme. Furthermore, these enzymes catalyze the insertion of molecular oxygen into polyunsaturated fatty acids, predominantly linoleic and arachidonic acids, with the initial formation of lipid hydroperoxides. In general, the overall biological activities of the lipoxygenase products are incompletely known and the significance of 12- and 15-lipoxygenase in man remains undefined.

#### 4.7.1. 12-Lipoxygenase
Two different enzymes termed 12-lipoxygenase (12-LO) catalyze the formation of 12-hydroperoxyeicosatetraenoic acid from arachidonic acid which is subsequently reduced to 12-hydroxyeicosatetraenoic acid (Fig. 12). The human platelet expresses one 12-LO

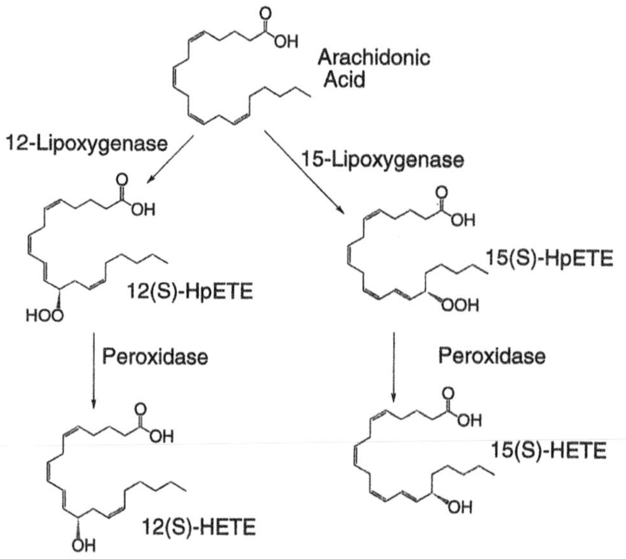

Fig. 12. Metabolism of arachidonic acid by 12- and 15-lipoxygenase pathways with corresponding stereospecific formation of hydroperoxyeicosatetraenoic acids (HpETE). Subsequent reduction of these hydroperoxides leads to the corresponding HETE at either carbon-12 or -15 which are thought to mediate biological activities of these enzymatic pathways.

type (EC 1.13.11.31) the cDNA of which has been cloned, sequenced, and found to encode a 662 amino acid protein with a molecular weight of 75,535 Da. A second 12-LO is observed in other mammalian systems including the mouse and rat and has been termed the leukocyte-type 12-LO. This latter lipoxygenase is very similar in many respects to a 15-lipoxygenase (15-LO) in terms of its substrate specificity and capability of forming both 12-HpETE and 15-HpETE from arachidonic acid. The human platelet 12-LO has approximately 65% identity in primary structure to that of 15-LO from human reticulocytes. In addition, there are other lipoxygenases less well studied including an epidermal lipoxygenase from newly differentiated keratinocytes and the lipoxygenase that oxygenates arachidonic acid at position C-15 and C-8. Both 12/15-LO types are suicide inactivated, but the leukocyte 12-LO undergoes autoinactivation at a much higher rate. Several lines of evidence suggest that the 12-LO pathway of arachidonate metabolism plays an important role in regulating cell survival and apoptosis (A.R. Brash, 1999).

### 4.7.2. 15-Lipoxygenase

The oxidation of arachidonic acid at carbon-15 is catalyzed by 15-LO, a soluble 661 amino acid containing protein with a molecular weight of 74,673 Da. Many cells express this enzyme which also efficiently oxidizes linoleic acid to 13-hydroperoxyocta-decadienoic acid and lesser extent 9-hydroperoxyoctadecadienoic acid because of broad substrate specificity as well as arachidonate to both 12-HpETE and 15-HpETE [28]. One distinguishing feature of this lipoxygenase is that it can oxidize arachidonic acid esterified to membrane phospholipids, thus forming esterified 15-HpETE. Expression of 15-LO is enhanced by several interleukins, suggesting a role of this enzyme in events such as atherosclerosis.

The X-ray crystal structure of mammalian 15-LO has revealed two domains, a catalytic domain and a β-barrel domain [29]. The β-barrel domain may be involved in the binding of this enzyme to phospholipid membranes, the source of either arachidonate or phospholipids in the oxidation process. The catalytic domain, which contains the histidine-coordinated Fe(III), holds the arachidonic acid assisted by an ionic bond between R403 and the ionized carboxyl group of arachidonic acid. The methyl terminus of arachidonic acid is thus placed deep within a hydrophobic binding pocket in an arrangement that is likely similar for other lipoxygenases.

Mammalian 15-LO is involved in the production of more complicated eicosanoids including the biologically active lipoxins [26]. Lipoxins are formed by the sequential reaction of both 15-LO and 5-LO acting on precursor arachidonate. For example, LTA$_4$ (the 5-LO product) can be converted to a lipoxin by action of 15-LO.

There are a host of biological activities initiated by these 15-LO dependent eicosanoids [28]. The unique activity of 15-LO in oxidizing intact phospholipids has been featured in several hypotheses linking the oxidation of phospholipids in atherosclerotic lesions to important role of these lipoxygenases. The recent availability of strains of mice which have a targeted deficiency in 5-, 12-, and 15-lipoxygenases (C. Funk, 2000) provides a powerful tool to ask specific questions concerning the role these lipoxygenases play in host defense reactions, cellular function, and perhaps disease processes.

366

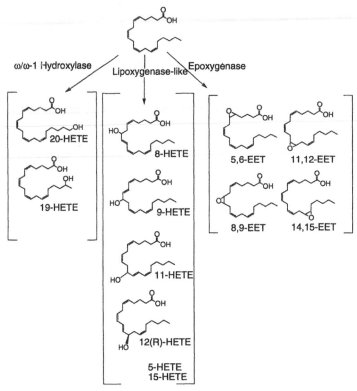

Fig. 13. Metabolism of arachidonic acid by cytochrome P-450 enzymes and the formation of three structurally distinct metabolite families. Omega-oxidation leads to a family of ω to ω-4 products of which 20-HETE (ω-oxidation) and 19-HETE (ω-1 oxidation) are indicated. The lipoxygenase-like mechanism of cytochrome P-450 metabolism leads to the formation of six different conjugated dienols, for which the structures of four are indicated. One unique biologically active lipoxygenase-like P-450 metabolite is 12(R)-HETE. The epoxygenase pathway leads to the formation of four regioisomeric epoxyeicosatrienoic acid (EETs) all of which are biologically active.

## 5. Cytochrome P-450s and epoxygenase pathways

Arachidonic acid can be metabolized to a series of products characterized by the introduction of a single oxygen atom from molecular oxygen and formation of three different types of initial products catalyzed by various cytochrome P-450 mixed function oxidases (Fig. 13) [30,31]. The three classes of products include a series of hydroxyeicosatetraenoic acids (HETEs) formed by an allylic oxidation mechanism resulting in a family of conjugated dienes isomeric to the reduced products of a lipoxygenase reaction. P-450 metabolites formed by this mechanism have been characterized as 5-, 8-, 9-, 11-, 12-, and 15-HETE, some of which are epimeric to the lipoxygenase catalyzed products, e.g., 12(R)-HETE. A second class of reactions involves oxidation of the terminal alkyl chain region of arachidonic acid with placement of a hydroxyl group between the terminal carbon atom (ω) through ω-4 position with formation of a family of ω-oxidized monohydroxyeicosatetraenoic acids. Insertion of oxygen into the carbon–carbon bond results in the formation of a family of regioisomeric *cis* epoxyeicosatetraenoic acids (EETs) from which the general pathway has been named, the epoxygenase pathway. These regioisomers include 14,15-, 11,12-, 8,9-, and 5,6-EETs which can be formed either as an *R,S*, or the *S,R* enantiomer (Fig. 13).

## 5.1. Epoxygenase P-450 isozymes

With the availability of recombinant P-450 isozymes, it has been possible to identify specific isozymes that can metabolize arachidonic acid. EET biosynthesis can be accomplished by CYP1A, CYP2B, CYP2C, CYP2D, CYP2G, CYP2J, CYP2N, and CYP4A subfamilies [31]. For each of these, unique EET regioisomers are formed. For example CYP2C8 produces 14(*R*),15(*S*)-EET and 11(*R*),12(*S*)-EET with optical purities of 86% and 81%, respectively. However, it is likely that more than a single P-450 contributes to EET biosynthesis within a specific cell or tissue and thus the individual arachidonate epoxygenase metabolite may depend upon expression of specific P-450 isoforms. It is thought that the majority of EET biosynthesis in human and rat kidney is a result of CYP2C expression in these tissues. However, the induction of specific P-450s can greatly alter the production of specific epoxygenase products.

## 5.2. Occurrence of EETs

Various EETs have been measured in tissues as well as physiological fluids such as urine (G. FitzGerald, 1990). Biologically active lipids originally defined as an endothelium-derived hyperpolarizing factor and an inhibitor of $Na^+/K^+$ ATPase found in the thick ascending loop of Henley cells were structurally characterized as 11(*R*),12(*S*)-EET and 20-hydroxyeicosatetraenoic acid, both derived from cytochrome P-450 mediated metabolism of arachidonic acid [30]. Interestingly, the EETs can readily form CoA esters and participate in reacylation of lysophospholipids which results in the reincorporation of these oxidized metabolites of arachidonic acid into phospholipid membranes, a biochemical feature not observed for prostaglandins, thromboxanes or leukotrienes. For example, human platelets have been found to contain 14,15-EET esterified within membrane phospholipids (Y. Zhu, 1995). It is felt that the majority of EETs produced within cells become reesterified to cellular glycerophospholipids.

## 5.3. Metabolism of EETs

A number of metabolic pathways operate on the primary epoxygenase metabolites of arachidonic acid. Some of the more abundant pathways include CoA-dependent reesterification as mentioned above as well as β-oxidation chain-shortening. A unique pathway involves epoxide hydrolase, a cytosolic enzyme that hydrates EETs to the corresponding vicinal dihydroxyeicosatrienoic acids [30]. There is also a microsomal epoxide hydrolase that can metabolize EETs, but at a somewhat lower rate. The soluble epoxide hydrolase does have substrate specificity, both in terms of the stereochemistry of the EET as well as its position in the arachidonic acid chain. As expected, there is nonenzymatic hydration of these epoxides especially under acidic conditions that can be accelerated during the isolation of these arachidonate metabolites. Therefore, it is sometimes difficult to distinguish between nonenzymatic and enzymatic hydration of EETs. The 5,6-EET is a poor substrate for cytosolic as well as microsomal epoxide hydrolase; however, it has been observed to be an efficient substrate for PGH synthase, leading to the formation of 5,6-epoxy-PGH$_1$. This reactive intermediate can subsequently be transformed into corresponding 5,6-epoxy-prostaglandins of the E, F, and I series or into an epoxy thromboxane analog. All of the EETs can also be substrates for lipoxygenases which would introduce molecular oxygen at any 1,4-*cis*-pentadienyl position not interrupted by the epoxide ring. The EETs can also be conjugated with reduced glutathione catalyzed by glutathione (*S*) transferases. Studies of the metabolism of EETs by rat or mouse liver microsomal P-450 revealed the formation of a series of diepoxyeicosadienoic acids as well as monohydroxyepoxyeicosatrienoic acids. Interestingly, the diepoxides were found to be further transformed into tetrahydrofurandiols mediated by intermediate diol epoxides formed by soluble epoxide hydrolase. The characterization of specific products of EET and PGH synthase metabolism of arachidonic acid have led to the observation of 5,6-epoxy-PGE$_1$ as a renal vasodilator with similar potency to that of PGE$_2$ and the metabolism of 8,9-EET by PGH synthase leading to 11-hydroxy-8,9-epoxyeicosatrienoic acid which is a mitogen for rat glomerular mesangial cells.

## 5.4. Biological actions of EETs

Metabolites of arachidonic acid derived from the epoxygenase P-450 pathway have been studied extensively in terms of their pharmacological properties. Potent effects have been observed in modulating various ion channels, membrane bound transport proteins, mitogenesis, PPARα agonists, and activators of tyrosine kinase cascades [30]. EETs likely play an important role in mediating Na$^+$/K$^+$-ATPase and inhibiting the hydroosmotic effect of arginine vasopressin in the kidney. A picture has emerged for an important role of EETs in regulating renal vascular tone and fluid/electrolyte transport placing the EETs in the pathogenesis of hypertension.

## 6. Future directions

There is currently a reasonable understanding of the structures of PGHS-1 and -2, but many of the relationships between structure and function remain to be identified. For example, it is known that the peroxidase activities of PGHSs preferentially utilize alkyl hydroperoxides such as PGG$_2$ versus hydrogen peroxide; the basis for this specificity is not evident from simple observation of the structures. It will also be important to characterize further the membrane binding domain of PGHSs in the context of the interaction of these domains with specific membrane lipids and the role of this domain in governing substrate entry into the cyclooxygenase site and product exit from this site.

Our understanding of the different biological roles of PGHS-1 and PGHS-2 is only beginning to emerge. The functions of these two isozymes in apoptosis, particularly as it relates to the development of a variety of cancers, angiogenesis, respiration, inflammation, pain and reproduction need further exploration. Of key importance is understanding the reason for the existence of the two PGHS isozymes and how coupling occurs between these enzymes, upstream lipases, and downstream synthases and receptors.

Considerable challenges remain in understanding the detailed biochemistry involved in the synthesis and release of biologically active leukotrienes. Little is known how these highly lipophilic molecules are released from cells; but even more curious is how the chemically reactive intermediate leukotriene $A_4$, made on the perinuclear membrane, can find its way into a neighboring cell in the process of transcellular biosynthesis. The mechanism of 5-lipoxygenase is still poorly understood; however, the detailed structure of 5-lipoxygenase (X-ray structure) may likely reveal important facets relevant to translocation events of 5-lipoxygenase as well as mechanism of suicide inactivation. Such information would be of great value in designing specific drugs as novel inhibitors of 5-lipoxygenase, a pharmacological approach highly successful for cyclooxygenase. Last, but not least, little is known concerning a potential intracellular role for leukotrienes. The currently known biological actions of leukotrienes all involve cell membrane G-protein-linked receptors, yet an understanding of why biosynthesis of these lipophilic molecules occurs deep within the cell remains a mystery.

## Abbreviations

| | |
|---|---|
| 2-AG | 2-arachidonyl-glycerol |
| BLT (1 or 2) | leukotriene $B_4$ receptor (subclass 1 or 2) |
| cysLT (1 or 2) | cysteinyl leukotriene receptor (subclass 1 or 2) for which leukotriene $C_4$, $D_4$, and $E_4$ are agonists |
| CYP4F(x) | cytochrome P-450 isozyme that carries out $\omega$-oxidation of leukotriene $B_4$ (subclass x) |
| EET | epoxyeicosatetraenoic acid |
| EP | prostaglandin E receptor |
| FLAP | 5-lipoxygenase activating protein |
| HETE | hydroxyeicosatetraenoic acid |
| HpETE | hydroperoxyeicosatetraenoic acid |
| LO | lipoxygenase |
| LT | leukotriene (followed by letter to designate structural type) |
| NSAID | nonsteroidal anti-inflammatory drug |
| PG | prostaglandin (followed by letter to designate structural type) |
| PGHS | prostaglandin endoperoxide H synthase |
| PLA$_2$ | phospholipase A$_2$ |

## References

1. Smith, W.L., DeWitt, D.L. and Garavito, R.M. (2000) Cyclooxygenases: structural, cellular and molecular biology. Annu. Rev. Biochem. 69, 149–182.

2. Reich, E.E., Zackert, W.E., Brame, C.J., Chen, Y., Roberts II, L.J., Hachey, D.L., Montine, T.J. and Morrow, J.D. (2000) Formation of novel D-ring and E-ring isoprostane-like compounds (D4/E4-neuroprostanes) in vivo from docosahexaenoic acid. Biochemistry 39, 2376–2383.

3. Gijon, M.A., Spencer, D.M., Siddiqi, A.R., Bonventre, J.V. and Leslie, C.C. (2000) Cytosolic phospholipase A2 is required for macrophage arachidonic acid release by agonists that Do and Do not mobilize calcium. Novel role of mitogen-activated protein kinase pathways in cytosolic phospholipase A2 regulation. J. Biol. Chem. 275, 20146–20156.

4. Murakami, M., Kambe, T., Shimbara, S. and Kudo, I. (1999) Functional coupling between various phospholipase A2s and cyclooxygenases in immediate and delayed prostanoid biosynthetic pathways. J. Biol. Chem. 274, 3103–3115.

5. Smith, W.L. and Langenbach, R. (2001) Why there are two cyclooxygenases. J. Clin. Invest. 107, 1491–1495.

6. Davletov, B., Perisic, O. and Williams, R.L. (1998) Calcium-dependent membrane penetration is a hallmark of the C2 domain of cytosolic phospholipase A2 whereas the C2A domain of synaptotagmin binds membranes electrostatically. J. Biol. Chem. 273, 19093–19096.

7. Murakami, M., Koduri, R.S., Enomoto, A., Shimbara, S., Seki, M., Yoshihara, K., Singer, A., Valentin, E., Ghomashchi, F., Lambeau, G., Gelb, M.H. and Kudo, I. (2001) Distinct arachidonate-releasing functions of mammalian secreted phospholipase A2s in human embryonic kidney 293 and rat mastocytoma RBL-2H3 cells through heparan sulfate shuttling and external plasma membrane mechanisms. J. Biol. Chem. 276, 10083–10096.

8. Kozak, K.R., Rowlinson, S.W. and Marnett, L.J. (2000) Oxygenation of the endocannabinoid, 2-arachidonylglycerol, to glyceryl prostaglandins by cyclooxygenase-2. J. Biol. Chem. 275, 33744–33749.

9. Thuresson, E.D., Lakkides, K.M., Rieke, C.J., Sun, Y., Wingerd, B.A., Micielli, R., Mulichak, A.M., Malkowski, M.G., Garavito, R.M. and Smith, W.L. (2001) Prostaglandin endoperoxide H synthase-1: the functions of cyclooxygenase active site residues in the binding, positioning, and oxygenation of arachidonic acid. J. Biol. Chem. 276, 10347–10357.

10. Kujubu, D.A., Fletcher, B.S., Varnum, B.C., Lim, R.W. and Herschman, H.R. (1991) TIS10, a phorbol ester tumor promoter inducible mRNA from Swiss 3T3 cells, encodes a novel prostaglandin synthase/cyclooxygenase homologue. J. Biol. Chem. 266, 12866–12872.

11. Patrono, C., Patrignani, P. and García Rodríguez, L.A. (2001) Cyclooxygenase-selective inhibition of prostanoid formation: transducing biochemical selectivity into clinical read-outs. J. Clin. Invest. 108, 7–13.

12. Marnett, L.J. and Kalgutkar, A.S. (1999) Cyclooxygenase 2 inhibitors: discovery, selectivity and the future. Trends Pharm. Sci. 20, 465–469.

13. Ueno, N., Murakami, M., Tanioka, T., Fujimori, K., Urade, Y. and Kudo, I. (2001) Coupling between cyclooxygenase, terminal prostanoid synthases and phospholipase A2s. J. Biol. Chem. 276, 34918–34927.

14. Smith, W.L. (1992) Prostanoid biosynthesis and mechanisms of action. Am. J. Physiol. 263, F181–F191.

15. Tilley, S.L., Coffman, T.M. and Koller, B.H. (2001) Mixed messages: modulation of inflammation and immune responses by prostaglandins and thromboxanes. J. Clin. Invest. 108, 15–23.

16. Narumiya, S. and GA, F. (2001) Genetic and pharmacological analysis of prostanoid receptor function. J. Clin. Invest. 108, 25–30.

17. Ford Hutchinson, A.W., Gresser, M. and Young, R.N. (1994) 5-Lipoxygenase. Annu. Rev. Biochem. 63, 383–417.

18. Radmark, O.P. (1999) 5-Lipoxygenase. In: G. Folco, B. Samuelsson and R.C. Murphy (Eds.), Novel Inhibitors of Leukotrienes. Birkhauser Verlag, Basel, pp. 1–22.

19. Evans, J. (1998) 5-Lipoxygenase and 5-lipoxygenase-activating protein. In: J.M. Drazen, S.-E. Dahlén and T.H. Lee (Eds.), Lung Biology in Health and Disease: Five-Lipoxygenase Products in Asthma. Marcel Dekker, New York, NY, pp. 11–32.

20. Thunnissen, M.M., Nordlund, P. and Haeggstrom, J.Z. (2001) Crystal structure of human leukotriene A(4) hydrolase, a bifunctional enzyme in inflammation. Nat. Struct. Biol. 8, 131–135.

21. Lam, B.K. and Frank Austen, K. (2000) Leukotriene $C_4$ synthase. A pivotal enzyme in the biosynthesis of the cysteinyl leukotrienes. Am. J. Respir. Crit. Care Med. 161, S16–19.

22. Silverman, E.S. and Drazen, J.M. (2000) Genetic variations in the 5-lipoxygenase core promoter. Description and functional implications. Am. J. Respir. Crit. Care Med. 161, S77–80.

23. Peters Golden, M. and Brock, T.G. (2000) Intracellular compartmentalization of leukotriene biosynthesis. Am. J. Respir. Crit. Care Med. 161, S36–40.

24. Murphy, R.C. and Wheelan, P. (1998) Pathways of leukotriene metabolism in isolated cell models and human subjects. In: J.M. Drazen, S.-E. Dahlén and T.H. Lee (Eds.), Lung Biology in Health and Disease: Five-Lipoxygenase Products in Asthma. Marcel Dekker, New York, NY, pp. 87–123.

25. Cui, X., Kawashima, H., Barclay, T.B., Peters, J.M., Gonzalez, F.J., Morgan, E.T. and Strobel, H.W. (2001) Molecular cloning and regulation of expression of two novel mouse CYP4F genes: expression in peroxisome proliferator-activated receptor alpha-deficient mice upon lipopolysaccharide and clofibrate challenges. J. Pharmacol. Exp. Ther. 296, 542–550.

26. Serhan, C.N. and Prescott, S.M. (2000) The scent of a phagocyte: Advances on leukotriene b(4) receptors. J. Exp. Med. 192, F5–8.

27. Lynch, K.R., O'Neill, G.P., Liu, Q., Im, D.S., Sawyer, N., Metters, K.M., Coulombe, N., Abramovitz, M., Figueroa, D.J., Zeng, Z., Connolly, B.M., Bai, C., Austin, C.P., Chateauneuf, A., Stocco, R., Greig, G.M., Kargman, S., Hooks, S.B., Hosfield, E., Williams, D.L., Ford Hutchinson, A.W., Caskey, C.T. and Evans, J.F. (1999) Characterization of the human cysteinyl leukotriene CysLT1 receptor. Nature 399, 789–793.

28. Conrad, D.J. (1999) The arachidonate 12/15 lipoxygenases. A review of tissue expression and biologic function. Clin. Rev. Allergy Immunol. 17, 71–89.

29. Gillmor, S.A., Villasenor, A., Fletterick, R., Sigal, E. and Browner, M.F. (1997) The structure of mammalian 15-lipoxygenase reveals similarity to the lipases and the determinants of substrate specificity. Nat. Struct. Biol. 4, 1003–1009.

30. Capdevila, J.H., Falck, J.R. and Harris, R.C. (2000) Cytochrome P450 and arachidonic acid bioactivation. Molecular and functional properties of the arachidonate monooxygenase. J. Lipid Res. 41, 163–181.

31. Zeldin, D.C. (2001) Epoxygenase pathways of arachidonic acid metabolism. J. Biol. Chem. 276, 36059–36062.

D.E. Vance and J.E. Vance (Eds.) *Biochemistry of Lipids, Lipoproteins and Membranes (4th Edn.)*

# CHAPTER 14

# Sphingolipids: metabolism and cell signaling

Alfred H. Merrill Jr. [1] and Konrad Sandhoff [2]

[1] *School of Biology, Petit Institute for Bioengineering and Biosciences, Georgia Institute of Technology,
Atlanta, GA 30332, USA, Tel.: +1 (404) 385-2842; Fax: +1 (404) 385-2917 or +1 (404) 894-0519;
E-mail: al.merrill@biology.gatech.edu*
[2] *Kekule-Institut für Organische Chemie und Biochemie der Rheinischen
Friedrich-Wilhelms-Universität Bonn, D-53121 Bonn, Germany*

## 1. Introduction

Sphingolipids were first described by Johann L.W. Thudichum in *A Treatise on the Chemical Constitution of Brain* (1884) [1]. Among the described compounds were sphingomyelin, cerebroside, and cerebrosulfatide (Fig. 1), which encompass the three categories of sphingolipids known today (phosphosphingolipids, neutral and acidic glycosphingolipids). Thudichum noted that hydrolysis of these lipids produced a compound that ". . . is of an alkaloidal nature, and to which, in commemoration of the many enigmas which it has presented to the inquirer, I have given the name of *Sphingosin*." Thus, this class of lipids became known as *sphingo*lipids.

Thudichum, a practicing physician throughout most of his life, was searching for a better understanding of disease, but appreciated that ". . . to reach this goal of complete knowledge. . . the medicinal chemist must. . . not. . . carry on research by a kind of fishing for supposed disease-poisons, of which, according to my view of the subject, the attempt of the boy to catch a whale in his mother's washing-tub is an appropriate parable." Later studies fulfilled Thudichum's faith in the value of basic research when

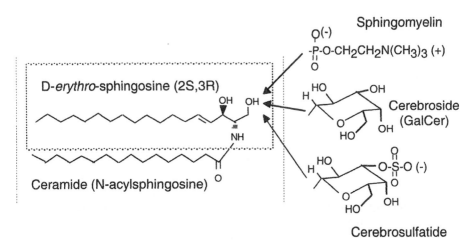

Fig. 1. Structures of sphingosine, ceramide, sphingomyelin, a cerebroside (galactosylceramide) and cerebrosulfatide from human brain.

several genetic diseases were found to have elevated amounts of sphingolipids (such as sphingomyelin in Niemann–Pick's disease and cerebrosides in Gaucher's disease) arising from defects in enzymes responsible for sphingolipid turnover, activator proteins for such enzymes, or lipid trafficking (for reviews, see Hakomori [2] and Schuette et al. [3]). This knowledge allowed development of methods for diagnosis of such sphingolipid storage diseases (or 'sphingolipidoses'), screening of families at risk, and, for at least Gaucher's disease, some degree of correction of the disorder by enzyme replacement. Progress is also being made using inhibitors of sphingolipid synthesis, and gene replacement offers promise for the future.

For many years, the only diseases associated conclusively with sphingolipids involved defective sphingolipid turnover. Disruption of sphingolipid biosynthesis is now known to be the major mechanism of action of mycotoxins (fumonisins and alternaria toxins) that cause a wide spectrum of diseases of plants and animals (A.H. Merrill, 2001). And in 2001, genetic defects in sphingolipid biosynthesis were shown to cause the most common hereditary disorder of peripheral sensory neurons (hereditary sensory neuropathy type I) (J.L. Dawkins, 2001; K. Bejaoui, 2001). It is certain that additional genetic diseases due to abnormal sphingolipid biosynthesis will be found, and knockout mice defective in the biosynthesis of glucosylceramide and other glycolipids have severe defects, especially in the developing nervous system (T. Kolter, 2000). There are also indications that sphingolipids and sphingolipid analogs may be useful for prevention and treatment of disease, e.g., gangliosides (R. McKallip, 1999) and α-galactosylceramide (M. Taniguchi, 1997 and 1998) have potent effects as modulators of the immune system, ceramide-coated balloon catheters limit neointimal hyperplasia after stretch injury in carotid arteries (R. Charles, 2000), and dietary sphingolipids protect against colon tumorigenesis (E.M. Schmelz, 2001). These probably reflect just a few of the ways sphingolipids are relevant to pathology, nutrition and medicinal chemistry.

## 1.1. Biological significance of sphingolipids

Sphingolipids are found in essentially all animals, plants, and fungi, as well as some prokaryotic organisms and viruses. They are mostly in membranes, but are also major constituents of lipoproteins. The functions of sphingolipids are still being discovered, but there are at least three, i.e., structure, recognition and signal transduction, which have been summarized diagrammatically in Fig. 2.

### 1.1.1. Biological structures
Some glycosphingolipids and sphingomyelins tend to cluster rather than behave like typical 'fluid' membrane lipids. This behavior arises from the mostly saturated alkyl sidechains, which allow strong van der Waals interactions, and the ceramide hydroxyls, amide bond and polar headgroups that are capable of hydrogen bonding and dipolar interactions (it is common for sphingolipids to have phase transition temperatures > 37°C). Sphingolipids contribute to the formation of regions of the plasma membrane termed 'rafts' and 'caveolae' ([5,6], Chapter 1), which are enriched in growth factor receptors, transporters and other proteins, especially proteins with a glycosylphos-phatidylinositiol-lipid anchor. Sphingolipids contribute to the stability of other types

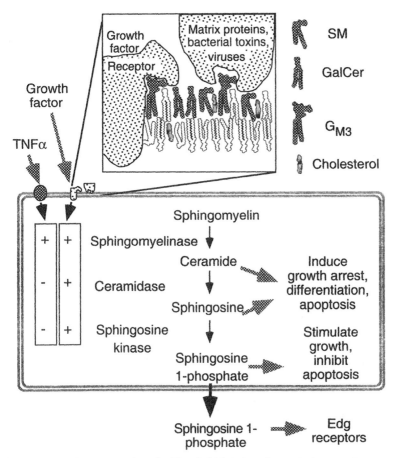

Fig. 2. Schematic representation of sphingolipid functions. Shown in the upper inset are sphingolipid- (and cholesterol-) enriched regions of the plasma membrane ('rafts'), and sphingolipids serving as ligands for extracellular proteins and receptors on the same cell. The lower diagram of a cell illustrates how agonists (such as tumor necrosis factor-α, TNF-α, and growth factors) can activate combinations of sphingolipid metabolizing enzymes to produce bioactive products that affect the shown cell behaviors. Sphingosine 1-phosphate is also secreted as an agonist for some members of the Edg family receptors (now named S1P).

of biological structures, such as the lamellar bodies that maintain the permeability barrier of skin (P. Wertz, 2000) and lipoproteins (S.L. Schissel, 1996). However, not all sphingolipids are so ordered, and some, such as sphingosine 1-phosphate and sphingosylphosphorylcholine (lysosphingomyelin), are sufficiently polar to exist in aqueous environments (Y. Yatomi, 1997).

### 1.1.2. Biological recognition
Membrane sphingolipids are located predominantly on the outer leaflet of the plasma membrane, the lumen of intracellular vesicles and organelles (endosomes, Golgi membranes, etc.), and in as yet undefined locations in mitochondria and nuclei. The complex

carbohydrate moieties are often signatures for particular cell types, and mediate inter-actions with complementary ligands, such as extracellular matrix proteins and receptors (S.I. Hakomori, 2000) (Fig. 2), including direct carbohydrate–carbohydrate binding with headgroups on neighboring cells (K. Handa, 2000). In some cases, sphingolipids interact with proteins on the same cell surface (Fig. 2). Such binding can be used to control the location of the protein (for example, in membrane rafts with other signaling proteins) as well as to modify the conformation of the receptor and its activity [7]. This is exemplified by the binding of ganglioside $G_{M3}$ by the epidermal growth factor receptor, which makes the receptor refractory to activation by this growth factor (E.J. Meuillet, 2000; A.R. Zurita, 2001). Sphingolipids are also recognized by viruses, bacteria and bacterial toxins as a means of both attachment and entry into the cell via membrane trafficking (K.A. Karlsson, 1992; C.A. Lingwood, 1999).

### 1.1.3. Signal transduction

The sphingolipid backbones are members of a signaling paradigm shown in Fig. 2, wherein receptor activation by agonists such as tumor necrosis factor-α and platelet-derived growth factor induce sphingomyelin turnover to elevate ceramide, or down-stream metabolites (sphingosine or sphingosine 1-phosphate) [8–10]. These products activate or inhibit multiple downstream targets (protein kinases, phosphoprotein phos-phatases and others) that control cell behaviors as complex as growth, differentia-tion and programmed cell death (apoptosis). Because ceramide and sphingosine 1-phosphate often have opposing signaling functions (e.g., induction versus inhibition of apoptosis; inhibition versus stimulation of growth), Sarah Spiegel has proposed that cells utilize a ceramide/sphingosine 1-phosphate 'rheostat' in deciding between growth arrest/apoptosis versus proliferation/survival (S. Spiegel, 1999). Sphingosine 1-phosphate can be released from cells and serve as an agonist for S1P receptors [10], hence, this compound serves as both a first and second messenger!

The field of sphingolipid signaling is still relatively young and has many new facets that reveal the biochemical 'logic' of using such complex molecules to control cell behavior. For example, sphingomyelin turnover not only produces 'signaling' metabo-lites, but also alters the structure of membrane domains that depend on the presence of this lipid (furthermore, when ceramide accumulates, its biophysical properties can profoundly affect membrane structure and the behavior of associated receptors and other proteins) [5,6]. A similar paradigm can be envisioned for glycosphingolipids. Thus, sphingolipid 'signaling' is an ensemble of changes in membrane structure and dynamics, the production (and removal) of bioactive metabolites, and the activation and/or inhibition of downstream targets.

### 1.2. Structures and nomenclature of sphingolipids

More than 300 different types of complex sphingolipids have been reported, and this does not include differences in the ceramide backbone. It has become necessary to develop a system of nomenclature for sphingolipids so that individual species can be referred to in a logical manner [11]. Nonetheless, there is still considerable variability in the names that are used for these compounds. For example, 'sphingosine' is still

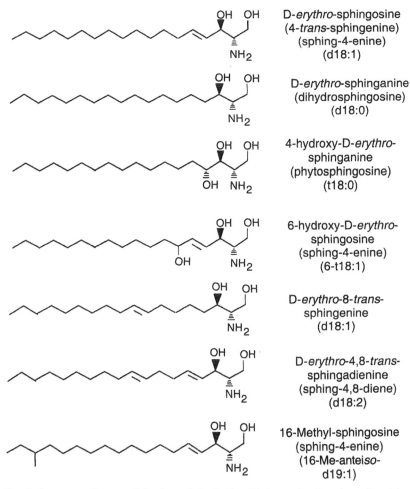

Fig. 3. Structures of some of the long-chain (sphingoid) bases that have been found in sphingolipids. Abbreviations for these compounds are shown in parentheses.

in common usage although the names recommended by the IUPAC are (*E*)-sphing-4-enine or (2*S*,3*R*,4*E*)-2–aminooctadec-4-ene-1,3-diol (by their recommendation, dihydro-'sphingosine' is sphinganine, and 4-hydroxysphinganine (also called phytosphingosine) is (2*S*,3*S*,4*R*)-2-aminooctadecane-1,3,4-triol). This chapter uses the most familiar names: sphingosine, sphinganine and 4-hydroxysphinganine.

Sphingosine is the prevalent backbone of most mammalian sphingolipids; however, over 60 different species of long-chain bases have been reported [12] and include compounds (Fig. 3) with (1) alkyl chain lengths from 14 to 22 carbon atoms, (2) different degrees of saturation at carbons 4 and 5, (3) a hydroxyl group at positions 4 or 6, (4) double bonds at other sites in the alkyl chain, and (5) branching (methyl groups) at the ω-1 (iso), ω-2 (anteiso), or other positions. Sphingoid bases are abbreviated by

378

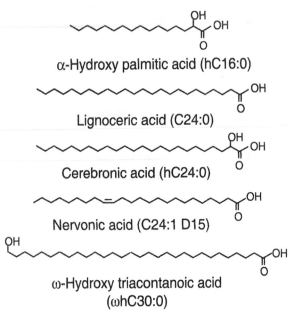

α-Hydroxy palmitic acid (hC16:0)

Lignoceric acid (C24:0)

Cerebronic acid (hC24:0)

Nervonic acid (C24:1 D15)

ω-Hydroxy triacontanoic acid
(ωhC30:0)

Fig. 4. Structures of representative fatty acids, including α- and ω-hydroxy fatty acids, that are found in mammalian sphingolipids. Abbreviations for these compounds are shown in parentheses.

citing (in order of appearance in the abbreviation) the number of hydroxyl groups (d and t for di- and tri-hydroxy, respectively), chain length and number of double bonds as shown in Fig. 3.

The majority of the sphingoid bases in cells are $N$-acylated with long-chain fatty acids to produce ceramides(s) (Fig. 1), although $O$-acylated (A. Abe, 1998), phosphorylated-(sphingosine 1-phosphate) and $N$-methylated-($N,N$-dimethylsphingosine) derivatives also exist. The fatty acids of ceramide vary in chain length (14 to 30 carbon atoms), degree of unsaturation (but are mostly saturated), and presence or absence of a hydroxyl group on the α- or ω-carbon atom. Structures and abbreviations for some fatty acids are shown in Fig. 4.

Most sphingolipids have a polar headgroup at position 1 (Figs. 1 and 5). Sphingolipids are often grouped based on the headgroups into the phosphosphingolipids and glycosphingolipids; however, these categories are not mutually exclusive: the major sphingolipids of yeast are ceramide phosphorylinositols. Glycosphingolipids are classified into broad types on the basis of carbohydrate composition. Neutral glycosphingolipids contain uncharged sugars such as glucose (Glc), galactose (Gal), $N$-acetylglucosamine (GlcNAc), $N$-acetylgalactosamine (GalNAc), and fucose (Fuc). Acidic glycosphingolipids contain ionized functional groups such as phosphate, sulfate (sulfatoglycosphingolipids), or charged sugar residues such as sialic acid ($N$-acetylneuraminic acid) in gangliosides or glucuronic acid in some plant glycosphingolipids. Further classification can be made on the basis of shared partial oligosaccharide sequences, sometimes referred to as 'root structures' as summarized in Table 1.

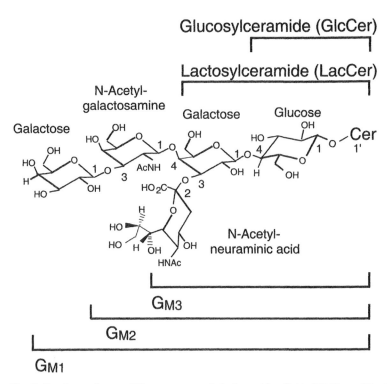

Fig. 5. Structures of some of the common neutral glycosphingolipids (GlcCer and LacCer) and gangliosides $G_{M1}$, $G_{M2}$ and $G_{M3}$. 'Cer' refers to the ceramide backbone [for another useful overview structure, for ganglioside $G_{D1_a}$, see G. van Echten-Deckert (1999)].

Table 1
Nomenclature for classification of glycosphingolipids

| Root name | Abbreviation | Partial structure [a] | | | |
|---|---|---|---|---|---|
| | | IV | III | II | I |
| Ganglio | Gg | Galβ1–3GalNacβ1–4Galβ1–4Glcβ1–1'Cer | | | |
| Lacto | Lc | Galβ1–3GlcNacβ1–3Galβ1–4Glcβ1–1'Cer | | | |
| Neolacto | nLc | Galβ1–4GlcNacβ1–3Galβ1–4Glcβ1–1'Cer | | | |
| Globo | Gb | GalNacβ1–3Galα1–4Galβ1–4Glcβ1–1'Cer | | | |
| Isoglobo | iGb | GalNacβ1–3Galα1–3Galβ1–4Glcβ1–1'Cer | | | |
| Mollu | Mu | GalNacβ1–2Manα1–3Manβ1–4Glcβ1–1'Cer | | | |
| Arthro | At | GalNacβ1–4GlcNacβ1–3Manβ1–4Glcβ1–1'Cer | | | |

[a] Roman numerals define sugar positions in the 'root' structure.

Gangliosides are often denoted by the 'Svennerholm' nomenclature [11] that is based on the number of sialic acid residues (e.g., $G_{M1}$ refers to a monosialo-ganglioside) and the relative position of the ganglioside upon thin-layer chromatography (thus, the order of migration of the series of monosialogangliosides in Fig. 5 is $G_{M3} > G_{M2} > G_{M1}$). By commonly used nomenclatures, the same compound might be called ganglioside $G_{M1}$, $II^3$-$\alpha$-$N$-acetylneuraminosyl-gangliotetraosylCer, $II^3$-$\alpha$-Neu5NacGg$_4$Cer, or depicted as:

Galβ1–3GalNacβ1–4Galβ1–4Glcβ1–1′Cer
|
Neu5Acα2–3

Note that the Roman numeral and Arabic superscript refer to the sugar in the root structure (cf. Table 1) that is substituted (counting from the ceramide toward the non-reducing end) and the position of that substitution, respectively.

A number of sphingolipids are referred to by their historic names as antigens and blood group structures, such as Forssman antigen ($IV^3$-$\alpha$-GalNAc-Gb$_4$Cer), a globo-pentosylceramide that is found in many mammals (but it is unclear if humans express this antigen) and the Lewis blood group antigens, which correspond to a family of α1–3-fucosylated glycan structures (Lewis ×, sialyl Lewis ×, etc.). For more information on these aspects of glycosphingolipidology see Varki et al. [13].

## 2. Chemistry and distribution

This section will summarize some of the properties of sphingolipids. More information is available in Merrill and Hannun [14].

### 2.1. Sphingoid bases

A distinctive feature of sphingoid bases is that they can bear a net positive charge at neutral pH, which is rare among naturally occurring lipids. Nonetheless, the p$K_a$ of the amino group is low for a simple amine (between 7 and 8) (A.H. Merrill, 1989), which means that a portion is uncharged at physiologic pH. This may help explain why sphingoid bases can readily move among membranes and across bilayers (in the uncharged state), unless transmembrane movement is impeded by acidic pH, such as in lysosomes.

Structural elucidation and quantitation of long-chain bases is possible using a variety of analytical techniques, including gas chromatography and high-performance liquid chromatography, mass spectroscopy, and nuclear magnetic resonance spectroscopy [13,14].

### 2.2. Ceramides

Ceramides per se are mostly found in small amounts in tissues, with the notable exception of the stratum corneum, where they are major determinates of the water permeability barrier of skin (P. Wertz, 2000). Many ceramides (even as part of complex sphingolipids) migrate on thin-layer chromatography as multiple bands due to the presence of at least several types of sphingoid bases and fatty acids. The molecular species can be analyzed by a number of techniques, such as gas chromatography, high-performance liquid chromatography (HPLC), or hydrolysis (or methanolysis) followed by analysis of the sphingoid bases and fatty acids [14]. However, the most information

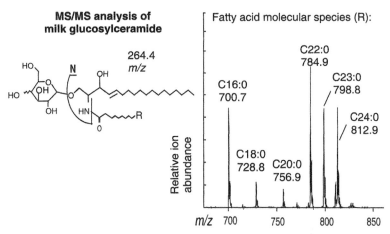

**MS/MS analysis of milk glucosylceramide**

264.4 m/z

Fatty acid molecular species (R):

Relative ion abundance

C16:0 700.7

C18:0 728.8  C20:0 756.9

C22:0 784.9

C23:0 798.8

C24:0 812.9

m/z  700    750    800    850

Fig. 6. Major fragmentation sites of a monohexosylceramide upon electrospray tandem mass spectrometry (left panel) and (right panel) a typical precursor ion spectrum (obtained with a bovine milk extract) monitoring $m/z$ 264.4 (a signature fragment obtained with sphingolipids with a sphingosine backbone) over the range $m/z$ 675–875. The labeled signals represent the various amide-linked fatty acids on the milk glucosylceramides. For more information see Sullards and Merrill [15].

is obtained by combining HPLC with electrospray tandem mass spectrometry (ESI MS/MS) [15]. In this method, the ceramides are separated as classes (free ceramides, sphingomyelins, glucosylceramides, etc.) by HPLC and the eluant is introduced directly into the ionizing chamber of the mass spectrometer, where the solvent is rapidly evaporated under high vacuum and the compounds are suspended in the gas phase as individual charged species. These 'parent' ions are separated by the first MS, then allowed to collide with a gas (such as $N_2$) to produce fragments that are separated by the second MS. Besides high sensitivity, the advantage of this instrumentation is the ability to focus on the compounds of interest in crude mixtures. For example, glycosylceramides containing sphingosine will fragment to $m/z$ 264.4 (Fig. 6); therefore, the second MS can be set to detect $m/z$ 264.4 and the first MS to identify the parent ions that produce this fragment. This is illustrated for the glucosylceramides in a milk lipid extract in Fig. 6. Quantitation is achieved by spiking the sample with an internal standard with a chemical composition sufficiently similar to the unknowns for them to fragment with similar efficiencies, as described in the legend to Fig. 7. More accurate and sensitive quantitation can be obtained using a specialized MS/MS technique known as multiple reaction monitoring (MRM), in which the mass spectrometer is programmed to maximize the time spent detecting specific precursor/product ion transitions, and the detection of each individual molecular species can be optimized with respect to ion formation and decomposition.

Use of such methods allows more accurate and facile quantitation of multiple sphingolipid species in cells and other biological materials. Fig. 7 gives a typical analysis of the sphingolipids of NIH 3T3 cells, which contain (nmol per $10^6$ cells): SM (2.7), GlcCer (0.31), Cer (0.082), sphingosine (0.017), sphingosine-1-phosphate (0.011), sphinganine (0.160), and sphinganine-1-phosphate (<0.001). For compari-

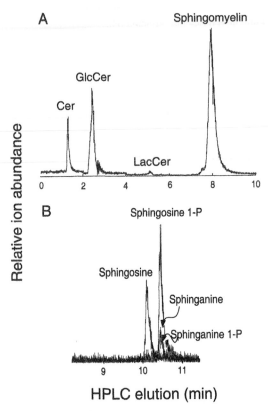

Fig. 7. A high-performance electrospray tandem mass spectrometry (HPLC–ESI–MS/MS) total ion chromatogram of endogenous levels of the complex sphingoid bases ceramide (Cer), glucosylceramide (GlcCer), lactosylceramide (LacCer), and sphingomyelin (SM) from NIH 3T3 cells (upper panel A). The cells were treated with base to remove glycerolipids and the organic solvent-soluble compounds were separated on a normal phase column to produce the profile shown. The amounts of each species can be quantified by comparison with spiked internal standards (Cer, GlcCer and SM with C12 fatty acide [15]). The elution profile in panel B is the extracted ion chromatogram for the free sphingoid bases from these cells (separated by reversed-phase chromatography prior to ESI–MS/MS) from NIH 3T3 cells (dC20:0, dC20:1 and dC17:1-1-phosphate are used as internal standards) [15]. Panel C demonstrates that this methodology can also be used to analyze lysosphingolipids (lysosphingomyelin and psychosine) and N-methyl sphingosines (these were not detected in NIH 3T3 cells so the data are for mixtures of standards).

son, human monocytes contain (nmol per $10^6$ cells): SM (1.2), GlcCer (0.032), Cer (0.027), sphingosine (0.024), sphingosine-1-phosphate (0.007), sphinganine (0.007), and sphinganine-1-phosphate (<0.001) (M.C. Sullards, unpublished). The amounts of the sphingolipid backbones (ceramides and sphingoid bases) are often small, but can be comparable to other sphingolipids in the cells (cf. GlcCer and Cer in monocytes).

Many studies add ceramides or ceramide analogs to cells in culture, enzyme assays, or other aqueous environments, but delivery is hampered by the hydrophobicity of long-chain ceramides, which have low critical micelle concentrations ($\leq 10^{-10}$ M). This is often circumvented by using short-chain ceramides (e.g., N-acetyl- or N-hexanoylsphingosine) or by delivering long-chain ceramides in detergents, liposomes, or organic solvent mixtures (C. Luberto, 2000). Fluorescent ceramide analogs, such as $N$-[6-(7-nitrobenz-2-oxa-1,3-diazol-4-yl)amino]hexanoylceramide (NBD-ceramide) and boron dipyrromethane difluoride ceramide (BODIPY-ceramide) are readily taken up by cells and have proven very useful in studies of sphingolipid transport and metabolism (A. Dagan, 2000; R.E. Pagano, 2000).

## 2.3. Phosphosphingolipids

Sphingomyelin is the major phosphosphingolipid in mammalian tissues. The biophysical properties of sphingomyelins have been described in Section 1.1.1 with respect to their role in rafts and caveolae [5] (see also Brugger, 1999 and 2000).

Animals additionally produce ceramide phosphorylethanolamines (M.N. Nikolova-Karakashian, 2000) and ceramide phosphate (S. Bajjalieh, 2000). Fungi such as *Saccharomyces cerevisiae* and *Sporothrix schenckii* contain ceramides derivatized with inositol phosphate with or without further addition of mannose and other carbohydrates (R.C. Dickson, 1999; C.V. Loureiro y Penha, 2001).

## 2.4. Glycosphingolipids

Plants and fungi often contain glycosphingolipids with relatively simple carbohydrate structures [16,17] whereas animals have a wide variety of simple to complex sphingolipids [18–20], including higher-order globosides, gangliosides and sulfatides.

### 2.4.1. Neutral glycosphingolipids

Glucosylceramide (Glcβ1–1′Cer, or GlcCer), galactosylceramide (Galβ1–1′Cer, or Gal-Cer), and lactosylceramide (Galβ1–4Glcβ1–1′Cer, or LacCer) are the most common neutral glycosphingolipids in higher organisms (Figs. 1 and 5). The glycosidic linkage to ceramide is of the β configuration in these lipids. The *gala* type, such as galabiosylceramide (Galα1–4Galβ1–1′Cer), is found primarily in kidney and pancreas. Other organisms utilize additional sugars in neutral glycosphingolipids; for example, the freshwater bivalve *Hyriopsis schlegelii* has Manβ1–1′Cer and Manβ1–2Manβ1–1′Cer, and Manβ1–4Glcβ1–1′Cer occurs in plants. More complex neutral glycosphingolipids are generally derived from LacCer or Manβ1–4Glcβ1–1′Cer. Numerous compounds also contain the oligosaccharide sequence –Galβ1–4GlcNAcβ1–3– as seen at the non-reducing end of the paragloboside Galβ1–4GlcNAcβ1–3Galβ1–4Glcβ1–1′Cer and a Lewis blood group-specific antigen from human adenocarcinoma, Galβ1–4[Fucα1–3]GlcNAcβ1–3Galβ1–4Glcβ1–1′Cer.

As discussed in Section 1.1.1, neutral glycolipids can form aggregates that may be important in establishing regions of the membrane with unique properties (see also T.E. Thompson, 1985).

### 2.4.2. Acidic glycosphingolipids

#### 2.4.2.1. Gangliosides.
Gangliosides are found in all cells of vertebrates, but in especially high amounts in the central nervous system. While diverse in structure, they

have in common one or more units of an acidic sugar called N-acetyl-neuraminic acid (more commonly called 'sialic acid') attached via α-glycosidic linkages to other sugars (Fig. 5). Sialic acids may have N-acetyl or N-glycolyl (i.e., hydroxyacetyl) groups at C5 (and less often C7, C8 or C9), and are distinguished by the names N-acetylneuraminic acid (Neu5Ac) and N-glycolylneuraminic acid (Neu5Gc). The simplest gangliosides contain one sialic acid and galactose or glucose, such as Neu5Gcα2-6Glcβ1–1'Cer (found in human brain). Most gangliosides are derivatives of LacCer (Fig. 5). The addition of one sialic acid gives $G_{M3}$ ganglioside (Neu5Acα2-3Galβ1–4Glcβ1–1'Cer), which is found in many biological sources. Gangliosides frequently contain a string of two or three sialic acid residues, attached to each other in α2–8 glycosidic linkages, examples of which are $G_{D3}$ (Neu5Acα2–8Neu5Acα2–3Galβ1–4Glcβ1–1'Cer) and $G_{T3}$ (Neu5Acα2–8Neu5Acα2–8Neu5Acα2–3Galβ1–4Glcβ1–1'Cer).

Gangliosides were initially classified into a few related series, such as the *ganglio* and *neolacto* types (Table 1), with the sialic acids on one or both root galactose residues of the *ganglio* type as, for example, in ganglioside $G_{D1a}$ (shown below) from human brain. However, additional kinds of naturally occurring structures (shown below) illustrate the structural diversity that can be obtained with glycosphingolipids of even the same carbohydrate composition.

$$
\begin{array}{cc}
\text{Galβ1–3GalNacβ1–4Galβ1–4Glcβ1–1'Cer} & G_{D1a} \\
| \qquad\qquad | & \\
\text{Neu5Acα2–3 \qquad Neu5Acα2–3} &
\end{array}
$$

$$
\begin{array}{cc}
\text{Neu5Acα2–6} & \\
| & \\
\text{Galβ1–3GalNacβ1–4Galβ1–4Glcβ1–1'Cer} & G_{D1α} \\
| & \\
\text{Neu5Acα2–3} &
\end{array}
$$

The extensive carbohydrate chains and (poly)anionic charge make many gangliosides highly amphiphilic (with critical micelle concentrations of ca. $10^{-8}$ M); many will partition into the aqueous phase when extracted with organic solvents.

*2.4.2.2. Phosphorus-containing glycosphingolipids.* As described in Section 2.3, fungi, plants, and protozoa contain sphingolipids in which ceramide is attached to an oligosaccharide via a phosphodiester linkage to *myo*-inositol and other carbohydrates.

*2.4.2.3. Sulfatoglycosphingolipids.* Over a dozen sulfated glycosphingolipids have been isolated from vertebrates echinoderms and microorganisms [21]. Cerebrosulfatide (3'-sulfo-Galβ1–1'Cer, or galactosylceramide-I³-sulfate) (Fig. 1) was the first such substance to be isolated, and is the major sulfoglycolipid of brain, kidney, the gastrointestinal tract and endometrium, and is a major glycolipid of mammalian male germ cells. Sulfatides are thought to be involved in neuronal cell differentiation, myelin formation and maintenance (F.B. Jungalwala, 1994; J.L. Dupree, 1999), and other processes (D. Mamelak, 2000). Two glucuronyl (GlcA) sulfatoglycolipids (3-O-SO₃-GlcAβ1–3Galβ1–4GlcNAcβ1–3Galβ1–4Glcβ1–1'Cer and 3-O-SO₃-GlcAβ1–3(Galβ1–4GlcNAcβ1–3)₂Galβ1–4Glcβ1–1'Cer) occur in the peripheral nervous system.

## 2.5. Lysosphingolipids

Lysosphingolipids lack the amide-linked fatty acid of the ceramide backbone, which makes them highly water-soluble. A lysoglycosphingolipid with one hexose (e.g., glucosyl- or galactosyl-sphingosine) is colloquially referred to as a 'psychosine'. It has long been suspected that the appearance of psychosines, which are toxic to cells in culture, in some sphingolipid storage diseases implicates them in the pathology [22]. The cause of the toxicity is not known, but lysosphingolipids tend to form micelles and disrupt membranes, as well as inhibit protein kinase C (Y.A. Hannun, 1987) and bind to a recently discovered family of psychosine receptors (D.S. Im, 2001). Novel lysosphingolipids with alkyl groups on the sugar have been discovered and termed plasmalopsychosines (K.K. Sadozai, 1993; T. Hikita, 2001).

## 2.6. Sphingolipids covalently linked to proteins

At least two examples of covalent attachments between sphingolipids and proteins have been found. Structural proteins of the cornified cell envelope of the skin permeability barrier are covalently attached to ω-hydroxyceramides and ω-hydroxyglucosylceramides through the ω-hydroxyl groups (T. Doering, 1999; M.E. Stewart, 2001). And for some organisms, such as yeast, ceramides replace diacylglycerols in the covalently attached phosphatidylinositolglycan-linkage (Chapter 2) that is used to attach the proteins to membranes (A. Conzelmann, 1995).

## 2.7. Sphingolipids in food

Unlike other lipids, for which dietary consumption has long been known, the per capita consumption in the United States has only recently been estimated to be 115–140 g/year, or 0.3–0.4 g/day [23]. The amounts in any given food vary considerably, from a few μmol/kg in fruits and some vegetables to ca. 2 mmol/kg (1–2 g/kg) in dairy products, egg and soybeans. Most foods of mammalian origin (beef, milk, poultry, etc.) have sphingomyelins, cerebrosides, globosides, gangliosides, sulfatides, etc., that are comprised of ceramide backbones with sphingosine (d18 : $1^{\Delta 4}$) or 4-hydroxysphinganine (t18 : 0). The complex sphingolipids of plants are mainly cerebrosides (mono- and oligohexosylceramides, with glucose the most common hexose), and ceramides that have relatively little d18 : $1^{\Delta 4}$ and mostly d18 : $1^{\Delta 8}$, d18 : $2^{\Delta 4,8}$, t18 : 0 and t18 : $1^{\Delta 8}$ sphingoid bases with α-hydroxy fatty acids.

# 3. Biosynthesis of sphingolipids

The general pathways for sphingolipid metabolism are reasonably well characterized [24] and their regulation is beginning to be understood through the development of specific inhibitors and identification of the genes for many of these enzymes.

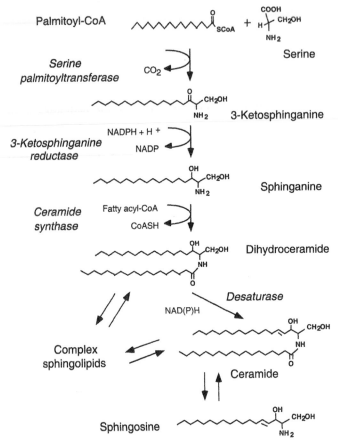

Fig. 8. Biosynthetic pathway for ceramide and sphingosine. Note that free sphingosine is formed after ceramide turnover, not as an intermediate of de novo biosynthesis.

## 3.1. Sphingoid bases and ceramide

### 3.1.1. Synthesis of the long-chain base backbone

Ceramide synthesis begins with the condensation of palmitoyl-CoA and L-serine (Fig. 8), catalyzed by the pyridoxal phosphate-dependent enzyme serine palmitoyltransferase. The reaction proceeds with overall retention of configuration of C2 of serine via the proposed mechanism shown in Fig. 9.

As would be predicted from this mechanism, serine palmitoyltransferase undergoes 'suicide' inhibition by β-halo-L-alanines and L-cycloserine. More potent and selective inhibitors have been isolated from microorganisms and are listed in Table 2. One is available commercially (myriocin or ISP-1) (Fig. 10), and is illustrated in Fig. 9 as the likely adduct that it forms with serine palmitoyltransferase.

For mammals and yeast, at least two gene products (termed SPTLC1 and SPTLC2, or sometimes SPT1 and SPT2) are necessary for activity (R.C. Dickson, 2000) and appear to be physically associated (K. Gable, 2000; K. Hanada, 2000). Human genetic

Table 2

Inhibitors of sphingolipid metabolism

| Enzyme | Inhibitor | Reference |
|---|---|---|
| Serine palmitoyltransferase | Cycloserine | M. Lev, 1984 |
| | β-Fluoroalanine | K.A. Medlock, 1988 |
| | Sphingofungin C | M.M. Zweerink, 1992 |
| | Lipoxamycins | S.M. Mandela, 1994 |
| | ISP-1/myriocin | Y. Miyake, 1995 |
| | Viridiofungin A | S.M. Mandela, 1997 |
| Ceramide synthase | Fumonisins $B_1$, $B_2$, etc. | E. Wang, 1991 |
| | Alternaria toxin | A.H. Merrill, 1993 |
| | Australifungins | S.M. Mandela, 1995 |
| | N-acylaminopentols | H. Humpf, 1998 |
| Glc : ceramide synthase | PDMP (D-*threo*-1-phenyl-2-decanoylamino-3-morpholino-1-propanol) and related analogs | N.S. Radin, 1990; A. Abe, 1995; J.S. Shayman, 2000 |
| | N-butyldeoxynojirimycon | F.M. Platt, 1997 |
| Lactosylceramide synthase | Epoxy-glucosylceramide | C. Zacharias, 1994 |
| Sphingomyelin synthase | PDMP (at very high concentrations) | A. Abe, 1995 |
| | D609 | C. Luberto, 1998 |
| Inositolphosphorylceramide synthase | Khafrefungin | S.M. Mandala, 1997 |
| | Aureobasidin A | M.M. Nagiec, 1997 |
| | Rustimicin | S.M. Mandala, 1998 |
| Sialidase | 2-Deoxy-2,3-dehydro-N-acetylneuraminic acid | P. Meindl, 1969; S. Usuki, 1988 |
| Glucocerebrosidase | Conduritol B-epoxide | G. Legler, 1977; S. Mahdiyoun, 1992 |
| Acidic sphingomyelinase | Phosphatidylinositol 4′,5′-bisphosphate, adenosine 3′,5′-diphosphate, adenine-9-beta-D-arabinofuranoside 5′-monophosphate | L.E. Quintern, 1987 |
| | SR33557 ((2-isopropyl-1-(4-[3-N-methyl-N-(3,4-dimethoxy-β-phenethyl) amino]propyloxy)-benzene sulfonyl))indolizine | J.P. Jaffrezou, 1992 |
| Neutral sphingomyelinase | 3-O-methyl-sphingomyelin | M.D. Lister, 1995 |
| | Glutathione | B. Liu, 1997 |
| | Difluoromethylene sphingomyelin | T. Yokomatsu, 2001 |
| | Scyphostatin and analogs | T. Izuhara, 2001; C. Arenz, 2001 |
| Ceramidase | N-Oleoyl-ethanolamine | M. Sugita, 1975 |
| | D-MAPP | A. Bielawska, 1996 |
| Sphingosine kinase | D- and L-*threo*-sphingosine (and sphinganine) | B.M. Buehrer, 1993 |
| | S-15183a and b | K. Kono, 2001 |
| Sphingosine 1-phosphate lyase | 4-Deoxypyridoxine-5′phosphate | W. Stoffel, 1969; P. Van Veldhoven, 1993 |

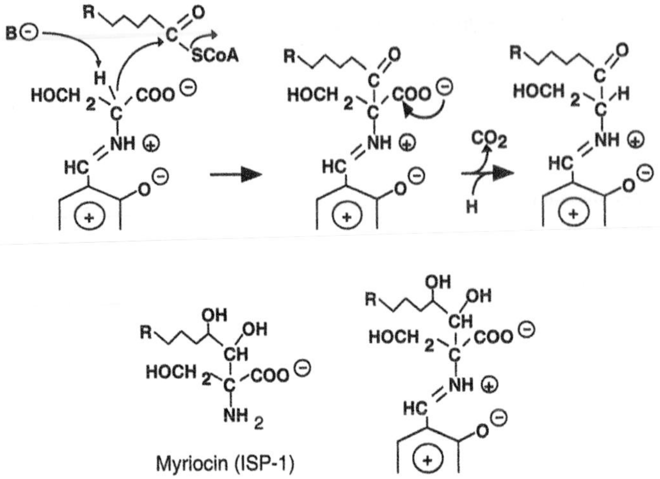

Fig. 9. A probable reaction mechanism for serine palmitoyltransferase (modified from K. Krisnangkura, 1976) and the adduct that presumably accounts for the high-affinity inhibition by myriocin (ISP1) (for the full structure of myriocin see Fig. 10).

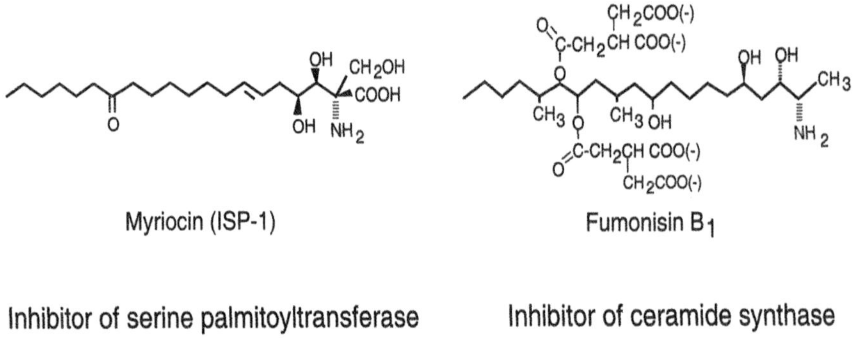

Myriocin (ISP-1)                                Fumonisin B$_1$

Inhibitor of serine palmitoyltransferase        Inhibitor of ceramide synthase

Fig. 10. Representative inhibitors of sphingolipid biosynthesis that have been isolated from microorganisms. For more inhibitors and references for these compounds, see Table 2.

defects in SPTLC1 have been shown to cause hereditary sensory neuropathy type I. In yeast, an 80-amino acid polypeptide has been shown also to affect activity (K. Gable, 2000). Relatively little is known about the structure of serine palmitoyltransferase due to its membrane association; however, a soluble, homodimeric enzyme is produced by *Sphingomonas* (H. Ikushiro, 2001) and should be amenable to structural and mechanistic studies.

Sphingoid base synthesis can be suppressed by addition of lipoproteins or free sphingoid bases to cells in culture, and studies with a phosphorylated but poorly degraded analog (*cis*-4-methylsphingosine) indicates that sphingoid base 1-phosphate(s) downregulate serine palmitoyltransferase activity (G. van Echten-Deckert, 1997). Regulation at a transcriptional level has been seen mostly with cells in culture (reviewed in Linn et al. [25]) but sometimes in vivo (R.A. Memon, 2001) and include endotoxin and cytokines, UV irradiation, retinoic acid, corticosteroids, and phorbol esters. Transcriptional regulation appears to involve mainly changes in SPTLC2 mRNA. There is also post-translational activation of serine palmitoyltransferase in response to etoposide in mammalian cells (D.K. Perry, 2000) and heat shock in yeast (G.M. Jenkins, 2001). The heat shock response is interesting, not only with respect to the downstream pathways that are affected, such as amino acid transport (M.S. Skrzypek, 1998) and ubiquitin-dependent proteolysis (N. Chung, 2000), but also because the long-chain bases that are elevated are predominantly eicosasphinganines (i.e., 20-carbon atoms in length) (R.C. Dickson, 1997).

The next step of sphingoid base synthesis, the reduction of 3-keto-sphinganine (Fig. 8), is apparently rapid in vivo because the 3-keto-intermediate is rarely detected. The gene for this reductase has been identified in yeast (T. Beeler, 1998).

### 3.1.2. Synthesis of the N-acyl-derivatives of sphingoid bases

As shown in Fig. 8, free sphinganine is acylated to dihydroceramides by ceramide synthase(s), which utilize a wide variety of sphingoid bases and fatty acyl-CoAs, and may be a family of isoenzymes (E. Wang, 2000). Ceramide can also be made by reversal of ceramidase (S. El Bawab, 2001); however, this reaction appears to account for relatively little ceramide synthesis under normal physiological conditions. Recent studies (I. Guillas, 2001) have identified yeast (Lag1P and Lac1P) and mammalian (K. Venkataraman, 2002) genes that encode ceramide synthase or are obligatory for ceramide synthase activity.

Microorganisms produce a number of inhibitors of ceramide synthase (Table 2). Fumonisins (Fig. 10), which are produced by some species of *Fusaria*, were discovered as causes of human esophageal cancer, equine leukoencephalomalacia, and porcine pulmonary edema (W.F.O. Marasas, 2001), and are now known to induce multiple pathologies, including liver and kidney toxicity and carcinogenicity, immunosuppression (and in some cases immunostimulation), and birth defects (A.H. Merrill, 2001). Fumonisins not only block complex sphingolipid formation, but also cause sphinganine to accumulate, which is both key to fumonisin toxicity (E.M. Schmelz, 1998) and provides a useful biomarker for exposure of organisms to this mycotoxin. There can also be elevation of sphinganine 1-phosphate [15], which might explain how fumonisins can be toxic and mitogenic for different cell types.

The last step of ceramide synthesis is the insertion of the 4,5-*trans*-double bond into the sphingoid base backbone, which occurs at the level of dihydroceramide (Fig. 8) (J. Rother, 1992). Therefore, free sphingosine per se is not an intermediate of sphingolipid biosynthesis de novo. The genes for desaturases have been characterized in plants (P. Sperling, 2000) and mammals (P. Ternes, 2002). For 4-hydroxysphinganines (phytosphingosine), insertion of the 4-hydroxyl group occurs at the level of sphinganine in yeast (M.M. Grilley, 2000) and plants (P. Sperling, 2000).

### 3.2. Sphingomyelin and ceramide phosphorylethanolamine

Sphingomyelin is synthesized by transfer of phosphorylcholine from phosphatidylcholine to ceramide, liberating diacylglycerol. This reaction links glycerolipid and sphingolipid signaling pathways, although it is not known if cells frequently capitalize on this relationship for signaling purposes. Most de novo sphingomyelin synthesis occurs in the Golgi apparatus (in liver); however, synthesis also occurs in the plasma membrane of many cell types and is probably a major site as well. Relatively little is known about the regulation of sphingomyelin biosynthesis, but it has interesting features such as stimulation by phorbol esters, 25-hydroxycholesterol, and Brefeldin A (G. Hatch, 1992; N. Ridgway, 1995), increases during development of the lung (C.A. Longo, 1997), decreases in aging (S.A. Lightle, 2001), and changes very early in the development of colon cancer (P.K. Dudeja, 1986).

Ceramide phosphorylethanolamine is synthesized from phosphatidylethanolamine and ceramide in a reaction analogous to sphingomyelin synthesis, and once formed can be methylated to sphingomyelin (M. Malgat, 1986; M.N. Nikolova-Karakashian, 2000). Inositolphosphoceramides are formed by transesterification (from phosphatidylinositol) (R.C. Dickson, 1999; A.S. Fischl, 2000) by transferases that share a conserved structural motif across yeast and pathogenic fungi, and which resembles somewhat a motif of lipid phosphatases (S.A. Heidler, 2000).

### 3.3. Neutral glycosphingolipids

Pathways for the biosynthesis of the different root glycosphingolipids (Figs. 11 and 12) appear complex, but are achieved by surprisingly few glycosyltransferases that commit precursors and intermediates to predictable products based on the specificities of the enzymes. The enzymes transfer a specific sugar from the appropriate sugar nucleotide (e.g., UDP-Glc, UDP-Gal, etc.) to ceramide or the non-reducing end of the growing carbohydrate chain attached to ceramide. The glycosyltransferases often recognize mainly the carbohydrate portion of the acceptor glycosphingolipid; however, ceramides with α-hydroxy-fatty acids are preferentially incorporated into GalCer (and sulfatides), whereas those with non-hydroxy-fatty acids are used to make GlcCer (I. van Genderen, 1995).

GalCer and GlcCer are synthesized with inversion of the configuration of the

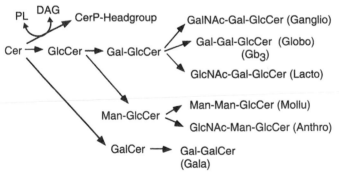

Fig. 11. Biosynthesis of phosphosphingolipids and 'root' glycosphingolipids. For phosphosphingolipids, the headgroup is transferred from a phosphoglycerolipid, such as phosphatidylcholine for sphingomyelin. For glycosphingolipids, the headgroup is transferred from a UDP-sugar.

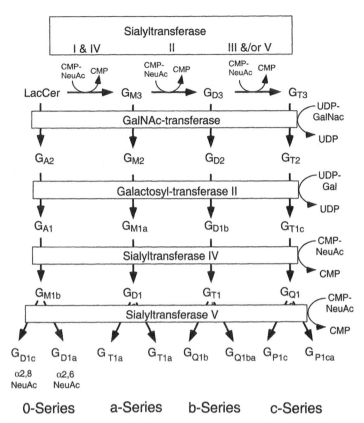

Fig. 12. Scheme for ganglioside biosynthesis from LacCer [24,26]. Due to cellular compartmentation sialyltransferases IV and V are probably not involved in $G_{M3}$ and $G_{T3}$ synthesis in vivo, respectively; however, they do so in vitro.

glycosidic bond ($\alpha$ to $\beta$) by UDP-Glc(or Gal) : ceramide glucosyltransferases, the genes for which have been identified in numerous organisms (S. Ichikawa, 1998; M. Leipelt, 2001; and reviewed in T. Tencomnao, 2001). Relatively little is yet known at a biochemical level about these enzymes; however, recent studies have identified amino acid residues of the glucosylceramide synthase active site that are essential for activity, and note that this enzyme and processive $\beta$-glycosyltransferases possess a conserved substrate-binding/catalytic domain (D.L. Marks, 2001).

A growing number of factors have been found to regulate expression of glucosylceramide synthase, including depletion of the amounts of GlcCer in the cell (I. Meivar-Levy, 1999), elevations in ceramide (H. Komori, 2000), endotoxin and acute phase response mediators (R.A. Memon, 2001), and basic fibroblast growth factor (although this appears to involve a post-translational mechanism) (S.A. Boldin, 2000). Glucosylceramide and protein-bound glucosylceramide play major roles in establishing the permeability barrier of skin, and GlcCer synthase is up-regulated during epidermal barrier development (K. Hanley, 1997; R. Watanabe, 1998; T. Doering, 1999).

The synthesis of GlcCer and GalCer can be inhibited by structural analogs of ceramide such as PDMP (Table 2), which decreases cellular levels of neutral glycosphingolipids and gangliosides and elevates ceramide, causing cell cycle arrest (C.S.S. Rani, 1995). GlcCer synthesis appears to be a major determinant of survival of tumor cells (by removal of ceramide and affecting multidrug resistance) (Y.Y. Liu, 2001; A. Senchenkov, 2001). But, the inability to make detectable glycolipids due to deficient GlcCer synthase is not lethal to some cell lines, although it slows growth and results in an elongated fibroblastic morphology (S. Ichikawa, 1994).

Additional glycosyltransferases are involved in the synthesis of the other neutral glycolipids as well as the addition of neutral sugars to gangliosides (Figs. 11 and 12), and many of the genes for these enzymes have been identified, including lactosylceramide synthase (T. Takizawa, 1999), globosylceramide (Gb3) synthases (Y. Kojima, 2000; J. Keusch, 2000), β-1,3-galactosyltransferases (M. Amado, 1999), β1,3 N-acetylglucosaminyltransferase (Lc3 synthase) (T. Henion, 2001) as well as numerous fucosyltransferases for Lewis × antigens. Labs are already beginning to engineer surface glycosphingolipids of desired composition by transfecting cells with combinations of these and other enzymes (E.G. Prati, 2000).

Synthesis of GlcCer occurs on the cytosolic aspect of the endoplasmic reticulum and/or early Golgi membranes (A.H. Futerman, 1991; D. Jeckel, 1992), whereas more complex neutral glycosphingolipids (beginning with LacCer) are made in the lumen of the Golgi apparatus (H. Lannert, 1994). Therefore, GlcCer and the sugar nucleotides must undergo transbilayer movement to the lumen of the Golgi for the synthesis of more complex sphingolipids. Sphingolipids are generally thought to reach their membrane locations via trafficking from Gogi; however, newly synthesized GlcCer can also be transported to the plasma membrane via a non-Golgi pathway (D.E. Warnock, 1994).

## 3.4. Gangliosides

The general pathway for the synthesis of gangliosides has evolved to the scheme depicted in Fig. 12 [24]. Gangliosides are synthesized by the stepwise transfer of neutral sugars and sialic acids by membrane-bound glycosyltransferases that are located in the regions of the Golgi apparatus that generally correspond to the order in which the sugars are added. For example, the sialyltransferase catalyzing the synthesis of $G_{M3}$ ganglioside is in the cis Golgi, whereas the enzymes involved in terminal steps in ganglioside synthesis are localized in the more distal trans Golgi network. Ganglioside biosynthesis can also involve the introduction of O-acetyl groups on sialic acid and N-deacetylation to produce a free amino group on position 5 of sialic acid. Gangliosides are incorporated into the outer leaflet of the plasma membrane by vesicle-mediated transport.

There has been considerable progress in identification of the genes responsible for the key reactions in Fig. 12, allowing these relationships to be tested by transfecting cells with the cDNA for enzymes of this pathway and determining the types of glycosphingolipids that are made; for example (a) transfection of GalNac-transferase cDNA into Chinese hamster ovary cells, which normally make mainly $G_{M3}$, produced cells that now synthesize mainly $G_{D2}$, whereas transfection of cells that are defective in sialylation yielded $G_{A2}$ (M.S. Lutz, 1994), (b) transfection of several cell lines with a GalNac-transferase yielded $G_{A2}$, $G_{M2}$ (the preferred product when both $G_{M3}$ and LacCer were available), $G_{D2}$, GalNAc sialylparagloboside and GalNAcG$_{D1a}$ (S. Yamashiro, 1995), and (c) transfection of the cDNA for $G_{D3}$ synthase (sialyltransferase II in Fig. 12) into Neuro2a cells increased $G_{D3}$ and $G_{Q1b}$ (N. Kojima, 1994). The findings to date support the view that ganglioside synthesis can be viewed as 'combinatorial' reactions (Fig. 12) catalyzed by a conservative number of key enzymes such that the ultimate composition is determined by the relative activities of these enzymes and the availability of their substrates [27].

Regulation of ganglioside biosynthesis involves both transcriptional and post-transcriptional factors. Transcriptional control of key glycosyltransferases appears to account for many of the developmentally regulated, tissue-selective variations in ganglioside amounts and types in mammalian organs, including large changes with oncogenic transformation (R.W. Ledeen, 1998; S. Hakomori, 1998). The biosynthesis of gangliosides is also controlled through post-translational modification of glycosyltransferases (R.K. Yu, 2001); several sialyltransferases (X. Gu, 1995) are down-regulated by protein kinase C in cell-free and intact cell systems and $N$-acetylgalactosaminyltransferase can be up-regulated by protein kinase A in cultured cells.

*3.5. Sulfatoglycosphingolipids*

Sulfatide (3′-sulfo-galactosylceramide) synthesis is catalyzed by GalCer sulfatotransferase (3′-phosphoadenylylsulfate : galactosylceramide 3′-sulfotransferase), which utilizes the activated sulfate donor 3′-phosphoadenosine-5′-phosphosulfate (PAPS). The cDNA encoding the sulfotransferase has been cloned (T. Honke, 1997). Regulation of sulfatide biosynthesis appears to reside in the activity of this sulfatotransferase.

# 4. Sphingolipid catabolism

Complex sphingolipids are lost from cells by (1) membrane internalization, recycling, and degradation, (2) hydrolysis to release bioactive products that participate in cell signaling, and (3) release from the cells by secretion or shedding.

In general, sphingolipids are internalized with endocytic vesicles, sorted in early endosomes, and recycled back to the plasma membrane (often with remodeling of the sphingolipid) or transported to lysosomes where they are degraded by specific acid hydrolases. Given that lysosomal membranes are rich in sphingolipids, it has been unclear why they, too, do not undergo hydrolysis if the endocytosed membranes and lysosomal membranes are simply fused. This dilemma was solved when it was shown (W. Furst, 1992; W. Mobius, 1999) that endocytosed sphingolipids (and presumably other components) become invaginated into intraendosomal vesicles that are delivered into the lumen of the lysosome. Thus, hydrolytic enzymes contact the sphingolipids to be digested in the lumen rather than as part of the lysosomal membrane, which is additionally protected by an elaborate glycocalyx that lines the inner leaflet.

The pathways for sphingolipid catabolism (Fig. 13) converge on ceramide; nonethe-

394

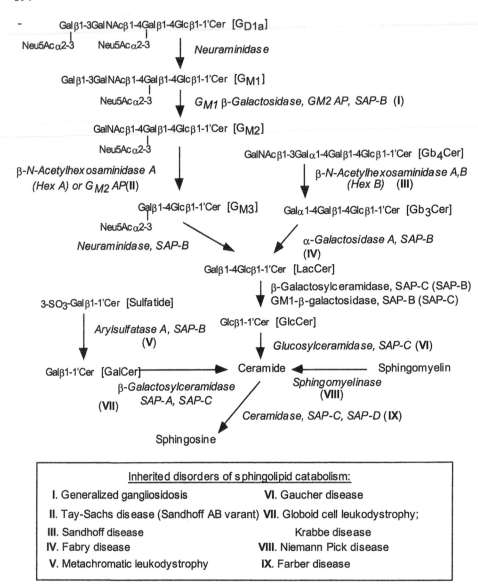

Fig. 13. Catabolism of complex sphingolipids and associated diseases [3,24].

less, cells contain at least small amounts of other lysosphingolipids, such as galactosylsphingosine (psychosine) and sphingosylphosphorylcholine, implying that such compounds may arise from as-yet-to-be-characterized enzymatic activities analogous to phospholipase $A_2$, or by de novo biosynthesis.

## 4.1. Sphingomyelin

The lysosomal hydrolysis of sphingomyelin to ceramide and phosphocholine is catalyzed by acid sphingomyelinase, a water-soluble, lysosomal glycoprotein that interacts with a sphingolipid activator protein (SAP-C) and anionic lipids such as bis(monoacylglycero)phosphate (T. Linke, 2001). Individuals may display a number of different molecular defects that cause insufficient lysosomal sphingomyelinase activity and result in accumulation of sphingomyelin in reticuloendothelial cells scattered throughout the spleen, bone marrow, lymph nodes, liver, and lungs. These are called Niemann–Pick disease Types A and B, and sphingomyelin also accumulates in Type C Niemann–Pick disease, but not due to genetic defects in sphingomyelinase per se. Screening for human Niemann–Pick disease is used to identify carriers and for prenatal diagnosis of affected fetuses. Acid sphingomyelinase-deficient mice have been generated (K. Horinouchi, 1995) as models for Niemann–Pick disease.

Sphingolipid turnover in other cellular compartments involves both acid sphingomyelinase and sphingomyelinases with neutral to alkaline pH optima. Acidic sphingomyelinase is also secreted, and the secreted form requires supplemental $Zn^{2+}$ for activity (I. Tabas, 1999). Neutral sphingomyelinases have been found in other cellular compartments, including the nuclear membrane, and at least one form resides in sphingolipid-enriched microdomains and is inhibited by the caveolin-scaffolding domain (R.J. Veldman, 2001). These are probably involved in cell signaling.

A sphingomyelinase D is found in the venom of brown recluse spiders, *Corynebacterium pseudotuberculosis* (which commonly infects sheep), *Vibrio damsela* (an aquatic bacterium that causes wound infections in humans) and the human pathogen *Arcanobacterium haemolyticum*. This enzyme, which yields ceramide 1-phosphate and choline, is able to produce much of the tissue damage caused by these organisms (A.P. Truett, 1993).

## 4.2. Glycosphingolipids

Glycosphingolipids are catabolized by the stepwise hydrolysis of the terminal monosaccharides through the concerted action of a series of specific exoglycosidases (Fig. 13). For the in vivo degradation of glycolipids with short oligosaccharide headgroups, i.e. of less than four carbohydrate residues, there is often a requirement for sphingolipid activator proteins (or saposins) SAP-A, -B, -C or -D. A number of inherited diseases are caused by mutations in the structural genes for these enzymes that result in reduced enzymatic activity, loss of the appropriate targeting signals for transport to lysosomes, or alteration of the domains that interact with other subunits of the enzyme and/or activator proteins (Fig. 13) [24].

That sphingolipidoses can result from several types of genetic defects is exemplified the $G_{M2}$ gangliosidoses, which arise from mutations of β-$N$-acetylhexosaminidase (Hex A or Hex B) (Fig. 13) or $G_{M2}$ activator proteins (R.A. Gravel, 1995) [24]. Hex A is an αβ heterodimer and degrades negatively charged and uncharged substrates, whereas Hex B is the ββ homodimer and cleaves mainly $N$-acetylgalactosamine residues from uncharged substrates such as $G_{A2}$, globotetraosylceramide and oligosaccharides.

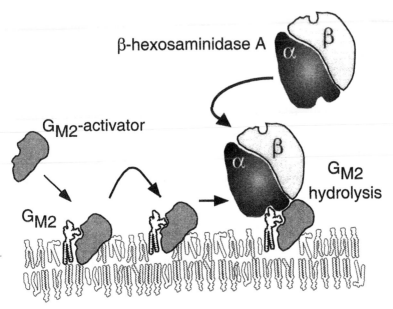

Fig. 14. A model for the role of $G_{M2}$ activator protein (SAP) in the hydrolysis of ganglioside $G_{M2}$ by β-hexosaminidase A (subunits α and β). For more information, see Schuette et al. [3] and Kolter and Sandhoff [24].

Therefore, mutations in the structural gene for the α subunit result in partial or complete loss of Hex A activity (Tay–Sachs disease), and mutations in the β subunit affect both Hex A and Hex B (Sandhoff disease). The phenotypes of these two gangliosidoses are similar, but they are easily distinguished by measuring Hex A and Hex B activities and by the accumulation of $Gb_4Cer$ in Sandhoff disease but not in Tay–Sachs disease ($G_{M2}$ accumulates in both disorders). The AB variant has normal levels of both Hex A and Hex B activity when measured in vitro, but there is a defective (or absent) $G_{M2}$ activator protein. The $G_{M2}$ activator protein is a membrane active protein that also binds ganglioside $G_{M2}$ as well as structurally related gangliosides forming complexes (usually in a 1 : 1 molar ratio) that present them to hexosaminidase A as illustrated in Fig. 14.

### 4.3. Ceramide

The major pathway for catabolism of the ceramide backbone is shown in Fig. 15.

In lysosomes, ceramides are hydrolyzed to free sphingoid bases and long-chain fatty acids by a ceramidase that has an acidic pH optimum. The lysosomal ceramidase is a water-soluble glycoprotein that hydrolyzes membrane-bound ceramide in an interfacial reaction that needs the stimulation by SAP-D and anionic phospholipids such as the lysosomal bis(monoacylglycero)phosphate (A. Klein, 1994; T. Linke, 2001). The human acidic ceramidase is a heterodimeric enzyme of 40 kDa and 13 kDa subunits synthesized as a single precursor polypeptide of approximately 53–55 kDa and targeted to the lysosome via the mannose 6-phosphate receptor (K. Ferlinz, 2001). Additional

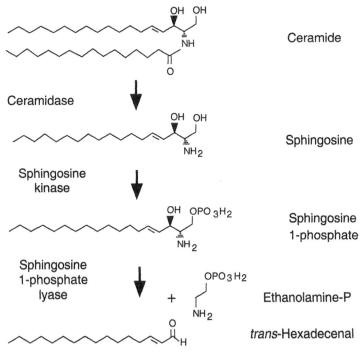

Fig. 15. Catabolism of ceramide and sphingosine [24].

ceramidases with neutral and alkaline pH optima have been found in various cell compartments (including mitochondria) and appear to be involved in signaling.

### 4.4. Sphingosine

Sphingosine undergoes reacylation by ceramide synthase (described in Section 3.2.2) or phosphorylation by sphingosine kinase. This family of ATP-dependent kinases is both cytosolic and membrane associated, and phosphorylates a wide range of sphingoid bases, although unnatural stereoisomers (e.g., L-*threo*-sphinganine) are inhibitors. The product can be dephosphorylated by specific sphingosine 1-phosphate phosphatases as well as more general lipid phosphatases, or cleaved to ethanolamine phosphate and *trans*-2-hexadecenal (Fig. 15) by a pyridoxal 5′-phosphate-dependent lyase (J. Zhou, 1998; P.P. Van Veldhoven, 2000). As shown first in the 1970s by W. Stoffel and co-workers, the phosphoethanolamine can be utilized for the synthesis of phosphatidyl-ethanolamine, and *trans*-2-hexadecenal reduced to the alcohol and incorporated into alkyl ether lipids (Chapters 8 and 9). Under certain conditions, degradation of sphingoid bases can account for as much as one third of the phosphoethanolamine in phosphatidylethanolamine (E. Smith, 1995). In addition to being an intermediate in the degradation of sphingoid bases, sphingosine 1-phosphate is packaged in some cells (e.g., platelets) for secretion as discussed in Section 6.2.3.

# 5. Regulation of sphingolipid metabolism

Sphingolipid metabolism is regulated at a number of levels: (1) the control of ceramide biosynthesis de novo and by the recycling of existing sphingolipids; (2) the partitioning of ceramide toward the major classes of sphingolipids (e.g., sphingomyelin versus GlcCer and GalCer); (3) the partitioning of intermediates to determine the complex glycolipid profiles of different cell types; (4) the trafficking of sphingolipids to the appropriate cellular membranes as well as to specialized regions of a given membrane; (5) the secretion (and shedding) of some categories of sphingolipids; (6) the internalization of sphingolipids during endocytosis and other membrane functions; (7) the turnover of sphingolipids for cell signaling; and (8) degradation.

These events vary in different cell types, and in a given cell type at different stages of development and in response to varying environmental conditions [29]. These issues are obviously too complex to deal with fully in this chapter; therefore, we present only a few examples.

## 5.1. De novo sphingolipid biosynthesis versus turnover in generating bioactive (signaling) metabolites

The traditional 'signaling' paradigm for sphingolipids is that agonists trigger the turnover of sphingomyelin to ceramide, sphingosine and sphingosine 1-phosphate (Fig. 2); nonetheless, the de novo biosynthetic pathway also produces bioactive products such as sphinganine, ceramide, and sometimes sphinganine 1-phosphate. That these intermediates of de novo synthesis also affect cell behavior was first shown by the role of sphinganine in the toxicity of fumonisins (E. Wang, 1991) and the involvement of de novo synthesized ceramide in daunorubicin-induced apoptosis (R. Bose, 1995). There are now a large number of natural agonists, drugs, toxins and toxicants, and even intermediates of common metabolic pathways (such as palmitoyl-CoA) that can alter cell behavior at least in part by affecting sphingolipid biosynthesis, as illustrated in Fig. 16 (Y.A. Hannun, 2001; S.C. Linn, 2001).

## 5.2. Complex sphingolipid formation in tissue development

Glycosphingolipids undergo quantitative and qualitative changes with development [2,28], such as the stage-specific expression of the antigen SSEA-1, which appears at the 8-cell stage, is maximally expressed at the morulae stage, and disappears at the blastocyte stage. This antigenic determinant is actually present on several glycosphingolipids, all of which have a terminal Galβ1–4[Fucα1–3]GlcNAcβ1–3Gal structure, as well as on cell surface glycoproteins. Globoside and Forssman antigen (IV$^3$-α-GalNAc-Gb$_4$Cer) also appear at the morulae and blastocyte stages, respectively. It is likely that some of these changes contribute to cell–cell and cell–matrix interactions that affect the migration of cells to target locations within the developing embryo, and cell differentiation when the correct location(s) are found.

The requirement for glycosphingolipids in development has been established by targeted disruption of the gene encoding glucosylceramide synthase (T. Yamashita,

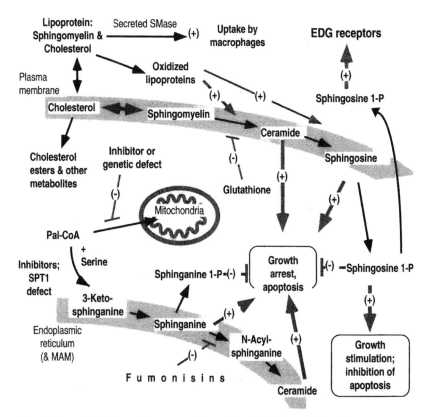

Fig. 16. Some of the factors that can modulate sphingolipid biosynthesis and turnover to impact cell behavior [23–25]. Agonist-induced changes in sphingolipid turnover were illustrated in Fig. 2; the additional pathways shown here are as follows (proceeding counter-clockwise from top left): (1) hydrolysis of lipoprotein sphingomyelin by secreted sphingomyelinase (SMase) to form particles that are taken up by macrophages; (2) induction of sphingomyelin turnover by oxidized lipoproteins to induce smooth muscle cell growth or death via sphingosine 1-phosphate versus ceramide, respectively; (3) associations between sphingomyelin and cholesterol that affect lipoprotein and membrane structure (as well as the efflux and metabolism of cholesterol, not shown); (4) perturbation of sphingolipid biosynthesis by factors that alter the amounts of the precursors (e.g., serine and palmitoyl-CoA) (Pal-CoA) including inhibitors and genetic defects of mitochondrial uptake of Pal-CoA; (5) inhibition of, or genetic defects in, serine palmitoyltransferase (SPT1); (6) inhibition of ceramide synthase (by fumonisins) to block complex sphingolipid formation and elevate sphinganine and sphinganine 1-phosphate; (7) induction of de novo sphingolipid biosynthesis at serine palmitoyltransferase and/or ceramide synthase by other stresses (irradiation, UV, cytokines, and heat shock, inter alia) (not shown); (8) the triggering of sphingomyelin hydrolysis by depletion of cytosolic glutathione, an inhibitor of neutral sphingomyelinase(s). More information about these factors is presented in the text.

1999). In the absence of glycosphingolipid synthesis, embryogenesis proceeded well into gastrulation with differentiation into primitive germ layers and patterning of the embryo, but was abruptly halted by a major apoptotic process. Embryonic stem cells deficient in glycosphingolipid synthesis were able to differentiate into endodermal, mesodermal, and ectodermal derivatives but were strikingly deficient in their ability to form well differentiated tissues.

## 5.3. Neural development and function

The ganglioside content of neuronal and glial cells changes quantitatively and qualitatively with development and aging (R.K. Yu, 1994) and it has been assumed that gangliosides help neuronal cells follow the appropriate path in development of neuronal networks, mediate cell–cell communication, and help regulate receptors and ion channels on neural cells, presumably by interactions with specific binding proteins (R.L. Schnaar, 1994).

These hypotheses have been confirmed by recent studies with mice lacking specific glycosyltransferases. Mice engineered to lack $G_{M2}/G_{D2}$ synthase express only the simple gangliosides ($G_{M3}$ and $G_{D3}$) and develop significant and progressive behavioral neuropathies (S. Chiavegatto, 2000). $G_{M2}/G_{D2}$ synthase knockout mice lack a calcium regulatory mechanism that is modulated by one or more of the deleted gangliosides (G. Wu, 2001). Mice with disruption of the gene encoding $G_{D3}$ synthase appear to undergo normal development and have a normal life span; however, when these mice are crossbred with mice carrying a disrupted gene encoding $\beta 1,4$-$N$-acetylgalactosaminyltransferase, the double mutants display a sudden death phenotype and are extremely susceptible to induction of lethal seizures by sound stimulus.

## 5.4. Physiology (and pathophysiology) of the intestinal tract

The intestinal epithelia undergo constant renewal through cell proliferation in the crypts at the base of the microvilli, differentiation and migration up the villi, and death as the cells are sloughed off. Crypt cells have more ceramide and $Gb_3Cer$, and less GlcCer and $G_{M3}$, than villus cells (J.-F. Bouhours, 1993), and differ also in their ceramide composition, with the crypt GlcCer having less hydroxy-fatty acids. All regions of the intestine contain relatively high levels of free ceramide (ca. 17% of the total sphingolipids). The glycosphingolipids of intestinal cells are part of the epithelial barrier, but are also used as attachment sites by microorganisms, viruses, and microbial toxins [23]; for example, *Candida albicans* binds to asialo-$G_{M1}$, HIV gp 120 to GalCer, cholera toxin to $G_{M1}$, and Shiga toxin and verotoxins to $Gb_3$.

## 5.5. Oncogenic transformation, tumor antigens, and immunomodulation

Changes in the glycosphingolipid composition of cells have long been associated with oncogenic transformation [2]. There are multiple strategies that might utilize sphingolipids to prevent or treat neoplasia, such as (a) administration of sphingolipids or sphingolipid analogs via diet or other means (E.M. Schmelz, 2001), (b) modulation of sphingolipid metabolism to trigger apoptosis via endogenous ceramide (S. Selzner, 2001), (c) altering ceramide metabolism to overcome drug resistance (A. Senchenkov, 2001), (d) controlling the immunomodulatory effects of shed glycolipids (S. Ladisch, 1995), and (e) using abnormally expressed glycolipids as tumor markers and/or to target chemotherapeutic agents more effectively [2].

# 6. Sphingolipids and signal transduction

Findings over the last decade have provided evidence that sphingolipids are involved in essentially all aspects of cell regulation (many of which are summarized in Figs. 2 and 16). (1) Sphingolipids (particularly sphingomyelin in association with cholesterol) help define 'microdomains' of the membrane that influence receptor signaling as well as the functions of important transport proteins. (2) Complex sphingolipids serve as ligands for receptors on neighboring cells or the extracellular matrix, and mediate changes in cell behavior in response to a cell's environment. (3) Complex sphingolipids modulate the properties of receptors on the same cell, thereby controlling the responsiveness of the cell to external factors. (4) Sphingolipids are extensively involved in membrane trafficking and, therefore, influence receptor internalization, sorting and recycling, as well as the movement and fusion of secretory vesicles in response to stimuli. (5) Ceramide, sphingosine and sphingosine 1-phosphate (and possibly others) are produced from agonist- or stress-induced turnover of sphingomyelin (and de novo synthesis) to participate in cell signaling. (6) Sphingosine 1-phosphate is released from some cells to serve as an extracellular agonist for S1P receptors. Moreover, because diverse factors can affect sphingolipid metabolism (availability of precursor substrates, genetic mutations in related pathways, naturally occurring inhibitors, etc.), these pathways may play a central role in many aspects of cell physiology and pathophysiology (Fig. 16).

## 6.1. Interactions between gangliosides and growth factor receptors

Gangliosides are bimodal regulators of cell growth: both inhibitory (often as the complex species) and stimulatory (as catabolites). This is exemplified by ganglioside $G_{M3}$, which inhibits growth through extension of the $G_1$ phase of the cell cycle, and makes cells refractory to stimulation by epidermal growth factor [7]. Upon removal of the fatty acid, lyso-$G_{M3}$ strongly inhibits cell proliferation whereas removal of the $N$-acetyl group from the sialic acid (producing de-$N$-acetyl-$G_{M3}$) enhances growth and epidermal growth factor receptor tyrosine kinase activity. The neutral glycolipid cores of $G_{M3}$, LacCer and GlcCer, also affect growth (K. Ogura, 1992).

## 6.2. Hydrolysis to bioactive lipid backbones

The finding that sphingosine is a potent inhibitor of protein kinase C (Y.A. Hannun, 1986) introduced the paradigm that the cellular functions of sphingolipids may reside not only in the complex species, but also in the lipid backbones. Soon thereafter, ceramide was found to be released from sphingomyelin in GH3 pituitary cells treated with diacylglycerol (R.N. Kolesnick, 1989) and HL-60 cells treated with 1α,25-dihydroxyvitamin $D_3$ (which induces these cells to differentiate) (T. Okazaki, 1989). Furthermore, treatment of HL-60 cells with a short-chain ceramide, or exogenous sphingomyelinase, could mimic the effects of 1α,25-dihydroxyvitamin $D_3$, which linked the agonist-induced turnover of a sphingolipid to a metabolite that met many of the criteria for an intracellular mediator. A large number of studies have now explored how the lipid backbones of sphingolipids serve as second messengers, but caution must be exercised in reading this literature because some of the effects of exogenous sphingolipids have not yet been linked to changes in endogenous mediators.

### 6.2.1. Ceramide

Sphingomyelin turnover to ceramide is now thought to mediate, at least in part, cellular responses to a wide spectrum of agents (cytokines, ionizing irradiation, corticosteroids, inter alia), many of which deal with some type of stress (R.N. Kolesnick, 1998; Y.A. Hannun, 2000; A Huwiler, 2000). Both neutral and acidic (including a secreted acidic) sphingomyelinases are involved. The regulatory mechanisms for these enzymes are involved, and their mechanisms of activation and inhibition are still being uncovered. As an example, CD95-triggers translocation of acid sphingomyelinase to the plasma membrane outer surface, where it releases ceramide from sphingomyelin and enables the clustering of CD95 in sphingolipid-rich membrane rafts triggering apoptosis (H. Grassme, 2001).

At least five direct targets for ceramide have been identified: ceramide-activated protein phosphatases, which are both type 1 and 2A phosphoprotein phosphatases (C.E. Chalfant, 2000); ceramide-activated protein kinase, which has been identified as the kinase suppressor of ras (Y.H. Zhang, 1997; D.B. Polk, 2001); cathepsin D (M. Heinrich, 2000); protein kinase C zeta (N.A. Bourbon, 1998); and cytosolic phospholipase $A_2$ (A. Huwiler, 2001]. In general, dihydroceramides (N-acylsphinganines) are ineffective in activating these targets, which makes them useful as controls. Elevations in cellular ceramide are usually associated with growth inhibition, differentiation and, in many cells, induction of apoptosis.

### 6.2.2. Sphingoid bases

A remarkable number of cellular systems are activated or inhibited by sphingoid bases, with the systems that are most likely to reflect direct activation being a 14-3-3 kinase (T. Megidish, 2000), and inhibition being protein kinase C (E.R. Smith, 2000), phosphatidic acid phosphohydrolase (D. Perry, 1992) and Akt kinase (H.C. Chang, 2001). There are also strong indications that sphingosine can affect ion transporters (C. Mathes, 1998; Y. Shin, 2000).

Relatively little is known about agonist-induced turnover of sphingolipids to sphingosine and the coupling of the sphingosine to intracellular responses. When added exogenously to cells, sphingoid bases are usually growth inhibitory and pro-apoptotic (V.L. Stevens, 1989; T. Shirahama, 1997). Agents affecting cellular levels of sphingosine include dexamethasone (R. Ricciolini, 1994), lipoproteins and phorbol esters (E. Wilson, 1988), platelet-derived growth factor (A. Olivera, 1993; E. Coroneos, 1995), Fas-induced apoptosis of type II Jurkat T cells (O. Cuvillier, 2000) and doxorubicin in MCF7 breast adenocarcinoma cells (O. Cuvillier, 2001). The in vivo correlates are few, but it has been recently reported that apoptosis in the skeletal muscle of rats with heart failure is associated with increased serum levels of tumor necrosis factor-α and sphingosine (L. Dalla Libera, 2001).

In addition, free sphingoid bases, sphingoid base 1-phosphates, and ceramides can be elevated by disruption of de novo sphingolipid biosynthesis by, for example, heat shock in yeast (R.C. Dickson, 1997; G.M. Jenkins. 1997) and fumonisins [4,15]. It should also be borne in mind when studying cells in culture that many cell lines undergo a 'burst' of sphinganine and sphingosine formation from both increased de novo sphingolipid biosynthesis and turnover when 'conditioned' medium is removed (E.R. Smith, 1995). This may have a physiologic significance because cells produce a novel factor (named betrachamine) that suppresses the burst (L. Warden, 1999).

### 6.2.3. Sphingosine 1-phosphate
Platelet-derived growth factor induces rapid increases in cellular sphingosine and sphingosine 1-phosphate, and several lines of evidence indicate that sphingosine 1-phosphate is a mediator of growth stimulation by this growth factor (A. Olivera, 1993; Y. Su, 1994). The downstream responses are release of calcium from intracellular compartment(s) (S. Kim, 1995) as well as activation of AP1 transcription factor. Sphingosine 1-phosphate is also mitogenic when added to cells exogenously, and is thought to signal through a pertussis toxin-sensitive G protein (K.A. Goodemote, 1995) that activates the mitogen-activated protein kinase pathway (J. Wu, 1995). Progress in characterizing these receptors, members of the EDG family (now called S1P) receptors, has been rapid [10], and sphingosine 1-phosphate clearly qualifies as both an extracellular and an intracellular messenger.

Whereas ceramide and sphingosine are often growth inhibitory and pro-apoptotic, sphingosine 1-phosphate is mitogenic and anti-apoptotic, which has been likened to a rheostat that determines cell survival versus death pathways (S. Spiegel, 1999). Ceramides and sphingosine induce release of cytochrome $c$ from mitochondria (an important event in apoptosis), and recent studies suggest that sphingosine 1-phosphate antagonizes apoptosis of human leukemia cells by inhibiting release of cytochrome $c$ and Smac/DIABLO from mitochondria (O. Cuvillier, 2001). A major determinant of the type(s) of sphingolipids that are formed is whether the agonists activate only sphingomyelinase or a combination of sphingomyelinase and ceramidase with or without sphingosine kinase (as shown in Fig. 2) (A. Olivera, 1993; E. Coroneos, 1995; M.N. Nikolova-Karakashian, 1997; N. Augé, 1999).

### 6.2.3.1. Other bioactive lysosphingolipids.
Considerable progress has been made in characterizing the occurrence and biological activities of other lysosphingolipids, but much work remains to elucidate their roles in cell regulation. For example, sphingosylphosphorylcholine (lysosphingomyelin) is a potent mitogen (T. Seufferlein, 1995), ceramide 1-phosphate is a potent calcium-mobilizing agent (S. Gijsbers, 1999), and psychosine (lyso-GlcCer or -GalCer), which has long been known to be highly cytotoxic, is the agonist for an orphan G protein-coupled receptor, T cell death-associated gene 8 (D.S. Im, 2001).

## 7. Bioactive sphingolipids appear to be at the heart of numerous aspects of cell regulation in normal and pathologic conditions

Fig. 16 illustrates interrelationships between sphingolipid metabolism and some of the factors that can alter the amounts of key intermediates and/or products to impact cell behavior. Beginning with the more familiar signaling paradigm:

- agonists, stress, oxidized lipoproteins and depletion of glutathione can induce sphingomyelinase activation to produce ceramide; and if ceramidase is also activated, sphingosine and/or sphingosine 1-phosphate (when sphingosine kinase is also active) (Section 6.2, and Fig. 2);
- a secreted acid sphingomyelinase hydrolyzes sphingomyelin associated with lipoproteins to form particles that are taken up by macrophages and promote foam cell formation (S. Marathe, 1998);
- hydrolysis of cellular sphingomyelin alters the structure of rafts and other membrane microdomains, as well as 'frees' cholesterol for efflux (Chapter 20), metabolism and down-regulation of de novo cholesterol synthesis (Chapter 15) [5,6];
- a number of factors can perturb sphingolipid biosynthesis by altering the amounts of the precursors (e.g., serine and palmitoyl-CoA) including inhibitors and genetic defects of mitochondrial uptake of palmitoyl-CoA (H. Vesper, 1999);
- genetic defects in serine palmitoyltransferase can, in an as yet to be elucidated way, result in sensory neuropathy (J.L. Dawkins, 2001; K. Bejaoui, 2001);
- a host of factors up-regulate de novo sphingolipid biosynthesis by transcriptional or post-translational activation of serine palmitoyltransferase and/or activation of ceramide synthase, often resulting in cell death due to elevations in pro-apoptotic intermediates (Section 5.1) [25];
- inhibition of ceramide synthase (by fumonisins) blocks complex sphingolipid formation and elevates sphinganine (to induce toxicity) and sometimes sphinganine 1-phosphate (which may induce a mitogenic response), whereas other factors (such as irradiation) can increase ceramide synthase and induce cell death (Section 3).

## 8. Future directions

Knowledge about basic 'sphingolipidology' — the physical properties of sphingolipids, sphingolipid metabolizing enzymes and targets regulated by sphingolipids — has at least doubled since the 1996 edition of this book. New ideas have also emerged about how sphingolipids are involved in disease etiology, and how naturally occurring and/or synthetic sphingolipids may be useful in disease prevention and treatment. Like other areas of biologic research, 'sphingolipidology' is rapidly evolving into 'sphingolipidomics', that is, a field where a rigorous evaluation of any one part requires a comprehensive analysis also of essentially all of the other components. For example, the 'sphingomyelin ceramide signaling pathway' was initially envisioned to involve agonist- (or stress-) induced turnover of sphingomyelin to ceramide to activate (or inhibit) intracellular target(s). As it is now understood, sphingomyelin turnover affects the organization (and intracellular trafficking) of important membrane microdomains (rafts and caveolae) and their associated receptors, transporters, and other bioactive lipids as well as the production of a cascade of products (ceramide, sphingosine and sphingosine 1-phosphate, etc.), each of which can interact with multiple intracellular (and sometimes extracellular) targets. Very often, a stimulus triggers not only sphingolipid turnover but also changes in de novo synthesis, including the formation of additional categories of bioactive sphingolipids (such as GlcCer). Hence, to understand completely how sphingolipids regulate a given biological process, one must analyze essentially all of the sphingolipids in (and around!) the cells. It is likely that many of the seeming contradictions in the current research literature occur because each study has evaluated only one or a few of these factors, which may be analogous to the parable of the blind men trying to describe an elephant.

New technologies such as electrospray tandem mass spectrometry, DNA and protein microarrays, etc., are making 'sphingolipidomic' analyses feasible. Nonetheless, additional methods are still needed to evaluate temporal, localized changes in subcellular compartments, and to be able to study integrated systems such as tissues, organs, and intact organisms.

With better understanding comes greater utility, and sphingolipids are impacting almost every translational research field, from bioengineering to nutrition. It is intriguing that Thudichum's interests included nutrition, and a century before 'nutraceuticals' and 'functional foods' were in vogue, he wrote in the preface to his last book, *Cookery. Its Art and Practice* (1895): "Physiologic deduction proves that . . . no cookery is rational which does not attain the utmost theoretically possible effect, namely, the production of the highest physiological force. . . . It is believed and hoped that the medical profession will find in this work many materials to assist them. . . ". He surely had sphingolipids in mind.

## *Abbreviations*

| | |
|---|---|
| Cer | ceramide |
| ESI–MS/MS | electrospray tanden mass spectrometry |
| Fuc | fucose |
| Gal | galactose |
| G, with subscript for the subclass | ganglioside |
| Gb, with subscript for the number of carbohydrates | globoside |
| Glc | glucose |
| GlcA | glucuronic acid |
| GM2-AP | $G_{M2}$ activator protein |
| Hex A or B | hexosaminidase A or B |
| Lac | lactose |
| Man | mannose |
| GalNAc | $N$-acetylgalactosamine |
| GlcNAc | $N$-acetylglucosamine |
| Neu5Ac | $N$-acetylneuraminic acid |
| Neu5Gc | $N$-glycolylneuraminic acid |
| Pal-CoA | palmitoyl-CoA |
| Ser | serine |
| SPT | serine palmitoyltransferase |
| SAP | sphingolipid activator protein |

| SM | sphingomyelin |
| SMase | sphingomyelinase |
| TNFα | tumor necrosis factor-α |
| UDP-sugar | uridine dinucleotide phosphate sugar |
| UV | ultraviolet |

## References

1. Thudichum, J.L.W. (1884) A Treatise on the Chemical Constitution of Brain. Bailliere, Tindall, and Cox, London.
2. Hakomori, S. (1983) Chemistry of glycosphingolipids. In: J.N. Kanfer and S. Hakomori (Eds.), Sphingolipid Biochemistry. Plenum, New York, NY, pp. 1–164.
3. Schuette, C.G., Doering, T., Kolter, T. and Sandhoff, K. (1999) The glycosphingolipidoses — from disease to basic principles of metabolism. Biol. Chem. 380, 759–766.
4. Merrill Jr., A.H., Sullards, M.C., Wang, E., Voss, K.A. and Riley, R.T. (2001) Sphingolipid metabolism: Roles in signal transduction and disruption by fumonisins. Environ. Health Perspect. 109(Suppl. 2), 283–289.
5. Brown, D.A. and London, E. (2000) Structure and function of sphingolipid- and cholesterol-rich membrane rafts. J. Biol. Chem. 27, 17221–17224.
6. Venkataraman, K. and Futerman, A.H. (2000) Ceramide as a second messenger: sticky solutions to sticky problems. Trends Cell Biol. 10, 408–412.
7. Hakomori, S., Yamamura, S. and Handa, A.K. (1998) Signal transduction through glyco(sphingo)lipids. Introduction and recent studies on glyco(sphingo)lipid-enriched microdomains. Ann. N.Y. Acad. Sci. 845, 1–10.
8. Hannun, Y.A., Luberto, C. and Argraves, K.M. (2001) Enzymes of sphingolipid metabolism: From modular to integrative signaling. Biochemistry 40, 4893–4903.
9. Kolesnick, R.N., Goni, F.M. and Alonso, A. (2000) Compartmentalization of ceramide signaling: physical foundations and biological effects. J. Cell Physiol. 184, 285–300.
10. Spiegel, S. and Milstien, S. (2000) Sphingosine-1-phosphate: signaling inside and out. FEBS Lett. 476, 55–57.
11. IUPAC-IUB Joint Commission on Biochemical Nomenclature (JCBN). Nomenclature of glycolipids. Recommendations 1997 (1998) Eur. J. Biochem. 257, 293–298.
12. Karlsson, K.-A. (1970) On the chemistry and occurrence of sphingolipid long-chain bases. Lipids 5, 6–43.
13. Varki, A., Cummings, R., Esko, J., Freeze, H., Hart, G. and Marth, J. (1999) Essentials of Glycobiology. Cold Spring Harbor Laboratory Press, Cold Spring Harbor, NY, p. 653.
14. Merrill, A.H. Jr. and Hannun, Y.A. (Eds.) (2000) Methods in Enzymology: Sphingolipid Metabolism and Cell Signaling. Part A, Vol. 311 and Part B, Vol. 312, Academic Press, San Diego, CA.
15. Sullards, M.C. and Merrill, A.H. Jr. (2000) Analysis of sphingosine 1-phosphate, ceramides and other bioactive sphingolipids by liquid chromatography-tandem mass spectrometry, Science Signal Transduction Environment (STKE) http://stke.sciencemag.org/cgi/content/full/OC_sigtrans;2001/67/pl1
16. Lynch, D.V. (1993) Sphingolipids. In: T.S. Moore Jr. (Ed.), Lipid Metabolism in Plants. CRC Press, Boca Raton, FL, pp. 285–308.
17. Merrill, A.H. Jr., Grant, A.M., Wang, E. and Bacon, C.W. (1996) Lipids and lipid-like compounds of Fusarium. In: R. Prasad and M.A. Ghannoun (Eds.), Lipids of Pathogenic Fungi. Ch. 9, CRC Press, Boca Raton, FL, pp. 199–217.
18. Wiegandt, H. (Ed.) (1985) Glycolipids. Elsevier, Amsterdam.
19. Yu, R.K. and Saito, M. (1989) Structure and localization of gangliosides. In: R.U. Margolis and R.K. Margolis (Eds.), Neurobiology of Glycoconjugates. Plenum, New York, NY, pp. 1–42.
20. Wiegandt, H. (1995) The chemical constitution of gangliosides of the vertebrate nervous system. Behav. Brain. Res. 66, 85–97.

21. Vos, J.P., Lopes-Cardozo, M. and Gadella, B.M. (1994) Metabolic and functional aspects of sulfo-galactolipids. Biochim. Biophys. Acta 1211, 125–149.

22. Suzuki, K. (1998) Twenty five years of the 'psychosine hypothesis': a personal perspective of its history and present status. Neurochem. Res. 23, 251–259.

23. Vesper, H., Schmelz, E.-M., Nikolova-Karakashian, M.N., Dillehay, D.L., Lynch, D.V. and Merrill Jr., A.H. (1999) Sphingolipids in food and the emerging importance of sphingolipids in nutrition. J. Nutr. 129, 1239–1250.

24. Kolter, T. and Sandhoff, K. (1999) Sphingolipids: Their metabolic pathways and the pathobiochemistry of neurodegenerative diseases. Angew. Chem. Int. Ed. 38, 1532–1568.

25. Linn, S.C., Kim, H.-S., Keane, E.M., Andras, L.M., Wang, E. and Merrill Jr., A.H. (2001) Regulation of *de novo* sphingolipid biosynthesis, and toxic consequences of its disruption. Bioc. Soc. Trans. 29, 831–835.

26. Kolter, T., Proia, R.L. and Sandhoff, K. (2002) Combinatorial ganglioside biosynthesis. J. Biol. Chem. (in press).

27. Muramatsu, T. (2000) Essential roles of carbohydrate signals in development, immune response and tissue functions, as revealed by gene targeting. J. Biochem. (Tokyo) 127, 171–176.

28. Nagai (1995) Essential roles of carbohydrate signals in development, immune response and tissue functions, as revealed by gene targeting. Behav. Brain Res. 66, 99–104.

29. Riboni, L., Viani, P., Bassi, R., Prinetti, A. and Tettamanti, G. (1997) The role of sphingolipids in the process of signal transduction. Prog. Lipid Res. 36, 153–195.

D.E. Vance and J.E. Vance (Eds.) *Biochemistry of Lipids, Lipoproteins and Membranes (4th Edn.)*
© 2002 Elsevier Science B.V. All rights reserved

# CHAPTER 15

# Cholesterol biosynthesis

## Laura Liscum

*Department of Physiology, Tufts University School of Medicine, 136 Harrison Avenue,
Boston, MA 02111, USA, Tel.: +1 (617) 636-6945; Fax: +1 (617) 636-0445;
E-mail: laura.liscum@tufts.edu*

## 1. Introduction

Cholesterol's structure, biosynthetic pathway and metabolic regulation have tested the ingenuity of chemists, biochemists and cell biologists for over 100 years. The last century began with the pioneering work of Heinrich Wieland, who deduced the structure of cholesterol and bile acids, for which Wieland was awarded the Nobel Prize in Chemistry in 1926. How was such a complex molecule synthesized by the cell? Investigation into the cholesterol biosynthetic pathway required the development of isotopic tracer methods in Rudi Schoenheimer's lab in the 1930s. Using these novel techniques, Konrad Bloch and David Rittenberg showed that the ring structure and side chain of cholesterol were derived from acetate, and they identified intermediates in the pathway. Subsequent work by Bloch, John Cornforth and George Popjak succeeded in establishing the biosynthetic origin of all 27 carbons of cholesterol. For his elegant work, Bloch was awarded the Nobel Prize in Chemistry in 1964.

By the 1980s, the cholesterol biosynthetic pathway was understood to be a complex pathway of over 40 cytosolic and membrane-bound enzymes, which was subject to feedback regulation by the end-product, cholesterol, and oxygenated forms (called oxysterols). Genes encoding the key enzymes were cloned, which subsequently revealed the transcriptional and post-translational control of these enzymes. Michael Brown and Joseph Goldstein were awarded the Nobel Prize in Physiology or Medicine in 1985 for their comprehensive work on feedback regulation of cholesterol metabolism. Today, the mechanisms of regulation have been elucidated on a molecular level, although it is still not clear how cholesterol elicits all of the regulation. Furthermore, the evidence is rapidly building that cholesterol's precursors and metabolites might serve as biologically active signaling molecules.

Fig. 1 is an overview of the metabolic and transport pathways that control cholesterol levels in mammalian cells (reviewed in Liscum and Munn [1]). Cholesterol is synthesized from acetyl-CoA via the isoprenoid pathway, and at least four enzymes in the biosynthetic pathway are regulated by cellular cholesterol levels. Essential non-steroidal isoprenoids, such as dolichol, prenylated proteins, heme A and isopentenyl adenosine-containing tRNAs are also synthesized by this pathway. In extrahepatic tissues, most cellular cholesterol is derived from de novo synthesis [2], whereas hepatocytes obtain most of their cholesterol via the receptor-mediated uptake of plasma lipoproteins, such as low-density lipoprotein (LDL). LDL is bound and internalized by the LDL receptor

410

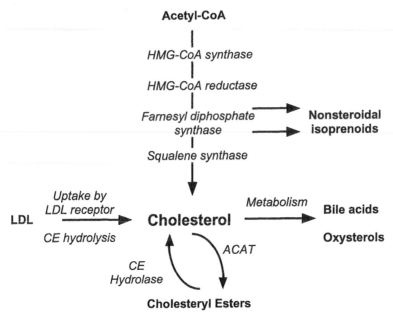

Fig. 1. Overview of the metabolic and transport pathways that control cholesterol levels in mammalian cells. Cholesterol is synthesized from acetyl-CoA and the four key enzymes that regulate cholesterol synthesis are indicated. Cells also obtain cholesterol by uptake and hydrolysis of LDL's cholesteryl esters (CE). End-products derived from cholesterol or intermediates in the pathway include bile acids, oxysterols, cholesteryl esters and non-steroidal isoprenoids. ACAT, acyl-CoA : cholesterol acyltransferase.

and delivered to the acidic late endosomes and lysosomes, where hydrolysis of the core cholesteryl esters occurs (discussed in Chapter 21). The cholesterol that is released is transported throughout the cell. Normal mammalian cells tightly regulate cholesterol synthesis and LDL uptake to maintain cellular cholesterol levels within narrow limits and supply sufficient isoprenoids to satisfy metabolic requirements of the cell. Regulation of cholesterol biosynthetic enzymes takes place at the level of gene transcription, mRNA stability, translation, enzyme phosphorylation and enzyme degradation. Cellular cholesterol levels are also modulated by a cycle of cholesterol esterification by acyl-CoA : cholesterol acyltransferase (ACAT) and hydrolysis of the cholesteryl esters, and by cholesterol metabolism to bile acids and oxysterols.

## 2. The cholesterol biosynthetic pathway

Fig. 2 takes a closer look at the cholesterol biosynthetic pathway, focusing on the enzymes that are regulated, sterol intermediates and the location of enzymes in the cell. Sterols are synthesized from the two-carbon building block, acetyl-CoA. The soluble enzyme acetoacetyl-CoA thiolase interconverts acetyl-CoA and acetoacetyl-CoA, which are then condensed by 3-hydroxy-3-methylglutaryl (HMG)-CoA synthase to form HMG-CoA. There are two forms of HMG-CoA synthase. A mitochondrial

411

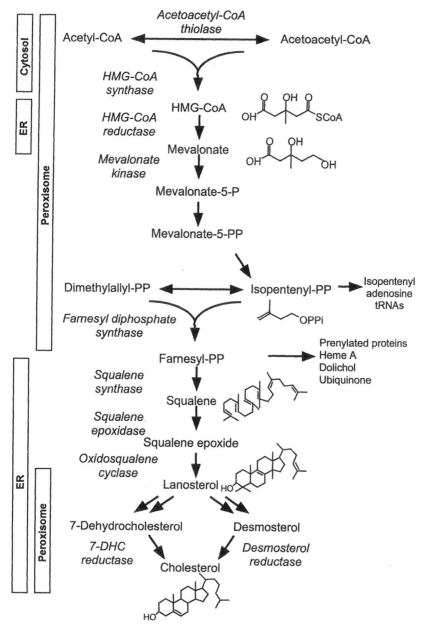

Fig. 2. The cholesterol biosynthetic pathway. Some of the major intermediates and end-products are indicated. Enzymes in the pathway are found in cytosol, endoplasmic reticulum (ER) and peroxisomes, as noted. Figure adapted from Olivier and Krisans [3]. HMG, 3-hydroxy-3-methylglutaryl; DHC, dehydrocholesterol.

form, involved in ketogenesis, predominates in the liver. In extrahepatic tissues, the most abundant form is a soluble enzyme of 53 kDa that is highly regulated by supply of cholesterol (G. Gil, 1986). Like acetoacetyl-CoA thiolase, HMG-CoA synthase has classically been described as a cytosolic enzyme because it is found in the $100,000 \times g$ supernatant of homogenized cells and tissues. However, both enzymes contain peroxisomal targeting sequences [3] and may reside in multiple cellular compartments.

HMG-CoA reductase catalyzes the reduction of HMG-CoA to mevalonate, utilizing two molecules of NADPH. HMG-CoA reductase is a 97-kDa glycoprotein of the endoplasmic reticulum (L. Liscum, 1985) and peroxisomes [3]. Analysis of the endoplasmic reticulum enzyme's domain structure revealed an N-terminal membrane domain with eight transmembrane spans (E.H. Olender, 1992), a short linker, and a C-terminal catalytic domain facing the cytosol (Fig. 3). Transmembrane spans 2–5 share a high degree of sequence similarity with several other key proteins in cholesterol metabolism; this region is termed the sterol-sensing domain (described in Section 3.5). Elucidation of the crystal structure of the HMG-CoA reductase catalytic domain indicated that the active protein is a tetramer [4], which is consistent with biochemical analysis. The monomers appear to be arranged in two dimers, with the active sites at the monomer–monomer interface. The dimer–dimer interface is predominantly hydrophobic.

HMG-CoA reductase is the rate-determining enzyme of the cholesterol biosynthetic pathway and, like HMG-CoA synthase, is highly regulated by supply of cholesterol. Thus, the enzyme has received intense scrutiny as a therapeutic target for treatment of hypercholesterolemia. The enzyme is inhibited by a class of pharmacological agents, generally called statins, which have an HMG-like moiety and a bulky hydrophobic group [5] (Fig. 4). Statins occupy the HMG-binding portion of the active site, preventing HMG-CoA from binding (E.S. Istvan, 2001). Also, the bulky hydrophobic group causes disordering of several catalytic residues. Thus, statins are potent, reversible competitive inhibitors of HMG-CoA reductase with $K_i$ values in the nanomolar range. Elevated plasma cholesterol levels are a primary risk factor for coronary artery disease, and statin inhibition of HMG-CoA reductase effectively reduces cholesterol levels and decreases overall mortality. However, complete inhibition of HMG-CoA reductase by statins will kill cells, even if exogenous cholesterol is supplied. That is because complete inhibition deprives cells of all mevalonate-derived products, including essential non-steroidal isoprenoids. To survive, cells must produce a small amount of mevalonate that, when limiting, is used preferentially by higher affinity pathways for non-steroidal isoprenoid production (S. Mosley, 1983).

Mevalonate is metabolized to farnesyl-diphosphate (-PP) by a series of enzymes localized in peroxisomes. First, mevalonate kinase phosphorylates the 5-hydroxy group of mevalonic acid. The enzyme is a homodimer of 40 kDa that is subject to feedback inhibition by several isoprenoid intermediates [6]. Mutations in the mevalonate kinase gene lead to the human genetic disease mevalonic aciduria (discussed in Section 2.2). The product of mevalonate kinase, mevalonate-5-P, is then phosphorylated to form mevalonic acid-5-PP, which is decarboxylated and dehydrated by mevalonate-PP decarboxylase to form isopentenyl-PP. Isopentenyl-PP is in equilibrium with its isomer, dimethylallyl-PP. Farnesyl-PP synthase catalyzes the head to tail condensations of two molecules of isopentenyl-PP with dimethylallyl-PP to form farnesyl-PP. The

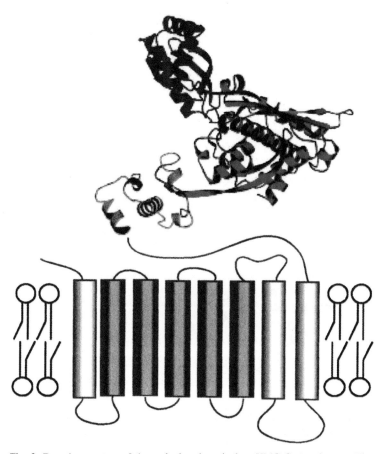

Fig. 3. Domain structure of the endoplasmic reticulum HMG-CoA reductase. The crystal structure of the catalytic domain has been determined and is depicted as a ribbon diagram (courtesy of Eva S. Istvan, Washington University School of Medicine). The catalytic domain consists of a small helical domain (green), a large central element resembling a prism (red), which contains the HMG-CoA-binding site, and a small domain to which NADPH binds (blue) [4]. The structure of the membrane domain has not been solved; however, it is known that eight transmembrane spans embed the protein into the endoplasmic reticulum membrane. Spans 2–5 (darker cylinders) are termed the sterol-sensing domain and mediate the regulated degradation of the enzyme.

enzyme is part of a large family of prenyltransferases that synthesize the backbones for all isoprenoids, including cholesterol, steroids, prenylated proteins, heme A, dolichol, ubiquinone, carotenoids, retinoids, chlorophyll and natural rubber (K.C. Wang, 2000).

Squalene synthase is a 47-kDa protein of the endoplasmic reticulum and catalyzes the first committed step in cholesterol synthesis. The enzyme condenses two molecules of farnesyl-PP and then reduces the presqualene-PP intermediate to form squalene. A large N-terminal catalytic domain faces the cytosol, anchored to the membrane by a C-terminal domain. This orientation may allow the enzyme to receive the hydrophilic substrates from the cytosol and release the hydrophobic product into the endoplasmic

414

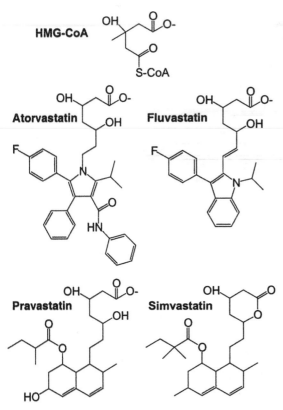

Fig. 4. Chemical structures of HMG-CoA and several statin inhibitors of HMG-CoA reductase. Atorvastatin (Lipitor), fluvastatin (Lescol), pravastatin (Pravachol) and simvastatin (Zocor) are widely prescribed cholesterol-lowering drugs.

reticulum membrane for further metabolism [7]. Squalene synthase is highly regulated by the cholesterol content of the cell. Thus, it plays an important role in directing the flow of farnesyl-PP into the sterol or non-sterol branches of the pathway (M.S. Brown, 1980) [7].

Squalene is converted into the first sterol, lanosterol, by the action of squalene epoxidase and oxidosqualene cyclase. Lanosterol is then converted to cholesterol by a series of oxidations, reductions, and demethylations. The required enzyme reactions have been defined and metabolic intermediates identified; however, the precise sequence of reactions between lanosterol and cholesterol remains to be established [8] (Fig. 5). There is evidence for two alternative pathways that differ in when the Δ24 double bond is reduced (discussed in Section 2.3). Both 7-dehydrocholesterol and desmosterol have been postulated to be the immediate precursor of cholesterol. One of the key enzymes in the latter part of the pathway is 7-dehydrocholesterol Δ7-reductase, a 55-kDa integral membrane protein. Mutations in the gene for 7-dehydrocholesterol Δ7-reductase cause the human genetic disease Smith–Lemli–Opitz syndrome (discussed in Section 2.3).

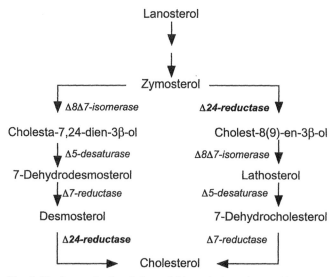

Fig. 5. Final steps in the cholesterol biosynthetic pathway. Alternate steps have been proposed for the conversion of zymosterol to cholesterol, which differ in when the Δ24-reductase reaction occurs. Figure adapted from Waterham and Wanders [8] and Kelly and Hennekam [11].

## 2.1. Enzyme compartmentalization

Where does cholesterol synthesis take place? All of the enzymes that convert acetyl-CoA to farnesyl-PP have classically been thought of as cytosolic enzymes, with the exception of HMG-CoA reductase, which is typically depicted as an endoplasmic reticulum enzyme with the catalytic site facing the cytosol. Enzymes that convert farnesyl-PP to cholesterol are classically described as microsomal. However, there is now strong evidence that all but one of these enzymes is also, or exclusively, peroxisomal [3]. The molecular cloning of cDNAs encoding many of these enzymes has revealed peroxisomal targeting sequences. The availability of antibodies has allowed immunocytochemical localization to peroxisomes. Together these data suggest that peroxisomes may play an active role in all steps in the cholesterol biosynthetic pathway except the conversion of farnesyl-PP to squalene, which is catalyzed by squalene synthase found solely in the endoplasmic reticulum.

HMG-CoA reductase is the one exception to the rule. Immunocytochemistry and immunoblotting have localized HMG-CoA reductase to both the endoplasmic reticulum and peroxisomes; however, no peroxisomal targeting motif has been found in the HMG-CoA reductase protein sequence. Furthermore, the peroxisomal HMG-CoA reductase has an apparent molecular weight of 90 kDa whereas the endoplasmic reticulum enzyme is 97 kDa (W.H. Engfelt, 1997). The peroxisomal enzyme exhibits other distinct properties: it is resistant to statin inhibition, the enzyme's activity is not regulated by phosphorylation, the protein's turnover is not regulated by mevalonate. Altogether, this evidence suggests that the endoplasmic reticulum and peroxisome enzymes are functionally and structurally distinct (N. Aboushadi, 2000).

Additional evidence for the involvement of peroxisomes in cholesterol biosynthesis comes from analysis of diseases of peroxisomal deficiency. Zellweger syndrome, neonatal adrenoleukodystrophy, and infantile Refsum's disease are all diseases of peroxisome biogenesis [9]. In most of these peroxisomal disorders, the peroxisomal matrix proteins are synthesized in the cytosol as normal, but they cannot be assembled into nascent peroxisomes due to mutations in one of at least 12 different genes encoding proteins necessary for peroxisomal protein targeting and import. Fibroblasts from individuals with peroxisome biogenesis disorders show reduced enzymatic activities of cholesterol biosynthetic enzymes, reduced levels of cholesterol synthesis and lower cholesterol content [3]. These data support the hypothesis that part of the cholesterol synthesis pathway is peroxisomal.

It is not clear why cholesterol synthesis is compartmentalized and requires intermediates to cycle between peroxisomes and the cytosol. It is also unclear why some of the enzymes are found in multiple compartments and others are solely in endoplasmic reticulum or peroxisomes. As noted, cholesterol synthesis is a very complex process and compartmentalization may represent another level of regulation [3].

## 2.2. Mevalonic aciduria

Cholesterol synthesis is essential for normal development and maintenance of tissues that cannot obtain cholesterol from plasma lipoproteins, such as brain. Furthermore, the biosynthetic pathway supplies non-steroidal isoprenoids that are required by all cells. Thus, it is not surprising that metabolic defects in the cholesterol biosynthetic pathway have devastating consequences.

The first recognized human metabolic defect in the biosynthesis of cholesterol and isoprenoids was mevalonic aciduria [10]. Mevalonic aciduria is an autosomal recessive disorder that is quite rare, with only 19 known patients. In normal individuals, a small amount of mevalonic acid diffuses into the plasma at levels proportional to the rate of cellular cholesterol formation. Patients with mild mevalonic aciduria excrete 3000–6000 times the normal amount of mevalonic acid and patients with the severe form of the disease excrete 10,000–200,000 times the normal amount. Enzyme assays using cell lysates showed that mevalonate kinase activity was markedly deficient in patient samples and genetic analysis has revealed nucleotide changes in the mevalonate kinase gene that lead to amino acid substitutions. Because of this enzyme deficiency, there is little to no feedback inhibition of HMG-CoA reductase and, thus, mevalonate is overproduced.

Clinical features of mevalonic aciduria include failure to thrive, anemia, gastroenteropathy, hepatosplenomegaly, psychomotor retardation, hypotonia, ataxia, cataracts, and dysmorphic features [10]. Surprisingly, patients with severe deficiencies in mevalonate kinase show normal plasma cholesterol levels and cultured mevalonic aciduria fibroblasts show rates of cholesterol synthesis half that of normal cells. Close examination of cholesterogenic enzymes in mevalonic aciduria fibroblasts has revealed a 6-fold increase in HMG-CoA reductase activity, which is postulated to compensate for the low mevalonate kinase activity.

## 2.3. Smith–Lemli–Opitz syndrome

A second metabolic defect in cholesterol synthesis leads to Smith–Lemli–Opitz syndrome (SLOS) (B.U. Fitzkey, 1999) [11]. SLOS is a relatively common autosomal recessive disorder, with estimates of incidence ranging from 1 in 10,000 to 1 in 60,000. Four lines of evidence pointed to the metabolic defect in SLOS patients. (1) Individuals with SLOS were found to have markedly elevated levels of plasma 7-dehydrocholesterol and low plasma cholesterol levels. (2) 7-Dehydrocholesterol $\Delta 7$-reductase activity was deficient in SLOS patient samples and the amount of residual activity could be correlated with severity of the disease. (3) Rodents treated with AY-9944, an inhibitor of 7-dehydrocholesterol $\Delta 7$-reductase, developed SLOS-like malformations [11]. (4) Cloning of the gene for 7-dehydrocholesterol $\Delta 7$-reductase led to identification of a splice-site mutation and amino acid substitutions in SLOS patients.

Severely reduced cholesterol synthesis is predicted to have severe consequences on development of the fetus because cholesterol is only obtained from the maternal circulation during the first trimester [11]. In addition, the brain is predicted to be severely affected because plasma lipoproteins cannot cross the blood–brain barrier and most, if not all, cholesterol needed for brain growth is synthesized locally (S.D. Turley, 1998) [2,12]. Indeed, severely affected SLOS infants who died soon after birth were found to have functionally null 7-dehydrocholesterol $\Delta 7$-reductase alleles [12], whereas typical affected individuals likely have some residual 7-dehydrocholesterol $\Delta 7$-reductase catalytic activity.

Patients with SLOS have mental retardation and microcephaly, which is consistent with cholesterol synthesis being required for normal brain development. Clinical features also include failure to thrive, and characteristic craniofacial, skeletal and genital anomalies. The clinical phenotype appears to be due to a lack of cholesterol rather than the cellular accumulation of 7-dehydrocholesterol (W. Gaoua, 2000). A recent multicenter clinical trial has shown that SLOS children fed a diet supplemented with cholesterol show improved growth and neurodevelopment (i.e. language and cognitive skills) (M. Irons, 1997; E.R. Elias, 1997). It is likely that the diet fulfilled the daily requirement for cholesterol and down-regulated endogenous 7-dehydrocholesterol synthesis.

What are the final steps in the cholesterol biosynthetic pathway? SLOS may provide an answer to that question. As noted above, there is evidence for two alternative pathways, which differ in when the $\Delta 24$ double bond is reduced [11]. In both pathways, lanosterol is demethylated to form zymosterol (Fig. 5). Then, zymosterol can be metabolized sequentially by a $\Delta 24$-reductase, $\Delta 8,\Delta 7$-isomerase, and $\Delta 5$-desaturase to form 7-dehydrocholesterol, which is reduced at the $\Delta 7$ position to form cholesterol. Alternatively, zymosterol can be metabolized by the $\Delta 8,\Delta 7$-isomerase and $\Delta 5$-desaturase, to form 7-dehydrodesmosterol. 7-Dehydrodesmosterol is metabolized by the $\Delta 7$-reductase to form desmosterol and then by the $\Delta 24$-reductase to form cholesterol. The fact that the SLOS deficiency in $\Delta 7$-reductase leads to a buildup of 7-dehydrocholesterol rather than 7-dehydrodesmosterol is interpreted to mean that the former pathway is the principal one. However, the latter pathway must also be used because desmosterol is an abundant cholesterol precursor in certain tissues. It has been suggested that the final steps in the biosynthetic pathway may be tissue specific.

Table 1
Inborn errors of sterol biosynthesis

| Syndrome | Metabolic defect |
| --- | --- |
| Mevalonic aciduria | Mevalonate kinase |
| Smith–Lemli–Opitz | Sterol $\Delta$7-reductase |
| Desmosterolosis | Sterol $\Delta$24-reductase |
| Rhizomelic chondrodysplasia punctata (CDP) | Pex7 peroxisomal enzyme import |
| CDP X-linked dominant (CDPX2) | Sterol $\Delta$8,$\Delta$7-isomerase |
| CHILD syndrome (congenital hemidysplasia with ichthyosis and limb defects) | Sterol $\Delta$8,$\Delta$7-isomerase Sterol C-4 demethylase |
| Greenberg skeletal dysplasia | Sterol $\Delta$14-reductase |

These syndromes and their corresponding metabolic defects are reviewed in Kelley [13].

Perhaps, in SLOS cells, any 7-dehydrodesmosterol that accumulates is metabolized by the available $\Delta$24-reductase to form 7-dehydrocholesterol.

*2.4. Other enzyme deficiencies*

Other inborn errors of sterol biosynthesis have been reviewed by Kelley [13] and are summarized in Table 1. Rhizomelic chondrodysplasia punctata, like Zellweger syndrome, exhibits defective sterol synthesis due to the lack of key peroxisomal enzymes of cholesterol biosynthesis. CDPX2, also known as Conradi–Hünermann syndrome, and most cases of CHILD syndrome are due to mutations in the sterol $\Delta$8,$\Delta$7-isomerase gene, which is located on the X chromosome. Mutations in a single gene may lead to different syndromes with similar, but distinct, pathologies due to the mosaicism of X-chromosome inactivation. A few cases of CHILD syndrome may be due to mutations in the sterol C-4 demethylase gene, also located on the X chromosome.

# 3. Regulation of cholesterol synthesis

Isoprenoid synthesis is regulated by the sterol end-product of the biosynthetic pathway, by non-sterol intermediates, and also by physiological factors. The cholesterol content of the cell controls several enzymes in the biosynthetic pathway, but the focus has been on the rate-limiting enzyme, HMG-CoA reductase. Different regulators have different mechanisms of action. For example, sterols have been shown to regulate at the level of HMG-CoA reductase transcription whereas non-sterols regulate HMG-CoA reductase mRNA translation. Both sterols and non-sterols are needed for regulation of HMG-CoA reductase protein degradation [14]. Physiological factors that influence cholesterol synthesis include diurnal rhythm, insulin and glucagon, thyroid hormone, glucocorticoids, estrogen and bile acids [15]. These factors regulate HMG-CoA reductase by transcriptional, translational and post-translational mechanisms.

## 3.1. Transcriptional regulation

HMG-CoA reductase is the rate-limiting enzyme in the cholesterol biosynthetic pathway and combined regulation of HMG-CoA reductase synthesis and turnover can alter steady state levels of the enzyme 200-fold. HMG-CoA reductase is regulated in parallel with at least three other enzymes in the cholesterol biosynthetic pathway, HMG-CoA synthase, farnesyl-PP synthase and squalene synthase, as well as the LDL receptor. This coordinate regulation is due to the fact that each gene has a similar sequence (*cis*-acting element) within the promoter that recognizes a common *trans*-acting transcription factor. Availability of the transcription factor to bind to the promoter sequence is influenced by the cellular cholesterol content.

Fig. 6 illustrates the current model of cholesterol-mediated transcriptional regulation [16–19]. The 5′ flanking regions of cholesterol-regulated genes have one to three copies of a 10-bp non-palindromic nucleotide sequence termed the sterol regulatory element (SRE). SREs are conditional positive elements that are required for gene transcription in cholesterol-depleted cells. The SRE sequence found in the LDL receptor gene is 5′-ATCACCCCAC-3′. SREs have been identified in the HMG-CoA synthase, HMG-CoA reductase, farnesyl-PP synthase and squalene synthase genes, as well as genes of fatty acid synthesis. However, there is not a strict SRE consensus sequence and identifying functional SREs has been difficult [17].

The transcription factor that binds the SRE is termed the SRE-binding protein (SREBP) [16,20,21]. The first SREBP to be identified was the protein that bound to the LDL receptor promoter (M.R. Briggs, 1990; X. Wang, 1990). Cloning of SREBP cDNAs (C. Yokoyama, 1993; X. Hua, 1993) revealed that there are two SREBP genes that produce three distinct proteins. SREBP-1a and -1c are derived from one gene that contains two promoters and differ in the length of the N-terminal transactivation domain. SREBP-2 is derived from a second gene and is 45% identical to SREBP-1a. SREBP-1c is the predominant isoform in liver and adipocytes. It was isolated independently and called adipocyte determination and differentiation-dependent factor 1 (P. Tontonoz, 1993). Here when the term SREBP is used, the information is relevant for all three isoforms.

SREBPs are cytosolic 68-kDa proteins with a canonical basic helix–loop–helix leucine zipper (bHLH-Zip) motif that is present in other transcription factors. Unlike other bHLH-Zip transcription factors, SREBPs have a tyrosine in place of a conserved arginine, which allows them to bind to the inverted E-box motif (5′-CANNTG-3′) in addition to SREs (J.B. Kim, 1995). By binding to SREs, SREBPs coordinately regulate multiple enzymes involved in fatty acid synthesis and lipogenesis [20,21].

An additional feature that distinguishes SREBPs from other bHLH-Zip transcription factors is that SREBP genes encode 125-kDa membrane proteins that are inserted into the endoplasmic reticulum and serve as precursors for the active transcription factors. SREBPs have three functional domains: an N-terminal 68-kDa fragment containing the bHLH-Zip transcription factor, two membrane-spanning segments, and a C-terminal regulatory domain. It is the sequential two-step cleavage of the full-length precursor SREBP and release of the 68-kDa N-terminal bHLH-Zip domain that is influenced by the cellular cholesterol content.

420

In cholesterol-depleted cells      In cholesterol-fed cells

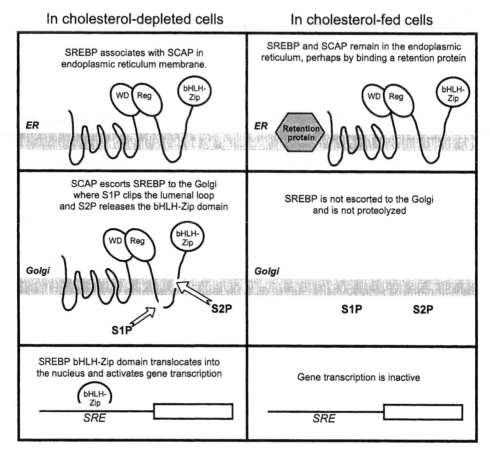

Fig. 6. Current model of cholesterol regulation of SREBP proteolysis. The sterol regulatory element-binding protein (SREBP) precursor is inserted into the endoplasmic reticulum (ER) membrane. The SREBP regulatory domain (Reg) interacts with the SREBP cleavage-activating protein (SCAP), likely through SCAP's WD repeats. When cholesterol levels are low, SCAP escorts SREBP to the Golgi where the bHLH-ZIP domain is released by site-1 protease (S1P) cleavage of a lumenal loop followed by site-2 protease (S2P) cleavage within a transmembrane span. The mature SREBP translocates into the nucleus and activates gene transcription. In cholesterol replete cells, the SREBP precursor and SCAP remain in the endoplasmic reticulum and the SREBP precursor is not proteolyzed to release the bHLH-Zip transcription factor.

Identification of the proteins required for regulated SREBP cleavage was accomplished using somatic cell genetic approaches. Mutant Chinese hamster ovary cells with abnormal regulation of cholesterol and fatty acid metabolism, which were selected over the past 20 years, proved invaluable for this goal [22]. Using expression cloning strategies, genes were isolated that restored SREBP-mediated transcription in each mutant. This work led to identification of two proteases and an escort protein required for SREBP precursor cleavage. Cleavage of the SREBP precursor at site 1 requires a subtilisin-like serine protease (J. Sakai, 1998), whereas cleavage at site 2 requires a zinc metalloprotease (R.B. Rawson, 1997). Transcription factor release is controlled by a chaperone, SREBP cleavage-activating protein (SCAP), which escorts the SREBP precursor to the Golgi where the proteases reside (X. Hua, 1996).

How is SREBP proteolysis controlled by cholesterol? A hint that this event may involve vesicle trafficking came from the finding that the SREBP precursor's $N$-linked carbohydrates are endoglycosidase-H-resistant (trimmed by Golgi mannosidases) when cellular cholesterol levels are low, and endoglycosidase-H-sensitive when cellular cholesterol levels are high (A. Norturfft, 1998, 1999). Thus, release of the mature SREBP transcription factor appeared to coincide with transport to the Golgi. The current model is as follows (Fig. 6). When cellular cholesterol levels are low, the SREBP precursor is synthesized and inserted into the endoplasmic reticulum membrane. The C-terminal regulatory domain of the SREBP precursor interacts with the C-terminus of SCAP, likely through SCAP's four WD repeats (J. Sakai, 1998). SCAP and the SREBP precursor are then transported to the Golgi where the lumenal loop of the SREBP precursor is clipped by the site-1 protease; however, the two halves of the protein remain membrane-anchored. Then the site-2 protease clips the N-terminal SREBP intermediate within the first membrane-spanning segment, releasing the soluble transcription factor, mature SREBP. Upon translocation to the nucleus, the mature SREBP binds SRE sequences within the promoters of target genes and enhances their transcription.

When cellular cholesterol levels rise to a threshold level, SCAP and the SREBP precursor no longer travel to the Golgi and the SREBP precursor is not proteolyzed to produce the mature SREBP. As a result, transcription of target genes declines to basal levels. Evidence for SCAP–SREBP transport is two-fold. The sterol-dependent movement of SCAP has been directly visualized in cultured cells transfected with a green fluorescent protein–SCAP fusion protein (A. Nohturfft, 2000). Furthermore, in vitro vesicle-budding assays have demonstrated that oxysterols suppress SREBP–SCAP complexes from entering into vesicles budding from the endoplasmic reticulum.

The mechanism by which SCAP senses cellular cholesterol is not known. Sensing is postulated to involve a five-transmembrane segment of SCAP with sequence similarity to a five-transmembrane segment of HMG-CoA reductase, called the sterol-sensing domain (discussed in Section 3.5). Consistent with this hypothesis, several constitutively active SCAP mutants have been isolated that have point mutations in the sterol-sensing domain (X. Hua, 1996). Also, there is recent evidence that SCAP interacts with a protein that is retained in the endoplasmic reticulum (T. Yang, 2000). Therefore, control of SCAP and SREBP transit to the Golgi could depend upon sterol-dependent binding of SCAP's sterol-sensing domain to a retention protein.

SREBPs control not only cholesterol synthesis, but also the synthesis of fatty acids (via fatty acid synthase, acetyl-CoA carboxylase, and stearoyl-CoA desaturase 2), triacylglycerols and phospholipids (glycerol-3-phosphate acyltransferase). This was illustrated most dramatically in transgenic mice overexpressing the nuclear forms of SREBP-1a, -1c and -2 in liver [20]. Overexpression of each isoform resulted in activation of a full spectrum of cholesterol and fatty acid biosynthetic enzymes; however, absolute levels of induction of each enzyme and the subsequent liver phenotype varied according to the isoform expressed. From these data, the following conclusions can be drawn. SREBP-1a is a strong activator of cholesterol and fatty acid synthesis and is likely to be important in rapidly dividing cells that require lipid for membrane production.

SREBP-1c predominates in the liver, where it primarily activates genes of fatty acid synthesis. SREBP-1c appears to be important for maintaining basal transcription levels of fatty acid and cholesterol biosynthetic enzymes during periods of fasting. It also plays a role in the insulin response [21]. SREBP-2 selectively activates cholesterol biosynthetic genes and the LDL receptor, and primarily responds when the liver's demand for cholesterol rises.

### 3.2. mRNA translation

HMG-CoA reductase is also subject to translational control by a mevalonate-derived non-sterol regulator (D. Peffley, 1985; M. Nakanishi, 1988). This component of the regulatory mechanism can only be observed when cultured cells are acutely incubated with statins, which block mevalonate formation. Under those conditions, sterols have no effect on HMG-CoA reductase mRNA translation; however, mevalonate reduces the HMG-CoA mRNA translation by 80% with no change in mRNA levels. Translational control of hepatic HMG-CoA reductase by dietary cholesterol was shown in an animal model in which polysome-associated HMG-CoA reductase mRNA was analyzed in cholesterol-fed rats (C.M. Chambers, 1997). It was found that cholesterol feeding increased the portion of mRNA associated with translationally inactive monosomes and decreased the portion of mRNA associated with translationally active polysomes. The mechanism of HMG-CoA reductase translational control has not been elucidated.

### 3.3. Phosphorylation

Many key metabolic enzymes are modulated by phosphorylation–dephosphorylation and it has long been known that HMG-CoA reductase catalytic activity is inhibited by phosphorylation (Z.H. Beg, 1973). Rodent HMG-CoA reductase is phosphorylated on Ser 871 by an AMP-activated protein kinase that uses ATP as a phosphate donor (P.R. Clarke, 1990). However, examination of HMG-CoA reductase activity in rat liver showed that phosphorylation–dephosphorylation could not account for the long-term regulation that occurred with diurnal light cycling, fasting, or cholesterol-supplemented diet (M.S. Brown, 1979). Approximately 75–90% of HMG-CoA reductase enzyme was found to be phosphorylated (inactive) under all physiological conditions. This reservoir of inactive enzyme may allow cells to respond transiently to short-term cholesterol needs.

The AMP-activated kinase that phosphorylates and inactivates HMG-CoA reductase also phosphorylates and inactivates acetyl-CoA carboxylase. It has been suggested that, when cellular ATP levels are depleted causing AMP levels to increase, the resultant activation of the kinase would inhibit cholesterol and fatty acid biosynthetic pathways, thus conserving energy (D.G. Hardie, 1992). Consistent with this hypothesis, cholesterol synthesis was reduced when ATP levels were depleted by incubation with 2-deoxy-D-glucose (R. Sato, 1993). However, cholesterol synthesis was not reduced when ATP levels declined in cells expressing a Ser 871 to Ala mutant form of HMG-CoA reductase, which is not phosphorylated (R. Sato, 1993). Therefore, HMG-CoA reductase phosphorylation appears to be important for preserving cellular energy stores rather than end-product feedback regulation.

## 3.4. Proteolysis

Raising the cellular cholesterol content not only stops transcription of the genes encoding the cholesterol biosynthetic enzymes, but it also leads to accelerated degradation of the rate-limiting enzyme, HMG-CoA reductase. In cholesterol-depleted cells, HMG-CoA reductase is a stable protein that is degraded slowly ($t_{1/2} = 13$ h) (J.R. Faust, 1982). If cholesterol repletion simply stopped transcription of the HMG-CoA reductase gene, that would lead to a slow decline in HMG-CoA reductase enzyme activity owing to stability of the protein; however, in the presence of excess sterols or mevalonate there is rapid ($t_{1/2} = 3.6$ h) (J.R. Faust, 1982) and selective degradation of the enzyme, which results in more precise control of cellular sterol synthesis.

The HMG-CoA reductase membrane domain is necessary and sufficient for regulated degradation. Expression of the cytosolic catalytic domain results in a stable protein that is not subject to regulated degradation (G. Gil, 1985), whereas expression of the N-terminal membrane domain linked to a reporter protein results in regulated degradation of the reporter (D. Skalnik, 1988). Despite the complex topology of HMG-CoA reductase, no proteolytic intermediates have ever been detected. How is HMG-CoA reductase proteolyzed? Possibilities include vesicular translocation to lysosomes or autophagy of HMG-CoA reductase-containing endoplasmic reticulum membranes; however, the degradation appears to be rapid and selective for HMG-CoA reductase. It is also unaffected by inhibitors of protein transport through the Golgi (K. Chun, 1990). Another possibility is HMG-CoA reductase digestion by resident endoplasmic reticulum proteases. Subcellular fractionation and use of specific protease inhibitors has revealed that HMG-CoA reductase degradation occurs in purified endoplasmic reticulum membranes and is inhibited by lactacystin, an inhibitor of the proteasome (T. McGee, 1996). Additional evidence that the proteosome is involved comes from study of the yeast ortholog, Hmg2p, which is subject to regulated degradation like the mammalian enzyme. A yeast genetic approach has led to the identification of three proteins that are involved in the reverse translocation of endoplasmic reticulum proteins and disposal by the proteosome (R.Y. Hampton, 1996) [23]. Evidence that HMG-CoA reductase is polyubiquitinated prior to proteolysis has been provided for the yeast (R.Y. Hampton, 1997) and mammalian (T. Ravid, 2000) enzymes.

What is the signal for accelerated HMG-CoA reductase degradation? In mammalian cells, both mevalonate-derived non-sterols and sterols are required. That is, in cholesterol-depleted cells, the addition of sterols leads to accelerated HMG-CoA reductase degradation only when the sterols are accompanied by a mevalonate-derived non-sterol signal (M. Nakanishi, 1988; J. Roitelman, 1992; T.E. Meigs, 1997). Regulated degradation has also been demonstrated through in vitro experiments. HMG-CoA reductase degradation is more rapid in endoplasmic reticulum membranes isolated from mevalonate- or sterol-treated cells (T. McGee, 1996) and in hepatic microsomes prepared from mevalonate-treated rats (C. Correll, 1994). Both the sterol and non-sterol signals are blocked by cycloheximide, indicating that regulated turnover of HMG-CoA reductase requires ongoing protein synthesis (K. Chun, 1990; J. Roitelman, 1992; T. Ravid, 2000).

There is strong evidence that the non-sterol isoprenoid signal for HMG-CoA reduc-

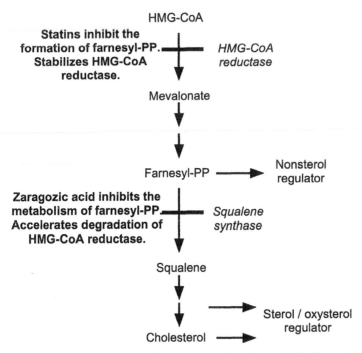

Fig. 7. Evidence that a farnesyl-PP-derived signal modulates HMG-CoA reductase degradation. In yeast, conditions that decrease farnesyl-PP levels stabilize HMG-CoA reductase levels whereas conditions that increase farnesyl-PP levels accelerate HMH-CoA reductase degradation.

tase degradation is derived from farnesyl-PP. This evidence has been recently reviewed (S.F. Petras, 2001) and includes the following. Intracellular farnesol levels increase significantly after mevalonate addition to cells (T.E. Meigs, 1996). Accelerated HMG-CoA reductase degradation can be induced in cells incubated with mevalonate, farnesyl-PP, or farnesol but not with a non-hydrolyzable analog of farnesyl-PP (C.C. Correll, 1994; M.D. Giron, 1994; T.E. Meigs, 1997). Furthermore, inhibition of the enzyme farnesyl pyrophosphatase blocks the mevalonate-dependent, sterol-accelerated degradation of the enzyme (T.E. Meigs, 1997).

Many, but not all, aspects of regulated HMG-CoA reductase degradation are conserved among eukaryotes. In yeast, modulation of Hmg2p stability by a farnesyl-PP-derived signal has been shown using pharmacologic and genetic approaches (R.G. Gardner, 1999). Conditions chosen to increase farnesyl-PP levels (inhibition of squalene synthase by zaragozic acid or down-regulation of squalene synthase) accelerated Hmg2p ubiquitination and degradation, whereas conditions chosen to decrease farnesyl-PP levels (inhibition of HMG-CoA reductase by statins, farnesyl-PP synthase down-regulation, or squalene synthase overexpression) stabilized Hmg2p (Fig. 7). One difference between the yeast and mammalian systems is the requirement for ongoing protein synthesis, which has not been shown in yeast (R.G. Gardner, 1999).

As mentioned above, regulation of mammalian HMG-CoA reductase turnover requires both sterols and non-sterol isoprenoids. In contrast, the yeast isozyme can be suppressed by farnesyl-PP alone, although sterols enhance the non-sterol regulation. Pharmacologic and genetic approaches have again been extremely informative, showing that endogenously produced oxysterols serve as a positive signal for Hmg2P degradation in yeast (R.G. Gardner, 2001). Oxysterols have long been known to accelerate HMG-CoA reductase degradation when added to the medium of cultured mammalian cells; however, an endogenously produced oxysterol regulator of mammalian HMG-CoA reductase degradation has not yet been identified.

How do sterol or non-sterol regulators influence HMG-CoA reductase degradation? They could have a direct effect on the enzyme itself, as an allosteric regulator or by binding to the membrane domain. Alternatively, they could have an effect on the biophysical properties of the endoplasmic reticulum membrane (i.e. fluidity, membrane organization) or affect the proteolytic machinery. These possible effects could be direct or through interaction with an effector protein (R.G. Gardner, 1999; S.F. Petras, 2001).

The half-life of HMG-CoA reductase is also influenced by the enzyme's oligomerization state and expression level (H.H. Cheng, 1999). Oligomerization of HMG-CoA reductase through its cytosolic domain appears to stabilize the protein, as does a higher expression level. Analysis of the crystal structure of the catalytic portion of human HMG-CoA reductase revealed that this domain forms tight tetramers (E.S. Istvan, 2000). It was suggested that oligomerization of the catalytic domain may induce association of the membrane domains, which would decrease accessibility of the enzyme to proteases. Sterols may induce dissociation of the HMG-CoA reductase tetramer, resulting in accelerated proteolysis.

*3.5. Sterol-sensing domain*

How do HMG-CoA reductase and SCAP detect rising cellular cholesterol levels? A key feature of these proteins was revealed when SCAP was cloned and the deduced protein sequence compared with that of HMG-CoA reductase (X. Hua, 1996). A five transmembrane domain was found in SCAP with 25% identity and 55% similarity with a corresponding region in HMG-CoA reductase. This domain, termed the sterol-sensing domain, appeared to be critical for SCAP's cholesterol-regulated SREBP escort function since certain amino acid substitutions in the sterol-sensing domain led to constitutive activity (X. Hua, 1996; A. Nohturfft, 1996). In HMG-CoA reductase, transmembrane span 2 (which is within the sterol-sensing domain) was shown to be necessary for regulated degradation (H. Kumagai, 1995). Chimeric mutants of HMG-CoA reductase were constructed that combined membrane-spanning domains from the hamster enzyme (which is subject to regulated degradation) with membrane-spanning domains from the sea urchin enzyme (which shares 62% amino acid sequence identity with the hamster enzyme in the membrane domain, but is not subject to regulated degradation). Analysis of regulated turnover of the chimeric molecules showed that hamster transmembrane span 2 was sufficient to confer regulated degradation upon the sea urchin enzyme (H. Kumagai, 1995).

Sterol-sensing domains with a high degree of sequence similarity are found in several other proteins with obvious connections to cholesterol homeostasis. One is the biosynthetic enzyme 7-dehydrocholesterol Δ7-reductase. The function of the sterol-sensing domain in 7-dehydrocholesterol Δ7-reductase is not clear; however, amino acid substitutions in that region cause the loss of 90% of catalytic activity (S.H. Bae, 1999). Another sterol-sensing domain-containing protein is NPC1, a 1278-amino-acid glycoprotein found in late endosomes and lysosomes (E.D. Carstea, 1997). NPC1 is hypothesized to play a role in trafficking of cholesterol, gangliosides and other cargo from late endosomes to destinations throughout the cell [24] (L. Liscum, 2000). Mutations in NPC1 lead to the predominant form of Niemann–Pick C disease, a human genetic disease characterized by progressive neurodegeneration. The biological function of NPC1 is still not clear, but structure/function analysis indicates that the sterol-sensing domain is important. Mutations that cause amino acid substitutions within the sterol-sensing domain lead to a rapidly progressing, infantile form of the disease, whereas amino acid substitutions throughout the rest of the protein cause the classical juvenile presentation [24]. NPC1L1 is a Golgi protein with 42% identity and 51% similarity with NPC1 (J.P. Davies, 2000). NPC1L1 has a sterol-sensing domain, but the function of this NPC1-like protein is unknown.

Other proteins with a sterol-sensing domain have a more tenuous link to cholesterol homeostasis. Two proteins, Patched and dispatched, are involved in developmental patterning (A.P. McMahon, 2000). Patched is the receptor for the morphogen, Sonic Hedgehog, which is the only known protein with a covalently attached cholesterol moiety. Dispatched is the plasma membrane protein required for secretion of cholesterol-modified Hedgehog. Hedgehog binding to Patched leads to a signal transduction cascade that activates transcription of specific genes. Mutations in Patched cause basal cell nevus syndrome, which is characterized by developmental abnormalities and basal cell carcinomas (R.L. Johnson, 1996). A role for Patched and dispatched in cholesterol metabolism has not been established; however, it is intriguing that exposure of embryos to inhibitors of cholesterol biosynthesis, such as Triparanol, AY-9944 or BM 15.766, cause profound developmental defects that resemble those in Sonic Hedgehog mutant embryos (J.A. Porter, 1996).

## 4. Metabolism of cholesterol

Cellular cholesterol levels are regulated, not only by feedback inhibition of cholesterol synthesis, but also by feedforward regulation of cholesterol metabolism. Excess cholesterol is metabolized to oxysterols. In addition to blocking SCAP-facilitated proteolysis of SREBP and thereby down-regulating endogenous cholesterol synthesis and LDL receptor levels, oxysterols also activate bile acid synthesis (discussed in Chapter 16) and cholesterol esterification, which further reduces the cellular content of unesterified cholesterol.

### 4.1. Oxysterols

Oxysterols are potent suppressors of cholesterol synthesis (A.A. Kandutsch, 1973, 1974). Their effectiveness has been attributed to their ability to diffuse into and through cells to activate regulatory processes, thus bypassing the need for receptor-mediated entry. It was long assumed that cholesterol was the natural regulator and that oxysterols were contaminants found in commercial supplies of cholesterol or formed upon storage of stock cholesterol solutions. Now it is known that there are many naturally occurring oxysterols that have diverse actions on cellular lipid metabolism (G.J. Schroepfer, Jr., 2000). 25-Hydroxycholesterol is the most studied oxysterol; however, other oxysterols are as, or more, physiologically important.

Oxysterols can be formed by the action of at least three distinct hydroxylases [25]. The mitochondrial sterol 27-hydroxylase participates in an alternative pathway of bile acid biosynthesis, hydroxylating cholesterol and several other intermediates in the bile acid synthetic pathway. 27-Hydroxycholesterol formed in peripheral tissues is a potent inhibitor of endogenous cholesterol synthesis. It is also thought to be secreted into the bloodstream and transported to the liver, where 27-hydroxycholesterol binds the liver X receptor (LXR) nuclear hormone receptor (described in Chapter 16). LXR forms a heterodimer with the retinoid X receptor (RXR) and activates transcription of genes encoding bile acid biosynthetic enzymes. The physiological significance of sterol 27-hydroxylase is illustrated by the genetic disease cerebrotendinous xanthomatosis, which is caused by mutations in the sterol 27-hydroxylase gene [26]. The absence of this critical hydroxylase activity precludes the mobilization of excess cholesterol from peripheral tissues and leads to cholesterol deposition and xanthoma development.

24-Hydroxylase is an endoplasmic reticulum enzyme predominantly expressed in brain. Bjorkhem and colleagues have provided strong evidence that 24-hydroxylase maintains cholesterol homeostasis in the brain, which cannot participate in high-density lipoprotein-mediated reverse cholesterol transport (I. Bjorkhem, 1999). 24-Hydroxycholesterol is secreted from the brain into the circulation, taken up by the liver and metabolized into bile acids.

25-Hydroxylase is an endoplasmic reticulum and Golgi enzyme with low-level expression in most tissues [25]. 25-Hydroxycholesterol is a potent regulator of SREBP proteolytic processing. Given that the enzyme resides in the same subcellular compartment as SREBP and SCAP, 25-hydroxycholesterol may be a physiological regulator of cholesterol synthesis.

Oxysterol binding to LXR activates transcription of several genes that play key roles in maintaining bodily cholesterol homeostasis. One is the gene encoding the ATP-binding cassette transporter ABCA1, a 2201-amino-acid plasma membrane protein that stimulates cholesterol and phospholipid efflux. Cholesterol effluxed from peripheral cells by the action of ABCA1 is transferred by plasma high-density lipoproteins to the liver in a process called reverse cholesterol transport [27] (discussed in Chapter 20). A second example is LXR-activated transcription of the gene encoding cholesteryl ester transfer protein, which promotes transfer of cholesteryl esters from high-density lipoproteins to very low-density lipoproteins for clearance by the liver. Finally, oxysterol binding to LXR also stimulates expression of SREBP-1c, but not SREBP-1a or SREBP-2 (J.J. Repa, 2000). Therefore, increased cellular cholesterol should lead to oxysterol formation, which would increase expression of SREBP-1c and increase fatty acid synthesis. Indeed, administration of an LXR selective agonist to mice led to increased lipogenesis and higher plasma triacylglycerol and phospholipid levels (J.R. Schultz, 2000).

Another protein that binds oxysterols with high affinity was first reported by A.A. Kandutsch et al. (1977) and called oxysterol-binding protein (OSBP). At the time the protein was purified and cDNA-cloned (F.R. Taylor, 1989; P.A. Dawson, 1989), OSBP was expected to be a cytosolic protein that translocated into the nucleus and repressed transcription of cholesterogenic genes when oxysterols were present. Given our current knowledge of transcriptional control, we might expect OSBP to bind to the SREBP precursor or SCAP in the endoplasmic reticulum or site-1 protease or site-2 protease in the Golgi to interfere with SREBP proteolytic processing. However, a direct role for OSBP in transcriptional control has not been demonstrated. OSBP is a high-affinity 25-hydroxycholesterol-binding protein ($K_d$ 10 nM) that translocates from cytosol and vesicles to the Golgi when ligand is bound [28].

How might OSBP transduce signals? One reasonable hypothesis is that when cellular cholesterol levels are high, cholesterol hydroxylation occurs. The resultant oxysterol then binds to OSBP, which translocates to the Golgi and signals suppression of cholesterol synthesis. However, OSBP responds paradoxically to cholesterol rather than oxysterols [28]. When cells are cholesterol replete, OSBP moves to the cytosol and vesicles, not to the Golgi. OSBP moves to the Golgi when cells are cholesterol-depleted. Thus, it has been difficult to establish the identity of OSBP's endogenous ligand and which downstream events are mediated by OSBP. The story is made more complex by the finding that the human OSBP family has at least five members, with sequence similarity in the C-terminal ligand-binding domain, whereas *Saccharomyces cerevisiae* has six related proteins [28].

## 4.2. Cholesteryl ester synthesis

Excess cholesterol can also be metabolized to cholesteryl esters. ACAT is the endoplasmic reticulum enzyme that catalyzes the esterification of cellular sterols with fatty acids. In vivo, ACAT plays an important physiological role in intestinal absorption of dietary cholesterol, in intestinal and hepatic lipoprotein assembly, in transformation of macrophages into cholesteryl ester laden foam cells, and in control of the cellular free cholesterol pool that serves as substrate for bile acid and steroid hormone formation. ACAT is an allosteric enzyme, thought to be regulated by an endoplasmic reticulum cholesterol pool that is in equilibrium with the pool that regulates cholesterol biosynthesis. ACAT is activated more effectively by oxysterols than by cholesterol itself, likely due to differences in their solubility. As the fatty acyl donor, ACAT prefers endogenously synthesized, monounsaturated fatty acyl-CoA.

The cloning of the human ACAT gene and its orthologs, as well as the subsequent generation of ACAT-deficient mice, led to the realization that two ACAT isozymes must contribute to the enzyme activity (reviewed in Farese [29], Rudel et al. [30] and Chang et al. [31]). Human ACAT-1 was cloned using an expression cloning strategy (C.C.Y. Chang, 1993). The gene encodes an integral membrane protein of 550 amino acids that is present in almost all cells and tissues examined. Orthologs were identified in other mammalian species, as well as *Drosophila melanogaster* and *Caenorrhabditis elegans*. The first indication of multiple ACATs came from the cloning of two ACAT-related enzymes (ARE1 and ARE2) from *S. cerevisiae* (H. Yang, 1996). The inactivation of both yeast genes was required to eliminate sterol esterification. In addition, ACAT-1-deficient mice showed the expected depletion of cholesteryl esters in adrenals, ovaries, testes and macrophages, but no changes in intestinal cholesterol absorption or hepatic cholesterol esterification (V.L. Meiner, 1996). This result indicated that a second ACAT must be present in those mouse tissues.

The cloning of ACAT-2 (R.A. Anderson, 1998; S. Cases, 1998; P. Oelkers, 1998) revealed a protein of similar size to ACAT-1, with a novel N-terminus but a C-terminus highly similar to ACAT-1. In adult humans, ACAT-2 is confined to the apical region of intestinal enterocytes, with low levels also expressed in hepatocytes. Disruption of the ACAT-2 gene in mice led to dramatic reduction in cholesterol absorption and prevention of hypercholesterolemia (A.K.K. Buhman, 2000). The data suggest that, in humans, ACAT-1 plays a critical role in foam-cell formation and cholesterol homeostasis in extrahepatic tissues, whereas ACAT-2 has an important role in absorption of dietary cholesterol [31]. ACAT-1 is the major isozyme in hepatocytes, although the total pool of cholesteryl esters produced by both enzymes regulates very low-density lipoprotein synthesis and assembly [31].

## 5. Future directions

Fifty years ago, it was recognized that hepatic cholesterol synthesis was subject to feedback regulation by dietary cholesterol (R.G. Gould, 1950). Only in the last decade have the mechanisms been elucidated for transcriptional and degradative regulation of the rate-limiting enzyme, HMG-CoA reductase. Both forms of regulation require that proteins sense the local cholesterol concentration. Rising cholesterol levels cause HMG-CoA reductase to be ubiquitinated and degraded by the proteosome. They cause SCAP to remain localized to the endoplasmic reticulum rather than translocating to the Golgi. The challenge ahead is to determine how HMG-CoA reductase and SCAP transduce the signal of increased cellular cholesterol content into action, i.e. protein degradation or movement to Golgi.

HMG-CoA reductase and SCAP are not the only cellular proteins equipped with a sterol-sensing domain. Does the sterol-sensing domain in 7-dehydrocholesterol $\Delta 7$-reductase confer cholesterol-mediated feedback regulation upon this last step in the biosynthetic pathway? What is the function of the sterol-sensing domain in NPC1, NPC1L1, Patched and dispatched? Is their subcellular location or their binding to another protein altered by cholesterol? Finally, is sterol-sensing domain-mediated regulation due to the action of cholesterol itself or another biologically active sterol?

## Abbreviations

| | |
|---|---|
| ACAT | acyl-CoA : cholesterol acyltransferase |
| bHLH-Zip | basic helix–loop–helix leucine zipper |
| CE | cholesteryl ester |
| DHC | dehydrocholesterol |

430

ER            endoplasmic reticulum
HMG           3-hydroxy-3-methylglutaryl
LDL           low-density lipoprotein
LXR           liver X receptor
OSBP          oxysterol-binding protein
PP            diphosphate
RXR           retinoid X receptor
S1P           site-1 protease
S2P           site-2 protease
SCAP          SREBP cleavage-activating protein
SLOS          Smith–Lemli–Opitz syndrome
SRE           sterol regulatory element
SREBP         SRE-binding protein

## References

1. Liscum, L. and Munn, N.J. (1999) Intracellular cholesterol transport. Biochim. Biophys. Acta 1438, 19–37.
2. Dietschy, J.M. and Turley, S.D. (2001) Cholesterol metabolism in the brain. Curr. Opin. Lipidol. 12, 105–112.
3. Olivier, L.M. and Krisans, S.K. (2000) Peroxisomal protein targeting and identification of peroxisomal targeting signals in cholesterol biosynthetic enzymes. Biochim. Biophys. Acta 1529, 89–102.
4. Istvan, E.S. and Deisenhofer, J. (2000) The structure of the catalytic portion of human HMG-CoA reductase. Biochim. Biophys. Acta 1529, 9–18.
5. Gotto, A. and Pownall, H. (1999) Manual of Lipid Disorders. 2nd ed., Williams and Wilkins, Baltimore, MD.
6. Houten, S.M., Wanders, R.J. and Waterham, H.R. (2000) Biochemical and genetic aspects of mevalonate kinase and its deficiency. Biochim. Biophys. Acta 1529, 19–32.
7. Tansey, T.R. and Shechter, I. (2000) Structure and regulation of mammalian squalene synthase. Biochim. Biophys. Acta 1529, 49–62.
8. Waterham, H.R. and Wanders, R.J. (2000) Biochemical and genetic aspects of 7-dehydrocholesterol reductase and Smith–Lemli–Opitz syndrome. Biochim. Biophys. Acta 1529, 340–356.
9. Gould, S.J., Raymond, G.V. and Valle, D. (2001) The peroxisome biogenesis disorders. In: C.R. Scriver, A.L. Beaudet, W.S. Sly and D. Valle (Eds.), The Metabolic and Molecular Bases of Inherited Disease. 8th ed., Vol. II, McGraw-Hill, New York, pp. 3181–3217.
10. Sweetman, L. and Williams, J.C. (2001) Branched chain organic acidurias. In: C.R. Scriver, A.L. Beaudet, W.S. Sly and D. Valle (Eds.), The Metabolic and Molecular Bases of Inherited Disease. 8th ed., Vol. II, McGraw-Hill, New York, pp. 2125–2163.
11. Kelley, R.I. and Hennekam, R.C.M. (2001) Smith–Lemli–Opitz syndrome. In: C.R. Scriver, A.L. Beaudet, W.S. Sly and D. Valle (Eds.), The Metabolic and Molecular Bases of Inherited Disease. 8th ed., Vol. IV, McGraw-Hill, New York, pp. 6183–6201.
12. Woollett, L.A. (2001) The origins and roles of cholesterol and fatty acids in the fetus. Curr. Opin. Lipidol. 12, 305–312.
13. Kelley, R.I. (2000) Inborn errors of cholesterol biosynthesis. In: L.A. Barness, D.C. DeVivo, M.M. Kaback, G. Morrow, A.M. Rudolph and W.W. Tunnessen (Eds.), Advances in Pediatrics. Vol. 47, Mosby, St. Louis, pp. 1–53.
14. Goldstein, J.L. and Brown, M.S. (1990) Regulation of the mevalonate pathway. Nature 343, 425–430.
15. Ness, G.C. and Chambers, C.M. (2000) Feedback and hormonal regulation of hepatic 3-hydroxy-3-methylglutaryl coenzyme A reductase: the concept of cholesterol buffering capacity. Proc. Soc. Exp. Biol. Med. 224, 8–19.

16. Brown, M.S. and Goldstein, J.L. (1997) The SREBP pathway: regulation of cholesterol metabolism by proteolysis of a membrane-bound transcription factor. Cell 89, 331–340.

17. Edwards, P.A., Tabor, D., Kast, H.R. and Venkateswaran, A. (2000) Regulation of gene expression by SREBP and SCAP. Biochim. Biophys. Acta 1529, 103–113.

18. Hampton, R.Y. (2000) Cholesterol homeostasis: ESCAPe from the ER. Curr. Biol. 10, R298–R301.

19. Sakai, J. and Rawson, R.B. (2001) The sterol regulatory element-binding protein pathway: control of lipid homeostasis through regulated intracellular transport. Curr. Opin. Lipidol. 12, 261–266.

20. Horton, J.D. and Shimomura, I. (1999) Sterol regulatory element-binding proteins: activators of cholesterol and fatty acid biosynthesis. Curr. Opin. Lipidol. 10, 143–150.

21. Osborne, T.F. (2000) Sterol regulatory element-binding proteins (SREBPs): key regulators of nutritional homeostasis and insulin action. J. Biol. Chem. 275, 32379–32382.

22. Chang, T.Y., Hasan, M.T., Chin, J., Chang, C.C., Spillane, D.M. and Chen, J. (1997) Chinese hamster ovary cell mutants affecting cholesterol metabolism. Curr. Opin. Lipidol. 8, 65–71.

23. Hampton, R.Y., Gardner, R.G. and Rine, J. (1996) Role of 26S proteasome and HRD genes in the degradation of 3-hydroxy-3-methylglutaryl coenzyme A reductase, an integral endoplasmic reticulum membrane protein. Mol. Biol. Cell 7, 2029–2044.

24. Patterson, M.C., Vanier, M.T., Suzuki, K., Morris, J.A., Carstea, E., Neufeld, E.B., Blanchette-Mackie, J.E. and Pentchev, P.G. (2001) Niemann–Pick disease type C: a lipid trafficking disorder. In: C.R. Scriver, A.L. Beaudet, W.S. Sly and D. Valle (Eds.), The Metabolic and Molecular Bases of Inherited Disease. 8th ed., Vol. III, McGraw-Hill, New York, pp. 3611–3633.

25. Russell, D.W. (2000) Oxysterol biosynthetic enzymes. Biochim. Biophys. Acta 1529, 126–135.

26. Bjorkhem, I., Boberg, K.M. and Leitersdorf, E. (2001) Inborn errors in bile acid biosynthesis and storage of sterols other than cholesterol. In: C.R. Scriver, A.L. Beaudet, W.S. Sly and D. Valle (Eds.), The Metabolic and Molecular Bases of Inherited Disease. 8th Ed., Vol. II, McGraw-Hill, New York, pp. 2961–2988.

27. Fayard, E., Schoonjans, K. and Auwerx, J. (2001) Xol INXS: role of the liver X and the farnesol X receptors. Curr. Opin. Lipidol. 12, 113–120.

28. Ridgway, N.D. (2000) Interactions between metabolism and intracellular distribution of cholesterol and sphingomyelin. Biochim. Biophys. Acta 1484, 129–141.

29. Farese Jr., R.V. (1998) Acyl-CoA : cholesterol acyltransferase genes and knockout mice. Curr. Opin. Lipidol. 9, 119–123.

30. Rudel, L.L., Lee, R.G. and Cockman, T.L. (2001) Acyl coenzyme A : cholesterol acyltransferase types 1 and 2: structure and function in atherosclerosis. Curr. Opin. Lipidol. 12, 121–127.

31. Chang, T.Y., Chang, C.C., Lin, S., Yu, C., Li, B.L. and Miyazaki, A. (2001) Roles of acyl-coenzyme A: cholesterol acyltransferase-1 and -2. Curr. Opin. Lipidol. 12, 289–296.

D.E. Vance and J.E. Vance (Eds.) *Biochemistry of Lipids, Lipoproteins and Membranes (4th Edn.)*

# CHAPTER 16

# Metabolism and function of bile acids

## Luis B. Agellon

*Canadian Institutes of Health Research Group in Molecular and Cell Biology of Lipids and Department of Biochemistry, University of Alberta, Edmonton, AB T6G 2S2, Canada*

## 1. Introduction

Bile acids make up a group of sterol-derived compounds that act as detergents in the intestine to facilitate the digestion and absorption of fats and fat-soluble molecules. In mammalian species, the cholesterol side chain is trimmed to yield C24-sterol derivatives. In other vertebrate species, the hydroxylation of the side chain does not lead to its removal and the products of the biosynthetic pathway are referred to as bile alcohols. Invertebrate species do not synthesize sterol bile acids. Over the last few years, much information has been gained about the function of bile acids and the mechanisms that regulate their synthesis. The focus of this chapter is to provide a general overview of bile acid biochemistry and to review recent discoveries that have advanced our understanding of bile acid metabolism and function in mammals.

The concept of bile was developed around the late 1600s to mid 1700s. It was early in the 1800s when bile solutes were crudely isolated. Among the components identified were the amino acid taurine (identified in ox bile, hence its name), cholesterol and a nitrogenous acid. The term 'cholic acid' was initially applied to the acidic component but this was changed to the generic term 'bile acid' shortly after. By the mid 1800s, taurine- and glycine-conjugated bile acids could be distinguished and it was also around this time that the idea that bile acids were responsible for solubilizing cholesterol in bile emerged. Nearly half a century ago, it became evident that bile acids are synthesized from cholesterol [1]. Bile acids are the major solutes in bile. The typical mammalian bile is comprised of about 82% water, 12% bile acids, 4% phospholipids (mostly phosphatidylcholines), 1% unesterified cholesterol and the remaining 1% as assorted solutes (including proteins).

## 2. Bile acid structure

The structure of bile acids holds the key for their ability to act as efficient detergents. In general, cholesterol is modified by epimerization of the pre-existing 3β-hydroxyl group, saturation and hydroxylation of the steroid nucleus and trimming of the side chain [2]. Fig. 1 shows the positions of the carbons in the steroid nucleus that are modified during bile acid biosynthesis. Under normal physiological conditions, one or two hydroxyl groups are added to the steroid nucleus. This modification renders the sterol less hydrophobic, enabling it to interact with an aqueous environment more efficiently. The hydroxyl groups of many bile acids are oriented towards one face of the

434

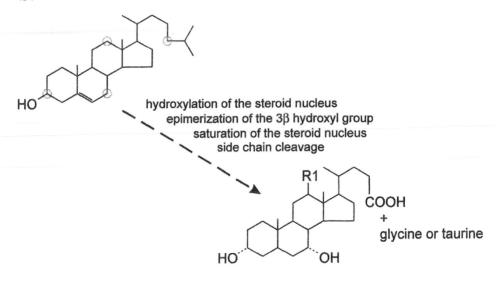

hydroxylation of the steroid nucleus
epimerization of the 3β hydroxyl group
saturation of the steroid nucleus
side chain cleavage

R1

COOH
+
glycine or taurine

R1=H, chenodeoxycholic acid
R1=OH, cholic acid

Fig. 1. Conversion of cholesterol into bile acids. The carbons in the cholesterol molecule that are modified during the conversion process are circled.

steroid nucleus giving the molecule an amphipathic character. After trimming 3 carbons from the side chain, the 'free' bile acids are covalently linked to one of two amino acids (either taurine or glycine) to form 'conjugated' bile acids. Conjugated bile acids readily ionize, allowing these polar molecules to efficiently interact with both hydrophobic and hydrophilic substances (Fig. 2).

The number and specific orientation of the hydroxyl groups added to the steroid nucleus vary according to animal species. Table 1 lists the bile acids that are commonly found in the bile of different mammalian species. Some bile acids, such as cholic and

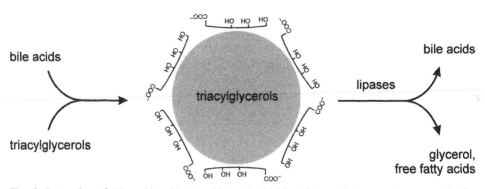

bile acids

triacylglycerols

triacylglycerols

bile acids

lipases

glycerol,
free fatty acids

Fig. 2. Interaction of bile acids with triacylglycerols. Lipid-soluble nutrients may be present in the triacylglycerol droplet. Lipases hydrolyze the triacylglycerols to liberate free fatty acids and glycerol.

Table 1
Abundant bile acids found in the bile of selected mammalian species

| Bile acid | Position and orientation of hydroxyl groups | Species |
|---|---|---|
| Chenodeoxycholic acid | 3α, 7α | bear, hamster, human, pig |
| Ursodeoxycholic acid | 3α, 7β | bear |
| Deoxycholic acid | 3α, 12α | cat, human, rabbit |
| Hyocholic acid | 3α, 6α, 7α | pig |
| β-muricholic acid | 3α, 6β, 7β | mouse, rat |
| Cholic acid | 3α, 7α, 12α | bear, cat, hamster, human, mouse, pig, rabbit, rat |

chenodeoxycholic acids, are common to many mammalian species whereas others are unique to certain species. Ursodeoxycholic acid, which is abundant in bear bile, has been found to be therapeutically useful for treating primary biliary cirrhosis and dissolving gallstones. It is chemically and biologically distinct from its isomer chenodeoxycholic acid, which differs only in the orientation of the hydroxyl group attached to carbon 7 of the steroid nucleus. Among the mammalian species that are commonly studied in the laboratory, the hamster is the only one that shows a biliary bile acid composition that is comparable to that of humans.

Fig. 3 shows the synthesis of taurine from cysteine via oxidation and decarboxylation reactions. Taurine is very rare in plants but is abundant in animal tissue, particularly in the brain. The bile acids of carnivores are mostly conjugated to taurine whereas those of herbivores are conjugated to glycine. Both taurine- and glycine-conjugated bile acids are found in the bile of omnivores. The bile of cats contains taurine-conjugated

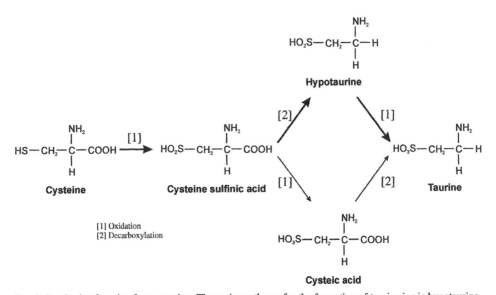

Fig. 3. Synthesis of taurine from cysteine. The major pathway for the formation of taurine is via hypotaurine.

bile acids exclusively. Cats appear to have a requirement for taurine as withdrawal of dietary taurine causes the degeneration of the retina leading ultimately to blindness. Interestingly, taurine deficiency does appear to have any significant consequences in other mammalian species. Conjugated bile acids are more acidic than unconjugated bile acids due to the additional carboxyl group contributed by the amino acid. Consequently, conjugated bile acids readily ionize and exist mainly as bile salts at physiological pH. The functional significance of the choice of amino acid used for conjugation is not clear. Cultured rat hepatoma cells show differential sensitivity to taurine- and glycine-conjugated bile acids [3]. In these cells, glycine-conjugated bile acids are toxic and induce cell death by apoptosis whereas taurine-conjugated bile acids are well-tolerated and even promote cell survival.

## 3. Biosynthesis of bile acids

Classical studies elucidated the major steps in the bile acid biosynthetic pathway mainly by analyzing the metabolites formed from labeled cholesterol and oxysterols [4]. At least 18 distinct reactions occurring in various subcellular compartments (cytosol, endoplasmic reticulum, mitochondria, and peroxisomes) are necessary to transform cholesterol into bile acids. Reactions involving modifications of the steroid nucleus occur in the endoplasmic reticulum and mitochondria. The removal of the cholesterol side chain involves peroxisomes. Many of the enzymes that catalyze these reactions have been purified, their cDNAs cloned and ectopically expressed in a variety of cultured cell lines. In addition, the impact of overactivity and deficiency of some of these enzymes in the formation of bile acids in vivo has been studied through the use of gene therapy, transgenic, and targeted gene disruption techniques.

### 3.1. The classical and alternative bile acid biosynthetic pathways

The classical pathway operates entirely in the liver (Fig. 4). It begins by α-hydroxylation of carbon 7 of the cholesterol steroid nucleus. This reaction is catalyzed by the microsomal cytochrome P-450 monooxygenase referred to as cholesterol 7α-hydroxylase (cyp7a) and is the rate-limiting step of the classical pathway. Several of the enzymes that participate in the transformation of cholesterol into bile acids belong to the cytochrome P-450 family. In general, this class of enzymes catalyzes the hydroxylation of various organic compounds using molecular oxygen as a cosubstrate. The heme-containing monooxygenases recognize specific molecules, or a group of related compounds, and work in concert with NADPH:cytochrome P-450 oxidoreductase which supplies electrons for the reactions. Cyp7a shows a high degree of selectively towards cholesterol. Bile acid output from the liver is correlated with the cyp7a activity, and it is generally considered that the classical pathway is the source of the bulk of the bile acids made by the liver.

The existence of an alternate pathway for the synthesis of bile acids was suspected because it was possible for oxysterols to be converted into bile acids (N. Wachtel, 1968). It is now recognized that a variety of oxysterols produced by an assortment of cell types

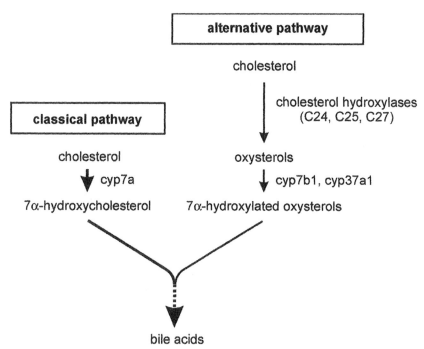

Fig. 4. The bile acid biosynthetic pathways. The classical pathway operates entirely in the liver. In other tissues, the entry of cholesterol into the alternate pathways is facilitated by cholesterol hydroxylases. The oxysterols generated by these enzymes are 7α-hydroxylated by oxysterol hydroxylases and the products enter the latter steps of the classical pathway.

can be converted into bile acids. The production of these oxysterols is catalyzed by several sterol hydroxylases: sterol 27-hydroxylase (cyp27) (J.J. Cali, 1991), cholesterol 25-hydroxylase (E.G. Lund, 1998) and cholesterol 24-hydroxylase (cyp46) (E.G. Lund, 1999). Cholesterol 25-hydroxylase is not a cytochrome $P$-450 monooxygenase, unlike the two other enzymes. Almost all of the 24-hydroxycholesterol that ends up in the liver originates from the brain, and it has been suggested that the production of this oxysterol is the major mechanism responsible for eliminating excess cholesterol from this organ (I. Bjorkhem, 2001). Cyp27 is also important in the latter stages of bile acid synthesis in the liver, as it is the major enzyme that catalyzes the hydroxylation of the side chain to facilitate the cleavage of the sterol side chain. The oxysterols generated outside the liver are 7α-hydroxylated, mainly by oxysterol hydroxylases distinct from cyp7a. The cyp7b1 oxysterol 7α-hydroxylase prefers 25-hydroxycholesterol and 27-hydroxycholesterol, while the cyp37a1 oxysterol 7α-hydroxylase is selective for 24-hydroxycholesterol (J. Li-Hawkins, 2000). Cyp7a can accept some oxysterols as a substrate, although it largely prefers cholesterol.

The latter steps required to complete the synthesis of bile acids occur only in the liver and are common to both the classical and alternative pathways. Consequently, the liver is the only organ in the body capable of producing bile acids. The isomerization of the 3β hydroxyl group and the saturation of the steroid nucleus involve 3β-hydroxy-

$\Delta^5$-$C_{27}$-steroid oxidoreductase (K. Wikvall, 1981), 3-oxo-$\Delta^4$-steroid 5β-reductase (O. Berseus, 1967) and 3α-hydroxysteroid dehydrogenase (A. Stolz, 1987). The activities of these enzymes are necessary for the formation of normal bile acids. The enzyme sterol 12α-hydroxylase (cyp8b1) catalyzes the addition of a hydroxyl group to carbon 12 of the steroid nucleus, and therefore controls the production of the cholic and chenodeoxycholic acids. Changes in the ratio of cholic to chenodeoxycholic acid affect the overall hydrophobicity of the bile acid pool.

The cDNA for an enzyme capable of catalyzing the 6α-hydroxylation of the steroid nucleus was cloned from pig liver (K. Lundell, 2001). This enzyme, named cyp4a21, is believed to be responsible for the formation of hyocholic acid, a bile acid typically found in porcine bile. The steps leading to the synthesis of β-muricholic acid (a 6β-hydroxylated bile acid) are less understood [5]. This bile acid appears only in rat and mouse bile. The conversion of lithocholic acid (a 3α-monohydroxylated bile acid) and chenodeoxycholic acid into β-muricholic acid has been observed, but the identities of the enzymes catalyzing the reactions are not known. Intestinal bacteria are thought to be responsible for modifying the steroid nucleus to form 7β-hydroxylated bile acids. However it was recently noted that bear liver has the capacity to produce ursodeoxycholic acid, indicating the existence of hepatic enzymes that can catalyze the direct 7β-hydroxylation of the steroid nucleus or epimerization of the 7α-hydroxyl group (L.R. Hagey, 1993).

### 3.2. Mutations affecting key enzymes involved in bile acid biosynthesis

Bile acid synthesis represents a major pathway for cholesterol catabolism. In humans, bile acid excretion can account for the disposal of up to ~0.5 g of cholesterol per day. In animal studies, direct stimulation of bile acid synthesis by increasing the abundance of cyp7a enzyme in the liver through gene therapy, reduces the concentration of cholesterol in the plasma (D.K. Spady, 1995, 1998; L.B. Agellon, 1997). It was reasonably expected that inhibiting bile acid synthesis by repression of cyp7a would impair cholesterol catabolism and lead to an increased concentration of plasma cholesterol. In mice, the complete loss of cyp7a function results in the high incidence of neonatal lethality due mainly to inefficient absorption of fats and fat-soluble vitamins [6]. Cyp7a-deficient mice that manage to survive beyond the weaning period synthesize bile acids via the alternative pathway [7] but these mice do not develop hypercholesterolemia [8]. In contrast, a recently discovered mutation in the human *CYP7A1* gene that causes cyp7a deficiency appears to cause hypercholesterolemia (J.P. Kane, 2002). It is not yet known if the loss of cyp7a activity has an effect on human neonatal survival.

Mutations in human cyp27 cause cerebrotendinous xanthomatosis (CTX) (J.J. Cali, 1991). This disorder, which is characterized by neurological defects and premature atherosclerosis, may well be the consequence of sterol accumulation in neural and other tissues [9]. CTX patients have reduced capacity for normal bile acid synthesis but produce large amounts of bile alcohols. This is consistent with the importance of cyp27 in the removal of the cholesterol side chain. Interestingly, deficiency of cyp27 in mice does not elaborate the full complement of defects observed in humans with CTX [10]. The basis for the difference is not completely understood. It has been suggested that other cytochrome *P*-450 enzymes, cyp3a4 in particular, partially compensate for the missing functions supplied by cyp27 in the murine species (A. Honda, 2001). Indeed, cyp27-deficient mice are still capable of producing normal C24 bile acids but overall bile acid synthesis is markedly diminished. This finding confirms that cyp27 activity is quantitatively important in side chain cleavage, but that hydroxylation of another carbon in the side chain can permit some side chain cleavage to proceed. McArdle RH-7777 rat hepatoma cells are deficient in both cyp7a and cyp27, and no longer possess the capacity to synthesize bile acids. Reinstatement of cyp7a activity enables these cells to synthesize C24 bile acids despite the absence of cyp27 activity (E.D. Labonté, 2000). Cyp27 deficiency causes hypertriglyceridemia and hepatomegaly in mice, indicating that cyp27 function affects other metabolic processes in this species (J.J. Repa, 2000).

The importance of the cyp7b1 oxysterol 7α-hydroxylase in bile acid synthesis has also been studied in mice. Mice lacking this microsomal enzyme are viable and do not exhibit obvious defects in cholesterol or bile acid metabolism [11]. The notable feature in cyp7b1-deficient mice is the accumulation of 25- and 27-hydroxycholesterol in plasma and cells, suggesting that cyp7b1 is important in the catabolism of these oxysterols into bile acids. In humans, mutations in cyp7b1 results in severe neonatal liver disease characterized by cholestasis (arrest of bile flow) and cirrhosis (damage and scarring of liver tissue resulting from chronic impaired liver function) [12]. Mutations in 3β-hydroxy-$\Delta^5$-$C_{27}$-steroid oxidoreductase and 3-oxo-$\Delta^4$-steroid 5β-reductase are also known to cause progressive intrahepatic cholestasis [13–15].

The importance of peroxisomes in the cleavage of the cholesterol side chain during bile acid synthesis is well illustrated in Zellweger syndrome (Chapter 9). This genetic disorder is characterized by peroxisome deficiency and accumulation of large amounts of bile alcohols in the plasma of afflicted patients (R.J. Wanders, 1987). Mice carrying an induced mutation in the *Scp2* gene also accumulate bile alcohols similar to those seen in Zellweger patients [16]. The *Scp2* gene codes for two proteins: the cytosolic sterol carrier protein-2 (SCP2) and the peroxisomal sterol carrier protein-x (SCPx) (Chapter 17). SCPx contains the entire SCP2 sequence plus an N-terminal domain that has a β-ketothiolase activity (U. Seedorf, 1994). The basis for the accumulation of bile alcohols in mice homozygous for a mutant *Scp2* gene is the deficiency in peroxisomal β-ketothiolase activity supplied by SCPx.

## 4. Transport of bile acids

### 4.1. Enterohepatic circulation

Bile acids circulate between the liver and intestines via bile and portal blood. The path traced by bile acids between these two organs is depicted in Fig. 5, and is referred to as the enterohepatic circulation [17]. A number of transporters involved in the transport of bile acids have been described [18]. Hepatocytes recover bile acids from portal blood by an active process involving sodium/taurocholate co-transporting polypeptide (ntcp) (B. Hagenbuch, 1991). The recovered bile acids, along with newly synthesized bile

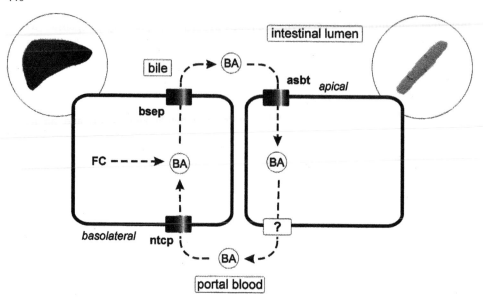

Fig. 5. Transport of bile acids in the enterohepatic circulation. The left and right sides of the figure depict a liver and intestinal cell, respectively. Note that the movement of bile acids in the enterohepatic circulation is vectorial. Abbreviations: asbt, apical/sodium bile acid cotransporter; BA, bile acids; bsep, bile salt export pump; FC, unesterified cholesterol; ntcp, sodium/taurocholate cotransporting polypeptide. The identity of the protein (depicted by '?') mediating the exit of bile acids from the basolateral pole of enterocytes is not yet known.

acids, are secreted into bile via the bile salt export pump (bsep, also known as sister of $p$-glycoprotein) (T. Gerloff, 1998). This protein belongs to the adenosine triphosphate binding (ABC) cassette family of transporters (M. Dean, 2001). Mutations in human bsep are known to cause progressive familial intrahepatic cholestasis type 2 [19]. However, targeted disruption of the murine gene encoding the bsep does not reproduce the human disease phenotype in mice [20]. This finding is another example illustrating a difference between human and murine bile acid metabolism.

The secreted bile acids are stored in the gallbladder prior to being released into the small intestine. An exception occurs in the rat (but not in the mouse), which lacks a gallbladder and thus continuously releases bile into the intestine. The primary bile acids (the products of bile acid biosynthesis in the liver) are metabolized by enteric bacteria to produce deconjugated (i.e., lacking taurine or glycine), and/or dehydroxylated derivatives referred to as secondary bile acids. The secondary bile acids may be further modified by sulfation and/or glucuronidation, but these modifications are not significant under normal physiological conditions. The deconjugated bile acids ('free' bile acids) are absorbed along the entire axis of the intestines. The majority of the conjugated bile acids are recovered in the terminal ileum via an active process involving the apical/sodium bile acid transporter (asbt) (M.H. Wong, 1994). In humans, mutations in asbt cause primary bile acid malabsorption (P. Oelkers, 1997). The identity of the intestinal bile acid exporter is not yet established.

As already mentioned, the major lipids found in bile are bile acids, phospholipids (mainly phosphatidylcholines, PC) and unesterified cholesterol. The solubility of cholesterol in bile is dependent upon the ratio of these lipids. The acyl chain composition of biliary PC (predominantly C16 : 0 at the *sn*-1 position and either C18 : 1 or C18 : 2 at the *sn*-2 position) differs from that normally found in bulk cell membranes (predominantly C18 : 0 at the *sn*-1 position and C20 : 4 at the *sn*-2 position). It is now known that the secretion of PC into bile requires a canalicular membrane protein referred to as mdr2 (an ABC-type transporter encoded by the *Abcb4* gene in mice). Mice that are deficient in mdr2 have a very low concentration of PC in bile (J.J.M. Smit, 1993). The secretion of bile acids into bile is not affected by mdr2 deficiency. However, cholesterol concentration in the bile of mdr2-deficient mice is diminished, indicating that the secretion of cholesterol into bile is dependent on biliary PC. Mutations in MDR3 (the human equivalent of mdr2) cause progressive familial intrahepatic cholestasis type 3 (J.M. De Vree, 1998). It was recently proposed that abca1 (Chapter 20), another ABC-type transporter, is involved in the cellular efflux of cholesterol. However, it is not yet clear if this transporter is directly responsible for mediating the transport cholesterol. Abca1 is found in a variety of organs including the liver. It remains to be determined if this protein is found in canalicular membrane of hepatocytes.

*4.2. Intracellular transport*

The mechanism for the intracellular transport of bile acids is less understood than the uptake and secretion of bile acids by liver and intestinal cells. Several intracellular proteins capable of binding bile acids have been identified but it is not yet clear if these proteins are involved in the transcellular transport of bile acids [18]. The best candidate protein in the intestine is the ileal lipid binding protein (ilbp). This protein, a member of the intracellular lipid binding protein family (A.V. Hertzel, 2000), is abundantly expressed in the distal portion of the small intestine where asbt is found. It has been suggested that ilbp and asbt interact to form a macromolecular bile acid transport system in intestinal cells [21]. The protein providing the equivalent function in liver cells is not known. The liver-fatty acid binding protein does not bind bile acids efficiently. High level expression of the human bile acid binder in hepatoma cells capable of active bile acid uptake does not appear to influence bile acid transport [3].

# 5. Molecular regulation of key enzymes in the bile acid biosynthetic pathways

Bile acid synthesis is modulated by a variety of hormonal and nutrient factors. Alterations in bile acid metabolism have been documented in response to thyroid hormones, glucocorticoids and insulin. It is also well known that cholesterol and bile acids have opposite effects on the activity of the bile acid biosynthetic pathway (Fig. 6). A major advance into the understanding of the mechanisms that regulate bile acid synthesis came with the cloning of the rat cyp7a cDNA, which permitted the expression of the cyp7a

442

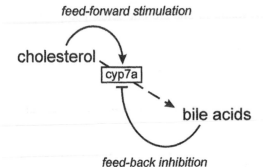

*feed-forward stimulation*

cholesterol

cyp7a

bile acids

*feed-back inhibition*

Fig. 6. Regulation of bile acid synthesis.

gene to be monitored at the molecular level [22]. Many of the details relating to the molecular mechanisms involved in regulating bile acid synthesis have been elucidated using both cultured cells and genetically modified mouse strains.

Feeding rats with a cholesterol-enriched diet induces bile acid synthesis. This increase is attributable to the rise in cyp7a activity, which catalyzes the rate-limiting step of the classical pathway (Fig. 6). Interrupting the return of bile acids to the liver, by diverting bile or by feeding a bile acid-binding resin, also stimulates the synthesis of bile acids. In contrast, reintroduction of bile acids into bile-diverted rats reverses the stimulatory effect, indicating that bile acid synthesis is subject to end-product inhibition. It was later discovered that cyp7a enzyme activity is closely correlated with cyp7a mRNA abundance, indicating that cyp7a gene transcription is the major determinant of cyp7a activity.

In many of the early studies, crystalline cholesterol was added directly to the standard rodent chow and fed to the animals. Although this experimental condition was useful in illustrating the stimulation of the cyp7a gene in response to dietary cholesterol, it does not normally exist in nature. The use of semi-purified diets has revealed that the composition of the fat in which cholesterol is presented has a marked influence on the ability of cholesterol to regulate cyp7a gene expression [23]. It has also become apparent that the fat component of the diet is capable of stimulating murine *Cyp7a1* gene expression, independent of exogenous cholesterol.

The regulation of the classical and alternative bile acid biosynthetic pathways has been studied mostly in mice and rats. The data indicate that the classical pathway is under stringent regulation, with much of the control exerted on the cyp7a gene. In contrast, the alternative pathway appears to operate constitutively. There is also emerging evidence indicating that the synthesis of bile acids in humans is only moderately regulated, unlike that in mice and rats.

### 5.1. Transcriptional control

Several transcription factor binding sites have been mapped in the cyp7a gene promoter, and many of these bind transcription factors that are members of the nuclear receptor superfamily (Table 2). Some of these receptors, notably the liver x receptor α (LXRα

Table 2

Transcription factors shown to have functional interaction with the cyp7a gene promoter

| Transcription factor | Cyp7a gene |
| --- | --- |
| BTEB | rat |
| C/EBPβ | rat |
| COUP-TFII (ARP-1) | rat |
| DBP | rat |
| HNF-1 | human |
| HNF-3 | human, hamster, rat |
| HNF-4 | human, hamster, rat |
| LRH-1 (also known as CPF, FTF) | human, rat |
| LXRα | rat |
| PPARα | mouse |
| TRα and TRβ | human |

The data in this table are compiled from studies published by academic (L.B. Agellon, J.Y. Chiang, A.D. Cooper, E. De Fabiani, G. Gil, D.J. Waxman, U. Schibler) and pharmaceutical (Tularik Inc., Glaxo-Wellcome Research and Development) laboratories.

[NR1H3]) and peroxisome proliferator-activated receptor α (PPARα [NR1C1]), bind to their target elements as heterodimers with retinoid x receptor α (RXRα [NR2B1]).

The stimulation of cyp7a gene expression by cholesterol involves LXRα (Fig. 7),

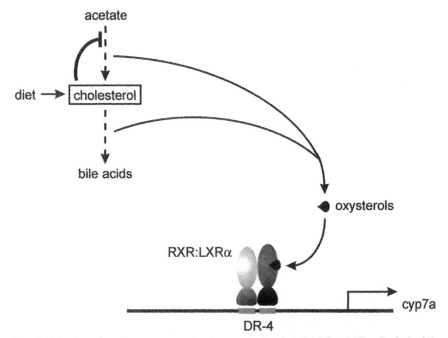

Fig. 7. Induction of cyp7a gene expression by oxysterol-activated LXRα : RXRα. Both the bile acid and cholesterol biosynthetic pathways generate oxysterols. The binding site of LXRα : RXRα in the cyp7a gene promoter is a DR-1 (a direct repeat of the hexanucleotide hormone response element separated by 4 nt).

an oxysterol-activated transcription factor (D.J. Mangelsdorf, 1996). In cultured cells, induction of the rat *Cyp7a1* gene promoter by oxysterols is dependent on LXRα (J.M. Lehman, 1997). In LXRα-deficient mice, the *Cyp7a1* gene is no longer induced by cholesterol feeding [24]. The oxysterols that serve as potent ligands for LXRα are likely generated by the early steps in the alternative bile acid biosynthetic pathway (i.e., 25-hydroxycholesterol and 27-hydroxycholesterol), and by the cholesterol biosynthetic pathway (i.e., 24(*S*),25-epoxycholesterol). It is notable that 24(*S*),25-epoxycholesterol is also capable of repressing 3-hydroxy-3-methyl-glutaryl coenzyme A reductase activity (T.A. Spencer, 1985).

Fatty acids and their metabolites can stimulate the murine *Cyp7a1* gene promoter via PPARα : RXRα in hepatoma cells [25]. Interestingly, LXRα : RXRα and PPARα : RXRα heterodimers bind to overlapping regions in the murine *CYP7a1* gene promoter. It is currently not known how these transcription factors interact with the cyp7a gene promoter when both are simultaneously activated, or whether this is relevant to the earlier finding that the type of fat in the diet influences the response of the murine *Cyp7a1* gene to dietary cholesterol. The corresponding region of the human *CYP7A1* gene promoter does not interact with either PPARα : RXRα or LXRα : RXRα. In transgenic mice, the human *CYP7A1* gene is not stimulated by cholesterol feeding (L.B. Agellon, 2002). A differential interaction of thyroid hormone receptors (TRα [NR1A1] and TRβ [NR1A2]) with the cyp7a gene promoters of different species has also been documented [26]. The human *CYP7A1* gene promoter binds and is inhibited by the thyroid hormone receptor. In contrast, the murine *Cyp7a1* gene promoter does not interact with the thyroid hormone receptor. These differences may indicate that the cyp7a gene promoters of different organisms are configured to respond to regulatory cues relevant to each species.

An indirect mechanism for the inhibition of cyp7a gene expression by bile acids has been proposed (Fig. 8). The liver receptor homolog protein-1 (LRH-1 [NR5A2]) is a monomeric orphan nuclear receptor bound to cyp7a gene promoter to enable expression in the liver (M. Nitta, 1999). It was recently discovered that bile acids are the physiological ligands of the farnesoid x receptor (FXR [NR1H4]), another transcription factor belonging to the nuclear receptor superfamily (M. Makashima, 1999; B.M. Forman, 1999). In the liver, FXR stimulates the expression of the gene encoding the nuclear factor known as small heterodimer partner (SHP; NR0B2) [27,28]. The interaction of SHP with LRH-1 renders the cyp7a gene promoter insensitive to stimulation by other transcription factors. However the proposed model (Fig. 8) cannot account for some observations. For example, the *Cyp7a1* gene is resistant to inhibition by bile acids in FXR-deficient mice (C.J. Sinal, 2000) even though it has been suggested that bile acids can stimulate SHP gene expression via an alternative mechanism involving the JNK/c-Jun pathway (J.H. Miyake, 2000; S. Gupta, 2001). Furthermore, feeding a diet containing both cholesterol and bile acids does not abolish cyp7a gene expression [29].

Bile acids inhibit the expression of the cyp8b1 gene (encodes the sterol 12α-hydroxylase) in parallel with the cyp7a gene (Z.R. Vlahcevic, 2000). Suppression of cyp8b1 gene expression is likely mediated through SHP, as LRH-1 is also required for cyp8b1 promoter activity (A. del Castillo-Olivares, 2000). The expression of the

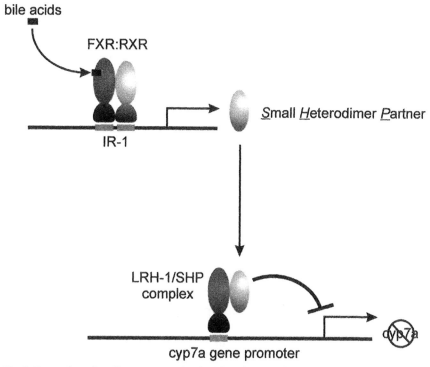

Fig. 8. Repression of cyp7a gene expression by bile acids. The liver receptor homolog-1 (LRH-1) binds to the cyp7a gene promoter to enable expression in the liver. A bile acid-activated FXR:RXR heterodimer binds to an IR-1 (an inverted repeat of the hexanucleodide hormone response element separated by 1 nt) in the promoter of the gene that encodes the small heterodimer partner (SHP) and stimulates its expression. Binding of SHP to LRH-1 arrests the expression of the cyp7a gene.

rat *Cyp8b1* gene is also inhibited by thyroid hormone, but it is not clear whether this effect involves the interaction of the thyroid hormone receptor with the rat cyp8b1 gene promoter (U. Andersson, 1999). Moreover thyroid hormones, unlike bile acids, exhibit opposite effects on the expression of *Cyp8b1* and *Cyp7a1* genes in rats.

## 5.2. Post-transcriptional control

The majority of the studies described in the literature dealing with the regulation of bile acid synthesis have focused on cyp7a and it is apparent that most of the control is exerted at the level of gene transcription. The cyp7a mRNA has a short half-life, and this is attributable to the existence of multiple copies of the AUUUA motif in its 3′-untranslated region. However, some bile acids can further accelerate the decay of chimeric mRNAs containing the 3′-untranslated region of the murine cyp7a mRNA in hepatoma cells, and this effect is independent of the AUUUA element [30]. The regulation of cyp7a enzyme activity by phosphorylation/dephosphorylation has been suggested. However the results obtained by several studies are conflicting and the topic remains controversial.

## 6. Future directions

The well-known function of bile acids is to aid in the digestion and absorption of lipids and lipid-soluble nutrients in the intestine. Since bile acids are synthesized from cholesterol, they are also regarded as the terminal products of cholesterol catabolism. Research in the past decade has provided a new understanding of the physiological importance of bile acids. Bile acids are now known be active regulators of cellular processes, such as signal transduction, by influencing the activity of proteins involved in signaling cascades, and gene expression by influencing the turnover of specific mRNA species as well as by serving as the natural ligand for the nuclear receptor FXR.

The synthesis of bile acids is under tight control, with both substrate and end-product actively participating in the regulatory process. The enzymes involved in bile acid synthesis are coordinately regulated with the proteins that transport bile acids. Much of the control is exerted at the level of transcription. It will be of interest to determine how bile acid metabolism is integrated into of other processes, such as reverse cholesterol transport and perhaps fat metabolism. The use of targeted gene disruption technology to generate specific mutations in the bile acid biosynthetic pathway has been highly useful in uncovering new components and evaluation of their relative importance. It should be realized that there are already recognized differences in the way standard mouse strains metabolize bile acids. Nevertheless, surprising new information has emerged from studies employing engineered mouse strains. It has become apparent that gender dimorphism exists with respect to several components of bile acid metabolism in the mouse. In addition, these studies reveal that the metabolism of bile acids in humans and mice may be more dissimilar than previously believed. It will be important to explain how these differences arise. Significant advances have also been gained in the area of bile acid transport. Proteins that mediate the passage of bile acids across cell membranes have been identified, although some still await discovery. The transport of bile acids within cells remains poorly understood but this should become clear in the coming years.

## Abbreviations

| | |
|---|---|
| cyp7a | cholesterol 7α-hydroxylase |
| cyp7b1 | oxysterol 7α-hydroxylase |
| cyp8b1 | sterol 12α-hydroxylase |
| cyp27 | sterol 27-hydroxylase |
| FXR | farnesoid x receptor |
| LXRα | liver x receptor α |
| LRH-1 | liver receptor homolog-1 |
| PC | phosphatidylcholine |
| PPARα | peroxisome proliferator-activated receptor α |
| SHP | small heterodimer partner |
| RXR | retinoid x receptor |

# References

1. Bloch, K., Berg, B.N. and Rittenberg, D. (1943) Biological conversion of cholesterol to cholic acid. J. Biol. Chem. 149, 511–517.

2. Hofmann, A.F. (1994). Bile acids. In: I.M. Arias, J.L. Boyer, N. Fausto, W.B. Jakoby, D.A. Schachter and D.A. Shafritz (Eds.), The Liver: Biology and Pathobiology. 3rd ed., Raven Press, New York, pp. 677–718.

3. Torchia, E.C., Stolz, A. and Agellon, L.B. (2001) Differential modulation of cellular death and survival pathways by conjugated bile acids. BMC Biochem. 2, 11.

4. Bjorkhem, I. (1985). Mechanism of bile acid biosynthesis in mammalian liver. In: H. Danielsson and J. Sjovall (Eds.), Sterols and Bile Acids. Elsevier, Amsterdam, pp. 231–278.

5. Elliott, W.H. (1985). Metabolism of bile acids in liver and extrahepatic tissues. In: H. Danielsson and J. Sjovall (Eds.), Sterols and Bile Acids. Elsevier, Amsterdam, pp. 303–329.

6. Ishibashi, S., Schwarz, M., Frykman, P.K., Herz, J. and Russell, D.W. (1996) Disruption of cholesterol 7α-hydroxylase gene in mice. I. Postnatal lethality reversed by bile acid and vitamin supplementation. J. Biol. Chem. 271, 18017–18023.

7. Schwarz, M., Lund, E.G., Setchell, K.D.R., Kayden, H.J., Zerwekh, J.E., Bjorkhem, I., Herz, J. and Russell, D.W. (1996) Disruption of cholesterol 7α-hydroxylase gene in mice. II. Bile acid deficiency is overcome by induction of oxysterol 7α-hydroxylase. J. Biol. Chem. 271, 18024–18031.

8. Schwarz, M., Russell, D.W., Dietschy, J.M. and Turley, S.D. (1998) Marked reduction in bile acid synthesis in cholesterol 7α-hydroxylase-deficient mice does not lead to diminished tissue cholesterol turnover or to hypercholesterolemia. J. Lipid Res. 39, 1833–1843.

9. Salen, G., Shefer, S. and Berginer, V. (1991) Biochemical abnormalities in cerebrotendinous xanthomatosis. Dev. Neurosci. 13, 363–370.

10. Rosen, H., Reshef, A., Maeda, N., Lippoldt, A., Shpizen, S., Triger, L., Eggertsen, G., Bjorkhem, I. and Leitersdorf, E. (1998) Markedly reduced bile acid synthesis but maintained levels of cholesterol and vitamin D metabolites in mice with disrupted sterol 27-hydroxylase gene. J. Biol. Chem. 273, 14805–14812.

11. Li-Hawkins, J., Lund, E.G., Turley, S.D. and Russell, D.W. (2000) Disruption of the oxysterol 7α-hydroxylase gene in mice. J. Biol. Chem. 275, 16536–16542.

12. Setchell, K.D., Schwarz, M., O'Connell, N.C., Lund, E.G., Davis, D.L., Lathe, R., Thompson, H.R., Weslie Tyson, R., Sokol, R.J. and Russell, D.W. (1998) Identification of a new inborn error in bile acid synthesis: mutation of the oxysterol 7α-hydroxylase gene causes severe neonatal liver disease. J. Clin. Invest. 102, 1690–1703.

13. Clayton, P.T., Leonard, J.V., Lawson, A.M., Setchell, K.D., Andersson, S., Egestad, B. and Sjovall, J. (1987) Familial giant cell hepatitis associated with synthesis of 3β,7α-dihydroxy- and 3β,7α,12α-trihydroxy-5-cholenoic acids. J. Clin. Invest. 79, 1031–1038.

14. Setchell, K.D., Suchy, F.J., Welsh, M.B., Zimmer-Nechemias, L., Heubi, J. and Balistreri, W.F. (1988) $\Delta^4$-3-oxosteroid 5β-reductase deficiency described in identical twins with neonatal hepatitis. A new inborn error in bile acid synthesis. J. Clin. Invest. 82, 2148–2157.

15. Schwarz, M., Wright, A.C., Davis, D.L., Nazer, H., Bjorkhem, I. and Russell, D.W. (2000) The bile acid synthetic gene 3β-hydroxy-$\Delta^5$-$C_{27}$-steroid oxidoreductase is mutated in progressive intrahepatic cholestasis. J. Clin. Invest. 106, 1175–1184.

16. Kannenberg, F., Ellinghaus, P., Assmann, G. and Seedorf, U. (1999) Aberrant oxidation of the cholesterol side chain in bile acid synthesis of sterol carrier protein-2/sterol carrier protein-x knockout mice. J. Biol. Chem. 274, 35455–35460.

17. Carey, M.C. and Duane, W.C. (1994). Enterohepatic circulation. In: I.M. Arias, J.L. Boyer, N. Fausto, W.B. Jakoby, D.A. Schachter and D.A. Shafritz (Eds.), The Liver: Biology and Pathobiology. 3rd ed., Raven Press, New York, pp. 719–767.

18. Agellon, L.B. and Torchia, E.C. (2000) Intracellular transport of bile acids. Biochim. Biophys. Acta 1486, 198–209.

19. Strautnieks, S.S., Bull, L.N., Knisely, A.S., Kocoshis, S.A., Dahl, N., Arnell, H., Sokal, E., Dahan, K., Childs, S., Ling, V., Tanner, M.S., Kagalwalla, A.F., Nemeth, A., Pawlowska, J., Baker, A., Mieli-Vergani, G., Freimer, N.B., Gardiner, R.M. and Thompson, R.J. (1998) A gene encoding a

liver-specific ABC transporter is mutated in progressive familial intrahepatic cholestasis. Nat. Genet. 20, 233–238.

20. Wang, R., Salem, M., Yousef, I.M., Tuchweber, B., Lam, P., Childs, S.J., Helgason, C.D., Ackerley, C., Phillips, M.J. and Ling, V. (2001) Targeted inactivation of sister of P-glycoprotein gene (spgp) in mice results in nonprogressive but persistent intrahepatic cholestasis. Proc. Natl. Acad. Sci. USA 98, 2011–2016.

21. Kramer, W., Girbig, F., Gutjahr, U., Kowalewski, S., Jouvenal, K., Muller, G., Tripier, D. and Wess, G. (1993) Intestinal bile acid absorption. Na$^+$-dependent bile acid transport activity in rabbit small intestine correlates with the coexpression of an integral 93-kDa and a peripheral 14-kDa bile acid-binding membrane protein along the duodenum-ileum axis. J. Biol. Chem. 268, 18035–18046.

22. Jelinek, D.F., Andersson, S., Slaughter, C.A. and Russell, D.W. (1990) Cloning and regulation of cholesterol 7α-hydroxylase, the rate-limiting enzyme in bile acid biosynthesis. J. Biol. Chem. 265, 8190–8197.

23. Cheema, S.K., Cikaluk, D. and Agellon, L.B. (1997) Dietary fats modulate the regulatory potential of dietary cholesterol on cholesterol 7α-hydroxylase gene expression. J. Lipid Res. 38, 157–165.

24. Peet, D.J., Turley, S.D., Ma, W., Janowski, B.A., Lobaccaro, J.-M.A., Hammer, R.E. and Mangelsdorf, D.J. (1998) Cholesterol and bile acid metabolism are impaired in mice lacking the nuclear oxysterol receptor LXRα. Cell 93, 693–704.

25. Cheema, S.K. and Agellon, L.B. (2000) The murine and human cholesterol 7α-hydroxylase gene promoters are differentially responsive to regulation by fatty acids via peroxisome proliferator-activated receptor α. J. Biol. Chem. 275, 12530–12536.

26. Drover, V.A.B., Wong, N.C.W. and Agellon, L.B. (2002) A distinct thyroid hormone response element mediates repression of the human cholesterol 7α-hydroxylase (CYP7A1) gene promoter. Mol. Endocrinol. 16, 14–23.

27. Lu, T.T., Makishima, M., Repa, J.J., Schoonjans, K., Kerr, T.A., Auwerx, J. and Mangelsdorf, D.J. (2000) Molecular basis for feedback regulation of bile acid synthesis by nuclear receptors. Mol. Cell 6, 507–515.

28. Goodwin, B., Jones, S.A., Price, R.R., Watson, M.A., McKee, D.D., Moore, L.B., Galardi, C., Wilson, J.G., Lewis, M.C., Roth, M.E., Maloney, P.R., Willson, T.M. and Kliewer, S.A. (2000) A regulatory cascade of the nuclear receptors FXR, SHP-1, and LRH-1 represses bile acid biosynthesis. Mol. Cell 6, 517–526.

29. Spady, D.K. and Cuthbert, J.A. (1992) Regulation of hepatic sterol metabolism in the rat. J. Biol. Chem. 267, 5584–5591.

30. Agellon, L.B. and Cheema, S.K. (1997) The 3'-untranslated region of the mouse cholesterol 7α-hydroxylase contains elements responsive to posttranscriptional regulation by bile acids. Biochem. J. 328, 393–399.

D.E. Vance and J.E. Vance (Eds.) *Biochemistry of Lipids, Lipoproteins and Membranes (4th Edn.)*
© 2002 Elsevier Science B.V. All rights reserved

# CHAPTER 17

# Lipid assembly into cell membranes

## Dennis R. Voelker

*The Lord and Taylor Laboratory for Lung Biochemistry, Program in Cell Biology, Department of Medicine,*
*The National Jewish Center for Immunology and Respiratory Medicine, Denver, CO 80206, USA,*
*Tel.: +1 (303) 398-1300; Fax: +1 (303) 398-1806; E-mail: voelkerd@njc.org*

## 1. Introduction

A fundamental problem of cell biology and biochemistry is the elucidation of the mechanisms by which the specific components of subcellular membranes are assembled into mature organelles. The major components of all cell membranes are lipids and proteins. The presence of discrete structural motifs contained in the primary sequence of proteins directs a large number of post-translational processes that enable their sorting among different membrane compartments [1]. The sorting process for proteins is essentially absolute such that plasma membrane proteins are never found in the mitochondria or vice versa. In contrast, lipid molecules do not contain discrete structural subdomains that exclusively direct their movement to specific membranes. The distribution of lipids among different organelles is heterogeneous, but (with a few exceptions) is not usually absolute. These observations about lipids indicate that specialized sorting and transport machinery must exist for their assembly into different membranes.

## 2. The diversity of lipids

A multiplicity of individual lipids can contribute to membrane formation. The biological role of this lipid heterogeneity is not completely defined and the list of significant actions continues to grow. Some of the diversity contributes to membrane fluidity. Other roles for lipid diversity are the storage of precursors that are metabolized to potent second messengers (e.g., diacylglycerol, ceramide, sphingosine, inositol trisphosphate and eicosanoids) (see Chapters 12, 13 and 14). In addition, many of the polyphosphoinositide (PI3P, PI4P, PI45P$_2$) function as membrane recognition and attachment sites for protein complexes involved in protein traffic and membrane fusion events [2]. Multiple anionic lipids (PS and polyphosphoinositides) can also regulate attachment of cytoskeletal proteins to membranes. Segregated domains of cholesterol and sphingolipid form microdomains or 'rafts' with unusual physical properties that contribute to protein sorting and are enriched in specific subsets of membrane proteins (see Chapter 1).

In addition to the large numbers of chemically distinct lipid species that occur within a prokaryotic or eukaryotic cell, there is another level of complexity, i.e., the asymmetric

450

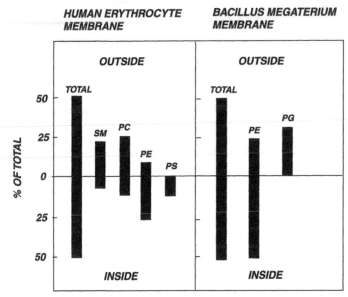

Fig. 1. The asymmetric distribution of lipids across the plane of the cell membrane of the human erythrocytes and *Bacillus megaterium*.

distribution of the lipids across the plane of the bilayer. Two striking examples of membrane lipid asymmetry are found in the red blood cell membrane [3], and the cytoplasmic membrane of *Bacillus megaterium* [4]. The data in Fig. 1 demonstrate that in the red cell membrane the outer leaflet of the lipid bilayer is composed primarily of sphingomyelin (SM) and phosphatidylcholine (PC), and the inner leaflet contains phosphatidylserine (PS) and phosphatidylethanolamine (PE), with lesser amounts of PC and SM. Relatively small amounts of phosphatidylinositol (PI) and its phosphorylated derivatives are also found in the erythrocyte membrane and these anionic lipids are distributed such that the majority is localized to the inner leaflet of the bilayer. In the prokaryote *B. megaterium* the distribution of PE has been shown to be asymmetric, with 30% of this lipid present on the outer leaflet of the bilayer and 70% on the inner leaflet. PE comprises about 70%, and phosphatidylglycerol (PG) about 30%, of the total phospholipid. Thus, nearly all the PG is in the outer leaflet of the bilayer.

Yet another level of complexity is found in cells that possess multiple membrane systems. The Gram-negative bacteria have both an inner and outer membrane system that differ in lipid composition (Chapters 1 and 3). In animal eukaryotes there are numerous membrane systems, the best characterized being endoplasmic reticulum (ER), Golgi membranes, plasma, mitochondrial, lysosomal, peroxisomal and nuclear membranes. In higher plants the eukaryotic organelle repertoire is expanded to include chloroplasts and other plastids, vacuoles and glyoxysomes (Chapter 4). Several of these membrane systems have dramatically different lipid compositions, as shown in Table 1 (A. Colbeau, 1971; T.W. Keenan, 1970). These differences in lipid content raise a variety of interesting questions: How are the different lipid compositions of different

Table 1
Lipid compositions of subcellular organelles from rat liver

| Phospholipid [a] | Endoplasmic reticulum | | Mitochondrial membranes | | Lysosomal membrane | Nuclear membrane | Golgi membrane | Plasma membrane |
|---|---|---|---|---|---|---|---|---|
| | Rough | Smooth | Inner | Outer | | | | |
| Lysophosphatidylcholine | 2.9 | 2.9 | 0.6 | – | 2.9 | – | 5.9 | 1.8 |
| Sphingomyelin | 2.4 | 6.3 | 2.0 | 2.2 | 16.0 | 6.3 | 12.3 | 23.1 |
| Phosphatidylcholine | 59.6 | 54.4 | 40.5 | 49.4 | 41.9 | 52.1 | 45.3 | 43.1 |
| Phosphatidylinositol | 10.1 | 8.0 | 1.7 | 9.2 | 5.9 | 4.1 | 8.7 | 6.5 |
| Phosphatidylserine | 3.5 | 3.9 | 1 | 1 | – | 5.6 | 4.2 | 3.7 |
| Phosphatidylethanolamine | 20.0 | 22.0 | 38.8 | 34.9 | 20.5 | 25.1 | 17.0 | 20.5 |
| Cardiolipin | 1.2 | 2.4 | 17.0 | 4.2 | – | – | – | – |
| Phospholipid/protein ($\mu$mole P/mg) | 0.33 | 0.47 | 0.34 | 0.46 | 0.21 | – | – | 0.37 |
| Cholesterol molar ratio Phospholipid | 0.07 | 0.24 | 0.06 | 0.12 | 0.49 | – | 0.152 | 0.76 |

Values for individual lipids are percentage of total phospholipid phosphorus.

organelles established? How are these differences maintained? Are the different lipid compositions essential for organelle function?

## 3. Methods to study intra- and inter-membrane lipid transport

### 3.1. Fluorescent probes

Pagano [5] and coworkers pioneered methods for the rapid insertion of fluorescent phospholipid analogs from liposomes into the plasma membranes of cultured cells. Virtually all of these analogs exhibit slight water solubility and high hydrophobic partitioning coefficients that enable them to be efficiently and reversibly transferred to cell membranes at low temperature from liposomes or albumin complexes containing the fluorescent lipid. A commonly used fluorochrome is nitrobenzoxadiazole (NBD) which is conjugated to short-chain fatty acids in the sn-2 position of glycerophospholipids or on the amine of sphingosine. Other analogs such as boron dipyrromethene difluoride (BODIPY) derivatized fatty acids have proved equally effective. Non-fluorescent molecules containing short-chain sphingosines or fatty acids such as diC8 SM or diC8 PC have similar physical properties to the fluorescent molecules, and have also proved to be important probes. The structures of some of these analogs are shown in Fig. 2. Subsequent to the insertion of these lipid analogs into cell membranes, the cells can be washed at low temperature to remove the donor liposomes or albumin complexes. In almost all cases this procedure results in the pulse labeling of the outer leaflet of the plasma membrane with the lipid analog. The lipid analogs can also be removed from the outer leaflet of the plasma membrane at reduced temperature by washing cells with a solution that contains liposomes (e.g., composed of dioleoyl-PC) or albumin. When the lipid analogs are fluorescent their intracellular movement can be observed

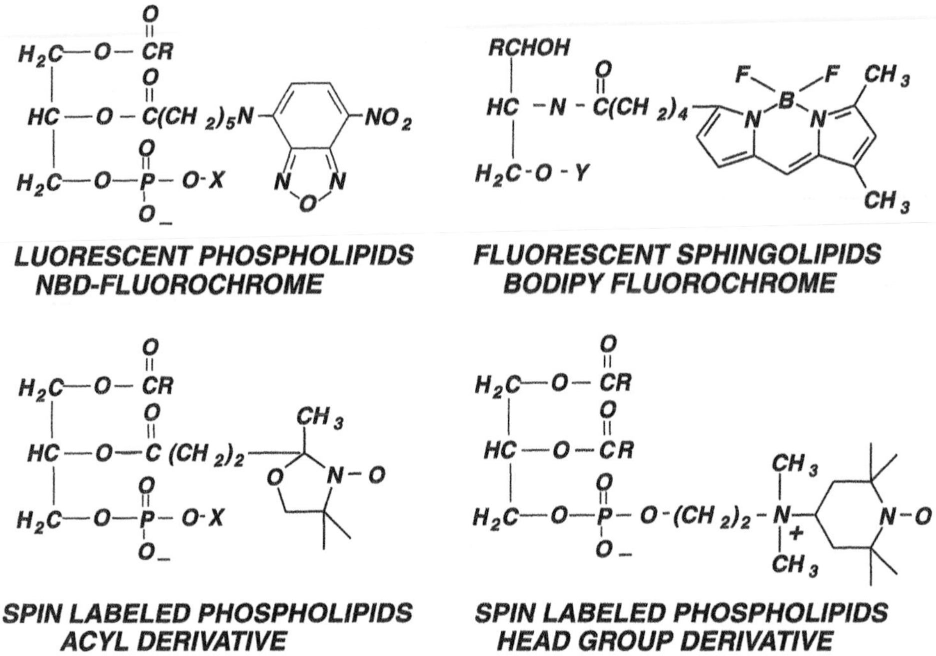

**LUORESCENT PHOSPHOLIPIDS**
**NBD-FLUOROCHROME**

**FLUORESCENT SPHINGOLIPIDS**
**BODIPY FLUOROCHROME**

**SPIN LABELED PHOSPHOLIPIDS**
**ACYL DERIVATIVE**

**SPIN LABELED PHOSPHOLIPIDS**
**HEAD GROUP DERIVATIVE**

Fig. 2. General structural features of fluorescent and spin-labeled lipid analogs. The fluorescent lipids contain a short-chain fatty acid, amino-caproic acid, that is derivatized with 4-nitrobenzo-2-oxa-1,3-diazole (NBD), or valeric acid that is derivatized with a boron dipyrromethene difluoride (BODIPY) moiety. For fluorescent phospholipids, X can be hydrogen, or the esterified forms of choline, ethanolamine, serine, or inositol. For fluorescent sphingolipids, Y can be hydrogen, or the esterified forms of phosphocholine or glucose. The spin-labeled lipids modified in the fatty acid portion contain a 4 doxylpentanoyl fatty acid in the *sn*-2 position. Those modified in the polar head group contain a tempocholine moiety in place of choline. The X substituent for the acyl spin-labeled lipids can be hydrogen, or the esterified forms of choline, ethanolamine or serine.

by fluorescence microscopy. A simplified outline of the use of these fluorescent lipids is shown in Fig. 3. In addition to their utility for examining the fluorescence pattern within cells, these lipids can be extracted from cells and their chemical metabolism analyzed using thin-layer chromatography or liquid chromatography/mass spectrometry. The fluorescent sterols dehydroergosterol and NBD-cholesterol are also available for in vivo studies. These sterols can be delivered to cells in reconstituted lipoproteins or from complexes with methyl β-cyclodextrin (E. Kilsdonk, 1995).

### 3.2. Spin-labeled analogs

Paramagnetic analogs of phospholipids have also been used to investigate lipid transport phenomena in model membrane systems (R.D. Kornberg, 1971) and in biological membranes. Representative structures are shown in Fig. 2. Several of these spin-labeled lipid analogs that are modified in the fatty acid chain can be readily and reversibly transferred from the bulk aqueous phase to biological membranes, in much the same way

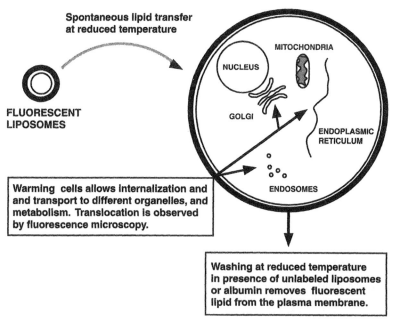

**Spontaneous lipid transfer at reduced temperature**

**FLUORESCENT LIPOSOMES**

MITOCHONDRIA

NUCLEUS

GOLGI

ENDOPLASMIC RETICULUM

ENDOSOMES

**Warming cells allows internalization and and transport to different organelles, and metabolism. Translocation is observed by fluorescence microscopy.**

**Washing at reduced temperature in presence of unlabeled liposomes or albumin removes fluorescent lipid from the plasma membrane.**

Fig. 3. Fluorescent labeling of living cells with lipid analogs. The heavy line represents the fluorescent phospholipid. Incubation of liposomes containing fluorescent lipid with eukaryotic cells at temperatures of 2–7°C results in the spontaneous transfer of fluorescence to the outer leaflet of the plasma membrane. The fluorescent lipid remains in the plasma membrane at low temperatures and can be reversibly removed by washing the cells with unlabeled liposomes or albumin solutions. Warming the cells to 37°C or intermediate temperature results in the internalization of phospholipid and subsequent labeling of intracellular organelles that can be monitored by fluorescence microscopy.

as the fluorescent lipid analogs. Since the amplitude of the ESR spectra is proportional to the amount of spin-labeled lipid present, these analogs can be used to measure the depletion or retention of the lipids (M. Seigneuret, 1984). In a typical experiment, an intact red cell is incubated with trace amounts of spin-labeled phospholipid at reduced temperature. This treatment effectively pulse labels the outer leaflet of the plasma membrane. Upon warming, the spin-labeled lipids can either remain in the outer leaflet of the plasma membrane or be internalized. If the cells are subsequently cooled and incubated in the presence of ascorbate, the ESR signal of lipid present in the outer leaflet (but not the inner leaflet) of the plasma membrane is quenched and the spectral difference can be used to determine both the rate and extent of transbilayer movement.

*3.3. Asymmetric chemical modification of membranes*

One method for ascertaining the distribution of lipids across the plane of the membrane bilayer is the use of membrane-impermeant reagents that react with the primary amines of PS and PE on only the external leaflet of the bilayer. Reagents such as trinitrobenzenesulfonate (TNBS) and isethionylacetimidate (IAA) (Fig. 4) are impermeant at reduced temperatures [6] and the chemically modified lipids can be readily identified

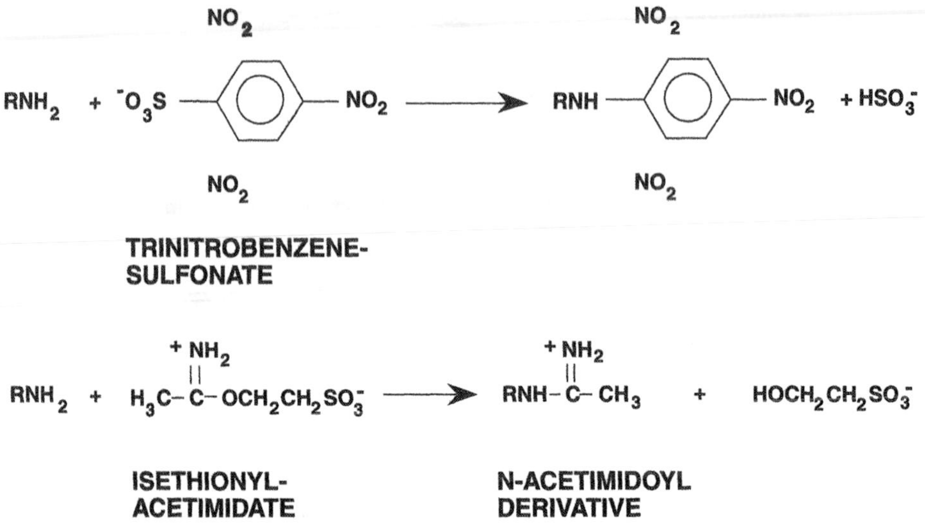

$$RNH_2 \ + \ ^-O_3S \text{—} \underset{\underset{NO_2}{\overset{NO_2}{\bigcirc}}}{} \text{—} NO_2 \ \longrightarrow \ RNH \text{—} \underset{\underset{NO_2}{\overset{NO_2}{\bigcirc}}}{} \text{—} NO_2 \ + HSO_3^-$$

**TRINITROBENZENE-
SULFONATE**

$$RNH_2 \ + \ H_3C\text{—}\overset{\overset{+NH_2}{\|}}{C}\text{—}OCH_2CH_2SO_3^- \ \longrightarrow \ RNH\text{—}\overset{\overset{+NH_2}{\|}}{C}\text{—}CH_3 \ + \ HOCH_2CH_2SO_3^-$$

**ISETHIONYL-
ACETIMIDATE**         **N-ACETIMIDOYL
DERIVATIVE**

Fig. 4. Primary amine modifying reagents. Phospholipids containing the primary amines PE and PS can be modified by treatment with either trinitrobenzenesulfonate or isethionylacetimidate yielding the *N*-trinitrophenyl derivative or the *N*-acetimidoyl derivative.

by thin-layer chromatography. When such reagents are used in conjunction with in vivo radiolabeling of the lipid, it is possible to discern the temporal and metabolic conditions required for the newly synthesized lipids to reach the compartment that is accessible to the chemical modifying reagents. A useful variation of this approach combines chemical reduction of NBD phospholipids with dithionite to eliminate fluorescence (J.C. McIntyre, 1991). When this latter technique is employed with fluorescence microscopy or spectrofluorometry, it can be extremely informative for resolving questions about transbilayer topology.

Specific pools of lipids on the external surface of cells can also be modified by the action of enzymes such as phospholipases, sphingomyelinases [3], and cholesterol oxidase (Y. Lange, 1985). These enzymes also generate characteristic derivatives of the parental lipids that can be readily identified by thin-layer chromatography and this approach provides another technique for identifying specific pools of lipid on the external surface of the cell membrane. Another means to sample lipids at membrane interfaces relies upon specific chemical desorption. The interaction between sterols and methyl β-cyclodextrin provides a high affinity interaction that can be used to selectively remove (or deliver) cholesterol and dehydroergosterol to membranes (E. Kilsdonk, 1995).

### 3.4. Phospholipid transfer proteins

In 1968, K.W. Wirtz identified a soluble intracellular protein derived from rat liver that was capable of binding PC and transferring it from one population of (donor) membranes to a second population of (acceptor) membranes (K. Wirtz, 1968). Since

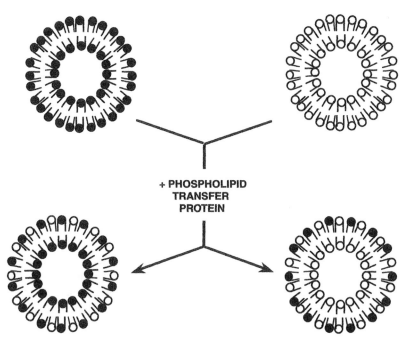

Fig. 5. The action of phospholipid transfer proteins. Mixing of equivalent populations of labeled (denoted in black) and unlabeled liposomes (denoted in white) with phospholipid transfer protein leads to the transfer of lipid between outer leaflets. In the absence of transbilayer movement of the lipid, only the outer leaflets equilibrate with each other.

this initial observation, many of these proteins have been identified in virtually all mammalian tissues, in plants, and in yeast and other microorganisms [7]. The well characterized phospholipid transfer proteins fall into three main categories: (1) those specific for PC; (2) those with high activity for PI and less, but significant, activity with PC, and in some cases SM (J. Westerman, 1995); and (3) those with transfer activity with most phospholipids and cholesterol (this latter protein is referred to as the nonspecific lipid transfer protein). In addition to the phospholipid transfer proteins, there are also intracellular proteins with high transfer activity for sphingolipids (T. Sasaki, 1990). The action of these proteins is typically a one for one exchange of lipid molecules between donor and acceptor membranes. As shown schematically in Fig. 5, the transfer proteins typically equilibrate the lipid present in the outer leaflets of liposomes. The ability of these proteins to transfer lipids from accessible membrane compartments has made them useful tools for inserting lipids into, or removing them from, membranes and probing the transbilayer movement of phospholipids. The role of these proteins in membrane biogenesis will be discussed in Section 4.3.2.6.

### 3.5. Rapid plasma membrane isolation

One approach to sampling the arrival of newly synthesized lipids at the cell surface utilizes a rapid plasma membrane isolation technique (R.F. DeGrella, 1982). In this approach, the intact cells are adsorbed onto cationic beads at reduced temperature. After adsorption, the cells are lysed by brief sonication, which liberates the majority of intracellular organelles. The density of the beads containing adsorbed plasma membrane allows them to be separated from the intracellular organelles by low-speed centrifugation. Subsequent to this isolation procedure, the lipids present in the membrane can be extracted and analyzed. When this method is used in combination with radiolabeling of the intracellular pool, the characteristics of the processes required for movement of the lipid from within the cell to the plasma membrane can be determined. A related approach has employed an iron-derivatized wheat germ agglutinin to bind the external surface of the plasma membrane (D. Warnock, 1993) that can be selectively isolated using high-intensity magnetic fields.

### 3.6. Organelle specific lipid metabolism

For a few lipids, distinct changes in structure also serve to define the arrival at certain organelles or their subcompartments. The enzyme PS decarboxylase is located at the inner mitochondrial membrane of mammalian cells (L.M.G. van Golde, 1974). The synthesis of PS, however, occurs primarily in the ER and related membranes. Thus the decarboxylation of PS can be used as an indicator of the transport of this lipid to the inner mitochondrial membrane [8]. Yeast also contain a mitochondrial PS decarboxylase (PSD1), and in addition, a second enzyme (PSD2) is found in the Golgi and vacuoles. Mutations in the *PSD1 or PSD2* genes of yeast make it possible to use PS metabolism to PE as an index of lipid transport to the locus of either the mitochondria or the Golgi–vacuolar compartment. In yeast, the movement of PE (derived from either PSD1 or PSD2) to the ER can also be followed by measuring its methylation to PC, since the methyltransferases are only present in the ER. Site-specific metabolism also occurs for sphingolipids (Chapter 14). Ceramide, the hydrophobic precursor for all sphingolipids, is synthesized in the ER. The formation of SM from ceramide (Cer) occurs at the luminal surface of the *cis*-medial Golgi (A.H. Futerman, 1990; D. Jeckel, 1990). Thus, SM synthesis from Cer can be used to follow ceramide transport from ER to Golgi. The synthesis of GlcCer also occurs at the Golgi but at the cytosolic side of the membrane. Subsequently, the GlcCer moves into the lumen of the organelle and is converted to LacCer and more complex glycosphingolipids. As with phospholipids, each metabolic step that occurs in a separate organelle or with different topology from the precursor, can serve as an indicator of lipid transport/translocation.

Important elements of sterol metabolism can also be used to elucidate where in the cell a particular precursor has moved [9]. The arrival of cholesteryl esters within lysosomes is revealed by cleavage of the fatty acid to yield free cholesterol. The subsequent transport of cholesterol to the ER can be monitored by the action of acyl CoA : cholesterol acyltransferase (Chapter 15) which results in the formation of new cholesteryl esters. In addition, sphingomyelinase treatment of the cell surface induces cholesterol movement from the plasma membrane to the ER where its arrival can likewise be monitored by acyl CoA : cholesterol acyltransferase action. Import of cholesterol into mitochondria (usually restricted to steroidogenic cells) can be followed by side-chain cleavage reactions that produce pregnenolone. Movement of pregnenolone out of mitochondria can also be followed by oxidations at positions 3, 17 and 21 which occur in the ER.

## 4. Lipid transport processes

The movement of lipids within the cell can be divided into two different general classes of transport: intramembrane transport, which entails the transbilayer movement of the lipid molecule; and intermembrane transport which is the movement of lipid molecules from one distinct membrane domain to another. Extensive reviews of these processes have been published [8–14].

### 4.1. Intramembrane lipid translocation and model membranes

The observation that biological membranes can be asymmetric with respect to transbilayer disposition of lipid components (Fig. 1) initially raised basic questions about how such asymmetry was established and maintained. An important issue that needed to be resolved on theoretical grounds, was whether lipids in model membranes could undergo spontaneous transbilayer movement. A simple consideration of the events that occur in the transbilayer movement of a zwitterionic molecule such as PC suggests that at least two energetically unfavorable events must occur. The first is desolvation of the molecule and the second is movement of the charged portion of the lipid through the hydrophobic portion of the bilayer.

Direct experiments to examine the transbilayer movement of phospholipids (R.D. Kornberg, 1971) made use of spin-labeled analogs of PC in which the choline moiety was replaced with the tempocholine probe, $N,N$-dimethyl-$N$-($1'$-oxyl-$2',2',6',6'$-tetramethyl-$4'$-piperidyl)-ethanolamine (Fig. 2). These workers found that only the ESR signal generated by molecules in the outer leaflet of unilamellar liposomes could be rapidly quenched by ascorbate. The ESR signal from lipid molecules initially residing at the inner leaflet of liposomes was accessible to ascorbate with a $t_{1/2}$ of >6.5 h, indicating slow transbilayer lipid movement (Fig. 6).

Additional evidence for slow transbilayer phospholipid movement in liposomes came from experiments using $^3$H-PC-labeled liposomes and PC-transfer protein. In the presence of excess unlabeled acceptor membranes, only the PC in the outer leaflet of the liposome membrane was rapidly transferred (J.E. Rothman, 1975). The $^3$H-PC initially present in the inner leaflet of the membrane moved to the outer leaflet with a $t_{1/2}$ of 11–15 days (Fig. 6). Further evidence demonstrating slow transbilayer movement of phospholipids was obtained from unilamellar liposomes containing 90% PC and 10% PE [6]. In these liposomes the PE initially residing in the outer leaflet of the membrane was rapidly modified by IAA. The PE at the inner leaflet remained refractory to modification by trinitrobenzenesulfonate (i.e., did not undergo transbilayer movement) with a $t_{1/2}$ of >80 days (Fig. 6).

In contrast to phospholipids, non-polar lipids such as diacylglycerol (DG) behave differently. B.R. Ganong (1984) synthesized a structural analog of DG in which the $sn$-3 hydroxyl group was replaced by an SH group that could be detected with dithio-

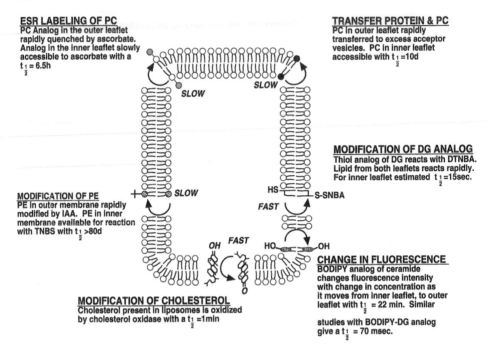

**ESR LABELING OF PC**
PC Analog in the outer leaflet rapidly quenched by ascorbate. Analog in the inner leaflet slowly accessible to ascorbate with a $t_{\frac{1}{2}} = 6.5h$

*SLOW*

*SLOW*

**TRANSFER PROTEIN & PC**
PC in outer leaflet rapidly transferred to excess acceptor vesicles. PC in inner leaflet accessible with $t_{\frac{1}{2}} = 10d$

**MODIFICATION OF DG ANALOG**
Thiol analog of DG reacts with DTNBA. Lipid from both leaflets reacts rapidly. For inner leaflet estimated $t_{\frac{1}{2}} = 15sec.$

**MODIFICATION OF PE**
PE in outer membrane rapidly modified by IAA. PE in inner membrane available for reaction with TNBS with $t_{\frac{1}{2}} > 80d$

*SLOW*

HS—S-SNBA

*FAST*

*OH  FAST*   HO—OH

**MODIFICATION OF CHOLESTEROL**
Cholesterol present in liposomes is oxidized by cholesterol oxidase with a $t_{\frac{1}{2}} = 1min$

**CHANGE IN FLUORESCENCE**
BODIPY analog of ceramide changes fluorescence intensity with change in concentration as it moves from inner leaflet, to outer leaflet with $t_{\frac{1}{2}} = 22 min.$ Similar studies with BODIPY-DG analog give a $t_{\frac{1}{2}} = 70 msec.$

Fig. 6. Summary of key experiments examining transbilayer lipid movement in liposomes. Abbreviations: IAA, isethionylacetimidate; TNBS, trinitrobenzenesulfonate; DTNBA, dithiobisnitrobenzoic acid (Ellman's reagent). Reaction of DTNBA with RSH gives R-S-SNBA. Each of the experiments was specifically designed to initially sample only the outer leaflet of the bilayer and then at subsequent periods detect the movement of lipid from the inner to the outer leaflet of the bilayer.

bisnitrobenzoic acid (Ellman's reagent). When liposomes containing the thiol analog of DG were reacted with Ellman's reagent, the $t_{1/2}$ for transmembrane movement was determined to be 15 s. Concentration-dependent changes in the fluorescent properties of BODIPY lipids have also been used to estimate transbilayer movement of DG and ceramide (J. Bai, 1997). By the fluorescence method the $t_{1/2}$ for transbilayer movement of BODIPY-DG is 70 ms and that of BODIPY-Cer is in 22 min (see Fig. 6). These results strongly suggest that the polar moiety of phospholipids is the portion of the molecule that greatly retards transbilayer movement of these molecules in model membranes. Cholesterol is another non-polar lipid whose transmembrane movement has been examined. Treatment of PC/cholesterol liposomes with cholesterol oxidase demonstrated that the entire cholesterol pool could be readily oxidized with a $t_{1/2}$ of 1 min at 37°C (J.M. Backer, 1981) (see Fig. 6).

Thus, studies with model membranes provide clear evidence that the transbilayer movement of phospholipids is a very slow process in this system, whereas the process appears to be rapid for non-polar lipids. The results imply that if transbilayer movement of phospholipids does occur in biological membranes, it must be a facilitated process.

## 4.2. Intramembrane lipid translocation and biological membranes

### 4.2.1. Prokaryotes

The primary consideration in the genesis of any biological membrane is the location of the synthetic apparatus that manufactures the subunits of the membrane and its relationship to the final distribution of its products. In *Escherichia coli*, substantial evidence indicates that the synthesis of phospholipids occurs at the inner (cytoplasmic) membrane by enzymes that have their active sites on the cytoplasmic surface of the inner membrane [13]. Such an orientation allows free access of water-soluble substrates and reaction products to the cytosol.

In experiments performed with *B. megaterium*, Rothman and Kennedy [15] used chemical modification with TNBS, under conditions where the probe did not enter the cell, to distinguish between PE molecules located on the outer and inner sides of the cell membrane. This technique was coupled with pulse-chase experiments with [$^{32}$P]inorganic phosphate and [$^{3}$H]glycerol and demonstrated that newly synthesized PE is initially found on the cytoplasmic surface of the cell membrane and is rapidly translocated to the outer leaflet of the membrane with a $t_{1/2}$ of 3 min at 37°C. Although the translocation is rapid, it does not occur coincident with synthesis, but rather, with a significant delay after the molecule is synthesized. In addition, the translocation can continue in the absence of PE synthesis. These findings indicate that lipid synthesis and translocation are two distinct events.

The energetic requirements for transmembrane movement of phospholipids have been investigated (K.E. Langley, 1979). Using *B. megaterium* and a TNBS probe, these studies demonstrated that the transbilayer movement of newly synthesized PE was unaffected by inhibitors of ATP synthesis and protein synthesis. Thus, the driving force for phospholipid translocation in *B. megaterium* is independent of metabolic energy, lipid synthesis, and protein assembly into cell membranes.

More recent work with closed vesicles derived from *B. megaterium* membranes demonstrates that NBD analogs of PE, PG, and PC can translocate across the membrane with a $t_{1/2}$ of 30 s at 37°C (S. Hraffnsdottir, 1997). Similar types of experiments conducted with closed vesicles isolated from *E. coli* inner membrane reveal that NBD phospholipids traverse the bilayer with a $t_{1/2}$ of 7 min at 37°C (R. Huijbregts, 1996). This latter process is insensitive to protease and *N*-ethylmaleimide treatments and does not require ATP. Collectively, the data indicate that transbilayer lipid movement is rapid and does not require metabolic energy in bacterial membranes. The basic characteristics of the lipid translocation in the intact cell appear to be retained in isolated membranes.

### 4.2.2. Eukaryotes

#### 4.2.2.1. Transbilayer movement of lipid at the endoplasmic reticulum.
In eukaryotic systems a detailed pattern of synthetic asymmetry has emerged with respect to the topology of the enzymes of phospholipid synthesis in rat liver microsomal membranes. Protease mapping experiments (R. Bell, 1981) have indicated that the active sites of the phospholipid synthetic enzymes are located on the cytosolic face of the ER. Thus, in both prokaryotic and eukaryotic systems it appears that the site of synthesis of the

bulk of cellular phospholipid is the cytosolic side of the membrane. This asymmetric localization of synthetic enzymes strongly implicates transbilayer movement of phospholipids as a necessary and important event in membrane assembly that is required for the equal expansion of both leaflets of the bilayer (reviewed in A. Zachowski, 1993).

The transbilayer movement of phospholipids in microsomal membranes has been measured using several different approaches. In preparations of liver microsomes that were first radiolabeled with lipid precursors in vivo, the transbilayer movement of lipids was examined using phospholipid transfer proteins (D.B. Zilversmit, 1977). The results from these experiments provided evidence that PC, PE, PS, and PI from both membrane leaflets were exchanged between labeled microsomes and excess acceptor membranes with a maximal $t_{1/2}$ of ~45 min. This value, was set as an upper limit because the amount of lipid transfer protein used could not exchange out the labeled phospholipid with a $t_{1/2}$ of less than 45 min.

In a different approach, a water-soluble, short-chain (dibutyroyl) analog of PC was used to measured the rate of uptake and luminal sequestration by isolated liver microsomes (W. Bishop, 1985). This PC analog was taken up in a time- and temperature-dependent manner. The kinetics of uptake were saturable with respect to substrate concentration and the transport activity was protease sensitive. The transporter was also shown to be stereospecific in its action and it was unaffected by the addition of ATP. Virtually identical properties have also been described for a microsomal transporter that utilizes butyroyl-lyso-PC (Y. Kawashima, 1987).

Additional studies utilized spin-labeled analogs of PC, PE, PS and SM, and the $t_{1/2}$ for the translocation of these lipid analogs from the cytosolic face to the luminal face of the microsomes was calculated to be 20 min. The transport process did not require ATP and the translocation of each class of lipid showed identical sensitivity to inhibition by $N$-ethylmaleimide (Fig. 7). Furthermore, different species of lipid showed transport kinetics that were consistent with mutual competition for a single transporter. These

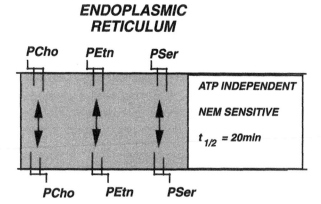

Fig. 7. Transbilayer movement of phospholipids in eukaryotic membranes. The general features of the transmembrane transporters of the ER are shown. The ER transporter does not require ATP and is inhibited by $N$-ethylmaleimide (NEM). The structure ($\llcorner_{\daleth}$) represents the DG portion of the lipids and PCho, PEtn and PSer are abbreviations for phosphocholine, phosphoethanolamine and phosphoserine, respectively.

results indicate that the ER has a relatively nonspecific, ATP-independent transporter that is capable of translocating multiple species of lipid across the bilayer.

Thus, the data from both bacteria and animal cells demonstrate that transbilayer movement of phospholipid occurs on a time scale of minutes, in an ATP-independent fashion in membranes that contain the majority of the enzymes involved in their biosynthesis. These intramembrane transport properties observed in the major biosynthetic membranes, however, are not generally true for other membrane systems. This is especially true of the plasma membrane.

*4.2.2.2. Transbilayer lipid movement at the eukaryotic plasma membrane.* The transbilayer movement of lipids at the cell surface of eukaryotes is being understood with increasing molecular and biochemical detail. Three fundamental classes of transport are now recognized and consist of the aminophospholipid translocases (also called flippases) the scramblases and the ATP binding cassette (ABC) pumps. A schematic summary of some of the properties of these proteins is shown in Fig. 8.

*Aminophospholipid translocase.* Several studies using either short-chain versions of PS, or spin-labeled or fluorescent analogs of PS and PE have established the properties of the translocases on the erythrocyte and nucleated cells [10,11]. The aminophospholipid translocase recognizes PS and PE at the external surface of the plasma membrane and translocates the lipids to the cytoplasmic side. This flipping process requires ATP on the cytosolic side of the membrane. The affinity for PS is about 30-fold higher than PE, and the equilibrium distribution is 95% PS and 90% PE in the inner membrane. The aminophospholipid translocases are susceptible to protease digestion and inactivation with $N$-ethylmaleimide. Additional susceptibilities to inhibition by $AlF_4$ and $Na_2VO_4$

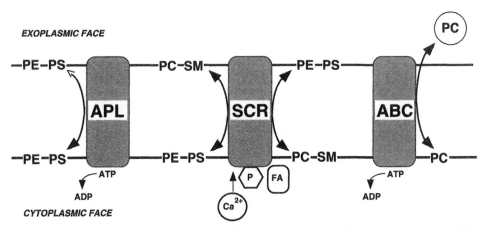

Fig. 8. Summary of the general features of transbilayer transporters found in plasma membranes. APL, amino phospholipid translocase; SCR, bidirectional transporter or scramblase; ABC, ATP binding cassette protein. The APL and ABC transporters utilize cytoplasmic ATP to drive transport. The APL is specific for PS and PE and the reaction greatly favors internalization. The ABC transporter pumps PC out of the plasma membrane. The scramblase is activated by high intracellular $Ca^{2+}$ levels and regulated by phosphorylation and acylation. The scramblase is nonspecific and randomizes the lipid distribution across the bilayer.

suggest that the translocases belong to the P-type ATPase family. The stoichiometry of ATP utilization per phospholipid translocation event is approximately one. All of the characteristics are consistent with the amino phospholipid translocase translocase being a specific plasma membrane ATPase that is activated by aminophospholipids.

An erythrocyte ATPase that is $Mg^{2+}$ and PS dependent, has been partially purified and reconstituted into vesicles. The reconstituted protein preparation is capable of transporting PS across the bilayer, albeit with low efficiency [11]. A second P-type ATPase, designated ATPase II, has been purified to homogeneity and its cDNA cloned. The deduced protein sequence reveals that it contains 3 P-type ATPase consensus sequences (X. Tang, 1996). Definitive reconstitution of the ATPase II into vesicles and demonstration of its aminophospholipid translocase activity has not yet been accomplished.

The cDNA sequence of the ATPase II was used to identify the yeast ortholog, which is the *DRS2* gene, that encodes the Drs2p protein, and for which *drs2* mutants are known, that have defects in ribosomal assembly. Some *drs2* mutant strains appear to have defects in their NBD-PS transport (X. Tang, 1996). However, *drs2Δ* mutants lacking any of the gene product have normal NBD-PS and NBD-PE transport (A. Siegmund, 1998). Thus, the aminophospholipid translocase has yet to be definitively identified in either mammalian or yeast systems.

*Bidirectional transporters.* The bidirectional transporters at the plasma membrane function to randomize the lipid distribution across the plane of the bilayer, and are commonly referred to as scramblases [10]. The action of the scramblase is similar to that of the previously described transbilayer transporter present in the ER. The protein was first functionally identified in erythrocytes, but is also present in nucleated cells. The scramblase shows no lipid specificity and essentially collapses the asymmetry of lipids at the cell surface. Phospholipids, SM and glycosphingolipids all serve as substrates. The randomizing function of this plasma membrane protein is activated by $Ca^{2+}$ and does not require ATP.

The scramblase protein was purified to homogeneity and its cDNA cloned [16]. The protein is oriented with its N-terminus in the cytosol and a short C-terminus at the external surface of the membrane. A $Ca^{2+}$ binding domain is found in the cytosolic region adjacent to the transmembrane domain. Post-translational modifications to the scramblase that alter activity include acylation and phosphorylation. There is currently much interest in the regulation of scramblase function, as it plays a critical role in the externalization of phosphatidylserine, a process that is important for the recognition of apoptotic cells by phagocytes. Four isoforms of the protein have now been identified (T. Wiedmer, 2000). In addition to regulation by $Ca^{2+}$ and oligomerization, the scramblase activity can be enhanced by phosphorylation directed by protein kinase C-delta (S. Frasch, 2000). The protein is also a substrate for the protein tyrosine kinase c-Abl (J. Sun, 2001). Detailed understanding of the mechanism of scramblase action is now likely to provide important fundamental information about the energetics and maintenance of lipid asymmetry.

*ABC transporters.* The ABC transporters are a large family of proteins involved in moving molecules across membranes in ATP-dependent reactions (I. Klein, 1999). A subset of this family transports molecules that include xenobiotics, bile acids, and hydrophobic compounds. In 1993, Smit and coworkers described a transgenic mouse with null allelles for the ABC transporter, *mdr2*, that exhibited a profound defect in transporting PC into bile. These findings led Ruetz and Gros [17] to examine the activity of the *mdr2* protein as a PC translocase. The heterologous expression of the *mdr2* cDNA in yeast leads to incorporation of *mdr2* protein into yeast secretory vesicles and acquisition of the ability to translocate NBD-PC across the bilayer. The translocation process is time-, temperature- and ATP-dependent. These findings indicate that the *mdr2* protein acts as a PC transporter. Unequivocal experiments in transgenic mice establish that the human MDR3 protein and mouse *mdr2* have identical function [18].

Further implication of ABC transporters in lipid translocation comes from work with human ABC A1, ABC R and ABC G5 plus ABC G8. The ABC A1 protein acts on cholesterol export and is described in Chapter 20. The ABC R protein is defective in Stargardt's macular dystrophy (R. Allikmets, 1997). In mice with null alleles for ABC R, retinylidene PE (a conjugate of retinaldehyde and PE) accumulates in the inner membrane of rod outer segment discs, due to failure of translocation by ABC R (J. Weng, 1999). These findings implicate ABC R as a translocase for retinylidene PE. The dimeric ABC G5 and ABC G8 transporter is defective in individuals with sitosterolemia (M. Lee, 2001). Under normal conditions, intestinal absorption of the plant sterol, sitosterol, appears to be minimized by an efflux pumping mechanism that continually translocates the sterol back to the lumen of the gut. In individuals with ABC G5 or ABC G8 defects, this pumping mechanism is lost, and blood and tissue levels of this deleterious sterol increase dramatically.

Data implicating ABC family transporters are growing, and it is likely that additional members will be added to the list. It is important to highlight one fundamental difference observed between the ABC transporters and other translocases (Fig. 8). In each case of the ABC family, the transported substrate either enters an environment that is a different phase from the membrane or the substrate is rapidly moved into another membrane or metabolized. In contrast, the substrate for the aminophospholipid translocases and scramblases remains within the bilayer across which it is transported.

### 4.3. Intermembrane lipid transport

From a theoretical perspective a number of processes could contribute to the intermembrane transport of lipids. These are outlined in Fig. 9 and include monomer solubility and diffusion (A), soluble carriers such as lipid transfer proteins (B), carrier vesicles (C), membrane apposition and transfer (D) and membrane fusion processes (E). Lipids such as free fatty acids, lysophosphatidic acid and CDP-DG may have sufficient solubility to allow for some monomeric transport but most other lipids are likely to require one of the other potential mechanisms due to their extremely low solubility.

### 4.3.1. Transport in prokaryotes
The presence of multiple membrane systems in organisms, such as Gram-negative bacteria, photosynthetic bacteria, and the eukaryotes, raise significant questions about the mechanisms of membrane biogenesis. In a 'simple' organism such as *E. coli* there are two membrane systems: the inner or cytoplasmic membrane, and the outer

464

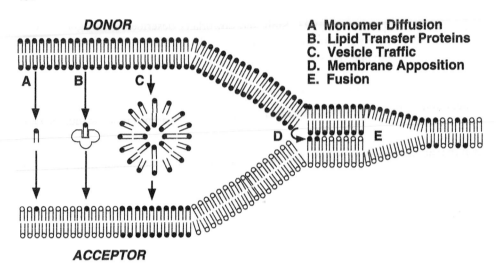

Fig. 9. Theoretical mechanisms for transporting lipids and altering membrane composition.

membrane (Chapter 3) [13]. The entire apparatus for phospholipid synthesis is located at the inner membrane. Consequently, there must exist a mechanism for exporting phospholipids from the inner membrane to the outer membrane.

In some of the earliest studies of phospholipid transport between the inner and outer membranes of *E. coli*, labeling of PE revealed that the specific activity of this lipid was fivefold higher in the inner membrane than the outer membrane, immediately following a 30-s pulse with [³H]glycerol (A.M. Donohue-Rolfe, 1980). During the chase period the specific activity of the outer membrane increased, while that of the inner membrane decreased. After several minutes the specific activities of both membranes asymptotically approached the same value, which indicated radioequilibration between the membranes. The $t_{1/2}$ for the translocation of PE was determined to be 2.8 min. The transport in these studies was independent of protein synthesis, lipid synthesis, and ATP synthesis. It appeared, however, to be dependent upon the cell's protonmotive force.

In recent work, Doerrler et al. [19] have examined the role of an ABC transporter in *E. coli* named *msbA*, in the phenomena of lipid transport between the inner and outer membrane. The *msbA* is an essential gene whose overexpression suppresses growth arrest in *E. coli* strains defective in the final steps of lipid A synthesis. Examination of the mode of rescue effected by high copy *msbA*, in the lipid A synthesis mutants, suggested that the protein acted to transport toxic lipid A precursors from the inner to the outer membrane. The function of *msbA* was examined further by creating a temperature-sensitive allele (*msbA^ts*). When cultures of strains with *msbA^ts* are shifted to the non-permissive temperature, cell growth arrests after about 60 min. In wild-type cells, subjected to the same treatments, cell growth is unaffected and 40–60% of the phospholipid, lipid A and protein, moved from the inner to the outer membrane. In the *msbA^ts* strain, <10% of the phospholipid and lipid A moved to the outer membrane, whereas protein traffic was unaffected. Furthermore, electron micrographs reveal significant invagination of the inner membrane consistent with excess lipid accumulation at this site. These experiments provide striking evidence for the role of *msbA* in the export of both phospholipid and lipid A from the inner membrane to the outer membrane of *E. coli*. These findings raise the possibility of other ABC transporters acting at sites of membrane contact or apposition, as a general mechanism for intermembrane lipid transport.

### 4.3.2. Transport in eukaryotes

Currently, there is a broad understanding of the elements of interorganelle transport for several different lipid classes. In many studies the questions have been narrowly focused to the movement of one lipid class between a donor and an acceptor compartment that are temporally, metabolically and geographically segregated within the cell. The discussion of these processes is organized by class of lipid and then by membrane systems examined.

### 4.3.2.1. Phosphatidylcholine.

*Transport of newly synthesized PC from the ER to the plasma membrane.* The principal site of PC synthesis is the ER (Chapter 8). The transport of PC from the ER to the plasma membrane has been examined using pulse-chase labeling with a [³H]choline precursor and rapid plasma membrane isolation with cationic beads (M. Kaplan, 1985). These studies reveal that PC transport is an extremely rapid process occurring with a $t_{1/2} \approx 1$ min (Fig. 10). This transport is unaffected by metabolic poisons that deplete cellular ATP levels, disrupt vesicle transport, or alter cytoskeletal arrangement. The mechanism of this transport is presently unknown, but the results are consistent with a soluble carrier mechanism such as PC transfer protein, or zones of apposition that facilitate rapid intermembrane transfer. Recent work by Pichler and colleagues has identified a subfraction of the ER that associates closely with the plasma membrane in yeast (H. Pichler, 2001). Future studies examining the effects of agents or mutations that disrupt these intracellular membrane associations will be critical for determining their role in lipid traffic.

*Transport of newly synthesized PC from the ER to the mitochondria.* Using conventional subcellular fractionation techniques, the transport of nascent PC to the mitochondria of BHK cells was examined by pulse-chase experiments with a [³H]choline precursor (M.P. Yaffe, 1983). These experiments show that the newly made PC pool equilibrates between the outer mitochondrial membrane and the ER in approximately 5 min (Fig. 10). Similar studies performed in yeast (G. Daum, 1986) revealed that the PC pool rapidly equilibrates between the ER and mitochondria. Addition of metabolic poisons did not eliminate the PC radioequilibration in yeast. Studies with isolated mitochondria demonstrate that PC loaded into the outer mitochondrial membrane can be transported to the inner membrane in an energy-independent manner (M. Lampl, 1994). Consistent with this finding is the observation that PC rapidly moves across the membrane of vesicles derived from mitochondrial outer membranes prepared from either mammalian cells or yeast (D. Dolis, 1996; M. Janssen, 1999).

*Transport of exogenous PC analogs from the cell surface to intracellular organelles.* Clear evidence for the movement of PC from the plasma membrane to intracellular

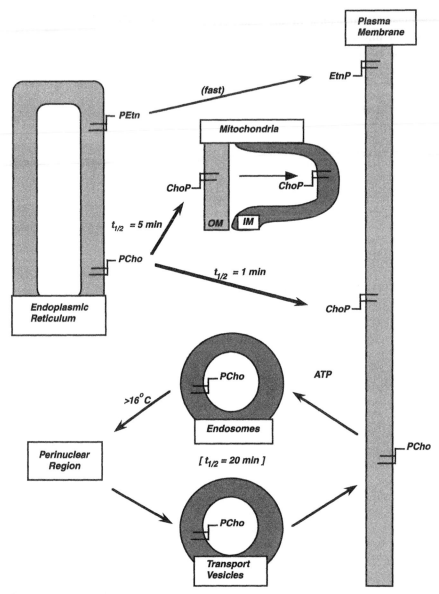

Fig. 10. Interorganelle transport of PC and PE within eukaryotic cells. The structure ( ⌐F) represents the DG portion of the phospholipid, and PCho and PEtn are the abbreviations for phosphocholine and phosphoethanolamine respectively. OM and IM are abbreviations for the outer and inner membranes of the mitochondria. The $t_{1/2}$ for PC transport from the perinuclear region of the cell to the plasma membrane is shown in brackets and estimated to be 20 min.

organelles has been obtained using the fluorescent lipid analog NBD-PC (R. Sleight, 1984). The fluorescent lipid can be pulse-labeled into the outer leaflet of the plasma membrane at 2°C. Upon warming the cells, the fluorescent lipid is transported from the plasma membrane to the perinuclear region of the cell in the proximity of the Golgi apparatus and the centrioles, via an ATP-dependent process (Fig. 10). The lipid transport occurs by endocytosis and the process can be disrupted by reducing the temperature to 16°C which causes the PC to accumulate in endosomal vesicles. The kinetics for endocytosed NBD-PC transport from intracellular membranes back to the plasma membrane occurs with $t_{1/2} = 20$ min. During the transit cycle the NBD-PC remains restricted to the non-cytosolic face of the respective membranes. The kinetics of this vesicle-based recycling of PC between the cell interior and the plasma membrane are markedly different from those for transport of newly synthesized PC to the cell surface. The disparity between the two transport processes suggests that there is restricted intermixing of nascent and recycling pools of PC.

### 4.3.2.2. Phosphatidylethanolamine.

*Transport of newly synthesized PE to the plasma membrane.* When an ethanolamine precursor is used, the primary site of PE synthesis is the ER (Chapter 8). The appearance of newly synthesized PE at the external leaflet of plasma membrane has been determined using chemical modification of the cell surface with TNBS at reduced temperature (R. Sleight, 1983). The results indicate that the initial rate of transport of PE is rapid and proceeds without a lag (Fig. 10). The transport process is insensitive to metabolic poisons that disrupt vesicle transport and cytoskeletal structure. The rapid transport kinetics occur at rates consistent with a soluble carrier-mediated process or transfer at zones of apposition between membranes. Analysis of the kinetics of the process is complicated since only PE at the outer leaflet of the plasma membrane is measured, and the ATP-dependent aminophospholipid transporter activity within the plasma membrane (P. Devaux, 1988; O. Martin, 1987) may be a required step for the lipid to arrive at this location. Despite these complications the results clearly indicate that the initial rate of arrival of PE at the plasma membrane occurs on a time scale that clearly distinguishes it from well characterized vesicle transport phenomena, and is independent of processes involved in protein transport to the cell surface.

PE derived from a PS precursor that is decarboxylated at the mitochondria is also transported to the plasma membrane (J. Vance, 1991) (Fig. 11). This mitochondrial PE is transported to the plasma membrane, with greater efficiency than PE synthesized from an ethanolamine precursor. The mechanism of this translocation remains to be elucidated but the process is unaffected by brefeldin A, a fungal metabolite that alters the structure and function of the Golgi apparatus.

*Transport of newly synthesized PE to the mitochondria.* Early studies examining the movement of newly synthesized PE from the ER to the mitochondria of hepatocytes demonstrated that the process was markedly slower ($t_{1/2} \approx 2$ h) than that observed for PC (M.P. Yaffe, 1983). These experiments used classical rate sedimentation to isolate the organelles. More recent studies indicate that such mitochondrial fractions are likely to contain another resolvable compartment, the mitochondria-associated membrane (MAM) (J. Vance, 1990). Evidence obtained using CHO-K1 cells [20] indicates that nascent PE (made via CDP-ethanolamine) is transported to the MAM but not to the inner mitochondrial membrane. It remains unclear whether some of this PE is transported to the outer mitochondrial membrane. The results are consistent with little import of PE

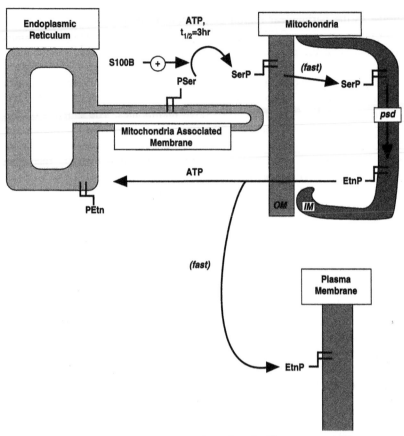

Fig. 11. Interorganelle transport of PS in eukaryotic cells. The structure ( ⌐F ) represents the DG portion of the phospholipid. PSer, PEtn and PCho are abbreviations for phosphoserine, phosphoethanolamine and phosphocholine. The term *psd* stands for PS decarboxylase. The rate for the transport of PS between the outer (OM) and inner (IM) mitochondrial membrane has not been determined but appears to be on the order of minutes.

derived from ethanolamine, into the mitochondria. Furthermore, yeast mutants lacking a functional allele for PS decarboxylase 1 are markedly deficient in mitochondrial PE (P.J. Trotter, 1995). The reduced PE in mitochondria cannot be fully restored by PE synthesized in the ER from an ethanolamine precursor, or that made in the Golgi or vacuole by PS decarboxylase 2 (R. Birner, 2001; M. Storey, 2001). These latter findings clearly demonstrate that there is compartmentation and restricted transport of different pools of PE within cells.

### 4.3.2.3. Phosphatidylserine.

*Transport of newly synthesized PS to the mitochondria.* The location of PS decarboxylase at the inner mitochondrial membrane [8] provides a convenient method for determining the arrival of PS at this cellular location. The extremely low steady state level of PS at the mitochondrial inner membrane (Table 1) coupled with kinetic considerations, indicates that PS is rapidly decarboxylated upon its arrival at the inner membrane. The general features of nascent PS transport are outlined in Fig. 11. The initial studies with intact mammalian cells that used PS decarboxylation as an indicator for lipid transport identified a clear ATP requirement for the transport process. Subsequent reconstitution studies, with isolated organelles, established that mitochondria could take up PS in an ATP-independent process. These findings indicated a requirement for ATP at a stage earlier than the presence of PS at the outer mitochondrial membrane. Additional studies with isolated organelles provided evidence for a tight association between specialized elements of the ER and the mitochondria. These in vitro associations were also shown to have in vivo counterparts evidenced by electron microscopy (D. Ardail, 1993). Successful isolation of these specialized ER structures, now called the mitochondria-associated membrane (MAM), indicated that they are selectively enriched in a subset of lipid synthetic enzymes especially PS synthase (J. Vance, 1990). A MAM structure has also been identified and isolated from yeast cells [21]. Pulse-chase experiments coupled with subcellular fractionation have now established the PS destined for the mitochondria must transit through the MAM and that exit from the MAM requires ATP [20].

The synthesis of PtdSer and transport to the mitochondria have been successfully reconstituted using permeabilized cells [8]. Permeabilized cells retain cellular organelles and cytoarchitecture but are depleted of soluble cellular components. The compromised plasma membrane enables the addition of defined soluble components under controlled conditions to reconstitute the transport processes. The transport of PS to the mitochondria in permeabilized cells occurs in the absence of cytosol, displays an absolute requirement for ATP and occurs with a $t_{1/2}$ of approximately 3 h at 37°C. This transport does not require ongoing synthesis of PS, and 45 fold dilution of the permeabilized cells does not alter the rate or extent of transport. These results are consistent with a membrane bound transport intermediate that utilizes zones of close membrane apposition or fusion between the ER and mitochondria. Although there is not an absolute requirement for cytosol in the transport reaction, a soluble 9 kDa $Ca^{2+}$ binding protein, named S100B, that is highly conserved across mammalian species can enhance the transport several fold (O. Kuge, 2001). Permeabilized yeast have also been used to examine PS transport (G. Achleitner, 1995). Unlike mammalian cells, the transport of PS to yeast mitochondria does not require ATP.

*Genetic approaches to identifying components involved in interorganelle aminophospholipid transport.* Genetic tools constitute a powerful approach for identifying the components involved in lipid transport. The ability to isolate mutant strains defective in a given transport step, and then clone the genes by complementation of the mutation, can lead to clear molecular and mechanistic resolution of complex processes. The genetic aspects of PS and PE transport in eukaryotic systems are shown in Fig. 12. Understanding the genetic approach to examining aminophospholipid transport requires an appreciation that the synthesis and decarboxylation of PS and the methylation of PE are all geographically separate events within a cell. A basic hypothesis of the genetic approach is that there are specific genes, designated *PST* for (*PS t*ransport), and *PEE* (for *PE e*xport), that either regulate or directly participate in the transfer process. Prominent methods for identifying mutant strains defective in these processes rely on

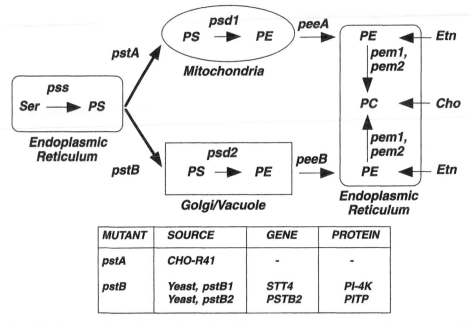

Fig. 12. Genetic analysis of aminoglycerophospholipid transport in eukaryotes. The transport of PS synthesized in the ER is hypothesized to be regulated by *PSTA* and *PSTB* genes. The acronym stands for *PS transport* (either A or B pathways). Likewise, the transport of PE synthesized in the mitochondria or Golgi/vacuole is proposed to be regulated by *PEEA* and *PEEB* genes. The acronyms stand for *PE export* (either A or B pathways). Both known and proposed mutations along the metabolic and transport pathways appear in lower case italics. The table summarizes the mutants, genes and proteins that have been identified. Other abbreviations: pss, PS synthase; psd, PS decarboxylase, pem, PE methyltransferase; PI-4K, PI-4-kinase; PITP, PI transfer protein; Cho, choline; Etn, ethanolamine; Ser, serine.

the identification of drug-resistant mammalian cells [22] or isolation of Etn auxotrophs (described in more detail in [8]) in suitable genetic backgrounds in yeast.

Studies with mammalian systems have identified a mutant line of CHO cells, designated R-41, that is resistant to an antibiotic that recognizes PE and causes cytolysis [22]. The cells have normal enzyme activity for PS synthases and decarboxylase, but labeling with [$^{14}$C]serine reveals a defect in PE formation. Further analyses with isolated mitochondria demonstrate that the rate of import of PS from the outer to the inner mitochondrial membrane occurs at approximately 40% of the rate found for wild-type cells. By the scheme shown in Fig. 12 this cell line belongs to the *pst A* class of mutants. Although the mutant line has a lesion in PS import into mitochondria, protein import into the inner membrane is unaffected. This cell line now provides an important tool for cloning the complementing cDNA that encodes an element required for PS import into the mitochondria.

In the yeast system, selection for Etn auxotrophs has yielded a number of mutant strains and genes that are involved in aminophospholipid transport [8]. Currently, the majority of the work has provided information about the transport of nascent PS from the ER to the Golgi/vacuole compartment in the *pstB* arm of the pathway. One mutant strain, *pstB1*, accumulates PS and shows reduced formation of PE. The gene complementing the defect, *STT4*, encodes a PI-4-kinase (P. Trotter, 1998). The results suggest that PI-4P or perhaps $PI4,5P_2$ may serve as a recognition motif on either the donor or the acceptor membranes participating in PS transport. Polyphosphoinositides are known to form recognition sites for the assembly of multiple protein components involved in membrane fusion machines [2].

A second mutant strain identified in the yeast system, *pstB2*, displays a 50% accumulation of PS and a 70% decrease in PE formation relative to wild-type strains when labeled with [$^3$H]serine. The gene that complements the defect, named *PSTB2*, is related to the PI transfer protein in yeast encoded by the *SEC14* gene. The protein encoded by the *PSTB2* gene, denoted PstB2p, is a PI transfer protein, but does not transfer PS (W. Wu, 2000). PstB2p is not a cofactor for the PS decarboxylase 2 but a protein that appears to regulate the access of PS to the enzyme. The site of action of PstB2p has now been shown to be at the acceptor membranes (W. Wu, 2001). The mechanism of action of PstB2p may be to dock donor and acceptor membranes, or it may act to regulate PI kinase activity in donor membranes.

### 4.3.2.4. Sphingolipids.

*Ceramide transport from ER to Golgi.* Ceramide is synthesized in the ER and the majority of this lipid is subsequently transported to the Golgi apparatus where it is metabolized to sphingomyelin and glycosphingolipids such as GlcCer and LacCer (Chapter 14) [14]. Measurement of conversion of ceramide to either sphingomyelin or glycosphingolipids can serve as an indicator of ceramide transport (Fig. 13). Hanada and coworkers have isolated mutant strains of CHO cells that are resistant to a toxin,

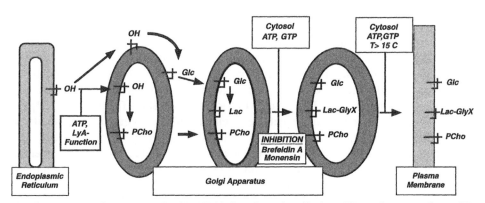

Fig. 13. Interorganelle transport of sphingolipids in eukaryotic cells from ER to plasma membrane. The structure (─┼─) represents the Cer portion of sphingolipids. PCho, Glc and Lac are the abbreviations for phosphocholine, glucose, and lactose. Cer is transported to the Golgi in an ATP-dependent reaction. GlcCer is synthesized on the cytosolic face of the Golgi. SM is synthesized on the luminal face. The LyA mutation selectively affects the access of Cer to the site of SM synthesis. GlcCer must reach the luminal face of the Golgi for conversion to LacCer and more complex glycosphingolipids, denoted by GlyX moieties. Movement of the sphingolipids through the Golgi requires cytosol, ATP, and GTP and is inhibited by brefeldin A, monensin, GTPγS and reduced temperature. The $t_{1/2}$ for sphingolipid transport from the Golgi to the plasma membrane is 20 min.

lysenin, that binds cell surface SM and causes cytolysis [23]. One class of mutants (LyA) is selectively defective in SM synthesis, but not glycosphingolipid synthesis, despite normal activity of SM synthase. These findings suggest that the routing of Cer to SM synthase and GlcCer synthases is different and regulated by different gene products.

Ceramide transport to the locus of SM synthase demonstrates that the process can be reconstituted in permeabilized cells (T. Funakoshi, 2000) and requires the addition of ATP and cytosol. Depletion of ATP in intact cells also yields arrest of ceramide (and BODIPY-ceramide) trafficking to SM synthase. When wild-type permeabilized cells are reconstituted with cytosol from the LyA cells, ceramide transport-dependent SM synthesis does not occur. Thus, the LyA lesion resides in a soluble protein that participates in transport of nascent Cer to the locus of SM synthase. In contrast to these findings the GlcCer synthesis failed to exhibit a clear requirement for cytosol for ceramide transport to the enzyme.

*Transport of newly synthesized sphingolipids from the Golgi to the plasma membrane.* The synthesis and intracellular trafficking of SM and glucosylceramide (GlcCer) has been examined using several different fluorescent ceramides and short-chain radiolabeled ceramides [14]. When fibroblasts are incubated with NBD-Cer at 2°C, it is rapidly taken up and distributed randomly among all cell membranes (N. Lipsky, 1985). Upon warming the cells to 37°C, the fluorescent lipid concentrates in the Golgi apparatus as it is converted to NBD-SM and NBD-GlcCer. These sphingolipids are subsequently exported from the Golgi apparatus to the plasma membrane by a process that is partially monensin-sensitive and brefeldin-A-sensitive in most cells and occurs with a $t_{1/2}$ of 20 min, a time similar to that required for the transport of proteins from the Golgi to the plasma membrane (Fig. 13). However, there appears to be a pool of Glc–Cer that can be transported by routes insensitive to inhibitors of vesicle trafficking [14]. Vesicle-based protein transport is arrested in mitotic cells as is the transport of newly synthesized NBD-SM and NBD-GlcCer (Kobayashi, 1989). Experiments using (non-fluorescent) short-chain analogs of Cer in permeabilized cells indicate that the export of nascent SM from the Golgi apparatus requires ATP and cytosol and occurs via a GTP-dependent mechanism that is also consistent with vesicle budding from the organelle (J.B. Helms, 1990). Export of nascent sphingomyelin from the Golgi is blocked at reduced temperatures such as 15°C and by the non-hydrolyzable GTP analog, GTPγS.

The movement of sphingolipids between elements of the Golgi has been monitored in reconstituted preparations from mutant Chinese hamster ovary cells defective in either the synthesis of lactosylceramide or the attachment of sialic acid to the latter (B. Wattenberg, 1990). In cell free systems, donor Golgi that accumulate lactosylceramide transfer this lipid to acceptor Golgi that are devoid of the substrate. The acceptor Golgi add sialic acid to the lactosyl ceramide to make $GM_3$. The lipid transfer reaction between Golgi compartments requires ATP and cytosol and is inhibited by GTPγS. The properties of glycosphingolipid transport between Golgi compartments are thus identical to those found for vesicular protein transport.

*Import of exogenous sphingolipids.* The NBD and BODIPY analogs of SM, GlcCer, and LacCer can be readily inserted into the outer leaflet of the plasma membrane of fibroblasts at reduced temperature (see Fig. 3). When fibroblasts treated in such a manner are warmed to 37°C the fluorescent sphingolipids are internalized and accumulate in the

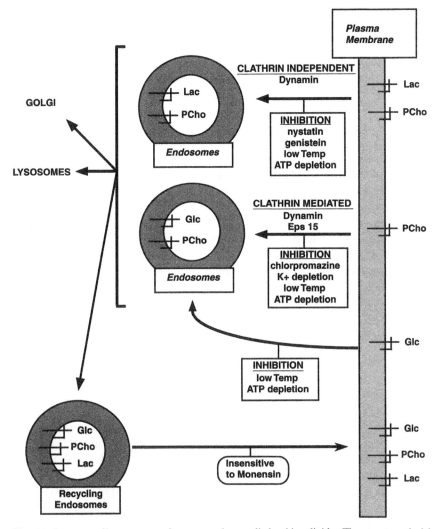

Fig. 14. Interorganelle transport of exogenously supplied sphingolipids. The structure ( ⊥ ) represents the Cer portion of sphingolipids. PCho, Lac and Glc are the abbreviations for phosphocholine lactose and glucose, respectively. Following insertion into the plasma membrane at reduced temperature, SM is internalized to the endosomal compartments by both clathrin-dependent and -independent pathways. LacCer is endocytosed primarily by the clathrin-independent pathways. The endocytic process can be generally inhibited by ATP depletion and reduced temperature. The endocytosed lipid can recycle back to the plasma membrane and this recycling is insensitive to monensin. The endocytosed lipid can also be transported to the Golgi apparatus or lysosomes.

endosomal compartments of the cell. General inhibition of endocytosis by ATP depletion, or maintenance at low temperature, effectively prevents any internalization of the polar sphingolipids. Internalized NBD-SM subsequently accumulates in the perinuclear region of the cell containing the centrioles (M. Koval, 1989) and the Golgi apparatus (see Fig. 14). The initial steps of BODIPY-SM internalization have been dissected using a variety of inhibitors and dominant negative structural variants of dynamin 2 (Dyn $2^{DN}$) and Eps 15 (Eps $15^{DN}$) [24]. The Eps 15 protein has a regulatory function in clathrin coated pit assembly. The internalization of BODIPY-SM is completely arrested by Dyn $2^{DN}$ which disrupts both clathrin-dependent and -independent endocytosis. The expression of Eps $15^{DN}$, or treatment of cells with chlorpromazine or $K^+$ depletion, inhibits clathrin-dependent endocytosis to a greater extent than SM endocytosis. Conversely, inhibition of clathrin-independent endocytosis with genistein or nystatin does not fully block SM endocytosis. These results indicate that BODIPY-SM is likely to be internalized by both clathrin-dependent and -independent pathways.

The movement of the endocytosed fluorescent SM from the internalized pool, back to the plasma membrane has also been examined in fibroblasts (M. Koval, 1989). This transport process occurs via vesicles. The properties of the recycling pool of NBD-SM are distinct from those observed for export of the newly synthesized SM out of the Golgi (Fig. 13). As stated above, monensin and brefeldin A arrest newly synthesized NBD-SM transport from the Golgi to the cell surface, but the recycling of endocytosed fluorescent SM is insensitive to monensin. The overall process of internalization of SM from the plasma membrane to the intracellular pool, and transport back to the cell surface occurs with a $t_{1/2}$ of approximately 40 min. These time constants are similar to those for membrane protein recycling processes from the plasma membrane.

The internalization of BODIPY-LacCer follows a route that partially overlaps with that for fluorescent SM [24]. The BODIPY-LacCer is internalized into endosomes and subsequently can be localized to the Golgi. The endocytosis of BODIPY-LacCer is inhibited by Dyn $2^{DN}$, nystatin and genistein but not Eps $15^{DN}$, chlorpromazine, or $K^+$ depletion. These latter results indicate that the fluorescent LacCer is endocytosed by a clathrin-independent mechanism.

The internalization and recycling of NBD-GlcCer from the cell surface is similar to that for SM and LacCer analogs [14,25]. The assignment of NBD-GlcCer endocytosis to either clathrin-dependent or -independent pathway has not yet been made. Following internalization the NBD-GlcCer is initially found in both early and late endosomal compartments and subsequently the Golgi apparatus. The lipid recycles back to the cell surface (Fig. 14) at a rate similar to that for SM. Transport of NBD-GlcCer from the endosomal compartment to the plasma membrane is insensitive to treatment of the cells with either monensin or brefeldin A.

### 4.3.2.5. Cholesterol.

*Transport of cholesterol to and from the plasma membrane.* Following its synthesis at the ER, cholesterol is transported throughout the cell and becomes enriched in the plasma membrane [9]. The transport of newly synthesized cholesterol to the plasma membrane has been examined in tissue culture cells using pulse-chase experiments with either the rapid plasma membrane isolation procedure (M. Kaplan, 1985), caveolae isolation (A. Uittenbogaard, 1998), oxidation of accessible cholesterol by cholesterol oxidase (Y. Lange, 1985), or desorption of newly labeled cholesterol with methyl-β-cyclodextrin (S. Heino, 2000). These lines of experimentation have revealed that the minimum transport time for cholesterol to the plasma membrane is 10 min at 37°C (Fig. 15). The transport process can be completely blocked by reducing the temperature

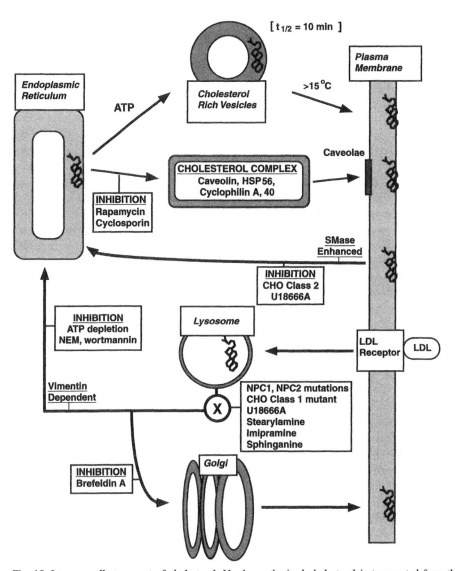

Fig. 15. Interorganelle transport of cholesterol. Newly synthesized cholesterol is transported from the ER to the plasma membrane in an ATP- and temperature-dependent process. One intermediate identified in this transport is a low-density cholesterol-rich fraction believed to be comprised of vesicles. A second proposed intermediate consists of a soluble cholesterol/protein complex. It is not clear if the vesicle fraction and soluble complex are the same. Cholesterol present in the plasma membrane can be induced to move to the ER by SMase treatment of the cell surface. This latter process is inhibited by hydrophobic amines and class 2 mutations in CHO cell lines. Low-density lipoprotein (LDL) derived cholesteryl ester enters the lysosome and is cleaved to form free cholesterol. The lysosomal cholesterol is exported from the lysosomes by a process regulated by NPC1 and NPC2 gene products that is susceptible to inhibition by hydrophobic amines. The cholesterol exported from the lysosome also traverses the Golgi en route to the plasma membrane, or travels directly to the ER via a process that exhibits partial dependence on intermediate filaments and requires ATP.

to 15°C or depleting cellular ATP levels with metabolic poisons. The transport of nascent cholesterol is unaffected by treatment of the cells with cytoskeletal poisons or monensin. When the translocation of cholesterol is inhibited by maintaining the cells at 15°C, this lipid accumulates in a low-density membrane fraction (M. Kaplan, 1985; Y. Lange, 1985). Intermediates in the transport of proteins between the ER and the Golgi apparatus accumulate at 15°C in vesicles of similar density to those containing cholesterol. However, the compartment containing the intermediates in protein transport is different from that containing cholesterol because the former is sensitive to brefeldin A treatment, whereas the latter is not [26]. This result demonstrates that cholesterol travels to the plasma membrane via intermediates that are distinct from those involved in protein transport. Collectively, these data suggest the presence of specialized machinery for cholesterol transport. One mechanism that has been proposed for this transport is non-vesicular and consists of a complex of caveolin with cholesterol and a heat shock protein, HSP 56, and the cyclophilins A and 40 (A. Uittenbogaard, 1998). This complex is believed to form a cytosolic cholesterol carrier that can transport the lipid from the ER to a caveolae-rich fraction of plasma membrane. Pulse-chase experiments with [$^3$H]acetate are consistent with caveolae serving as an entry point for cholesterol at the plasma membrane (Chapter 11). Both cyclosporin A and rapamycin are predicted to disrupt the interactions of the cyclophilins and HSP 56 with caveolin. Treatment of cells with cyclosporin A and rapamycin markedly inhibited the appearance of nascent [$^3$H]cholesterol in caveolae and the total plasma membrane. However, there is still uncertainty as to the relationship between this cytosolic complex and the low-density fraction that accumulates at 15°C.

In addition to the outward movement of cholesterol to the plasma membrane, cells display a retrieval system for recovering the sterol from the plasma membrane. When the plasma membrane of mammalian cells is rapidly depleted of SM by sphingomyelinase treatment, a significant fraction of the cholesterol is transported to the ER and esterified by acyl CoA : cholesterol acyltransferase [27]. The cholesterol retrieval is blocked by hydrophobic amines including U18666A, and sphingosine, and steroids such as proges-terone [27] but is insensitive to ATP depletion. Liscum and coworkers have isolated a cell line denoted CHO 3–6 (N. Jacobs, 1997) (also described as a Class 2 mutant) that is defective in recovering plasma membrane cholesterol after sphingomyelinase treatment, that should prove most useful for dissecting the transport mechanism.

*Recycling of exogenous cholesterol.* Exogenous cholesterol imported into the cell via the low-density lipoprotein (LDL) receptor can be utilized for membrane biogenesis and regulation of sterol metabolism (Chapter 15). The mechanisms whereby lipoprotein-derived cholesterol (generated from cholesteryl esters within lysosomes) is disseminated throughout the cell is being understood with increasing detail (Liscum, 1999; E. Blanchette-Mackie, 2000). The current view indicates that approximately 70% of the lysosomal cholesterol pool is directed to the plasma membrane, whereas 30% is directed to the ER by a separate pathway. One set of inhibitors or mutations appears to affect the export of cholesterol from the lysosomes before the bifurcation of the transport between the plasma membrane and the ER. These early acting conditions include U18666A, imipramine, Niemann–Pick C mutations and CHO class 1 mutants. Subsequent to the bifurcation in the pathway the routing to the plasma membrane is sensitive to brefeldin

A disruption of the Golgi. After the bifurcation the transport to the ER is sensitive to disruption of intermediate filaments, ATP depletion, $N$-ethylmaleimide treatment, and (weakly) wortmannin intoxication.

Important insights into the mechanism of cholesterol transport have come from LDL metabolism in cells from individuals with Niemann–Pick type C (NPC) disease. In NPC fibroblasts, cholesterol transport from the lysosomal compartment to the plasma membrane is markedly retarded compared to that in normal fibroblasts (Liscum, 1999; E. Blanchette-Mackie, 2000). The NPC cells also have impaired regulation of acyl CoA : cholesterol acyl transferase, 3 hydroxy-3-methyl glutaryl CoA reductase and LDL receptor levels, in response to LDL. In addition, free cholesterol accumulates in the lysosomal compartment. In contrast, the transport of newly synthesized cholesterol from the endoplasmic reticulum to the plasma membrane of NPC fibroblasts is essentially identical to that found for normal cells. These findings localize one abnormality of NPC disease to cholesterol export from the lysosomes to other organelles.

Two genes, NPC1 and NPC2 are now known to be responsible for the NPC phenotype. The NPC1 gene product shows significant homology to the morphogen receptor, *Patched*, to a protein of unknown function, NPCL1, and to members of the RND (resistance–nodulation–division) family of prokaryotic permeases. NPC1 has sequence homology to proteins containing sterol sensing domains (3-hydroxy-3-methylglutaryl CoA reductase and SREBP cleavage activating protein, Chapter 15). NPC1 is also closely related to bacterial permeases that transport hydrophobic compounds including acriflavine [28,29]. Consistent with the proposed permease function, NPC1-defective cells cannot efficiently export acriflavine out of their lysosomes. The characteristics of NPC1 protein structure most closely resemble those of fatty acid permeases, and expression of human NPC1 protein in *E. coli* markedly enhances the uptake of oleic acid by the bacteria. Thus a specific biochemical function assignable to NPC1 is modulation of fatty acid transport. The relationship between this defined function and the export of cholesterol remains to be clarified.

The NPC2 gene has also been identified [29]. The gene product is a soluble lysosomal protein that was identified in a global proteomics screen of lysosomal constituents. The protein is found in secreted and intralysosomal forms, with retrieval of the secreted protein mediated by the mannose-6-P receptor. Incubation of NPC2-defective cells with medium containing the secreted form of wild-type NPC2 leads to protein uptake and rectification of the cholesterol accumulation seen in the mutant cells. In addition, individuals with the NPC2 phenotype (which is identical to that for NPC1) show specific mutations in the NPC2 gene, thereby confirming the gene-mutation–disease relationship. The emerging picture is that soluble NPC2 and membrane-bound NPC1 must cooperate in the recognition and translocation of cholesterol out of the lysosome.

*Cholesterol import into mitochondria.* In steroidogenic tissues cholesteryl esters are hydrolyzed in response to hormonal stimuli, and cholesterol is imported into mitochondria for the synthesis of pregnenolone, the precursor of all steroid hormones. The transit of cholesterol between the outer and inner mitochondrial membrane is regulated by steroidogenic acute regulatory protein (StAR) [30]. StAR is rapidly synthesized in response to hormonal stimuli and targeted to the mitochondria by N-terminal sequences. The StAR protein is imported into the mitochondria and becomes associated with the matrix and inner membrane. Initially, transit intermediates of StAR were proposed to be the cholesterol carriers between the outer and inner membranes. However, current data demonstrate that import of StAR is not essential and that the association of the C-terminus of StAR with the outer mitochondrial membrane may be the critical interaction required for promoting cholesterol transfer. The mechanism of StAR action is not understood, but interaction of the C-terminus with cholesterol and outer membrane proteins may play a role in assembling a transport complex that moves the lipid to the inner membrane. Individuals with lipid congenital adrenal hyperplasia lack functional StAR and are unable to make pregnenolone. Analysis of the StAR sequences in these individuals reveals that the mutations accumulate in the carboxy terminal region of the protein. The carboxy terminus of StAR and a structural homolog MLN-64 that binds cholesterol are now recognized to define a large protein family capable of binding hydrophobic molecules (L.M. Iyer, 2001).

*4.3.2.6. Phospholipid transfer proteins and membrane biogenesis.* Since their discovery in the early 1960s, the PLTPs have been attractive candidates for soluble lipid carriers between membranes in vivo (K. Wirtz, 1968). However, there have been two major points of debate about a lipid trafficking role for the PLTPs. The first point has remained that the proteins effect exchange of lipid between model membranes and do not yield a net transfer of mass. Such an action by the proteins in vivo might be able to change lipid composition of membranes but would not be able to cause net synthesis and accumulation of new lipid mass. However, to be circumspect one cannot rule out the possibility that either protein modification or protein–protein interactions in vivo, may enable the proteins to accomplish net transfer. A second point of concern has been whether the lipid binding and exchange of the proteins simply reflects a lipid binding property that has other functions.

Genetic approaches addressing the role of PLTPs have provided significant insight into their function [31]. The identification of the *SEC14* gene and its corresponding Sec14p protein, as the major PI/PC transfer protein in yeast enabled critical tests of in vivo function. The *SEC14* gene was shown to be essential and play an important role in regulating protein traffic through the Golgi. Tests of the required function of Sec14p demonstrate that the ability of the protein to effect transfer of PI is completely dispensable (S. Phillips, 1999). The PC bound form of Sec14p is now thought to be crucial to its function. In the PC bound form, Sec14p is proposed to be a negative regulator of phosphocholine cytidylyltransferase (Chapter 8) (H.B. Skinner, 1995). The consequences of this regulation of the cytidylyltransferase are thought to be steady state maintenance of a Golgi diacylglycerol pool required for secretory vesicle formation. Other functions of Sec14p have also been proposed including the regulation of polyphosphoinositide pools that are required for operation of the secretory pathway. This latter idea finds support from data in other systems that show a requirement for PI transfer proteins in either exocytosis or signal transduction that appear related to modulation of polyphosphoinositide pools [32].

Another prominent test of in vivo lipid transfer protein function has focused upon the nonspecific lipid transfer protein known as SCP2. Early studies suggested that SCP2 played an important role in intracellular cholesterol traffic. However, antibodies raised to the protein localized the antigen primarily in peroxisomes, rather than the predicted cytosolic compartment. More detailed analysis revealed that SCP2 arose as a proteolytic fragment of the peroxisomal enzyme 3-ketoacyl-CoA thiolase (Chapter 5). These findings were more consistent with a role for SCP2 in peroxisomal β-oxidation than cholesterol traffic. The development of gene-targeted mice with defects in SCP2 expression revealed a lesion in pristanic acid oxidation and in cholesterol side-chain oxidation [33] but not in intracellular lipid traffic. Collectively, these data fail to define any role for SCP2 in cholesterol transport.

The in vivo role of PC transfer protein has also been examined by genetic deletion in mice (A. van Helvoort, 1999). Animals with homozygous null alleles for the PC transfer protein had no identifiable phenotype. Examination of tissues with high rates of synthesis and secretion of PC, such as liver and lung, also failed to show any perturbation in lipid traffic or metabolism, indicating that the PC transfer protein cannot be essential for these processes.

Thus far, critical biochemical and genetic studies of lipid transfer proteins do not demonstrate the proteins act as lipid carriers that directly function in the transport reactions for membrane biogenesis. However, the issue of a role for such proteins remains open, in light of the observations that some lipids (e.g., PE) move with surprising rapidity between membranes in reactions that are not demonstrably ATP-dependent. Certainly, other mechanisms are possible for such rapid lipid transport reactions, but additional data are required before a role lipid for transfer proteins can be ruled out.

## 5. Future directions

Lipid transport is a fundamental process essential to all cell growth, division and differentiation. Our understanding of lipid transport has changed markedly in the last five years, and the pace of change is now increasing. Most notably, the identification of mammalian cell lines, yeast, and bacterial strains with defects in lipid traffic is a major advance. The identification of human diseases with lesions in lipid traffic makes additional important tools, and in some cases cell lines, available. Advances in reconstitution of lipid traffic in permeabilized cells now allow for more precise and critical tests of protein function in the processes. The application of fluorescent probes continues to provide new insights and real time images of selected aspects of lipid transport. As the examination of these processes now begins to enter the realm of the manipulation of mutant cells, genes, and gene products, there remains much to be accomplished. Future studies need continued focus on the development of new genetic tools. For many of the lipid trafficking processes described in this chapter there are still no mutants available, and a concerted effort must be made to develop novel selections and screens that attack the voids in our understanding. The current expansion in genomic information and the ease of manipulating genes in heterologous systems now also allows for approaches in which educated guesses can be used for targeting candidate sequences.

Any candidate sequence that appears in yeast can now be obtained in a hemizygous null strain from commercial sources. Straightforward manipulation allows for the recovery of strains harboring null alleles that are covered by a plasmid-borne copy of the wild-type gene under inducible or repressible promoters. Such tools allow for rapid critical testing of gene product function in lipid traffic. In addition, phylogenetic jumping among databases can also permit the rapid identification, isolation, and testing of specific gene products in suitable in vivo and in vitro assay systems. The mechanisms of intracellular lipid traffic in membrane assembly have thus far been difficult to elucidate, but recent advances are grounds for much optimism. The current molecular tools, combined with new genetic strategies are likely to provide both new research opportunities and rewards for those who tackle this long-standing problem of cell biology.

## Abbreviations

| | |
|---|---|
| BODIPY | boron dipyrromethene difluoride |
| Cer | ceramide |
| DG | diacylglycerol |
| diC8- | dioctanoyl- |
| ER | endoplasmic reticulum |
| ESR | electron spin resonance |
| GlcCer | glucosylceramide |
| GTP | guanosine triphosphate |
| GTPγS | guanosine 5′-$O$-(-3-thiotriphosphate) |
| IAA | isethionylacetimidate |
| LDL | low-density lipoprotein |
| PC | phosphatidylcholine |
| PE | phosphatidylethanolamine |
| PG | phosphatidylglycerol |
| PI | phosphatidylinositol |
| PS | phosphatidylserine |
| PSD | phosphatidylserine decarboxylase |
| MAM | mitochondria-associated membrane |
| mdr | multidrug resistance |
| NBD | $N$-[7-(4-nitrobenzo-2-oxa-1,3-diazole)]-6-aminocaproyl |
| NPC | Niemann–Pick type C |
| SM | sphingomyelin |
| TNBS | trinitrobenzenesulfonate |

## References

1. Mellman, I. and Warren, G. (2000) The road taken: past and future foundations of membrane traffic. Cell 100(1), 99–112.
2. Simonsen, A., Wurmser, A.E., Emr, S.D. and Stenmark, H. (2001) The role of phosphoinositides in membrane transport. Curr. Opin. Cell Biol. 13(4), 485–492.
3. Verkleij, A.J., Zwaal, R.F.A., Roelofsen, B., Comfurius, P., Kastelijn, D. and van Deenen, L.L.M.

(1973) The asymmetric distribution of phospholipids in the human red cell membrane. A combined study using phospholipases and freeze-etching electron microscopy. Biochim. Biophys. Acta 323, 178–193.

4. Rothman, J.E. and Kennedy, E.P. (1977) Asymmetrical distribution of phospholipids in the membrane of *Bacillus megaterium*. J. Mol. Biol. 110, 603–618.

5. Pagano, R.E. and Sleight, R.G. (1985) Defining lipid transport pathways in animal cells. Science 229, 1051–1057.

6. Roseman, M., Litman, B.J. and Thompson, T.E. (1975) Transbilayer exchange of phosphatidylethanolamine for phosphatidylcholine and *N*-acetimidoyl phosphatidylethanolamine in single-walled bilayer vesicles. Biochemistry 14, 4826–4830.

7. Wirtz, K.W.A. (1991) Phospholipid transfer proteins. Annu. Rev. Biochem. 60, 73–99.

8. Voelker, D.R. (2000) Interorganelle transport of aminoglycerophospholipids. Biochim. Biophys. Acta 1486(1), 97–107.

9. Liscum, L. and Munn, N.J. (1999) Intracellular cholesterol transport. Biochim. Biophys. Acta 1438(1), 19–37.

10. Bevers, E.M., Comfurius, P., Dekkers, D.W. and Zwaal, R.F. (1999) Lipid translocation across the plasma membrane of mammalian cells. Biochim. Biophys. Acta 1439(3), 317–330.

11. Daleke, D.L. and Lyles, J.V. (2000) Identification and purification of aminophospholipid flippases. Biochim. Biophys. Acta 1486(1), 108–127.

12. Trotter, P.J. and Voelker, D.R. (1994) Lipid transport processes in eukaryotic cells. Biochim. Biophys. Acta 1213, 241–262.

13. Huijbregts, R.P., de Kroon, A.I. and de Kruijff, B. (2000) Topology and transport of membrane lipids in bacteria. Biochim. Biophys. Acta 1469(1), 43–61.

14. Van Meer, G. and Holthuis, J.C. (2000) Sphingolipid transport in eukaryotic cells. Biochim. Biophys. Acta 1486(1), 145–170.

15. Rothman, J.E. and Kennedy, E.P. (1977) Rapid transmembrane movement of newly synthesized phospholipids during membrane assembly. Proc. Natl. Acad. Sci. USA 74, 1821–1825.

16. Zhou, Q., Zhao, J., Stout, J.G., Luhm, R.A., Wiedmer, T. and Sims, P.J. (1997) Molecular cloning of human plasma membrane phospholipid scramblase. A protein mediating transbilayer movement of plasma membrane phospholipids. J. Biol. Chem. 272(29), 18240–18244.

17. Ruetz, S. and Gros, P. (1994) Phosphatidylcholine translocase: a physiological role for the *mdr2* gene. Cell 77, 1071–1081.

18. Borst, P., Zelcer, N. and van Helvoort, A. (1999) ABC transporters in lipid transport. Biochim. Biophys. Acta 1486, 128–144.

19. Doerrler, W.T., Reedy, M.C. and Raetz, C.R. (2001) An *Escherichia coli* mutant defective in lipid export. J. Biol. Chem. 276(15), 11461–11464.

20. Shiao, Y.J., Lupo, G. and Vance, J.E. (1995) Evidence that phosphatidylserine is imported into mitochondria via a mitochondria-associated membrane and that the majority of phosphatidylethanolamine is derived from decarboxylation of phosphatidylserine. J. Biol. Chem. 270, 11190–11198.

21. Daum, G. and Vance, J.E. (1997) Import of lipids into mitochondria. Prog. Lipid Res. 36, 103–130.

22. Emoto, K., Kuge, O., Nishijima, M. and Umeda, M. (1999) Isolation of a Chinese hamster ovary cell mutant defective in intramitochondrial transport of phosphatidylserine. Proc. Natl. Acad. Sci. USA 96(22), 12400–12405.

23. Fukasawa, M., Nishijima, M. and Hanada, K. (1999) Genetic evidence for ATP-dependent endoplasmic reticulum-to-Golgi apparatus trafficking of ceramide for sphingomyelin synthesis in Chinese hamster ovary cells. J. Cell Biol. 144(4), 673–685.

24. Puri, V., Watanabe, R., Singh, R.D., Dominguez, M., Brown, J.C. and Wheatley C.L. et al. (2001) Clathrin-dependent and -independent internalization of plasma membrane sphingolipids initiates two Golgi targeting pathways. J. Cell Biol. 154(3), 535–547.

25. Hoekstra, D. and Kok, J.W. (1992) Trafficking of glycosphingolipids in eukaryotic cells; sorting and recycling of lipids. Biochim. Biophys. Acta 1113, 277–294.

26. Urbani, L. and Simoni, R.D. (1990) Cholesterol and vesicular stomatitis virus G protein take separate routes from the endoplasmic reticulum to the plasma membrane. J. Biol. Chem. 265, 1919–1923.

27. Liscum, L. and Munn, N.J. (1999) Intracellular cholesterol transport. Biochim. Biophys. Acta 1438(1), 19–37.

28. Davies, J.P., Chen, F.W. and Ioannou, Y.A. (2000) Transmembrane molecular pump activity of Niemann–Pick C1 protein. Science 290(5500), 2295–2298.

29. Naureckiene, S., Sleat, D.E., Lackland, H., Fensom, A., Vanier, M.T. and Wattiaux R. et al. (2000) Identification of HE1 as the second gene of Niemann–Pick C disease. Science 290(5500), 2298–2301.

30. Stocco, D.M. (2000) Intramitochondrial cholesterol transfer. Biochim. Biophys. Acta 1486(1), 184–197.

31. Li, X., Xie, Z. and Bankaitus, V.A. (1999) Phosphatidylinositol/phosphatidylcholine transfer proteins in yeast. Biochim. Biophys. Acta 1486, 55–71.

32. Cockcroft, S. (2001) Phosphatidylinositol transfer proteins couple lipid transport to phosphoinositide synthesis. Semin. Cell Dev. Biol. 12(2), 183–191.

33. Seedorf, U., Ellinghaus, P. and Nofer, J.R. (2000) Sterol carrier protein-2. Biochim. Biophys. Acta 1486, 45–54.

D.E. Vance and J.E. Vance (Eds.) *Biochemistry of Lipids, Lipoproteins and Membranes (4th Edn.)*

# CHAPTER 18

# Lipoprotein structure

Ana Jonas

*Department of Biochemistry, College of Medicine, University of Illinois at Urbana-Champaign,*
*506 South Mathews Avenue, Urbana, IL 61801, USA, Tel.: +1 (217) 333-0452;*
*Fax: +1 (217) 333-8868; E-mail: a-jonas@uiuc.edu*

## 1. Introduction

Lipoproteins are soluble complexes of proteins (apolipoproteins) and lipids that transport lipids in the circulation of all vertebrates and even insects. Lipoproteins are synthesized in the liver, in the intestines, arise from metabolic changes of precursor lipoproteins, or are assembled at the cell membranes from cellular lipids and exogenous lipoproteins or apolipoproteins. In the circulation, lipoproteins are highly dynamic. They undergo enzymatic reactions of their lipid components, facilitated and spontaneous lipid transfers, transfers of soluble apolipoproteins, and conformational changes of the apolipoproteins in response to the compositional changes. Finally, lipoproteins are taken up and catabolized in the liver, kidney, and peripheral tissues via receptor-mediated and other mechanisms. This chapter deals almost exclusively with the human lipoproteins.

### 1.1. Main lipoprotein classes

Although the assembly, structure, metabolism, and receptor interactions of lipoproteins are determined by their apolipoprotein components, the most common classifications of lipoproteins are based on their hydrated density or mobility on agarose gel-electrophoresis.

The classification into chylomicrons (CM), very low-density (VLDL), low-density (LDL), and high-density (HDL) lipoproteins is based on their relative contents of protein and lipid that determine the densities of these lipoprotein classes. Chylomicrons have only 1–2% protein while HDL have about 50% protein by weight. The diameters of lipoproteins are inversely correlated with their densities and range from about 6000 Å for CM down to 70 Å for the smallest HDL (Fig. 1).

The general structural organization is similar for all the lipoprotein classes: the apolipoproteins and amphipathic lipids (mostly phospholipids and unesterified cholesterol) form a 20-Å shell on the surface of spherical particles. This shell encloses a core of neutral lipids (triacylglycerols, cholesteryl esters, and small amounts of unesterified cholesterol and other dissolved lipids, e.g., lipid-soluble vitamins). The main protein components are characteristic of each lipoprotein class; they are indicated in Fig. 1, and will be described in detail in Section 3 of this chapter.

The principal functions of the lipoprotein classes are determined by their apolipoprotein (apo) and lipid components. The CM are synthesized in the intestines for the transport of dietary triacylglycerols to various tissues. VLDL are synthesized in the

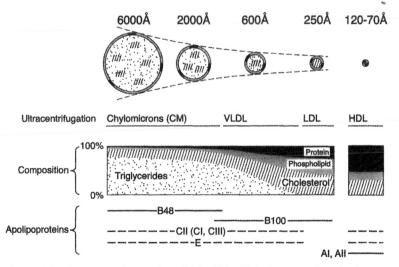

Fig. 1. Major lipoprotein classes (CM, VLDL, LDL, HDL) based on their density. Lipoprotein diameters range from about 6000 Å for CM to 70 Å for HDL. The outer shell (~20 Å) of all lipoproteins consists of apolipoproteins, unesterified cholesterol, and phospholipids; the spherical core contains triacylglycerols and cholesteryl esters. CM and VLDL have the highest contents of triacylglycerols, and 1–10% apolipoproteins by weight; LDL and HDL contain mostly cholesteryl esters in their cores, and 20–50% of apolipoproteins. The major apolipoprotein components of the various classes of lipoproteins are indicated with the solid lines; secondary or minor apolipoprotein components are indicated with the dashed lines. In this figure, 'cholesterol' refers to both esterified and unesterified cholesterol; triglycerides = triacylglycerols.

liver for the export of endogenous triacylglycerols, while LDL arise from the metabolic transformation of VLDL in circulation. The function of LDL is to deliver cholesteryl esters to peripheral tissues and to the liver. HDL are synthesized and assembled in the liver and intestine or are formed from metabolic transformations of other lipoproteins in circulation, and from cellular lipids at the cell membranes (see Chapter 20). HDL remove excess cholesterol from cells and transport it to liver and steroidogenic tissues for metabolism and excretion.

Lipoproteins are also classified by their electrophoretic mobility on agarose gels into α, preβ, and β lipoproteins, corresponding to HDL, VLDL, and LDL density classes respectively; CM, when present, remain at the electrophoretic origin.

Although lipoprotein concentrations in blood plasma are highly variable, depending on age, sex, feeding state, metabolic/hormonal state, and disease state of individuals, a representative lipoprotein distribution for a fasting, healthy, adult male in plasma is approximately 0 mg/dl for CM, 150 mg/dl for VLDL, 410 mg/dl for LDL and 280 mg/dl for HDL [1].

## 1.2. Lipoprotein subclasses

The lipoproteins within each class are heterogeneous in terms of their density, size, and lipid and apolipoprotein contents and compositions, as well as in their functional properties. They can be separated into subclasses by ultracentrifugation, gel filtration, electrophoresis, or affinity chromatography methods.

Based on ultracentrifugal and gel-filtration separations, HDL have been subdivided into HDL$_1$, HDL$_2$, and HDL$_3$ subclasses, from the largest and least dense to the smallest and most dense particles. The HDL$_1$ subclass is enriched in apo E and is least abundant and often disregarded. Non-denaturing gel electrophoresis has further separated the main HDL$_2$ and HDL$_3$ subclasses into HDL$_{2a}$, HDL$_{2b}$, HDL$_{3a}$, HDL$_{3b}$, and HDL$_{3c}$ species spanning a density range from 1.085 to 1.171 g/ml and size (diameter) range from 106 to 76 Å, respectively [2].

Using anti-apolipoprotein immunoaffinity columns two major subclasses of HDL have been separated: one containing apo A1 but no apo A2 (LpA-I) and another containing both apo A1 and apo A2 (LpA-I/A-II). Minor proteins (apo E, apo Cs) may or may not be present in significant amounts in these HDL subclasses [3]. On average, human HDL contain about 70% by weight of apo A1, 20% of apo A2, and 10% of the minor apolipoproteins.

Two-dimensional separations of HDL (agarose gel-electrophoresis in one dimension and non-denaturing polyacrylamide gel-electrophoresis in the second dimension) have yielded α-migrating, preβ-migrating, and γ-migrating HDL subclasses of various sizes and compositions. The preβ-migrating subclasses are present in low concentrations in plasma (in contrast to the abundant α species), and in somewhat higher concentrations in interstitial fluid, but are metabolically very important as they represent the nascent forms of HDL that are especially active in cholesterol uptake from cells (preβ$_1$-HDL) and cholesterol esterification by lecithin cholesterol acyltransferase (LCAT) (preβ$_2$- and preβ$_3$-HDL) [4]. Because of their key role in cholesterol uptake from cells, the preβ$_1$-HDL have been studied quite intensively (J.P. Kane, 1985). They contain 2 molecules of apo A1 per particle, and no apo A2 nor other proteins. Their mass ranges from 60 to 80 kDa, with a content of about 90% protein, 7% phospholipids, 0.3% unesterified cholesterol, and 1.8% esterified cholesterol. The preβ$_1$-HDL represent only 7% of total apo A1 in plasma and have an apo A1 conformation distinct from apo A1 in α-migrating HDL. Much less is known about the other preβ-HDL subclasses, except that they are discoidal in shape, contain large proportions of phospholipid, and have masses around 300 kDa.

Other lipoprotein classes can also be separated into subclasses of varying density and size by the same separation methods. Subclasses of LDL in the density range from 1.027 to 1.060 g/ml and size range from 270 to 210 Å have been obtained and shown to have different metabolic properties [5]. Small dense LDL, containing high amounts of triacylglycerols, appear to be the most proatherogenic LDL species. Chylomicrons and VLDL undergo continuous density, size, and composition changes due to the hydrolysis of their triacylglycerols by lipoprotein lipase and exchanges of soluble apolipoproteins; therefore, these lipoprotein classes consist of a continuous spectrum of particles.

## 2. Lipid components

### 2.1. Lipid composition

The lipid content and composition of the major lipoprotein classes are listed in Table 1 [6], which shows that the total lipid content is inversely correlated with the density of the lipoproteins. Glycerolipids, mainly triacylglycerols, are the major lipid components of CM and VLDL, but constitute less than 11% of the lipids of LDL and HDL, which are enriched in cholesteryl esters (24–51%). Unesterified cholesterol is found in all the lipoprotein classes in relatively low proportions because it is actively esterified by LCAT on HDL and then redistributed to LDL and VLDL by cholesterol ester transfer protein. The total phospholipid content of lipoproteins increases with increasing density and is directly related to the surface area of the lipoprotein particles, as the surface of lipoproteins is covered by a monolayer of phospholipids (PL) and apolipoproteins. By far, the main phospholipid constituents are phosphatidylcholines (PC) (57–81% wt PL) followed by sphingomyelins (12–26% wt PL). Lyso PC and other phospholipids (phosphatidylethanolamine, phosphatidylserine, and phosphatidylinositol) constitute 5–15% wt PL. Glycolipids are present only in trace amounts and free fatty acids contribute less than 3% to the total mass of lipids.

### 2.2. Fatty acid composition

In the fasted state, the fatty acid composition of lipoprotein lipids reflects their biosynthetic origins and metabolic transformations in circulation (Table 2) [6].

The glycerolipids (predominantly triacylglycerols) have high proportions of 18 : 1 and 16 : 0 fatty acids, reflecting synthesis in the liver. The bulk of cholesteryl esters of human lipoproteins is formed in circulation by the action of LCAT. This enzyme acts on HDL

Table 1
Lipid composition of lipoprotein classes [a]

|  | CM [b] | VLDL [c] | LDL [c] | HDL [c] |
|---|---|---|---|---|
| Density (g/ml) | <0.94 | 0.94–1.006 | 1.006–1.063 | 1.063–1.210 |
| Total lipid (% wt) | 98–99 | 90–92 | 75–80 | 40–48 |
| Glycerolipids (% wt lipid) [d] | 81–89 | 50–58 | 7–11 | 6–7 |
| Cholesteryl esters (% wt lipid) | 2–4 | 15–23 | 47–51 | 24–45 |
| Unesterified cholesterol (% wt lipid) | 1–3 | 4–9 | 10–12 | 6–8 |
| Phospholipids (% wt lipid) [e] | 7–9 | 19–21 | 28–30 | 42–51 |
| PC (% wt PL) | 57–80 | 60–74 | 64–69 | 70–81 |
| SM (% wt PL) | 12–26 | 15–23 | 25–26 | 12–14 |
| Lyso PC (% wt PL) | 4–10 | ~5 | 3–4 | ~3 |
| Other (% wt PL) | 6–7 | 6–10 | 2–10 | 5–10 |

[a] Adapted from Skipski [6].
[b] Chylomicrons (CM) were isolated during absorption of fat meals from plasma or lymph.
[c] VLDL, LDL, and HDL were isolated from fasting plasma or serum.
[d] Most glycerolipids are triacylglycerols, only about 4% are diacylglycerols or monoacylglycerols.
[e] Phospholipid (PL), phosphatidylcholine (PC), sphingomyelin (SM).

Table 2

Fatty acid composition of glycerolipids, cholesteryl esters and phospholipids present in lipoprotein classes [a]

| Fatty acid | VLDL | LDL | HDL |
|---|---|---|---|
| Glycerolipids: | | | |
| 16:0 | 26.9 | 23.1 | 23.4 |
| 18:0 | 2.9 | 3.3 | 3.5 |
| 18:1 | 45.4 | 46.6 | 43.8 |
| 18:2 | 15.7 | 15.7 | 15.9 |
| 20:4 | 2.5 | 5.4 | 7.6 |
| Cholesteryl esters: | | | |
| 16:0 | 11.5 | 11.3 | 11.4 |
| 18:0 | 1.4 | 1.0 | 1.1 |
| 18:1 | 25.6 | 22.4 | 22.4 |
| 18:2 | 51.8 | 59.9 | 54.8 |
| 20:4 | 6.1 | 6.9 | 6.3 |
| Phospholipids: | | | |
| 16:0 | 33.8 | 36.0 | 31.8 |
| 18:0 | 14.7 | 14.3 | 14.5 |
| 18:1 | 12.2 | 11.6 | 12.3 |
| 18:2 | 20.3 | 18.9 | 20.6 |
| 20:4 | 13.6 | 13.2 | 15.7 |

[a] Adapted from Skipski [6]. Lipoproteins were isolated from fasting plasma. Composition is in wt% of total fatty acids.

phosphatidylcholines and unesterified cholesterol to form lyso PC and the cholesteryl ester products. In this reaction, LCAT uses preferentially PC species with 18:2 or 18:1 fatty acids in the $sn$-2 position, thus enriching cholesteryl esters in these fatty acids. In contrast, PC containing 18:0 or 20:4 fatty acids is a poor substrate for LCAT, explaining the decreased contents of these fatty acids in the cholesteryl esters. The fatty acid composition of the phospholipids found in lipoproteins is similar to the phospholipid fatty acid composition of cell membranes, especially of hepatic and intestinal cells, including relatively high contents of the essential fatty acids 18:2 and 20:4.

In general, in the fasting state, the fatty acid compositions across the lipoprotein classes for specific types of lipids are fairly similar due to the transfers of lipids among all the lipoproteins by cholesterol ester transfer protein and the phospholipid transfer protein. Postprandially the fatty acid composition of VLDL glycerolipids reflects to some extent the fatty acid composition of dietary fat, but the fatty acid compositions of LDL and HDL are hardly affected. In chylomicrons, the fatty acid compositions do reflect the fatty acid composition of the meal, especially 8–10 h after the meal when CM concentrations in lymph and plasma are maximal.

## 2.3. Lipid organization

The surface of all lipoproteins consists of a lipid monolayer containing all the phospholipids, and about 2/3 of all unesterified cholesterol, plus the corresponding apolipoproteins.

The dynamic properties of the lipid monolayers depend largely on the nature of the constituent lipids. For example, the surface lipids of LDL are more condensed and rigid than those of HDL due to the presence of more saturated fatty acids in the phospholipids of LDL, a higher sphingomyelin-to-PC ratio, and a higher unesterified cholesterol-to-phospholipid ratio in LDL (M.C. Phillips, 1989). The surface monolayer of VLDL is even more fluid than that of HDL due to differences in the monolayer lipids. Apolipoproteins exert little or no effect on the average dynamic properties of the surface lipids especially in LDL and VLDL. In fact, isolated surface lipids, reconstituted into vesicles or microemulsions, have similar fluidity, diffusion rates, and mobility as the lipid monolayer components in the intact lipoproteins. The core lipids (triacylglycerols and cholesteryl esters) partition poorly into the surface lipid monolayer, accounting for approximately 3 and 1 mol%, respectively, of the surface lipids [7].

Under physiological conditions, the interior of lipoproteins is a fluid spherical droplet of the neutral lipids, including small amounts of dissolved unesterified cholesterol (about 1/3 of the total) and other lipophilic molecules. In HDL and VLDL, the core lipids do not appear to be organized because of the small volume available in HDL, and because of the high content of fluid triacylglycerols in VLDL. However, the core of LDL particles is known to have the cholesteryl esters organized into two interdigitated concentric layers that may undergo cooperative phase transitions at temperatures between 19–32°C. This organization is optimal at a specific cholesteryl ester/triacylglycerol ratio of 7/1 where a separate fluid phase of triacylglycerol is present in the center. At different core lipid ratios there is mixing of the neutral lipids [8]. Phase transitions in the core lipids of LDL apparently result in changes in the secondary structures of apolipoprotein B100 on the surface. This indicates some coupling of hydrophobic regions of the apolipoprotein with adjacent neutral lipids.

## 3. Apolipoproteins

### 3.1. Classes and general properties

The apolipoproteins found in plasma are classified into two broad types: the non-exchangeable and the exchangeable (or soluble) apolipoproteins. Apolipoprotein B100 (apo B100) and apo B48, the principal protein components of LDL, VLDL, Lp(a) and CM are non-exchangeable apolipoproteins. They are very large and water-insoluble proteins that are assembled with lipids at their site of synthesis in the endoplasmic reticulum of liver or intestinal cells (Chapter 19). These non-exchangeable apolipoproteins circulate bound to the same lipoprotein particle through various metabolic transformations in plasma, until they are cleared, as lipoproteins, via specific receptors (Chapter 21). In contrast, the exchangeable apolipoproteins (e.g., apo A1, apo A2, apo Cs, apo E) have much smaller molecular masses than apo B100 or apo B48, are more or less soluble in water in their delipidated states, can transfer between lipoprotein particles, and can acquire lipids while in circulation (see Table 3) [9].

The common function of all apolipoproteins is to help solubilize neutral lipids in the circulation. The apolipoproteins bind readily to phospholipid/water interfaces

Table 3

Major human apolipoproteins

| Apolipoprotein [a] | Molecular weight [b] | Lipoprotein class [c] | Concentration in plasma (mg/dl) |
|---|---|---|---|
| Apo A1 | 28,100 | **HDL**, *CM* | 130 |
| Apo A2 [d] | 17,400 | **HDL** | 40 |
| Apo A4 [e] | 44,500 | **CM** | 15 |
| Apo(a) [f] | $3–8 \times 10^5$ | **Lp(a)** | 0.1–40 |
| Apo B100 [g] | 512,000 | **LDL, VLDL** | 250 |
| Apo B48 [g] | 242,000 | **CM** | |
| Apo C1 | 6,600 | **VLDL, CM**, *HDL* | 3 |
| Apo C2 | 9,000 | **VLDL, CM**, *HDL* | 12 |
| Apo C3 | 9,000 | **VLDL, CM**, *HDL* | 12 |
| Apo D [h] | 22,000 | **HDL** | 12 |
| Apo E [i] | 34,200 | **VLDL, CM, HDL** | 7 |

[a] Other, minor apolipoproteins isolated from lipoprotein fractions include apo F, apo H, apo J, apo L, and apo M. They are present in plasma in low concentrations and do not have well-defined functions in lipoprotein metabolism.

[b] Polypeptide molecular weights do not include carbohydrate contributions.

[c] In bold are the lipoprotein classes containing the highest proportion of the apolipoprotein; in italics are secondary lipoprotein classes.

[d] Human apo A2 is a disulfide-linked dimer of two identical monomers of 8.7 kDa; other mammalian apo A2s are monomeric.

[e] Apo A4 is a glycoprotein found in lymph CM and is mostly (90%) lipid-free in plasma. Exists in several polymorphic states.

[f] Apo(a) is a highly polymorphic protein, containing variable numbers of kringle structural units and glycan chains. Homologous to plasminogen, apo(a) is bound to apo B100 in LDL by a disulfide linkage forming Lp(a).

[g] Apo B48 contains 48% of the N-terminal sequence of apo B100. Both are glycoproteins. Apo B48 is present in variable amounts depending on feeding state and CM concentration.

[h] Apo D belongs to the lipocalin protein family, it may bind progesterone, but its role in lipoprotein metabolism is unknown.

[i] Apo E is glycosylated and consists of three polymorphic forms (apo E2, E3, and E4).

and, under appropriate conditions, can spontaneously form discrete particles with phospholipids. In vivo, the assembly of apolipoproteins with lipids to form lipoproteins may require the assistance of cell proteins such as the microsomal lipid transfer protein or the ABC1 transporter.

Intimately related to their ability to bind phospholipids, and to solubilize neutral lipids within lipoprotein particles, the apolipoproteins have the ability to change conformation to adjust to changing lipid contents, compositions, and metabolic states of the lipoproteins. In apo A1, the adaptable structural regions have been called the mobile or hinge domains. Analogous domains likely exist in other apolipoproteins, as epitope recognition by specific antibodies and exposure to proteolytic digestion change with changing lipid contents and compositions of various lipoproteins.

Several of the exchangeable apolipoproteins in their respective lipoproteins are known to activate or inhibit plasma enzymes. Apo A1, and to a lesser extent apo E, apo A4, and apo C1, activate LCAT, apo C2 activates lipoprotein lipase, and apo C3 and apo A2 may act as inhibitors of hepatic lipase.

Apolipoproteins also have roles in receptor recognition. The best characterized interaction is the recognition of the LDL receptor by apo B100- and apo E-containing lipoproteins. Recently, a receptor for HDL (SR-BI) has been described which binds various apolipoproteins, including apo A1 and apo A2.

In addition to the well-known functions of apolipoproteins in lipid binding and solubilization, modulation of enzymatic activities, and receptor recognition, other functions have been described for apolipoproteins. For example, apo E has been implicated in in situ nerve repair and regeneration, as well as in plaque formation in Alzheimer's disease. Apo(a) may have a function in the clotting process, while apo A4, produced in the intestine and the hypothalamus, may have a role in signaling satiety in the fed state.

Evidently, the multiple functions of apolipoproteins are determined by their unique, modular structures encoded by families of genes.

### 3.2. Gene organization

The gene structures of the major exchangeable apolipoproteins are very similar to each other, whereas the apo B gene is distinct, as are the gene structures of apo(a) and apo D. Most of the genes of the exchangeable apolipoproteins contain four exons and three introns, with similar locations of intron/exon boundaries, and similar intron and exon lengths for the first three exons (see Fig. 2) [10]. The differences in the total length of the mRNA are due to the differences in the length of the 4th exon. The mRNAs encompass a 5′-untranslated region, a region encoding the signal sequence, a short pro-segment, the mature sequence of the protein, and a 3′-untranslated region. Exons 3 and 4 of the genes encode the entire mature sequence. For the apo A4 gene, the only difference from the other exchangeable apolipoprotein genes is the absence of the first exon and intron in the 5′-untranslated region. The homologies in the gene and protein sequences of the exchangeable apolipoproteins indicate that these genes evolved by gene duplication from a primordial gene resembling the gene of apo C1.

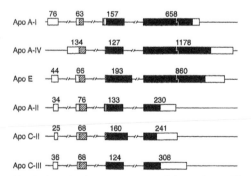

Fig. 2. Organization of the genes of the exchangeable apolipoproteins [10]. The boxes represent the exons joined by introns (broken lines). The open portions of the boxes correspond to the 5′- and 3′-untranslated regions, the hatched portions are the signal peptide regions, and the filled parts represent the regions that code the mature protein sequences. The narrow open portions in exon 3 of the apo A1 and apo A2 genes represent their pro-segments. Numbers above the exons indicate the number of nucleotides in each exon.

In contrast to the genes of the exchangeable apolipoproteins, the apo B gene is short with respect to the length of its 29 exons. It is also quite asymmetric: 19 introns are concentrated within the first 1000 codons, and two very long exons (exons 26 and 29) occur in the 3′-third of the gene sequence. No homology is found between the apo B gene and other known genes [9]. It encodes one of the longest known polypeptide chains containing 4536 amino acid residues. Apo B48 is encoded by the same gene as apo B100, but the mRNA is edited by a specific enzyme in human intestine. The enzyme changes codon 2153 from a Gln to a stop codon, resulting in the production of apo B48, a protein of 242 kDa, lacking the LDL receptor-binding region of apo B100 (D. Driscoll, 1990).

### 3.3. Primary sequences

The exchangeable apolipoproteins, in addition to having similar genes, have similar primary amino acid sequences. Their sequences contain 11 and 22 amino acid homologous repeats, the latter consisting of two 11-mers, as shown in Fig. 3 [11]. The last 33 amino acids encoded by exon 3 can be aligned into three 11-mers, while the sequences encoded by exon 4 fit, in general, into 22-mer segments that often start with

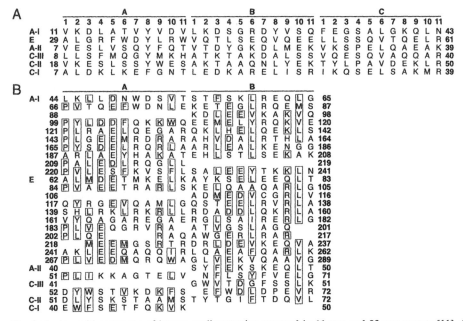

Fig. 3. Amino acid sequences of human apolipoproteins arranged in 11-mer and 22-mer repeats [11]. (A) The last 33 amino acids encoded by exon 3 are divided into three (A, B, C) 11-mer repeats. (B) The sequences encoded by exon 4 contain 22-mer repeats each consisting of two 11-mers (A and B), and some 11-mer repeats. If there are 10 or more amino acids in the same column with equivalent physical properties (i.e., hydrophobic, basic, acidic), they are boxed. Numbers at the beginning and end of a sequence give the amino acid locations in the mature apolipoprotein.

a Pro (J.I. Gordon, 1986). The significance of these repeated sequences is that they are predicted to form amphipathic α-helices. These helices in apolipoproteins and synthetic peptides have been shown experimentally to bind to phospholipid surfaces and to be effective in solubilizing lipids. Thus, the 22 amino acid repeats in α-helical organization are the lipid-binding units of the exchangeable apolipoproteins. Furthermore, specific repeats in the sequences encoded by exon 4 have other functional roles such as LCAT activation by the 143–165 repeat of apo A1, LDL receptor-binding by the 136–150 region of apo E, and lipoprotein lipase activation by residues 44–79 of apo C2.

In contrast, the apo B100 sequence contains many internally repeated sequences that bear little resemblance to the sequences of the exchangeable apolipoproteins [12]. There are unique, Pro-rich repeats of 25 and 52 residues that contain high proportions of hydrophobic amino acids. These regions could be in contact with the core lipids in LDL and VLDL. Other features of the apo B100 sequence include (1) 25 cysteine residues (16 in disulfide linkages), (2) at least 4 heparin-binding regions, and (3) 19 potential glycosylation sites (16 of which are occupied by glycan chains, contributing about 9% of the molecular mass of apo B100). Most of the disulfide linkages and free Cys residues cluster in the amino-terminal region of the apolipoprotein. The receptor-binding domain of apo B100, between residues 3353 and 3371, does have some homology with the corresponding receptor-binding region of apo E.

Apo(a) is a distinct, highly polymorphic glycoprotein that is covalently linked to apo B100 of LDL by a disulfide linkage, to form Lp(a). Apo(a) is homologous to plasminogen, containing variable kinds and numbers of kringle sequences. Each kringle repeat contains about 80 amino acids and three internal disulfide bridges. While apo(a) has sequences homologous to those forming the plasminogen catalytic, serine protease site, the peptide sequence corresponding to the cleavage site required for activation of plasminogen is modified in apo(a), so that the serine protease activity is not expressed. In fact, the function of apo(a) in plasma is not known in the 30% of the population that expresses this apolipoprotein in significant levels (A.M. Scanu, 1988).

## 3.4. Secondary structures

The exchangeable apolipoproteins, devoid of lipids, have substantial amounts of α-helical structure in physiological aqueous solutions and their α-helix content increases markedly upon lipid binding. For example, apo A1 is about 50% helical in the lipid-free state, and becomes 60–85% helical when bound to lipids, as measured by circular dichroism spectroscopy.

Various computer algorithms have predicted the existence of amphipathic α-helix segments, coinciding quite well with the 22-mer repeats of the exchangeable apolipoproteins. The predicted helical structures have a non-polar face, and a larger polar face. Depending on the relative distribution of the charged residues on the polar face, the amphipathic helices of the exchangeable apolipoproteins fall into three classes: A, G$^*$, and Y helices [13].

In the A-type helices (see Fig. 4) the basic residues are found at the boundary between the polar and non-polar helix faces. The hydrophobic chains of the basic amino acids, in fact, contribute to the hydrophobicity of the non-polar face of the helix. The

Ala • Leu • A̅s̅p̅ • L̇ẏṡ • Leu • L̇ẏṡ • G̅l̅u̅ • Phe • Gly • Asn • Thr • Leu • G̅l̅u̅ • A̅s̅p̅ • L̇ẏṡ • Ala • Ȧṙġ • G̅l̅u̅
              11                              16                           21

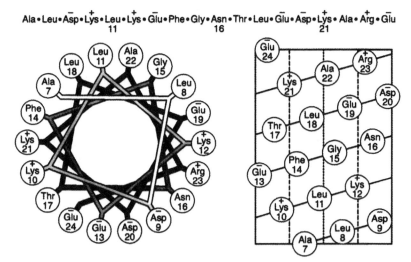

Fig. 4. Type-A amphipathic helix corresponding to residues 7–24 of apo C1. (Left) Helical wheel representation with the α-helix axis in the center of the wheel and consecutive amino acids at 100° from each other. The non-polar residues are at the top of the wheel, basic residues occur at the interface between the non-polar and polar sides of the helix, and negatively charged residues appear in the middle of the polar face (at the bottom of the wheel). (Right) The helix is represented as a flattened cylinder cut along the center of the polar face.

acidic residues, on the other hand, are located along the middle ridge of the polar face and are fully exposed to solvent. Such A helices in specifically designed synthetic peptides bind effectively to phospholipids and lipoproteins, and readily solubilize lipids. The A-type helices are indeed the fundamental lipid-binding units of apolipoproteins. Two 22 amino acid synthetic A helices in tandem, joined by a Pro, are even more effective in binding lipids, suggesting inter-helix cooperativity.

The G* helices are amphipathic helices typically found in globular proteins. Their basic and acidic residues are randomly distributed on the polar face of the helix. The G* helices usually participate in protein–protein interactions rather than protein–lipid interactions. G*-type helices are mostly found in the N-terminal regions of exchangeable apolipoproteins, encompassing the 11-mer repeats encoded by exon 3.

Amphipathic helices of the Y-type have alternating clusters of basic and acidic residues on the polar face. Their distinction from the A-type helices in terms of lipid binding is not clear, as synthetic peptides representing A- and Y-types of helical segments of apo A1 have comparable lipid-binding properties.

Because of the large size and insolubility in water of apo B100, secondary structure measurements have been conducted on intact LDL particles and proteolytic fragments solubilized in detergents or reassembled with lipids. Reported α-helix contents in LDL range from 20 to 43%, and for β-structure from 12 to 41%, as determined by circular dichroism and infrared spectroscopic measurements. Computer analysis of the sequence of apo B100 suggests the presence of five regions of clustered secondary structures: (1) an N-terminal sequence containing G-type helices (residues 58–476); followed by

(2), an amphipathic β-strand region (residues 827–1961); (3) a region with α-helices resembling A- and Y-type amphipathic helices and including other kinds of helices (residues 2103–2560); (4) a second β-strand region (residues 2611–3867); and (5) a C-terminal, third region of amphipathic α-helices (residues 4061–4338) [14]. The N-terminal region of apo B100, containing the G-helices and high contents of Cys, consists of one or more globular domains (residues 22–303 and 512–721) that are involved in interactions with the microsomal triglyceride transfer protein, a protein required for the assembly of nascent apo B100 with lipids. The central and C-terminal regions of α-helices may be involved in reversible lipid binding by apo B100, and the β-strand regions may participate in irreversible interactions with lipids, possibly involving core lipids.

### 3.5. Three-dimensional structures in solution

While most of the known functions of apolipoproteins are associated with their lipid-bound states, lipid-free or lipid-poor exchangeable apolipoproteins do exist in plasma and interstitial fluid, and have important metabolic roles in lipid uptake from cells, transfers between lipoproteins, structural remodeling of lipoproteins, and apolipoprotein catabolism.

Fragments of two apolipoproteins, apo E (22 kDa N-terminal fragment) and apo A1 (C-terminal fragment missing 43 N-terminal amino acid residues) have been crystallized and their structures determined by X-ray methods (Fig. 5). The apo E structure [15] (Fig. 5A) consists of an elongated (65 Å) four-helix bundle and a connecting short helix. The structure is stabilized by hydrophobic interactions, salt-bridges, and leucine-zipper

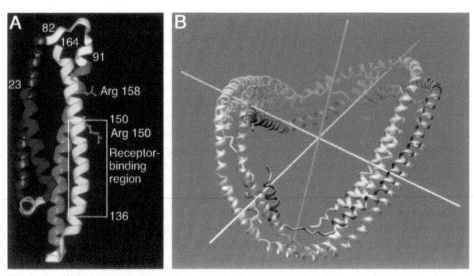

Fig. 5. (A) Ribbon X-ray structure of the apo E 22 kDa N-terminal fragment, showing the elongated 4-helix bundle of the lipid-free protein monomer [15]. The LDL receptor-binding region is boxed. (B) X-ray structure of a tetramer of the apo A-I fragment lacking the N-terminal 43 amino acids [16].

interactions. The receptor-binding region contains a cluster of basic residues on the surface of one long helix. This structure confirms the importance of amphipathic α-helices as a fundamental structural motif of apolipoproteins, even in the lipid-free state. Indeed, the four main helices contain 19, 28, 36, and 35 amino acids and encompass several of the predicted 11-mer and 22-mer repeats. The first two helices and their connecting short helix consist of the three 11-mer repeats encoded by exon 3 plus the first 22-mer repeat of exon 4, while the last two helices each consist of two 22-mer repeats. However, as important as this partial structure of apo E is for our understanding of apolipoprotein folding in solution, it does not represent accurately the configuration of the receptor-binding region because the lipid-free form of this apo E fragment binds to the LDL receptor with very poor affinity. Only in the lipidated form does apo E adopt the correct structure for high-affinity binding to the receptor.

It is hypothesized that when apo E binds to lipid surfaces the four-helix bundle opens at a hinge region to expose the hydrophobic faces of the helices to the lipid. In addition, the C-terminal 10-kDa fragment, missing from the crystal structure, is a major lipid-binding domain of apo E, but its contribution to the intact structure of apo E is not known.

A fragment of apo A1 lacking 43 amino acids from the N-terminus has been crystallized into 2 or 3 distinct conformations. The best studied one is a tetramer where each monomer polypeptide chain forms continuous amphipathic α-helices that curve into a horseshoe with a diameter of about 100 Å [16]. The monomers are organized into dimer pairs of antiparallel overlapping sequences giving a closed, distorted circular structure (Fig. 5B). This structure again illustrates the importance of amphipathic α-helices, shows that they can form long continuous structures, and suggests models for interaction with lipids; however, it does not agree with known hydrodynamic and spectroscopic properties of lipid-free apo A1 monomers under physiologic conditions. The best current information for intact monomeric apo A1 in solution indicates that it has an elongated and rather rigid structure ($125 \times 25$ Å) about two helices thick. The N-terminal Trp residues cluster near each other, and the compact and structured N- and C-terminal regions of the protein are within 35 Å of each other (A. Jonas, 2002). In fact, one of the crystallized forms of the apo A1 fragment suggests a monomeric protein in a hair-pin fold of extended helices, that is quite compatible with the proposed fold of the intact monomeric apo A1.

The three-dimensional structures of two insect apolipoproteins, apolipophorin-III from locust and Sphinx moth, have been solved by X-ray and NMR methods (M.A. Wells, 1991; R.O. Ryan, 1997). Both of these proteins bind reversibly to insect lipoprotein (lipophorin) surfaces and undergo dramatic structural changes in the process. The structures of the insect apolipophorins, similar to that of the apo E fragment, consist of elongated helix bundles of 5 amphipathic helices (Fig. 6).

It is important to note that exchangeable apolipoproteins have an extraordinary capacity for structural adaptation in response to solution conditions or to different lipid environments. In solution, high salt conditions, pH changes, or inclusion of organic solvents can lead to distinct structural states, with abnormally high α-helix contents, and specific degrees of oligomerization, as seen for the crystallized tetramer of apo A1. Under physiologic conditions, the transformation from a lipid-free to a lipid-bound state often involves a marked increase in α-helix content, and transfer of hydrophobic amino

496

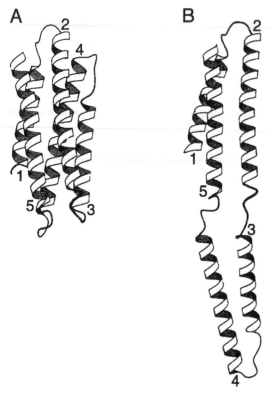

Fig. 6. Ribbon diagrams of the structure of moth apolipophorin III: (A) the helix-bundle configuration determined by X-ray analysis of the lipid-free protein; (B) proposed 'open' configuration of the apolipophorin as it binds to the surface of the insect lipoproteins. The helices are numbered 1 through 5, starting at the N-terminus (D.R. Breiter, 1991).

acid residues from a buried protein environment to a lipid environment. In lipid-bound states, changes in the content and composition of the lipid — surface phospholipids, cholesterol, or core lipids — can induce structural changes in the apolipoproteins. Also, binding of other apolipoproteins to the lipoprotein surface can affect the conformation of the resident apolipoproteins. Thus, there are potentially many distinct conformational states accessible to each exchangeable apolipoprotein. This is likely the case also for lipid-bound apo B100 and apo B48.

## 4. Complexes of apolipoproteins with lipids

### 4.1. Binding of apolipoproteins to phospholipid surfaces

Interactions of apolipoproteins with phospholipids are essential for the assembly of lipoproteins, stabilization of lipoprotein structures, and expression and modulation of apolipoprotein functions. The main experimental approaches for the study of apolipoprotein interactions with phospholipids have used isolated, exchangeable apolipoproteins in conjunction with aggregated lipids dispersed in water or spread at the air/water interface. The aggregated lipid states include lipid monolayers, various types of liposomes (small unilamellar vesicles, large unilamellar vesicles, multilamellar vesicles), and emulsions. All these lipid systems consist of or include phospholipids, especially phosphatidylcholines.

Apolipoproteins bind or adsorb readily to the phospholipid surfaces [17]. For fluid egg PC surfaces, in vesicle or emulsion form, the binding affinities ($K_d$s) for exchangeable apolipoproteins range from 0.2 to 9 $\mu$M, depending on the apolipoprotein, presence or absence of cholesterol in the surface, and curvature of the surface. Apo A4 has the lowest affinity for egg PC surfaces, in agreement with the observation that most of the apo A4 in plasma is lipid-free. The other exchangeable apolipoproteins have affinities comparable to one another, and similar calculated stoichiometries of amino acids bound per PC, averaging $0.70 \pm 0.35$. The percent of phospholipid surface occupied by the apolipoproteins, at saturation, is in the range from 8 to 12%. In the absence of cholesterol, this corresponds to the maximal compression of the surface PC, required to accommodate the amphipathic helices of the apolipoproteins. Cholesterol added to the surface initially increases the available surface area for apolipoprotein binding, but then decreases it at higher cholesterol contents (e.g., 30 mol%) [17].

The rates of interaction with phospholipid surfaces are highest for lower molecular weight apolipoproteins or peptide analogs that are monomeric and unfolded [18]. However, upon binding to phospholipid surfaces most apolipoproteins and peptides become highly $\alpha$-helical. In fact, helix formation largely accounts for the negative enthalpy of binding. At the phospholipid surface, the amphipathic helices lie with their axes parallel to the surface. The polar and charged residues of the $\alpha$-helices are at the level of the polar head groups of the phospholipids, but do not appear to interact electrostatically with them. The prevalence and importance of intra or intermolecular helix-to-helix interactions is not yet elucidated.

Desorption of exchangeable apolipoproteins from phospholipid surfaces occurs in two steps: unfolding and loss of $\alpha$-helical structure, followed by a slow desorption step whose rate depends on the length of the apolipoprotein and its affinity constant for the surface. Generally, the smaller molecular weight apolipoproteins desorb more rapidly and exchange between phospholipid surfaces more readily than larger apolipoproteins. One exception is apo A2, which has a higher affinity for PC than apo A1, and can displace apo A1 from model PC surfaces and from native lipoproteins.

As already noted, delipidated apo B100 and apo B48 are not soluble in water; therefore, their investigation has been limited to the proteins solubilized with detergent or reconstituted with selected lipids.

The properties of lipoprotein-like particles reconstituted with defined lipids will be described in the next section.

## 4.2. Lipoprotein-like complexes

Although phospholipid liposomes are favored systems for the study of apolipoprotein binding to phospholipid surfaces, vesicle–apolipoprotein complexes are not ideal models for lipoproteins. Vesicles have an interior water compartment not present in lipoproteins, are incapable of solubilizing large amounts of neutral lipids within the phospholipid bilayer, and are too large to mimic the surface curvature of HDL. Thus, methods have been developed to prepare small, micellar complexes of exchangeable apolipoproteins (in particular apo A1) with lipids that mimic discoidal and spherical HDL in shape, composition, and functional properties. For LDL and VLDL, microemulsions and emulsions of lipids of selected diameter and composition, with added apo B100, make good models of the native lipoproteins.

Several methods are known for the reconstitution of HDL-like complexes from pure components: (1) spontaneous formation of HDL discs from dimyristoylphosphatidylcholine liposomes; (2) detergent mediated reconstitution of HDL discs with various phospholipids; and (3) co-sonication of apolipoproteins and lipids to form either discoidal or spherical HDL analogs [19].

Dimyristoyl-PC liposomes can bind apolipoproteins reversibly, as described in the preceding section; however, at the transition temperature ($T_m$) of the lipid (24°C) and at sufficiently high proportions of apolipoprotein to dimyristoyl-PC (1/3 or greater, wt/wt), the apolipoproteins can lyse the liposomes to give rise to small discs analogous to nascent HDL. The rate of the liposome disruption and solubilization depends on the temperature of the reaction. It is highest at the onset of the main phase transition of dimyristoyl-PC and decreases a thousand-fold on either side of $T_m$. While the same reaction does occur with dipalmitoyl-PC liposomes at the $T_m$ of 41°C, the rate is much slower than for dimyristoyl-PC. For long-chain, unsaturated PC, such as palmitoyloleoyl-PC, the rates of reaction are too slow to be measured at accessible temperatures. Therefore, for practical purposes, under ordinary conditions, only dimyristoyl-PC can be used effectively to reconstitute HDL discs by this method. In this system, smaller apolipoproteins and peptide analogs disrupt dimyristoyl-PC liposomes at higher rates than larger apolipoproteins. Thus under some circumstances synthetic peptide mimics of A-type amphipathic helices may lyse even egg PC or palmitoyloleoyl-PC vesicles.

The mechanism of the disruption of dimyristoyl-PC liposomes into discoidal reconstituted HDL particles requires binding of the apolipoprotein to the liposome surface, followed by protein–protein interactions that lead to penetration of the bilayer by apolipoprotein, and breakdown of the liposome into bilayer discs. The discs are surrounded and stabilized on the periphery by the amphipathic helices of the apolipoproteins [20].

A more universal method for producing discoidal reconstituted HDL particles uses detergent (usually Na cholate) to solubilize the phospholipid into mixed micelles, followed by addition of exchangeable apolipoprotein and removal of detergent by dialysis or column chromatography. Depending on the proportions of PC to apolipoprotein, in the range from 1/1 to 4/1 (wt/wt), reconstituted HDL discs of different diameters containing different numbers of apolipoprotein molecules with varied conformations can be produced. Cholesterol (up to 15 mol%) or small amounts of other lipids can be readily incorporated during the lipid solubilization step.

The third method for making HDL-like particles is extensive co-sonication of apolipoproteins with phospholipids, in the absence or presence of neutral lipids. Usually, the components are mixed in the proportions found in native HDL and yield discoidal or spherical analogs of HDL depending on the absence or presence of neutral lipids. While the ability to synthesize spherical reconstituted HDL particles by this method is very attractive, controlling particle homogeneity and yield is difficult.

To produce models of LDL, VLDL, and CM, lipid mixtures of PC and a neutral lipid are first sonicated to give metastable emulsion particles of the general size of the desired lipoproteins, and then the apo B component is added from a detergent dispersion, or exchangeable apolipoproteins are added in solution [17].

The advantages of reconstituted lipoproteins over the native lipoproteins are that the model particles can be made with a single apolipoprotein and one or a few defined lipid components to study each component individually. In addition, particles of uniform size can be isolated where high-resolution structural analysis is potentially possible. Also the reconstituted lipoproteins lend themselves for the study of structure–function relationships. In fact, reconstituted lipoproteins display all the known functions ascribed to native lipoproteins, including enzyme activation, receptor binding, and uptake and transfer of lipids.

## 4.3. Reconstituted HDL

The best studied reconstituted HDL are the discoidal particles containing apo A1 and a single type of PC, with or without added small amounts (<20 mol%) of other phospholipids or cholesterol [20]. The discoidal shape of the particles has been confirmed by negative-stain electron microscopy, small-angle X-ray scattering, and atomic force microscopy methods. All methods indicate a disc thickness of 45–55 Å, that corresponds to the phospholipid bilayer thickness, and diameters ranging from 70 to 180 Å. The main phase transition of the PC is preserved in the particles, but $T_m$ is shifted about 3°C to higher temperatures, indicating a greater ordering and restriction of the PC in the particles than in corresponding liposomes.

Analyzed by non-denaturing gradient gel electrophoresis the particles display discrete diameters and size distributions that vary depending on the initial PC/apo A1 ratios of the preparations. Isolated particles of a specific size have reproducible, characteristic physical and functional properties. For example, some particles containing palmitoyloleoyl-PC (78 and 109 Å) are poor substrates for LCAT while others, particularly the 96 Å particles, are the best known LCAT substrates (A. Jonas, 1989). In addition, there is evidence of different binding affinities of the particles for SR-BI receptors (D.R. van der Westhuyzen, 2001). Thus reconstituted HDL subclasses, with different lipid contents, have different apolipoprotein conformations that result in dramatically altered functional properties. In vivo, this conformational adaptability of apolipoproteins probably leads to metabolic switching for the diverse functions of HDL subclasses. Apo A1 conformation is also regulated by the saturation or unsaturation of the PC acyl chains and by high contents of cholesterol (>15 mol %) or polyun-

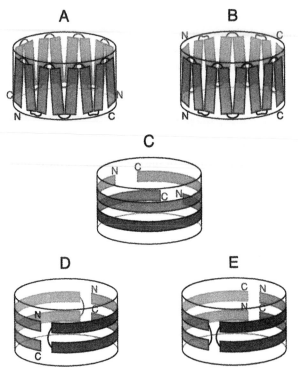

Fig. 7. Diagrams of the proposed arrangement of two apo A1 molecules (residues 44–243) on the periphery of a discoidal reconstituted HDL particle (M.A. Tricerri, 2001). N, denotes the N-terminus and C, the C-terminus of each protein molecule. (A) 'Picket-fence' model in head-to-tail arrangement. (B) 'Picket-fence' in a head-to-head arrangement. (C) 'Belt' model. (D) 'Hair-pin' fold in a head-to-tail arrangement. (E) 'Hair-pin' fold in a head-to-head arrangement.

saturated PC. Other exchangeable apolipoproteins (apo A2, apo E, apo A4) also form discoidal reconstituted HDL with diameters that roughly correspond to their contents of amphipathic α-helices and molecular weights.

Reconstituted spherical HDL can be made by co-sonication of selected HDL components (e.g., apo A1, PC, cholesteryl ester) or by extensively reacting discoidal reconstituted HDL with LCAT in the presence of an exogenous source of cholesterol. The products are spheroidal and contain 2, 3, or 4 apo A1 molecules per particle, PC, cholesterol, and a cholesteryl ester core. The diameters of the particles range from 80 to 120 Å. Although not well studied, the conformation of apo A1 appears distinct from that in the discoidal particles and variable depending on the particle diameter (A. Jonas, 1990).

In reconstituted discoidal HDL, the apolipoproteins form a protective shell, one helix thick, around the periphery of the discs. According to current experimental evidence and models [21] (Segrest, 1999; Jonas, 2001), the helix axes are predominantly perpendicular to the acyl chains of the PC bilayer (see Fig. 7C,D). Earlier models, however, oriented the helical repeats almost parallel to the acyl chains, with the antiparallel helical segments joined by tight β-turns (Fig. 7A,B) (Rosseneu, 1990; Schulten, 1997). The spatial organization of the amphipathic helices in spherical reconstituted HDL is not known and has not been modeled.

## 4.4. Structures of native lipoproteins

Electron microscopic images of lipoproteins show predominantly spherical shapes. Only nascent HDL appear as stacks of discs by negative-stain transmission electron microscopy. The stacks are artifacts of the method, since in solution nascent HDL and their reconstituted HDL analogs are free-standing discs. Regarding the structure of native HDL, their electron microscopic images do not have enough resolution to show any significant surface features, and other methods have not yet been successful in providing dependable structural information. The computer models generated for the reconstituted HDL discs are probably applicable for the larger nascent HDL particles, but no credible computer models exist for the spherical HDL. Based on the dimensions of HDL$_3$ particles, and the molecular volumes and surface areas of its component phospholipids, cholesterol, cholesteryl esters and amino acids, a space-filling model was constructed by Edelstein et al. in 1979 [22]. Their model is still valid today, with the exception that more apolipoprotein helix contacts are probably present, and the space under the helices is probably filled with cholesteryl ester (or triacylglycerol) molecules rather than with unesterified cholesterol, as shown in Fig. 8.

The N-terminal domain of apo B may have structural homology with lipovitellin. Lipovitellin is a soluble oocyte lipoprotein that contains 16% by weight of lipid and has sequence homology with the N-terminal 670 residues of apo B and with the microsomal lipid transfer protein. The X-ray structure [23] of lipovitellin contains a funnel-shaped cavity formed by two β-sheets lined with hydrophobic residues that are presumed to bind

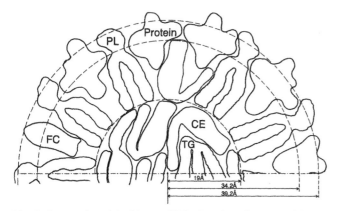

Fig. 8. Structural model of human HDL$_3$ based on the composition and dimensions of the particles, and the molecular volumes and surface areas of the individual components. PL, phospholipid; FC, unesterified cholesterol; CE, cholesteryl ester; TG, triacylglycerol. Modified form of the model presented in Edelstein et al. [22] showing a cholesteryl ester molecule in the space under the apolipoprotein helices, and unesterified cholesterol (FC) in the surface monolayer.

502

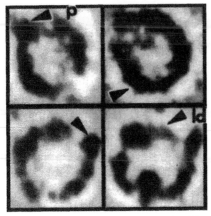

Fig. 9. Individual images of LDL obtained by cryoelectron microscopy [22]. The box sizes are 270 × 270 Å. The unlabeled arrow indicates a region of high density that corresponds to a domain of apo B100; arrow with a 'p' indicates the pointed, N-terminal domain of apo B100; and an arrow with 'ld' points to a region of low density in apo B100.

phospholipids. A similar structure may be present in apo B and in the microsomal lipid transfer protein. Higher-resolution cryoelectron microscopy and negative-stain electron microscopy combined with monoclonal antibody tagging have revealed some structural features of apo B100 on LDL particles of relatively homogeneous sizes [24]. Images of LDL appear slightly ovoid and with average diameters of 233 Å. Individual LDL images show four or five high-density regions that correspond to the single molecule of apo B100. The dense regions are joined by lower-density connections. Although the entire protein appears in many images as a ring on the surface of the LDL (Fig. 9), the actual distribution of the protein domains is probably more dispersed and complex. The N-terminal domain of apo B100 can be identified as a high-density pointed end that protrudes more from the particle surface than any of the other domains. Some of the putative domains have approximately circular projections but the largest one, the one corresponding to residues 1700–3070, is elongated and is located at 50–140° latitude from the N-terminus.

## 5. Future directions

In spite of the great advances made during the last two decades in the elucidation of the structures and structure–function relationships of lipoproteins, much remains to be accomplished in this field of research.

Key areas of future research:
(1) Structural and functional studies of the roles of minor lipid and apolipoprotein components of lipoproteins, including the products of oxidative reactions that occur in vivo and in vitro.
(2) Investigation of the conformational adaptability of apolipoproteins during assembly

with lipids and during metabolic transformations of lipoproteins in circulation. Elucidation of the conformational changes at the atomic level will be a major challenge dependent on the success of high-resolution analysis of apolipoprotein and lipoprotein structures.

(3) High-resolution determination of the three-dimensional structure of intact apolipoproteins and lipoproteins, under physiologic conditions. Accomplishing this goal will require not only major advances in the crystallization, X-ray, and NMR methodologies, but also the isolation or preparation of highly homogeneous reconstituted or native lipoproteins.

(4) Further developments in computer modeling of reconstituted or native lipoproteins will require powerful molecular dynamics algorithms, and much more extensive experimental information on the folding and topology of the apolipoproteins. Especially promising are mass spectrometric methods combined with specific or random chemical cross-linking of apolipoproteins to establish spatial proximity.

## Abbreviations

| | |
|---|---|
| apo | apolipoprotein |
| CM | chylomicrons |
| HDL | high-density lipoproteins |
| $K_d$ | equilibrium dissociation constant |
| LCAT | lecithin cholesterol acyltransferase |
| LDL | low-density lipoproteins |
| Lp(a) | lipoprotein(a) |
| PC | phosphatidylcholine |
| PL | phospholipid |
| SR-BI receptor | scavenger receptor BI |
| $T_m$ | main phase transition temperature |
| VLDL | very low-density lipoproteins |

## References

1. Barklay, M. (1972) Lipoprotein class distribution in normal and disease states. In: G.J. Nelson (Ed.), Blood Lipids and Lipoproteins: Quantitation, Composition, and Metabolism. Wiley-Interscience, New York, pp. 587–603.
2. Blanche, P.J., Gong, E.L., Forte, T.M. and Nichols, A.V. (1981) Characterization of human high-density lipoproteins by gradient gel electrophoresis. Biochim. Biophys. Acta 665, 408–419.
3. Cheung, M.C. and Albers, J.J. (1984) Characterization of lipoprotein particles isolated by immunoaffinity chromatography. J. Biol. Chem. 259, 12201–12209.
4. Fielding, C.J. and Fielding, P.E. (1995) Molecular physiology of reverse cholesterol transport. J. Lipid Res. 36, 211–228.
5. Shen, M.M.S., Krauss, R.M., Lindgren, F.T. and Forte, T.M. (1981) Heterogeneity of serum low density lipoproteins in normal human subjects. J. Lipid Res. 22, 236–244.
6. Skipski, V.P. (1972) Lipid composition of lipoproteins in normal and diseased states. In: G.J. Nelson

504

(Ed.), Blood Lipids and Lipoproteins: Quantitation Composition, and Metabolism. Wiley-Interscience, New York, pp. 471–583.

7. Miller, K.W. and Small, D.M. (1983) Surface-to-core and interparticle equilibrium distribution of triglyceride-rich lipoprotein lipids. J. Biol. Chem. 258, 13772–13784.

8. Pregetter, M., Prassl, R., Schuster, B., Kriechbaum, M., Nigon, F., Chapman, J. and Laggner, P. (1999) Microphase separation in low density lipoproteins. Evidence for a fluid triglyceride core below the lipid melting transition. J. Biol. Chem. 274, 1334–1341.

9. Pownall, H.J. and Gotto, A.M. Jr. (1992) Human plasma apolipoproteins in biology and medicine. In: M. Rosseneu (Ed.), Structure and Function of Apolipoproteins. CRC Press, Boca Raton, FL, pp. 1–32.

10. Li, W.-H., Tanimura, M., Luo, C.-C., Datta, S. and Chan, L. (1988) The apolipoprotein multigene family: biosynthesis, structure, structure–function relationships, and evolution. J. Lipid Res. 29, 245–271.

11. Luo, C.-C., Li, W.-H., Moore, M.N. and Chan, L. (1986) Structure and evolution of the apolipoprotein multigene family. J. Mol. Biol. 187, 325–340.

12. Yang, C.-Y., Chen, S.-H., Gianturco, S.H., Bradley, W.A., Sparrow, J.T., Tanimura, M., Li, W.-H., Sparrow, D.A., DeLoof, H., Rosseneu, M., Lee, F.-S., Gu, Z.W., Gotto Jr., A.M. and Chan, L. (1986) Sequence, structure, receptor-binding domains and internal repeats of human apolipoprotein B-100. Nature 323, 738–742.

13. Segrest, J.P., Jones, M.K., DeLoof, H., Brouillette, C.G., Venkatachalapathi, Y.V. and Anantharamaiah, G.M. (1992) The amphipathic helix in the exchangeable apolipoproteins: a review of secondary structure and function. J. Lipid Res. 33, 141–166.

14. Segrest, J.P., Jones, M.K., Mishra, V.K., Anantharamaiah, G.M. and Garber, D.W. (1994) ApoB-100 has a pentapartite structure composed of three amphipathic α-helical domains alternating with two amphipathic β-strand domains. Arterioscler. Thromb. 14, 1674–1685.

15. Wilson, C., Wardell, M.R., Weisgraber, K.H., Mahley, R.W. and Agard, D.A. (1991) Three-dimensional structure of the LDL receptor-binding domain of human apolipoprotein E. Science 252, 1817–1822.

16. Borhani, D.W., Rogers, D.P., Engler, J.A. and Brouillette, C.G. (1997) Crystal structure of truncated human apolipoprotein A-I suggests a lipid-bound conformation. Proc. Natl. Acad. Sci. USA 94, 12291–12296.

17. Atkinson, D. and Small, D.M. (1986) Recombinant lipoproteins: implications for structure and assembly of native lipoproteins. Annu. Rev. Biophys. Biophys. Chem. 15, 403–456.

18. Pownall, H.J., Massey, J.B., Sparrow, J.T. and Gotto, A.M. Jr. (1987) Lipid–protein interactions and lipoprotein reassembly. In: A.M. Gotto Jr. (Ed.), Plasma Lipoproteins. Elsevier, Amsterdam, pp. 95–127.

19. Jonas, A. (1986) Reconstitution of high-density lipoproteins. Methods Enzymol. 128, 553–582.

20. Jonas, A. (1992) Lipid-binding properties of apolipoproteins. In: M. Rosseneu (Ed.), Structure and Function of Apolipoproteins. CRC Press, Boca Raton, FL, pp. 217–250.

21. Brouillette, C.G., Anantharamaiah, G.M., Engler, J.A. and Borhani, D.W. (2001) Structural models of human apolipoprotein A-I: a critical analysis and review. Biochim. Biophys. Acta 1531, 4–46.

22. Edelstein, C., Kézdy, F.J., Scanu, A.M. and Shen, B.W. (1979) Apolipoproteins and the structural organization of plasma lipoproteins: human plasma high density lipoprotein-3. J. Lipid Res. 20, 143–153.

23. Anderson, T.A., Levitt, D.G. and Banaszak, L.J. (1998) The structural basis of lipid interactions in lipovitellin, a soluble lipoprotein. Structure 6, 895–909.

24. Spin, J. and Atkinson, D. (1995) Cryoelectron microscopy of low density lipoprotein in vitreous ice. Biophys. J. 68, 2115–2123.

D.E. Vance and J.E. Vance (Eds.) *Biochemistry of Lipids, Lipoproteins and Membranes (4th Edn.)*

# CHAPTER 19

# Assembly and secretion of lipoproteins

Jean E. Vance

*CIHR Group on Molecular and Cell Biology of Lipids and Department of Medicine,*
*328 Heritage Medical Research Centre, University of Alberta, Edmonton, AB T6G 2S2, Canada,*
*Tel.: +1 (780) 492-7250; Fax: +1 (780) 492-3383; E-mail: jean.vance@ualberta.ca*

## 1. Overview of lipoprotein secretion into the circulation

Lipoproteins are secreted into the circulation from hepatocytes of the liver and enterocytes of the intestine. All plasma lipoproteins share a common structure consisting of a neutral lipid core of triacylglycerols (TGs) and cholesteryl esters surrounded by a surface monolayer of phospholipids, unesterified cholesterol and specific proteins called apolipoproteins (Fig. 1 and Chapter 18). The primary function of plasma lipoproteins is to transport hydrophobic, water-insoluble lipids in the circulation. The TG-rich lipoproteins deliver TGs made in the liver and intestine to other tissues in the body for either storage or utilization as an energy source. High-density lipoproteins are thought to remove cholesterol from peripheral tissues (i.e. tissues other than liver) and deliver it to the liver for excretion in bile in a process referred to as 'reverse cholesterol transport' (Chapters 16, 20). A high level of plasma lipids is a strong risk factor for development of cardiovascular disease (Chapter 22). The steady state level of lipoproteins in the cir-

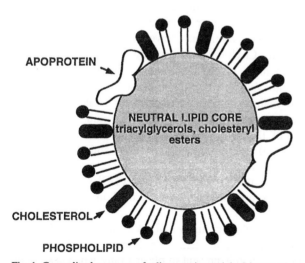

Fig. 1. Generalized structure of a lipoprotein particle. Lipoproteins are approximately spherical particles that consist of a neutral lipid core (triacylglycerols and cholesteryl esters) surrounded by a surface monolayer of amphipathic lipids (unesterified cholesterol and phospholipids) and specific apoproteins. Courtesy of Dr. J. Segrest, University of Alabama, with permission.

506

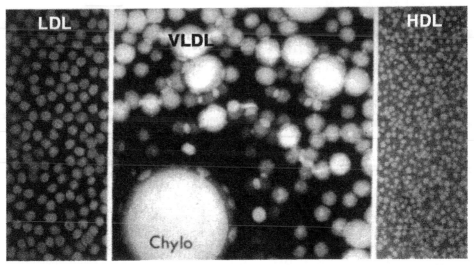

Fig. 2. Negative staining electron micrographs of human plasma lipoproteins (diameters 10–1000 nm). The largest particles [chylomicrons (Chylo) and VLDLs] contain a higher ratio of lipid to protein, and are therefore less dense, than low-density lipoproteins (LDLs) and high-density lipoproteins (HDLs) which contain relatively more protein. Photograph courtesy of Dr. R. Hamilton, University of California at San Francisco, with permission.

culation results from a balance between their rates of secretion into, and removal from, plasma. This chapter focuses on the mechanisms by which lipoproteins are assembled and secreted from the liver and intestine. The synthesis of the constituent lipids of the lipoproteins is described in Chapters 8 (phospholipids), 15 (cholesterol) and 10 (TGs).

The major plasma lipoproteins (Fig. 2) are usually classified according to density (table 1 in Chapter 18). Since lipids have lower buoyant densities than proteins, lipoproteins that have a high ratio of lipid to protein have lower density than lipoproteins having a lower ratio of lipid to protein. Chylomicrons, the largest and most lipid-rich particles, whose major lipid component is TG, are secreted by the intestine and are abundant in plasma only after a meal. Very low-density lipoproteins (VLDLs), which are also rich in TG, are secreted mainly by the liver, although some are also of intestinal origin. The size and lipid composition of chylomicrons and VLDLs vary according to the nutritional status of the animal. This chapter discusses how lipids are assembled with apolipoprotein (apo) B and secreted as VLDLs and chylomicrons. Intermediate-density lipoproteins and low-density lipoproteins (LDLs) are generated in the circulation by lipolysis of TGs within chylomicrons and VLDLs. High-density lipoproteins are particles of diverse composition that are formed in the circulation (Chapter 20).

Apo B is an essential component of chylomicrons, VLDLs and low-density lipoproteins. Unlike the 'exchangeable' plasma apolipoproteins (e.g. apos E, A1, A2, C), apo B does not exchange among lipoproteins and is present in plasma only when associated with lipids. In addition to apo B, VLDLs and chylomicrons contain apo E and apos C, and chylomicrons also contain some apos A1 and A2.

In many respects, the secretion of apo B from cells follows a pathway identical to that of typical secretory proteins. The mRNA encoding apo B is translated on ribosomes bound to the endoplasmic reticulum (ER) and the protein is translocated across the ER membrane into the lumen. A unique difference between apo B and other secretory proteins is that apo B must be non-covalently associated with lipids before secretion. Newly formed apo B-containing lipoproteins undergo vesicular transport from the ER through the Golgi stacks. Secretory vesicles, containing lipoproteins, are formed in the *trans*-Golgi network and subsequently fuse with the plasma membrane releasing apo B-containing lipoproteins from the cell. Evidence that apo B transits the same secretory route as other secretory proteins was provided by electron microscopy studies of rat liver in which lipid droplets the size of VLDL were detected in the lumen of the ER and Golgi [1]. In cultured rat hepatocytes, synthesis of an apo B molecule takes 7 to 15 min and apo B appears in the culture medium 30 to 40 min later. Pulse-labeling experiments indicate that the slow step in passage of apo B through the secretory pathway is movement of the protein out of the ER (R.A. Davis, 1987).

## 2. Structural features of apolipoprotein B

Full-length human apo B, designated as apo B100, is one of the largest single polypeptide chains known with 4536 amino acids ($M_r$ 513,000) [2]. In humans, the liver secretes exclusively apo B100 whereas the intestine secretes primarily a truncated form of apo B100, called apo B48 ($M_r$ 264,000), along with some apo B100 (J.P. Kane, 1983). Rodent livers, in contrast, secrete both apo B100 and apo B48. A convenient 'centile' nomenclature for apo B variants is commonly used in which apo B species are designated by a number indicating the size of the molecule relative to full-length apo B100. Therefore, in this shorthand scheme, apo B48 refers to the N-terminal 48% of apo B100 and apo B15 refers to the N-terminal 15% of apo B100.

Apo B100 is an amphipathic protein. Some of the complex molecular interactions required for assembly of this protein with lipids can be inferred from its unique domain structure. Indeed, apo B is the only protein known to require association with a lipid microemulsion for secretion. Each VLDL particle contains a single molecule of apo B (J. Elovson, 1988). Computer-based structural analysis of apo B100 predicts a pentapartite structure (Fig. 3) (R. Nolte, 1994) [3] consisting of a globular N-terminal domain followed by alternating amphipathic β-sheets and α-helices. As much as 40–50% of apo B100 is β-sheet structure, which is concentrated in two large regions. At least 25% of the molecule consists of α-helices which are too short to span a membrane. It is likely that the surface of a VLDL particle does not contain sufficient phospholipid molecules to completely cover the surface and, consequently, apo B is probably in direct contact with some TG in the particle core. The β-strands are thought to be the major motif mediating association with lipids. Since the amphipathic β-sheets of apo B have both polar and non-polar faces, the prediction is that the non-polar face interacts with neutral lipids, such as TGs, whereas the polar face is exposed to the aqueous environment.

A general correlation exists between the length of the apo B molecule and its ability

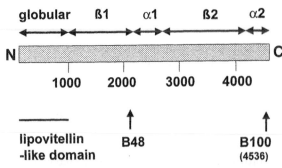

Fig. 3. Schematic representation of the pentapartite structure of apo B100. The N-terminal globular domain and the alternating α and β domains are indicated above the representation of the polypeptide. Numerical values below the polypeptide indicate an amino acid residue scale. Human apo B100 contains 4536 amino acids; the size of apo B48 is indicated. The N-terminal 1000 amino acid residues of apo B100 are homologous with the lipid-binding pocket of lamprey lipovitellin.

to associate with lipids. Naturally occurring mutations in the human apo B gene, in the group of diseases known as hypobetalipoproteinemia, show that secretion of some truncated forms of apo B as lipoprotein particles is severely impaired [4] (S.G. Young, 1993). In cultured rat hepatoma cells transfected with cDNAs encoding C-terminally truncated variants of human apo B100 (from apo B15 to B94) the size and density of the secreted apo B-containing particles are dependent upon the length of the apo B. For example, apo B18 is secreted essentially without neutral lipids whereas apo B28 is secreted in particles the size and density of high-density lipoproteins. As the size of the apo B molecule increases, from B37 to B48 to B53 to B72 to B94, the apo B associates with progressively more lipid in larger and less dense particles [5]. Thus, the sequence of apo B between apo B17 and B41 is necessary for assembly of TG-rich lipoproteins. Sequences in the C-terminus of apo B29 bind phospholipids and diacylglycerols but only a little TG. TGs, however, associate with the region between apo B29 and B32.5, and a massive accumulation of TGs occurs between B32.5 and B41 where ~1 molecule of TG binds for each 2 amino acids of apo B (D.M. Small, 2000).

The four alternating α-helices and β-sheets (Fig. 3) are common features of apo B100 in all vertebrate species examined. Amphipathic β-strands are also found in the egg yolk protein, lipovitellin. The X-ray crystal structure of lamprey lipovitellin reveals a 'lipid pocket' joined by three anti-parallel β-sheets. Based on the prediction that human apo B100 contains a similar lipid-binding pocket at the N-terminal domain of the molecule, a model has been proposed in which this pocket provides the initial site of association of TGs with apo B. Sequential accumulation of TGs within this cavity, in concert with expansion of the pocket, is thought to occur by unfolding of the amphipathic β-strands of the β1-domain (J. Segrest, 1999). Not explained by these studies or models, however, is the reason why apo B48 assembles into chylomicrons, the largest and most TG-rich particles, whereas apo B100, which is approximately twice as large as apo B48, forms the smaller VLDLs.

# 3. Transcriptional regulation of apo B synthesis

## 3.1. DNA elements regulating apo B transcription

Under most metabolic conditions, changes in apo B mRNA levels do not influence how much apo B is secreted. Moreover, changes in apo B secretion, over at least a 7-fold range, occur without alteration in the amount of apo B mRNA (J. Scott, 1989). Consequently, it is generally accepted that apo B secretion is primarily regulated co- and post-translationally.

The liver and intestine produce the majority of apo B, but in mice the production of apo B in the yolk sac is crucial during fetal development. It is thought that apo B plays a role in transferring lipid nutrients to the developing mouse embryo (R.V. Farese, 1999).

Interestingly, apo B is also expressed at low levels in the hearts of humans and mice (S.G. Young, 1998). The presence of apo B and microsomal triacylglycerol transfer protein (which is also required for lipoprotein assembly, Section 6.4) in the heart presumably leads to the secretion of apo B100-containing lipoproteins by cardiac myocytes. Recent experiments suggest that secretion of TG-rich lipoproteins by the heart is a mechanism for regulating the amount of TGs stored in the heart and thereby protects the heart from a detrimental accumulation of lipids [6].

The human apo B gene spans 43 kb, is located on chromosome 2p and consists of 28 introns and 29 exons. Over half of the apo B coding sequence is encoded by one exceptionally large exon, exon 26, which contains 7572 bp. The apo B mRNA transcript is 14 kb and is remarkably stable, having a half-life of 16 h (J. Scott, 1989). Some elements responsible for expression of the human apo B gene have been defined. The gene promoter (bp −898 to +1) contains a TATA box. Within the promoter region are sequences that bind the liver-specific transcription factors HNF-3 and NF-1. Positive and negative cis-acting DNA sequences required for transcription of apo B in the liver were identified using transient transfection assays of promoter–reporter constructs in cultured human hepatoma (HepG2) cells and intestinal (CaCo2) cells. The majority of enhancer activity was localized to 443 bp within intron 2 [7]. By footprint analysis this enhancer region was shown to bind two nuclear transcription factors, HNF-1 and C/EBP. An additional, weaker, enhancer was found within intron 3, and a reducer element was identified in the 5′-flanking region between bp −3211 and −1802. However, when these promoter elements were linked to reporter genes and expressed in transgenic mice, a more complex picture emerged. In mice, both the proximal promoter (bp −898 to +1) and intron 2 enhancer were required for expression of a β-galactosidase reporter in liver, yet surprisingly no intestinal expression occurred even though the same constructs induced expression in the intestinal cell line, CaCo2. Moreover, the reducer element (bp −3200 to −1802) that was identified in transient transfection studies in HepG2 cells, did not decrease expression in livers of transgenic mice. Consequently, while CaCo2 and HepG2 cells are frequently used as experimental models of the intestine and liver, respectively, it is important to verify in intact animals results obtained from cultured cells.

Recently, factors controlling the intestinal expression of apo B have been described.

A high level of intestinal expression of human apo B in transgenic mice was achieved with a construct containing ~80 kb of the 5′-flanking sequence [8]. The region of the gene driving expression of apo B transgenes in mouse intestine was subsequently localized to a 3 kb segment over 55 kb upstream of the transcriptional start site [9,10]. This region contains putative binding sites for the transcription factors HNF-3β, HNF-4 and C/EBPβ; the corresponding segment of the murine apo B gene exhibits a high degree of sequence conservation. Interestingly, elements of the apo B gene that had been previously demonstrated in vivo to confer expression in the liver (i.e. intron 2 enhancer and 5′ upstream liver enhancer) do not play a role in intestinal expression of apo B transgenes in mice.

## 3.2. Apo B mRNA editing

Apo B100 contains 4536 amino acid residues and in humans is the only form of apo B produced by the liver. In contrast, the small intestine of all mammals, and the liver of some species, synthesize apo B48, the N-terminal 2152 residues of apo B100. Apo B100 and apo B48 are generated from a single structural gene. Subsequently, apo B48 is produced by an unusual RNA editing process in which a single cytidine at position 6666 of human apo B mRNA is deaminated to a uridine [11]. A stop codon is thereby introduced and the truncated variant, apo B48, is generated. The cis-acting elements directing this site-specific deamination are located within approximately 25 nucleotides in an AU-rich region flanking the target cytidine. A region of 11 nucleotides 5 bases downstream of this cytidine, is particularly important for editing and has been called the 'mooring sequence'.

A 27 kDa protein from enterocytes is required for apo B RNA editing [12]. This protein, called apobec-1 (for 'apo B editing complex-1'), is highly homologous to other cytidine deaminases but does not, by itself, edit the RNA. Rather, apobec-1 is the catalytic subunit of a multi-protein complex containing auxiliary proteins. Apobec-1 is a homodimer that is present in both the cytosol and nucleus and apo B48 is formed only in tissues that express apobec-1 mRNA (i.e. human and rodent intestine and rodent liver). However, apobec-1 is also expressed in some tissues that do not make apo B, suggesting that this enzyme also acts on other mRNAs in addition to apo B mRNA. This idea is supported by the finding that transgenic mice over-expressing apobec-1 develop hepatocellular carcinomas and liver dysplasia (T.L. Innerarity, 1995).

Apo B48-containing lipoproteins are removed from plasma more rapidly than apo B100-containing lipoproteins since apo E, which is added to apo B48-containing lipoproteins after secretion, is a high-affinity ligand for the LDL receptor (Chapter 21). It was reasoned, therefore, that if apo B100 secretion were blocked, by inducing the complete editing of apo B mRNA so that only apo B48 was produced, fewer circulating apo B-containing lipoproteins would be present and, consequently, the animals would be less susceptible than normal to diet-induced hypercholesterolemia. As predicted, when the apobec-1 gene was over-expressed in livers of atherosclerosis-prone mice via adenovirus-mediated gene transfer, plasma apo B100-containing lipoproteins and cholesterol were greatly decreased and the extent of atherosclerosis was reduced (B. Teng, 1994).

On the other hand, targeted disruption of the murine apobec-1 gene abolished apo B editing in all tissues and resulted in a complete deficiency of apo B48 in serum (N.O. Davidson, 1996) demonstrating that there is no functional duplication of the apobec-1 gene. Apobec-1$^{-/-}$ mice are fertile, healthy, and serum levels of cholesterol and TG are normal showing that the apobec-1 gene is not essential for normal life. Moreover, mice that express only apo B48 (by replacement of the apo B48-editing codon with a stop codon) show no obvious defects in growth, reproduction or function (S.G. Young, 1996). A question that remains is: why is apo B100 produced at all?

## 4. Models used for studying apo B and VLDL secretion

The most commonly used models for studying the assembly and secretion of apo B-containing lipoproteins are hepatoma cell lines (HepG2 human hepatoma cells and McArdle 7777 rat hepatoma cells) and primary hepatocytes from rats, mice or hamsters. Each model has its limitations. In addition, several transgenic and gene-targeted mouse models have been developed.

Hepatoma cell lines are convenient laboratory models since they are easily maintained and manipulated under defined conditions. However, although hepatoma cells have retained many properties of hepatocytes, several key liver functions, including some important aspects of lipid metabolism, have been altered. One advantage of HepG2 cells is that they are of human origin and, like human liver, secrete apo B100 but not apo B48. However, the basal rate of apo B secretion in HepG2 cells is low, and the apo B-containing lipoproteins are not true VLDLs but contain less lipid than VLDLs. Thus, the use of HepG2 cells for studying VLDL assembly is not ideal. McArdle 7777 hepatoma cells are of rat origin and, like rat liver, secrete both apo B100 and B48. When incubated with oleate some of the secreted apo B is in VLDLs, but the majority of apo B is secreted in more dense, partially lipidated, lipoproteins. One advantage of McArdle cells is that they are excellent cells in which to express heterologous cDNAs, such as those encoding truncated and mutated forms of apo B.

Primary hepatocytes isolated from livers of rats, mice and hamsters are also commonly used for studying VLDL assembly and secretion. Although these cells must be freshly isolated for each experiment, they retain most properties of native hepatocytes, at least for 24 h. Unlike human hepatocytes, primary rat and mouse hepatocytes secrete both apo B48 and apo B100 but these particles are the size, composition and density of VLDLs. Hamster primary hepatocytes are also useful since they secrete VLDLs that contain apo B100, but not apo B48 (K. Adeli, 2000). However, stable transfection of primary hepatocytes with cDNAs is not possible.

Several gene-targeted mice have been effectively used to study apo B metabolism [4]. Expression of human apo B in mice fed a high-fat diet resulted in a 20-fold increase in the plasma content of apo B-containing lipoproteins and extensive atherosclerosis. Since the secretion of apo B from hepatocytes of these mice was only slightly higher than in wild-type mice the lipoprotein accumulation and atherosclerosis were attributed primarily to defective lipoprotein clearance. Another line of transgenic mice was produced in which Arg-3500 of apo B (the proposed site of binding to the LDL receptor) was mutated. These mice were unable to clear apo B-containing lipoproteins from the plasma and developed severe hypercholesterolemia.

Generation of apo B knock-out mice was problematic since most apo $B^{-/-}$ embryos died during development. Nevertheless, mice lacking the murine apo B gene, but expressing human apo B, were generated. During the suckling period the offspring were indistinguishable from wild-type mice but because the human apo B gene was not expressed in the intestine (Section 3.1) the mice did not produce chylomicrons and suffered severe fat malabsorption and retarded growth [8].

The disease hypobetalipoproteinemia is caused by point mutations in the apo B gene that result in production of low levels of poorly lipidated, truncated apos B. Several mouse models of this disorder have been generated. For example, transgenic mice expressing apos B70, B81 or B83 have low levels of plasma apo B-containing lipoproteins and develop severe neurodevelopmental abnormalities [4].

## 5. Covalent modifications of apo B

Several co- and post-translational modifications of apo B occur. Human apo B100 contains 25 cysteine residues of which 16 exist as intramolecular disulfide bonds; 14 of these are clustered in the N-terminal region. Disulfide bond formation likely occurs co-translationally by the action of protein disulfide isomerase, a lumenal ER protein that coordinates disulfide bond formation and protein folding. As discussed in Section 6.4, protein disulfide isomerase is one subunit of the microsomal TG transfer protein heterodimer that is required for efficient secretion of apo B-containing lipoproteins. To determine if disulfide bond formation in the N-terminus of apo B were required for assembly of lipid with apo B, individual cysteine residues were replaced with serines or alanines. When either or two cysteine pairs in this region was eliminated, lipid assembly and secretion of apo B were greatly reduced (G.S. Shelness, 1997; Z. Yao, 1998) implying that disulfide bonds in this region are critical for the correct folding and secretion of apo B.

Plasma apo B100 contains 8–10% by weight carbohydrate and at least 20 potential N-linked glycosylation sites. Both N- and O-linked glycosylations occur in apo B100. N-linked glycosylation begins co-translationally in the ER lumen but it is not yet clear whether or not glycosylation is required for apo B secretion. Treatment of chicken hepatocytes with tunicamycin, which inhibits formation of N-linked oligosaccharide chains, did not reduce the amount of apo B secreted (D.M. Lane, 1982). However, when HepG2 cells were incubated with tunicamycin, apo B100 secretion was reduced and the intracellular degradation of apo B (Section 7) was increased. Moreover, the apo B100-containing particles that were secreted had a normal buoyant density suggesting that N-linked glycosylation is not required for assembly of lipids with apo B100 (L. Chan, 2001). An alternative explanation, however, is that the decreased secretion of apo B100 resulted from inadequate lipidation of apo B so that defective particles underwent pre-secretory degradation.

Several studies have shown that secreted apo B contains multiple phosphorylated serine residues (R.A. Davis, 1984) and that the phosphorylation is increased in diabetic, compared to non-diabetic, rats (J.D. Sparks, 1990). The physiological relevance of this phosphorylation is not known. Apo B phosphorylation has been suggested to occur in the Golgi apparatus in agreement with the finding that the Golgi contains kinases that phosphorylate secretory proteins (L.L. Swift, 1996).

Human apo B secreted by HepG2 cells is also covalently acylated with palmitic acid on at least one cysteine residue (J.M. Hoeg, 1988). Mutation of Cys-1085 in human apo B29 expressed in McArdle hepatoma cells abolished palmitoylation at this site. Interestingly, the secreted lipoproteins containing the mutated apo B29 were associated with less TG than those containing apo B29 with palmitoylated Cys-1085 [13]. These studies suggest that palmitoylation of apo B might play a role in assembling neutral lipids with apo B. However, no information is available on how many cysteines of apo B are palmitoylated, which palmitoyltransferase is involved or the topology of the palmitoylation reaction.

## 6. Regulation of apo B secretion by lipid supply

### 6.1. Fatty acids and triacylglycerols

In most instances, apo B is synthesized constitutively and in excess of the amount secreted so that the rate of apo B synthesis does not regulate how much apo B is secreted. The assembly of apo B-containing lipoproteins requires the synthesis of apo B, TGs, cholesterol, cholesteryl esters and phospholipids. When insufficient lipid is available for assembly of apo B into stable lipoprotein particles, excess apo B is not secreted but is degraded intracellularly, primarily by the proteasome and also by undefined proteases within the lumen of the secretory pathway (Section 7).

When the fatty acid oleate is added to HepG2 cells the synthesis of TG and phospholipids is stimulated and the amounts of apo B and TG secreted are increased [14]. Thus, an increased lipid supply enables a larger proportion of newly synthesized apo B to be translocated across the ER membrane and enter the secretory pathway. In agreement with the idea that lipid supply can regulate apo B secretion, the oleic-acid-induced increase in apo B secretion is blocked by Triacsin D, an inhibitor of TG synthesis. The rate of apo B secretion is not, however, solely a function of the rate of TG synthesis since glucose stimulates both TG synthesis and secretion, but does not increase apo B secretion. Furthermore, when oleic acid is supplied to primary rat hepatocytes, in contrast to HepG2 cells, both the synthesis and secretion of TG are increased whereas apo B secretion is unaltered. Therefore, the difference in response of apo B secretion to oleate in HepG2 cells versus primary hepatocytes cannot be ascribed to differences in TG synthesis.

Early work suggested that two pools of TG exist in hepatocytes: a large cytosolic storage pool that turns over slowly, and a small microsomal pool that turns over rapidly. The source of TG used for assembly with apo B appears to be primarily (perhaps ~70%) the cytosolic storage pool rather than the pool made from de novo synthesis [15]. Recent data support a model in which cytosolic TG is hydrolyzed, perhaps by the recently discovered TG hydrolase in the ER lumen [16]. The mechanism by which an ER lumenal lipase accesses cytosolic TG is not clear but the small amount of TG (−5% of total membrane lipids) in the ER membrane is a potential source of this TG. The lipolysis products (diacylglycerols and monoacylglycerols) are thought to be subsequently re-esterified within the ER lumen to produce TG that is assembled with apo B.

Operation of this lipolysis–re-esterification cycle also requires fatty acids, as well as a mono- or di-acylglycerol acyltransferase activity, in the ER lumen for re-synthesis of TG. Until recently, all TG was thought to be synthesized by diacylglycerol acyltransferase-1, a microsomal enzyme (Chapter 10). Surprisingly, however, diacylglycerol acyltransferase-1 knock-out mice retain the ability to synthesize TG and have normal fasting serum TG levels [17]. These data indicate that secretion of TG-rich lipoproteins does not require diacylglycerol acyltransferase-1. Consequently, another gene product has been implicated in synthesizing TG for VLDL assembly: diacylglycerol acyltransferase-2 [18]. Consistent with this idea, diacylglycerol acyltransferase-2 is highly expressed in liver. Confirmation of this hypothesis awaits generation of diacylglycerol acyltransferase-2-deficient mice.

The type of fatty acid supplied to hepatocytes also influences the secretion of apo B-containing lipoproteins. Compared to oleic acid, the (n-3) fatty acids eicosapentaenoic acid (20 : 5) and docosahexaenoic acid (22 : 6), found in fish oils, decrease plasma TG levels in humans and decrease the secretion of apo B-containing lipoproteins from rat hepatocytes and hepatoma cells. Consistent with these findings, (n-3) fatty acids also stimulate apo B degradation (E.A. Fisher, 1993).

### 6.2. Phospholipids

Phosphatidylcholine is the major phospholipid on the surface monolayer of all lipoproteins including VLDLs. In the liver, phosphatidylcholine is synthesized by two biosynthetic pathways: the CDP-choline pathway and the phosphatidylethanolamine N-methylation pathway (Chapter 8). Choline is an essential precursor of phosphatidylcholine synthesis via the CDP-choline pathway. When cells or animals are deprived of choline, this biosynthetic pathway is specifically inhibited and secretion of TGs and apo B, but not other proteins, is reduced by ~70% even though the phosphatidylcholine content of ER and Golgi membranes is essentially normal [19]. Thus, phosphatidylcholine synthesis is required for VLDL secretion. Interestingly, choline deficiency inhibits the secretion of only apo B variants larger than apo B28 (i.e. only those that assemble a neutral lipid core) indicating that association of apo B with neutral lipids and/or VLDL secretion depend on phosphatidylcholine synthesis. The mechanism by which decreased synthesis of phosphatidylcholine inhibits apo B secretion is by pre-secretory degradation. Translocation of apo B into the ER lumen is not impaired since the number of apo B-containing particles within the ER lumen is normal. In contrast, the number of apo B-containing particles in the Golgi lumen is decreased by ~70% implying that defective VLDL particles are degraded in a post-ER compartment of the secretory route (Section 7).

An alternative pathway for phosphatidylcholine synthesis, which is quantitatively significant only in the liver, is the methylation of phosphatidylethanolamine to phosphatidylcholine via phosphatidylethanolamine $N$-methyltransferase. Male mice in which the methyltransferase gene was disrupted (D.E. Vance, 1997) have greatly reduced plasma levels of TG and apo B100, but not apo B48. Moreover, the secretion of apo B100 and TG from hepatocytes derived from these mice is similarly impaired (A. Noga, 2002). Thus, phosphatidylcholine synthesis is required for normal apo B secretion.

Phospholipid turnover also appears to play a role in VLDL assembly (G.F. Gibbons, 1996). A significant proportion of TG in VLDLs contains fatty acids derived from the deacylation of phospholipids. Moreover, the assembly of VLDLs has been proposed to require the calcium-independent phospholipase A2 (Z. Yao, 2000).

In addition to phosphatidylcholine, VLDLs contain smaller amounts of other phospholipids (e.g. phosphatidylethanolamine, phosphatidylserine, phosphatidylinositol) as well as sphingomyelin and ceramide. Inhibition of sphingolipid synthesis by >90% by addition of fumonisin B to rat hepatocytes (Chapter 14), does not, however, inhibit VLDL secretion. Therefore, normal amounts of sphingolipids are not required for VLDL assembly/secretion (A.H. Merrill, 1995).

### 6.3. Cholesterol and cholesteryl esters

Sterol response element binding proteins (SREBPs) are master regulators of fatty acid and cholesterol synthesis at the level of transcription (Chapters 6, 7, 15). The transcriptionally active forms of SREBPs are produced from precursor proteins by a sterol-dependent proteolytic cleavage [20]. SREBPs co-ordinately increase the expression of genes involved in synthesizing fatty acids (fatty acid synthase, acetyl-CoA carboxylase, stearoyl-CoA desaturase) and cholesterol (3-hydroxy-3-methylglutaryl-CoA reductase), as well as the microsomal TG transfer protein (Section 6.4). All these components are required for VLDL assembly. The ability of the liver to secrete VLDLs is also linked via SREBP to the cholesterol/bile acid pathway through changes in the hepatic level and metabolism of cholesterol (R.A. Davis, 2001).

Cholesteryl esters are relatively minor constituents (5–15% of total lipids) of VLDLs but the amount of cholesteryl esters relative to TGs increases when rats are fed a high-cholesterol diet (R.A. Davis, 1982). There are conflicting data on whether or not the availability of cholesterol and/or cholesteryl esters directly influences apo B secretion. Inhibition of cholesteryl ester formation in hepatocytes decreases apo B secretion is some studies but not others. A severe reduction in the cholesteryl ester content of hepatoma cells reduces apo B secretion whereas an increased cellular content of cholesteryl esters above normal does not stimulate apo B secretion. Cholesterol is esterified by acyl-CoA:cholesterol acyltransferase (ACAT) which is encoded by at least two different genes. In mouse liver and intestine the majority of cholesteryl esters are made via ACAT-2. In animal studies, ACAT inhibitors reduce plasma cholesterol probably primarily by decreasing intestinal cholesterol absorption rather than by reducing VLDL secretion. In support of this idea, targeted disruption of the ACAT-2 gene in mice greatly reduces intestinal cholesterol absorption and results in complete resistance to diet-induced hypercholesterolemia (R.V. Farese, 2000). However, apo B-containing lipoproteins are made in normal amounts in ACAT-2$^{-/-}$ mice despite the absence of essentially all hepatic ACAT activity. Interestingly, plasma VLDLs of these mice have smaller diameters than those of wild-type mice. Thus, normal amounts of cholesteryl esters might be required for assembling large VLDLs. In mice with a targeted disruption of the ACAT-1 gene, hepatic cholesterol esterification activity and plasma cholesterol levels are normal [21].

Based on experiments with ACAT$^{-/-}$ mice it has been suggested that ACAT-1 functions primarily in providing cholesteryl esters for storage in cytosolic droplets whereas ACAT-2 is linked to secretion of cholesteryl esters into lipoproteins. In agreement with this concept, the two ACATs have been proposed to have different membrane topologies with the active site of ACAT-1 facing the cytosol and that of ACAT-2 facing the ER lumen (L. Rudel, 2000). It is not yet clear, however, how these findings apply to humans since ACAT-1 accounts for the majority of ACAT activity in *human* hepatocytes whereas ACAT-2 is the major ACAT in human intestine.

### 6.4. Microsomal triacylglycerol transfer protein

Lipids are assembled with apo B into VLDLs in the ER lumen (J.E. Vance, 1993). During this process TGs are concentrated into the core of the particle. A microsomal lumenal protein, the 97 kDa microsomal triacylglycerol transfer protein (MTP), that has the ability to transfer TGs between membranes in vitro, is proposed to transfer TG to nascent apo B particles [22]. MTP is present in liver and intestine as a soluble heterodimer with protein disulfide isomerase (55 kDa) (J.R. Wetterau, 1990). The latter is a ubiquitous ER lumenal protein that catalyzes formation of disulfide bonds during protein folding. The 97 kDa MTP subunit confers all lipid transfer activity to the heterodimer. In in vitro assays, MTP catalyzes lipid transfer from donor to acceptor liposomes with substrate specificity of: TG > cholesteryl esters > diacylglycerols > phosphatidylcholine. The lipid transfer reaction displays ping-pong, bi–bi kinetics implying that MTP transfers lipids by a 'shuttle' mechanism. Although MTP contains no C-terminal KDEL sequence (this sequence retains soluble proteins within the ER lumen), association of MTP with protein disulfide isomerase, which does contain a KDEL sequence, is thought to retain MTP within the ER lumen. The tissue and subcellular location of MTP, and its preference for transporting neutral lipids, suggest that MTP is involved in loading nascent apo B particles with TG.

### 6.4.1. MTP deficiency in humans and mice

The hypothesis that MTP is required for secretion of apo B-containing lipoproteins from the liver and intestine was strengthened by the discovery that individuals with the rare genetic disease abetalipoproteinemia lack detectable MTP protein and intestinal MTP lipid transfer activity (J.R. Wetterau, 1992). These patients suffer from fat malabsorption and despite a normal apo B gene, their plasma apo B is barely detectable. Specific inhibitors of MTP lipid transfer activity have been developed based on the premise that MTP might be a target for reducing atherosclerosis by inhibiting production of apo B-containing lipoproteins. MTP inhibitors lower plasma cholesterol levels by up to 80% in rats, hamsters and rabbits (J.R. Wetterau, 1998). However, heterozygosity for MTP deficiency in humans does not result in altered plasma lipid or lipoprotein levels suggesting that MTP is not normally rate-limiting for lipoprotein production unless MTP activity is greatly reduced.

The function of MTP has been investigated in genetically modified mice [23,24]. Complete elimination of the murine MTP gene is embryonically lethal, probably because apo B-containing lipoproteins are required for transferring lipids from the yolk sac to the developing mouse embryo. However, the murine MTP gene was inactivated specifically in the liver by generating mice harboring a 'floxed' MTP gene and subsequent Cre-mediated recombination. In these mice, plasma apo B100 levels were reduced by >90% but, surprisingly, apo B48 levels were reduced only slightly. Moreover, in hepatocytes derived from these mice apo B100 secretion was eliminated whereas secretion of apo B48 was almost unaltered. However, the apo B48-containing lipoproteins contained less lipid than those from wild-type mice. In addition, ultrastructural analyses revealed that, compared to wild-type hepatocytes, MTP$^{-/-}$ hepatocytes contained very few VLDL-sized, lipid-staining particles within the ER or Golgi lumina. Moreover, MTP$^{-/-}$ mouse livers contained numerous lipid droplets in the cytosol, a hallmark of impaired TG secretion. These observations suggest that (i) MTP is required for transferring TG into the lumen of the ER, and (ii) apo B100 secretion requires MTP whereas secretion of apo B48 is either less sensitive to MTP deficiency or does not require MTP. On the other hand, over-expression of MTP in mouse liver via recombinant adenoviruses increased the secretion and plasma levels of apo B100 and B48, as well as TG (D.J. Rader, 1999).

It should be noted that although MTP is assumed to function as a lipid transfer protein, based on its ability to transfer lipids between membranes in vitro, a lipid transfer function of MTP in providing lipid for VLDL assembly has not been directly demonstrated in intact cells or animals.

### 6.4.2. Studies on the function of MTP in cultured cells

Recently, investigators have extensively studied the role of MTP in the assembly of apo B-containing lipoproteins in hepatocyte-derived cells, as well as non-hepatic cells that have been genetically modified to express apo B and MTP. MTP and apo B have been shown by co-immunoprecipitation experiments to interact physically during lipoprotein assembly via specific sites that have been identified on MTP and apo B (C.C. Shoulders, 1999; G.S. Shelness, 1998; T. Grand-Perret, 1999). The association between MTP and apo B is transient and increases when lipoprotein assembly is stimulated upon addition of oleic acid, and decreased when TG synthesis is inhibited by Triacsin D (H.N. Ginsberg, 1996). Apo B also co-immunoprecipitates with other ER lumenal chaperone proteins such as calnexin, calreticulin, GRP94, Erp72 and BiP which presumably facilitate the correct folding of apo B (H. Herscovitz, 1998). MTP itself has also been implicated as a chaperone for apo B during VLDL assembly (H.N. Ginsberg, 2001).

There is considerable controversy regarding the precise role of MTP in assembling lipids with apo B [25]. MTP has been suggested to play a role in several steps along the VLDL assembly pathway: translocation of apo B across the ER membrane and into the ER lumen; co-translational assembly of lipids with apo B; transfer of the bulk of TG into the core of VLDLs; movement of TG into the ER lumen. Experimental data both support and refute the participation of MTP at each of these steps. Some of the conflicting data likely arise from different experimental models used, e.g. hepatoma cells versus primary hepatocytes, cultured cells versus genetically modified mice, human versus rodent models.

A general consensus is that lipid addition to apo B begins during translation and translocation of apo B across the ER membrane (E.A. Fisher, 1998; D.A. Gordon, 1996; S.-O. Olofsson, 1998). Experiments with cultured cells indicate that MTP stimulates, but is not absolutely required for, co-translational lipidation of apo B. Short apo B variants, such as apo B18, are secreted from hepatoma cells in poorly lipidated form independent of MTP, whereas secretion of longer (i.e. >apo B23), more highly lipidated apo B variants is stimulated by MTP. These findings suggest that the TG-transferring property of MTP is important for VLDL assembly. Moreover, the requirement of MTP is much greater for the secretion of apo B100 than for apo B48 suggesting that there might be different mechanisms involved in assembly of these two types of lipoproteins. In spite of a suggested chaperone function for MTP, this protein is apparently not *required* for translocation of apo B across the ER membrane since (i) apo B41-containing lipoproteins are secreted from mammary-derived cells lacking MTP activity (D.M. Small, 1995), (ii) apo B48 is lipidated and translocated equally well into the lumen of dog pancreatic microsomes (which lack MTP) and liver microsomes (which contain MTP) (A.E. Rusiñol, 1997), and (iii) an almost normal amount of apo B48, albeit poorly lipidated, is secreted from MTP$^{-/-}$ hepatocytes (S.G. Young, 1998).

Whether or not MTP plays a role in supplying the bulk of TG for VLDLs is controversial. Several studies indicate that MTP is not directly involved in this process. One suggestion is that MTP transports TGs into the ER lumen to form lumenal TG droplets which are subsequently incorporated into apo B-containing lipoproteins. This proposed role for MTP is derived from observations that (i) MTP$^{-/-}$ livers of mice lack VLDL-sized, lipid-staining particles that are normally present in the ER lumen of wild-type mice, and (ii) ER lumenal lipid droplets, the size of chylomicrons, accumulate in enterocytes of mice lacking intestinal apo B (R.L. Hamilton, 1998). However, a role for MTP in transferring TG to apo B, either by shuttling TG monomers to apo B or by mediating the fusion between partially lipidated apo B particles and a lumenal TG droplet, has not been unequivocally demonstrated. Nor is it clear how the proposed 'shuttle mechanism' for MTP can be reconciled with a role in transferring such large amounts of TG to apo B.

Based on available experimental evidence several models have been proposed for the assembly of apo B with lipids. In one popular model, called the 'two-step' model (Fig. 4A), MTP transfers lipid to apo B during translocation of the protein into the ER lumen to form partially lipidated apo B. These small particles then fuse with a TG droplet in the ER lumen to produce VLDLs. As discussed above, MTP might be involved in forming the lumenal TG droplet and/or in fusion of the droplet with the small apo B-containing particle. However, the biophysical mechanism by which a TG droplet would fuse with a small apo B-containing lipoprotein is not clear. A second model of assembly of apo B-containing lipoproteins is one in which lipid is sequentially added to apo B that is either undergoing translocation across the membrane, or that is associated with the lumenal face of the ER membrane (Fig. 4B). During times when lipid supply is low, small, lipid-poor apo B particles would be formed and secreted, but when lipid supply is abundant, fully lipidated VLDLs would be produced.

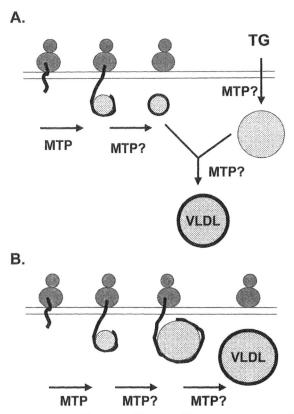

Fig. 4. Two models proposed for assembly of apo B-containing lipoproteins in hepatocytes. (Panel A) The 'two-step' model. Shown from left to right is the co-translational translocation of apo B into the ER lumen during which small amounts of lipid are assembled with apo B in a MTP-dependent step. The partially lipidated apo B particle is then released into the lumen and fuses with a lumenal triacylglycerol (TG) droplet to form VLDL. Alternatively, MTP might 'shuttle' TG monomers to the particle. It is proposed that MTP mediates formation of a TG droplet in the ER lumen for assembly with apo B. (Panel B) The concerted assembly model. From left to right is shown the co-translational translocation of apo B into the ER lumen. Lipid is sequentially added to apo B during translocation until a VLDL-sized particle is formed. In the event of limited lipid supply, small, partially lipidated apo B particles would be secreted. The bulk of TG might be transferred either from TG within the ER membrane or from a lumenal TG droplet, as indicated in panel A. The role of MTP in some steps of both schemes (indicated by ?) is still controversial.

Many issues remain concerning the role of MTP in the assembly of apo B-containing lipoproteins. (i) The existence of ER lumenal TG droplets has not been unambiguously demonstrated by isolation of these droplets from hepatocytes. (ii) The mechanism of formation and stabilization of TG droplets in the ER lumen is not understood. (iii) It is not clear why some lumenal TG droplets lacking apo B would not be secreted by default from hepatocytes. (iv) The partially lipidated apo B-containing particles that are formed co-translationally have not been directly shown to be precursors of fully lipidated VLDLs. (v) It is not clear why MTP deficiency abrogates the secretion of apo B100 but not apo B48.

### 6.4.3. MTP gene expression

The promoter region of the MTP gene contains elements predicted to bind transcription factors (e.g. HNF-1, HNF-4 and AP-1) that might dictate the cell type-specific expression of MTP. Hepatic MTP mRNA levels are increased in hamsters fed a high-fat diet. The activity of the human promoter is increased by cholesterol, most likely via the SREBP pathway (Chapter 15), and expression of the MTP gene is inhibited by insulin. However, it is not yet known if these alterations in gene expression regulate the secretion of apo B-containing lipoproteins.

## 7. Translocation and intracellular degradation of apo B

Typical secretory proteins co-translationally translocate across the ER membrane without interacting with the membrane bilayer or being exposed to the cytosol. However, translocation of apo B into the ER lumen is unusual in several respects. First, complete translocation of apo B occurs only when apo B is associated with lipid. Second, an unusual translocational pausing occurs during which translation continues and domains of newly synthesized apo B become exposed to the cytosol (V.R. Lingappa, 1990). Translocational pausing of apo B has been proposed to be mediated by multiple 'pause–transfer sequences' which are similar, but non-identical, sequences of ~10 amino acids that direct stopping and starting during translocation [26]. The translocating chain-associated membrane protein, TRAM, in the ER membrane appears to mediate translocational pausing (V.R. Lingappa, 1998). The signal for resumption of translocation after pausing is not known but has been suggested to be association of lipid with apo B. Others have argued that translational, rather than translocational, pausing occurs.

### 7.1. Proteasomal degradation of apo B

A third unusual property of apo B is that its translocation across the ER membrane is linked to proteasomal degradation (Fig. 5A). Pulse-chase studies in hepatocytes show that 40 to 60% of newly synthesized apo B is not secreted but is degraded intracellularly [27]. When lipid supply is restricted, or when MTP is inhibited, apo B does not efficiently translocate into the ER lumen but becomes exposed to the cytosol and is degraded. However, this degradation does not determine how much apo B is secreted because when degradation is inhibited by N-acetyl-Leu-Leu-norleucinal, apo B accumulates in microsomes whereas apo B secretion is not increased (R.A. Davis, 1994). This arrest in translocation and cytosolic exposure of apo B result in apo B being ubiquitinylated and targeted for degradation by cytosolic proteasomes [28] in a process facilitated by the cytosolic chaperone hsp70 (E.A. Fisher, 2001). Several ER lumenal and membrane proteins are known to undergo proteasomal degradation by a pathway in which the completely translocated protein undergoes retrotranslocation into the cytosol and is subsequently ubiquitinylated and degraded by proteasomes. In contrast, the ubiquitinylation and proteasomal degradation of apo B are thought to be initiated while apo B is attached to the ribosome and still within the translocation channel (E.A. Fisher, 2001).

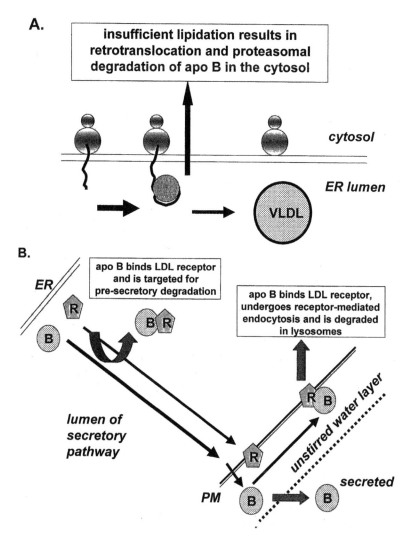

Fig. 5. Intracellular degradation of apo B. (Panel A) When lipid supply is limited, or when MTP is inhibited, lipid-poor apo B particles are formed and apo B is exposed to the cytosol, becomes ubiquitinylated and degraded by the proteasome. Sufficient supply of lipids results in VLDL being secreted. (Panel B) Apo B is also degraded non-proteasomally. Nascent apo B particles, indicated as 'B' are transported via the secretory route to, and across, the plasma membrane (PM) and enter the 'unstirred water layer' adjacent to the cell. Some particles are fully secreted whereas others bind to cell surface LDL receptors, indicated by 'R', are internalized by receptor-mediated endocytosis and undergo lysosomal degradation. The LDL receptor is also proposed to mediate the pre-secretory degradation of apo B by binding to apo B100 within the lumen of the secretory pathway and targeting apo B100 for degradation by as yet unidentified lumenal proteases.

## 7.2. Degradation of apo B within the secretory pathway

Apo B is also degraded within the secretory pathway. This site of degradation was demonstrated in HepG2 cells in which apo B was degraded by a lumenal protease after the protein had completely translocated into the ER lumen and had therefore escaped proteasomal degradation. In primary rat hepatocytes apo B is also degraded within the secretory pathway when phosphatidylcholine biosynthesis is inhibited (D.E. Vance, 1993). The ER lumenal protease ER60 has been suggested to mediate pre-secretory apo B degradation (K. Adeli, 1997). Degradation of apo B within the secretory pathway is thought to provide a quality control mechanism for eliminating defective particles before their secretion. In contrast to the proteasomal degradation of apo B, which occurs in *response* to inefficient translocation across the ER membrane but does not control how much apo B is secreted, the post-translocational degradation of apo B can directly control how much apo B is secreted.

A clue to the mechanism of non-proteasomal degradation of apo B came from studies in which hepatocytes lacking functional LDL receptors were found to secrete more apo B than did wild-type hepatocytes. The LDL receptor was subsequently implicated in two facets of apo B degradation [29] (Fig. 5B). First, when fully assembled apo B-containing particles are exported across the plasma membrane of hepatocytes, the nascent lipoproteins enter the unstirred water layer in the vicinity of the plasma membrane. These particles do not immediately diffuse away from the cell (E.A. Fisher, 1990) but instead can bind LDL receptors on the hepatocyte surface. Consequently, the lipoproteins can be taken up by receptor-mediated endocytosis and degraded in lysosomes (Chapter 21). Second, LDL receptors and apo B both transit the secretory pathway en route to the plasma membrane. The ligand-binding domain of the LDL receptor is oriented towards the ER lumen and therefore has the opportunity to bind apo B intralumenally and target it for degradation before it can be secreted. Approximately equal amounts of apo B100 are degraded via the LDL receptor using the re-uptake and pre-secretory mechanisms [29]. Apo B100 is more susceptible than apo B48 to LDL receptor-mediated degradation because apo B100 contains the LDL receptor-binding domain that is absent from apo B48. Small amounts of apo B48 are also degraded because these particles associate with apo E which binds to the LDL receptor.

This mode of presecretory degradation of nascent apo B was also invoked as a possible explanation for why apo B secretion is impaired in apo E$^{-/-}$ mice, whereas apo B secretion is increased in transgenic mice and hepatoma cells that over-express apo E. The basis for this conjecture was that apo E binds avidly to the LDL receptor and thus might compete with apo B for receptor binding, both within the secretory pathway and on the cell surface. The prediction, however, was found not to be correct because when apo E production was eliminated in LDL receptor$^{-/-}$ mice, VLDL secretion was still 50% less than in LDL receptor$^{+/+}$ mice (L.M. Havekes, 2001).

## 8. Assembly and secretion of chylomicrons

When dietary fat enters the intestine TGs are hydrolyzed to monoacylglycerols and fatty acids which diffuse across the microvillus membrane. TGs are subsequently re-synthesized in the enterocyte and large amounts of TGs are packaged with other lipids and apo B48 into chylomicrons that are secreted into mesenteric lymph. The wide variation in sizes of chylomicrons (35 to >250 nm) reflects the supply of TGs to the intestine and affords a mechanism by which TG secretion can be readily altered in response to diet.

A cell culture model frequently used to study intestinal apo B secretion is the human colon carcinoma cell line, CaCo2, which secretes both apo B100 and apo B48 (E.J. Schaefer, 1987). In many respects, the intestinal assembly of chylomicrons is similar to the hepatic assembly of VLDLs, although there are some differences. Both apo B48 and MTP are required for normal chylomicron assembly. ER lumenal TG droplets, the size of chylomicrons, but lacking apo B, were observed in enterocytes of mice lacking intestinal apo B synthesis, suggesting that chylomicron assembly might utilize these apo B-free, ER lumenal TG droplets (R.L. Hamilton, 1998). Interestingly, in CaCo2 cells almost no newly synthesized apo B is degraded intracellularly. Moreover, ubiquitinylation of intracellular apo B is not detectable although CaCo2 cells do possess the machinery for adding ubiquitin to proteins for proteasomal degradation (L. Chan, 2000).

One unique feature of chylomicron assembly is revealed in individuals with chylomicron retention disease (Anderson's disease). These people secrete apo B100-containing VLDLs from the liver but do not secrete apo B48-containing chylomicrons from the intestine. This rare disease is characterized by hypocholesterolemia and lipid malabsorption [30]. Since the intestinal apo B editing mechanism is intact and apo B48 is synthesized normally, a defective gene, different from genes involved in VLDL assembly, appears to be responsible for this disease.

## 9. Assembly of lipoprotein(a)

In addition to its presence on VLDLs and LDLs, plasma apo B is also found in humans, some primates and hedgehogs (but not rodents) covalently bound via a single disulfide linkage to a glycoprotein called apo(a). Apo(a) is synthesized in significant quantities only in the liver and associates with apo B of LDL to form lipoprotein(a). In human plasma, the concentration of lipoprotein(a) varies from <1 to >100 mg/dl. In general, a high level of lipoprotein(a) is an independent risk factor for development of coronary artery disease (A.M. Scanu, 1992).

Apo(a) contains tandem repeats of sequences that are very similar to the 'kringle' motifs of plasminogen suggesting that apo(a) arose by duplication of the plasminogen gene. A cysteine residue (Cys-4057) in one kringle domain forms a disulfide linkage with Cys-4326 of apo B100. Apo B100 of many mammals cannot form a disulfide linkage with human apo(a). For example, in transgenic mice expressing human apo(a), this apo(a) circulates in plasma free of apo B-containing lipoproteins. A step-wise model for assembly of lipoprotein(a) has been proposed. Initially, a non-covalent interaction occurs between a lysine residue (Lys-680) in the N-terminus of apo B100 and a lysine-binding site in apo(a) (M.L. Koschinsky, 2001). Next, a non-covalent interaction between the C-terminal region of apo B100 and amino acids 4330–4397 of apo(a) occurs, followed by formation of the disulfide bond between apo B and apo(a) (S.P.A. McCormick, 2000). Most data support a model of lipoprotein(a) assembly in which apo(a) associates with apo B of LDL in the plasma after secretion, rather than within hepatocytes prior to secretion.

## 10. Future directions

The past few years have been marked by an explosion of data on lipoprotein metabolism obtained from genetically modified mice. These studies have been extremely useful for elucidating mechanisms of assembly and secretion of lipoproteins, particularly because there are limitations to studying this process in cultured cells. However, one must also acknowledge the limitations of using mice as models for studying human lipoprotein assembly since lipoprotein metabolism in mice is distinct in several respects from that in humans.

Several important questions remain regarding the mechanisms of lipoprotein assembly. It is surprising that although the MTP gene and protein have been so extensively studied, and although we know that an absence of MTP severely impairs the secretion of apo B-containing proteins, we still do not know exactly how this protein participates in lipoprotein assembly. In in vitro assays, MTP transfers lipids between membranes. Consequently, it is assumed that MTP transfers TGs to apo B. However, this lipid transfer function of MTP has not yet been demonstrated in cells. Another proposed role for MTP is in the transfer of TGs across the ER membrane to form lumenal TG droplets, but how this process occurs and the composition of the droplets, are not known. There are currently no data explaining why animals make apo B100 in addition to apo B48 since apo B48-only mice appear normal. Nor do we understand why apo B is made in large excess of its requirement for lipoprotein secretion or why hepatocytes degrade a large fraction of newly synthesized apo B intracellularly in an apparently wasteful process. In addition, proteases involved in apo B degradation within the lumen of the secretory pathway remain to be identified. Several post-translational covalent modifications of apo B occur (glycosylation, phosphorylation and palmitoylation) yet we do not understand the biological relevance of these modifications. The mechanisms that regulate chylomicron assembly are poorly understood. For example, the defective gene responsible for Anderson's disease, in which intestinal apo B48/chylomicron secretion is severely impaired while secretion of apo B100 from the liver is normal, has not yet been isolated nor has its function been identified. The stage is now set for resolution of these issues within the next few years.

## Abbreviations

| ACAT | acyl-Co A : cholesterol acyltransferase |
|------|------------------------------------------|
| apo | apolipoprotein |
| ER | endoplasmic reticulum |
| LDL | low-density lipoprotein |
| MTP | microsomal triacylglycerol transfer protein |
| SREBP | sterol response element binding protein |
| TG | triacylglycerol |
| VLDL | very low-density lipoprotein |

# References

1. Alexander, C.A., Hamilton, R.L. and Havel, R.J. (1976) Subcellular localization of B apoprotein of plasma lipoproteins in rat liver. J. Cell Biol. 69, 241–263.
2. Yang, C.-Y., Chen, S.-H., Gianturco, S.H., Bradley, W.A., Sparrow, J.T., Tanimura, M., Li, W.-H., Sparrow, D.A., DeLoof, H., Rosseneu, M., Lee, F.-S., Gu, Z.-W., Gotto, A.M. and Chan, L. (1986) Sequence, structure, receptor-binding domains and internal repeats of human apolipoprotein B-100. Nature 323, 738–742.
3. Segrest, J.P., Jones, M.K., Mishra, V.K., Anantharamaiah, G.M. and Garber, D.W. (1994) ApoB-100 has a pentapartite structure composed of three amphiphilic a-helical domains alternating with two amphipathic β-strand domains. Arterioscler. Thromb. 14, 1674–1685.
4. Kim, E. and Young, S.G. (1998) Genetically modified mice for the study of apolipoprotein B. J. Lipid Res. 39, 703–723.
5. Yao, Z., Blackhart, B.D., Linton, M.F., Taylor, S., Young, S.G. and McCarthy, B.J. (1991) Expression of carboxyl-terminally truncated forms of human apolipoprotein B in rat hepatoma cells. Evidence that the length of apolipoprotein B has a major effect on the buoyant density of the secreted lipoproteins. J. Biol. Chem. 266, 3300–3308.
6. Bjorkegren, J., Veniant, M., Kim, S.K., Withycombe, S.K., Wood, P.A., Hellerstein, M.K., Neese, R.A. and Young, S.G. (2001) Lipoprotein secretion and triglyceride stores in the heart. J. Biol. Chem. 276, 38511–38517.
7. Brooks, A.R., Blackhart, B.D., Haubold, K. and Levy-Wilson, B. (1991) Characterization of tissue-specific enhancer elements in the second intron of the human apolipoprotein B gene. J. Biol. Chem. 266, 7848–7859.
8. Young, S.G., Cham, C.M., Pitas, R.E., Burri, B.J., Connolly, A., Flynn, A., Pappu, A.S., Wong, J.S., Hamilton, R.L. and Farese, R.V. (1995) A genetic model for absent chylomicron formation: mice producing apolipoprotein B in the liver but not in the intestine. J. Clin. Invest. 96, 2932–2946.
9. Antes, T.J., Goodart, S.A., Chen, W. and Levy-Wilson, B. (2001) Human apolipoprotein B gene intestinal control region. Biochemistry 40, 6720–6730.
10. Antes, T.J., Goodart, S.A., Huynh, C., Sullivan, M., Young, S.G. and Levy-Wilson, B. (2000) Identification and characterization of a 315-base pair enhancer located more than 55 kilobases 5′ of the apolipoprotein B gene, that confers expression in the intestine. J. Biol. Chem. 275, 26637–26648.
11. Powell, L.M., Wallis, S.C., Pease, R.J., Edwards, Y.H., Knott, T.J. and Scott, J. (1987) A novel form of tissue-specific RNA processing produces apolipoprotein-B48 in intestine. Cell 50, 831–840.
12. Navaratnam, N., Morrison, J.R., Bhattacharya, S., Patel, D., Funahashi, B., Giannoni, F., Teng, B.-B., Davidson, N.O. and Scott, J. (1993) The p27 catalytic subunit of the apolipoprotein B mRNA editing enzyme is a cytidine deaminase. J. Biol. Chem. 268, 20709–20712.
13. Zhao, Y., McCabe, J.B., Vance, J. and Berthiaume, L.G. (2000) Palmitoylation of apolipoprotein B is required for proper intracellular sorting and transport of cholesteryl esters and triglycerides. Mol. Biol. Cell 11, 721–734.
14. Dixon, J.L., Furukawa, S. and Ginsberg, H.N. (1991) Oleate stimulates secretion of apolipoprotein B-containing lipoproteins from Hep G2 cells by inhibiting early intracellular degradation of apolipoprotein B. J. Biol. Chem. 266, 5080–5086.
15. Wiggins, D. and Gibbons, G.F. (1992) The lipolysis cycle of hepatic triacylglycerol. Biochem. J. 284, 457–462.
16. Lehner, R. and Vance, D.E. (1999) Cloning and expression of a cDNA encoding a hepatic microsomal lipase that mobilizes stored triacylglycerol. Biochem. J. 343, 1–10.
17. Smith, S.J., Cases, S., Jensen, D.R., Chen, H.C., Sande, E., Tow, B., Sanan, D.A., Raber, J., Eckel, R.H. and Farese, R.V. (2000) Obesity resistance and multiple mechanisms of triglyceride synthesis in mice lacking Dgat. Nat. Genet. 25, 87–90.
18. Cases, S., Stone, J., Zhou, P., Yen, E., Tow, B., Lardizabal, K.D., Voelker, T. and Farese Jr., R.V. (2001) Cloning of DGAT2, a second mammalian diacylglycerol acyltransferase, and related family members. J. Biol. Chem. 42, 38870–38876.
19. Verkade, H.J., Fast, D.G., Rusiñol, A.E., Scraba, D.G. and Vance, D.E. (1993) Impaired biosynthesis

of phosphatidylcholine causes a decrease in the number of very low density lipoprotein particles in the Golgi but not in the endoplasmic reticulum of rat liver. J. Biol. Chem. 268, 24990–24996.

20. Brown, M.S. and Goldstein, J.L. (1997) The SREBP pathway: regulation of cholesterol metabolism by proteolysis of a membrane-bound transcription factor. Cell 89, 331–340.

21. Buhman, K.K., Chen, H.C. and Farese, R.V. (2001) The enzymes of neutral lipid synthesis. J. Biol. Chem. 276, 40369–40372.

22. Berriot-Varoqueaux, N., Aggerbeck, L.P., Samson-Bouma, M. and Wetterau, J.R. (2000) The role of the microsomal triglyceride transfer protein in abetalipoproteinemia. Annu. Rev. Nutr. 20, 663–697.

23. Raabe, M., Flynn, L.M., Zlot, C.H., Wong, J.S., Veniant, M.M., Hamilton, R.L. and Young, S.G. (1998) Knockout of the abetalipoproteinemia gene in mice: reduced lipoprotein secretion in heterozygotes and embryonic lethality in homozygotes. Proc. Natl. Acad. Sci. USA 95, 8686–8691.

24. Chang, B.H.-J., Liao, W., Nakamuta, M., Mack, D. and Chan, L. (1999) Liver-specific inactivation of the abetalipoproteinemia gene completely abrogates very low density lipoprotein/low density lipoprotein production in a viable conditional knockout mouse. J. Biol. Chem. 274, 6051–6055.

25. Gordon, D.A. and Jamil, H. (2000) Progress towards understanding the role of microsomal triglyceride transfer protein in apolipoprotein-B lipoprotein assembly. Biochim. Biophys. Acta 1486, 72–83.

26. Kivlen, M.H., Dorsey, C.A., Lingappa, V.R. and Hegde, R.S. (1997) Asymmetric distribution of pause transfer sequences In apolipoprotein B 100. J. Lipid Res. 38, 1149–1162.

27. Borchardt, R.A. and Davis, R.A. (1987) Intrahepatic assembly of very low density lipoproteins. Rate of transport out of the endoplasmic reticulum determines rate of secretion. J. Biol. Chem. 262, 16394–16402.

28. Yeung, S.J., Chen, S.H. and Chan, L. (1996) Ubiquitin–proteasome pathway mediates intracellular degradation of apolipoprotein B. Biochemistry 35, 13843–13848.

29. Twisk, J., Gillian-Daniel, D.L., Tabon, A., Wang, L., Barrett, P.H.R. and Attie, A.D. (2000) The role of the LDL receptor in apolipoprotein B secretion. J. Clin. Invest. 105, 521–532.

30. Dannoura, A.H., Berriot-Varoqueaux, N., Amati, P., Abadie, V., Verthier, N., Schmitz, J., Wetterau, J.R., Samson-Bouma, M.-E. and Aggerbeck, L.P. (1999) Anderson's disease: exclusion of apolipoprotein and intracellular lipid transport genes. Arterioscler. Thromb. Vasc. Biol. 19., 2494–2508.

D.E. Vance and J.E. Vance (Eds.) *Biochemistry of Lipids, Lipoproteins and Membranes (4th Edn.)*
© 2002 Elsevier Science B.V. All rights reserved

# CHAPTER 20

# Dynamics of lipoprotein transport in the human circulatory system

Phoebe E. Fielding and Christopher J. Fielding

*Cardiovascular Research Institute and Departments of Medicine and Physiology, University of California,
San Francisco, CA 94143-0130, USA*

## 1. Overview

### 1.1. Functions of the major lipoproteins

Plasma lipoproteins are soluble complexes of lipids with specialized proteins (apolipoproteins). Their function is to deliver lipids from the tissues where they are synthesized (mainly the liver and intestine) to those that utilize or store them. The apolipoproteins solubilize and stabilize the insoluble lipids of the lipoprotein particles, and prevent the formation of aggregates. Many apolipoproteins have additional functions in plasma lipid metabolism. Some are ligands for cell surface receptors, and determine the tissue-specific delivery of lipids. Some are cofactors for plasma lipases. Others regulate lipid reactions in the plasma, as competitive inhibitors of lipid uptake or metabolism (see table 3 in Chapter 18). It is the protein composition of lipoproteins that in large part specifies their metabolism in the plasma compartment. Conversely, the apolipoprotein content of lipoprotein particles alters during recirculation, as changes in the lipid composition of the particles modify the affinity of apolipoproteins for their surface. The interaction of these processes largely specifies the delivery of lipids to different tissues. Lipids delivered via plasma lipoprotein particles, in addition to neutral acylglycerols, phospholipids, and free and esterified cholesterol, include fat-soluble vitamins and antioxidants. Lipid binding to apolipoproteins is in most cases via the hydrophobic faces of amphipathic helical domains (J.A. Gazzara, 1997). Apolipoproteins have little tertiary structure, which gives them flexibility on the surface of the lipoprotein, as the diameter of the particle responds to the loading or unloading of lipids. Amino acid sequences functional in receptor binding or enzyme activation usually include clusters of charged residues.

While blood plasma contains the highest levels of lipoprotein particles, most lipoproteins, with the exception of the largest, triacylglycerol-rich particles, can cross the vascular bed, though their concentrations in the extracellular space are significantly lower. Interstitial lipoprotein particles interact directly with the surface of peripheral cells, delivering and receiving lipids. This recirculation is completed when interstitial fluid is collected into the main trunk lymph ducts, and returned to the plasma.

## 1.2. 'Forward' lipid transport

Functionally there are two main classes of lipoproteins. The first consists of particles whose main role is to deliver lipids (mainly triacylglycerols) from the liver or intestine to peripheral, extrahepatic tissues. These particles contain apolipoprotein B (apo B) together with a changing admixture of other lipids and proteins. In chylomicrons, which are secreted from the small intestine, this triacylglycerol originates from dietary long-chain fatty acids. These are re-esterified in the intestinal mucosa before being incorporated into lipoproteins. They contain a single molecule of a truncated form of apo B (apo B48). After loss of most of their triacylglycerol during recirculation in the plasma compartment, the chylomicron 'remnants' are cleared by the liver. Very low-density lipoproteins (VLDLs) secreted from the liver, contain one molecule of the full-length form of apo B (apo B100). Following the loss of most of their triacylglycerol to peripheral tissues, some VLDLs are returned to the liver, endocytosed and catabolized. Others remain in the circulation as intermediate density lipoprotein particles (IDLs). These still contain significant amounts of triacylglycerol and most of their original content of cholesteryl ester and free cholesterol, together with apolipoproteins B and E. After further modification by plasma lipases, most of the apo B100 particles remain in the circulation in the form of low-density lipoproteins (LDLs). After a plasma half-life of about 2 days, LDL are endocytosed, mainly by the liver. Their protein is degraded; their sterol content can be recycled into newly secreted lipoprotein particles, or degraded to bile acids.

Functionally, VLDL (density < 1.006 g/ml), IDL (density 1.006–1.019 g/ml) and LDL (density 1.019–1.063 g/ml) particles represent a continuum of decreasing size and increasing density created by the lipolysis of triacylglycerol. The traditional density limits of these fractions, shown in fig. 1 of Chapter 18, reflect this continuum. The density of each fraction depends mainly on its weight ratio of triacylglycerol (density 0.91 g/ml) to protein (density 1.33 g/ml). In terms of apolipoprotein composition, VLDL particles contain apo B100 and apo Cs with or without apo E, IDLs contain apo B100 and apo E but not apo Cs, and LDLs contain only apo B100.

## 1.3. 'Reverse' lipid transport

The second major class of lipoprotein particles carries lipids (mainly free and esterified cholesterol) from peripheral tissues to the liver. These high-density lipoprotein particles (HDLs) contain 1–4 molecules of apolipoprotein A1 (apo A1), together with other apolipoproteins that specify the metabolism and delivery of these lipids. HDL-dependent lipid transport is often defined as *Reverse cholesterol transport (RCT)* (C.J. Fielding, 1995).

Several sources of cellular cholesterol contribute to RCT. A part reflects peripheral sterol synthesis, despite the downregulation of this pathway by cholesterol in circulating plasma lipoproteins, mainly LDL. A second part represents the recycling of lipoprotein cholesteryl esters, mainly in LDL, internalized via endocytosis at peripheral LDL receptors; these are also highly downregulated, under physiological conditions, by LDL. Probably the major part of RCT responds to the selective cellular uptake of preformed lipoprotein free cholesterol, independent of LDL receptors. This enters recycling endosomes returning to the cell surface. Cholesterol from all these sources transfers to HDL for further metabolism, including esterification, outside the cell. Some free cholesterol transfers within the circulation to HDL from other plasma lipoproteins.

Part of the HDL cholesteryl ester formed is transferred to apo B lipoproteins prior to their uptake by the liver. The remainder, mostly cholesteryl ester, is selectively internalized (that is, without the rest of the lipoprotein particle) from HDL by hepatocytes, and by steroidogenic tissues.

HDLs accumulate lipids from the peripheral tissues, and return them to the liver. Newly formed HDLs have high density and little lipid. Their density decreases as they accumulate lipid in the circulation. The classical subfractions of HDL [HDL-3 (density 1.12–1.219 g/ml), HDL-2 (density 1.063–1.12 g/ml), HDL-1 (density < 1.063 g/ml)] reflect this functional and structural continuum.

## 2. Lipoprotein triglyceride and lipolysis

### 2.1. Initial events

The structure of newly synthesized intestinal apo B-containing lipoprotein particles (chylomicra) is described in Chapter 19. Each consists of a triacylglycerol core containing a small proportion of cholesteryl esters, stabilized by a surface film made up mainly of phospholipid, some free cholesterol, and one molecule of apo B. Triacylglycerol and cholesteryl and retinyl esters in chylomicrons are derived almost entirely from dietary cholesterol, vitamin A and unesterified fatty acids; chylomicron phospholipids and free cholesterol are made in the enterocyte. Editing of full-length apo B transcripts (see Chapter 19) generates apo B48, which contains only the terminal 2152 residues of full-length apo B100. Since apo B does not exchange between lipoprotein particles during recirculation, apo B48 is an effective marker for chylomicron particles, and dietary triacylglycerol (E. Campos, 1992). Mice in which the editing enzyme was knocked out, when fed the same triacylglycerol load as control mice, were significantly less efficient in secreting chylomicron particles [1]. Dietary triacylglycerol accumulated in the intestinal mucosal cells. This finding indicates that apo B editing may have evolved along with dietary fat consumption to optimize the synthesis of large triacylglycerol-rich particles.

Chylomicrons are co-secreted with apo A1 (the intestine is the major source of this apolipoprotein in human subjects). This apo A1 is lost spontaneously to HDL as soon as chylomicrons reach the circulation. The transfer is independent of triacylglycerol lipolysis. At the same time, apo E and apo C proteins move to the surface of chylomicrons from reservoirs within the plasma population of large spherical HDL particles.

VLDLs secreted from the liver include a single molecule of full-length apo B100 containing 4536 amino acids (Chapter 19). The triacylglycerol-rich core of VLDLs contains significant levels of hepatic cholesteryl esters. Studies with isolated rat livers indicate that the incorporation of cholesteryl esters into VLDLs is necessary for their

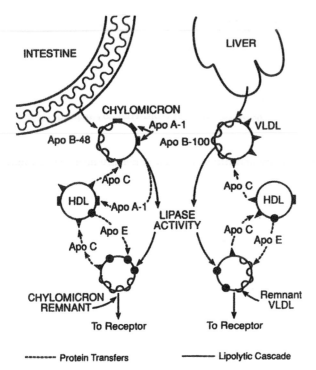

Fig. 1. Transfer of plasma apolipoproteins to newly secreted chylomicrons and VLDL. These small proteins play a critical role in optimizing the reaction rate of these triacylglycerol-rich particles with peripheral lipases and receptor proteins.

successful secretion into the perfusate. The phospholipid moiety of newly synthesized VLDL is enriched in phosphatidylethanolamine, in comparison with circulating VLDL. Newly synthesized plasma VLDLs contain apo C apoproteins but little apo E. As in the case of chylomicrons, enrichment of VLDL with apo E and additional apo Cs takes place in the plasma compartment.

These preliminary events in the circulatory system occupy about 5 min. In the case of both chylomicrons and VLDLs, the product is a triacylglycerol-rich apo B-particle functional to deliver triacylglycerol fatty acids to the peripheral tissues (Fig. 1). The purpose of this time lag is probably to allow these lipoproteins to distribute through the plasma compartment, prior to the inception of hydrolysis. A chylomicron or VLDL fully activated for lipolysis contains 10–20 molecules of apo C2, the cofactor of lipoprotein lipase (LPL). Titration of apo C2 content vs the rate of lipolysis indicates that 2–3 apo C2 molecules per chylomicron or VLDL are needed for maximal activity. Apo C2 and other apo Cs leave VLDL and chylomicrons as lipolysis proceeds, the triacylglycerol core shrinks, and surface phospholipid and proteins are transferred away to other lipoproteins, particularly HDL. Because apo C2 is present in initial excess, lipolysis rates are maintained until a major part (~80%) of initial triacylglycerol content of the particles has been lost.

*2.2. The structure and activation of lipoprotein lipase (LPL)*

LPL hydrolyzes the 1(3)-ester linkages of triacylglycerol of chylomicrons and VLDLs whose surface contains apo C2. The primary product of LPL-mediated lipolysis is 2-monoacylglycerol. After spontaneous isomerization of this lipid, LPL has activity against the 1-monoacylglycerol formed. Limited further lipolysis by plasma and platelet monoacylglycerol hydrolases also takes place. Monoacylglycerol is also readily internalized by vascular cells. As a result the end-products of LPL-mediated triacylglycerol hydrolysis are unesterified fatty acids, monoacylglycerol and glycerol. Fatty acids originating from LPL activity are cleared by adipose tissue and re-esterified under postprandial conditions and stored. Under fasting conditions, hormone-sensitive lipase promotes the release of unesterified fatty acids from adipocyte triacylglycerol back into the circulation (Chapter 10). Fatty acids from LPL-mediated lipolysis in muscle tissue are mainly catabolized to two-carbon subunits as part of oxidative metabolism.

LPL gene transcription is stimulated by sterol response element binding protein-1 (SREBP-1) (Chapter 15) and by Sp-1, and inhibited by Sp-3. The regulation of LPL expression is tissue-specific. During fasting, adipocyte LPL expression is reduced, while expression in muscle cells is increased. Postprandially, expression is upregulated in adipocytes, and decreased in muscle cells. Tissue-specific expression of LPL is mediated by the transcription factor PPARγ via its PPAR/RXR heterodimer [2]. This mechanism is related to the need to supply fatty acids to muscle for oxidative metabolism under conditions of scarcity, and to direct excess fatty acids to adipose tissue for storage postprandially, conditions where circulating glucose levels provide alternative substrate for muscle cells.

The secreted human LPL protein has 448 amino acids. It is functional as a dimer. LPL is a member of a triacylglycerol lipase protein family (Chapter 10) others of which include hepatic lipase, which like LPL is released into the plasma by heparin, and pancreatic lipase. Pancreatic lipase and several related fungal lipases have been crystallized. LPL is ~30% homologous in primary sequence to pancreatic lipase, whose X-ray coordinates have been used to model LPL structure. Other information on structure–function relationships in LPL has been obtained from the site-directed mutagenesis of key amino acids of receptor- and heparin-binding sites that are absent from pancreatic lipase (Fig. 2). LPL is a serine hydrolase whose active site triad is made up of the $S_{132}$, $D_{156}$ and $H_{241}$ residues. Consistent with other lipases in this family, the primary sequence of LPL predicts a polypeptide 'lid' (residues 239–264) which opens when LPL binds to its lipoprotein substrate. A short sequence of hydrophobic amino acids in the C-terminus (residues 387–394) has also been implicated in LPL binding to triacylglycerol-rich lipoproteins. Other data implicate residues 415–438 in both substrate interaction and dimer stability. Heparin binding by LPL was thought earlier to be mediated mainly via five basic residues in two adjacent clusters ($R_{279}$, $K_{280}$, $R_{282}$, $K_{296}$, $R_{297}$). Additional basic residues ($K_{403}$, $R_{405}$, $K_{407}$) were recently implicated (R.A. Sendak, 1998) The involvement of additional sequences at the C-terminus (residues 390–393, 439–448) has also been described (Y. Ma, 1994). These, while not directly heparin-binding, amplify binding by the other domains.

LPL is present within intracellular pools in adipocytes and muscle cells, but the

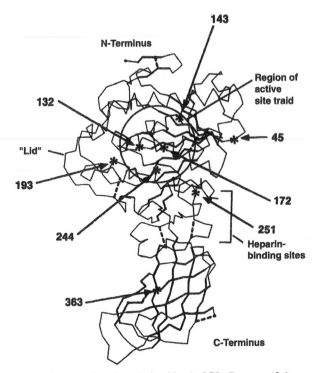

Fig. 2. Structure–function relationships in LPL. Because of the overall sequence similarity between LPL and pancreatic lipase, structural features and the locations of selected amino acids in LPL (which has not been crystallized) have been superimposed on the three dimensional structure of pancreatic lipase. (Modified from Faustinella et al. (1991) J. Biol. Chem. 266, 9481–9485, with permission).

functional fraction of LPL is at the vascular endothelial surface, where it is bound by heparin-like glycosaminoglycans. The products of the reaction of chylomicrons and VLDL with endothelial LPL, lipoprotein remnants, continue to circulate in the plasma compartment (see below). Small amounts of LPL are also present in the circulation, especially postprandially. LPL binds to several members of the LDL receptor protein family (specifically, LDL receptor-like proteins-1 and -2, and VLDL receptor protein) to induce receptor-mediated lipoprotein catabolism [3]. Some triacylglycerol clearance may occur via this endocytic route, but the predominant role of receptor binding by LPL is likely to lie in the endocytosis and degradation of LPL itself. Consistent with this, mice in which genes encoding the VLDL receptor, or both VLDL and LDL receptor proteins, were knocked out, had normal levels of plasma triacylglycerol. The receptor-binding domain in the LPL primary sequence lies within the C-terminal region of the protein. It includes $K_{407}$, but is distinct from the lipid-binding domain, which includes $W_{393}$ and $W_{394}$.

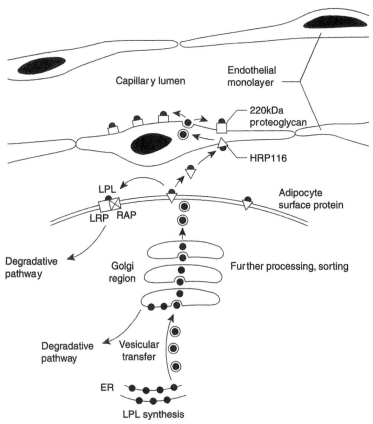

Fig. 3. Synthesis, secretion and transport of LPL from the adipocyte to the vascular endothelial surface. ER, endoplasmic reticulum. Degradative pathways from the Golgi compartment and cell surface are illustrated. RAP, receptor-associated protein; LRP, LDL-receptor-like protein. The suggested roles of HRP (heparin-released protein)-116, and 220 kDa proteoglycan are also shown.

## 2.3. Transport of LPL to its endothelial site

LPL, synthesized in adipocytes and myocytes, is transported out of the parenchymal cells, through the pericyte layer, and across the endothelium, before binding to functional sites on the vascular endothelial surface (Fig. 3). Mature LPL contains several polysaccharide chains, which are required for effective LPL secretion. Unesterified fatty acids and lysophosphatidylcholine increase the rate of LPL secretion from adipocytes. Adipocytes can degrade secreted LPL. Two distinct pathways are involved. One requires the 39 kDa receptor-associated protein (RAP) which binds to the LDL receptor-related protein [4]. LPL is also internalized via a proteoglycan-dependent pathway.

Subsequent stages of the activation of newly secreted LPL involve its transendothelial migration to specific binding site on the capillary vascular surface (Fig. 3). Transcytosis was recently shown to involve both the VLDL receptor protein, and proteoglycans

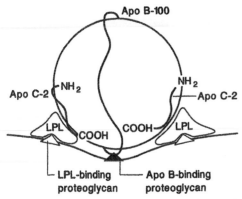

Fig. 4. Model to show interactions between the endothelial cell surface, triacylglycerol-rich lipoproteins, apo C2 and LPL. Two LPL molecules are shown reacting with the same VLDL particle. These are representative of the multiple LPLs probably reactive with each triacylglycerol-rich lipoprotein particle.

[3]. Earlier, a 116-kDa LPL-binding protein, released by heparin, had been implicated in LPL binding to endothelial cells. Microsequencing of peptides from this protein indicated it to be a fragment of apo B. The exact role of this fragment in the migration of LPL remains to be established. LPL is bound to the endothelial vascular surface via a 220 kDa proteoglycan whose functional site is probably a highly sulfated decasaccharide [5].

### 2.4. Structure of the LPL–substrate complex at the vascular surface

It seems likely that LPL and its large triacylglycerol-rich lipoprotein substrates both establish multiple interactions with each other and with the capillary wall to anchor the enzyme–substrate complex to the vascular surface. Components of such a multi-protein functional complex would include LPL itself, apo C2 and apo B on the VLDL or chylomicron, the 220 kDa proteoglycan, and possibly VLDL receptor protein or another member of this family (Fig. 4).

Apo B100 has a length sufficient to make only a single circumference of VLDL. This was estimated from electron microscopic studies of apo B in the smaller lipolysis product, LDL [6]. It follows that contact between apo B and the endothelial cell must be restricted to a relatively small fraction of the primary sequence. The same considerations apply to the interaction of the larger chylomicron particles, which are stabilized by the shorter apo B variant, apo B48. Since triacylglycerols in VLDL and chylomicrons are competitive substrates for LPL in mixtures of these lipoproteins, the same LPL binding sites must accommodate either lipoprotein particle.

Kinetic data suggest that several molecules of LPL simultaneously catabolize the triacylglycerol of each VLDL or chylomicron particle. The turnover number of LPL under physiological conditions is about $10$ $s^{-1}$. For a chylomicron containing $3 \times 10^5$ molecules of triacylglycerol, catabolism of 50% of this lipid by a single LPL molecule would take about 3 h, yet the measured $t_{1/2}$ is 10–15 min. These data suggest that several molecules of LPL become attached to the circumference of each chylomicron

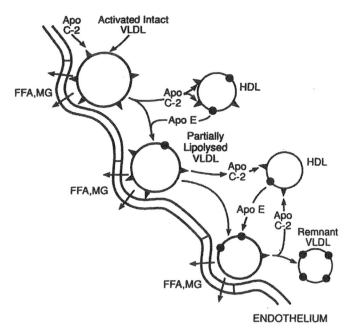

Fig. 5. Mechanism of remnant lipoprotein formation at the endothelial surface. Apo B is not illustrated. FFA, free fatty acid; MG, monoacylglycerol; apo C2, closed triangles; closed circles, apo E. This model reflects the appearance of partially lipolyzed lipoprotein particles in the circulation during LPL-mediated lipolysis of triacylglycerol-rich lipoproteins.

or VLDL particle during lipolysis, with each LPL activated by one molecule of apo C2. In Fig. 4, apo B is illustrated binding directly to the endothelial cell surface, while individual proteoglycan anchors bind LPL to the endothelium. Each apo C2 would link one LPL to the surface of the lipoprotein. If this model were correct, LPL binding sites on the capillary endothelium must be fluid yet highly organized, and able to adapt rapidly to substrates with different diameters and apo C2 contents.

### 2.5. Kinetics of the LPL reaction and the role of albumin

As VLDL and chylomicrons pass down their delipidation cascade, partially catabolized intermediates formed as a result of LPL activity are detected in the circulation (Fig. 5). This observation makes it likely that lipolysis does not result from a single binding event. Rather, there must be repeated dissociation and rebinding, during which lipoprotein triacylglycerol is catabolized, apo C2 is gradually lost, and LPL catalytic rate is decreased while remnant end-products are formed.

There has been considerable discussion of mechanisms by which triacylglycerol-rich lipoproteins could be reversibly displaced from the endothelial surface. The most likely would involve the transient accumulation of lysogenic lipolysis products at the lipoprotein surface within the LPL-binding surface microdomain. After dissociation of the lipoprotein particle, these lipids would diffuse away, leaving the partially lipolyzed particle once more competent to bind to the lipase site. There are three candidate lipids for such a role: unesterified fatty acids, monoacylglycerols, and lysophosphatidylcholine.

Only a portion of the fatty acids generated by LPL are cleared locally. The rest remain in the circulation, after transfer from the surface of the substrate lipoprotein to albumin, part of which remains bound to the lipoprotein surface. Under physiological conditions, fatty acids are largely converted to their sodium and potassium salts, and can act as detergents. Monoacylglycerols are effective lysogens. Even at concentrations of 1–2 μM, they inhibit LPL activity in the isolated perfused rat heart. Monoacylglycerols do not bind to albumin, but are taken up rapidly by cells. LPL also generates lysophospholipids. These are effective lysogens but unlike monoacylglycerols, they form stable complexes with albumin. Although further research is needed, monoacylglycerols seem the most likely contributors to the transient displacement of triacylglycerol-rich lipoproteins from the vascular surface.

LPL has an important role in directing VLDL triacylglycerol to muscle tissues during fasting, and VLDL and chylomicron triacylglycerol to adipose tissue postprandially. In addition to the transcriptional regulation described above, there is evidence of posttranslational and kinetic differences between LPL sites in adipose and muscle tissues that contribute to the distribution of lipolysis products between tissues.

In adipocytes, fasting is associated with the synthesis of LPL molecules whose N-linked polysaccharide chains retain an unmodified high-mannose structure. In the fed state, these chains are modified by mannose trimming, and the addition of glucose, hexosamine and sialic acid units. The high mannose form of LPL has low specific activity and is retained within the adipocyte. The modified form is actively secreted. Insulin levels are an important determinant of LPL processing.

The apparent $K_m$ of endothelial LPL in adipose tissue is relatively high, compared to that in muscle tissues such as the heart (C.J. Fielding, 1976). This means that the hydrolysis of lipoprotein triacylglycerol by LPL in adipose tissue remains proportional to substrate concentration. In contrast, LPL at the surface of muscle capillaries is saturated, even at the low circulating levels of triacylglycerol-rich lipoproteins characteristic of the fasting state.

### 2.6. Later metabolism of chylomicron and VLDL triacylglycerol

Chylomicrons recirculate until about 80% of initial triacylglycerol content has been catabolized in peripheral tissues. The chylomicron remnant is then endocytosed by hepatic receptors (Chapter 21). Chylomicron remnants retain almost the whole of their original cholesteryl and retinyl ester content. This is cleared by the liver along with remnant triacylglycerol.

The metabolism of VLDL remnants is more complex. In humans some VLDL remnants (IDL) are cleared by the liver via the LDL receptor; but a significant proportion (estimated at 50–70%) is further modified within the circulation to generate LDL. Essentially the whole of circulating LDL is formed in this way. Comparison of the composition of IDL with that of LDL indicates that this conversion involves the loss of 80–90% of IDL triacylglycerol, some phospholipid, and the dissociation of remaining apo E. In contrast, IDL free cholesterol content is the same as that in VLDL while cholesteryl ester is increased in LDL, as a result of the activity of cholesteryl ester transfer protein (CETP) that exchanges apo B-associated triacylglycerol for apo A1-associated cholesteryl esters (see Section 4.3).

It was formerly considered that the loss of lipids from IDL was mediated mainly via the activity of hepatic lipase, a triacylglycerol hydrolase with a role in the generation of small HDL (see Section 3). Recent data make this explanation less likely. Mice in which hepatic lipase was inactivated, and mice and rabbits overexpressing hepatic lipase, had similar, normal levels of circulating total and remnant triacylgycerol, even postprandially. In contrast, IDL is an optimal substrate for CETP. While VLDL and IDL contain similar numbers of cholesteryl ester molecules per apo B, LDLs have about 50% more cholesteryl ester molecules per apo B than either. These data suggest that in normal metabolism, the conversion of IDL to LDL is driven mainly by CETP, and that the role of hepatic lipase is to hydrolyze triacylglycerol on HDL, not on IDL. If this model is correct, the loss of apo E which is part of the IDL-to-LDL conversion is probably passive, and reflects the changing surface lipid composition of IDL.

### 2.7. Congenital deficiencies of lipoprotein triacylglycerol metabolism

The functional pools of LPL and hepatic lipase are quantitatively released into plasma by heparin (post-heparin plasma). Genetic deficiency of LPL is associated with a massive increase in the circulating levels of chylomicrons, and an absence of LPL activity from post-heparin plasma. However, there is less increase in VLDL levels than would be predicted from the role of LPL in VLDL catabolism. An alternative, low-capacity pathway probably exists for the clearance of intact VLDL particles by the liver. Numerous mutations within the human LPL gene have now been identified. Their effects on LPL function were discussed in Section 2.2. Congenital hepatic lipase deficiency is associated with increased levels of plasma triacylglycerol compared to controls.

Because of the dominating role of apo C2 as cofactor for LPL activity, the effects of congenital apo C2 deficiency in human plasma mimic those of LPL deficiency. Mice overexpressing or deficient in LPL have been developed. Their plasma lipoprotein patterns resemble those of the corresponding human genetic deficiency, and confirm the roles of LPL and hepatic lipase in plasma lipid metabolism described above. Mice with muscle-specific overexpression of LPL developed insulin resistance along with the expected increase in muscle triacylglycerol levels [7]. This effect was associated with a decrease in insulin-stimulated glucose uptake. These findings show the power of transgenic mouse models in studying complex metabolic diseases.

## 3. HDL and plasma cholesterol metabolism

### 3.1. The origin of HDL

Unlike apo B-containing lipoproteins, HDL-containing apo A1 are formed in the extracellular space. This process involves the association of lipid-poor apo A1 with cell-derived phospholipids and cholesterol. The association of apo A1 and phospholipid is thermodynamically favorable; phospholipid-free apo A1 has not been detected in biological fluids. Nevertheless isolated, lipid-free apo A1 is often used as a convenient surrogate for lipid-poor apo A1 in the analysis of lipid transfers from the cell surface.

Newly synthesized apo A1 made by the liver and (particularly in humans) by the small intestine is recovered loosely associated with the surface of lymphatic triacylglycerol-rich lipoproteins. The apo A1, probably in association with small amounts of phospholipid, dissociates spontaneously after entering the plasma compartment, in a reaction independent of lipolysis. Lipid-poor apo A1 can also be generated via the action of lipid transfer proteins and/or hepatic lipase (Sections 4.2 and 4.3), when these reduce the core size of mature, spherical HDL.

Both lipid-poor and lipid-free apo A1 demonstrate preβ-migration when plasma is fractionated by nondenaturing agarose gel electrophoresis. Under these conditions, the bulk of HDL, made up of spherical lipid-rich particles, has more rapid, α-migration, while LDL migrates more slowly in a β-position. This technique has proven to be useful for discriminating 'early' or lipid poor HDL from mature, lipid-rich particles in the RCT pathway. The major preβ-HDL of human plasma (preβ$_1$-HDL) has a molecular weight of about 70 kDa An increase in preβ-HDL levels has been correlated with an impairment of RCT and an increased risk of coronary artery disease in human patients.

There is little information yet on the physical structure of prebeta-HDL, although at least two inter-convertible forms, containing one and two molecules of apo A1, may be present in plasma.

### 3.2. Role of the ABCA1 transporter in HDL genesis

Studies in vitro have shown that the prebeta-HDL population includes avid acceptors of cell-derived cholesterol and phospholipids. These lipoprotein complexes are precursors of mature HDL. In human Tangier Disease, there is an almost complete deficiency of mature HDL. The low levels of apo A1 present (1–2% of normal) have preβ-mobility. Tangier Disease patients also have localized patches of orange, lipid-laden macrophages, classically in the tonsils. LDL levels are very low. Cultured Tangier Disease fibroblasts lack significant ability to transfer either phospholipid or free cholesterol to lipid-free apo A1, though transfer of cellular lipids to mature HDL is almost normal (G. Rogler, 1995). These data have led to the conclusion that Tangier Disease patients inherit a defect in the ability of peripheral cells to build normal mature HDL from lipid-poor, apo A1-containing precursors.

Genetic analysis of Tangier Disease families recently led to the identification of one of the key factors for HDL assembly. The DNA of these patients was found to contain deletions or other defects in the ABCA1 gene. ABCA1 is an ATP-binding cassette (ABC) transporter protein closely related to the multidrug resistance transporter, to several hepatic bile acid transporters, and other transporter proteins active in the transmembrane movement of small amphipathic solutes [8].

ABCA1 mRNA is widely expressed among tissues and in cultured cells. It has been studied most intensively in fibroblasts and macrophages, where its activity has been linked to the efflux of cholesterol and phospholipids to extracellular apo A1.

The regulation of ABCA1 expression is complex, and incompletely understood. At least three classes of mRNA transcripts have been identified, corresponding to different transcriptional start sites. These are under the regulation of different promoter sequences. In macrophages and hepatocytes, ABCA1 mRNA levels are strongly upregulated by oxysterols and retinoic acid via a LXR/RXR tandem transcription site (see Chapter 16) [9]. In a few rodent transformed macrophage cell lines, ABCA1 expression is regulated by cAMP but human cells generally are unaffected (A.E. Bortnick, 2000). ABCA1 mRNA levels are also upregulated by cholesterol itself. The molecular mechanism of the response to cholesterol has not yet been clarified. It may involve oxysterols generated intracellularly (X. Fu, 2001).

Despite the stimulus that the identification of ABCA1 has given to HDL studies, a number of key questions on its role in HDL formation remain unresolved. The first is the mechanism by which ABCA1 promotes lipid efflux, and the identity of the lipids transported. In ABCA1$^{-/-}$ mouse embryos, a defect was identified in the catabolism of apoptotic cell bodies. At the same time, ABCA1$^{-/-}$ cells were found to be defective in annexin V binding, an assay for exofacial phosphatidylserine in apoptotic cells (Y. Hamon, 2000). At the present time, the primary substrate for ABCA1 has not been identified. Phospholipid leaving the cell surface under the influence of ABCA1 is almost entirely phosphatidylcholine. One current hypothesis is that ABCA1 could modify the phospholipid composition, and possibly charge, of the exofacial leaflet of the membrane bilayer, thereby secondarily reducing the activation energy for efflux of phosphatidylcholine (Fig. 6). Another hypothesis is that ABCA1 directly transports phosphatidylcholine, and possibly free cholesterol. This possibility is consistent with the loss of both free cholesterol and phospholipid efflux from Tangier cells, and the restoration of both activities in cells transfected with ABCA1 cDNA. However, several recent reports suggest that ABCA1 might not play a direct role in cholesterol transport. These studies each showed that phospholipid efflux by ABCA1 was regulated independently of free cholesterol efflux [10]. These data appear more consistent with a two-step process: (1) addition of phospholipid; followed by (2) addition of FC.

### 3.3. The role of caveolae in HDL genesis

Another area of active investigation is the origin of the cellular cholesterol transferred to preβ-migrating (lipid-poor) HDL. There is general agreement that unesterified cholesterol transferred to lipid-poor apo A1 originates mainly from the plasma membrane. Caveolae (see Chapter 1) are microdomains of the cell surface implicated in cholesterol homeostasis and transport as well as signal transduction. These functions are probably related, because unesterified cholesterol levels in caveolae regulate the efficiency of signal transmission. In primary cells including fibroblasts, smooth muscle cells and endothelial cells, caveolae are implicated as the direct precursors of cholesterol in lipid-poor HDL. While caveolae contain several proteins involved in cellular cholesterol homeostasis, such as the scavenger receptor BI (see Section 4.4) there is no convincing evidence at present that the ABCA1 transporter is located there, consistent with the phospholipids and unesterified cholesterol on apo A1 originating from different membrane microdomains.

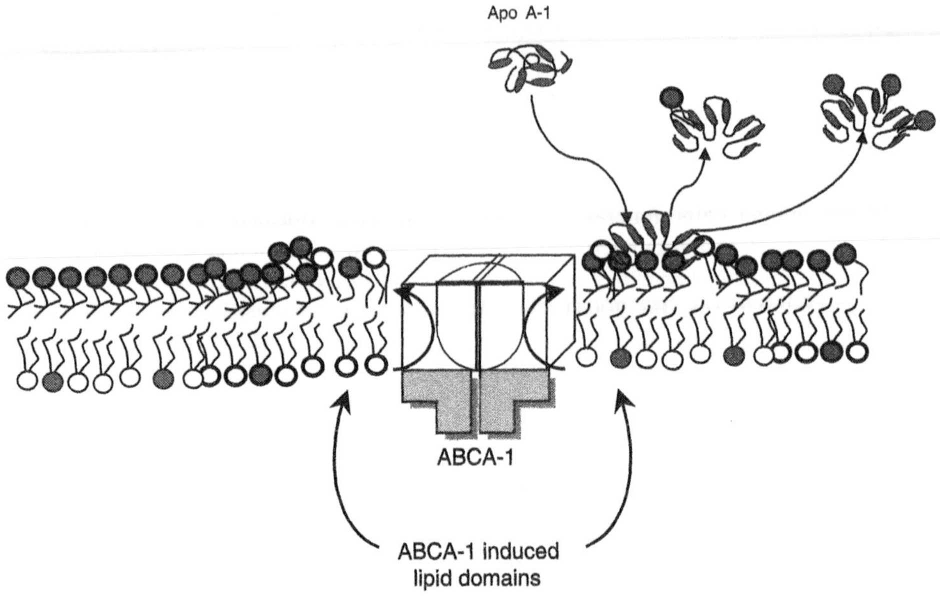

Fig. 6. The ABCA1 transporter and the formation of lipid-poor HDL reactive as acceptors of cell-derived unesterified cholesterol. The figure illustrates the role proposed for ABCA1 in the distribution of phospholipids between the exo- and cyto-facial leaflets of the membrane bilayer. The exofacial leaflet is rich in phosphatidylcholine, substrate for apo A1 at ABCA1 transporter sites. Closed circles, phospholipid; open circles, free cholesterol. Modified from G. Chimini (2002) by permission.

## 3.4. The role of LCAT in HDL genesis

Further growth and maturation of apo A1-HDL depend on the activity of lecithin : cholesterol acyltransferase (LCAT):

Unesterified cholesterol + phosphatidylcholine

$\longrightarrow$ cholesteryl ester + lysophosphatidylcholine

Though present in lymph, LCAT is active mainly in the plasma compartment. It had been thought until recently that the primary substrates of LCAT were phospholipid-rich, discoidal apo A1-containing particles which support maximal acyltransferase rates, and accumulate in the plasma of LCAT-deficient subjects. Recently, it was reported that LCAT could be directly reactive with lipid-poor HDL [11]. This could indicate that more than one pathway can convert lipid-poor to mature HDL.

LCAT consumes unesterified cholesterol and phospholipids to produce insoluble cholesteryl ester (Fig. 7). This is retained in the HDL core, while the water-soluble lysophosphatidylcholine formed at the same time is transferred away to albumin. In this way, LCAT maintains concentration gradients of cholesterol and phosphatidylcholine between cell and lipoprotein surfaces and the growing HDL particle. The later stages of HDL genesis probably depend entirely on diffusion of lipids from the surface of other lipoprotein particles that is independent of ABCA1 activity.

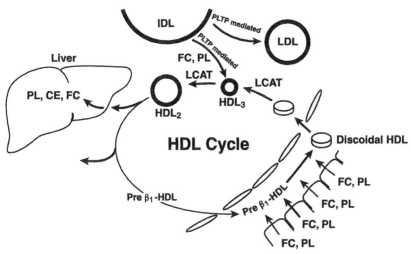

Fig. 7. The HDL cycle, showing: lipid-poor, prebeta-migrating (preβ-1) particles incorporating cell- and lipoprotein-derived unesterified cholesterol and phospholipid; the formation of discoidal HDL; the role of the LCAT reaction in generating spheroidal, alpha-migrating HDL; and the generation of a new cycle of lipid-poor particles at the surface of liver cells. The transfer of excess phospholipid from VLDL and IDL as a result of lipolysis, is catalyzed by phospholipid transfer protein (PLTP).

### 3.5. Regeneration of prebeta-migrating HDL

Lipid-poor, preβ-migrating HDL are formed when the size of the central lipid core of α-migrating HDL, which is mainly cholesteryl ester, is decreased, or when HDL surface lipids are increased (Fig. 7). The surface area occupied by HDL protein, cholesterol and phospholipids then exceeds the surface area of the core. Apo A1 dissociates from the particle in the form of lipid-poor, preβ-migrating HDL. Once released, these prebeta HDL are available as acceptors of additional cell-derived lipids. They may be sufficiently primed with phospholipid to participate in a new cycle of cholesterol efflux. Preβ-HDL particles in plasma are sphingomyelin-rich, while particles newly formed as the result of ABCA1 transporter activity contain mainly phosphatidylcholine molecules. This finding is consistent with the hypothesis that once formed, preβ-HDL can recycle via LCAT, losing phosphatidylcholine to transesterification but retaining sphingomyelin, which is not a LCAT substrate.

Three pathways have been identified for preβ-HDL formation from alpha-HDL:
(1) phospholipid transfer protein (PLTP) activity;
(2) exchange of HDL cholesteryl esters for triacylglycerol in VLDL and LDL catalyzed by cholesteryl ester transfer protein (Section 4.3) concomitant with hepatic lipase-mediated lipolysis of HDL triacylglycerol.
(3) selective uptake of cholesteryl ester from HDL catalyzed by the cell-surface scavenger receptor SR-BI.

It is not known if the preβ-HDL formed by these different pathways have the same kinetic properties as acceptors of cell-derived lipids, though it seems likely. Because triacylglycerol molecules have a similar volume to that of cholesteryl esters, CETP alone would seem unlikely to promote preβ-HDL formation. The relative contribution of the different pathways towards the recycling of apo A1 is likely to differ significantly under different physiological conditions.

Several other plasma apolipoproteins (particularly apo A4 and apo E) have marked sequence similarity to apo A1. Lipid-poor HDL with these proteins in place of apo A1 have been identified. Their concentration is much lower than those of apo A1 particles. Also, it is not clear if apo A4 and apo E particles, two of those identified, can recycle between lipid-rich and lipid-poor populations in the way described for apo A1. As a result, apo A1 is likely to play the predominant role in transporting peripheral cell cholesterol through the plasma compartment to the liver, at least in normal metabolism.

*3.6. Regulation of gene expression and structure of apo A1*

The apo A1 gene codes for a 287-aa preproprotein. Following the loss of its leader sequence, and the removal of a 6-aa pro-sequence in plasma, mature apo A1 circulates as a 243-aa polypeptide. Apo A1 gene transcription rates are not highly regulated, compared to those of ABCA1 and other catalytic factors of the HDL cycle. Sp1, a 'housekeeping' transcription factor, plays a major role in regulating apo A1 transcription rates. PPARs which regulate phospholipid efflux to apo A1, are reported to have little effect on the expression of apo A1 itself. Regulation of the cholesterol transporting activities of apo A1 in plasma is probably determined for the most part by its distribution within its three structural forms, i.e. the amorphous (lipid-poor), discoidal and spheroidal HDL species.

Apo A1, like other phospholipid-binding plasma apolipoproteins, is largely made up a series of amphipathic helical segments, typically 22 amino acids in length [12] (Chapter 18). These are separated by helix-breaking proline or glycine residues. Synthetic amphipathic helical segments whose primary sequence is unrelated to that of native apo A1 can be effective mimics of native apolipoprotein in binding phospholipid, promoting cholesterol efflux from cells, and activating the formation of cholesteryl esters by the LCAT reaction. In spite of this, some repeats in native apo A1 are clearly of more significance than others. This has been seen in experiments where the position of repeats within the primary sequence was systematically varied, though the amino acid sequence of each repeat was unchanged (M. Sorci-Thomas, 1997). The biological activity of such mutant apo A1 species varied widely. These data indicate that apo A1 retains significant tertiary structure. Some generalizations are now possible. The central ('hinge') region of the polypeptide (residues 143–164) appears to be of particular significance in promoting the LCAT reaction [13]. The same domain is important in promoting cellular cholesterol efflux, in the presence or absence of LCAT activity. A C-terminal domain has been implicated in phospholipid binding.

Most of the information on apo A1 function has been obtained from synthetic discoidal recombinants of apo A1 and pure phospholipids, with a molar ratio of 1 : 200 to 1 : 500. The size and molecular properties of these particles, produced by sonication or dialysis from detergent solution, are quite similar to those of discoidal lipoproteins found in the plasma of LCAT-deficient human subjects. The particles have been shown to consist of a planar phospholipid bilayer. The edges of the bilayer are sealed from the

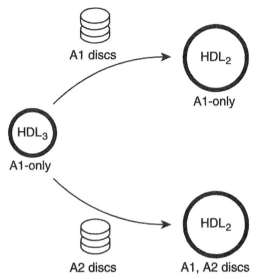

Fig. 8. HDL fusion and the formation of apo A1, apo A2 HDL. The role proposed for PLTP in the promotion of fusion is illustrated, together with the formation of apo A1, apo A2 products from alpha-HDL.

aqueous medium by apo A1. It was thought earlier that a 'picket fence' model in which the repeats were at right angles to the lipid bilayer, accurately reflected the structure of these particles (Chapter 18). A more recent 'belt' model has the repeats aligned circumferentially parallel to the bilayer [14]. The balance of evidence still suggests that discoidal HDL are the normal intermediate of the conversion of lipid-poor, prebeta HDL to mature, spherical particles. The presence of discoidal HDL in lymph is consistent with this interpretation. In any case, the end products of the action of LCAT on apo A1 complexes rich in cholesterol and phospholipid are alpha-migrating, spherical HDL particles rich in cholesteryl esters.

Most alpha HDL particles, unlike preβ- and discoidal HDL, include apo A2 as well as apo A1. Evidence recently obtained suggests that these are a product of the fusion of apo A1-only and apo A2-only HDL particles [15] (Fig. 8). This fusion could be mediated locally by the lysophosphatidylcholine formed in the LCAT reaction. In LCAT deficient plasma, apo A1 and apo A2 form distinct populations of HDL particles. Apo A2 has been considered an inhibitor of the LCAT reaction, and thus indirectly, of reverse cholesterol transport. Apo A2 might thus limit the size reached by spherical HDL. Mice transgenic for apo A2 were atherosclerosis-prone compared to normal animals of the same strain, but this effect has not been seen in mice of all genetic backgrounds.

The formation of α-HDL is accompanied by large changes in the conformation of apo A1. This was made clear by studies with monoclonal antibodies, as well as a variety of sensitive physical techniques. The unique properties of apo A1 in lipid binding and the promotion of reverse cholesterol transport reflect this elasticity.

## 3.7. Structure and properties of LCAT

Plasma LCAT originates mainly from hepatocytes. Hepatic levels of LCAT mRNA are determined mainly by the interplay of Sp1 and Sp3 promoter binding sites. The rate of the LCAT reaction in plasma is regulated for the most part not by changes in circulating LCAT protein levels, but by differences in its catalytic rate with the different HDL particles. Postprandially, LCAT rates are increased as unesterified cholesterol and phospholipid are transferred to HDL from triacylglycerol-rich lipoproteins; the level of LCAT protein in the circulation is unchanged.

There is enough LCAT in plasma for only about 1% of HDL particles to contain one molecule of enzyme. Either LCAT must move rapidly between HDL particles or, more likely, its substrates and products must be transferred effectively from a metabolically active HDL subfraction containing LCAT to other HDL particles. The spontaneous transfer of free cholesterol and lysophosphatidylcholine in plasma is rapid. That of phosphatidylcholine and cholesteryl esters is much slower. These transfers are stimulated by dedicated plasma lipid transfer proteins (Sections 4.2 and 4.3).

LCAT is a 416-amino acid serine hydrolase [16]. It has only limited sequence homology (<5%) to other lipases (LPL, hepatic lipase, pancreatic lipase). The amino acid residues which make up its active site triad have been identified. Several carbohydrate chains modify the reaction rate and substrate specificity of the enzyme. LCAT has not been crystallized. Efforts have been made to explain its three-dimensional structure using the coordinates obtained from X-ray diffraction analysis of triacylglycerol lipases. LCAT, like these lipases, probably has a mobile 'lid' responsive to the lipid interface of HDL. A helical domain, adjacent to the active site serine residue and partly homologous to a sequence in apo E, may be involved in lipid binding. To date, these insights have been insufficient to explain the unique selectively of LCAT for unesterified cholesterol, rather than the hydroxyl group of water, as acyl acceptor. In the complete absence of cholesterol, LCAT is an efficient phospholipase.

Apo A1 is required for both the acyltransferase and phospholipase activities of LCAT. Three arginine residues within the 143–164 repeat of apo A1 ($R_{149}$, $R_{153}$, and $R_{160}$) are essential for its activation by apo A1, suggesting a possible role for salt-bridges between these residues and either negatively charged amino acids in LCAT, and/or phosphate groups within phosphatidylcholine [17]. In reaction with LDL, LCAT catalyzes phosphatidylcholine acyl exchange. LCAT can also hydrolyze short-chain lipid esters. This reaction is independent of the presence of apo A1, consistent with the view that the apoprotein may be needed to align the enzyme and its substrates at a phospholipid–water interface.

## 3.8. Congenital deficiencies of LCAT and HDL

Two variants of LCAT deficiency are recognized. In the first, LCAT synthesizes no cholesteryl esters in plasma. Cholesterol accumulates as droplets in peripheral tissues. Apo-E-rich particles accumulate in LCAT deficient plasma, indicating this alternative cholesterol transport pathway, though upregulated, cannot fully substitute for that catalyzed by LCAT. Only lipid-poor and discoidal HDL particles are present under these conditions. In the second type of LCAT deficiency (Fish-Eye Disease) LCAT can transesterify cholesterol from VLDL and LDL, but not from exogenous HDL. LCAT deficiency and Fish-Eye Disease are the result of different mutations in the primary sequence of the LCAT protein (J.A. Kuivenhoven, 1997).

Normal HDL are absent from plasma in congenital apo A1 deficiency, and also in ABCA1 deficiency (Tangier Disease). In apo A1 deficiency, no HDL is present. In Tangier Disease, HDL present is all in the form of prebeta-HDL. Whether prebeta-HDL in Tangier Disease have the same composition as those in normal plasma has apparently not been reported.

Epidemological studies consistently show that low HDL cholesterol is correlated with an increased risk of atherosclerotic vascular disease [18]. The relationship is usually stronger than that between the same disease and LDL. The evidence that heart disease is systematically increased in LCAT deficiency, apo A1 deficiency and Tangier Disease is equivocal at best, in spite of the fact that LCAT, apo A1 and ABCA1 play key roles in regulating cellular cholesterol content. This paradox has several possible explanations. The first is that these HDL deficiency diseases are also characterized by low levels of circulating LDL (25–50% of normal). This reduces the delivery of cholesterol to peripheral cells to partially offset reduced reverse cholesterol transport. A second possibility is that HDL cholesterol concentrations may not reflect the rate of reverse cholesterol transport. For example, mice transgenic for SR-BI (Section 4.4) increase cholesterol clearance to bile but decrease HDL cholesterol levels (K.F. Kozarsky, 1997). Other pathways, such as that involving apo E, may be able to assume part of the function of apo A1. A third possibility is that it is not HDL cholesterol as such, but a metabolically active subfraction of HDL, that is antiatherogenic. Changes in its composition could be less extreme than those of HDL cholesterol levels. Trace HDL proteins that could play such a role are the antioxidant proteins paraoxonase and platelet activating factor which are responsible for the protective role of HDL in neutralizing oxidized phospholipids in LDL (M. Navab, 2001).

## 4. Reactions linking the metabolism of apo A1 and apo B lipoproteins

### 4.1. Metabolic implications

Triacylglycerol carried by apo B lipoproteins is mainly catabolized in peripheral (that is, non-hepatic) tissues. Its fatty acids are used for oxidative metabolism or storage. In contrast, very little cholesterol is needed for growth or repair in peripheral tissues. Nevertheless there is a continuous 'forward' delivery of cholesterol to peripheral cells. Two main reasons can be suggested. Cholesteryl ester is needed for triacylglycerol-rich particles to be successfully secreted from the liver. Second, the recycling of free cholesterol between the liver and peripheral cells suppresses local cholesterol synthesis and the expression of lipoprotein receptors. These receptors would otherwise promote the futile uptake up large amounts of lipoprotein cholesteryl ester.

Despite the different roles of the apo A1- and apo B-lipoproteins systems, exchange reactions in plasma, catalyzed by lipid transfer proteins, have been identified. These promote the movement of lipids between the major transport pathways. Transfer proteins are ATP-independent. Their reactions (i) are reversible; and (ii) proceed only down preexisting concentration gradients.

## 4.2. Phospholipid transfer protein (PLTP)

PLTP is a 476 amino acid protein showing ~20% sequence similarity to several other lipid-binding proteins, which include cholesteryl ester transfer protein and bacterial permeability inducing protein. Short, highly hydrophobic sequences in conserved regions of the primary sequence may represent the strands of a hydrophobic basket or cleft involved in lipid binding. The expression of PLTP is PPAR-gamma dependent, and may involve the LXR/RXR orphan receptor heterodimeric complex, the same factors that regulate ABCA1 expression, and phospholipid efflux from cells.

In plasma, PLTP catalyzes the transfer of phospholipids, particularly phosphatidylcholine, between lipoprotein classes [19]. The generation of excess surface phosphatidylcholine as a result of triacylglycerol lipolysis, and the consumption of phosphatidylcholine by LCAT, both ensure that a phospholipid gradient is maintained from VLDL and LDL to HDL. PLTP activity is reported to be present in all mammalian plasmas. Human genetic PLTP deficiency has not yet been unequivocally identified.

PLTP activity is needed for maximal LCAT activity because the rate of transfer of phospholipids from cells to plasma via ABCA1 transporter activity, and the spontaneous transfer of phospholipids from other lipoproteins, are both much slower than that of cholesterol. Without PLTP, reverse cholesterol transport might otherwise be limited. PLTP also plays a major role in generating prebeta-HDL, the major acceptor of cellular cholesterol. An additional role for PLTP recently identified is in the secretion of apo B lipoproteins from the liver [20]. Finally, PLTP can promote phospholipid efflux from the surface of fibroblast monolayers to preformed HDL, though not to lipid-free apo A1.

## 4.3. Cholesteryl ester transfer protein (CETP)

CETP is a 476-amino acid plasma protein structurally related to PLTP (Section 4.2). Like PLTP, CETP expression in hepatocytes is PPAR-dependent [21]. The C-terminus of CETP, a domain absent in PLTP, plays a key role in the transfer of both triacylglycerols and cholesteryl esters. Neither the tertiary structure nor the detailed mechanism of CETP-mediated lipid transfer are yet fully established. A model of CETP tertiary structure based on X-ray coordinates established for bacterial permeability-increasing protein has been described. Like LCAT, the activity of CETP is regulated more by the composition of substrate lipoproteins than by the circulating level of CETP protein. For example, increased CETP activity observed postprandially appears to be almost completely the consequence of increased triacylglycerol/cholesteryl ester ratios in triacylglycerol-rich dietary lipoproteins.

Like PLTP, the CETP reaction transfers lipids down a preexisting concentration gradient maintained by LCAT. CETP normally promotes transfer of CE to VLDL and LDL, at a rate typically ~50% that of LCAT. This means that much of the

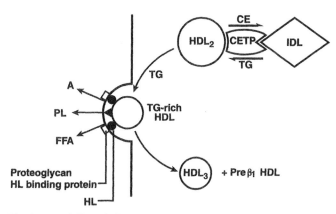

Fig. 9. Remodelling of HDL by hepatic lipase (HL). The hydrolysis of triacylglycerol (TG) transferred from VLDL and IDL via the activity of cholesteryl ester transfer protein (CETP) is shown, together with the displacement of lipid-poor (prebeta$_1$) HDL from the diminished surface of the spherical HDL particle. FFA, free fatty acids.

cholesteryl ester generated by LCAT is cleared directly from HDL, not from LDL after CETP-mediated lipid transfer (Section 4.4).

The net effect of CETP activity is to reduce HDL CE and increase LDL CE. In normal plasma, CE transfer is complemented by a similar and opposite transfer of triacylglycerol from VLDL and LDL to HDL. Under conditions where there is no cholesteryl ester concentration gradient between lipoproteins (for example, if VLDL secreted from the liver contains as much cholesteryl ester as HDL) CETP can still catalyze the unproductive exchange of cholesteryl esters between lipoprotein particles.

There has been considerable debate whether CETP should be considered a 'proatherogenic', pathologically neutral, or 'antiatherogenic' factor. The activities of CETP and hepatic lipase contribute to the recycling of lipid-poor apo A1, and the formation of prebeta-HDL (Fig. 9). Whether or not increased CETP activity leads to an increase in circulating LDL cholesterol levels depends on the capacity of hepatic LDL receptors. A study of Japanese subjects expressing partial CETP deficiency did not suggest any increased resistance to atherosclerosis. On the other hand, reduced CETP levels in hemodialysis patients have been linked to an increased incidence of heart disease (although other relevant activities, including LCAT, were also reduced). Several groups of thiol reagents have been described which inhibit CETP activity. As of the time of writing, CETP inhibitors have not been shown to reduce atherosclerosis in human populations, though beneficial effects in rabbits have been reported.

### 4.4. Scavenger receptor BI (SR-BI)

SR-BI (the human protein is also known as CL-A1) is a 409-amino acid transmembrane protein member of the scavenger B-family. Its primary sequence contains no consensus ATP binding site. As a result, lipid transfers mediated by SR-BI, like those supported by CETP and PLTP, are driven by established concentration gradients. (Fig. 10). SR-BI

548

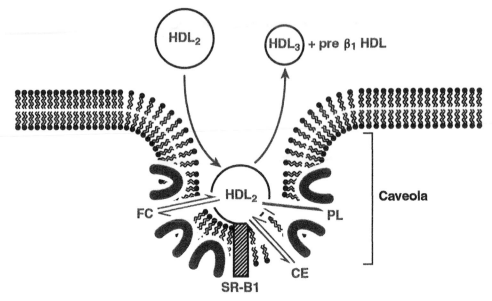

Fig. 10. Lipid transfers catalyzed by SR-BI. SR-BI is shown localized to caveolae. The selective transfers of cholesteryl ester (CE), free cholesterol (FC) and phospholipids (PL) are shown between HDL and the cell surface. The model suggests that SR-BI acts to promote facilitated (that is, protein-mediated) diffusion of lipoprotein lipids down their physiological concentration gradients, under conditions where FC efflux is driven by the LCAT reaction.

catalyzes the selective uptake of lipids, particularly CE, from lipoprotein particles, particularly HDL, though it is active with a wide array of lipids and lipoproteins. SR-BI also promotes efflux of unesterified cholesterol from cells [22]. Transcriptional regulation of SR-BI expression by transcription factor SF-1, by sterol regulatory element binding protein, and by Sp proteins 1 and 3 has been described. While there are significant species differences in the tissue specificity of SR-BI expression, SR-BI is usually expressed at high levels in tissues forming steroid hormones (such as adrenal and gonadal cells). Expression is high in mouse liver but low in human liver. This pattern is the inverse of that seen with CETP and reflects the specialization of humans and mice as 'LDL' and 'HDL' animals respectively. SR-BI has been localized to caveolae, the cholesterol-rich microdomains that are abundant in many peripheral cells.

The selective uptake from HDL requires SR-BI binding. Two positively charged residues ($R_{402}$, $R_{418}$) are important for efficient uptake of cholesteryl esters into the cell. The mechanism by which SR-BI promotes transport of unesterified cholesterol out of the plasma membrane is not yet clear. From a study of mutant SR-BI species and anti-SR-BI antibodies, it was suggested that HDL binding was essential for cholesterol efflux. In contrast, kinetic studies suggested that much of the effect of SR-BI on cholesterol efflux was indirect, possibly the result of induced local modifications in the distribution of lipids within the membrane bilayer. It is unclear to what extent cholesterol efflux from caveolae depends on the presence of SR-BI. CD-36, a second scavenger protein, also found in caveolae, has a weaker effect.

SR-BI may facilitate uptake of unesterified cholesterol from the intestinal lumen. ABCA1 has been implicated in the transport of unesterified cholesterol out of the intestine into lymph. While outside the scope of this chapter, these findings indicate that mechanisms parallel to those described in this chapter may be in place to regulate cholesterol transport to plasma from other extracellular spaces.

## 4.5. Animal models of human plasma cholesterol metabolism

The availability of technology to over-express or delete individual genes in mice has had a wide impact in this field. The effects of modulating the levels of a single enzyme or transport protein can be studied in vivo against the background of interacting factors. Many of these studies were initiated to estimate the role of each gene product in promoting or inhibiting atherogenesis. A major problem of the transgenic/knockout approach has been that in several respects plasma cholesterol metabolism and transport in mice differ significantly from that in humans. As a result, despite successful efforts to create mouse models of human lipid diseases, the quantitative role played by individual factors has sometimes been difficult to establish.

Identification of ABCA1 as the defective protein in human Tangier Disease led rapidly to the production of ABCA1$^{-/-}$ mice [23]. The plasma lipoprotein pattern in these animals, in particular the almost complete absence of HDL, mirrors that of human ABCA1 deficiency. In the developing mouse fetus, disposal of apoptotic cell bodies was inhibited. Cellular cholesterol accumulation in ABCA1$^{-/-}$ mice, particularly in the lungs, was dramatic. These abnormalities have not been reported in human Tangier patients.

Apo A1$^{-/-}$ mice showed decreased HDL cholesterol levels (about 25% of normal). Apo E levels in HDL were increased, and probably as a result, atherosclerosis susceptibility was not increased. Plasma LCAT activity was almost completely inhibited in these animals. This suggests that other apolipoproteins are unable to replace apo A1 in vivo as LCAT cofactors. In mice transgenic for human apo A1, the human protein displaced mouse apo A1 almost completely from HDL. These animals were protected against diet-induced atherosclerosis.

LCAT$^{-/-}$ knockout mice have a marked phenotypic resemblance to LCAT deficient human subjects. Discoidal and lipid-poor HDL species accumulate in the plasma, consistent with the identification of these as precursors of mature, alpha-migrating HDL. LCAT transgenic mice were not protected from atherosclerosis, in contrast to LCAT transgenic rabbits [24].

PLTP$^{-/-}$ mice had reduced HDL levels, consistent with the role proposed for this transfer protein in supplying phospholipids to the LCAT reaction, but atherosclerosis susceptibility was not increased [20]. This may be linked to a concomitant reduction in the secretion of apo B lipoproteins (Section 4.2). Overexpression of PLTP in mice was associated with atherosclerosis resistance, and the appearance of increased concentrations of prebeta-HDL, consistent with the role predicted for these particles in the early steps of reverse cholesterol transport.

CETP is absent from mouse plasma. The human CETP gene was expressed in mice in a number of independent studies. Its effects were complex. In one study, the expression of human CETP together with human LCAT, reduced atherosclerosis susceptibility. In a second study, overexpression of human CETP alone in mice led to increased levels of cholesteryl ester in apo B lipoproteins, and induction of atherosclerosis. The molecular basis of these differences is still not clear.

SR-B1 deficiency was associated with reduction in the bilary cholesterol content, and an increase in circulating HDL levels. Intestinal cholesterol absorption was not increased in SR-BI$^{-/-}$ mice. Mice transgenic for human SR-BI showed significantly lower plasma HDL levels. Paradoxically, atherosclerosis susceptibility was reduced [25].

## 5. Summary and future directions

Since the last edition of this volume, there have been many advances in understanding plasma lipid metabolism. Major developments in basic mechanisms include the identification of ABCA1 and SR-BI. In the field of triacylglycerol transport, the roles of new members of the LDL receptor family have been established, and new insights into the regulation and transport of LPL identified. The efflux of cholesterol from cells to plasma lipoproteins, previously considered the result of passive diffusion, is now recognized to be highly regulated by cell membrane proteins, and on a par with influx as a key determinant of cellular cholesterol homeostasis. The availability of mice overexpressing or deficient in almost all known factors of plasma lipid transport has provided key insights into regulatory pathways.

In other areas, much remains to be done. Knowledge of the tertiary structure of key proteins of plasma lipid metabolism remains very incomplete. Our understanding of the transcriptional regulation of newly identified lipid transport proteins is, not surprisingly, still in its infancy. Efforts to generate mouse models better reflective of human vascular biochemistry and pathology continue. Plasma lipoprotein metabolism has shown itself again to be much more complex, and much more highly regulated than previously thought, and still a target for further intensive research.

## Abbreviations

| | |
|---|---|
| ABCA1 | ATP-binding cassette transporter A1 |
| Apo- | apolipoprotein |
| CETP | cholesteryl ester transfer protein |
| HDL | high-density lipoprotein |
| IDL | intermediate-density lipoprotein |
| LCAT | lecithin : cholesterol acyltransferase |
| LDL | low-density lipoprotein |
| LPL | lipoprotein lipase; |
| PLPT | phospholipid transfer protein |
| RCT | reverse cholesterol transport |
| SR-BI | scavenger receptor BI; |
| VLDL | very low-density lipoprotein |

## Acknowledgements

Research by the authors cited in this chapter was supported by the National Institutes of Health via HL 14237, HL 57976 and HL 67294.

## References

1. Kendrick, J.S., Chan, L. and Higgins, J.A. (2001) Superior role of apolipoprotein B48 over apolipoprotein B100 in chylomicron assembly and fat absorption: an investigation of apobec-1 knockout and wild type mice. Biochem. J. 356, 821–827.
2. Carroll, R. and Severson, D.L. (2001) Peroxisome proliferator-activated receptor-alpha ligands inhibit cardiac lipoprotein lipase activity. Am. J. Physiol. 281, H888–H894.
3. Obunike, J.C., Lutz, E.P., Li, Z.H., Paka, L., Katopodis, T., Strickland, D.K., Kozarsky, K.F., Pillarisetti, S. and Goldberg, I.J. (2001) Transcytosis of lipoprotein lipase across cultured endothelial cells requires both heparan sulfate proteoglycans and the very low density lipoprotein receptor. J. Biol. Chem. 276, 8934–8941.
4. Obunike, J.C., Sivaram, P., Paka, L., Low, M.G. and Goldberg, I.J. (1996) Lipoprotein lipase degradation by adipocytes: Receptor-associated protein (RAP)-sensitive and proteoglycan-mediated pathways. J. Lipid Res. 37, 2439–2449.
5. Parthasarathy, N., Goldberg, I.J., Sivaram, P., Mulloy, B., Flory, D.M. and Wagner, W.D. (1994) Oligosaccharide sequences of endothelial cell surface heparan sulfate proteoglycan with affinity for lipoprotein lipase. J. Biol. Chem. 269, 22391–22396.
6. Chatterton, J.E., Phillips, M.L., Curtiss, L.K., Milne, R., Fruchart, J.-C. and Schumaker, V.N. (1995) Immunoelectron microscopy of low density lipoproteins yields a ribbon and bow model for the conformation of apolipoprotein B on the lipoprotein surface. J. Lipid Res. 36, 2027–2037.
7. Kim, J.K., Fillmore, J.J., Chen, Y., Yu, C., Moore, I.K., Pypaert, M., Lutz, E.P., Kako, Y., Velez-Carrasco, W., Goldberg, I.J., Breslow, J.L. and Shulman, G.I. (2001) Tissue-specific overexpression of lipoprotein lipase causes tissue-specific insulin resistance. Proc. Natl. Acad. Sci. USA 98, 7522–7527.
8. Dean, M., Hamon, Y. and Chimini, G. (2001) The human ATP-binding cassette (ABC) transporter superfamily. J. Lipid Res. 42, 1007–1017.
9. Costet, P., Luo, Y., Wang, N. and Tall, A.R. (2000) Sterol-dependent transactivation of the ABC1 promoter by the liver X receptor/retinoid X receptor. J. Biol. Chem. 275, 28240–28245.
10. Fielding, C.J. and Fielding, P.E. (2001) Cellular cholesterol efflux. Biochim. Biophys. Acta 1533, 175–189.
11. Sparks, D.L., Frank, P.G., Braschi, S., Neville, T.A. and Marcel, Y.L. (1999) Effect of apolipoprotein A-1 lipidation on the formation and function of prebeta- and alpha-migrating LpA-1 particles. Biochemistry 38, 1727–1735.
12. Frank, P.G. and Marcel, Y.L. (2000) Apolipoprotein A-1: structure–function relationships. J. Lipid Res. 41, 853–872.
13. Sorci-Thomas, M.G., Thomas, M., Curtiss, L. and Landrum, M. (2000) Single repeat deletion in apo A-1 blocks cholesterol esterification and results in catabolism of delta-6 and wild type apo A-1 in transgenic mice. J. Biol. Chem. 275, 12156–12163.
14. Segrest, J.P., Jones, M.K., Klon, A.E., Sheldahl, C.J., Hellinger, M., De Loof, H. and Harvey, S.C. (1999) A detailed molecular belt model for apolipoprotein A-1 in discoidal high density lipoprotein. J. Biol. Chem. 274, 31755–31758.
15. Clay, M.A., Pyle, D.H., Rye, K.A. and Barter, P.J. (2000) Formation of spherical, reconstituted high density lipoproteins containing both apolipoproteins A-I and A-II is mediated by lecithin : cholesterol acyltransferase. J. Biol. Chem. 275, 9019–9025.
16. Jonas, A. (2000) Lecithin cholesterol acyltransferase. Biochim. Biophys. Acta 1529, 245–256.
17. Roosbeek, S., Vanloo, B., Duverger, N., Caster, H., Breyne, J., De Beun, I., Patel, H., Vanderkerckhove, J., Shoulders, C., Rosseneu, M. and Peelman, F. (2001) Three arginine residues in apolipoprotein A-1 are critical for activation of lecithin : cholesterol acyltransferase. J. Lipid Res. 42, 31–40.

18. Von Eckardstein, A., Nofer, J.R. and Assmann, G. (2001) High density lipoproteins and arteriosclerosis. Role of cholesterol efflux and cholesterol transport. Arterioscler. Thromb. Vasc. Biol. 21, 13–27.

19. Huuskonen, J., Olkonnen, V.M., Jauhiainen, M. and Ehnholm, C. (2001) The impact of phospholipid transfer protein (PLTP) on HDL metabolism. Atherosclerosis 155, 269–281.

20. Jiang, X.C., Qin, S., Qiao, C., Kawano, K., Lin, M., Skold, A., Xiao, X. and Tall, A.R. (2001) Apolipoprotein B secretion and atherosclerosis are decreased in mice with phospholipid-transfer protein deficiency. Nat. Med. 7, 847–852.

21. Luo, Y., Liang, C.P. and Tall, A.R. (2001) The orphan nuclear receptor LRH-1 potentiates the sterol-mediated induction of the human CETP gene by liver X receptor. J. Biol. Chem. 276, 24767–24773.

22. Gu, X., Kozarsky, K. and Krieger, M. (2000) Scavenger receptor, class B, type I-mediated $^{3}$H-cholesterol efflux to high and low density lipoproteins is dependent on lipoprotein binding to the receptor. J. Biol. Chem. 275, 29993–30001.

23. McNeish, J., Aiello, R.J., Guyot, D., Turi, T., Gabel, C., Aldinger, C., Hoppe, K.L., Roach, M.L., Royer, L.J., de Wet, J., Broccardo, C., Chimini, G. and Francone, O.L. (2000) High density lipoprotein deficiency and foam cell accumulation in mice with targeted disruption of ATP-binding cassette transporter-1. Proc. Natl. Acad. Sci. USA 97, 4245–4250.

24. Sakai, N., Vaisman, B.L., Koch, C.A., Hoyt, R.F., Meyn, S.M., Talley, G.D., Paiz, J.A., Brewer, H.B. and Santamarina-Fojo, S. (1997) Targeted disruption of the mouse lecithin : cholesterol acyltransferase (LCAT) gene. Generation of a new animal model for human LCAT deficiency. J. Biol. Chem. 272, 7506–7510.

25. Mardones, P., Quinones, V., Amigo, L., Moreno, M., Miquel, J.F., Schwartz, M., Miettenen, H.E., Krieger, M., VanPatten, S., Cohen, D.E. and Rigotti, A. (2001) Hepatic cholesterol and bile acid metabolism and intestinal cholesterol absorption in scavenger receptor class B type I-deficient mice. J. Lipid Res. 42, 170–180.

D.E. Vance and J.E. Vance (Eds.) *Biochemistry of Lipids, Lipoproteins and Membranes (4th Edn.)*

# CHAPTER 21

# Lipoprotein receptors

Wolfgang J. Schneider

*Institute of Medical Biochemistry, Department of Molecular Genetics, Dr. Bohr Gasse 9/2,
A-1030 Vienna, Austria, Tel.: +43 (1) 4277-61803; Fax: +43 (1) 4277-61804;
E-mail: wjs@mol.univie.ac.at*

## 1. Introduction

From a systemic lipoprotein metabolism point of view, the main task of lipoprotein receptors is the clearance of lipoproteins from the circulation, body fluids, and interstitial spaces. One can envision several reasons why lipoproteins need to be cleared from extracellular fluids into cellular compartments, for instance: (1) they are transport vehicles for components that are vital to the target cells; (2) their uptake serves signalling and/or regulatory roles in cellular metabolism; (3) they have done their job and have become dispensable; and (4) they might have deleterious effects if allowed to remain extracellular for prolonged time. Chapters 18, 19, and 20 have described the structures, syntheses, and interconversion pathways of lipoprotein particles in the circulation. Here, the mechanisms of lipoprotein transport from the plasma compartment to various types of cells of the body, which is one of the best understood aspects of receptor biology, is described. In addition, newly discovered functions of receptors thus far thought to be specialized exclusively for lipoprotein transport are described.

In the context of lipoprotein transport via cell surface receptors, it helps to recall that the two major transported lipid components of lipoproteins, triacylglycerols and cholesterol, have quite different fates. Triacylglycerols are delivered primarily to adipose tissue and muscle where their fatty acids are stored or oxidized for production of energy, respectively. Cholesterol, in contrast, is continuously shuttled among the liver, intestine, and other extra-hepatic tissues. Actually, the major transport form of cholesterol is its esterified form; within cells the cholesteryl esters are hydrolyzed and the unesterified sterol has multiple uses. Among their many functions, sterols serve as structural components of cellular membranes, as substrates for the synthesis of steroid hormones and bile acids, and they perform several regulatory functions (a classical example is the low density lipoprotein (LDL) receptor pathway, see Section 2.2). For correct targeting of lipoproteins to sites of metabolism and removal, the lipoproteins rely heavily on the apolipoproteins (apos) associated with their surface coat. Apos mediate the interaction of lipoprotein particles with enzymes, transfer proteins, and with cell surface receptors, the main topic of this chapter.

Key features of human lipoprotein metabolic pathways are schematically summarized in Fig. 1; this outline necessarily omits many of the details which are less significant for aspects of receptor-mediated removal of lipoproteins. The interwoven complex pathways can be divided into exogenous and endogenous branches, concerned with the transport of dietary and liver-derived lipids, respectively. Both metabolic sequences start with the

554

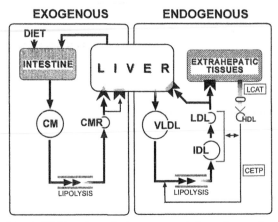

Fig. 1. Highly schematized summary of the pathways for receptor-mediated lipoprotein metabolism in humans. The liver is the crossing point between the exogenous pathway (left-hand side) dealing with dietary lipids, and the endogenous pathway (right-hand side) that begins with hepatically synthesized lipoproteins. The exogenous metabolic branch starts with the production of chylomicrons (CM) in the intestine, whereas the liver synthesizes very low density lipoprotein particles (VLDL). Abbreviations: CMR, chylomicron remnants; IDL, intermediate density lipoproteins; LDL, low density lipoproteins; HDL, high density lipoproteins; LCAT, lecithin : cholesterol acyltransferase; CETP, cholesteryl ester transfer protein; ▲, LDL receptor-related protein (LRP); and ◥, LDL receptor; 'Lipolysis' denotes lipoprotein lipase-catalyzed triacylglycerol lipolysis in the capillary bed.

production and secretion of triacylglycerol-rich lipoproteins (Chapter 19). Intestinally derived chylomicrons (CM in Fig. 1) are secreted into the lymph and from there enter the bloodstream, where they function as energy carriers by providing triacylglycerol-derived fatty acids to peripheral tissues. This lipolytic extraction of fatty acids from the triacylglycerol core of the lipoprotein particles ('Lipolysis' in Fig. 1) is achieved mainly by the enzyme lipoprotein lipase, which is bound to the lumenal surface of the endothelial cells lining the capillary bed. Removal of triacylglycerol in extrahepatic tissues results in decreased size of the chylomicrons and produces cholesteryl ester-rich lipoprotein particles termed chylomicron remnants. During this conversion, apoCs are lost from the surface of the particles; the remnants, having finished their task, are destined for catabolism by the liver, which occurs almost exclusively by receptor-mediated processes. Both the so-called LDL receptor-related protein (▲ in Fig. 1, and see Section 3.1) and the LDL receptor (◥ in Fig. 1, and see Section 2) mediate their removal via recognition of apoE. The apoB48, which resides on chylomicrons throughout their life span, is not recognized by these receptors.

In analogy to the exogenous lipid transport branch, the endogenous pathway begins with the production and secretion of triacylglycerol-rich lipoproteins by the liver, here termed very low density lipoprotein (VLDL). In significant contrast to chylomicrons, VLDL contains apoB100, in addition to the apoCs and apoE. Lipoprotein lipase in the capillary bed hydrolyzes triacylglycerol of secreted VLDL, but less efficiently than from chylomicrons, which is likely one of the reasons for slower plasma clearance of VLDL ($t_{1/2}$, days) compared to chylomicrons ($t_{1/2}$, minutes to a few hours). Lipolysis during the prolonged residency of VLDL particles in the plasma compartment generates intermediate density lipoproteins (IDL) (Fig. 1) and finally, LDL. In parallel, apos and surface components (mostly phospholipids and unesterified cholesterol), but also cholesteryl esters and triacylglycerol, are subject to transfer and exchange between particles in the VLDL lipolysis pathway and certain species of high density lipoproteins (HDL). In addition, it appears that another receptor akin to the LDL receptor, the so-called VLDL receptor (see Section 3.2), might act in concert with lipoprotein lipase in delivering fatty acids to a limited set of peripheral tissues. Enzymes involved in cholesterol loading and esterification, and in interparticle-transfer reactions are, e.g., lecithin–cholesterol acyltransferase (LCAT) and cholesteryl ester transfer protein (CETP) (see Chapter 20). IDL particles (which still harbor some apoE) to a variable degree, and LDL as the end product of VLDL catabolism in the plasma (and at least in man, free of apoE), are catabolized via the LDL receptor. This receptor is found in the liver (which harbors 60–70% of all the LDL receptors in the body) as well as in extrahepatic tissues, and is a key regulatory element of systemic cholesterol homeostasis (see Section 2.2).

Thus, steady-state plasma LDL levels are not only the result of lipoprotein receptor numbers, but also are influenced by the rate of VLDL synthesis, the activity of lipoprotein lipase and other lipases, the VLDL receptor, and by other metabolic processes. As far as HDL levels are concerned, one result of the LCAT- and CETP-catalyzed reactions is the production of a dynamic spectrum of particles with a wide range of sizes and lipid compositions. The further metabolic fates of these HDL fractions are described in Chapter 20.

In addition to receptor-mediated metabolism of lipoproteins, which clearly is the predominant mechanism for removal from the plasma of intact lipoproteins, individual components of lipoproteins, in particular unesterified cholesterol, might diffuse into cells across the plasma membrane. Other minor uptake processes may include so-called fluid-phase endocytosis, which does not involve binding of lipoproteins to specific cell surface proteins, and phagocytosis, in which lipoproteins are thought to attach to the cell surface via more or less specific forces, and are engulfed by the plasma membrane.

## 2. Removal of LDL from the circulation

The supply of cells with cholesterol via receptor-mediated endocytosis of LDL is one of the best characterized processes of macromolecular transport across the plasma membrane of eukaryotic cells. The following sections describe this process, provide an overview of the biochemical and physiological properties of the LDL receptor, and discuss the molecular basis for the genetic disease, familial hypercholesterolemia.

### 2.1. Receptor-mediated endocytosis

This multi-step process, originally defined as a distinct mechanism for the cellular uptake of macromolecules, emerged from studies by M.S. Brown, J.L. Goldstein and their colleagues which they performed in the mid-1970s in order to elucidate the

556

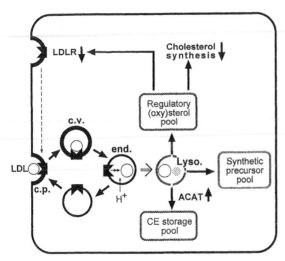

Fig. 2. The LDL receptor pathway, regulation of cellular cholesterol homeostasis. LDL receptors are synthe-sized in the endoplasmic reticulum and undergo post-translational modification in the Golgi compartment; from there they travel to the cell surface and collect in coated pits (c.p.). LDL particles bound to LDL receptors (▶) are internalized in coated vesicles (c.v.) which become uncoated and acidified by protons (H⁺) being pumped into their lumen, resulting in endosomes (end.), in which the LDL particles dissociate from the receptors due to the low pH. From there, the LDL are delivered to lysosomes (Lyso.), but almost all of the receptors travel back to the cell surface (where they become incorporated into c.p. again) within a recycling vesicle. Lysosomal degradation of LDL results in complete breakdown of apoB100 and liberation of cholesterol via hydrolysis of cholesteryl esters. The LDL-derived cholesterol has three main fates: (a) it is reconverted to cholesteryl esters via stimulation of acyl-CoA cholesterol acyltransferase (ACAT) for storage in droplets (CE storage pool; bottom); (b) it is used as biosynthetic precursor for bile acids, steroid hormones, membranes, etc. (synthetic precursor pool; right-hand side); and (c) it serves, especially if converted to oxysterols (top), several regulatory functions. The most important of these are suppression of cholesterol synthetic enzymes, and decreasing the production of LDL receptors.

normal function of LDL. The salient features of the itinerary of a LDL particle (mean diameter, ∼22 nm) from the plasma into a normal human fibroblast are summarized in Fig. 2. First, the lipoprotein particle binds to one of the approximately 15,000 LDL receptors on the surface of the cell. LDL receptors are not evenly distributed on the cell surface; rather, up to 80% are localized to specialized regions of the plasma membrane comprising only 2% of the cell surface. These regions form pits and are lined on their cytoplasmic side with material that in electron micrographs has the appearance of a fuzzy coat. Each of these so-called 'coated pits' contains several kinds of endocytic receptors in addition to LDL receptors, but LDL particles bind only to 'their' receptors, due to their extremely high affinity and specificity. Next, the receptor/LDL complex undergoes rapid invagination of the coated pit, which eventually culminates in the release of the coated pit into the interior of the cell. At this point, the coated pit has been transformed into an endocytic 'coated vesicle', a membrane-enclosed organelle that is coated on its exterior (cytoplasmic) surface with a polygonal network of fibrous protein(s), the main structural component of which is a fascinating protein called clathrin [1]. Subsequently, the coat is rapidly removed, in concert with acidification of the vesicles' interior and fusion with other uncoated endocytic vesicles. Transiently, LDL and the receptor are found in smooth vesicles in which the lipoprotein particle dissociates from the receptor due to the acidic environment. LDL is then delivered to lysosomes, where it is degraded, while the receptor escapes this fate and recycles back to the cell surface, homes in on a coated pit and is ready to bind and internalize new ligand molecules [1].

There are variations to the theme; not in all systems of receptor-mediated endocytosis are ligand degradation and receptor recycling coupled; however, all have in common the initial steps leading to the formation of endosomes. Then, the receptors are either degraded, recycled back to the cell surface, or are transported (for example, across polarized cells); their respective ligands can follow the same or divergent routes [1]. The reutilization of the LDL receptor via recycling constitutes an economical way to ensure efficient removal of LDL from the extracellular space.

## 2.2. The LDL receptor pathway

The LDL receptor is the key component in the feedback-regulated maintenance of cholesterol homeostasis in the body [1]. In fact, as an active interface between extra- and intra-cellular cholesterol pools, it is itself subject to regulation at the cellular level (cf. Fig. 2). LDL-derived cholesterol (generated by hydrolysis of LDL-borne cholesteryl esters) and its intracellularly generated oxidized derivatives mediate a complex series of feedback control mechanisms that protect the cell from over-accumulation of cholesterol. First, (oxy)sterols suppress the activities of 3-hydroxy-3-methylglutaryl-CoA (HMG-CoA) synthase, and HMG-CoA reductase, two key enzymes in cellular cholesterol biosynthesis. Second, the cholesterol activates the cytoplasmic enzyme acyl-CoA : cholesterol acyltransferase (ACAT; E.C. 2.3.1.26) which allows the cells to store excess cholesterol in re-esterified form. Third, the synthesis of new LDL receptors is suppressed, preventing further cellular entry of LDL and thus cholesterol overloading. The coordinated regulation of LDL receptors and cholesterol synthetic enzymes relies on the sterol-modulated proteolysis of a membrane-bound transcription factor, SREBP, as described in Chapter 15.

The overall benefits from, and consequences of, this LDL receptor-mediated regulatory system are the coordination of the utilization of intra- and extra-cellular sources of cholesterol at the systemic level. Mammalian cells are able to subsist in the absence of lipoproteins because they can synthesize cholesterol from acetyl-CoA. When LDL is available, however, most cells primarily use the LDL receptor to import LDL cholesterol and keep their own synthetic activity suppressed. Thus, a constant level of cholesterol is maintained within the cell while the external supply in the form of lipoproteins can undergo large fluctuation.

Most of these concepts have arisen from detailed studies on cultured fibroblasts from normal subjects and from patients with the disease, familial hypercholesterolemia (FH). Lack of the above described regulatory features in FH fibroblasts led to the conclusion that the abnormal phenotype is caused by lack of LDL receptor function, and thus, disruption of the LDL receptor pathway. In particular, the balance between extracellular and intracellular cholesterol pools is disturbed. Clinically, the most important effect of

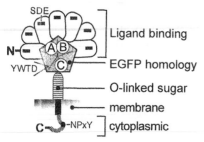

Fig. 3. Domain model of the LDL receptor. The five domains of the mature protein, from the N-terminus (bold N) to the carboxy-terminus (bold C) are as follows. (1) The ligand binding domain, characterized by seven cysteine-rich repeats, each with a cluster of negatively charged amino acids whose core consists of Ser–Asp–Glu ('SDE'); repeats 2–7 cooperatively bind apoB100 and apoE. (2) The EGF-precursor (EGFP) homology region, consisting of 400 amino acid residues (central pentagon); adjacent to the ligand binding domain and at the carboxy-terminus of this region, respectively, are located three repeats with high homology to repeat motifs found in the precursor to epidermal growth factor (encircled letters A, B, and C). The remaining portion of this domain consists of five internally homologous stretches of approximately 50 amino acid residues each of which contains the sequence Tyr–Trp–Thr–Asp (YWTD). (3) The O-linked sugar region, consisting of 58 amino acids with 18 serine and threonine residues containing O-linked carbohydrate chains. (4) A single membrane-spanning domain. (5) The cytoplasmic tail with 50 amino acid residues containing the internalization sequence Asn–Pro–Val–Tyr (NPxY; the Val is not absolutely conserved in all species).

LDL receptor deficiency is hypercholesterolemia with ensuing accelerated development of atherosclerosis and its complications (Chapter 22). In the following sections, a detailed description of the LDL receptor is provided, with emphasis on the impact of mutations on its structure and function.

### 2.3. Relationships between structure and function of the LDL receptor

Studies at the levels of protein chemistry, molecular biology, and cell biology have led to a detailed understanding of the biology of the LDL receptor. The mature receptor is a highly conserved integral membrane glycoprotein consisting of five domains (Fig. 3). In order of appearance from the amino terminus these domains are: (1) the ligand binding domain; (2) a domain that has a high degree of homology with the epidermal growth factor precursor (EGFP); (3) a domain that contains a cluster of O-linked carbohydrate chains; (4) a transmembrane domain; and (5) a short cytoplasmic region. Until direct information on the three-dimensional structure of the 839-residue receptor becomes available, an arrangement of these domains as presented in Fig. 3 may serve as a useful model.

### 2.3.1. The ligand binding domain

This domain mediates the interaction between the receptor and lipoproteins containing apoB100 and/or apoE [2]. The function is localized to a region at the amino terminus of the receptor, comprised of seven repeats of approximately 40 residues each. These seven repeats have six cysteines each, which presumably mediate the folding of the domain into a rigid structure with clusters of negatively charged residues on its surface (with the signature tripeptide Ser–Asp–Glu). These clusters are thought to participate in the binding of lipoprotein(s) via positively charged residues on apoB100 or apoE.

### 2.3.2. The epidermal growth factor precursor homology domain
This region of the LDL receptor lies adjacent to the ligand binding site and is comprised of approximately 400 amino acids; the outstanding feature is the sequence similarity of this region to parts of the EGFP, i.e., three regions termed 'growth factor repeats'. Two of these repeats (A, B in Fig. 3) are located in tandem at the amino terminus, and the other (C21) is at the carboxy-terminus of the precursor homology region of the LDL receptor. The remainder consists of five ~50-residue stretches that contain tetrapeptide sequences with a consensus of Tyr–Trp–Thr–Asp. Experimental evidence suggests an involvement of this region in the receptor's acid-dependent dissociation from LDL and its subsequent recycling (cf. Fig. 2).

### 2.3.3. The O-linked sugar domain
The O-linked sugar domain of the human LDL receptor is a 58-amino acid stretch highly enriched in serine and threonine residues, located just outside the plasma membrane. Most, if not all, of the 18 hydroxylated amino acid side chains are glycosylated. The O-linked oligosaccharides undergo elongation in the course of receptor synthesis and maturation: when leaving the endoplasmic reticulum, N-acetylgalactosamine is the sole O-linked sugar present, and upon processing in the Golgi, galactosyl and sialyl residues are added. Despite the detailed knowledge about the structure of this region, its functional importance remains unclear.

### 2.3.4. The membrane anchoring domain
The membrane anchoring domain lies carboxy-terminally to the O-linked carbohydrate cluster. It consists of 22–25 hydrophobic amino acids; as expected, the deletion of this domain in certain naturally occurring mutations, or by site-directed mutagenesis, leads to secretion of truncated receptors from the cells.

### 2.3.5. The cytoplasmic tail
The cytoplasmic tail of the LDL receptor constitutes a short stretch of 50 amino acid residues involved in the targeting of LDL receptors to coated pits. Naturally occurring mutations and site-specific mutagenesis [3] have identified an 'internalization signal', Asn–Pro–Xxx–Tyr (NPxY; where x denotes any amino acid). Recently, the cytoplasmic domains of the LDL receptor and structural relatives have come into new focus, since they hold the key to the involvement of these receptors in signal transduction as indicated below (Section 5).

### 2.4. The human LDL receptor gene — organization and naturally occurring mutations

The ~48 kb human LDL receptor gene contains 18 exons and is localized on the distal short arm of chromosome 19. There is a strong correlation between the functional domains of the protein and the exon organization in the gene. For instance, the seven cysteine-rich repeats of the ligand binding domain are encoded by exons 2 (repeat 1), 3

560

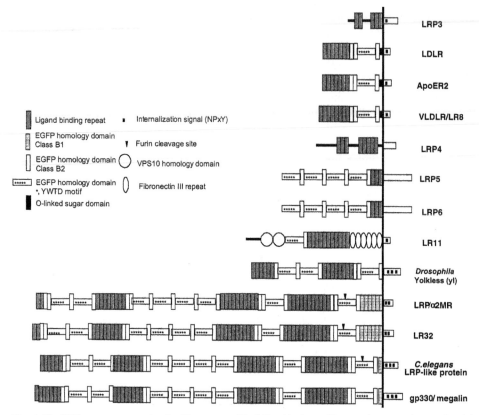

Fig. 4. The LDL receptor gene family. The structural building blocks making up these proteins are listed in the left-hand top part (for more details, see Fig. 3); presumed extracellular domains are depicted to the left of the plasma membrane (black vertical line). The standard modules are: negatively charged ligand binding repeats with six cysteines each; EGFP homology repeats (in the entire family, two subclasses with slightly different consensus sequences are distinguished, termed B1 and B2; these repeats also contain six cysteines each); the five 'YWTD' motifs within the 5-times-50 residues structure of EGFP homology domains; the O-linked sugar domains, just outside the plasma membrane, typical for LDL receptor, apoER2 and VLDL receptor/LR8; and the consensus or presumed internalization signals, NPxY. Several large members of the gene family harbor consensus furin cleavage sites; LR11 (Section 6.2) contains domains not found in other relatives, i.e., two VPS10 domains and six fibronectin type III repeats. Except for the yolk protein transport receptor of *Drosophila melanogaster* (yl), all receptors are discussed in the text.

(repeat 2), 4 (repeats 3, 4, and 5), 5 (repeat 6) and 6 (repeat 7). The EGFP homology domain is encoded by eight exons, organized in a manner very similar to the gene for the EGFP itself. The third domain is translated from a single exon between introns 14 and 15. Thus, the LDL receptor gene is a compound of shared coding sequences; in fact, many more molecules containing all or some of these elements have been discovered and likely will continue to be found. Membrane proteins with clusters of ligand binding repeats in their extracellular domains are now recognized as relatives of the so-called LDL receptor gene family (Fig. 4).

Molecular genetic studies in FH patients have identified over 600 different mutations in the LDL receptor gene. A listing of all mutations with their original literature citations can be found at http://www.ucl.ac.uk/fh/. In order to gain some insight into the nature of these mutations, they are grouped into five classes according to their effects on the protein [4] as follows.

### 2.4.1. Class 1: null alleles — no detectable receptor
These mutant alleles fail to produce receptor proteins as determined by immunoprecipitation with many anti-receptor antibodies, and thus, cells carrying these mutations do not bind any LDL. The spectrum of these mutations includes point mutations causing premature termination codons early in the protein coding region, mutations in the promoter region that block transcription, mutations that lead to abnormal splicing and/or instability of the mRNA, and large deletions.

### 2.4.2. Class 2: slow or absent processing of the precursor
These alleles, probably accounting for at least half of all mutant LDL receptor alleles, specify transport-deficient receptor precursors which fail to move with normal rates from the endoplasmic reticulum to and through the Golgi compartment(s) and on to the cell surface. As a consequence, the typical sudden increase in apparent $M_r$ observed during biosynthesis of the normal receptor is lacking. While some mutations only slow down processing, most of these mutations are complete in that there is total absence of transport from the endoplasmic reticulum, and the mutant receptors never reach the cell surface.

### 2.4.3. Class 3: defective ligand binding
These receptors in general reach the cell surface at normal rates, but are unable to bind LDL efficiently due to subtle structural changes near the ligand binding domain. By definition, these mutant receptors undergo the normal maturation process.

### 2.4.4. Class 4: internalization-defective
Here, one of the prerequisites for effective ligand internalization — localization of LDL receptors to coated pits — is not met. The failure of these 'internalization-defective' receptors to localize to coated structures results from mutations that directly or indirectly disrupt the carboxy-terminal domain of the receptor. Variants of class 4 mutations have been identified in which large deletions lead to a lack of both the cytoplasmic and transmembrane domains; the majority of these mutant truncated proteins are, as expected, secreted.

### 2.4.5. Class 5: recycling-defective
The classification of these mutations into a separate class [4] is based on the observation that upon deletion of the first two EGFP domains of the human LDLR, the truncated receptor binds and internalizes ligand, but fails to release it in the acidic environment of the endosome [5]. The mutant receptor is rapidly degraded without returning to the surface in an unoccupied state. Class 5 mutants all affect the EGFP homology domain, and most often the YWTD regions thereof.

In summary, to a large extent through the delineation of natural mutations in the LDL receptor gene, structural as well as regulatory features of receptor-mediated metabolism of the major cholesterol-carrying lipoprotein in human plasma are now well understood. In the following sections, our still somewhat limited knowledge about the events involved in receptor-mediated plasma clearance of triacylglycerol-rich lipoproteins is outlined.

## 3. Removal of triacylglycerol-rich lipoproteins from the plasma

### 3.1. Catabolism of chylomicrons

Chylomicrons are too large to cross the endothelial barrier; thus, their prior lipolysis to remnants serves a dual function: transport of energy to tissues, and decrease in size to facilitate terminal catabolism. Classical experiments in hepatocytes, perfused rat livers, and transgenic and knockout mice studies have shown that chylomicron remnant transport into the liver is mediated by cell surface receptors. Although apoE, which is recognized by the LDL receptor, is the surface component that targets chylomicron remnants to their site of uptake, studies in LDL receptor-deficient model systems predicted that chylomicron remnant removal would be LDL receptor independent. Individuals with homozygous FH, who lack functional LDL receptors, show no signs of delayed clearance of chylomicron remnants. Furthermore, evidence for a separate hepatic chylomicron remnant removal mechanism came from studies in which dietary, drug, and hormonal factors were shown to regulate the number of hepatic LDL receptors without greatly affecting the clearance rate for chylomicron remnants.

Since the LDL receptor and the proposed chylomicron remnant receptor share at least one property, namely apoE binding, attempts to isolate this receptor were based on the presumed similarity of its ligand binding region to that of the LDL receptor. Indeed, homology cloning resulted in the characterization of an unusually large membrane protein, composed exclusively of structural elements found in the LDL receptor molecule; it has therefore been termed LDL receptor-related protein, or LRP [6]. As shown in Fig. 4, LRP, a 4526-amino acid integral membrane glycoprotein, contains (among other structural elements found in the LDL receptor) 31 repeats of the type forming the ligand binding domain in the LDL receptor and 22 repeats of the growth factor type. The unusually large membrane protein binds lipoproteins in apoE-dependent fashion.

Soon after its cloning, LRP was shown to be identical to the receptor for $\alpha_2$-macroglobulin ($\alpha_2$MR), a major plasma protein that functions in 'trapping', and thereby inactivating, cellular proteinases that have entered the plasma compartment. Since then, many more plasma proteins and protein complexes have been identified, which at least in vitro bind to LRP [7]. Importantly, $\alpha_2$-macroglobulin–proteinase complexes are cleared rapidly by the liver (with the same kinetics as chylomicron remnants), indicating that LRP may indeed perform multiple functions in the removal of spent vehicles of intestinal lipid transport and of potentially harmful proteinases. Another LDL receptor relative with an even broader range of functions is introduced in the following section, and further properties of this protein are described in detail in Sections 4 and 5.

### 3.2. The so-called VLDL receptor: a role in catabolism of VLDL?

The name VLDL receptor was coined for a protein discovered in 1992 by Takahashi et al. [8]. The overall modular structure of the VLDL receptor is virtually superimposable with that of the LDL receptor, except that the ligand recognition domain contains an additional binding repeat located at the amino terminus (Fig. 4). The VLDL receptor shows an amazing degree of conservation among different species; within mammals, there is 95% identity between the corresponding proteins. Even the VLDL receptor homologs of more distant species such as the chicken and frog share 84% and 73%, respectively, of identical residues with the human VLDL receptor. In addition, VLDL receptors exist in variant forms, arising from differential splicing of exon 16 which specifies an $O$-linked sugar domain.

However, uptake of VLDL as such into tissues in vivo has not been conclusively shown to involve the eight ligand binding repeat receptor to a significant extent. This is despite the fact that its tissue distribution is highly suggestive of a role in triacylglycerol transport into metabolically active tissues, such as heart, skeletal muscle, and adipose tissue. In contrast to the LDL receptor, and as expected from a receptor implicated in triacylglycerol transport, the VLDL receptor is not regulated by cellular sterols, but its level appears to be influenced by hormones such as estrogen and thyroid hormone. On the other hand, its expression pattern is not congruent with that of lipoprotein lipase with which the VLDL receptor would be expected to act in concert. Nevertheless, numerous studies suggest that the VLDL receptor is, at least in part, involved in the delivery of fatty acids derived from VLDL-triacylglycerols to peripheral tissues, such as adipose tissue [9].

Subsequent to the revelation that despite the proposed and/or intuitive physiological function, the VLDL receptor likely has only a limited role in lipoprotein metabolism, our view of VLDL receptor function has dramatically changed. In 1999, it was discovered by elegant experimentation that it clearly plays a role in neuronal migration in the developing brain, via binding of a ligand quite distinct from lipoproteins. These important results are described in Section 5. However, there is a homologue of the VLDL receptor which indeed does deserve this name, since it has a very well defined function in lipoprotein metabolism, as described in the following section.

## 4. Multifunctional receptors in the chicken

A particularly interesting VLDL receptor homolog is that of the chicken (termed LDL receptor relative with eight binding repeats, or LR8; Fig. 4), as its functions are documented by both biochemical and genetic evidence: it mediates a key step in the reproductive effort of the hen, i.e., oocyte growth via deposition of yolk lipoproteins [10,11]. This conclusion is based on studies of a non-laying chicken strain carrying a single mutation at the *lr8* locus that disrupts LR8 function (the 'restricted ovulator',

R/O, strain) [12]. As a consequence of the mutation, the hens fail to deposit into their oocytes VLDL and the lipophosphoglycoprotein vitellogenin, which are produced at normal levels in the liver, and the mutant females develop severe hyperlipidemia and features of atherosclerosis. The phenotypic consequences of the single-gene mutation in R/O hens revealed the extraordinary multifunctionality of LR8, i.e., that the receptor recognizes in essence over 98% of all the yolk precursors that eventually constitute the mass of the fully grown oocyte [11]. Obviously, R/O hens, which represent a unique animal model for an oocyte-specific receptor defect leading to familial hypercholesterolemia (Section 2.2), are sterile due to non-laying.

Those tissues which express the VLDL receptor in mammals, i.e., heart, skeletal muscle, brain, and adipose tissue, but not the liver, also express this receptor in chicken, albeit at very low levels compared to the oocytes [13]. One difference between the structures of the major VLDL receptors in mammals and chicken LR8 is the presence (in mammalian tissues) and absence (in chicken oocytes) of the $O$-linked sugar domain, respectively [13]. Here, the larger form is termed LR8$^+$, and the smaller one, LR8$^-$. It was found that in chicken, the somatic cells and tissues, in particular the granulosa cells surrounding the oocytes, heart, and skeletal muscle express predominantly LR8$^+$ (at very low levels, as indicated above), while the oocyte is by far the major site of LR8-expression. In the male gonad, the same expression dichotomy exists in that somatic cells express the larger, and spermatocytes the shorter, form of LR8 [14].

The properties of LR8 and its central role in reproduction strengthen the hypothesis that the avian receptor is the product of an ancient gene with the ability to interact with many, if not all, ligands of more recent additions to the LDL receptor gene family (Fig. 4). In this context, vitellogenin, absent from mammals, and apoE, not found in birds, possess certain common biochemical properties and regions of sequence similarities, and have been suggested to be functional analogues [15]. Even high density lipophorin, an abundant lipoprotein in the circulatory compartment of insects, is likely endocytosed in a variety of tissues via an LR8 homologue with very high similarity to the VLDLR/LR8 group [16]. Presumably, binding of lipophorin to this receptor is mediated by apolipophorins I and II, which share sequence homology with mammalian apoB, and thus may behave similarly to the major yolk precursor proteins.

Furthermore, studies in the chicken have revealed that members of the LDL receptor gene family from different animal kingdoms have common structures, and share a growing list of physiological roles, including the most recently discovered function(s) in signal transduction, as described in the following section.

## 5. VLDL receptor and apoE receptor type 2 (apoER2) as signal transducers

### 5.1. ApoER2 — a close relative of the VLDL receptor

The structure of apoER2, which was discovered by homology cloning, is highly reminiscent of that of the VLDL receptor (Fig. 4). However, the produced proteins are now known to harbor a cluster of either three, four, five, seven, or eight binding repeats, dependent on the species and organ expressing apoER2 [17]. Murine apoER2 mRNA may contain an additional small exon following that encoding repeat 8, giving rise to a variant containing a furin consensus cleavage site at the carboxy-terminal end of the ligand binding domain. This may lead to the secretion of a soluble receptor fragment constituting the ligand binding domain, which could act as a dominant negative ligand-trapping receptor. Most interestingly, this variant is the only one detectable in the placenta, showing that some of the splice events are tissue-specific.

ApoER2 is predominantly found in brain, placenta, and testis. This is in contrast to other members of the LDL receptor family, which are all expressed to a small extent in the brain, but most prominently in a variety of other organs and cells. Besides the liver, the brain is also the most prevalent site of apoE expression in mammals, and it is widely believed that apoE serves a role in local lipid transport in the central nervous system [18]. Ligand binding studies with the human apoER2 demonstrated high affinity of the receptor for β-VLDL, indicating that the receptor might be involved in apoE-mediated transport processes in the brain. In addition, at least the apoER2 splice variant containing eight binding repeats can act as receptor for $\alpha_2$-macroglobulin (also a ligand of LRP, see Section 3.1) in brain, which suggests a role in the clearance of $\alpha_2$-macroglobulin–proteinase complexes from the cerebrospinal fluid and from the surface of neurons. In turn, proteinases may play a role in synaptic plasticity [19], and the balance between proteolytic activity and its inhibition might be controlled by proteinase inhibitors and their receptors. Despite these intriguing possibilities, an even more important, recently delineated, function of apoER2 — and of the VLDL receptor — is described below.

## 5.2. Genetic models reveal new roles for apoER2 and VLDL receptor

In addition to their potential to bind and endocytose ligands in a variety of metabolic and cellular contexts, these receptors are now known to be involved in signal transduction [20], which likely is independent of ligand internalization. Surprisingly, targeted disruption of both the VLDL receptor and the apoER2 genes in mice (so-called double-knockout mice) display a dramatic phenotype, essentially identical to that of mice lacking the extracellular matrix glycoprotein reelin [20]; single-knockout mice of either receptor gene show only very subtle phenotypes. Reelin is secreted by Cajal–Retzius cells in the outermost layer of the developing cerebral cortex and orchestrates the migration of neurons along radial fibers, thus forming distinct cortical layers in the cerebrum.

The reason for the grossly abnormal phenotype of the double-knockout mice, i.e., disturbed foliation of the cortical layers, is that reelin normally interacts with the extracellular domains of both the VLDL receptor and apoER2 [21,22], but the ensuing vital signal cascade remains inactivated when the receptors are missing. The current model suggests that reelin binding to the VLDL receptor and apoER2 leads to phosphorylation of the cytoplasmic adapter protein Disabled-1 (mDab-1), which is associated with the NPxY-motifs present in the receptors' tails. Reelin-triggered tyrosine-phosphorylation of mDab-1 may then start kinase cascade(s) controlling cellular motility and shape by acting on the neuronal cytoskeleton. The specificity of the reelin signalling via apoER2 and VLDL receptor seems to be achieved by selective binding of reelin to these receptors and not to other members of the LDL receptor family. This is an important aspect, since mDab-1 not only binds to the VLDL receptor and apoER2, but also to LRP and the LDL receptor [23].

## 6. Other relatives of the LDL receptor family

In the past few years, several additional LDL receptor gene family members have been identified at the molecular level. Since all of these, by definition, contain LDL receptor ligand binding repeat clusters and may therefore play roles in lipid-related metabolism, they are also described in this chapter. For simplified schematized structures of these membrane proteins, please refer to Fig. 4.

### 6.1. Small and mid-sized LDL receptor relatives: LRP 3, 4, 5, and 6

These rather new additions to the LR gene family were discovered more or less serendipitously. Degenerative probes corresponding to the highly conserved amino acid sequence WRCDGD, found in LDL receptor ligand repeats, were used to screen a rat liver cDNA library, resulting first in the cloning of LRP3, and then of its human homologue from a HepG2 cDNA library [24]. The same approach, using a murine heart cDNA library, resulted in cloning of LRP4. LRP3 is a 770 residue membrane protein with clusters of two and three binding repeats, respectively. Murine LRP4 contains two clusters of LDL receptor binding repeats with three and five modules, respectively. Future studies will help to clarify whether these proteins are capable of binding ligands that have been shown to interact with other LDL receptor relatives, whether they are endocytotically competent, and of course, what their relevant in-vivo functions are.

Two other new members of the LDLR family, LRP5 [25] and LRP6 [26] have been discovered (Fig. 4), quite surprisingly, in the course of attempts to identify the nature of the insulin-dependent diabetes mellitus locus IDDM4 on chromosome 11q13. Human and mouse LRP5 and LRP6 are type I membrane proteins, approximately 1600 residues long (about twice as large as the LDL receptor), and their extracellular domains are organized exactly as a portion of LRP (Fig. 4). The cytoplasmic domains of LRP5 and LRP6 do contain motifs (dileucine, and aromatic-X-X-aromatic/large hydrophobic) similar to those known to be functional in endocytosis of other receptors.

Probably more importantly, they harbor serine- and proline-rich stretches that may serve as ligands for Src homology 3 (SH3) and WW (a variant of SH3) domains, properties that relate these receptors to signal transduction pathways, albeit different ones from those of apoER2 and the VLDL receptor described above. Indeed, LRP6 has been shown to be a co-receptor for proteins of the Wnt family, which trigger signalling pathways important for correct development of anterior structures. Most recently, LRP6 has been demonstrated to interact with proteins called Dickkopf and Axin, respectively [27]. Dickkopf inhibits Wnt signalling by releasing receptor-bound Wnt. Axin is one of the components in the cascade that regulates the activation of gene expression in the nucleus of target cells. The interplay of Wnt-, Dickkopf- and Axin-binding to LRP5/6 may hold the key to important developmental signals, similar to the role of VLDL receptor and apoER2 in neuronal migration (Section 5.2).

## 6.2. The unusual one: LR11

Also by homology cloning, LR11, a novel and unusually complex member of the LDL receptor gene family (Fig. 4) has been discovered first in rabbit, and subsequently in man, mouse and chicken [28]. Significantly, overall sequence identities between the ∼250 kDa proteins range from 80% (man vs. chicken) to 94% (man vs. rabbit). The predicted amino acid sequence of LR11 suggests that the polypeptide is made up of seven distinct domains (Fig. 4), among which is a cluster of eleven LDL receptor ligand repeats. Unusual features are a large domain (∼400 residues) highly homologous to a yeast receptor for vacuolar protein sorting, VPS10p, and modules found in cellular adhesion molecules (six tandem fibronectin type III repeats). The membrane spanning and cytoplasmic domains are extremely highly conserved; e.g., the presumed internalization signal is FANSHY in all LR11s known to date.

LR11 is found mainly in the nervous system, and depending on the species, also in testis, ovary, adrenal glands, and kidney. It appears that LR11 is developmentally regulated, and several studies have demonstrated induced expression during morphogenetic processes. For instance, LR11 levels are increased during proliferation, but become downregulated following differentiation of neuroblastoma cells. LR11 is also markedly increased in arterial intimal smooth muscle cells during atherogenesis and in the proliferative phase of smooth muscle cells in culture. Thus, to date, the limited information available suggests that LR11 is involved in cellular proliferation during development and possibly, in pathological processes.

However, knowledge about the cell biology of LR11 is still highly rudimentary. Only 10% of the receptors are found on the cell surface in transfected cells, and thus far identified ligands of LR11, such as apoE-containing lipoproteins, have not offered significant insights into LR11 functions [29]; finally, endocytic competence of the receptor has not been demonstrated unambiguously to date.

## 6.3. Large LDL receptor relatives: megalin and LR32

### 6.3.1. Megalin, a true lipid transport receptor

Encoded by a different gene from that for LRP, this 600-kDa protein is another large member of the LDL receptor gene family containing four LDL receptor ligand binding repeat clusters. Although many proteins which bind to LRP are also ligands for megalin (with the exception of apoJ, also known as clusterin), its expression pattern and specificity for certain ligands account for physiological roles distinct from those of LRP. Megalin is essential for development of the forebrain by taking up apoB-containing lipoproteins into the embryonic neuroepithelium (reviewed in Willnow et al. [7]). Another important function is its involvement in the metabolism of certain lipophilic vitamins. For instance, in the kidney, vitamin B12/transcobalamin complexes are recaptured from the ultrafiltrate directly by binding to megalin expressed on proximal tubule cells [30]. Furthermore, megalin mediates the reabsorption from the proximal tubules of 25-(OH) vitamin D3/vitamin D binding protein complexes, which constitutes a key step in converting the precursor into active vitamin D3 in the kidney [7].

### 6.3.2. LR32 (LRP1B)

This 4599-residue type I membrane protein contains 32 LDL receptor ligand binding repeats in its extracellular portion (Fig. 4). Among all of its relatives, LR32 shows the highest homology to LRP, containing one additional ligand binding repeat and an insertion of 33 amino acids in the cytoplasmic domain compared to the sequence of LRP (cf. Fig. 4). This newly discovered receptor molecule may thus have a role in lipoprotein metabolism.

However, homology searches revealed that LR32 is identical to the product of a candidate tumor suppressor gene, *lrp1b* [31]. The human LR32 gene locus was mapped to chromosome 2q21 by fluorescence in-situ hybridization. The receptor is expressed mainly in brain and skeletal muscle, and its regulation has been studied in smooth muscle cells derived from rabbit arteries and in an established smooth muscle cell-line. In both systems, LR32 expression is induced during the exponential phases of cellular proliferation. Peaks of expression seem to occur at later time points than those observed for LR11 induction (see Section 6.2), consistent with different roles of LR11 and LR32. The future will tell us more about the true physiological role(s) of this close relative of LRP.

## 7. Scavenger receptors

In addition to the type of scavenger receptors (SRs) described in Chapter 20, there is also a growing list of hepatic and extrahepatic SRs with potential disease-related functions. For many of these, the criterion to be considered as SR is their broad spectrum of ligands which includes diverse polyanionic compounds and, importantly, modified lipoproteins, such as oxidized LDL. The more than a dozen currently known SRs are classified according to their primary structure, tissue distribution, and proposed function(s) into six groups (SR-A to -F). The most prominent and probably best understood SRs in the context of lipoprotein metabolism are the SR class A (SR-A) and a class E SR, LOX-1 (for a concise review of SRs, see Terpstra et al. [32]).

### 7.1. Class A SRs

SR-As are trimeric membrane proteins characterized structurally by the presence of a small amino terminal intracellular region, an extracellular coiled-coil collagen-like stalk, and a cysteine-rich carboxy-terminal domain [33]. Three isoforms of this receptor are produced from the same gene by differential splicing. SR-A isoforms are expressed at different levels in tissue macrophages, Kupffer cells, and various extrahepatic endothelial cells. SR-A expression is induced by some of its ligands, which include, in addition to modified lipoproteins and polyanions, Gram-positive bacteria, heparin, lipoteichoic acid, and a precursor of lipid A from LPS of Gram-negative bacteria.

The potential role of SR-A in atherosclerotic plaque development was demonstrated in a study on apoE- and SR-A double-knockout mice. Mice deficient in apoE develop severe plaques, but simultaneous absence of SR-A leads to a reduction in plaque size by 58%. This reduction may be related to the greatly reduced uptake of acetylated LDL and oxidized LDL that can be observed in in-vitro uptake studies using macrophages and liver cells of SR-A knockout mice.

## 7.2. Lectin-like oxidized LDL receptor (LOX)-1

This 50-kDa transmembrane protein shows no structural similarity to other SRs. It belongs to the C-type lectin family of molecules, and its carboxy-terminal cytoplasmic tail contains several potential phosphorylation sites [34]. It can act as endocytic receptor for atherogenic oxidized LDL, but in contrast to SR-A, it interacts only weakly, if at all, with acetylated LDL. Thus, LOX-1 differs from other SRs in that binding of oxidized LDL is inhibited by polyinosinic acid and delipidated oxidized LDL, but not by acetylated LDL, maleylated bovine serum albumin, or fucoidin. LOX-1 is found in thoracic and carotid vessels, and highly vascularized tissues such as placenta, lungs, brain, and liver. Its expression is apparently not constitutive, but can be induced by proinflammatory stimuli; it is then detectable in cultured macrophages and in activated smooth muscle cells. LOX-1 also mediates the recognition of aged red blood cells and apoptotic cells. However, little is known about whether LOX-1 has similar functions in clearance of damaged cell in vivo.

In summary, SRs are a widely expressed and highly diverse group of proteins that are appropriately named for their recognition of a broad array of ligands. At least some of this intriguing group of receptors, as indicated here, may play roles in the metabolism of modified lipoproteins, and thus may well be related to lipid anomaly disorders.

## 8. Future directions

It is exciting to realize that multifunctionality of LDL receptor gene family members can no longer be viewed as being limited to their extracellular moiety. The originally established concepts regarding their functions in lipoprotein metabolism remain valid, but have been extended significantly by the discovery that these membrane proteins also play a role in important signalling pathways. Given the wealth of signalling pathways, future efforts will be directed towards developing concepts that define the contributions of the individual intracellular domains of LDL receptor relatives. These efforts can also be expected to add alternative aspects to the ongoing quest to delineate the evolutionary history of the LDL receptor family. In fact, based on the most recently gained knowledge, an evolutionary theory will have to consider the combinatorial events arising from the multitude of both intra- and extra-cellular domains of these proteins in the generation of new signalling and/or transport units.

Another point to consider is the functional redundancy of receptors involved in a multitude of metabolic pathways and events. This aspect has been addressed here not only for the LDL receptor gene family, but also for the growing group of scavenger receptors, which show broad overlapping ligand specifities (Section 7). In the case of the well understood LDL receptor gene family, when need arises (e.g., when one or more receptors are unfunctional), certain members can substitute for others, or are at least sufficiently active for preserving life. For example, VLDLR$^{-/-}$ mice, and apoER2$^{-/-}$ mice, have phenotypes that are at first sight indistinguishable from normal; but double-knockout mice are grossly abnormal, as described in Section 5.2. Moreover, (1) LRP can physiologically substitute for the LDL receptor in hepatic clearance of chylomicron remnants (which also bind to the LDL receptor in vitro), (2) VLDLR, not expressed in the liver and normally not significantly involved in lipoprotein metabolism, can, when expressed in the liver, substitute for the LDL receptor, and (3) in chickens, an oocyte-specific LRP mediates growth of oocytes, although not sufficient for egg laying, in the LR8-deficient R/O hens (see Section 4). Thus, functional redundancy of LDL receptor relatives can be due to their simultaneous expression on the same cells in a given organ. In turn, different functions despite overlapping ligand spectra may arise from their expression in different cell types within a tissue or organism. LDL receptor family members presumably have in-vivo access to different ligands in different environments, and in addition, their cytoplasmic domains interact with cell-type-specific adaptor proteins in order to mediate a spectrum of signal transduction pathways.

As a consequence, specific functions of the gene family must be defined at two levels: the cellular level, in order to delineate molecular events, and in physiological studies including state-of-the-art genetic manipulations, which should reveal the functional relevance of receptor redundancy.

## Abbreviations

| | |
|---|---|
| $\alpha_2$MR | $\alpha_2$-macroglobulin receptor |
| ACAT | acyl-CoA : cholesterol acyltransferase |
| Apo | apolipoprotein |
| CETP | cholesteryl ester transfer protein |
| CM(R) | chylomicron (remnants) |
| EGFP | epidermal growth factor precursor |
| FH | familial hypercholesterolemia |
| HDL | high density lipoprotein |
| HMG- | 3-hydroxy-3-methylglutaryl- |
| LCAT | lecithin–cholesterol acyltransferase |
| LOX-1 | lectin-like oxidized LDL receptor |
| LR | LDL receptor relative |
| LRP | LDL receptor-related protein |
| (V)LDL | (very) low density lipoprotein |

## References

1. Goldstein, J.L., Brown, M.S., Anderson, R.G., Russell, D.W. and Schneider, W.J. (1985) Receptor-mediated endocytosis: concepts emerging from the LDL receptor system. Annu. Rev. Cell Biol. 1, 1–39.

2. Esser, V., Limbird, L.E., Brown, M.S., Goldstein, J.L. and Russell, D.W. (1988) Mutational analysis of the ligand binding domain of the low density lipoprotein receptor. J. Biol. Chem. 263, 13282–13290.

3. Lehrman, M.A., Schneider, W.J., Brown, M.S., Davis, C.G., Elhammer, A., Russell, D.W. and Goldstein, J.L. (1987) The Lebanese allele at the low density lipoprotein receptor locus. Nonsense mutation produces truncated receptor that is retained in endoplasmic reticulum. J. Biol. Chem. 262, 401–410.

4. Hobbs, H.H., Russell, D.W., Brown, M.S. and Goldstein, J.L. (1990) The LDL receptor locus in familial hypercholesterolemia: mutational analysis of a membrane protein. Annu. Rev. Genet. 24, 133–170.

5. Davis, C.G., Goldstein, J.L., Südhof, T.C., Anderson, R.G.W., Russell, D.W. and Brown, M.S. (1987) Acid-dependent ligand dissociation and recycling of LDL receptor mediated by growth factor homology region. Nature 326, 760–765.

6. Herz, J., Hamann, U., Rogne, S., Myklebost, O., Gausepohl, H. and Stanley, K.K. (1988) Surface location and high affinity for calcium of a 500-kDa liver membrane protein closely related to the LDL-receptor suggest a physiological role as lipoprotein receptor. EMBO J. 7, 4119–4127.

7. Willnow, T.E., Nykjaer, A. and Herz, J. (1999) Lipoprotein receptors: new roles for ancient proteins. Nat. Cell. Biol. 1, E157–E162.

8. Takahashi, S., Kawarabayasi, Y., Nakai, T., Sakai, J. and Yamamoto, T. (1992) Rabbit very low density lipoprotein receptor: A low density lipoprotein receptor-like protein with distinct ligand specificity. Proc. Natl. Acad. Sci. USA 89, 9252–9256.

9. Tacken, P.J., Hofker, M.H., Havekes, L.M. and van Dijk, K.W. (2001) Living up to a name: the role of the VLDL receptor in lipid metabolism. Curr. Opin. Lipidol. 12, 275–279.

10. Bujo, H., Hermann, M., Kaderli, M.O., Jacobsen, L., Sugawara, S., Nimpf, J., Yamamoto, T. and Schneider, W.J. (1994) Chicken oocyte growth is mediated by an eight ligand binding repeat member of the LDL receptor family. EMBO J. 13, 5165–5175.

11. Schneider, W.J., Osanger, A., Waclawek, M. and Nimpf, J. (1998) Oocyte growth in the chicken: receptors and more. Biol. Chem. 379, 965–971.

12. Bujo, H., Yamamoto, T., Hayashi, K., Hermann, M., Nimpf, J. and Schneider, W.J. (1995) Mutant oocytic low density lipoprotein receptor gene family member causes atherosclerosis and female sterility. Proc. Natl. Acad. Sci. USA 92, 9905–9909.

13. Bujo, H., Lindstedt, K.A., Hermann, M., Dalmau, L.M., Nimpf, J. and Schneider, W.J. (1995) Chicken oocytes and somatic cells express different splice variants of a multifunctional receptor. J. Biol. Chem. 270, 23546–23551.

14. Lindstedt, K.A., Bujo, H., Mahon, M.G., Nimpf, J. and Schneider, W.J. (1997) Germ cell — somatic cell dichotomy of a low density lipoprotein receptor gene family member in testis. DNA Cell Biol. 16, 35–43.

15. Steyrer, E., Barber, D.L. and Schneider, W.J. (1990) Evolution of lipoprotein receptors. The chicken oocyte receptor for very low density lipoprotein and vitellogenin binds the mammalian ligand apolipoprotein E. J. Biol. Chem. 265, 19575–19581.

16. Dantuma, N.P., Potters, M., De Winther, M.P., Tensen, C.P., Kooiman, F.P., Bogerd, J. and Van der Horst, D.J. (1999) An insect homolog of the vertebrate very low density lipoprotein receptor mediates endocytosis of lipophorins. J. Lipid Res. 40, 973–978.

17. Nimpf, J. and Schneider, W.J. (2000) From cholesterol transport to signal transduction: low density lipoprotein receptor, very low density lipoprotein receptor, and apolipoprotein E receptor-2. Biochim. Biophys. Acta 1529, 287–298.

18. Mahley, R.W. and Huang, Y. (1999) Apolipoprotein E: from atherosclerosis to Alzheimer's disease and beyond. Curr. Opin. Lipidol. 10, 207–217.

19. Seeds, N.W., Williams, B.L. and Bickford, P.C. (1995) Tissue plasminogen activator induction in Purkinje neurons after cerebellar motor learning. Science 270, 1992–1994.

20. Trommsdorff, M., Gotthardt, M., Hiesberger, T., Shelton, J., Stockinger, W., Nimpf, J., Hammer, R., Richardson, J.A. and Herz, J. (1999) Reeler/Disabled-like disruption of neuronal migration in knock out mice lacking the VLDL receptor and apoE receptor-2. Cell 97, 689–701.

21. Hiesberger, T., Trommsdorff, M., Howell, B.W., Goffinet, A., Mumby, M.C., Cooper, J.A. and Herz, J.

572

(1999) Direct binding of reelin to VLDL receptor and ApoE receptor 2 induces tyrosin phosphorylation of the adaptor protein disabled-1 and modulates tau phosphorylation. Neuron 24, 481–489.

22. D'Arcangelo, G., Homayoundi, R., Keshvara, L., Rice, D.S., Sheldon, M. and Curran, T. (1999) Reelin is a ligand for lipoprotein receptors. Neuron 24, 471–479.

23. Trommsdorff, M., Borg, J.-P., Margolis, B. and Herz, J. (1998) Interaction of cytosolic adaptor proteins with neuronal apoE receptors and the amyloid precursor proteins. J. Biol. Chem. 273, 33556–33565.

24. Ishii, H., Kim, D.H., Fujita, T., Endo, Y., Saeki, S. and Yamamoto, T. (1998) cDNA cloning of a new low-density lipoprotein receptor-related protein and mapping of its gene (lrp3) to chromosome bands 19q12–q13.2. Genomics 51, 132–135.

25. Hey, P.J., Twells, R.C.J., Phillips, M.S., Nakagawa, Y., Brown, S.D. and Kawaguchi Y. et al. (1998) Cloning of a new member of the low-density lipoprotein receptor gene family. Gene 216, 103–111.

26. Brown, S.D., Twells, R.C.J., Hey, P.J., Cox, R.D., Levy, E.R. and Sodermann, A.R. (1998) Isolation and characterization of LRP6, a novel member of the low density lipoprotein receptor gene family. Biochim. Biophys. Res. Commun. 248, 879–888.

27. Nusse, R. (2001) Developmental biology. Making head or tail of Dickkopf. Nature 411, 255–256.

28. Moerwald, S., Yamazaki, H., Bujo, H., Kusunoki, J., Kanaki, T., Seimiya, K., Morisaki, N., Nimpf, J., Schneider, W.J. and Saito, Y. (1997) A novel mosaic protein containing LDL receptor elements is highly conserved in humans and chickens. Arterioscler. Thromb. Vasc. Biol. 17, 996–1002.

29. Jacobsen, L., Madsen, P., Jacobsen, C., Nielsen, M.S., Gliemann, J. and Petersen, C.M. (2001) Activation and functional characterization of the mosaic receptor SorLA/LR11. J. Biol. Chem. 276, 22788–22796.

30. Moestrup, S.K., Kozyraki, R., Kristiansen, M., Kaysen, J.H., Rasmussen, H.H., Brault, D., Pontillon, F., Goda, F.O., Christensen, E.I., Hammond, T.G. and Verroust, P.J. (1998) The intrinsic factor-vitamin B12 receptor and target of teratogenic antibodies is a megalin-binding peripheral membrane protein with homology to developmental proteins. J. Biol. Chem. 273, 5235–5242.

31. Liu, C.X., Musco, S., Lisitsina, N.M., Forgacs, E., Minna, J.D. and Lisitsyn, N.A. (2000) LRP-DIT, a putative endocytic receptor gene, is frequently inactivated in non-small cell lung cancer cell lines. Cancer Res. 60, 1961–1967.

32. Terpstra, V., van Amersfoort, E.S., van Velzen, A.G., Kuiper, J. and van Berkel, T.J. (2000) Hepatic and extrahepatic scavenger receptors: function in relation to disease. Arterioscler. Thromb. Vasc. Biol. 20, 1860–1872.

33. Krieger, M. (1997) The other side of scavenger receptors: pattern recognition for host defense. Curr. Opin. Lipidol. 8, 275–280.

34. Kume, N. and Kita, T. (2001) Lectin-like oxidized low-density lipoprotein receptor-1 (LOX-1) in atherogenesis. Trends Cardiovasc. Med. 11, 22–25.

D.E. Vance and J.E. Vance (Eds.) *Biochemistry of Lipids, Lipoproteins and Membranes (4th Edn.)*

# CHAPTER 22

# Lipids and atherosclerosis

## Ira Tabas

*Departments of Medicine and Anatomy and Cell Biology, Columbia University, 630 West 168th Street,*
*New York, NY 10032, USA, Tel.: +1 (212) 305-9430; Fax: +1 (212) 305-4834;*
*E-mail: iat1@columbia.edu*

## 1. Introduction

Atherosclerotic vascular disease is the cause of heart attacks, stroke, aortic aneurysms, and peripheral vascular disease, which together represent the most frequent causes of death in the industrialized world. Indeed, the aging of the population and the 'westernization' of world diet is predicted to increase the impact of atherosclerosis worldwide over the next few decades despite continuing advances in plasma lipid-lowering therapy (E. Braunwald, 1997).

Atherosclerosis progresses in a series of stages, although some lesions at each stage may not progress further or may even regress if inciting events, such as hypercholesterolemia, diabetes, smoking, or hypertension, are controlled [1–4] (Fig. 1). The initial

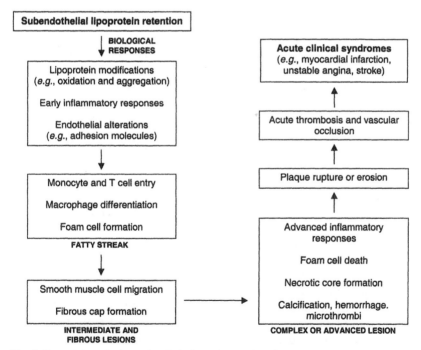

Fig. 1. Progression of atherosclerotic lesions. As noted in the text, only a portion of lesions at any stage progress and, under the proper conditions, lesion regression may occur.

stage involves the accumulation of subendothelial lipoproteins in focal areas of the arterial tree, usually at branch points with disturbed laminar flow. In response to this retention, a series of biological responses ensue, including (1) lipoprotein oxidation, (2) endothelial alterations, (3) inflammatory responses including T cell recruitment, cytokine secretion, monocyte chemotaxis, and subendothelial macrophage accumulation, (4) and intracellular cholesterol accumulation in macrophages. Much of the cholesterol is stored as cholesteryl fatty acid esters (CE) in cytoplasmic lipid droplets surrounded by a monolayer of phospholipid. These cytoplasmic droplets give the macrophages a foamy appearance when viewed by microscopy, and thus these cells are referred to as 'foam cells' (Fig. 2C). The presence of macrophage foam cells defines the earliest pathological lesion, referred to as the 'fatty streak'. Interestingly, fetuses of hypercholesterolemic mothers have been observed to have fatty streaks (W. Palinski, 1997). These fatty streaks disappear soon after birth, but virtually all Westerners have fatty streaks by the teenage years.

Although sensitive tests of endothelial function show abnormalities in vasodilatation in the very earliest phases of atherosclerosis (R.A. Vogel, 1998), fatty streaks are not occlusive and cause no overt symptoms. However, some fatty streaks may progress over years to more complex lesions that can give rise to chronic symptoms or, more importantly, acute events. An important event in the progression of fatty streaks involves the migration of smooth muscle cells from the media to the intima and the secretion of large amounts of collagen and other matrix proteins by these cells. In addition, macrophages proliferate and continue to accumulate more lipid. Smooth muscle cells can also accumulate lipid and become foam cells. These events give rise to so-called fibrous lesions, which are eccentric lesions consisting of lipid-loaded macrophages and smooth muscle cells covered by a fibrous cap. Further progression to complex lesions involves the accumulation of extracellular lipid, which results from a combination of aggregation and fusion of matrix-retained lipoproteins and release of lipid droplets from dying foam cells. Calcification, hemorrhage, and microthrombi can also be observed in these complex lesions [1].

At this stage, several fates of the lesion are possible [1,5,6]. Plasma cholesterol lowering can result in lesion regression, particularly of the foam cells. In this case, the macrophages begin to lose their cholesterol via the process of cholesterol efflux, and the number of macrophages decreases, probably through a combination of decreased monocyte entry, decreased macrophage proliferation, and increased macrophage egress and apoptosis. Alternatively, the complex lesions can progress. If arterial occlusion increases gradually, the patient may experience exercise-induced ischemia, but collateral vessel formation often prevents additional clinical symptoms. However, if the lesions rupture or erode before they become large and occlusive, acute vascular events such as unstable angina, heart attacks, sudden death, or strokes can occur. Rupture involves the abrupt disruption of the fibrous cap, followed by exposure of thrombogenic material and acute thrombosis. Importantly, rupture mostly occurs in lipid-rich and macrophage-rich 'shoulder' regions of the plaque and is probably triggered by the degradation of the fibrous cap by proteases secreted by macrophages or released from dying foam cells. Physical stresses related to pools of soft lipid underneath a thin fibrous cap also contribute to plaque rupture. These pools of lipid and cellular debris, often referred

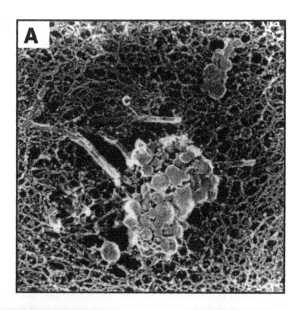

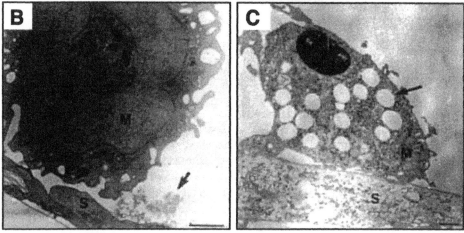

Fig. 2. Lipoprotein aggregation and macrophage foam cell formation. (A) Freeze-etch replica-plated electron micrograph of rabbit aorta subendothelium 2 hours after intravenous injection of LDL (from Nievelstein et al. (1991) Arterio. Thromb. 11: 1795–1805). (B) A J774 murine macrophage (M) immediately after plating on sphingomyelinase-induced LDL aggregates (arrow) formed on the surface matrix of smooth muscle cells (S). (C) After 24 hours of incubation, the aggregates have been internalized by the macrophage, which now has large cytoplasmic neutral lipid droplets consisting mostly of cholesteryl ester (arrow). The cytoplasmic droplets are characteristic of lesional foam cells. Bars in B and C, 1 μm. Panels B and C are from Tabas et al. (1993) J. Biol. Chem. 268: 20419–20432.

to as 'necrotic' or 'lipid' cores, result from the death of macrophage foam cells (M.J. Mitchinson, 1995).

As is evident from this overview, lipids are the sine qua non of atherosclerosis. Indeed, the 'athero' of 'atherosclerosis' is derived from the Greek word for 'gruel',

Table 1

Percent weight of major lesion lipids in four progressive stages of atherosclerotic lesions

| Lipid | Fatty streak | Intermediate lesion | Fibrous lesion | Advanced lesion |
|---|---|---|---|---|
| Cholesterol | 9.6 | 21.1 | 22.5 | 31.5 |
| Triglyceride | 2.8 | 4.4 | 5.2 | 6.0 |
| Cholesteryl ester | 77.0 | 55.0 | 55.5 | 47.2 |
| Phospholipid | 10.1 | 19.6 | 16.8 | 15.3 |
| Phosphatidylcholine | 4.8 | 7.6 | 4.5 | 4.3 |
| Sphingomyelin | 5.6 | 11.0 | 11.7 | 10.1 |
| Lysophosphatidylcholine | 0.3 | 1.0 | 0.6 | 0.9 |

Adapted from Katz et al. [16].

which refers to the massive accumulation of lipids in these vascular lesions. The major types of lipids that accumulate during the various stages of atherosclerosis are shown in Table 1. In addition, there are many lipids that are minor in quantity but, because of their biological activities, are thought to have a major impact on atherogenesis. This chapter will cover the properties and activities of many of the lipids that occur in atherosclerotic lesions, with an emphasis on their roles in lesion development and progression.

## 2. Cholesterol and atherosclerosis

### 2.1. Cholesterol deposition in the arterial wall

As alluded to in the Introduction, the primary event in atherogenesis is cholesterol deposition in the arterial wall. The cholesterol originates from circulating plasma lipoproteins, which contain both unesterified cholesterol ('free' cholesterol, or FC) and cholesteryl ester (CE) (see Chapter 18). The two classes of lipoproteins that contribute most to atherogenesis are low density lipoprotein (LDL) and so-called remnant lipoproteins, which are the lipolytic products of chylomicrons and very low density lipoprotein (VLDL). Plasma lipoproteins continually enter the subendothelial space of vessels via 'leakage' through transient gaps between endothelial cells and probably also via endothelial transcytosis. Under normal conditions, lipoproteins are not retained in the subendothelium and simply re-enter the circulation. In certain focal areas of the arterial tree, however, lipoprotein retention by subendothelial extracellular matrix is increased, leading to their net accumulation in the arterial wall. This retained material elicits a series of biological responses, leading to the cellular and extracellular processes that constitute atherosclerotic lesion formation ([4,7]; see also Section 1). Because a high concentration of circulating atherogenic lipoproteins promotes the accumulation of these lipoproteins in the arterial wall, this model explains the well-established relationship between plasma cholesterol levels and atherosclerosis in both experimental animal models and humans.

The fate of the FC and CE moieties of retained lipoproteins includes both extracellular and intracellular processes. Extracellular matrix-retained lipoproteins are modified by lipases, proteases, and oxidation reactions (P.T. Kovanen, 2000) [8]. These reactions can lead to the generation of lipid vesicles that are rich in FC but poor in protein and CE (H. Kruth, 1985). The biological or pathological significance of these FC-rich vesicles is not known. Other reactions lead to the generation of modified lipoproteins that act as extracellular signaling molecules on lesional cells or that are avidly internalized by macrophages and smooth muscle cells. Thus, these modified lipoproteins are responsible for foam cell formation and a variety of cell signaling events.

## 2.2. Cholesterol accumulation in lesional macrophages: lipoprotein internalization

The major cell type that internalizes subendothelial lipoproteins is the macrophage [2,3]. Lesional macrophages are derived from circulating monocytes that enter the arterial wall in response to chemokines; the chemokines are secreted by endothelial cells in response to both underlying retained lipoproteins and T cell-derived cytokines. Under the influence of other molecules secreted by endothelial cells, notably macrophage colony stimulating factor, subendothelial monocytes differentiate into macrophages. The differentiated macrophages then engage and internalize subendothelial lipoproteins and thus accumulate lipoprotein-derived cholesterol in the form of intracellular cholesteryl ester droplets (foam cell formation). As outlined in the Section 1, this cellular event is the hallmark of early lesion development and also contributes to late lesional complications.

Two key issues in the area of macrophage foam cell formation include the cell-surface processes and receptors involved in lipoprotein internalization and the metabolic fate of lipoprotein-derived cholesterol following internalization [9]. Most studies examining macrophage–lipoprotein interactions use an experimental system in which monolayers of cultured macrophages are incubated with soluble, monomeric lipoproteins dissolved in tissue culture media. These studies have revealed that native LDL is poorly internalized by macrophages, suggesting that LDL undergoes modification in the arterial wall. While a variety of LDL modifications have been proposed, two types, namely oxidation and aggregation, have received the most attention [8,10,11].

LDL particles with oxidative modifications of both its protein and lipid moieties are known to exist in atherosclerotic lesions and are readily internalized by macrophages. A number of receptors have been implicated in oxidized LDL uptake by macrophages, including class A and B scavenger receptors (e.g., CD36) and lectin-like oxidized LDL receptor-1 (LOX-1) (Chapter 21). While internalization of oxidized LDL by macrophages may have important implications in atherogenesis, it is unlikely that all of the hallmarks of macrophage intracellular cholesterol metabolism that are known occur in lesions can be explained by this process alone [8].

As stated above, lipoproteins in the subendothelium are also known to be aggregated and fused, which may result from oxidation, lipolysis, or proteolysis [8,12] (Fig. 2A). For example, hydrolysis of the sphingomyelin on LDL particles to ceramide by sphingomyelinase leads to LDL aggregates that appear similar to those that exist in lesions, and there is evidence that LDL in the arterial wall is hydrolyzed by a form of sphingomyelinase secreted by arterial-wall cells [13] (Fig. 2B). Aggregated lipoproteins, like oxidized LDL, are readily internalized by macrophages. When aggregated LDL is added in tissue culture medium to monolayers of cultured macrophages, the LDL receptor (Chapter 21) seems to participate in a phagocytic-like process to internalize these particles. In vivo, however, most of the aggregated lipoproteins are bound to extracellular matrix, and newer experimental systems that attempt to mimic the uptake of retained and aggregated LDL have revealed that multiple receptors in addition to or instead of the LDL receptor are involved. Most importantly, macrophage internalization of aggregated lipoproteins leads to massive CE accumulation, which is the key intracellular cholesterol metabolic event that is known to occur in macrophages in lesions [9] (Fig. 2C).

Remnant lipoproteins are also important in atherogenesis (R.W. Mahley, 1985; R.J. Havel, 2000). These particles can be internalized by macrophages in their native form, although both oxidation and aggregation of these particles occur and probably further enhance macrophage uptake. The receptor or receptors involved in the uptake of remnant particles is not definitively known, but the likely candidates are the LDL receptor and the LDL receptor-related protein (LRP), which interact with the apolipoprotein E moiety of the remnant lipoproteins (Chapter 21). Remnant lipoproteins, like aggregated lipoproteins, lead to massive cholesteryl ester accumulation in macrophages. Finally, it is worth mentioning that another lipoprotein called lipoprotein(a), in which a large glycoprotein called apolipoprotein(a) is covalently attached to the apolipoprotein B100 moiety of LDL, has been implicated in atherogenesis (A.M. Scanu, 1998). Although macrophage receptors for lipoprotein(a) have been described, neither the mechanism of atherogenicity nor the role of lipoprotein(a) lipids in macrophage cholesterol loading and lesion development are known.

### 2.3. Cholesterol accumulation in lesional macrophages: intracellular trafficking of lipoprotein-derived cholesterol

The fate of lipoprotein cholesterol after internalization is a key issue in understanding the biology and pathology of lesional macrophages. After internalization by receptor-mediated endocytosis or phagocytosis, the lipoproteins are delivered to late endosomes or lysosomes, where hydrolysis of proteins and lipids occurs. Most importantly, the large lipoprotein-CE stores are hydrolyzed by a lysosomal enzyme called lysosomal acid lipase. The liberated FC then trafficks to the plasma membrane and other cellular sites [9].

The trafficking of lipoprotein-derived cholesterol from lysosomes has been a major area of focus in the field of intracellular cholesterol metabolism, and many of the cellular and molecular events are not known (Chapter 17). By analyzing cells with mutations in cholesterol transport, investigators have identified roles for two proteins, called npc1 and npc2 (HE1), in lysosomal and/or endosomal cholesterol transport (E.J. Blanchette-Mackie, 2000; P. Lobel, 2000). In addition, the lipid lysobisphosphatidic acid may also play a role in these processes (J. Gruenberg, 1999). The mechanisms by which these molecules are involved in cholesterol transport, however, are poorly understood. One current model suggests that there is an initial npc1-independent phase consisting of rapid cholesterol transport from late endosomes or lysosomes to the plasma membrane, probably by vesicular transport (T.Y. Chang, 2000; Y. Lange, 2000).

According to this model, the cholesterol is then internalized into a 'sorting organelle', from which cholesterol is distributed to peripheral cellular sites in an npc1-dependent manner. This transport process probably also occurs via vesicular transport. It must be emphasized, however, that until the mechanism of action of the molecules mentioned above and other molecules are elucidated and the cellular sites in the itinerary identified, this model must be considered hypothetical. It is also likely that different cell types and different conditions in the same cell type may result in different cholesterol trafficking patterns.

From the point of view of atherosclerosis, the two most important peripheral trafficking pathways are those to the endoplasmic reticulum, where cholesterol is esterified by acyl-CoA : cholesterol acyltransferase (ACAT), and to the plasma membrane, where cholesterol can be transferred to extracellular acceptors in a process known as cholesterol efflux (Chapter 20). The former process leads to the massive cholesteryl ester accumulation seen in foam cells [9,14,15]. The ACAT reaction utilizes primarily oleoyl-CoA, and so ACAT-derived CE is rich in oleate. In contrast, plasma lipoprotein-CE tends to be rich in linoleate. As expected, therefore, the cholesteryl oleate : cholesteryl linoleate ratio in foam cell-rich fatty streak lesions is relatively high (1.9) [16]. However, the ratio in advanced lesions is only 1.1, suggesting an increase in lipoprotein-CE in advanced atheromata due to poor cellular uptake of lipoproteins or to defective lysosomal hydrolysis following uptake by lesional cells. Further discussion of the cholesterol esterification pathway appears in Chapter 15, and cholesterol efflux, which is an important mechanism that may prevent or reverse foam cell formation, is covered in Chapter 20.

## 2.4. Accumulation of free cholesterol in lesional macrophages

Interestingly, foam cells in advanced atherosclerotic lesions accumulate large amounts of FC [16], some of which is in crystalline form and may be deposited in the extracellular space when foam cells die (Fig. 3). For example, while 2 of 13 abdominal aortic and femoral artery fatty streak lesions contained cholesterol crystals, all of 24 advanced lesions had these structures [16]. The mechanism of FC accumulation is not known, but could involve either defects in cholesterol trafficking to ACAT or a decrease in ACAT activity itself. Because much of the FC accumulating in the cells appears to be associated with lysosomes, it is tempting to speculate that defects in lysosomal cholesterol transport arise in advanced foam cells. In this context, macrophages exposed to oxidized LDL can internalize a substantial amount of cholesterol, but there is relatively little stimulation of ACAT-mediated cholesterol esterification [8]. According to one model, oxysterol-induced inhibition of lysosomal sphingomyelinase leads to accumulation of lysosomal sphingomyelin, which binds cholesterol and thus inhibits transport of the cholesterol out of lysosomes (M. Aviram, 1995).

Free cholesterol accumulation in macrophages may be an important cause of macrophage death in advanced atherosclerotic lesions [17] (Fig. 4). Death induced by intracellular free cholesterol excess probably involves both necrosis and apoptosis. Necrotic death may result from the malfunction of critical plasma membrane proteins exposed to a microenvironment with a high cholesterol : phospholipid ratio. Intracellular cholesterol crystal accumulation may also contribute to this form of death.

580

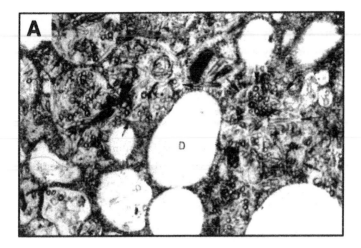

Fig. 3. (A) Intracellular free cholesterol accumulation in a lesional foam cell. Electron micrograph of the cytoplasm of a foam cell isolated from an advanced aortic atherosclerotic lesion in a cholesterol-fed rabbit. The cell was treated with filipin, which forms spicules with unesterified cholesterol. Multiple spicules are observed in vesicles, shown to be lysosomes (depicted by arrows). Bar, 0.5 μm. (From Shio et al. (1979) Lab. Invest. 41: 160–167.) (B) Extracellular cholesterol crystals in an advanced atherosclerotic lesion. The section is from the proximal aorta of a fat-fed apolipoprotein E knockout mouse. This mouse model is often used to study atherosclerosis in vivo because the high plasma levels of remnant lipoproteins resulting from absence of apolipoprotein E leads to a much greater degree of atherosclerosis lesion development than observed in wild-type mice. The arrows depict the areas of cholesterol crystals.

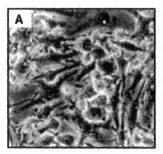
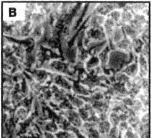
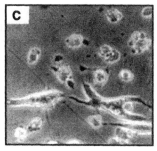

Fig. 4. FC-induced death in macrophages. Cultured macrophages were cultured under control conditions (A) or conditions leading primarily to CE loading (B) or FC loading (C). Note that FC-loaded macrophages, but not those loaded mostly with CE, have obvious signs of cytotoxicity, including detachment from the dish and altered morphology. (From Tabas et al. (1996) J. Biol. Chem. 271: 22773–22781.)

Free cholesterol-induced apoptosis in cultured macrophages involves both activation of Fas ligand and release of cytochrome $c$ from mitochondria, which is associated with increased levels of Bax in these cells (I. Tabas, 2000 and 2001).

### 2.5. Cholesterol accumulation in lesional smooth muscle cells

Smooth muscle cells in atherosclerotic lesions also accumulate large amounts of cholesteryl ester, although the mechanisms involved are poorly understood [18]. As with macrophages, native LDL is a poor inducer of foam cell formation, but substantial cholesterol accumulation has been induced in cultured smooth muscle cells by aggregated LDL (L. Badimon, 1998). Cytokine treatment of cultured smooth muscle cells leads to the induction of the type A scavenger receptor, but there are no data specifically showing that oxidized LDL can cause foam cell formation in smooth muscle cells either in vitro or in vivo. Finally, remnant lipoproteins, including β-VLDL, cationized LDL, and cholesteryl ester emulsions, can induce cholesterol accumulation in cultured smooth muscle cells, but their roles in vivo are not known.

### 2.6. The fate of foam cell cholesterol in atheromata

Cholesteryl esters, which exist in membrane-bound droplets in the macrophage cytoplasm, undergo a continuous cycle of hydrolysis by neutral cholesteryl ester hydrolase and re-esterification by ACAT (M.S. Brown and J.L. Goldstein, 1980). If extracellular cholesterol acceptors, like HDL or apolipoprotein A-I, are available, some of this cholesterol can leave the cell, enter the circulation, and be transported to the liver in a process known as reverse cholesterol transport (Chapter 20). The fatty acyl and neutral lipid composition of foam cell droplets may influence this process by affecting the fluidity of the droplets. It is also possible that the foam cells themselves can leave lesions, although this event has been difficult to document. Finally, as described above, foam cells in atherosclerotic lesions die, and thus cellular stores of cholesteryl ester and free cholesterol, including cholesterol crystals, can be released into the lesions. This process undoubtedly contributes to the formation of the necrotic, or lipid, core of advanced atheromata, because such areas contain macrophage debris (M.J. Mitchinson, 1995). As described in the Introduction, necrotic cores have important pathophysiologic significance because they predispose lesions to plaque rupture, the proximate cause of acute vascular clinical syndromes.

## 3. Oxysterols and atherosclerosis

### 3.1. Origins of oxysterols

Oxysterols arise from dietary sources, non-enzymatic oxidation, and enzymatic oxidation reactions [19]. The structure of some of the oxysterols that may be involved in atherosclerosis are shown in Fig. 5. Dietary oxysterols are incorporated into chylomicrons and include 7-ketocholesterol (7K), 7α- and 7β-hydroxycholesterol (7OH), and α- and β-5,6-epoxycholesterol (EPOX). 7OH, 7K, 24-hydroxycholesterol (24OH), 25-hydroxycholesterol (25OH), and 27-hydroxycholesterol (27OH) can be formed in vivo, but it is not clear whether non-enzymatic or enzymatic mechanism are involved. Specific enzymatic reactions include the formation of 7αOH by cholesterol 7α-hydroxylase in liver (Chapter 16) and 27OH and 3β-hydroxy-5-cholestenoic acid by 27-hydroxylase in liver and macrophages.

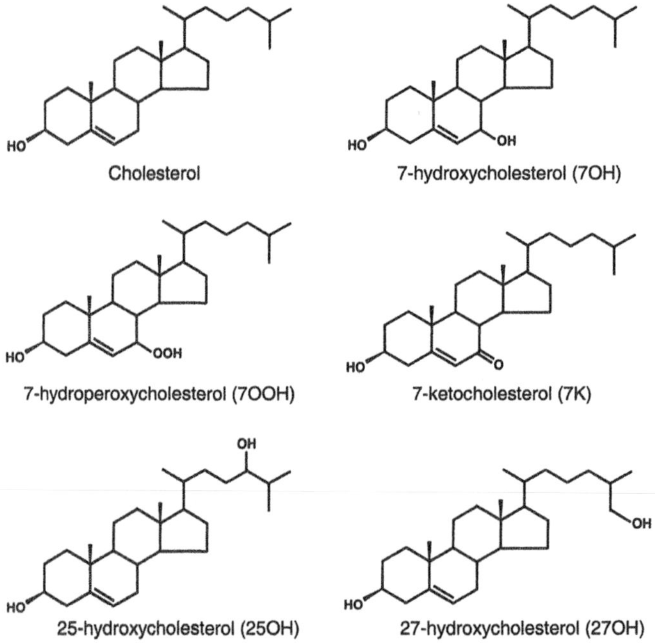

Fig. 5. Structures of some oxysterols that have been implicated in atherogenesis. (Adapted from Brown and Jessup [19].)

## 3.2. Oxysterols in plasma, lipoproteins, and atherosclerotic lesions

The most abundant oxysterols in human plasma are 27OH, 24OH, and 7αOH (Fig. 5), and most of these are esterified to fatty acids at the 3β position by lecithin:cholesterol acyltransferase [19]. Both free and esterified oxysterols partition in lipoproteins similar to cholesterol and cholesteryl ester, respectively, although 27OH is generally not found in VLDL, and unesterified 25OH can be associated with albumin. With the possible exception of 7βOH, there is no clear relationship between plasma levels of oxysterols and atherosclerosis.

Oxysterols are also found in copper-oxidized LDL and consist predominantly of 7K, 7-hydroperoxy-cholesterol (7OOH), 7OH, and EPOX. LDL oxidized by more physiologic means, such as during contact with macrophages or by incubation with lipoxygenase, accumulates 7OOH (Fig. 5). Myeloperoxidase-treated cholesterol leads to the formation of unique chlorinated cholesterol derivatives, which can give rise to cholesterol epoxides [19].

The predominant oxysterols in atherosclerotic lesions include 27OH and 7K, where the levels are approximately 1% of cholesterol, with smaller amounts of 7OH; as in plasma, the vast majority are in the esterified form [19]. These oxysterols are found mostly in macrophage foam cells, which probably reflects the abundance of 27OH (Section 3.3). Moreover, 27-hydroxylated 7K is found in human atherosclerotic lesions, probably via macrophage sterol 27-hydroxylase acting on 7K internalized by macrophages (W. Jessup, 2000). When cultured macrophages are incubated with copper-oxidized LDL, 50% of the accumulated sterols are oxysterols, many of which accumulate in lysosomes as non-ACAT-derived oxidized fatty acid esters of oxysterols (W. Jessup, 2000). Although copper-oxidized LDL contains substantial amounts of 7OOH, macrophages accumulate very little of this lipid probably due to conversion to 7OH by phospholipid hydroperoxide glutathione peroxidase [19].

## 3.3. Physiologic significance of oxysterols in atherosclerosis

The proposed roles of oxysterols in atherosclerosis are based primarily on the results of cell-culture experiments. There are a number of in vivo studies in which investigators have exposed animals to oxysterols through diet or injection, but the overall results are not conclusive. For example, in a review by Brown and Jessup of 13 oxysterol dietary studies, six showed an increase in atherosclerosis, but four demonstrated a decrease in lesion size and three showed no effect [19]. Differences in animal models and types and doses of oxysterols undoubtedly account for some of these differences.

There are four major areas of oxysterol biology that have emerged from cell-culture studies. These include oxysterol effects on the regulation of intracellular cholesterol metabolism, cellular cytotoxicity, sterol efflux from macrophages, and activation of nuclear transcription factors. Issues related to oxysterols and intracellular cholesterol metabolism are covered in detail in Chapters 15 and 16. In brief, both cholesterol and certain oxysterols, such as 25OH and 7K, suppress the proteolytic activation of sterol response element binding protein. This, in turn, leads to transcriptional down-regulation of the LDL receptor and certain enzymes in the cholesterol biosynthetic and fatty acid synthesis and metabolism pathways. 25OH and 7K, like cholesterol, can also promote the degradation of 3-hydroxyglutaryl-3-methylglutaryl CoA reductase, a rate-limiting enzyme in isoprenoid and cholesterol biosynthesis, and 25OH can activate ACAT and suppress neutral cholesteryl ester hydrolase activity. Cells possess a protein that binds oxysterols, called oxysterol binding protein, but its role in cellular responses to oxysterols is not known. Importantly, the physiologic role of oxysterols in cellular cholesterol metabolism as it relates to atherosclerosis is far from certain, particularly because the concentrations of oxysterols used in most macrophage cell-culture studies far exceeds those found in macrophage foam cells in vivo [19].

Oxysterols have diverse roles in cholesterol efflux, a critical topic in foam cell biology. On the one hand, cells incubated with 7K and 25OH have decreased cholesterol efflux. Possible mechanisms include inhibition of membrane desorption of cholesterol or phospholipids or, as mentioned above, inhibition of lysosomal sphingomyelinase leading to lysosomal sequestration of cholesterol (M. Aviram, 1995). On the other hand, the conversion of cholesterol by macrophage sterol 27-hydroxylase to 27OH and 3β-hydroxy-5-cholestenoic acid, which are efficiently effluxed from cells, has been proposed to promote sterol efflux from foam cells (I. Björkhem, 1994). Indeed, 27OH is a prominent oxysterol found in lesions, and some studies have shown an inverse correlation between 27-hydroxylase levels and atherosclerosis (N.R. Cary, 2001). Future studies with 27-hydroxylase-deficient mice in general, and mice with a macrophage-specific deficiency of 27-hydroxylase in particular, should shed further light on the physiologic importance of this pathway.

A major area of investigation has been on the cytotoxic effects of certain oxysterols on cultured endothelial cells, macrophages, and smooth muscle cells [19]. The most potent cytotoxic oxysterols include 27OH, 7βOOH, 7αOH, and 7K (G.M. Chisolm, 1996). These sterols damage cells through a variety of mechanisms including cholesterol starvation, membrane perturbation, cellular lipid peroxidation, and activation of apoptotic pathways. Although death of endothelial cells, macrophages, and smooth muscle cells does occur in atherosclerotic lesions and might be expected to promote lesion development and complications, the role of isolated oxysterols or oxysterols in oxidized LDL in these events is far from certain. The conditions used in many cell-culture studies, notably high concentrations of oxysterols and/or serum-free medium, may not reflect the situation in vivo.

Exciting recent work has revealed that certain oxysterols are activators of nuclear transcription factors [20]. In particular, four oxysterols found in vivo, 24,25 EPOX, 24OH, 22OH, and 27OH, but not cholesterol, activate LXRα, LXRβ, and FXR (Chapter 16). Once activated, these molecules heterodimerize with activated RXR, forming active transcription factors which translocate to the nucleus and induce several genes important in atherosclerosis. In particular, a set of genes important in the reverse cholesterol transport pathway is activated by this pathway [21]. The proteins encoded by these genes include (1) macrophage ABCA1 and apolipoprotein E, which promote cholesterol efflux from foam cells, (2) plasma cholesteryl ester transfer protein, which transfers HDL-cholesterol to lipoproteins that can be internalized by hepatocytes, (3) and liver cholesterol 7α-hydroxylase, which is the key enzyme that converts hepatocyte cholesterol into bile acids for excretion. Recent studies with genetically manipulated mice have demonstrated the importance of the LXR pathway in vivo. For example, activation of RXR reduces atherosclerosis in apolipoprotein E knockout mice (J. Auwerx, 2001), and the livers of cholesterol-fed LXRα knockout mice accumulate very large amounts of cholesterol (D.J. Mangelsdorf, 1998).

In summary, oxysterols are known to exist in atherosclerotic lesions and have been demonstrated in cell-culture experiments to have profound cellular effects that could influence the development, progression, and reversal of atherosclerosis. The key question in this field of research, however, is whether the concentrations of oxysterols in vivo are high enough to influence atherogenesis. Thus far, only the oxysterol-activated nuclear transcription pathway has been directly supported by in vivo data, and even in this case the precise roles and identification of the activating oxysterols in vivo have not yet been elucidated.

## 4. Triglycerides and atherosclerosis

There are two major issues that arise when considering the role of triglycerides in atherosclerosis: the effect of triglyceride-containing lipoproteins in the plasma on atherosclerotic lesion development, and the direct role of arterial-wall triglycerides in atherogenesis. The association between triglyceride-rich lipoproteins and atherosclerotic vascular disease is often difficult to assess due to complex metabolic relationships between these lipoproteins and other risk factors for atherosclerosis, including low plasma HDL and hyperfibrinogenemia [22]. Certainly, those triglyceride-containing lipoproteins that also have a high content of cholesterol, such as remnant lipoproteins, have been shown to be associated with atherosclerotic disease and probably function largely by delivering large amounts of cholesterol to the subendothelial space. The possibility that triglycerides and triglyceride-derived fatty acids also contribute to the atherogenicity of these lipoproteins, however, must also be considered (Section 5). Interestingly, metabolic disorders resulting in severe increases in plasma triglyceride, such as lipoprotein lipase deficiency, are not associated with increased risk of atherosclerosis. In these disorders, the triglyceride-rich lipoproteins are so large that they cannot enter the arterial wall (D.B. Zilversmit, 1989).

Triglycerides constitute a measurable proportion of the lipid content of atherosclerotic lesions, although considerably less than that of cholesterol. In one study, for example, the weight-percentages of triglycerides in fatty streaks and advanced lipid-rich lesions were 2.8 and 6.0, respectively; the corresponding total cholesterol percentages were 9.6 and 31.5, respectively [16]. However, both extracellular and intracellular triglycerides could play an important role in atherogenesis by serving as a source of free fatty acids following hydrolysis by extracellular and intracellular lipases. Indeed, the relatively low content of lesional triglyceride may be partially due to this process. As discussed in Section 5, free fatty acids are precursors of potentially important bioactive lipids.

Lesional cells in general, and macrophages in particular, can accumulate triglycerides via the uptake of triglyceride-rich lipoproteins and by intracellular triglyceride synthesis. Although triglycerides are a relatively minor component of neutral lipid droplets in lesional foam cells, even small percentages can lower the melting temperature of these droplets [16]. Polyunsaturated fatty acids in the cholesteryl esters and triglycerides of lipid droplets also lower their melting temperature. Liquid neutral lipid droplets in foam cells are hydrolyzed at a more rapid rate than liquid crystalline droplets, and thus foam cells with a higher content of triglycerides may have an increased rate of cholesterol efflux (J.M. Glick, 1989).

## 5. Fatty acids and atherosclerosis

### 5.1. Direct effects of fatty acids

Fatty acids may have direct effects on atherogenesis and are also the precursors of specific bioactive lipids that may have important roles in lesion development. Lesions contain up to 0.4 mg of free fatty acids per gram of wet tissue (L. Robert, 1976). In terms of direct effects, high concentrations of extracellular free fatty acids could, in theory, lower the pH of focal areas in lesions, thus enabling the action of certain enzymes, such as lysosomal hydrolases that are secreted or leak from cells. When taken up by cells, fatty acids stimulate cholesteryl ester, phospholipid, and triglyceride syntheses. Individual types of fatty acids can have specific effects. For example, oleate, but not linoleate, is a potent stimulator of the ACAT reaction [15], and neutral lipids esterified to polyunsaturated fatty acids have a lower melting temperature, which tends to promote neutral lipid hydrolysis and lipid efflux.

### 5.2. Oxidation of long-chain polyunsaturated fatty acids: introduction

The major bioactive products of free fatty acid metabolism relevant to atherosclerosis are those that result from enzymatic or non-enzymatic oxidation of polyunsaturated long chain fatty acids. In most cases, these fatty acids are derived from phospholipase $A_2$-mediated hydrolysis of phospholipids (Chapter 11) in cellular membranes or lipoproteins, or from lysosomal hydrolysis of lipoproteins after internalization by lesional cells. In particular, arachidonic acid is released from cellular membrane phospholipids by arachidonic acid-selective cytosolic phospholipase $A_2$. In addition, there is evidence that group II secretory phospholipase $A_2$ (Chapter 11) hydrolyzes extracellular lesional lipoproteins, and lysosomal phospholipases and cholesterol esterase release fatty acids from the phospholipids and cholesteryl esters of internalized lipoproteins. Indeed, Goldstein and Brown recently surmised that at least one aspect of the atherogenicity of LDL may lie in its ability to deliver unsaturated fatty acids, in the from of phospholipids and cholesteryl esters, to lesions (J.L. Goldstein and M.S. Brown, 2001).

### 5.3. Oxidative metabolites of arachidonic acid

An important fate of arachidonic acid is enzymatic conversion to prostaglandins by one of two prostaglandin G/H synthases [23]. As described Chapter 13, these enzymes have both cyclooxygenase and hydroperoxidase activities, and are often referred to as COX-1 and COX-2. Although atherosclerotic lesions express both isoforms, mature

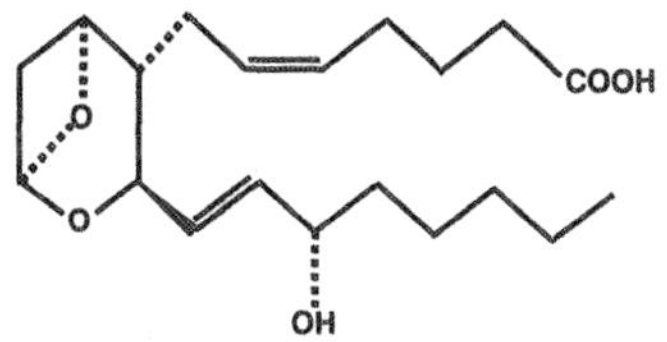

Thromboxane A$_2$

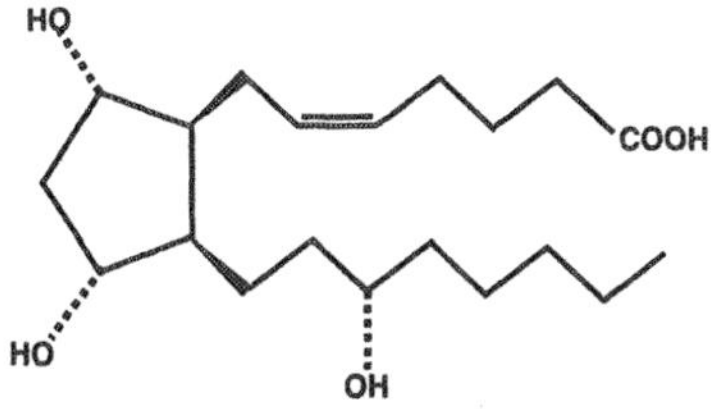

8-*iso*-PGF$_{2\alpha}$

Fig. 6. Structures of two arachidonic acid derivatives proposed to play important roles in thrombosis and atherogenesis. Thromboxane A$_2$ is a potent inducer of platelet aggregation that contributes to acute thrombosis in advanced atherosclerosis in vivo. The isoprostane 8-*iso*-PGF$_{2\alpha}$ is being investigated as a marker of oxidative stress in atherosclerosis and may also have direct atherogenic effects on platelets and smooth muscle cells.

human platelets express only COX-1. In this regard, the most well-documented role of a cyclooxygenase product on atherothrombotic vascular disease is platelet-derived thromboxane A$_2$ (Fig. 6). Thromboxane A$_2$ is a potent inducer of platelet aggregation and vasoconstriction, and aspirin-induced inhibition of platelet COX-1 accounts for its benefit in the secondary prevention of strokes and myocardial infarction. The roles of other prostaglandins synthesized in lesions are uncertain. For example, vascular endothelial cells synthesize prostacyclin, which blocks platelet aggregation, cellular interactions, and vascular smooth muscle cell proliferation in vitro and in vivo. However, there is no evidence using drug-induced inhibition of prostacyclin synthesis or prostacyclin receptor-deficient mice that prostacyclin has a net effect on atherosclerotic lesion development in vivo (G.A. Fitzgerald, 2001). A third COX-derived prostaglandin that has received recent interest is 15-deoxy-D12,14-prostaglandin J$_2$ (15d-PGJ$_2$), which, at least in vitro, is an agonist of peroxisomal proliferator-activated receptor-$\gamma$ (PPAR$\gamma$) (C.K. Glass, 1998). PPAR$\gamma$ is expressed in atherosclerotic lesions and in cultured endothelial cells, vascular smooth muscle cells, and monocyte/macrophages (B. Staels, 2000). Although cell-culture studies have revealed a variety of biological effects that may be pro- or anti-atherogenic, recent in vivo studies suggest that PPAR$\gamma$ plays an anti-atherogenic role (C.K. Glass, 2000). The physiologic significance of 15d-PGJ$_2$, which also acts on I$\kappa$B kinase, in PPAR$\gamma$ biology remains to be determined.

A relatively new class of oxidized arachidonic acid derivatives with potential relevance to atherosclerosis are F$_2$ isoprostanes [24] (Fig. 6). These compounds form as a result of non-enzymatic, free-radical attack of the fatty acid moieties of cellular or lipoprotein phospholipids, followed by release of the isoprostanes from the phospholipids by a phospholipase. 8-*iso*-PGF$_2$ may also be formed by the action of COX-1 or COX-2 in platelets or monocytes, respectively, but the significance of COX-dependent 8-*iso*-PGF$_2$ formation in vivo is unproven. F$_2$ isoprostanes circulate in the plasma and appear in the urine as free compounds or esterified to phospholipids, and 8-*iso*-PGF$_2$ is found in atherosclerotic lesions in association with macrophages and smooth muscle cells. The potential significance of isoprostanes to atherosclerosis are their effects on platelets and vascular cells, as demonstrated in cell-culture studies, and their potential usefulness as a non-invasive marker of oxidant stress. 8-*iso*-PGF$_2$ induces platelet aggregation, DNA synthesis in vascular smooth muscle cells, and vasoconstriction. Moreover, elevated levels of F$_2$ isoprostanes are found in cigarette smokers, diabetics, and subjects with hypercholesterolemia, where they may serve as an indicator of increased lipid peroxidation.

Arachidonic acid can also be oxidized by 5-, 12-, and/or 15-lipoxygenases to various mono-, di-, and trihydroxyderivatives called leukotrienes (Chapter 13), some of which are present in atherosclerotic lesions [25]. The monohydroxylated leukotrienes 12(*S*)- and 15(*S*)-HETEs can be produced by human arterial endothelial cells and can promote monocyte adherence, an important early event in atherogenesis. Although 15(*S*)-HETE is found at relatively high levels in human atherosclerotic lesions, its role in vivo is not known. Atheromatous tissue has the capacity to synthesize the dihydroxylated leukotrienes LTC$_4$ and LTB$_4$ (R. De Caterina, 1988; C. Patrono, 1992). LTC$_4$ can be made by monocytes, macrophages, and endothelial cells, and LTB$_4$ is synthesized by activated monocytes. In theory, LTC$_4$ could promote vasoconstriction, and LTB$_4$ could contribute to atherosclerosis-related endothelial alterations, such as increased permeability and adhesiveness. Moreover, LTB$_4$ is also an activator of PPARα, which appears to promote atherogenesis in vivo (C.F. Semenkovich, 2001). Finally, the trihydroxylated derivatives of arachidonic acid, the lipoxins, possess some anti-inflammatory properties, but it has been speculated that their ability to induce monocyte chemotaxis might promote atherogenesis [25].

Another fate of arachidonic acid with potential relevance to atherosclerosis is cytochrome P450 monooxygenase-derived metabolism to epoxyeicosatrienoic acids (EETs), which may also be formed nonenzymatically by the interaction of arachidonic acid with free radicals [25]. EET synthesis in cultured endothelial cells can be induced by LDL, and EETs are found both in LDL and in human atherosclerotic lesions. Biological effects of EETs include potentially anti-atherogenic effects, such as vasodilatation and prevention of platelet aggregation, and atherogenic responses, such as increased monocyte adhesion.

## 5.4. Atherogenic and anti-atherogenic effects of other long-chain polyunsaturated acids

Dietary intake of n-6 fatty acids such as linoleic acid, and n-3 fatty acids, such as the fish oils eicosapentanoic acid and docosahexaenoic acid, lower plasma cholesterol and antagonize platelet activation, but the fish oils are much more potent in this regard [26]. In particular, n-3 fatty acids competitively inhibit thromboxane synthesis in platelets but not prostacyclin synthesis in endothelial cells. These fatty acids have also been shown to have other potentially anti-atherogenic effects, such as inhibition of monocyte cytokine synthesis, smooth muscle cell proliferation, and monocyte adhesion to endothelial cells. While dietary intake of n-3 fatty acid-rich fish oils appears to be atheroprotective, human and animal dietary studies with the n-6 fatty acid linoleic acid have yielded conflicting results in terms of effects on both plasma lipoproteins and atherosclerosis. Indeed, excess amounts of both n-3 and n-6 fatty acids may actually promote oxidation, inflammation, and possibly atherogenesis (M. Toberek, 1998). In this context, enzymatic and non-enzymatic oxidation of linoleic acid in the *sn*-2 position of LDL phospholipids to 9- and 13-hydroxy derivatives is a key event in LDL oxidation (Section 6.2).

# 6. Phospholipids and related lipids

## 6.1. Introduction

Phospholipids compose the outer monolayer of lesional lipoproteins and the membranes of lesional cells. In lipoproteins, the phospholipid monolayer provides an amphipathic interface between the neutral lipid core and the aqueous external environment, and it provides the structural foundation for the various apolipoproteins (Chapter 18). In the specific context of atherosclerosis, the phospholipids of lesional lipoproteins are modified by various oxidative reactions that could have important pathological consequences. In lesional cells, membrane phospholipids not only play structural roles but also are precursors to important phospholipase-generated signaling molecules that may play important roles in atherogenesis.

## 6.2. Oxidative modification of phosphatidylcholine in lesional lipoproteins

Oxidation of LDL and probably other lesional lipoproteins occurs in atherosclerotic lesions and may contribute to lesion pathology at various stages of atherogenesis [10,11]. Based on in-vitro studies and, in some cases, genetically altered mutant mouse models, LDL oxidation may be triggered by oxidative enzymes secreted by lesional cells, including myeloperoxidase, inducible nitric oxide synthase, and, 15-lipoxygenase. In the above sections, oxidative modifications of cholesterol, cholesteryl ester, and free fatty acids in LDL were discussed. The phospholipids of LDL also undergo oxidative modification, and the products of these reactions are found in atherosclerotic lesions and have potentially important atherogenic effects on lesional cells.

The most abundant and important oxidative changes in LDL phospholipids are those that occur to the unsaturated fatty acids in the *sn*-2 position. Witztum and Berliner described several products of phospholipid oxidation that result from this process [27] (Fig. 7). An early event is the addition of oxygens to these fatty acids, resulting in the generation of hydroxy fatty acids, hydroperoxy fatty acids, and isoprostanes. In one model, lipoxygenase-generated hydroperoxyoctadecadienoic acid (HPODE) and hydroperoxyeicosatetraenoic acid (HPETE) in LDL surface phospholipids act as 'seeding' molecules. These hydroperoxy fatty acids then trigger the oxidation of arachidonate-containing phospholipids, notably, 1-palmitoyl-2-arachidonyl-*sn*-glycero-3-phosphorylcholine. This early series of events occurs before any oxidative

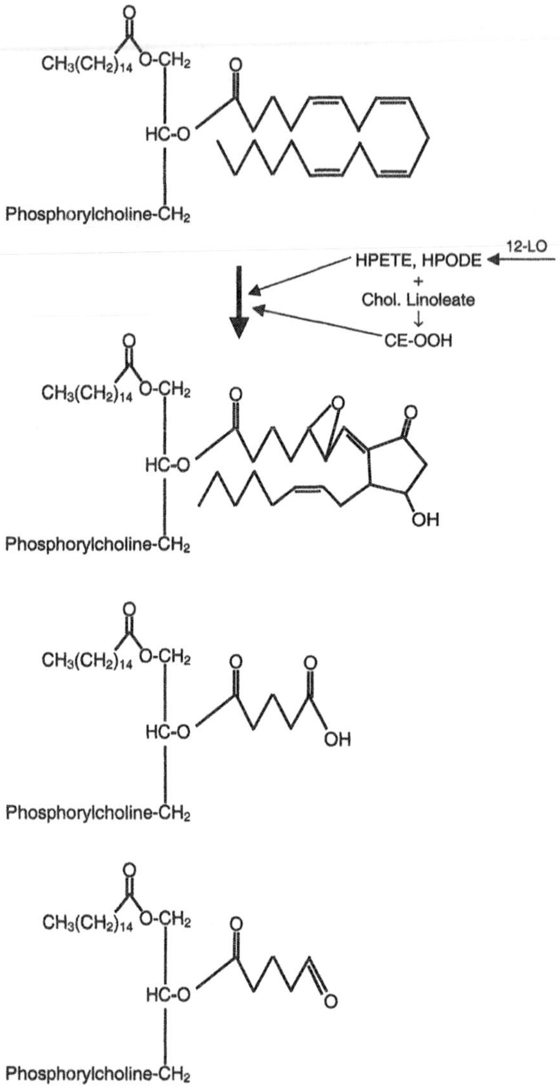

Fig. 7. Oxidation of LDL phospholipids in the generation of minimally modified LDL. 'Seeding' molecules like HPETE, HPODE, and cholesteryl linoleate hydroperoxide (CE-OOH) are proposed to trigger the oxidation of 1-palmitoyl-2-arachadonoyl phosphatidylcholine in LDL. This leads to the generation of three oxidized phosphatidylcholine species that confer atherogenic activity to minimally modified LDL. (Adapted from Navab et al. [28].)

modification of apo B100 and result in the generation of so-called 'minimally modified' LDL. Minimally modified LDL is found in atherosclerotic lesions and promotes monocyte binding to endothelial cells and monocyte chemotaxis in cultured cell studies [28]. The oxidized arachidonate-containing phospholipids, which can account for much of the biological activity of minimally modified LDL, include those in which the arachidonyl group is modified to 5-oxovaleroyl, glutaroyl, or 5,6-epoxyisoprostane phosphatidylcholine. The potentially atherogenic molecules induced in endothelial cells and smooth muscle cells by these oxidized phospholipids include E-selectin, VCAM-1, monocyte chemoattractant protein-1, macrophage colony stimulating factor, P-selectin, and interleukin-8. These cellular effects, like those of $LTB_4$ (Section 5.3), likely involve the activation of PPARα.

Another possible consequence of oxidation of sn-2 unsaturated fatty acids in oxidized LDL is fragmentation of the fatty acid, resulting in a phospholipid with a short acyl group in the sn-2 position [27]. If the sn-2 fatty acid were arachidonate, the truncated acyl group would be a 5-carbon aldehyde or a 5-carbon carboxylic acid. If the sn-2 group were linoleate, the products would be a 9-carbon aldehyde or carboxylic acid. Phospholipids containing these shortened sn-2 acyl chains have biological activity similar to that of platelet-activating factors, where an ether-linked fatty acyl group occupies the sn-1 position and an acetyl group is in the sn-2 position. By interaction with a G-protein-coupled receptor on a variety of cell types, platelet activating factor activates both platelets and leukocytes and increases vascular permeability. In terms of atherosclerosis, studies with PAF receptor antagonists have suggested that the chemotactic activity of minimally modified LDL may be mediated through platelet activating factor-like phospholipids acting directly on monocytes (P.D. Reaven, 1997). Similarly, the smooth muscle cell mitogenic activity of oxidized LDL can be mimicked by PC containing a 5-carbon carboxylic acid and can be blocked by a platelet activating factor receptor antagonist (S.M. Prescott, 1995).

Phospholipase $A_2$-mediated hydrolysis of oxidized phospholipids can result in the release of either intact oxygenated free fatty acids by phospholipase $A_2$, discussed in Section 5, or fragmented fatty acids, such as malondialdehyde, which can lead to protein modification. The other product of this reaction is lysophosphatidylcholine (D. Steinberg, 1988) [29]. In-vitro studies have revealed multiple effects of lysophosphatidylcholine on lesional cells, including expression of adhesion molecules on endothelial cells, monocyte chemotaxis and macrophage scavenger receptor expression, growth factor expression by smooth muscle cells and macrophages, cellular cytotoxicity, and inhibition of T cell activation. Although it is not known whether enough lysophosphatidylcholine exists in lesions to cause these effects in vivo, a receptor for lysosphospholipids, called G2A, was recently identified, and studies with G2A-deficient mice suggest a possible role for lysophospholipids in the regulation of T cell activation in vivo (O.N. Witte, 2001). Finally, adducts of oxidized phospholipids and apo B100 can result from LDL oxidation. These adducts are recognized by macrophages and thus can mediate cellular uptake of oxidized LDL, and they form a potent epitope that can elicit cellular and humoral immune responses that may play important roles in atherogenesis (J.L. Witztum, 1999).

Navab et al. have suggested that one of the anti-atherogenic mechanisms of HDL may be its ability to prevent LDL oxidation or reduce its atherogenic activity [28]. Multiple mechanisms may be involved, including removal of the 'seeding' molecules HPODE and HPETE and degradation of the oxidized phospholipids themselves by paraoxonase, an esterase/peroxidase, and platelet-activating factor-acetylhydrolase, a

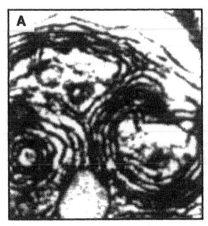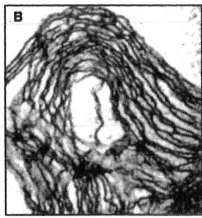

Fig. 8. Phospholipids whorls in cholesterol loaded macrophages. (A) Electron micrograph showing a membrane whorl in the cytoplasm of a foam cell. The cell was isolated from an advanced aortic atherosclerotic lesion in a cholesterol-fed rabbit. (From Shio et al. (1979) Lab. Invest. 41: 160–167.) (B) Electron micrograph showing a membrane whorl in the cytoplasm of an FC-loaded J774 macrophage. The cell was in the 'adaptive' stage, i.e., before the onset of FC-induced death. (From Shiratori et al. (1994) J. Biol. Chem. 269: 11337–11348.)

lipoprotein-bound phospholipase $A_2$-like enzyme that can cleave oxidized acyl groups from the *sn*-2 position of oxidized phospholipids. In apo E knockout mice, targeted disruption of the paraoxonase gene increases lipoprotein oxidation and atherosclerosis (A.J. Lusis, 2000).

### 6.3. The phospholipids of lesional cells

Phosphatidylcholine is the major phospholipid of lesional cells and, as mentioned above, serves both structural and signaling functions. In terms of cellular membrane structure, the cholesterol:phospholipid ratio in lesional cells must be kept within a certain limit in order for the proper functioning of membrane proteins [9]. Cholesterol-rich foam cells isolated from atherosclerotic lesions have intracellular phospholipid whorl-like structures, and PC biosynthesis is increased in lesional areas of the arterial wall (Fig. 8). Cell culture studies have revealed that free cholesterol (FC) loading of macrophages directly leads to the activation of CTP:phosphocholine cytidylyltransferase and an increase in phosphatidylcholine biosynthesis and mass. Proof that this is an adaptive response to FC excess comes from a study in which the cytidylyltransferaseα gene was disrupted in macrophages, which resulted in accelerated FC-induced death (I. Tabas, 2000). Thus, activation of phosphatidylcholine biosynthesis in cholesterol-loaded lesional macrophages may help to protect these cells from the toxicity of FC excess [9].

  Cellular phospholipids, particularly phosphatidylinositol and phospholipids containing unsaturated fatty acids in the *sn*-2 position, are precursors to a variety of signaling molecules (Chapter 12). These include (1) diacylglycerol and inositol trisphosphate by phosphatidylinositol-specific phospholipase C-induced hydrolysis of phosphatidylinositol, (2) phosphatidic acid by phospholipase D, (3) fatty acids and lysophosphatidylcholine by phospholipase $A_2$, and (4) platelet activating factor-like molecules by oxidation. Diacylglycerol activates protein kinase C, and inositol triphosphate leads to intracellular calcium release. Both of these reactions are involved in a variety of signaling processes that occur in lesional smooth muscle cells, macrophages, and endothelial cells, including responses to cytokines and growth factors. Oxidized LDL has been shown to activate phospholipase D in cultured vascular smooth muscle cells by a tyrosine kinase-mediated mechanism, and phosphatidic acid could mimic the proliferative effects of oxidized LDL in these cells (S. Parthasarathy, 1995). Finally, given the potential importance of apoptosis of macrophages and smooth muscle cells in atherosclerosis (above), phosphatidylserine is an important phospholipid in lesional cells. Phosphatidylserine is normally a component of the inner leaflet of the plasma membrane, but it becomes externalized during apoptosis and acts as a recognition signal and ligand for phagocytes. Interestingly, both CD36 and scavenger receptor B-1 on macrophages are receptors for phosphatidylserine on apoptotic cells as well as for oxidized LDL. Thus, it is possible that phagocytosis and clearance of apoptotic cells in lesions may be competitively inhibited by oxidized LDL (D. Steinberg, 1999).

## 6.4. Sphingomyelin and ceramide

Sphingomyelin is an important component of both the phospholipid monolayer of lesional lipoproteins and of membranes of lesional cells. In atherogenic lipoproteins like LDL, hydrolysis of sphingomyelin to ceramide results in lipoprotein aggregation and fusion, resulting in the formation of large aggregates that appear similar to those that occur in extracellular regions of the subendothelium of atherosclerotic lesions [8]. The mechanism of sphingomyelinase-induced aggregation and fusion, which is dependent on lipoprotein ceramide content, probably lies in both the physical effects of ceramide on lipoprotein structure and on hydrogen bonding between ceramide on one particle and phospholipids on a neighboring particle. Tabas and coworkers have provided evidence that extracellular hydrolysis of LDL-sphingomyelin by sphingomyelinase occurs in the subendothelium of atherosclerotic lesions and is catalyzed by a form of acid sphingomyelinase, called S-sphingomyelinase, that is secreted by endothelial cells and macrophages [13]. Although the overall importance of this reaction in vivo remains to be determined, its potential importance is that subendothelial lipoprotein retention promotes lipoprotein retention in the arterial wall and is a potent substrate for macrophage and possibly smooth muscle cell foam cell formation. Lipoproteins with a high sphingomyelin:phospholipid ratio are particularly good substrates for S-sphingomyelinase. In this context, lipoproteins isolated from atherosclerotic lesions have a very high sphingomyelin (as well as ceramide) content. Moreover, a recent analysis of plasma samples from a case:control study showed that a high sphingomyelin:phospholipid ratio in plasma lipoproteins was an independent risk factor for coronary artery disease in humans (X.C. Jiang, 2000).

HDL also contains sphingomyelin. Because sphingomyelin avidly binds cholesterol, sphingomyelin may increase the ability of HDL to act as an extracellular acceptor for cholesterol effluxed from cells (G. Rothblat, 1997). However, HDL-sphingomyelin has also been shown to inhibit the binding of lecithin:cholesterol acyltransferase to the lipoprotein, and so this effect may balance the effect of HDL-sphingomyelin-induced cholesterol efflux on reverse cholesterol transport (A. Jonas, 1996).

Sphingomyelin in cellular membranes may have several important roles related to atherogenesis. Because sphingomyelin interacts strongly with cholesterol, accumulation of sphingomyelin in cellular sites that are involved in cholesterol trafficking may influence cellular cholesterol distribution. For example, the defective intracellular trafficking and ACAT-mediated esterification of oxidized LDL-derived cholesterol in macrophages may be due to the inhibition of acid sphingomyelinase by oxidized LDL lipids (M. Aviram, 1995). Moreover, the sphingomyelin accumulation that occurs in acid sphingomyelinase-deficient macrophages leads to defective cholesterol trafficking and efflux (I. Tabas, 2001). On the other side of this issue is the response of sphingomyelin biosynthesis to FC loading of cells. Investigators have shown that the synthesis and mass of cellular sphingomyelin is increased in advanced atherosclerotic lesions [16]. In cultured FC-loaded macrophages, sphingomyelin biosynthesis is increased, suggesting that sphingomyelin, like phosphatidylcholine, plays a role in the adaptation of cells to FC excess.

Another area of sphingomyelin biology with potential relevance to atherosclerosis is related to cell signaling [30]. Hydrolysis of cellular sphingomyelin by either neutral or acid sphingomyelinases results in the generation of intracellular ceramide, which is involved in a variety of cell-signaling reactions (Chapter 14). In terms of atherosclerosis, ceramide-mediated signaling may play roles in smooth muscle cell proliferation and apoptosis and macrophage apoptosis [30]. Alterations in ceramide synthesis and ceramide hydrolysis by cellular ceramidases may also influence these events. In this context, ceramidase-generated sphingosine can be phosphorylated to sphingosine-1-phosphate, which is another signaling molecule. For example, Ross and colleagues showed that sphingosine-1-phosphate blocks the migration of vascular smooth muscle cells induced by platelet-derived growth factor (R. Ross, 1995).

*6.5. Glycosphingolipids*

Sugar transferases can convert ceramide to a variety of glycosphingolipids, including neutral glycosphingolipids such as glucosylceramide and lactosylceramide, and polar glycosphingolipids such as gangliosides, which contain ceramide, sugars, and sialic acid and/or $N$-glycolylneuraminic acid (Chapter 14). Glycosphingolipids are found both in plasma lipoproteins and in the cells and extracellular regions of atherosclerotic lesions [30]. Chatterjee and colleagues have proposed that lactosylceramide, synthesized from glucosylceramide by the enzyme UDP-galactose:glucosylceramide-$\beta$1–4 galactosyltransferase activity (GalT-2), is a lipid second messenger that is involved in the proliferation of vascular smooth muscle cells by oxidized LDL [30]. In cultured smooth muscle cells, oxidized LDL stimulated GalT-2 activity and lactosylceramide synthesis. Proliferation induced by oxidized LDL in these cells was blocked by an inhibitor of GalT-2, and exogenous lactosylceramide was able to stimulate proliferation in the absence of oxidized LDL. The mechanism may involve 5-oxovaleroyl phosphatidylcholine-mediated stimulation of NADPH oxidase by lactosylceramide, leading to signaling cascade triggered by superoxide radicals and involving Ras activation and p44-mitogen-activated protein kinase. Interestingly, native LDL was shown to actually decrease GalT-2 activity and lactosylceramide synthesis in smooth muscle cells in an LDL receptor-dependent manner. While these cell-culture studies have provided a potentially interesting role for GalT-2 activity and lactosylceramide in atherosclerosis, the physiologic significance of these findings overall must await future in-vivo studies.

Table 2

Summary of proposed roles of lesional lipids in atherosclerosis

| Lipid | Overall effects in atherosclerosis | Specific examples |
|---|---|---|
| Free cholesterol | Accumulation in and alteration of macrophages and smooth muscle cells, including gene regulation and, in excess, death | Stimulation of ACAT Repress transcription of LDL receptor gene FC-induced macrophage death |
| Cholesteryl ester | Accumulation in macrophages and smooth muscle cells Substrate for oxidation | Foam cell formation Cholesterol linoleate hydroperoxide as a pro-oxidant |
| Oxysterols | Regulation of cellular cholesterol metabolism Cytotoxicity Sterol efflux pathways Activation of nuclear transcription factors | Stimulation of ACAT 7K-induced macrophage death Efflux of 27OH Activation of LXR by 22OH |
| Triglycerides | Source of fatty acids Affect neutral lipid droplet fluidity in foam cells | Liquid crystalline $\rightarrow$ liquid neutral transformation of foam cell droplets |
| Fatty acids | Stimulate CE, triglyceride, and phospholipid synthesis Polyunsaturated fatty acids are sources of bioactive eicosanoids | Thromboxane $A_2 \rightarrow$ platelet aggregation Isoprostanes $\rightarrow$ smooth muscle cell proliferation |
| Phospholipids (other than sphingolipids) | Structural roles in lipoproteins and lesional cells Source of signaling molecules Substrate for oxidative modification into bioactive molecules | Part of adaptive response to FC-induced cytotoxicity Lysophosphatidylcholine $\rightarrow$ monocyte chemotaxis Oxidized phospholipids $\rightarrow$ induction of endothelial adhesion molecules |
| Sphingolipids | Source of signaling molecules Involve on lipoprotein aggregation Influence intracellular cholesterol trafficking | Ceramide $\rightarrow$ lesional cell death and proliferation Sphingomyelinase-induced LDL aggregation Lactosylceramide $\rightarrow$ smooth muscle cell proliferation |

## 7. Future directions

The potential roles of the many types of lesional and lipoprotein lipids in atherogenesis is staggering (Table 2). Not surprisingly, most studies investigating these roles have been conducted on cultured cells where the concentrations of the lipids and the overall state of cells may be very different from that occurring in atherosclerotic lesions. Thus, one of the most important, and difficult, areas in future studies will be to sort out these effects in vivo through the use of inhibitory compounds or genetic manipulations in mice. In some cases, such studies have provided impressive results, such as the decrease in atherosclerosis observed in 15-lipoxygenase knockout mice (C.D. Funk, 1999). On the other hand, the effects of anti-oxidants in both humans and animal models have yielded conflicting results (R. Stocker, 2001). Further understanding of the molecular basis of lipid synthesis and catabolism, and of the action of bioactive lipids in cells, will help in the design of improved in-vivo models. The most important of these lipids include cholesterol, oxidized phospholipids and fatty acids, oxysterols, and sphingolipid derivatives. Key areas for investigating the cellular effects of bioactive lipids include (1) inflammatory responses in endothelial cells, T cells, and macrophages, (2) secretion of atherogenic and anti-atherogenic molecules by lesional cells, (3) proliferation of macrophages and smooth muscle cells, (4) and apoptotic and necrotic death in lesional cells. Moreover, the mechanism and consequences of macrophage and smooth muscle cell lipid accumulation, particular cholesteryl ester and free cholesterol accumulation, represent fundamental areas in lesional cell biology that require further investigation. New advances in genomics and proteomics have already begun to aid in these effort and will increasingly do so. Ultimately, the goal of these studies is to elucidate novel targets for drug or gene therapy that can complement plasma cholesterol-lowering therapy in the fight against the leading cause of mortality worldwide.

## References

1. Stary, H.C. (2000) Natural history and histological classification of atherosclerotic lesions: an update. Arterioscler. Thromb. Vasc. Biol. 20, 1177–1178.
2. Ross, R. (1995) Cell biology of atherosclerosis. Annu. Rev. Physiol. 57, 791–804.
3. Glass, C.K. and Witztum, J.L. (2001) Atherosclerosis. The road ahead. Cell 104, 503–516.
4. Williams, K.J. and Tabas, I. (1995) The response-to-retention hypothesis of early atherogenesis. Arterioscler. Thromb. Vasc. Biol. 15, 551–561.
5. Small, D.M. (1988) Progression and regression of atherosclerotic lesions. Arteriosclerosis 8, 103–129.
6. Newby, A.C., Libby, P. and van der Wal, A.C. (1999) Plaque instability — the real challenge for atherosclerosis research in the next decade? Cardiovasc. Res. 41, 321–322.
7. Williams, K.J. and Tabas, I. (1998) The response-to-retention hypothesis of atherogenesis, reinforced. Curr. Opin. Lipidol. 9, 471–474.
8. Tabas, I. (1999) Nonoxidative modifications of lipoproteins in atherogenesis. Annu. Rev. Nutr. 19, 123–139.
9. Tabas, I. (2000) Cholesterol and phospholipid metabolism in macrophages. Biochim. Biophys. Acta 1529, 164–174.
10. Chisolm, G.M. and Steinberg, D. (2000) The oxidative modification hypothesis of atherogenesis: an overview. Free Radic. Biol. Med. 28, 1815–1826.
11. Navab, M., Berliner, J.A., Watson, A.D., Hama, S.Y., Territo, M.C., Lusis, A.J., Shih, D.M., Van

Lenten, B.J., Frank, J.S., Demer, L.L., Edwards, P.A. and Fogelman, A.M. (1996) The Yin and Yang of oxidation in the development of the fatty streak. A review based on the 1994 George Lyman Duff Memorial Lecture. Arterioscler. Thromb. Vasc. Biol. 16, 831–842.

12. Oorni, K., Pentikainen, M.O., Ala-Korpela, M. and Kovanen, P.T. (2000) Aggregation, fusion, and vesicle formation of modified low density lipoprotein particles: molecular mechanisms and effects on matrix interactions. J. Lipid Res. 41, 1703–1714.

13. Tabas, I. (1999) Secretory sphingomyelinase. Chem. Phys. Lipids 102, 131–139.

14. Brown, M.S. and Goldstein, J.L. (1983) Lipoprotein metabolism in the macrophage: Implications for cholesterol deposition in atherosclerosis. Annu. Rev. Biochem. 52, 223–261.

15. Chang, T.Y., Chang, C.C.Y. and Cheng, D. (1997) Acyl-coenzyme A : cholesterol acyltransferase. Annu. Rev. Biochem. 66, 613–638.

16. Katz, S.S., Shipley, G.G. and Small, D.M. (1976) Physical chemistry of the lipids of human atherosclerotic lesions. Demonstration of a lesion intermediate between fatty streaks and advanced plaques. J. Clin. Invest. 58, 200–211.

17. Tabas, I. (1997) Free cholesterol-induced cytotoxicity. A possible contributing factor to macrophage foam cell necrosis in advanced atherosclerotic lesions. Trends Cardiovasc. Med. 7, 256–263.

18. Tabas, I. and Krieger, M. (1999) Lipoprotein receptors and cellular cholesterol metabolism in health and disease. In: K. Chien (Ed.), Molecular Basis of Heart Disease. W.B. Sanders, New York, pp. 428–457.

19. Brown, A.J. and Jessup, W. (1999) Oxysterols and atherosclerosis. Atherosclerosis 142, 1–28.

20. Peet, D.J., Janowski, B.A. and Mangelsdorf, D.J. (1998) The LXRs: a new class of oxysterol receptors. Curr. Opin. Genet. Dev. 8, 571–575.

21. Tall, A.R., Costet, P. and Luo, Y. (2000) 'Orphans' meet cholesterol. Nat. Med. 6, 1104–1105.

22. Durrington, P.N. (1998) Triglycerides are more important in atherosclerosis than epidemiology has suggested. Atherosclerosis 141(Suppl. 1), S57–S62.

23. FitzGerald, G.A., Austin, S., Egan, K., Cheng, Y. and Pratico, D. (2000) Cyclo-oxygenase products and atherothrombosis. Ann. Med. 32(Suppl. 1), 21–26.

24. Patrono, C. and FitzGerald, G.A. (1997) Isoprostanes: potential markers of oxidant stress in atherothrombotic disease. Arterioscler. Thromb. Vasc. Biol. 17, 2309–2315.

25. Sellmayer, A., Hrboticky, N. and Weber, P.C. (1999) Lipids in vascular function. Lipids 34(Suppl.), S13–S18.

26. Sanders, T.A. (1990) Polyunsaturated fatty acids and coronary heart disease. Baillieres Clin. Endocrinol. Metab. 4, 877–894.

27. Witztum, J.L. and Berliner, J.A. (1998) Oxidized phospholipids and isoprostanes in atherosclerosis. Curr. Opin. Lipidol. 9, 441–448.

28. Navab, M., Berliner, J.A., Subbanagounder, G., Hama, S., Lusis, A.J., Castellani, L.W., Reddy, S., Shih, D., Shi, W., Watson, A.D., Van Lenten, B.J., Vora, D. and Fogelman, A.M. (2001) HDL and the inflammatory response induced by LDL-derived oxidized phospholipids. Arterioscler. Thromb. Vasc. Biol. 21, 481–488.

29. Hurt-Camejo, E. and Camejo, G. (1997) Potential involvement of type II phospholipase A2 in atherosclerosis. Atherosclerosis 132, 1–8.

30. Chatterjee, S. (1998) Sphingolipids in atherosclerosis and vascular biology. Arterioscler. Thromb. Vasc. Biol. 18, 1523–1533.

# Subject Index